CAMBRIDGE LIBRARY COLLECTION

Books of enduring scholarly value

Botany and Horticulture

Until the nineteenth century, the investigation of natural phenomena, plants and animals was considered either the preserve of elite scholars or a pastime for the leisured upper classes. As increasing academic rigour and systematisation was brought to the study of 'natural history', its subdisciplines were adopted into university curricula, and learned societies (such as the Royal Horticultural Society, founded in 1804) were established to support research in these areas. A related development was strong enthusiasm for exotic garden plants, which resulted in plant collecting expeditions to every corner of the globe, sometimes with tragic consequences. This series includes accounts of some of those expeditions, detailed reference works on the flora of different regions, and practical advice for amateur and professional gardeners.

A History of Garden Art

Marie Luise Gothein (1863–1931) published this scholarly two-volume history of garden design in German in 1913. Its second edition of 1925 was translated into English by Laura Archer-Hind, edited by gardening author Walter P. Wright (1864–1940), and published in 1928. The highly illustrated work is still regarded as among the most thorough and important surveys of its kind. It begins by examining evidence from both archaeology and literature, as well as climate and soil conditions, to discuss the gardens of ancient Egypt and Assyria, and continues to survey developments worldwide until the twentieth century. Individual gardens, technical innovations, and fashions in horticulture are all discussed in detail. Volume 2 considers northern European gardens of the Renaissance, the cultural importance of Louis XIV's France, the impact of the introduction of foreign plants, and gardening in Europe, the Far East and North America up to the early twentieth century.

Cambridge University Press has long been a pioneer in the reissuing of out-of-print titles from its own backlist, producing digital reprints of books that are still sought after by scholars and students but could not be reprinted economically using traditional technology. The Cambridge Library Collection extends this activity to a wider range of books which are still of importance to researchers and professionals, either for the source material they contain, or as landmarks in the history of their academic discipline.

Drawing from the world-renowned collections in the Cambridge University Library and other partner libraries, and guided by the advice of experts in each subject area, Cambridge University Press is using state-of-the-art scanning machines in its own Printing House to capture the content of each book selected for inclusion. The files are processed to give a consistently clear, crisp image, and the books finished to the high quality standard for which the Press is recognised around the world. The latest print-on-demand technology ensures that the books will remain available indefinitely, and that orders for single or multiple copies can quickly be supplied.

The Cambridge Library Collection brings back to life books of enduring scholarly value (including out-of-copyright works originally issued by other publishers) across a wide range of disciplines in the humanities and social sciences and in science and technology.

A History of Garden Art

From the Earliest Times to the Present Day

VOLUME 2

MARIE LUISE GOTHEIN
EDITED BY WALTER P. WRIGHT
TRANSLATED BY LAURA ARCHER-HIND

CAMBRIDGE
UNIVERSITY PRESS

CAMBRIDGE
UNIVERSITY PRESS

University Printing House, Cambridge, CB2 8BS, United Kingdom

Cambridge University Press is part of the University of Cambridge.

It furthers the University's mission by disseminating knowledge in the pursuit of
education, learning and research at the highest international levels of excellence.

www.cambridge.org
Information on this title: www.cambridge.org/9781108076159

© in this compilation Cambridge University Press 2014

This edition first published 1928
This digitally printed version 2014

ISBN 978-1-108-07615-9 Paperback

A HISTORY OF
GARDEN ART

FROM THE EARLIEST
TIMES TO THE PRESENT DAY
IN TWO VOLUMES

WATER-LILIES IN BLOOM

MARIE LUISE GOTHEIN

A
HISTORY OF GARDEN ART

Edited by
WALTER P. WRIGHT

Translated from the German by
MRS. ARCHER HIND, M.A.

With over Six Hundred Illustrations

VOLUME II
LONDON & TORONTO
J. M. DENT & SONS LIMITED
NEW YORK: E. P. DUTTON & CO. LTD.

PRINTED IN GREAT BRITAIN

CONTENTS OF VOLUME II

LIST OF ILLUSTRATIONS

CHAPTER XIII. THE FRENCH GARDEN IN EUROPEAN COUNTRIES

CHAPTER XIV. CHINA AND JAPAN

CHAPTER XV. THE ENGLISH LANDSCAPE GARDEN

CHAPTER XVI. TENDENCIES OF GARDEN ART IN THE NINETEENTH CENTURY

CHAPTER XVII. MODERN ENGLISH GARDENING

CHAPTER XVIII. LANDSCAPE ARCHITECTURE IN NORTH AMERICA (UNITED STATES AND CANADA)

CHAPTER XI

GERMANY AND THE NETHERLANDS IN THE TIME OF THE RENAISSANCE

CHAPTER XI

GERMANY AND THE NETHERLANDS IN THE TIME OF THE RENAISSANCE

WHEREVER the German language is spoken the history of garden art has been varied and irregular, but interesting. There was a miscellaneous collection of workers in this field, just as there was in all the other departments of study and culture that flourished at the time of the Renaissance. Side by side were princes, burghers and learned scholars. In many ways the division of labour and interests was a check on the development of large standard types. Hitherto we have found this development working outward from individual important centres. In Italy one great town handed over the leadership, so to speak, to another; in France, England and Spain everything was concentrated at one royal court and in the nobility that was so closely connected therewith; but in Germany a peaceful uniform progress was out of the question. True, the stream of Italian art breaks through with much force at the end of the fifteenth century, but it is at once dissipated in a multitude of different channels whose fructifying work disappears only too often from our sight. Therefore we get the surprise of some work of art suddenly appearing, apparently in no connection with anything else that is going on; and we cannot feel (as we so strongly felt in France, and even in England) that foreign influence is simply food for what is indigenous to the soil; because these separate creations for the most part seem to fade away, leaving no real successors.

Almost at the same time that Charles VIII. took his adventurous journey through Italy to Naples, a German prince, Duke Eberhard von Würtemberg, made his way across the Alps. He travelled modestly, like a tourist, with only a small retinue, among whom was the learned Reuchlin. Lorenzo de' Medici received him as his guest in Florence, and showed him his garden, among his other treasures, with great pride. The duke, like his learned friend, admired everything immensely, but his own estate was too small to allow of his making anything similar at home. His journey must be counted among those numerous travels for purposes of study that were taken by learned men of his day. The result was not so much that they gave a strong impetus to effective art at home as that they imbibed at the great universities that interest in botany which was now flourishing in Italy, and caused it to spread rapidly in Germany from the first years of the sixteenth century. An accurate knowledge of medicinal virtues was in great demand, and almost all the plants then known were used in medicine. From this point of view most foreign plants were introduced, even in the first half of the sixteenth century, and the earliest botanists were physicians: they were the men who made botanic gardens in their own country when they came back from their travels in Italy. For the most part they were burghers of wealthy flourishing towns, and they had the means for indulging their tastes. These learned students maintained intercourse eagerly with one another. Their gardens

3

were worth seeing and attracted visitors from abroad. There is no doubt that in the first intention they were laid out for botanical purposes; but as soon as educated persons had returned from Italy, where they generally spent years, and had assimilated the art and skill they desired, it was natural that, in those early days when utility and ornament had gone hand in hand, the chief honour should be given to utility.

All the important scientific workers at botany in Germany were the owners of large gardens. As early as 1525 Henricus Cordus started a garden in Erfurt after he had taken a doctor's degree at Ferrara, and therewith added a new title of honour to the town: at that time it was already called the Garden of the Holy Roman Empire because of the extensive gardens there. When, five years later, Cordus left for Marburg as professor, one of his first acts was to lay out a garden. His son Valerius, like himself a learned botanist, was a close friend of Conrad Gesner, the well-known physician and man of varied learning at Zurich, who devoted his entire life to science, and died in 1565 at the age of forty-nine of the plague, which he was attempting to combat a second time in his native city. Gesner was the centre of all botanical study in Germany and far beyond. At Zurich he owned a lovely garden, he travelled a great deal, and in his writings he has left behind the names of the well-known gardens of his time, with information about them. These gardens, belonging at first to private scholars, soon became attached as academic properties to the universities. After the Italian botanic gardens founded at Padua, Bologna, and Pisa about the middle of the century had shown how helpful they were for medicine, for plant production, and for the acclimatisation of foreign herbs, Germany and the Netherlands were the first to follow the lead. In 1577 there was founded a botanic garden at Leyden, in 1580 another at Leipzig, and in 1597 another at Heidelberg, to mention the most famous; but soon there were smaller ones in nearly all the universities.

The interest in botany in German countries was now so much in the foreground that it gave a certain scientific character even to the private gardens of important and educated owners. Erasmus, in his *Convivium Religiosum*, makes the guests walk before their meal into a well-kept garden, a square with a wall round it. "The place is dedicated to the honourable pleasures of rejoicing the eye, refreshing the nose, and renewing the spirit." Only sweet-scented herbs grow therein; the plants are set out in the most perfect order; every species has its own place, and each one has its own *vexillum* with its name and special virtues. Thus for example speaks the marjoram: "Keep away from me, swine; my scent is not for you," for although this plant has a sweet smell, swine cannot endure its scent. Thus the owner has talking plants, not dumb ones. The individual beds are enclosed with palisades, in one place striped with green, and in another with the complementary red. The whole garden is divided in two by a brook, which flows into a basin of stucco on whose base one sees beasts of all kinds and colours, and plants mirrored in the clear water. Round about the place there are covered halls, two stories high, making a shady border to the house. The lower story, made with stucco columns, is painted all over, and the pictures portray a second garden with animals and flowers.

Special excuses are given to justify these pictures, and here also the scientific interest in foreign fauna and flora preponderates, in contradistinction to the upper halls, which, lighted from windows on the outside, are painted with serious religious pictures. One gets there directly from the gallery which is beside the house in front of the library. At the end of these corridors there are little summer-houses where one can rest and look out over the kitchen-garden. This comprises the vegetable-garden, which is called "the woman's kingdom"; and the medicinal-garden, which contains the useful herbs for home consumption. On the left is the playground, a meadow with a quickset hedge, and a summer-house in one corner where people can take meals, and which can also be used as an isolation house in case of infectious disease. On the right of the vegetable-garden is the orchard, where more foreign trees are grown; at one end is the apiary, and by the pillared walk of the flower-garden a bird-house is approached by a flying bridge, which perhaps had water under it for the sake of water-fowl.

All the features that characterise the garden of Erasmus are entirely of the northern

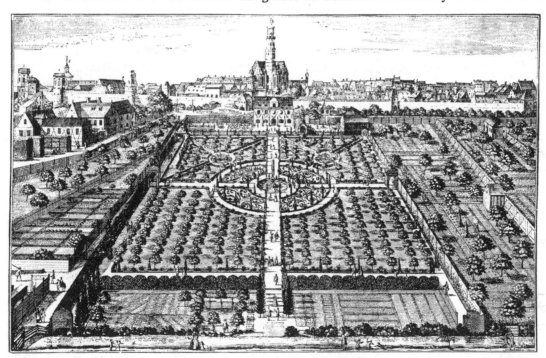

FIG. 356. THE TOWN GARDENS AT AUGSBURG

type—the two-storied corridor in the flower-garden which is laid out with a botanical intention, the bright stream enclosed in a basin, the special meadow playground with hedges round it—all showing the style of the North, one might say the style of the towns-man's garden, belonging to the well-to-do scholar; and it is quite likely that Erasmus had in his mind some particular garden of a gentleman's family at Basle.

Germany succeeded earlier than any country of the North in winning a place for the town garden. As a fact, the flourishing towns of Upper Germany, and Augsburg ahead of all, felt the influence of the new art sooner than princes at their homes, because of the trade carried on with Italy. The gardens of the Fugger family play a great part in the history of the town, and when Charles V. visited them in 1539 he was astonished at their splendour. A year later Beatus Rhenanus writes enthusiastically about them, and sets them above the French gardens at Blois. Towards the end of the century these gardens have become so extensive that townspeople complain in 1584 that they are encroaching on their living space. When Montaigne sees them on his journey to Italy, he has much to say about their water-tricks and devices. Drawings of a somewhat later date (Fig. 356) show a great number of notable gardens at Augsburg, founded in the sixteenth century, as many inside the town as outside the walls. There is as a rule an oblong strip, with leafy paths all round, and beautiful flower-beds. But the botanical interest always reigned supreme. In the year 1560 the famous botanist Clusius went on a travelling expedition with the heir of Count Anton Fugger to collect new plants for his gardens, because everyone showed the utmost eagerness to be the first in introducing a new plant into his own garden. It was a wonderful claim to honour and glory when Councillor Johann Heinrich Herward of Augsburg flowered the first tulip in 1559, its bulbs having been sent to Augsburg by the hand of Busbecq the imperial messenger. Thither went Conrad Gesner, and had a woodcut made for his book *De Hortis Germaniæ*. These flowers were destined to develop to such effect as to influence greatly the history of Dutch trade.

Other towns were not much behind Augsburg, for travelling scholars carried the fashion to all parts. As early as 1489 a garden belonging to the Canon Mariensüss on the cathedral island at Breslau was talked about; and this one, as well as the garden of the physician Woysel (which was flourishing between the years 1540 and 1560), belongs naturally to the class of private botanical gardens, though the descriptions of Erasmus have taught us that such places were not wanting in style and artistic skill. In the last third of that century there was another doctor at Breslau, Laurentius Scholz, and a picture of his garden reminds us strongly of what Erasmus says. Scholz had studied in Padua, leaving there in 1579. Six years later he returned to his native town, and as his means increased so also did his delight in the garden: he felt that the care of it was a patriotic duty, and its reputation soon spread outside Breslau. Like the garden of Erasmus, that of Scholz was a formal square, and was divided into four sections by crossing paths. A Latin inscription was chiselled on the chief gate: "To the praise and honour of Almighty God, to the glory of my native town, for the use of friends and students of botany, also for my own delight, I have established this garden, long neglected heretofore, at my own expense, and have furnished it with indigenous and foreign plants."

The first section, reached from the main gate, was the flower-garden, which was laid out in beds, perhaps enclosed with a palisade, and planted with flowers which were used for wreaths and nosegays. Doctor Scholz took great pains that we should know what the plants were; not only was he a useful medical writer who was always pleased to go beyond his own garden, but after the fashion of his day he had his plants faithfully drawn by a nature artist of Breslau. The chief constituents of his garden were still the old native plants: in spring, snowdrops, violets, crocuses, primulas, auriculas, and crown imperials; in summer, columbines, snapdragons, cornflowers, poppies, and lilies, but during the last thirty years tulips had come from the East, and were shown with great pride in this garden. To the doctor and botanist, however, the second section, the real medicinal-garden, was more important. Here there were 385 kinds, and among them many foreign plants, which the doctor had procured through his connection with Spain, Italy, and Austria. They were planted in beds, and here also for each kind there was a separate bed. By the side of the medicinal herbs (just as we know them by the *Capitulare* and the cloister plan of St. Gall) there are found the aromatic plants of Italian gardens, such as basil, marjoram, balm, hyssop, rosemary, and dittany. But certain novelties also flowered here,

FIG. 357. THE GARDEN OF CHRISTOPHER PELLER, NUREMBERG

which Portuguese seafarers had brought from India, such as canna and balsam, and best of all the hitherto unknown potato.

Next to the flower-garden was the tree-plantation, and then the orchard; and there grew flowering shrubs, such as laburnum, snowball, and Turkish elder. In the shade of covered walks there were many kinds of amusements. In the last section was the labyrinth, its winding ways overgrown with espaliers and all kinds of climbing plants, and farther on the rose-garden with its nine sorts of roses brought from the East, and various vineyards. The middle of the sections is adorned with fountains, and one is overshadowed by a tree of life, the largest, finest, and also the oldest that Silesia can boast of. The Arbor Vitæ was brought by Francis I. from Canada to Paris, and thence spread very rapidly over Europe. On the west the garden was bounded by a winter-house for bays, pomegranates, oleanders and myrtles, and its walls were presumably painted with Italian scenes. There were two aviaries, and a decorated ice chamber, and in a grotto among other pieces there was a Polyphemus hurling his rock, who was much admired. In the middle of the garden was a summer-house which was open on four sides, and contained pictures, works of art, and musical instruments. It was reserved for merry parties.

This famous travelled physician, who also owned a room full of "excellent rarities" of art, held festivals inspired by the true antique spirit. His friends, men and women, were assembled here to a cheerful feast, for song, for recitations, or for conversation, and here they crowned themselves and their goblets, just as people had done on the shores of the Mediterranean. And the people of Silesia, always fond of poetry, were grateful to their fellow-townsman for the joys of garden and feast: Scholz collected no fewer than seventy poems praising his work. What a ray of sunshine we have here in a country whose flowers are so soon to be crushed under the iron heel of the Thirty Years' War!

After Breslau, Nuremberg and Frankfort could boast of fine gardens at an early date. Eoban Hesse sang the praises of Nuremberg gardens as early as 1532 in a Latin poem, and Hans Sachs speaks of them in his smooth words. Towards the end of the century the garden of Camerarius, the doctor and botanist, won widespread fame. A picture by Sandrart shows the garden of a gentleman of Nuremberg, the wealthy Christopher Peller, in the state it was in about the middle of the seventeenth century, when it was still almost unaltered. The main fabric stands round a court, which is enclosed with a very fine balustrade ending in two gates flanked with obelisks. In the court people amuse themselves with all manner of ball-games and ninepins. The garden, of which only half appears in the drawing (Fig. 357), is in four rows of beds edged with stone, each bed made in the old way to contain only one kind of plant. Round the beds, which are joined together in groups of three, there are lower stone borders with ornamental pots set on them: these contain plants of many kinds, with orange-trees and other costly foreign plants that have to pass the winter in a hothouse. To right and left one sees into the tree-gardens, which are separated off by pretty wooden palisades.

Finer than this, which is really a simple garden, is one of the same date belonging to Johannes Schwindt (Fig. 358), a burgomaster of Frankfort who loved display. In this case the enclosure is made of green lattice-work with pillars, windows, and gates. In the windows are pots of flowers; the pillars have little obelisks on them, and busts. One walks into a fine parterre with geometrical patterns marked out in box, and little trees at the corners; round the encompassing hedges there are again benches with flower-pots.

The wide middle path leads into the flower-garden behind, introduced by two huge statues, a Hercules and a Hermes, and flanked by two obelisks at the end, where there is another parterre. Round the second and third sections alleys covered with green lattice and foliage follow the line, with entrance gates and windows. At the sides there are fountains and statues, and the eye passes over the scene into other gardens.

A very charming picture is shown of the garden at Ulm made by the architect and private gentleman Joseph Furttenbach at the side of his pretty house (Fig. 359) after his return from Italy. The garden is certainly small, "but so arranged that an ordinary

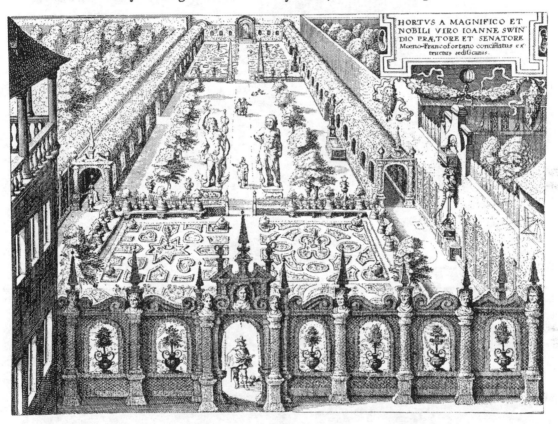

FIG. 358. THE GARDEN OF JOHANNES SCHWINDT, FRANKFORT

private person can get all the pleasures he desires." Because of its situation, which looking to the south enjoys "the blessed sun," it was able to produce an abundance of flowers. The order of the beds as described in 1638 recalls, with their simple walled divisions, the same idea as the garden of Erasmus at the beginning of the sixteenth century, and also is like the garden of Scholz at Breslau. But the flowering bulbs give a greater diversity of colour. The little garden is enclosed by arboured walks on the side of the house, and between these stands the chief pride of the architect educated in Italy—the grotto, a small erection in *rustica* with pretty ingenious devices. Also behind this grotto, quite cutting off the garden from the house, there is a summer-house, thought of as a dining-room, "where the master, if ever he is tired and weary from his daily work, can enjoy his slice of bread with his companions *in bona caritate*, and has a good opportunity of there thanking God for it."

FIG. 359. JOSEPH FURTTENBACH'S GARDEN AT ULM

All the gardens that we have observed so far are burgher gardens really, however well they are carried out. But for the gardens of a prince something special was desired at that period. In 1560 Gesner divides up the different kinds, by no means scientifically, but with a view to doing justice to all, in this fashion: "1. Ordinary gardens for the household, with vegetables, vines, orchard and grass for the nourishment of man and beast. 2. Medicinal gardens, containing in addition to these things various healing plants, foreign and native. 3. Miscellaneous gardens, with not only healing herbs but also

FIG. 360. CASTLE GARDEN AND LABYRINTH

peculiar plants that attract attention and admiration. 4. Elegant gardens only meant for ornament, with arbours, pleasure-houses, and places to stroll about in, with fine ever-green trees, and all the various designs that can be made by curving and weaving the branches. Such are the gardens of wealthy ladies and all well-to-do people, especially monks. 5. Show gardens, such as learned men and princes or the state itself may possess, with splendid buildings, ponds, and water-works, artificial mounds, squares for tourna-ments or tennis." A picture from the Werth collection (Fig. 360) gives an illustration of these requirements for the garden of a prince. Close to the castle is the spacious play-ground, and behind is the labyrinth with arcades round it and a magnificent fountain, also an animal park with a bath and other features.

Among the princely houses Hapsburg was early addicted to the culture of gardens, and Maximilian I. himself wrote works on the subject. The relation of Charles V. with Spain we have already studied. He made use of his huge domain, as did his brother Ferdinand, to introduce rare plants into Vienna. At that time this city, the court capital since the eleventh century, was the flourishing centre of culture, before the first invasion of the Turks. In the fifteenth century we find Bonfini praising its marvellous situation, lying like a palace in the midst of suburbs, several of which are comparable with itself for beauty and size, so that the whole of Vienna might be one enormous garden, adorned with vineyards and orchards. In the picture are pleasant cheerful hills, with charming country houses and fish-ponds. But this beauty of the olden time was almost all destroyed in the first cruel year (1529) of the Turkish invasion.

It is not Vienna, however, that gives us the earliest gardens due to a Hapsburg prince, but Tyrol, at the Castle of Ambras near Innsbruck. The Archduke Ferdinand, second son of the Emperor Ferdinand I., gave this castle in 1564 to his wife Philippina Welser. By rebuilding and laying out the garden skilfully he had made an exceedingly beautiful princely seat out of a mediaeval fortress. There lived the fair woman, who had been exalted to the rank of a fully recognised wife by her knightly and (in the truest sense) princely husband. This was in spite of the disfavour of the emperor his father, and in the face of many difficulties. Here she lived a life of romantic love that became almost legendary, in surroundings that were truly worthy of her. In the year 1574, when the castle and its inhabitants were at the very summit of their happiness and glory, they received a visit from the learned Jesuit and jurist Stephanus Pighius, who was conducting the young prince Carl von Jülichberg, nephew of the archduke, on his travels through Tyrol to Italy. He describes how they rode out to the summer dwelling, the *villa suburbana* of the archduke.

The castle was built on the hill, "in magnificence excelling the finest villas of the ancients." In the women's part the visitor first saw hanging gardens and wired aviaries, but it is not clear whether by this is meant real roof-gardens or high terraces, for the gardens proper are at the foot of the hill. Below they saw "paradises"—probably garden parterres with pillared corridors—labyrinths, grottoes dedicated to various nymphs, and grand fountains; the numerous springs received their supply of water from the natural brooks. Arbours, open and covered, where meals are taken, are decked with the finest topiary, especially one round arbour with a table for guests to sit at in the middle; this table suddenly begins to move, spinning round quicker and quicker till they are quite giddy. In the eyes of the visitors the *chef-d'œuvre* is an underground wine cellar, where they are led through several grottoes into the sanctuary of Bacchus. Round these well-kept gardens lies the park, well supplied with thickets, fish-ponds, cages for animals, and enclosures for game. Near the house are large playgrounds for knightly sport, racecourses, stadiums, and a tennis-house, where the archduke played with his guests.

At the beginning of the seventeenth century the elder Merian produced an etching (Fig. 361) which exhibits the garden of the castle, though its size is somewhat cut down, and of the parts on high ground near the women's quarters only one corner can be seen. But at the foot of the hill lies the large square intended for pomps and games, surrounded by buildings among which on the left front the Chamber of Art is especially interesting with its famous collection. Near the Chamber of Art is the granary, and on the other side the library. Above is the part where the master's sons live, with a flower-garden next to it; on one side this is bordered by an airy walk, and the lower building is very likely the armoury. Perhaps this little garden was one of the "paradises" seen by the strangers. Close by the castle and in the wall is built the great festival hall, whose windows look out on the ornamental garden below. The second side of this square parterre encloses the covered tennis-court, and there is a low wall protecting the third side, while the fourth appears to be open towards the park, but shut off by a row of tall trees, which follow the line of the house.

The parterre is divided into nine geometrical squares, which are apparently encircled by low fencing. In the centre there is a small round open pavilion on pillars, probably to take meals in, while in the park, which goes all about this side of the castle hill, there

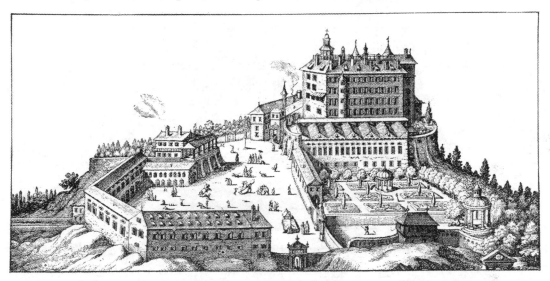

FIG. 361. CASTLE AMBRAS, TYROL

stands, raised high up on the wall, the great round pavilion with the revolving table. A long narrow piece between walls and near the parterre may perhaps be the stadium, and beside it is the cellar with its underground delights. This place corresponds to what Gesner demanded for a princely garden. It makes us think of early French gardens, such as Gaillon or Blois. In the copious supply of playgrounds and their buildings, and in the art collection, we find the personality of this prince, whose knightly nature and love for art were renowned far and wide. At his beautiful seat were held the wonderful Shrove-Tuesday feast, the shows, and the tournaments of the year 1580, when Ferdinand was able to take his delight in splendour and happiness and in his wife's love. But a few weeks later the mistress of the castle died quite suddenly, and Philippina Welser took with her to the grave the finest flowering of the princely home.

Only a short time after Ferdinand had made Ambras the glory of his Tyrolese land, his elder brother Maximilian II. created in the neighbourhood of Vienna a castle of pleasure that not only caused admiration in his contemporaries and surprised wonder in succeeding generations, but has offered many a riddle to ourselves. On the south-east of the town there is still, on a height, a square place surrounded by battlemented walls. These are topped by towers at regular intervals, but in the front, facing the Danube, where the main building is, the towers are wanting. This gives a warlike appearance, so that to-day it seems only natural that there should be a powder magazine there guarded by a row of sentinels in front of their sentry-boxes. But as a fact within what looks like a girdle of fortresses a "Tusculum" had been erected by one of the most peace-loving of princes. The period of Maximilian's reign brought to the noble-minded ruler many conflicts and sorrows, which saddened his life and left him unsettled; through all his troubles there remained the longing for peace, which he found best in his architectural and gardening interests, as he repeatedly writes to his friend Veit von Dornburg, the Venetian ambassador.

About the year 1569, when we first hear of the emperor's villa, the works at the

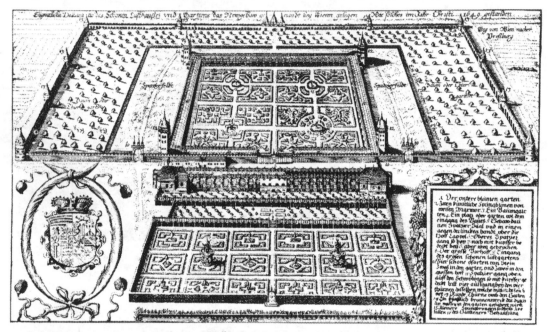

FIG. 362. NEUGEBÄUDE, VIENNA—PLAN OF THE GARDENS

"Pheasantry," as it was then called, were in full swing. The name probably came from an earlier pheasantry in the same place, but years later the dull name of "New Building" was used instead, which points to the fact that Maximilian's successor, Rudolph II., completed the main building. What Maximilian had done himself became, two generations later, a very unusual sort of place, which rightly appeared to a good Viennese to have no connection with anything outside; and from the middle of the seventeenth century there was a common legend that could not be got rid of—that the emperor had converted an old camp of Sultan Soliman into a country house. When it was finished the New Building must certainly have presented a far more warlike appearance than it does to-day, for inside the tower-crowned circle of walls there was a second barrier (Fig. 362). There was a square exactly the width of the villa, which was enclosed by a wide wall with arcades: at its four corners it was topped by high two-storied turrets, whose dome-shaped roofs cased in copper commanded a distant view over the outer walls, while on the arcade roof itself there was a fine promenade. This square included the upper flower-garden, divided into sixteen beds in a geometrical pattern, two of which showed the Austrian arms, the double eagle, and were provided with many fountains. The arcades and the pavilions in the corner turrets were finely painted.

From the upper garden one passed next into a narrow court dividing the garden from the villa. This was enclosed on the valley side by a loggia in the real Italian style. Hence one could see over the lower gardens, which fell away towards the Danube in four terraces, first two narrow ones planted with trees, next a flower-garden in size and situation almost the same as the upper ones, and on the lowest of all a large pond—so at any rate it appears to be in the engravings. Bongarsius, a Swiss visitor, who saw the gardens when they were completed in 1585, speaks of two ponds, but this may not be quite accurate, for he makes no mention of the beautiful fountains adorned with statues which (according to the account of expenses) were set up in the lower gardens. Round the flower-garden of the highest terrace stretched the park within the great battlemented and tower-crowned walls: it was approached through two immense gates on either side. Bongarsius thus describes it: "Round the flower-garden there is a park of fruit-trees planted well, and a fine labyrinth; in the middle of the whole park there is a trench three or four feet wide, marked out by stones, and getting its water from a hill a mile and a half away."

To what class does this place belong as a whole, so separate as it is from any German natural development? The builder of it is unknown, and the court of Hapsburg at that time was quite cosmopolitan. Maximilian himself, great as was his personal inclination towards German Protestantism, employed only foreigners for his works of art, and Italian sculptors and painters were working on the New Buildings. Also for the fortified parts, which Austria at that time constantly had to keep up with a view to the danger threatening from Turkey, the emperor always employed Italian architects. The house at New Building, with its Italian features, leaves no doubt as to the builder's nationality, and so it seems quite possible that one of the architects for fortifications also designed the whole of the garden. We have Italian examples in Vignola's work at Caprarola, and San Michele's charming villa San Vigilio on Lake Garda; and in common with this last we have the swallow-tails on the outside walls; and if San Michele used a fortified tongue of land as a garden-motive, why should not a Venetian architect adopt the tent of an encampment as the ground-plan for a villa?

It remains, however, a strange thing, that the artist should have made anything very anti-Italian in the terrace arrangement of these gardens. The strict axial line has something bald and matter-of-fact in its character that is always avoided in Italian gardens, whether it is by breaking off the set plan, or by introducing important crossways, or by dividing the water-works, or by getting some picturesque grouping of the terraces, more or less as at San Vigilio. On the other hand, to find in Italy an axial order carried out with steps or through an exaggerated emphasis laid on the middle line, we need only think of Villa d'Este. In this place there is nothing of that kind, and however severe the axial line of the gardens in relation to one another, they are each individually treated and separated off in the true northern style. The castellated wall cuts off the upper garden with its arcades from court and house alike. The lower terraces have as a rule no intercommunication, and one would be unable to get into the separate gardens if there were not now a friendly corridor leading as a covered way to the lower ones. The water is only applied as shell fountains, canals, and ponds—ideas and arrangements that we have long been familiar with in France. The upper garden with its arcades may be more or less likened to Bury, and the corner turrets in both places are turned into pavilions by being painted. The same treatment of arcades is found in nearly all the French gardens, and still more do we feel the resemblance in the wide canals ending in tanks.

The relations between the Hapsburg court and the French were much strained at this period, but later King Henry III. was received at Vienna with great pomp when he was on his journey to take the throne of Poland. Artists of all countries were at that time

FIG. 363. BELVEDERE, HRADSCHIN, PRAGUE

a wandering folk, and at the Viennese court there was always a great deal doing. We do not hear much about actual plantation, but in 1537 Maximilian succeeded in getting the famous botanist Clusius as Inspector of Gardens, and his presence vouches for the cultivation of unusual trees and plants in these gardens. The happy time for the villa was of short duration, for by the end of the century lamentations began about its decay, and in the middle of the seventeenth century buildings and gardens both had lost not only their beauty but also their festive character. Instead of merry guests, wild beasts were entertained there; buildings and gardens became a mere menagerie. At this time was started the fairy-tale about the Turkish camp, and when in 1683 the dreaded Turks really invaded, the legend was so far dressed up that the foreign soldiers were reported to have wept at the sight of the ancient camp of Sultan Soliman. For us, however, this New Building remains as a unique document of German garden art, just because it stands alone and unconnected. The stream of art presses forward from outside, and is greedily absorbed, but without steadfastness, with no national feeling, which ought properly to accept what is foreign merely as an incentive to its own fruitful development.

Rudolph the Second's increasing estrangement from Vienna, his slackening interest in his possessions there, had brought about the downfall of New Building. More and more did the emperor confine himself to the circle he had himself drawn, till he never went outside his own residential town, Prague. But it was only in his last years that he withdrew, melancholy and hating mankind, from all the outside world. Before that time his interest was centred in art. His Chamber of Art at the Hradschin had a world-wide reputation because of its fabulous treasures. He also felt a wonderful affection for his garden. At the Hradschin he found a particularly fine castle, called the Belvedere (Fig. 363), which was built by a pupil of Sansovino, and was unquestionably the greatest building of the Italian Renaissance on this side of the Alps. From the pillars round the rooms one saw a long garden which had been laid out by an Italian gardener, just like the pleasure place of Ferdinand I. and his son who founded Ambras. Unfortunately, there is no picture or description of it in its early state; we only hear of whimsical festivities, such as the theatre performance at night on the entrance of Ferdinand in 1558, which he saw from the corridor of the Belvedere. It was one of the masques, when the fireworks

FIG. 364. BELVEDERE, HRADSCHIN, PRAGUE--THE SINGING WELL

and automatic arts of this period won astonishing triumphs. But Rudolph soon turned his whole heart to the garden, which was separated from the castle by rather deep trenches where deer were kept. There were wooden bridges leading to the main castle, but the trenches themselves were not filled with water in Rudolph's time, they were changed into a deer-park. It was a wonderful pleasure to watch the wild creatures from the high covered bridges, "especially in the breeding season, when they rushed about over the hill and wall."

The garden itself is almost level, and is orientated with the castle; the main axis is indicated by fountains, one of which, with two shells and ornamental figures, is still preserved before the garden front of the Belvedere (Fig. 364). It bears the name of "the Singing Well," from a bagpiper on the top, who must have blown the water out of his pipe with a singing sound. It looks as though an arboured path from this spot marked out the middle. The grottoes were especially famous, and the emperor in his later, sadder days liked to be there, and took pleasure in their wonderful mirrors and invisible music. But visitors liked the wild animals best, and at the other end of the garden there were lions, camels, and other beasts in wooden cages. Later on, towards the middle of the seventeenth century, the noise seemed inconveniently near to the castle, and the animals were moved farther off; for apparently New Building had been turned into a menagerie.

FIG. 365. A FLEMISH STUDY OF SUMMER WORK IN THE GARDEN

Rudolph had plants brought from Italy, Spain, and Asia; pomegranates, oranges, citrons, and lemons were imported, according to an account of the year 1632. And the honour of having been the first to grow tulips in Europe (which so many gardeners tried to do) was accorded to Rudolph by the travellers' books. Rudolph also adorned his garden with beautiful statues, and by the castle there were set up two tennis-courts, which, like the greater part of the garden, were destroyed in the sieges during the Seven Years' War. In this garden the shy prince passed the last sick and sad years of his life; and strangers who wanted to see the emperor came disguised as gardeners or stablemen to satisfy their desire.

Among the painters who experienced the kindness of this lover of art the Dutch stood first. Because of their alliance with Spain, who ruled the world, the provinces of Flanders at that time provided a favourable soil for the development of the art which from the very beginning had made a great effort in Holland to surpass its natural limitations. With an insatiable hunger, Flemish artists seized upon the material of the whole world. Dutch garden art lives for us still in an abundant supply of pictures, both paintings and woodcuts, and above all in those copper engravings that the Flemish people loved so dearly. It may be that with this mass of material individuality was sometimes wiped out; for in the immense number of pictures of the months and seasons we find spring and summer, April, May, and generally some autumn months as well, depicted as garden scenes; and these in the faithful reproductions of this most realistic kind of art give many significant details of the style of that day; yet these gardens are not to be taken as individual portraiture, for the same scene is often found transferred from one picture to another. For instance, a painting by Breughel in the Museum at Lille (Fig. 365) and another by Abel Grimmer at Antwerp (Fig. 366) show exactly the same garden with scarcely any change in the accessories. Again, an Italian goldsmith, finding on two drawings by Hans Boll certain garden scenes, copies them on an embossed plaque, adding details of ornament in the Italian style (Fig. 367). It is still harder to discover what is truly Flemish in the pictures produced by those Dutch artists who went to earn their living at foreign courts. There is no doubt that the beautiful garden landscape by

FIG. 366. A FLEMISH STUDY OF SPRING WORK IN THE GARDEN

Valckenborch (Fig. 368) in the museum at Vienna really depicts the seat of an Austrian prince, with labyrinth, pond, and different gardens.

The architectural painter Vredemann de Vries lived and died at the court of Rudolph II. His numerous garden sketches, which were published in 1568 and 1583 in a series of engravings called *Hortorum Viridariorumque Formæ*, have more of the town character; and although he chiefly took his examples from the gardens of his own home, they are best ranked as specimens of the German style. De Vries was a zealous student of Vitruvius, and thought that his gardens could have no higher recommendation than that they were divided as Vitruvius demanded into Doric, Ionic, and Corinthian. But it is not the fact that there is any real difference in style. The gardens are very much alike, even in so far as they try in individual parts to obtain a great variety. At any one site there are always different gardens divided by hedges or barriers, and seldom with any axial arrangement. For the most part they have a tree in the middle, now and then a pavilion (Fig. 369) with a fountain, and occasionally are varied with a sunk basin. Round the grander gardens run pergolas with doors, windows, and domes (Fig. 370); the arbours are often supported on pretty pillars with Herms, but there are very few real statues. The beds are always laid out geometrically, and often have small trees at the corners or in the middle, and borders of stone or box. Frequently the whole garden is enclosed with galleries. This sort of parterre is not as a rule next to the house, which usually has a lawn by it, meant

FIG. 367. THE GARDEN IN APRIL; FROM AN EMBOSSED PLAQUE

for a playground. There is not much water, and this is generally treated as a fountain in the middle, never as a canal unless that is just a large basin—the sort so commonly found in French gardens of the same date.

If we compare the drawings of de Vries with gardens that were really carried out at this time, such as the Kielmann gardens at Vienna (Fig. 371), the relationship is easy to see. This important place, in spite of its coherence in general plan, is divided up into a whole series of separate gardens, each dominated by a summer-house with which

FIG. 368. VIEW OF A GARDEN IN FRONT OF AN AUSTRIAN TOWN

FIG. 369. A DUTCH STUDY OF A GERMAN GARDEN WITH ARBOURS

are ranged twin fountains as chief ornament of the geometrical beds. The bordering
of arbour walks and the entrance gates recall the Schwindt garden, to which this one is
superior in size and magnificence, but not in the feeling for style.

To this period belongs the garden full of a proud and artistic charm that was
laid out by Rubens at his fine house at Antwerp (Fig. 372). A triple "gate of triumph"
leads out of a great court beside the house, and gives a view of the garden, which has
a pretty summer-house at the end of it, set up in that peculiar mixture of Flemish and
Italian styles which is characteristic of the whole house. If we pass into the garden, the
large parterre decorated with vases is at our side in front of the chief room of the house.
On this follows a curiously divided plot of garden, where a very fine pavilion of lattice-
work between pergolas divides the place either from the neighbouring garden or
from a final tree-garden of its own. Although the laying-out of the parterre as shown in
the engraving of 1692 points to a later time, the actual site, the complete absence of water,
the gaiety of it all, the garden houses, and also the simplicity of the carved work,
indicate that spirit which pervaded wealthy Antwerp at the turn of the century, when
its greatest artist could make himself this lovely home, he being also one of its most
important citizens.

The flat country of Holland hardly comes under consideration at this time in respect
of garden formation. Dutch families were still living outside the towns, mostly in
their strong water-castles. In the last decades of the century Dutch trade first made its

marvellously rapid start, and from the beginning concentrated on flowers and the importation of foreign plants to Holland. It is obvious how great a difference the bulbs, and especially tulips, must make in the appearance of European gardens; but in spite of this the development of French gardens and parterres was to become more influential, and to grow less and less like the gay variegated show of Holland. It came about that the tulip trade when finally established diverted men's minds from the true art of gardening, and became only an affair of bargaining, for which the innocent bulbs served as material. All the same, the interest in botany bore timely fruit in Holland, and by 1557 the Botanical Garden at Leyden was founded, which for some time was at the head of all the scientific gardens. But the real importance of the Dutch garden begins at a later period, when in the seventeenth century this country of rich tradesmen set up flourishing villas on the outskirts of the growing towns, and along the course of the canals.

In Germany garden interest reached its highest point at the turn of the century, before the outbreak of the Thirty Years' War. In Vienna a garden was shown on Shrove Tuesday, 1613, by the barons Georg Wilhelm Jörger and Wolf Tonradtel, "decked with lovely trees, citrons and others, and with music playing." French and Italian books on the subject now began to be translated into German. In 1597 appeared the first German garden book, by Johann Peschel, which (in its long title) proposes to give instruction in all garden matters. For the lay-out he demands before all things that it shall be thought out and drawn on paper first and properly squared, then the beds are to be put

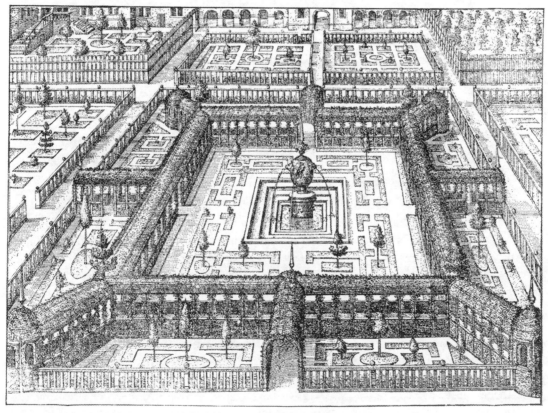

FIG. 370. A DUTCH STUDY OF A GARDEN WITH PERGOLAS AND DOME

in. But there must also be covered walks, and these he calls *Stackete*, or *Gelender*, or *Khemerer*. Only after all has been thought out, set on paper, and strictly measured, can the plan be transferred to the land. This demand shows the German spirit in the tendency to theorise. Peschel's book passed through many editions, and its first rival in success was the garden book of Doctor Peter Laurenberg of Rostock: this work appeared in a German edition in 1671. But the numerous copperplates of parterres taken from French examples were influential also.

About this time there was for the most part no lack of examples of any kind of art, and in particular the copper engravers followed the lead of the Dutch by using their

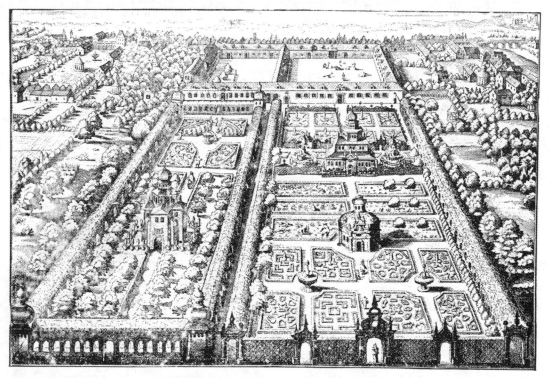

FIG. 371. THE KIELMANN GARDENS AT VIENNA

skill in the service of the garden. Right in the front stood Joseph Furttenbach of Ulm, who lived ten years in Italy and pursued his architectural studies there. He resided for the most part in Northern Italy, but had travelled about the other districts, and especially had studied eagerly at Caprarola. His plans for gardens show a strong inclination for fortress surroundings. When he went home he published a book of travels which was much read at the time, and then published his studies and sketches from 1628 onward. But although the small garden at his own house did not go beyond the ordinary town garden of the German Renaissance period, his plans clearly show the signs of a foreign influence. There are architectural plans which bear the stamp of a school in their severe regularity and axial arrangement, while the gardens are often surrounded with moats treated in the style of a fortress, and with little cannons, in the fashion of the time, stuck up on the projections.

This man, so industrious and so full of imagination, was also much concerned for the

FIG. 372. RUBENS' HOUSE AND GARDEN AT ANTWERP

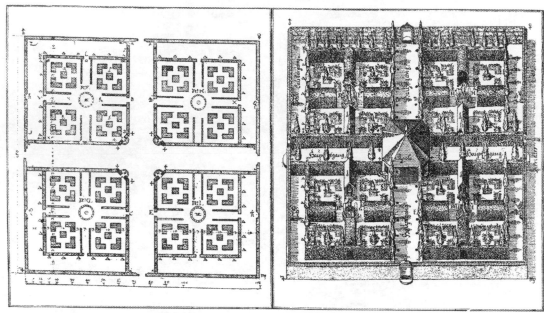

FIG. 373. A LITTLE GARDEN OF PARADISE—A DESIGN FOR A SCHOOL GARDEN

welfare of his native town, and thought out a careful social scheme which was very far in advance of his time, and has only nowadays pressed into the circle of modern interests. For not only did he plan school buildings, where seats and tables were carefully designed for the children's health, but he wished to have a school garden in front, which he named the "little garden of Paradise" (Fig. 373), "so as to awaken good thoughts in the children, of walking in Paradise, and so practising them in the Christian religion and other good, useful and honourable studies." Thither the teachers were to take their scholars, and there a public examination was to be held. As a room for trials, he designed a large domed place in the middle of the garden, with four chairs in it, where there should be children, boys and girls together, holding their little disputations, and on the walls there were to be hung up the things they had made. Four doors led into the quarters of a large garden square that was traversed by wide walks, and these divisions were cut up by little arboured paths into four flower-gardens, where the young examinees were allowed to gather the flowers in the beds as a reward; in the middle was a large fountain. In the first section there was a model of Adam and Eve, with the mother of mankind plucking an apple from a real tree of Paradise and offering it to her spouse. Below this group the children could read the words, cut in stone:

> Through Adam's fall, on garden ground,
> Mankind, alas, his ruin found.

But consolation was at hand in the garden section on the right, where on a "very charming mound in the middle there was a figure carved in stone of Our Lord and Only Saviour Jesus Christ rising out of his grave in the garden," with the following inscription below:

> In garden ground, where Christ lay dead,
> Mankind is now deliverèd.

When there was to be no examination, the children were to run about in the garden, and enjoy its beauty; they might pick flowers and fruit, and each child was to receive a cake baked for himself. It does not appear whether the people of Ulm ever carried out a scheme so kindly and so good for children, or whether this too was only an architect's dream.

At the same time as Furttenbach, the elder Merian began his useful work, though perhaps he laid less stress on the garden side of it. He gives a great number of illustrations of German castles and the gardens attached to them, and these have the advantage of having been actually carried out. But the pictures of foreign gardens, Italian and French as well as German, must have had a great influence on his contemporaries.

Among those that were immediately affected by Italy, Hellbrunn, at Salzburg, stands supreme. In the years 1613–19 Marcus Sittich was bishop of Salzburg. He was a member of the Hohenem family, who because of their origin on the borders of Latin territory had a very close connection with Italy. Since the time of Pius IV., the family had continued to live in a similar fashion to "nephews" in Italy (where we have often found them as predecessors of the Borghesi) with the same delight in buildings and gardens as they showed in Germany. Marcus Sittich first completed the Mirabellschloss built by his predecessors (Fig. 374). The ornamentation of the garden, the design of the parterre, belong to the French period, but the whole lay-out has preserved its Renaissance character. It is not only that any connection with the castle is entirely absent, and the chief garden separated from the sides by high walls that are only broken by narrow doors, but the separate beds are encircled by balustrades which interfere with the *ensemble* of the design. This whole scheme differs little from the town gardens we know so well at Augsburg, Nuremberg, and Vienna.

The most original creation of Marcus Sittich is the charming little castle of Hell-

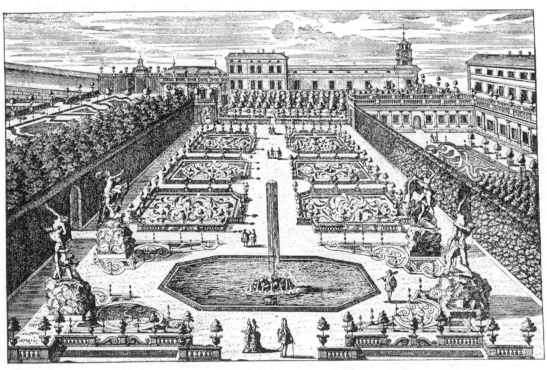

FIG. 374. MIRABELL CASTLE, SALZBURG—THE GARDENS

brunn before the gates of Salzburg (Fig. 375). He started it as soon as his rule began, and finished the whole thing in fifteen months. Although the garden in some parts was changed into an English park, and in others was enlarged under French influence, it still preserves in a great many respects the character that Sittich gave it. In the castle itself grottoes were introduced on the ground floor, not only on the entrance side (1) but also on the garden side. By the garden below the terraces there is a great grotto construction; here we find the abundance of statues, water-plays, and automata, which we recognise as like those in Italian gardens, and which are still mostly intact. There is the

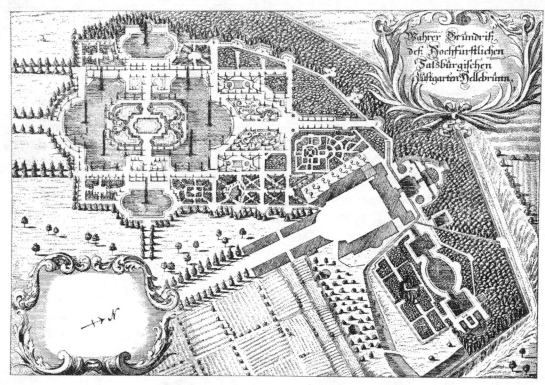

FIG. 375. HELLBRUNN, SALZBURG—GENERAL PLAN

rain grotto, the mirror grotto, and another with a dragon coming out of a hole in a rock, who drinks from a fountain and vanishes again. There are all sorts of birds singing, and actually the beloved "vault of ruins" is not wanting, with its stones threatening to fall.

In front of this garden façade extends the large, deeply-sunken star-pond (2), into which falls a cascade of three tiers, ending above in a semicircular theatre. The ibex, the Hohenem crest, is to be seen everywhere, and is a sure guide to the estates of Marcus Sittich. A long narrow canal passes out from this pond on both sides, beset with an inexhaustible supply of grottoes and little water-plays. Certain regions at the end of the canal hail from a later day, as for example the mechanical theatre, a costly toy with marionettes doing all sorts of things. The passion for grottoes is at its highest in this garden. An engraving by Merian shows a great number of small grotto-houses, which enliven the garden, sometimes open at the top and sometimes shut.

FIG. 376. HELLBRUNN, SALZBURG—THE WATER ON THE EAST SIDE

On the other side of the walk of grottoes there stood near the house a small summer pavilion, which was orientated as centre piece with four corner turrets and a tank in the shape of a trefoil. People were beginning to imitate Italian casinos by setting up unusual summer-houses in their German gardens. We find them still more varied in the seats of princes of Middle Germany, and it looks as though a peculiar style of central erection came about with this feature much emphasised. The chief parterre at Hellbrunn was a water parterre: there were four basins on small lawns, the middle one having a summer-house on a round hill which was approached by thirty steps. On the eastern façade of the house there was another water site (3), which, unlike the chief garden, is still for the most part preserved (Fig. 376). It has three basins, one behind the other, the middle one oval, and the two others square; these are connected by narrow canals, and with each is a fountain group with a stone table and ten stone seats. In the middle water poured forth to fill the drinking-glasses; but woe betide those who sat down on the seats. They were driven off by water spurting out, and if they tried to escape back to the steps that led to the semicircular theatre at the grotto, or to the pretty balustraded galleries on the side, they were met by a fresh shower of rain from many little pipes. From pedestals Democritus and Heracleitus—allegorical figures of tragedy and comedy —looked down upon this merry play; while above was Rome, placed in the broken gable of the semicircle, which was adorned with blue stones and shells, above the arms of Marcus Sittich. On the right near this part there was similarly a small grotto-house, which contained Orpheus and a sleeping Eurydice and all sorts of animals; near by was a charming little menagerie.

The archbishop made something else that was very remarkable at that time, in his park up on the Waldberg, and named it Hohenems. He built a little casino, called the Castle of the Month, because it was said that, to gratify the wish of a Bavarian prince who was passing through, he surprised him on his way back, after one month, with the place quite finished. Nothing is preserved of the garden, but at the back of the small castle there is a very interesting wide road. From an opening we walk down into an antique theatre cut in the living rock, with seats all round and entrances and exits. Perhaps a stage was also set up for the occasion. Marcus Sittich had pastoral plays and operas performed, as for example on 31 October, 1617, before the retinue of a prince who was on the return journey after a hunting expedition to Berchstoldsgaden.

This theatre is important, not only because of its attempt to revive Palladio's great effort to have an antique theatre at Vicenza, but also because of its position in a lonely park, set in the cleft of a rock, all of which creates a feeling that we often find in the gardens of the later eighteenth century. Also it is the first permanent theatre in the open air about which we have any information; for the use of natural stages, with hedges as side scenes, certainly came in quite a hundred years later. We have already noticed, in France and Spain, how closely related in Renaissance days were the feelings for pleasure, games, and piety; and Marcus Sittich was wont to lead his guests from the

FIG. 377. THE "PRETTY GARDEN," MUNICH

theatre in the rock—through the deer-park, where white stags were kept—for a few minutes into his hermitage. He entertained them in the little castle of Belvedere, in the room with the pictures, from whose windows they looked over the Salzach towards Hallein, and led them to the eight hermits' cells which he had put there with six small chapels. In one of these lived a French brother, called Antonius the Fifth, whose tombstone at the parish church of Anif hard by thus told of his life:

> Nicholas Mudet was my name,
> From Lyon city first I came.
> With fear of God, by lonely ways,
> At Hellenbrunn I spent my days.
> Often at Rome the time I passed,
> But here I found my grave at last.

The garden owed this abundance of ideas, and its fanciful arrangement, to the owner's connection with Italy.

In the rest of Germany also the intent eagerness of the many smaller princes was aroused, and everyone jealously observed the progress of his neighbour. On the other hand, there was a good feeling that prompted them to share all new discoveries. The book of travel by a gentleman of Augsburg called Philip Hainhofer, written between 1611 and 1613, gives us an amusing view of this exchange, which was of course concerned first and foremost with special botanical wonders. Hainhofer was an art-dealer at court, perhaps one of the earliest of his kind, a man of great learning, fine taste, and intelligent eye, all of which helped him to that knowledge of men which he needed: he was vain

FIG. 378. THE RESIDENCE GARDEN, MUNICH

enough and snobbish enough to let princes feel that he enjoyed their condescension and confidence; and clever and unconscientious enough to help himself with time-serving, if it was to his own advantage. He became so indispensable to princes that they also employed him on their lesser diplomatic errands, but he never abandoned his chief aim of being an art agent. Into chambers of art and houses of pleasure he was constantly seeking an entrance, and also into their gardens, so that we owe many happy descriptions to him.

He seems to have cemented his earliest alliance with the Pomeranian Duke Philip II., who desired to build a pleasure-castle, in 1611; for this was acquired by means of Hainhofer's duplicates of plants, and his drawings and sketches of other German castles. The old Duke William of Bavaria, who devoted most of his leisure to art after he abdicated in favour of his son Maximilian, wanted to show his gratitude to the Pomeranian lord, and commissioned Hainhofer, during a visit to Augsburg, to make a journey to Eichstätt,

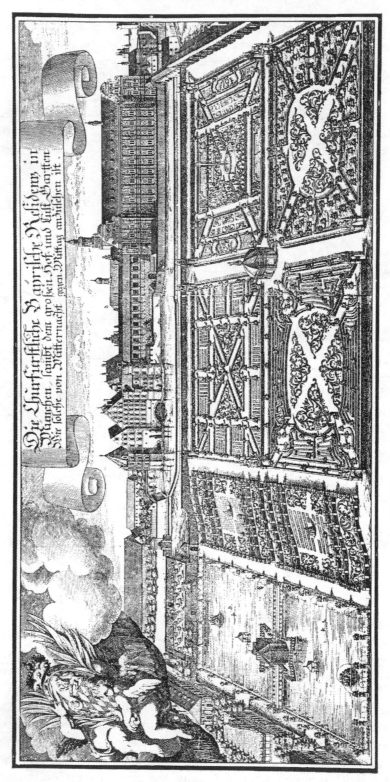

FIG. 379. COURT GARDEN AT THE RESIDENCE, MUNICH

whither all botanical interest was directed. The learned, high-minded, though frail and delicate Johann Conrad von Gemmingen, a prince of the Church, had built a fine castle, Willibaldsburg, close to the bishop's see, Eichstätt, and had laid out the gardens afresh. Hainhofer enumerates eight different ones, "which are variously adorned in area, in divisions, in order of flowers—and wonderful with their roses, lilies, and other plants." They are all, as at Ambras, at the foot of the hill, with the castle at the top, which according to his account has a moat round it. The bishop was at that time very busy with the rebuilding of the castle, and wanted to "make the garden turn the other way and (going down the hill) unite together castle and mountain." Whether the plan was carried out is doubtful, for the bishop died in 1612, and in 1634 the Swedes razed the castle to the ground. The prelate, however, had made a monumental work for his contemporaries and one of lasting renown for a later world. He had all his plants drawn and then engraved in copper. Week by week a messenger on horseback with a box of fresh flowers was sent to Nuremberg, where the chemist Basilius Beseler made drawings of them and arranged them according to their time of flowering. This work was valued at 3000 florins, and first came out after the bishop's death in 1613 under the title of *Hortus Eychstedtensis*. All the princes added flower pictures of the same kind, though not in such a costly get-up as this, and made exchanges by way of polite greetings.

Soon after Hainhofer's journey to Eichstätt had been accomplished to the satisfaction of his patron, he went to Munich to report to Duke William. Munich was already, under the rule of the duke's father, Albert V., embarked upon its first great period of building; and he had at his residence, on the east, the other side of the city trenches, enlarged the pleasure-garden and laid it out in the Italian style. In this garden, called "Rosengart," at the fête held in honour of Charles V., the emperor had led the dance with the duke's consort. Hainhofer, who was always well received, obtained admission, and saw this garden, whose days were numbered, and he describes a very pretty pergola, and a pleasure-house that was handsomely painted, and from the back looked into a deer-park. But far more important in his opinion were the two gardens which William laid out during his residence in the new part on the south. The smaller one of these, called the "Pretty Garden" (Fig. 377), is now only known under the name of the grotto court, but it is the most attractive feature of the residence. The grotto still remains, fantastically made of stalactites, shells, and many half-precious stones; it contains a golden Mercury reminiscent of Giovanni da Bologna, and other fountain figures. The painting of the walls is for the most part restored. In the garden, scarcely thirty by twenty metres in size, only the beautiful middle fountain, a copy of Benvenuto Cellini's Perseus, which was intended as a fountain figure, has been preserved. "The water runs out from head and neck, like blood from human veins and arteries," says Hainhofer. But in his time "the garden was divided into four compartments, the plots with the beds marked out in white marble, in each division a stone trough with running water for irrigating the plots." The walls were adorned with statues, and in front of the grotto was a mosaic of blue stone worked in the Italian manner. This pretty scene was viewed from a balcony with a gilded parapet.

Turning from this look-out to the opposite side, one saw the other garden (Fig. 378), which had for chief feature an open room with inside decorations of fountains and statues. The garden itself was a long oblong, with an open loggia on one side, and on the other a trellis covered with greenery; six of the divisions are bounded by hedges, and

II—D

two of them by white stone. Small trees stand at the corners, and inside it are "all sorts of pretty flowers." But the chief piece is a great tank at the end of the garden with a fountain and many figures, Neptune being the most imposing one; and opposite is a grotto bearing a life-size Bavaria at the top, which now ornaments the rotunda in the chief garden. Finally, there is a round temple, with a Pegasus. There is a surprising number of fountains for a garden that is not large. In this ornamentation Hainhofer forgets his interest in botany.

Duke William has here made a masterpiece of a Residence garden, a place to live in the open, where artistic ornament was of chief moment. But at that time he was not living at the Residence, which his reigning son improved very finely, but had his own private place, now Maxburg, where Hainhofer certainly found no garden, but tells of a hermitage instead; and this proves that places of the sort were not only set up in remote parks as at Gaillon and Hellbrunn, but also in the middle of the town, at a Residence. "All this grotto is made in one piece, just as we see them copied in paintings and copper-plates of fathers and hermits." So says the learned Hainhofer. "It is made out of the actual rock with cells cut in it, and there are firs and wild trees all about it, and water gushes out of the rock, making a stream and a little pond; therein, made of lead, are snakes, lizards, toads, crabs, etc. In this grotto everything is woven of bass, straw, and sticks, and the altar is made of rock. In the little room in the winter there is only a poor oven, and it is all dark, melancholy, gloomy, and even frightening. On the wall St. Francis in the Wilderness is painted; and the ceiling is thatched with straw and sticks as huts are. On the wall there is a tree with a stopper in it; and when you take out the stopper you see through the tree out to the tower in the city, and its clock, and thus know what hour has struck; and this is the peculiarity of the grotto. It also has a little loggia above the water, and in it a long plank on trestles, and there are twelve low stools of straw and thatch, made for the use of the princes . . . when they take a meal with the Carthusians in the grotto. There are two of them here, a priest and a lay brother. I asked the priest if the time seemed long and he said 'No,' for he was always meditating as to '*quid Deus fecerit pro se, quid Deus faciat in se, quid Deus facturus sit de se.*'" This was quite in accordance with William's way of thinking, who always went about in coarse clothing like a monk, and dressed all his servants in black; he had private paths made from his castle to the Jesuits and the Capuchins, although his whole time and inclination were devoted to the art and pride of this world.

But the main garden of the Residence was not even started at the time of Hainhofer's first visit, for Maximilian laid it out soon after in the north at the other side of the town trenches, as the last grand work of his otherwise completed Residence, after the old pleasure-garden had fallen a victim, as we have related, to the extension buildings. This garden was absorbed into town fortifications at the time of the Thirty Years' War, but at the outset of this fatal epoch it was already finished. Maximilian had made a journey to Italy as heir to the throne, and there his taste for art was much enlightened, which made a great difference to the internal ornament of his Residence. The garden (Fig. 379) certainly has a fine Italian casino with a flat balustraded roof and an open hall as chief feature, but in its *ensemble* it shows far more leaning towards French examples. In front of the casino there are two ponds separated with a balustraded path down the middle, where there are fountains. Half-way down this middle walk, which widens out into a

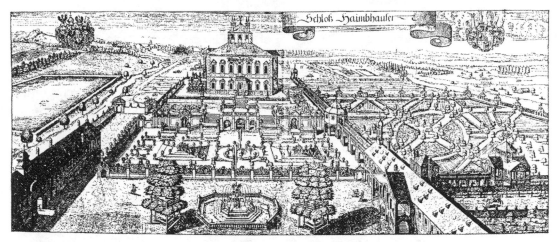

FIG. 380. HAIMHAUSEN CASTLE

rotunda, a green pavilion was erected later. But from the beginning the large centre pavilion, which connects the separate parts of the garden, appears to have been ornamented with the "Bavaria" from the Residence garden. This great place is higher up than the ponds, and leading to it is an inclined slope in six divisions. There are special gates covered with greenery leading into all the four parts of the garden. The plantation shown in the picture belongs to a later date, to the end of the seventeenth century, and the engraving by Merian has preserved the far simpler arrangement of beds.

Maximilian's brother Albert also practised a fine style of gardening; his handsome place was at the so-called Sailor's Gate, on the far side of the town prison. There Albert used to go with his friend and protégé, the Jesuit Jacob Balde, who was the last and best of the neo-Latin poets, and sang the praises of his patron in Horatian verse. These have passed into the German language in Herder's translation. Balde speaks of hanging gardens; leaning over the columns he gazes with his royal friend into the depth of the prison, which was laid out as a garden in times of peace. At the entrance stood the charming figure of a boy in stone, with a wonderful song of praise upon it, and the words: "Did Flora give thee life, when like a mother she had ordered all this garden?" The

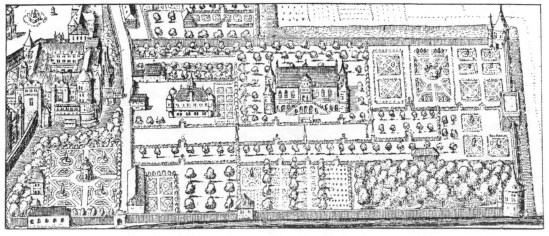

FIG. 381. STUTTGART—THE CASTLE GARDEN

poet knows not how he can praise enough the exuberance of the flowers in this starry meadow, as he calls it. This was in the middle of the Thirty Years' War, which hindered building activities even in Bavaria. But at any rate the gardens were kept up as protection in the precincts of the town until a more peaceful time should arrive.

The period before the war when people were interested produced other charming works in Bavaria, and not only at Munich. The castle of Haimhausen (Fig. 380) shows a

FIG. 382. HEIDELBERG CASTLE—PLAN OF THE GARDEN

feeling for style that is almost Italian in its fine array of terraces with a grotto, and an approach by steps. At the end of the garden the kitchen court is decorated with a fine fountain, flanked by two tall trees, in which are fixed rooms and seats in different tiers: this is a custom inherited in the Middle Ages from antiquity, but in Germany carried particularly far.

In rivalry with the princes at Munich, Frederick of Würtemberg made a new castle at Stuttgart (Fig. 381). Here too there is no immediate connection between the noble residence and the flower-garden, for they are separated by a wide walk. The garden has certainly a summer-house as architectural centre, but this idea is overmastered by the northern spirit which prevails throughout. This fine pleasure-house, with its wonderful mixture of Gothic and Renaissance forms, has artistic charm on a large scale: the wide, airy balcony that runs round the place with corner turrets and lofty gable is a capital room for a cheerful party, and here we might perhaps find an immediate link between house and garden; but one must needs compare with it the severe axial lay-out of those gardens at Munich that really feel French influence. There appear in these any number of distinct ideas, and every part is treated individually, with no reference to its neighbours. The chief feature, the large flower-garden, certainly is at the side of the house, but is not connected with it: thus the middle division, a circular hill with steps, corresponds to the style of the house because the pavilion at the top is treated as a little castle. Among the other parts of the garden, playgrounds and tree-gardens, a special ornament is the orangery, one of the oldest and most famous in Germany.

In almost every garden, or group of gardens, we have been able to detect a new style and its influence, but in these busy times no unity of style is found in Germany, in spite of close communication and eager exchanges. We pass from one German individuality to another, always to find a new picture, and that a different one. France and Italy begin to contend for supremacy in Germany. The originators of the castle of Heidelberg near Stuttgart took a high aim in their rivalry. There is scarcely a building in the

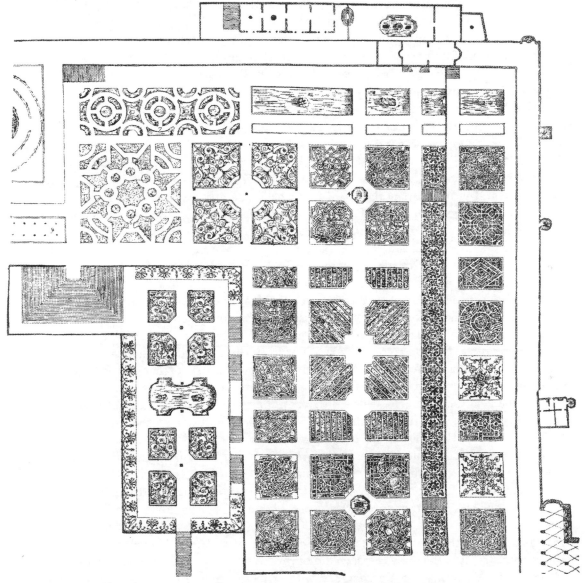

FIG. 382. HEIDELBERG CASTLE—PLAN OF THE GARDEN

whole world which has excited admiration of so varied and so critical a kind as this castle in the pleasant valley of the Neckar—as much in the days of its greatest pride and glory as now, when, in ruins, it has acquired a new charm. The steep projecting spur of the "Königsstuhl" would add to the difficulty of making a garden in any sort of larger style, and accordingly the Counts Palatine were satisfied for a long time with keeping a garden on the flat land in front of the walls of their little town, and this was called the court or the lord's garden. It cannot have been without importance, for it contained beautiful fountains, and most of all its orangery enjoyed a world-wide reputation as early as the end of the sixteenth century. Olivier de Serres in his book praises the orangery at Heidelberg as a marvellous example. At that period these noble trees, and figs also, were not planted in pots, but straight into the earth, and in winter a wooden house was built over them, so that in summer people could stroll among the trees as in an orange grove in the South. For long enough this kind of protection for southern trees had been known in Germany. Here also the princes sent their specimens, and as early as 1559 Joachim II. of Brandenburg received one from Prince August of Saxony.

It is not only Olivier de Serres who speaks of the size and beauty of the Heidelberg trees with admiration, but also the curious adventurer of those parts, Michael Heberer, who was nicknamed Robinson of the Palatinate because of his sea voyages, his imprisonment in Turkey, and his wonderful escape. This great traveller says he "never found the like in Italy or Egypt." He continues: "His lordship has also a large, lovely garden next to the town, below the mountain [Gaisberg], which is enclosed by walls and partly by noble vines. In this garden there are often held knightly exercises by the gentlefolk, and also meetings of the rest of the rural population." But it gradually came to be felt that the long way from the castle to this garden was inconvenient. When Frederick V. of the Palatinate married Elizabeth, the daughter of James I. of England, in 1613, and soon after became the ruler of his own country, he made a plan for a large show garden (Fig. 382) beside his castle, such as Germany had never yet seen.

Elizabeth knew the right man to carry out this plan, and summoned her own tutor, Salomon de Caus, who since her brother's death and her own departure from England had lost his occupation. Elizabeth had been very fond of her brother Henry Frederick, whose intention it had been to accompany his sister to Germany for the wedding festivities; and now, at Heidelberg, she was glad to keep a living memorial of her lost brother in this teacher whom they had shared; thus it came about that Salomon de Caus became architect to the Palatine court. He made a little book of a collection of drawings of fountains and grottoes, and dedicated it to Elizabeth in 1615 in memory of her brother, to whom he had already dedicated an earlier work in 1612. It was destined to serve him as a real support and storehouse of ideas in aid of the great work that he now began without delay and with the utmost eagerness—the construction of the gardens of Heidelberg.

At that time he found outside the walls by the castle nothing besides a little level plot made in 1508, about two hundred feet square, with a wall. This ground was called the hare garden; and the few drawings show nothing at all of real garden design, though we may assume that there were vegetables or something of the sort. Above this place the mountain rose steeply, broken by the deep dip of the Friesenberg valley. The configuration could not have been worse for a garden site with terraces, such as de Caus desired; and it is indeed wonderful what the architect accomplished by hewing away the mountain side and filling up the valley. Even at the Villa d'Este there were no greater difficulties to contend with; in many places the walling amounts to seventy or eighty feet, so that the colossal niches which are visible in the great terraces are not by any means the least important part of this earthwork. And in two years the completion of four and in some parts five terraces was accomplished, so that it could support the famous garden of Heidelberg Castle (Fig. 383).

FIG. 383. THE CASTLE AND GARDEN, HEIDELBERG—GENERAL VIEW

Although de Caus proved in these foundation works that he was indeed a master of the soil, he was nevertheless unable, in spite of his Italian travel, and of his studies at Villa d'Este and Pratolino, to make his own the idea of artistic unity, of proportion, of the subordination of all the parts to the whole. He was a person of many-sided intellect, and not without considerable artistic gifts, as is proved by the number of his ideas, more plentiful in this garden than in any other of the time. He also had propounded to him here an extraordinarily difficult task, for the garden was situated outside any possible connection with the castle. It was a collection of gardens, which very different centuries had thrown together, each in its own style, each picturesque, but without any uniformity. The irregular trenches separated castle from hill, and to make anything really satisfactory the count would have had to create some quite new middle point in a summer-house on

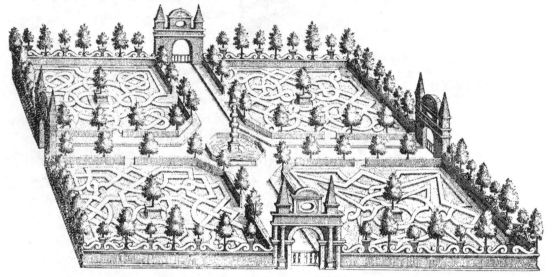

FIG. 384. HEIDELBERG—PARTERRE IN THE CASTLE GARDEN

the heights; but there was no need for this. Because architectural co-ordination could not be ensured, de Caus renounced all axial order, but this was owing chiefly to the fact that making steps was not really feasible. The garden at Heidelberg is the best argument on the negative side of the immense support given to the structure of a terrace garden by well-proportioned steps of harmonious design. Here all artistic plan is wanting in the steps, and they are nothing but steep breakneck connections between the terraces.

The Italians, above all the Romans, had long learned what a good background is provided by ornamental and convenient steps for any ceremonious occasion or large party. On the steps of such delightful gardens one imagines groups of lovely smiling ladies and proud nobles, moving up and down, whereas on the steep steps of the Heidelberg garden we see at best nothing but the young Lieselotte [Princess of the Palatinate, and afterwards Duchess of Orleans] jumping and scrambling down, tired with a romp on the upper terrace at ball or ninepins. But independently of the steps difficulty, an axial arrangement was outside any possible artistic idea for de Caus. He was compelled by the Friesenthal to bend the garden in two at a right angle, and so he got two sets of terraces. In each case he equipped the upper one with a fantastic but effective top part with grottoes. But on no side was the view arranged so as to get a middle point; both groups of grottoes are set—it would appear to be intentionally—to one side, while the eye seems to be straining to look out at the end. In the same fashion the water is on several sides, but not treated systematically. The architect had too deeply engrained in his nature the mediaeval custom of treating every part separately, and never once is a terrace combined with anything else.

Salomon de Caus described the garden exactly two years later, in words and by illustrations, after the building had been rudely interrupted in 1618 when Frederick V. was called to the throne of Bohemia, and the Thirty Years' War broke out. Thus we have before us not only the work that was completed with amazing quickness up to the year 1618, but also the plan of the whole garden as the architect designed it. After this publication the Fouquière picture was painted, which again served as a copy for the engraving of Merian.

FIG. 385. HEIDELBERG—FOUNTAIN STATUES IN THE CASTLE GARDEN

On the wide main terrace, which is on the same level as the castle entrance, we walk through a building at the end, a kind of aviary, and then stroll through five different parterres, each treated separately, and often cut off by special little entrance gates (Fig. 384). There is a fountain or a statue in the centre, and the paths are bordered with hedges or with pergolas and pavilions. The beds are laid out in different patterns so as to hold flowers, or little orange-trees, or tiny lawns. Behind, close to the wall of the terrace at the top, is a water-garden. From the basin at the corner of the higher side-terrace the water flows into a fountain, and a basin lower down transfers it into two receptacles adorned with statues (Fig. 385). The place ends with a pretty and effective water-parterre (Fig. 386), but oddly enough this is not arranged axially with the basins.

In the wall of the higher terraces, almost at the corner of the garden, is the entrance to the great grotto, which receives its waters from a reservoir that is again bordered with balustrades and is turned aside and adorned with a Venus fountain. The water plunges down on a step inside the grotto, and thence throws out streams to various fountains (Fig. 387). From the Venus fountain a peculiar convex double stairway leads to the pretty structure carved in greenery on the highest step. The long terrace at the side, to which the two steep stairways lead, has insignificant parterre beds, and above lies the long narrow tennis path, which is cut off on the mountain side by a great alcove, still to be seen, crowned with the portrait of Frederick V. A similar construction should mark the end at the other side also. The intention of the architect was to carry on this road

FIG. 386. HEIDELBERG CASTLE—THE WATER PARTERRE

at right angles as far as the end of the great terrace. Opposite the entrance to the castle, on an ascending terrace, more large grottoes were planned, wonderfully decorated on the façades, and with fountains and statues inside, but this part was never quite finished. The great terrace also, which stands opposite the castle on powerful arches above the Friesenthal, has only in part kept its early form.

At first the famous old orange-trees had been brought out of the garden up the mountain side with incredible pains and difficulty, and had been replanted in a long narrow garden—a thing that rightly caused universal admiration. De Caus wanted to replace the wooden winter-house with a stone one, whose roof and window could be taken out in summer, so that the supports might act as a broken wall. Above this garden, which was 280 feet in length, a labyrinth was to have been put, as a crown to the garden and also a protection, but it was never finished. Behind lay the medicinal garden, in a pretty part with pavilions at the corners, and lastly was to come a great square tower, with near it a small room cut out of the hedge; the tower also was never made, except its foundations. In the corner of this large middle terrace there was a three-sided stairway on a large scale, but very clumsy, leading to the lowest of the gardens, which, as it was so small, was treated as one whole. On both sides of the large basin adorned with figures which stood in the centre, there were beds held together in fours by a statue, while steps in sets of two, at the front dividing wall and with a slight edging of fountain-work, led to the higher terrace. All this must have made a separate garden, pretty and characteristic.

It is curious that de Caus says nothing in his description about the special piece in front of the new building which he made for his young mistress Elizabeth, and yet this garden on the level on the terrace next to the town was no doubt actually laid out. The decorative entrance-gate, which still stands, shows the same type of architecture that de Caus proposed for his stone orangery. There is an inscription which says that the Count Palatine erected this gate in honour of his wife. It was let into a wall, which had at its other end an aviary against the wall of the terrace. There is no picture giving the interior divisions of this garden, which was united by bridges with the new part of the castle. It was remade after the ravages of the Thirty Years' War, and little alterations were made here and there, before it came utterly to grief, castle buildings and all. At last in 1805 a terrace garden was laid out in the English style in a most unfortunate attempt to accommodate the picturesque grouping of shrubs and trees to the mighty terrace skeleton.

To-day one seeks with great difficulty for specimens of wrecked remains of grottoes and alcoves, to build up the old scene.

Andreas Harten, a deeply religious man and an enthusiastic Protestant, published a curious pamphlet about gardens in 1648, in which in 233 pages he compares the Bible to a pleasure-garden. The title of his book, which is full of superstition and witch-lore, is *Worldly and Heavenly Gardens*. In spite of everything, including his start in life as a taverner, he was himself a clever gardener, and at that time in the employ of Christian von Schönburg-Glauchau-Waldenburg, at Rochsburg in Saxony. Harten describes this garden (which he brought to great beauty), with its hedged-in paths, furnished with domes, towers, doors and windows, symmetrically laid-out parterres, again broken up into separate divisions with box edging, and each of these containing only one species of plant. He says in his book that since the Reformation (to which he attributes every good thing) "useful and necessary buildings for gardens and herbs have again reached so flourishing a state, though after great expense, that there is hardly a townsman who keeps anything in a town, but spends his all on garden building, to say nothing of potentates, lords, and nobles."

Harten contrasts this happy time before the Thirty Years' War with the gloomy days he was then living in. "To hinder the delightful garden-building day by day, the devil is always at work, and seeks out the right places; that is, on account of our sins, he incites the great potentates against one another, so that they lose sight of all the peaceful pleasures of eye and heart (which aforetime they took in their gardens), and he makes them

FIG. 387. HEIDELBERG CASTLE—THE INTERIOR OF THE GROTTO

go forth and spend all they have on unspeakable dissensions and wars (wherefrom they suffer not only pain, want, and danger, but all manner of adversity), though before they had spent their substance on beautiful pleasure-gardens, whence they got all that they needed and enjoyed and which they might still enjoy." But in the midst of the storms of war which are hostile to all culture and especially to the gardener's art, there were many exceptions. Harten himself boasts of his master "that he felt a remarkable kindness and affection for fine garden-making." And in the same year that Harten wrote, another prince's gardener gives a detailed description of the garden of his patron, the Duke of Brunswick. The beautiful garden of Hessem at Wolfenbüttel surrounded by water

FIG. 388. CLIPPED HEDGES IN THE CASTLE GARDEN, HESSEM, BRUNSWICK

has its different parterres richly supplied with every kind of figure cut in the hedges (Fig. 388). The parterres are most elaborately laid out with stars and armorial bearings. One of these contains a masterpiece, a magnificent fountain that was once bought from merchants at Augsburg for 8000 gulden.

Out of the middle of the war-time, when the heart of the country was beating most wildly and restlessly, we hear a voice which is never tired of calling people to the peaceful art of garden-making, from the greatest sites to the smallest. Wallenstein never lost sight of home-life, although he himself stood over a dizzy abyss. When he took for his residence Gitschin in Bohemia, it was a wretched patch of 198 shingle-roofed houses. With unwearying trouble, warnings, threats, and most of all with money support, he brought the townsmen to the point of building better houses, and introduced orderliness and sobriety. He built a great castle for himself, and untiringly superintended its building. The garden, which lay behind the castle, appears to have been like the one he planned afterwards at Prague. In June 1630, shortly after his fall (in consequence of the assembly of princes at Ratisbon and just before he took up his residence at Gitschin), he writes as follows: "If I am correct, there is no fountain in the garden plan exactly in front of the loggia. Tell the architect that there must be a large fountain put right in the middle of the square before the loggia; all the water must run into it and then out, dividing into two streams right and left, making the other fountains in the courts run in the same way. Send me the design of the garden, with not less than one of each set of

fountains marked with numbers, and write what is needed for each of them."

Gitschin had an imposing park suitable for a country house, whose area amounted to 12,000 metres; and a wide avenue of limes with four rows of trees led to the town. The prince ordered guards to be set there "so that the limes should not be spoiled by the numbers of people who come from the town." In the park he grew uncommon trees and shrubs. The water in the garden, six fountains and a pond for swans, was conducted by eight artificial canals; there was a pheasantry and a garden for animals. We can get no clearer picture of his garden at the town house in Prague, "the Duke's house," with which he was concerned during the last years before he was murdered. He had about twenty town houses and other buildings taken down so as to obtain a good site here. As at Gitschin, the garden front of this large, irregularly built house had a loggia in the pure Italian style (Fig. 389). On one side was the prince's bathroom, treated

FIG. 389. WALLENSTEIN'S GARDEN AT PRAGUE

in the grotto style. A winding stairway led from here to his private workroom. In front of the loggia was a wide square of garden with a beautiful fountain in the centre, and round this four beds making a parterre.

The unfavourable nature of the place does not allow the garden to stretch out as far as it should in length, and thus neither the noble loggia nor the garden can make a really finished picture. To get a view of the length axis, it is necessary to go to the aviary at the side of the hall, and here also the parterres are connected with fountains, and the garden is bounded along its whole breadth by a large water-mirror; in the centre is an island, with very probably a fountain group to make a *point de vue*. In the arcade-like alcoves, statues used to stand, for the letters speak of their being set up there. Among others there was a bronze Hercules standing even as late as 1793. The aviary is a great wired building, the walls adorned with all sorts of shells and stalactites, and this is still there. In the aviary there were, in the Italian style, hedges, bushes, and trees, and the birds nested in them, as at the Palazzo Doria at Genoa. Italian influence is supreme in all these minor details, and at this period is very conspicuous. Baccio di Bianco, the gardener who afterwards migrated to Spain, was active here for a time. But the whole picture of the Kleinseite district at Prague must have produced an Italian effect in its buildings. The great pile of the seventeenth century, the Belvedere, was on the high part, the palaces of Lubkowitz and Fürstenburg on the slopes, with their terrace gardens: of these one can still see certain traces in isolated remains of steps, grottoes and pavilions, in what seem to be their former places, and lastly on the level ground the house of Wallenstein.

A man similar in many respects to Wallenstein, one whom the long war had moulded as a hero, and yet one who had added to the life of adventure a keen delight in the works of peace, such as building and gardening, was Maurice of Nassau. He certainly out-lived the war, and his building activities in Germany belong to the period that followed it; still he had shown beforehand how well he knew how to combine the arts of peace and war. Fabulous stories are told of his buildings in Brazil, where he was sent in the service of the Dutch State. In the short space of seven years he had built no fewer than three great places in what is now Pernambuco, of which two, Freiburg and Boa Vista (Bellevue), were palaces; and these were castles with trenches round them and flanked with towers, according to superficial, untrustworthy accounts. There were bridges that led into fine pleasure-gardens, where probably French influence will have predominated, as also in the buildings, though the tall vegetation of the South must always have given a peculiar effect.

After Maurice came home to Europe in 1644 his work went to pieces. But he created a new field of activity, when in 1647 he entered the service of the "Great Elector." His lord wanted, like himself, to make the country more beautiful. With this intention he furthered the plans of a stadtholder, who, though in his service and domain, began to build in the spirit of an independent prince. Just as Wallenstein converted Gitschin, so did Maurice convert Cleves from a wretched place into a flourishing residential town, and that in a very short time. He put avenues everywhere, and when he found a view but not enough height, he piled up artificial mounds, and built a series of country houses. These, according to the accounts we have, showed a purely Italian influence, especially the ornamental part of the so-called new animal-garden, which was built in terraces, one above the other, adorned with fountains. The lowest of these splashed its water up to a height of twenty-four feet from the beak of a black eagle which stood in the middle of a basin, of which the back wall was covered with rock-work and masks. The end at the lowest level was defined by two carved heraldic lions, a present from the Council of Amsterdam, and a fountain, in the shape of a star, sprinkled water above. On the third ascending terrace was the figure of a boy blowing in a shell and sitting on a dolphin. Finally the whole place was crowned by a Minerva in white marble, also a present from Holland: she stood in an amphitheatre adorned with vases, urns, and basins. Maurice was a travelled man, and acquainted with the gardens of the South; so although his eyes were often directed towards Holland, which was near at hand, a place like this did not originate in Holland; and he was always able to tell his chief, afresh and justifiably, that his guests from Holland were full of admiration and astonishment for his work.

But a flourishing creation like this, coming immediately in the war and out of it, was a strange exception, for only slowly and little by little did people begin to recover tracts of land that had been entirely laid waste, and to build and plant anew. Therefore the Thirty Years' War completely put an end to the renaissance of garden art in Germany. The German princes once more had power and sufficient substance. Ornament and luxury in their homes became an increasing need. And a new star which gave an irresistible direction to culture of every kind, but especially influenced garden art, had arisen: this was Louis XIV.

CHAPTER XII
THE TIME OF LOUIS XIV

CHAPTER XII

THE TIME OF LOUIS XIV

IN the second half of the seventeenth century France held the leadership, obtained after long efforts and severe struggles, over the whole of Europe. She was incontestably at the highest summit of political power, but was still more obviously of first-rate importance in her culture, which was fortified by foreign examples and yet flourished independently on her own soil. Italy was compelled to resign the sceptre in many departments of art, but especially in gardening, to this rival in the North, after fulfilling her mission during a century and a half on the other side of the Alps.

The growth of native art in France was greatly favoured by the fact that there were no outside disturbances to hinder it. At the beginning of the seventeenth century garden art had received a strong impetus everywhere in the North, while in Germany the Thirty Years' War had left few flowers untrampled, and in England the rule of the Puritans, with their hatred for luxury, had perceptibly interfered with garden tradition. In France, however, progress had not suffered; for the civil disturbances of the Fronde were really only a trial of strength—skirmishing which could be put a stop to by reconciliations and persuasions, not war to the knife and therefore hostile to art. After the death of Henry IV., the king, as such, was in a less commanding position for a time, and the long regencies and the great power of the Church had made it easy for proud nobles to raise their pretensions to an equal height with his. The waves of political life communicated movement even to garden art. After the building of Saint-Germain-en-Laye, Henry the Fourth's stately home, there was no important new royal house, and the old ones like Fontainebleau added few new features to castle and garden. On the other hand the Luxembourg, the residence of the queen regent, grew; and Richelieu's country house at Ruel attracted foreign visitors. With this a nobleman's castle like Liancourt might compete, or indeed that of a foreign adventurer, like Saint-Cloud. And if Fontainebleau and Saint-Germain may be included, they are after all only acquired properties, and at best the gardens and castles stand *inter pares*.

About the middle of the century, on the threshold of a new age, there stood out conspicuously a man who by virtue of his personality seemed to exhibit the spirit of his time—both the arrogance and the autocratic defiance characterising that period: this was Fouquet, the Finance Minister of Mazarin. Fouquet's delusion, and the snare that led to his fall, was that he—although having the finest instinct for what the future held in matters of art and science—failed entirely to realise what the embryo monarch would prove himself to be; and so in blind security rushed to his fate. There was an element of noble tragedy in his career. A shining example to Louis XIV., who knew himself to be his pupil but never acknowledged it, he was nevertheless overthrown by the king.

Fouquet was in the prime of life and at the zenith of his powers, when at the beginning of 1650 he arranged a contract with the architect Le Vau to build a castle called Vaux in his own viscounty, at Melun. He was still under forty, and he had no reason to doubt that by favour of Mazarin he would soon be appointed First Minister of

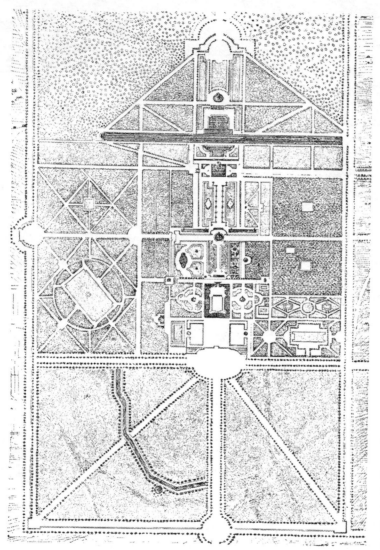

FIG. 390. VAUX-LE-VICOMTE—GROUND-PLAN

Finance, and he supposed that the rivalry of Colbert might be easily ignored. The young king, carelessly enmeshed in his love affairs and other pleasures, did not appear to feel the least desire to take the reins of state into his own hands. In order to clear the ground for castle, garden, and a proper open space round them, it was necessary for Fouquet to buy three villages, and to pull them down. The place grew with astonishing quickness over its foundations. The powerful financier had inexhaustible wealth at his disposal, so he pressed on the work eagerly. It is said that at times as many as eighteen thousand labourers were employed together, and the cost was computed at sixteen million livres.

Vaux-le-Vicomte is a fine building with pavilions about it, and has a wide moat all round (Fig. 390)—the natural thing for any country place, for almost without exception the neighbouring castles of recent erection were still made as water-castles, for instance Ruel, Liancourt, and Louis the Thirteenth's palace at Versailles. In front of the castle was the wide entrance court, the *cour d'honneur*, cut off by a handsome semicircular balustrade and fine trellis barriers, to which the broad carriage roads of the park led. The stabling on both sides of the court covered kitchen places and vegetable plots. On the other side of the house, erected on terraces that were built up artificially, there was a vestibule inside the moat, which was bordered by a balustrade. From here there was a very good view of the greater part

FIG. 391. VAUX-LE-VICOMTE—VIEW OF THE GARDEN FROM THE CASTLE TERRACE.

of the garden, which had perhaps been begun before the foundation stones of the castle were laid.

As early as 1625 the editor of Claude Mollet's work, *Le Théâtre des Plans et Jardinage*, dedicated it to Fouquet, and in it speaks of the wonderful garden at Vaux-le-Vicomte, where "they very delightfully allow art to strive with nature, and every day bring fresh beauties and new treasures." The work at the garden must at that time have been carried on with much eagerness, and it is certain that it was in great beauty when the castle was finished. Fouquet secured the then famous painter Charles Le Brun to decorate his castle and the other rooms at the villa. Le Brun then recommended to the Minister of Finance his young friend André Le Nôtre, who had, like himself, studied painting. He had got to know and like him through their common teacher, Simon Vouet, and he admired his imaginative fancy in decoration and his knowledge of garden art. Le Nôtre owed his scientific knowledge to his own home. His father was superintendent of the Tuileries Gardens. Here in Vaux-le-Vicomte André was to win that rank as an expert which was recognised half a century later all over Europe.

Mademoiselle de Scudéry, who describes Vaux in her novel *Clélie*, says soon after the place was completed: "The most wonderful thing is how this garden lies between two shrubberies, which agreeably break the view." This fine observer has here caught the peculiarity of the garden. We can stand with her on the terrace of the castle, whence a drawbridge leads over the moat (Fig. 391). On both sides of the castle there are parterres decked with fountains, intentionally made in simple lines so as not to distract the eye too much from the fine *parterre de broderie* that lies below the house. To right and left there are flower-gardens, which are worked out with much play of fountains (Fig. 392). The greatest admiration is caused by "la fontaine de la couronne," which balances on the shining waters a crown of water-spouts: there is a round fountain at the end, and the parterre is finished off with two small, narrow canals. Hence one goes down by the main path through a watery road. On both sides there are lawns edged with flowers and decked with fountains, and the water finds its way into a great square basin. If one stays to look from the terrace, the bright ribbon of the canal can be seen glittering from the depths below. This canal ends the garden abruptly, and has no bridge: it widens into a large basin, behind which the natural hill rises somewhat steeply. A triangle, regular but ending in a semicircle, occupies a space in the park ground.

The hill is marked out and ornamented with grotto-work, fountains, and water-beasts right up to the top, where the imposing scene ends with a huge figure of Hercules and a great column of water. This limitation of the background corresponds to the confinement of the sides with shrubberies. On the left of the castle the ground rises; and this is made use of for a terrace, approached in the side-garden axis by a small piece of ground with a cascade and a fine ascent of steps. On the other side are shrubberies with trellis bordering the walks, and here are found either flower-lawns or gardens, or paths with fountains, or what is called a water-theatre. Leaving the castle terrace and going down through the water road to the terrace of the great canal, the visitor is met with a surprise. At the dividing wall of the castle, and so not visible from there, a cascade appears, a contrast in life and movement to the peaceful gleaming line of the canal. But such cascades are not only the adornment of the lowest of the terraces, which is sunk

FIG. 392. VAUX-LE-VICOMTE—PARTERRE AT THE SIDE OF THE CASTLE

between two high places, they also enliven the scene from the hill where the great Hercules stands with a view of garden and castle (Fig. 393).

Two important currents of thought that in every field were dominating men's minds at this period were now adopted by Le Nôtre, who saw how to use them in his gardens from the very beginning and how to blend them. The one represented the spirit of discipline, of firm, distinct, defined rule, of proportion: this idea found expression in literature in the work of Boileau, in politics in the ever-increasing monarchical sentiment, in fashionable life in etiquette cultivated to the very extreme of refinement. In opposition to this, however, there was the unrestrained and constantly growing desire for variety, for change. The society which was subjected willingly and consciously to this spirit of discipline, and found therein the expression of its highest culture, the fixed form and rule of life, would have grown old before its time, indeed would have died of sheer boredom, had it not been for its constant search—so often perplexing to us—for novelty and change, which rendered its votaries continually breathless and excited. Vaux-le-Vicomte was the first attempt to combine these two requirements.

French gardens before then had the severe axial order, even from the time of Du Cerceau, but the terraces in his day presented the same or similar pictures, each separately; the parterres again were exactly alike, and symmetry was marked in the repetition of the same lines. When then in the seventeenth century Italian influence began to be felt in a new direction, there was no real mastery of the novel idea of the whole and its parts. The cascades at Saint-Cloud and Ruel lie at the side in their own special axis, with no relation to that of the house, and the famous "variété" at Ruel is, as a whole, unrestful and scattered. But Le Nôtre perceived that the most important thing was to create a magnificent scene which could be viewed as a whole from the house, and demanded this character of being visible before anything else, although its main lines might show all sorts of variety in parterres and in water: for this garden was intended to serve for royal fêtes and the display of magnificent costumes, exercises, and fireworks; everything and everybody must be seen. But for a picture a frame was needed, and this was provided by shrubberies that each separately constituted a private garden, in which more and more variety would be found to answer the most extravagant demands.

Fouquet had before all things in his mind, at the making of Vaux-le-Vicomte, the holding of great fêtes, such as France had never yet heard of. He could scarcely wait till everything was quite ready, but in the summer of 1661 he gave his first great fête in honour of the young, much admired lady, Henrietta, wife of the Duke of Orleans, the king's brother. Molière, at that time in Monsieur's company as actor and poet, first introduced his *Ecole des Maris* at this entertainment. Fouquet seemed overwhelmed with every sort of good luck, and in the year 1659 he became sole "General Intendant." His party seemed extraordinarily strong, though certainly for a few months past storm-clouds had been rising, and on 9 March Mazarin had died. In the conferences at his minister's death-bed the monarchical feelings of Louis seem to have matured and hardened to decision. Soon after Mazarin's death he summoned his cabinet, and announced that henceforth he was going to be his own prime minister.

Fouquet still did not believe the signs, or the warnings of his friends, while Colbert was steadily working towards his downfall. Nothing could show up the perilous system of Fouquet to hostile eyes more glaringly than the hitherto unheard-of magnificence of

FIG. 393. VAUX-LE-VICOMTE—VIEW TOWARDS THE CASTLE FROM THE CANAL

his new castle. It was said that Colbert visited it secretly while it was still in course of building, and informed the king of the immense number of workmen and the enormous expenses. The brilliant fête, at which the king was not present, excited his anger and also his greed, and he offered to appear at another one on 17 August. But Fouquet's fall was a prearranged affair; indeed the king had purposed to arrest his host at his own house while the fête was going on, and it was only his mother who dissuaded him from this. The king arrived, and was received with pomp; but he could scarcely restrain his anger at the luxury which he could not equal. In the grand procession which inaugurated the fête, he saw everywhere the armorial bearings of his minister, an ibex with the proud but ill-starred motto, *Quo non ascendet.*

After the great feast came the performance, in the garden, of Molière's piece, *Les Fâcheux*, at the theatre erected at the end of an avenue of pines. This had been written in fifteen days especially for the festival. Le Brun had painted the decorations; and Pélisson, the well-known prose writer, and Fouquet's secretary, had composed a prologue for it, which was recited by the best actress of the day, La Béjart. A ballet, suited to the persons of the comedy, was conducted by Giacomo Torelli of Urbino, whose cleverness at decoration and machinery had gained him the name of "Le grand Jongleur." After the acting a firework display caused the greatest enthusiasm. At this there occurred an unfortunate accident: two horses belonging to the queen-mother's carriage shied and were drowned in the great canal. Lafontaine, who wrote an eloquent description of this gala to his friend Mancroi, ends his letter with the words: "I did not imagine that my account would have such a sad ending." The poet did not know what a gloomy import these words held; for he and the host of admirers never suspected that, one short month later, their Mæcenas, their friend, and their protector, in strict custody, and accused of high treason, would only just escape a sentence of death, and that his doom of exile for life would only be commuted to imprisonment for life by the king who hated and would not forgive him.

The wonderful beauty of Vaux-le-Vicomte flowered quickly but soon faded and was gone, leaving behind only loneliness and neglect. But Fouquet's name and character never shone so brightly in the days of glory as now in the season of misfortune. Though a strong opposing party might have oppressed him, though his own ambition might have so far misled him that he had become a dangerous servant to the king, and though the new course of the ship of state might have baffled him, he did as a fact acquire a glorious halo, arising from the love, the unwearying faithfulness, and the active help of friends, poets, artists, and men of letters. Fouquet was no ordinary patron; he knew how to make these people his friends. He was really something of an artist himself, and understood what he gave other people to do. It is related of him that, when he was coming out of the sessions room during his own trial and was being led through a court outside, he saw some workmen busy at a well; he stopped and went up to them, and forgetting his own troubles, gave them advice as to how they could start it better, "for I have some knowledge of these matters."

The kind thought that he had constantly for those about him was shown in various ways. Corneille, who wrote his *Œdipus* at Fouquet's request, says in his preface that Fouquet, who at that time was living at the fine castle of Saint-Mandé, had opened his library in the town (as well as the one in the country) as a waiting-room. Scarcely was the news of his downfall known when a general cry of lamentation arose. Pélisson, who had been confined in the Bastille, wrote from there one pamphlet after another against Colbert. Loret, a journalist who had glorified Fouquet's life and his festivities in his paper, both in poetry and prose, published such a violent article against Colbert that he was deprived of his pension. As soon as Fouquet in his prison heard of this, he wrote to Mademoiselle de Scudéry asking her to send an indemnity to his faithful friend. Lafontaine, who had spent some happy years in Fouquet's circle of artists, composed a moving poem praying the king to grant mercy. "The air in your deep grotto is full of laments, the nymphs of Vaux are weeping." Lafontaine had already begun a small book, in Fouquet's day of good fortune, wherein he was to sing of the beauties of Vaux in allegorical style. This remained long a mere unpublished fragment, but ten years later he issued it under the title of *The Dream of Vaux*, and in his introductory verses musically laments the fate of the unhappy man "who displeased his king and lost his friends, yet to whom in spite of his fall I dedicate these tears." In honour of Fouquet he makes four fairies appear, Architecture, Painting, Garden Art, and the Art of Poetry; they approach the Judge's seat, where Fouquet presides, and plead for the foremost rank. After Architecture and Painting, who advance proudly and certain of victory, Garden Art appears, so fair, quiet, and lovable, that the judges are at once struck by her modest charm. Then when she simply and sweetly spoke of her beauties, all hearts were disposed towards her; and had not Painting brought forward a picture the truth of which Garden Art sadly admitted, showing what she was like in winter, the prize of victory, ultimately carried off by Poetry, would have been awarded to her.

Thus high did Garden Art stand among her sisters, whose best representatives all assembled around Fouquet. "He was called *Le cœur le plus magnifique du royaume*, but this meant wounding Louis XIV. in his tenderest spot and incurring his spite." So writes Sainte-Beuve in his essay. And yet perhaps it was the greatest reward that the prisoner won within his dungeon walls, that Louis could do no better than take into his own service (being as he was the real pupil of his hated minister) that same circle of artists which Fouquet had collected about him, to educate them and fill them with his own spirit—so much so that they never lost their longing and their compassion for him. When in the king's mind the plan ripened of building at Versailles a royal palace which should eclipse every other, he took Le Vau as his architect, Le Brun as his painter, and Le Nôtre as the designer of his garden; while it was Lafontaine and Molière who contributed to the palace half its glory and splendid shows.

Ever since the year 1624 Louis XIII., whose only passion was the chase, had had a small shooting-box in the wide marshy ground at Versailles. He first hunted there when he was a boy of six, and had gone to Versailles since with an ever-increasing affection. The little castle, made of brick and rough-cast, built round a court with pavilions at the four corners, and surrounded by a wide moat, was pretty. The garden was quite in the style of the first third of the seventeenth century, and was laid out by Jacques Boyceau, who was superintendent of it as long as he lived. In his work there are two drawings still to be seen of parterres at Versailles, one a *parterre de broderie*, near the front of the castle on the garden side, the other a *parterre de pelouse*, with larger strips of lawn; and as this one is inscribed "from the park at Versailles," it was probably farther away. There is no reason to suppose that the garden was extended so early as the time of Louis XIII., nor, as many have assumed, that it already showed the main lines of its later state.

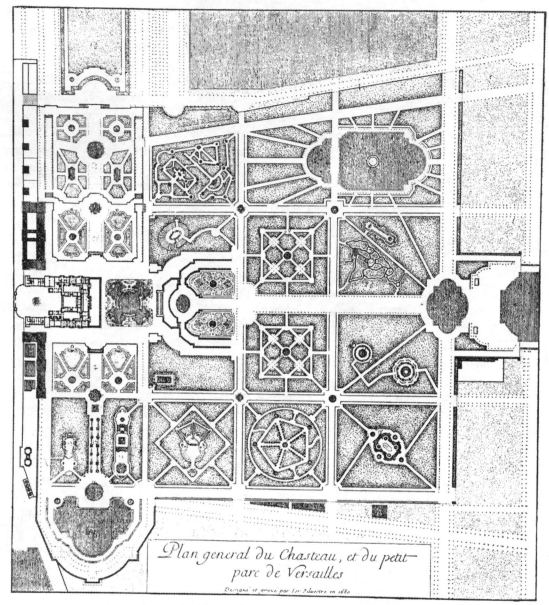

Plan general du Chasteau, et du petit
parc de Versailles

FIG. 394. VERSAILLES—PLAN OF THE PETIT PARC

Also the plan drawn by Gomboust in 1652 of Paris and its environs shows Versailles with only a few parterres and surrounded by an immense park.

Louis XIV. seems to have shared his father's affection for the place from the beginning. Its interest turned on the hunting; and after he had his first hunt there as a boy of twelve, he grew so fond of the simple, pretty little castle that it took him a long time to decide on a new building. He never did make up his mind to pull the old one down, but instead gave the builders the hard task of making new shells, one after another, round the old kernel. And so it was the garden and not the beauty of the castle that first excited his envy and sense of rivalry at Vaux-le-Vicomte.

The main lines of the gardens of Versailles must have been planned immediately after the fall of Fouquet—in 1662–3 (Fig. 394). In these days we have accustomed ourselves to bring the immense castle buildings—showing a garden front of 415 metres—into harmony with the powerful lines of the garden, and to understand one by way of the other. But when the king gave orders to Le Nôtre to design a royal garden such as the world had never yet seen, there was nothing but the little hunting-box; and since he carried out the garden in its main lines and huge dimensions in those first six years before Louis could make up his mind that Le Vau should undertake the first extension of the buildings, we must imagine that Le Nôtre had in his mind some sort of vision of a palace of immense size, for he could hardly have ventured to combine this little moated castle with a far-reaching garden scheme of such magnitude.

As a fact, Louis did help him with his demands and desires. He often came out to Versailles, and the castle was itself changed and made magnificent inside. Here Le Brun was quite in his proper place, with his inexhaustible fancy in decoration. Fouquet had already started a Gobelin factory, and Le Brun was at the head of it, and had his own ideas carried out there. In this too the king had copied him, or rather he had had the factory moved to Versailles. But if he did have his rooms set out more tastefully, his hunting-box could not extend beyond the moat, and Louis planned fêtes that should even surpass those of the Spanish court in style and magnitude. Therefore, the garden must serve as a theatre, since the castle could not; and Le Nôtre knew very well what he was about when he brought down the horseshoe paths from the great upper terrace, and there laid out a second large parterre, and followed on with a very wide avenue, 335 metres long. This avenue passed from the parterre through thick, park-like shrubberies, of which at first only two were completed with their inside decoration, to a great oval basin, beyond which a wide cross-road formed the end to the garden. In 1664 the king was at the height

FIG. 395. LOUIS THE THIRTEENTH'S OLD CASTLE AT VERSAILLES, WITH THE ORANGERY

of his passion for Madame La Vallière, with whom he frequently had a rendezvous in
the hunting-box at Versailles. It is reported that it was in honour of the birth of her child
that the king designed the grand fête which from the seventh to the fourteenth of May
converted the lower part of the garden into the fairy palace of Ariosto's *Alcina*. The
palace itself was built in the middle of the great pond at the end of an island; the four
walks in the avenues leading to the rotunda ended in triumphal arches, and between
there were rows of seats rising like an amphitheatre, where the spectators were to sit
and look on at performances, exercises, and sports, and as a finish at the wonderful

FIG. 396. VERSAILLES—WATER-AVENUE AND DRAGON FOUNTAIN

firework piece, wherein after the disenchantment of Ruggiero the magic castle goes up in
flames. On another day, a hundred paces farther along, there was an improvised theatre,
covered with a white linen cloth, and here Molière's *La Princesse d'Elide* was produced.
There were water-sports in the moat, and a royal lottery came at the close of this most
varied fête. As it was blessed with the finest possible weather, it was a brilliant success.
In spite of all, the courtiers were angry, as Madame de Sévigné says, because the king
had invited six hundred people and had not paid the least attention to their accommodation,
"so that the gentlemen of Switzerland and Elbeuf had not a hole to hide in."

Meanwhile the king did not contemplate any enlargement of his castle, though the
works in the garden were pressed forward eagerly. The orangery under the south parterre
was made before 1664 (Fig. 395); it was only half as wide as it is now, thus corresponding
to the small building, and it had twelve narrow arches with the parterre and its basin in
front, where the orange-trees were placed in summer. Louis had 190 trees transplanted

from Vaux to Versailles, getting them brought there by Fouquet's former gardener, La Quinteny, whom he also took into his service. Above the orangery was the flower parterre, which must have existed in Louis the Thirteenth's time, for Boyceau had made the plan of it. This parterre was shut off towards the castle garden by a wooden lattice, and remained of the same size when the castle was enlarged the first time; it was only much later that it was made twice as big. From 1678 it looked over the new orangery made by Mansart, and beyond it to the huge *pièce d'eau des Suisses*, at the side of which were the

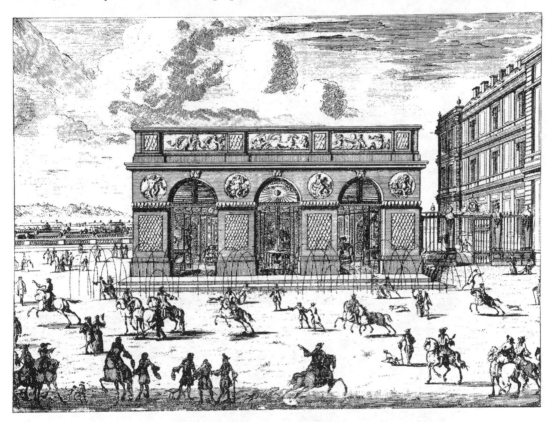

FIG. 397. VERSAILLES—GROTTO OF THETIS

tree- and kitchen-gardens. But on the north side the corresponding part towards the castle already had in 1664 its large parterre and wonderful water-devices whence the avenue with the pretty groups of children slopes gently down to the Basin of Neptune between two boskets that were often altered (Fig. 396).

The finest piece of ornamental work, however, on the northern side was one of the earliest that Louis made, the Grotto of Thetis (Fig. 397). It was situated where the chapel is now, and later it had to give place to the extension buildings of the castle's north wing, carried out by Mansart. The maker of the grotto was a certain Francini, no doubt one of the family of architects that Maria de' Medici summoned to France. We hear of two brothers Francini who did their utmost as architects in water, and utilised every drop they could find in the neighbourhood, which was poor in springs. The chief reservoirs were put above the grotto, in the three chambers of which the water was dispersed into thousands

FIG. 398. VERSAILLES—INTERIOR OF THE GROTTO OF THETIS

of the devices so popular at the time, and into fountains. The approach to the grotto was closed by three gigantic arched gates with gilded iron openings, upon which rays starting from a sun disk streamed down to six maps of the world in the form of medallions. Here started the worship of the sun, which the king himself personified in many and various ways at Versailles, as "Le Roi Soleil." Above these gates was seen a relief carving of Helios coming down to Thetis and hailed by nymphs and tritons. The interior of the grotto with its endless decorations of shells did not receive its finest ornamentation till 1675, when three groups of statuary were put up: a reclining Apollo

FIG. 399. VERSAILLES—THE LATONA FOUNTAIN

with nymphs around him, and to right and left the horses of the sun-god being watered by tritons (Fig. 398).

The great terrace immediately before the castle was a child of many sorrows for Le Nôtre, because no other part of the place was changed so many times. First of all there was a way across the drawbridge over the moat, with balustrades here and there, leading into a formal carpet-patterned parterre. This is what Boyceau designed, and what Le Nôtre found. He does not appear to have altered it at all for some time, because the first water-parterre is seen in the pictures that give the castle as enlarged by Le Vau. Originally only gently inclined walks of semicircular shape led from the great terrace to the Latona parterre, which from its form got the name of the Horseshoe. It was in 1666 that Le Nôtre laid down the imposing steps which gave the feature of immense size to the show garden, which again derives much of its importance and justification from the new buildings of Le Vau. In these years also the ornamentation was carried through of the two

II—F

basins at the beginning and end of the great King's Avenue, intended to give the proper rhythm to the middle axis of the garden. If the cult by courtiers of the "Roi Soleil" had already begun in the grotto, the brotherhood of earthly and heavenly Sun-Kings had purposely united, so that the whole garden could be converted into a veritable temple of the sun.

The decoration of the water in the Horseshoe depicts the birth of Latona's son. On an island the goddess, with the twins at her side (Fig. 399), is praying Jupiter to

FIG. 400. VERSAILLES—THE BASIN OF APOLLO AND "TAPIS VERT"

pour his anger upon the rough people who, scattered about in the water below, are spurting jets upon her, that rise out of human bodies with mouths of frogs or from actual frogs. The original place was like this, but later on the noisy horrid creatures were quite close to the goddess, and had been reared up so that when the jets were playing the whole place looked like a water-mountain. But at the end of the avenue, in the centre of the gigantic pond, which once had supported Alcina's palace of enchantments, there arose from the waves the young god with his four fiery steeds, just surmounting the water with half their bodies (Fig. 400). Tritons trumpeting at the side announced the new light of day. But behind the Apollo fountain men were eagerly digging out the great canal. Le Nôtre had never felt the smallest doubt as to constructing this work on so immense a scale and with such incomparably imposing effect; for it was started at the trench of the crossways, and was gradually extended farther to the four points at the ends. This work satisfied by one ingenious effort the practical necessity of draining the marshy ground of the low-lying park, and of collecting the many separate canals and ponds into one, so making the garden an enclosure and giving it width at the same time, as only a moving mirror of water on a large scale can do. Possibly the canal at Fontainebleau had led the way to this, and the first attempt at Vaux had also been happy in garden and ground; but the wide canal at Versailles, 1560 metres by 120 metres, whose cross-arms (Fig. 401) extended to the length of 1013 metres, achieves the finest realisation of this idea of a French canal garden; and from then till now it has never been surpassed, but only imitated.

Louis had done a great deal for the garden, but still could not make up his mind to enlarge the little house. At the beginning Colbert was very hostile to Versailles, and he could not bear to see Le Nôtre and Le Vau always coming with new plans, so as to keep in the king's good graces—plans that Louis accepted unhesitatingly and with real pleasure. There is a story that Le Nôtre was one day walking through the garden with the king, who gave the same cheerful answer to every proposal of his, "Le Nôtre, I will allow you 10,000 francs for that," when Le Nôtre suddenly stopped and exclaimed, "I will not say another word, or I shall be the ruin of your Majesty." It was far more in accordance with Colbert's ideas to enlarge the Louvre in a grand way, and so to fix the king in Paris. In 1664 he wrote him an outspoken, pressing letter: "Your Majesty is aware that, except for brilliant deeds of arms, nothing shows the greatness and the spirit of a prince so much as monumental buildings. Posterity esteems a prince according to the measure of those noble buildings that he puts up in his reign. How deplorable will it be, if the greatest and mightiest of kings, who shows excellence in the highest degree that a prince can attain to, should be measured by the standard of Versailles!—and such a danger is likely to befall."

As events proved, Colbert's fears were justified up to a point that he could never have dreamed of.

Louis did certainly build at the Louvre, but with little interest, for he did not care for Paris, and it was at Versailles that he celebrated his festivals and his victories. In the year 1668 the Peace of Aix was made the occasion of festivities far more splendid and fairylike than anything that took place four years earlier. Madame La Vallière was nearing the end of her power, for the king had just come under the influence of Madame de Montespan, who, like himself, was in the full tide of youth. Louis had carried out the fateful ideas of monarchic rule, the Fronde affair was over, and its supporters were now his faithful servants; he had also been victorious over the foreign foe. The lady was beautiful and full of spirit and caprice. Madame La Vallière, who was modest and affectionate and inclined to be sad, hesitated a little while and then retired to a cloister.

In 1668 the king would only have about him what was entirely pleasurable. The garden was now, as we have said, quite ready as seen from the castle, with its great lines, and its main ornament—the parts for show. But the twelve shrubberies made in the three main walks and the three crossways were at that time for the most part only simple *massifs*, as they were called, consisting of groups of trees with paths among them. There is only one of these thickets that appears to have been turned into a decorative garden inside, perhaps with a view to this fête; it was called the Water Hill, on account of the rock-like fountains springing from a round basin; or else the Star, because of the five paths which approached these fountains and were bordered by trellis in a clever way. Here the "grand folk" took their breakfast, where the place was ornamented with vases, arcades of cypresses, and sculptures. From here they proceeded to the theatre at one of the crossings, and soon came to the basin where the Saturn fountain was set up, with foliage overhead and costly tapestries intended to imitate marble. Here Molière's comedy *Georges Dandin* was performed, with music by Lulli. As early as 1665 Molière had become the king's court poet; and Louis (who had at first detested him because of his satire) treated him later on as protégé and friend, as long as Molière lived; and it is well known that the greater number of his comedies were written for the king's fêtes and were often directly commissioned.

In French gardens permanent theatres of greenery, without which later on in the eighteenth century no garden was complete, were still unknown. Pastoral plays and masques, which originated at these fêtes and because of them, were by preference played with a garden background. Spain had done things of the sort, as we know, that were quite astonishing. The theatre at Buen Retiro, with the stage opening into the park, was a peculiar blend of permanent theatre and garden background. When the party stayed in the garden, the stage was placed here or there in a suitable place. It was the business of mechanicians and stage-managers to produce surprising effects and wonders of decoration in the least possible time. The theatrical performance was for Louis XIV., as for Philip IV., the centre point of the splendid fête.

A place specially beloved by the king was the grotto, which he also used as a background for all kinds of musical performances. The other festivities on these May days of the year 1668, the supper and the tennis, took place at the various crossways of the avenues, or at the basins of Flora and Ceres, which were converted by the artists into wonderful rooms, all green and adorned with fountains, sculpture and draperies, so that people could say (comparing them with the rooms in the castle which had their own fountains and flowers at these fêtes) that "palaces have turned into gardens, and gardens into palaces." Two displays of fireworks given in the chief walk came at the end of these joyful days: the cost was computed at nearly 120,000 livres. A picture by the artist Patel has preserved the scene of castle and garden, as the king arrived at the fête in his chariot with six horses.

Shortly after this, when the remains of the fête were still ornamenting the garden, Lafontaine paid a visit to Versailles accompanied by three friends, Racine, Boileau, and Molière, with a view to seeing the king's new arrangements. Lafontaine made use of this visit in a kind of prologue to his poem *Psyche*, and to him we owe the first description of the garden. The friends came in a carriage, and made their first halt at the menagerie. This perhaps belongs to the works of Louis XIII., and is situated in the park at the place where the cross-road of the great canal ends, and where the Trianon stands on the opposite side (Fig. 401). But Louis XIV. was incessantly building there: the central piece which comprises the chief animal-house in the middle, and the little ones grouped round it like the spokes of a wheel, was designed by Mansart, and has supplied the pattern for many other menageries in parks. Next to the menagerie comes the beautiful orangery, and this draws spirited verses from one of the three.

At table the writers talk about the owner of all these lovely things. "It is only Jupiter who can without rest rule the world; human beings need to pause. Alexander gave his leisure to debaucheries, Augustus to games, Scipio and Lælius often amused themselves with throwing flat stones over the water, but our king is content to build

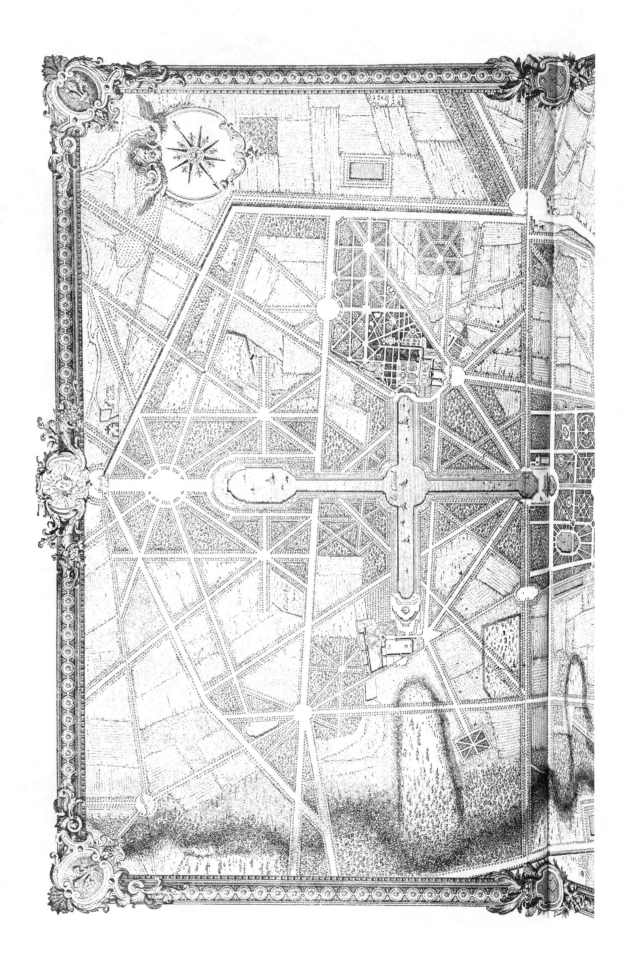

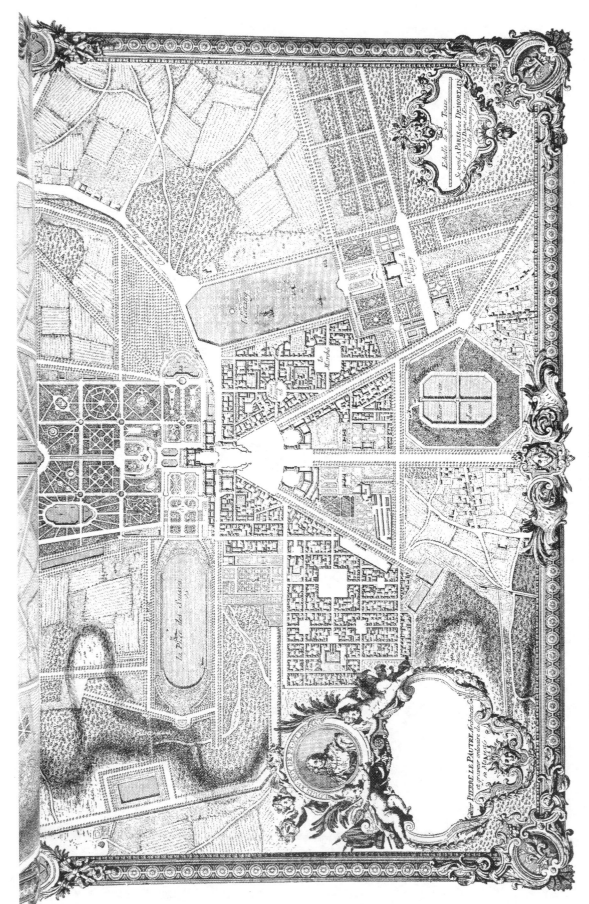

FIG. 401. VERSAILLES—GENERAL PLAN, INCLUDING CLAGNY, THE TRIANON AND THE MENAGERIE, AND SHOWING THE GREAT CANAL.

FIG. 402. VERSAILLES—THE BOSKET OF THE THREE FOUNTAINS

palaces, and that is worthy of him . . . gardens so lovely and buildings so glorious are an honour to our country." This kind of flattery was what Louis liked best to hear. The end and aim of the friends' visit was, however, the grotto, and this is described in detail with its ornament and water-plays. But when the man on guard wants them to take part in the fun of these tricks, they cry off, and he can "keep them for townspeople and Germans." They ask for a dry place, and then Lafontaine reads his friends the first book of his poem. Between the reading of the first and second books they look at the other garden, and especially enjoy the charm of the basin of Latona, revealed to them as a whole from the stairs above it. This Lafontaine describes in his melodious verse. A little later Mademoiselle de Scudéry visits the garden, and her description is accurate to the minutest details, in her own eloquent and picturesque style. In the grotto she brings before our minds the concert of artificial birds and water-organs.

After this fête of 1668 the want of room in the small castle made itself over-whelmingly felt, and Louis could not again venture to house his guests so badly. Therefore he decided to build, but only on the strict understanding that the old castle must be preserved, whereupon Le Vau put the first shell round it. But it is characteristic that the king kept, to the end, his own private bedroom and sitting-room in this old central kernel of the castle. Le Vau's building was set up over the filled-in moat and the terrace in front of it, so that the parterre at the side had not to be removed; and the disappearance of the moat is significant architecturally, for it was a final farewell to the Renaissance feeling in France. A new age needed more space than a ring of water would allow. It is probable that from this date no castle of importance in France had a canal entirely round it, although, as we shall see, the founders did not like to be quite without water-trenches.

The fête had a greater effect on the garden of the future than on the actual castle, for both Le Nôtre and other artists had observed, when they put up the perishable green rooms in their splendid variety, that it would be much better to make such private fête-apartments with out-of-door decoration that would last, and so would be at hand always, and could be quickly got ready; for now, when they had to realise, as Colbert did at last, that the court would be living in the open, they must be continually inventing new surprises and new fêtes. For this purpose they could best make use of the shrubberies in the *Petit Parc* (Fig. 394), as the garden reaching to the great canal was now called. Some of these were already started. Now, however, in rivalry with Le Vau, who was very ambitious, a most prolific period set in for Le Nôtre. In the years 1669–74 all the boskets on the northern side of the great avenue came into being. First, after the making of the *étoile*, the two at the side of the Children's Walk that led to Neptune's basin were laid out. In the green *massif* of trees on the left—for the most part confined by a trellis—there appeared a water-arbour (*berceau d'eau*) where people could walk under the arches and keep dry; on the other side there was a water-pavilion, that consisted of a great series of jets of water and dolphins. Later on there was made in this place the beautiful Bosket of the Three Fountains (Fig. 402).

Le Nôtre and the other artists who were working together during these feverishly busy and productive years knew well that they could only hope to please a court that was *blasé*, and easily bored, by the production of more and ever more startling novelties. The artists in water had the most trouble, for they had to produce Le Nôtre's new surprises. An immense supply of water was wanted to feed the endless fountains that were made

FIG. 403. VERSAILLES—THE BOSKET OF THE WATER-THEATRE

in these years, and new ones were always in request. Wonderful fountains arose at every crossing-point of walks, after the temporary buildings for theatre and fête were put up in 1668. On the top of the basins were Flora, Ceres, Bacchus, and Saturn. These are some of the few statues still in existence. But any number more were needed in the groves.

The water-theatre (Fig. 403) in the northern group is one of the most artistic and most admired of these; at any rate it is designed as a permanent garden theatre artistically decorated, and it was so used. There is a semicircular stage with three paths rising gently from it in the form of a star, and in each of these there is a water-stairway with several sets of jets spouting high; then on both sides of the water-stairway there are narrow

paths, crossed by jets that form arches and rise from between pyramidal yews. Where the paths meet there are smaller places with fountains; and the tall trees, which throw the whole into deep shadow, are held back with low trellis-work. A narrow space with fountains divides the stage from the spectators, who are in a large apartment with seats of grassy lawn raised up as an amphitheatre. The picture is completed with vases of flowers and fine statue decoration.

One of the most charming ideas of Le Nôtre was the labyrinth scheme, which was carried out in 1674. Lafontaine had brought the name of Æsop into everybody's mouth with his *Fables*, and had also made French people familiar with Æsop's appearance in the beautiful prose preface to the work. So Le Nôtre set up the Greek fable-monger opposite a figure of Eros at the entrance to the labyrinth. Anyone strolling in, enticed by Eros, will find that Æsop will be his guide, with his fables. For at every complication of the paths, which are very intricate, there stands a fountain adorned with animals taken from

FIG. 404. VERSAILLES—FOUNTAIN IN THE LABYRINTH

one of the fables (Fig. 404). There are thirty-nine fountains as different surprises, all very pretty and humorous. At that time it was of the first importance to find some new device instead of the well-known designs for the labyrinth. And here there was a veritable garden of enchantment; the stranger had the loveliest pictures before his eyes, whichever way he turned. The animals were made of lead, and painted in natural colours, but the fountains were of coloured stones and shells, and their background mostly of lattice or smoothly-cut beech hedges. Nothing more charming could be imagined than these animal groups designed by the best artists.

The king and his court took great part in such developments, and there is a whole series of drawings and engravings showing Louis and his retinue inspecting one of the new groves. To make plans for fresh boskets became a sort of game at court. Le Nôtre certainly did not smile on amateur efforts, but when the powerful mistress, Madame de Montespan, wanted one of her own ideas carried out, all hands had to set to work. Madame little knew what an old scheme she was introducing into French gardens with her bronze tree that spurted water from the tips of its leaves. Perhaps her idea came straight from Spain. The tree was in the centre of a square basin crowned with bronze bulrushes, from which rose jets that crossed with those starting out of the tree, while swans at the corners sent their own jets into the pond. Other fountains completed this pretty artificial picture, as for example a pond with a

FIG. 405. VERSAILLES—BOSQUET DE L'ARC DE TRIOMPHE (*see page* 79)

bowl of fruits and water spurting out of them, and a so-called water-buffet, a very popular structure. At the time when Madame de Montespan was in power, this bosket, called Le Marais (Fig. 406), was one of the most admired, but the king's taste in ladies changed and with it his taste for art; and so the designer was fated to see it changed in her lifetime, in the year 1705. All the groves, indeed the whole garden, contained a host of statues at that time, some of them copies of antiques; but far more were original, and one may easily suppose that from the year 1669 onwards every important sculptor in the kingdom was busy on Versailles. The garden to-day gives only a feeble idea of the original decoration. The statues have mostly fallen down, and in the revolution some were destroyed, though a few have been kept in museums, such as Puget's Milo of Crotona and the Perseus and Andromeda, both in the Louvre. But one must mentally put these colossal groups back into their park surroundings, by the green walls of the King's Avenue, to get any notion of their effect.

The whole terrace was greatly changed at the same time as Le Vau's building. Le Nôtre thought there were far too many parterres for flowers: at the side of the castle, only slightly sunken, were the two large flower-gardens, and on the lower side of the old *parterre de broderie* on the chief terrace was the one behind the basin of Latona. The

FIG. 406. VERSAILLES—LE MARAIS BOSQUET

old one did not seem to Le Nôtre conspicuous enough, so he conceived the idea of having a handsome water-mirror in this part. At first he seems to have been a little irresolute about it, and he made one large basin and four smaller ones; but afterwards combined them all—water, flowers, and grass—very cleverly into a fine water-parterre. According to the plans of Le Brun, the water-parterre was filled with an incredible number of marble statues and groups of allegorical nature, and between these stood great vases: towards the stairs of the Latona there were two sphinxes with children on their backs, which were afterwards moved into the northern parterre. New statues were continually being brought into the Latona Horseshoe, and in the King's Avenue (now

FIG. 407. VERSAILLES—THE BOSKET OF LA GALERIE D'EAU

the *Tapis Vert*, Fig. 400) rows of vases and statues stood on both sides of the way. At the end of this avenue, where the chariot of the sun and the god appeared, rising out of the waves, there was also the canal, now extended to its immense length. It was generally covered with boats, manned by Dutch sailors; later on there were even Venetian gondoliers, who settled in the park, in a small colony still called "Little Venice."

All of this was ready, though part of the statuary was as yet only in model form, awaiting for completion the royal command and a full state purse, when the king, who, for months past, had chosen to stay at Versailles, resolved to give a splendid fête in honour of the second conquest of Franche-Comté in 1674. "One thing specially remarkable about the king's fêtes," says Félibien, to whom we owe the description of it all, "is the great speed with which all this glory appeared. His commands were so carefully and diligently carried out that it seemed like enchantment; almost in one moment, before you observe it, you are amazed to find theatres erected, groves with fountains and figures, refreshments carried about, and thousands of things going on that it would seem impossible to get done without a great many workmen and in a long time." It was in order to achieve with promptitude what the king required that Le Nôtre made so many groves (Fig. 407). They stood ready to supply the best of fanciful backgrounds at the six-days' fête. The courtiers began by paying Madame de Montespan the compliment of taking their morning meal in her newly finished bosket, the Marais, the beauty of which was now enhanced with garlands of flowers and many porcelain vases. Operas, plays, and fireworks now rang the changes once more. In front of the grotto Molière's *Malade Imaginaire* was performed, and on the canal there were naval shows, while over all a fairy-like illumination shone into the black night, where suddenly all the fountains leapt into view, lighted by flames of many colours.

The year 1674 was the crowning stage of Louis' life in every way. He was thirty-six years old, and felt in full possession of his powers. He was not unjustifiably proud in his conviction that he had become the central point and focus of art and culture in his own country, then incontestably at the head of all Europe. Every objection, even from the Church, was silenced in these days; for Louis was undoubtedly monarch and Mæcenas in one. His love for Madame de Montespan gave the inward stimulus to his being; she was the woman who supplied what in these years he wanted. She was beautiful, proud, self-willed, and full of spirit and fun, a nature which wanted to rule and gave in to him alone. She was the kind of queen he needed for his fêtes, and one who inspired him with a desire for peaceful days after the campaigns of war. When the fête of 1674 was over, it was an understood thing that the king would make Versailles his proper residence, since he cared for Paris less and less. Then when his passion for the marquise knew no bounds, he determined to make a fine place for her near Versailles, inferior to nothing but Versailles itself. She would be near enough at Clagny, a royal property lying north-east of Versailles, and yet not so admittedly by his side. On 22 May of the same year Colbert's son submitted to him a plan designed by a young architect called Mansart, as yet little known. The king's answer was: "I can say nothing yet; I will first hear what Madame de Montespan thinks." On 12 June the answer was: "We both agree; they are to carry out the plans at once and begin work without a moment's delay. Madame de Montespan is very anxious to see the garden so far advanced that it can be planted this autumn." Colbert knew that the king was one who expected things done that to other people would seem impossible, not only in the case of buildings of this kind, but also at his fêtes.

When Madame de Sévigné visits Clagny in the next August, 1675, she writes to her daughter: "We have been at Clagny, and what shall I tell you about it? It is a palace of Armida; the building rises *à vue d'œil*, the gardens are already made. You know what Le Nôtre is. He has left standing a little dark wood which is very nice; and next comes a little wood of oranges in great tubs: you can stroll in this wood, which has shady avenues, and there are hedges on both sides cut breast-high, so as to conceal the tubs, and these are full of tuberoses, roses, jessamine, and pinks. This novelty is certainly the prettiest, most surprising and ravishing that one could imagine, and the little wood is greatly liked." Naturally at that time the castle and gardens were not yet finished: before everything was complete in the way of marble sculptures, paintings, gildings inside the house, orangery, and gardens—costing the sum of about 17,000,000 francs—the year 1680 had arrived, and the star of Madame de Montespan was rapidly waning. She no doubt did live for a time in the beautiful castle, and in 1683, after his marriage to Madame de Maintenon, Louis did present it to his former mistress. But the pain caused to the lady by her humiliations was too much for her pride in the long run, and in June 1692 she left everything to her son, the Duc de Maine, and entered a nunnery.

At the end of the eighteenth century the ruined castle was pulled down, and to-day there is no trace of it. Mansart made his first experiment at Clagny (Fig. 4c8). As at

FIG. 408. CLAGNY—GROUND-PLAN

Versailles, the buildings were round an inner court with a large one in front, at the sides of which kitchen-gardens and stables were cleverly concealed; and this front court was enclosed by a moat, showing how hard it was to part from the traditional idea that a castle must have trenches as a protection. The main part of the building was set forward into the garden by side wings, a plan carried out on a small scale by Mansart, serving as an example for the future enlargement of the castle at Versailles. The garden, which Le Nôtre laid out "in his own style" as Madame de Sévigné says, and which has been so greatly praised, could be no more than the private garden of a small house. But the style of Le Nôtre is here fully exemplified, and the middle show-garden gently sloping in terraces from the castle (Fig. 409) ends in a large water-mirror, the great pond which

FIG. 409. CHÂTEAU DE CLAGNY—THE PRINCIPAL PARTERRE

the master found to his hand and made into the shape of the long canal; there were groves bordering the middle garden, planted with hornbeam and silver fir. In front of one wing of the castle, whose ground floor served as an orangery, there was that little orange wood so charmingly described by Madame de Sévigné, with a dark grove at the side that made a pleasing contrast to it.

The works of Mansart at Clagny recommended him to the favour of the king. When Louis was living at Versailles more and more uninterruptedly, it became inevitable that the castle must be enlarged; for though he might receive his court for summer entertainment, he could not possibly receive the Government. Le Vau had died by 1670, and his pupils had finished the building; but now the king, who could never brook delay, needed a young, energetic person, and so Mansart began the gigantic structure of the two side

FIG. 410. VERSAILLES—BOSKET OF L'ÎLE ROYALE

wings, which was started and carried through with a really superhuman speed. As many as 22,000 or even 36,000 workmen were employed. The king would hear of no interruptions—even an epidemic must not stop the workmen. Madame de Sévigné writes that in October 1678 there were "cartloads of dead bodies brought out of a sick-house every night." Le Nôtre worked on unweariedly at the garden, for the king and his court had to have something new—and again something new. In the official *Court News* would constantly appear the statement that the king had been to visit a successful alteration. And so new groves were made as often as possible, and in due course the best one of all, called the King's Island. This was a very large basin, comprising two boskets, with an island in the middle, made lively with an endless supply of water-jets (Fig. 410).

Greater work was needed in the alterations that from the year 1677 gradually affected almost all the groves; indeed in Louis' long reign some of them were altered five times. First the architect renewed the two groves at the side of the Avenue of Children, and there appeared the Triumphal Arch bosket (Fig. 405) that Rigaud has engraved. Le Nôtre got longer leave in 1678, so that he could go to Italy and find inspiration. He was no longer young, but he was greatly esteemed by the king, who liked his open, careless, even childlike, enthusiasm, and was pleased to have him about. New legends

II—G

grew about his name, now well known all over Europe because of his doings at Versailles, especially on the occasion of this visit to Italy, and every beautiful garden must
needs enjoy the reputation of having been designed by him. This was asserted over and
over again about Villa Ludovisi, which as a fact had been in existence more than thirty
years. Though Le Nôtre had, it is true, studied in Italy when he was a young man, such
a work as that would hardly have been entrusted to an unknown artist. We have an
example of the kind of impression his childlike, frank character produced in an anecdote
told about him. They say that during an audience of Innocent XI., Le Nôtre was so

FIG. 411. VERSAILLES—BOSKET OF THE FONTAINE DE LA RENOMMÉE

charmed with the Pope that he kissed him, and when this was reported at the French
court the people about the king refused to believe it; but Louis said, "I can very well
credit that, for he has often kissed me when he was excited." Le Nôtre kept his eyes
open, and made friends with Italian artists, so that he came back with a great many new
ideas, which he at once carried out at home.

The garden at Versailles had now to bear a loss, for the beautiful grotto, which had
been its chief ornament and a favourite resting-place of the king for twenty years or so,
fell a victim to the building of the north wing. There was no thought of transferring it
to another place, because the taste for this kind of ornament and many-coloured shell-
work began to fade in the eighties. The king had become more serious, and his taste
changed to a love for stricter, simpler, and larger lines. But at any rate some pretty
groups were kept; they were first moved into another grove, whose fountain bore the
statue of Fame (Fig. 411). They remained there till 1704, and were then transferred to

FIG. 412. VERSAILLES—COLONNADE, WITH BOSKET BEHIND

a new bosket that was made in the place of Madame de Montespan's weeping tree. For their present unsuitable position in three niches of a great semi-artificial rock above an irregularly made basin, the era of English taste, with its sham romanticism, which fortunately has not done much harm elsewhere at Versailles, is responsible.

The king lost with his grotto a kind of concert-room in the open, for he had greatly enjoyed having musical performances in front of this grotto or inside it. Not till two years later was the charming pillared bosket set up (Fig. 412), which was mostly built by Mansart. Thirty-two marble pillars coupled with pilasters were set in an elliptical curve, connected with balustrades and arches, with fountains between, and in the centre there was a beautiful group representing the Rape of the Sabines. The king often had the concerts held in this grove, which was an early example of the new simpler taste. The new orangery

FIG. 413. VERSAILLES—BRONZE STATUE BY WATER-MIRROR

was made by Mansart at the same time that the grotto was broken up, also the enlargement of the south parterre to make it the same size as the one on the north, and again the final extension of the great water called the Swiss Lake, which now made a suitable ending to a large crossway that issued from the basin of Neptune, at that time called the Dragon fountain, because of its decoration. The central feature of this up-and-down cross-road, the great terrace in front of the castle, was also completed in 1684. The eye grew somewhat weary with all the prettiness, and the water-parterre blocked the straight way to the castle, so that people had to go round; therefore a large wide avenue was made in the middle where all the fine processions could be displayed, having on both sides great water-mirrors—which can still be seen—with their incomparable decoration of bronze statues made by Tuby, Keller, Coysevox, Le Hongre (Fig. 413), and others.

Two wonderful groups were designed for the middle of this water: the Birth of Venus, and the Thetis group; but it was only for a short time and only as models that they adorned the place, which now looks rather bare. These severe waters with their paucity of ornament (as compared with the host of marble statues) made exactly the right prelude for the garden, suiting the character of the Versailles court, as it developed in these years with ever statelier ceremonial. We must imagine all the accessories of pomp and splendour, when the king was carried for a promenade in the park in his chair, accompanied by dignitaries in order of precedence. These journeys he always took in small specially made carriages, or with some guest, to show him the beauty of his park, as he loved to do. He himself wrote the first "Guide" to the park and gardens, and his servants had to follow this in conducting people round. It was arranged precisely how they were to pause on the

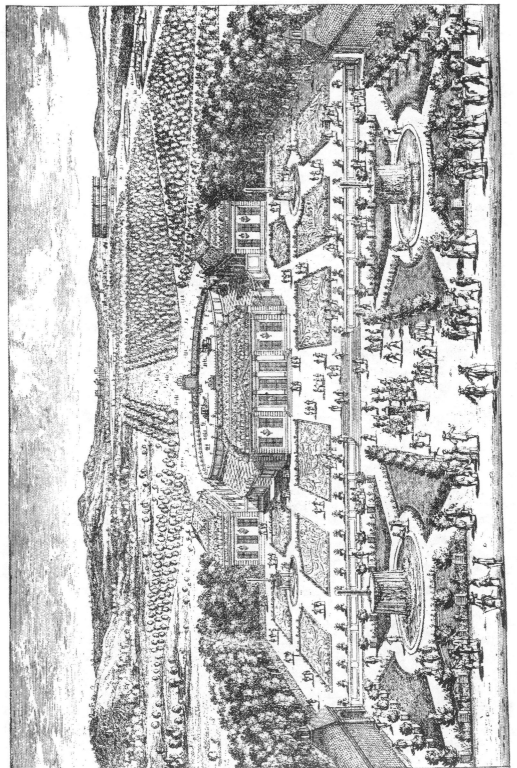

FIG. 414. TRIANON DE PORCELAINE—VIEW OF THE GARDENS

steps of the Latona parterre, so as to get the orientation of the whole of the terraces, and then pass down to the Latona basin. At the foot, a point of view is particularly recommended, which allows all the most noteworthy of the water arrangements to be seen at a glance; and this is even now known as the *point de vue*. Then they are to go on through the main avenue as far as the great canal, and then look back, so as to get the entire castle as a complete view seen over the garden. Then come the groves on the left of the canal, next the orangery, and finally the various places on the right side. This is a method of conducting which is used by every modern guide.

The king's passion for building and altering was not long satisfied with Versailles;

FIG. 415. TRIANON DE PORCELAINE, SHOWING THE SIDE PAVILIONS

and beside the mighty works there grew up a whole series of castles. Even before Clagny was begun, there sprang up in a few months one small costly erection: the Trianon de Porcelaine. This happened in 1670, and it was on the northern end of the crossway of the canal. "It was regarded by everyone as a marvel," says Félibien, "for it was only started at the end of the winter, and by the spring there it stood, as though it had grown out of the earth with all the flowers about it" (Fig. 414). Louis wanted to please Madame de Montespan with the little house: it was only a tea-house where you might take refreshments at midday in the heat of the summer. It was about this time that reports had arrived from the French missionaries in China, which were destined in a short time to play such an important part in the story of the changes of taste in gardens; and these reports had astonished the world. People began eagerly to collect porcelain, stuffs, and paintings from China, and the porcelain tower of Nanking was the eighth wonder of the world. Louis desired to have something of the same kind, and so he must have the small Trianon tea-house adorned *à la chinoise*. Owing to lack of porcelain they used faience in the Dutch manner, but this was produced at a factory of faience newly founded at Trianon itself. On the façade faience plaques were put everywhere, and great blue vases on the cornices and on the steps which led to the canal, with white marble busts on plinths of faience. The inside was to correspond; and you passed through one large room in the middle with a separate apartment on each side of it. The room was all white, with blue figures as ornament, and the floor was paved with faience tiles in the same colouring.

Since each of the side rooms was only one chamber and a cupboard from which issued a bird-house, they added on either side of the house, which flanked an oval court, side pavilions, each set up with the same decoration as a *petit palais* (Fig. 415). Here we meet for the first time with the detached buildings of the dwelling-pavilions, which group themselves as attendants to the chief house, an arrangement that was carried out at the completion of Marly, and from now onwards is repeated many thousands of times, particularly in Germany. To this taste in interiors the garden laid out by Le Bouteux corresponded. On the terraces there had to be flowers, brilliant in colour and beauty, to suit the blue and white pottery; both the seats and the flower-boxes were painted blue and white. Because the flowers in the other gardens were kept back by strict lines of water and trellis, here too this new *variété* must be introduced. "On the parterre opposite the rooms," says Félibien, "you find four waterspouts, which leap up high into the air from four basins raised on pedestals." [The engraving, Fig. 414, shows only two, and this is probably correct.] "From this parterre you mount into another garden which might justly be called the everlasting abode of Spring, for every time you go to it, you find it full of all sorts of flowers, and the air is sweet with the pleasant scents of jessamine and orange as you stroll among them."

A large house made of wood was set up over a winter garden, in which oranges, citrons and other trees were kept rooted in the earth, with surrounds of myrtle and jessamine. The scent of the flowers, tuberoses, hyacinths, and pinks, was often so strong that it was intolerable on the terraces after a while to some people; all the same the odours were, so to speak, tuned to the right pitch, and inside the place there was a *cabinet de parfum* as its finest ornament, and the sweet-smelling flowers of many kinds combined to make a real harmony. It is a very noticeable thing, that the ambassadors from Siam, who were received magnificently at Versailles in 1686, admired this *cabinet de parfum* beyond everything, "for they loved strong scents, and were delighted with this way of making perfume from flowers."

Versailles and all the other places of the king were regarded with veneration in France, and also in Europe, where all eyes were fixed upon France: of this people were very well aware. In 1686, the *Mercure Galant* writes: "The Trianon at Versailles aroused in all private persons the wish to have something of the same kind, and almost all great lords who possessed country houses had one built in the park, with smaller ones at the end of the garden. The burghers, who could not go to the expense of these buildings, dressed up some old booth, or perhaps a sentry-box, as a Trianon or at least as a kind of cabinet in their houses." And all through the eighteenth century this custom grew. The Encyclopædists followed the lead of the *Mercure Galant*, which had only been in jest with the catchword "Trianon." Trianon and Hermitage became synonymous. In the great *Universal Lexicon* of Zeller, which appeared in 1734, we find "Hermitage, a retreat, a low-built pleasure-house, in a shrubbery or a garden, furnished with rough stones, or poor woodwork, and left practically wild, so that one may cultivate solitude and live in the fresh air. It is also called a Trianon." A building like this has gone far afield from the pretty porcelain Trianon—and into a new world. True, Louis himself had aimed at a certain kind of solitude at the Trianon. In his fine barge on the canal he often made his way in the afternoons to this place, whence, looking back, he enjoyed the *ensemble* of his garden and house. But he soon felt that he was still too near the pomp and show from

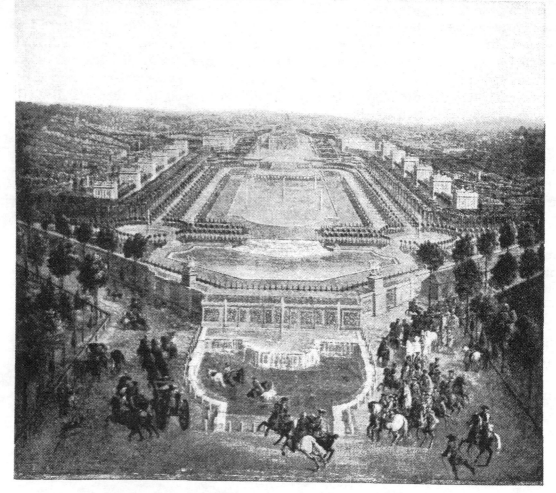

FIG. 416. MARLY-LE-ROI—VIEW WITH THE HORSE-POND IN FRONT

which he wanted to escape. He needed something that was far distant from the castle, and from the ceremonial that he had made himself, and by which he was confined. Hence came Marly.

Saint-Simon thus writes of the beginnings of Marly in his *Memoirs*:

Louis the Fourteenth, *lassé du beau* and tired at last of the swarm of courtiers, persuaded himself that from time to time he needed a small place and solitude. He searched near Versailles for somewhere to satisfy this new taste, inspecting several sites and exploring the lovely banks of the Seine. At last he found a narrow valley behind Louveciennes, unapproachable because of its marshes, shut in by hills on every side, and very narrow. with a miserable village on one slope, which was called Marly.

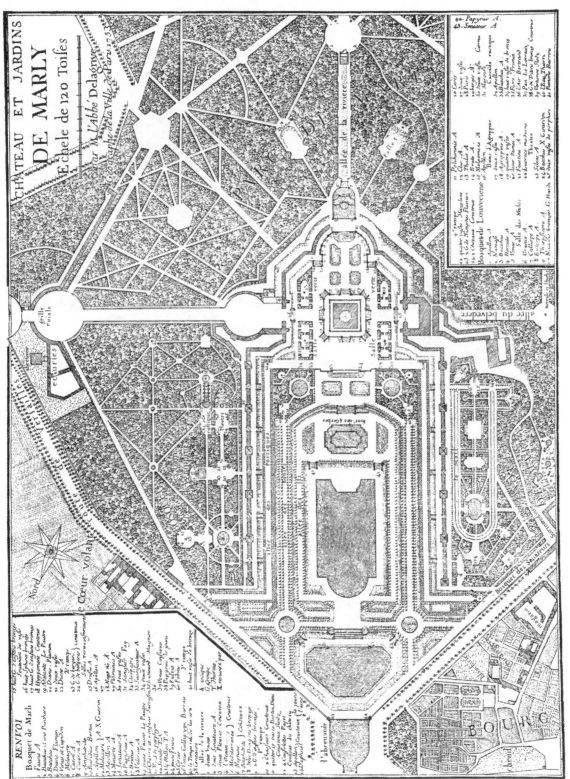

FIG. 417. MARLY-LE-ROI—GENERAL PLAN, 1753

FIG. 418. MARLY-LE-ROI—THE PRINCIPAL GARDEN

So the hermitage was built. The plan was to spend there three nights only, from a Wednesday to a Saturday, two or three times a year, with a dozen or so courtiers for necessary attendance. But what actually happened [Saint-Simon continues], was that the hermitage was enlarged, building after building sprang up, hills were removed, and water-works and gardens were put in (Fig. 416). It is no exaggeration to say that Versailles, as we saw it, did not cost as much as Marly. . . . It was the king's bad taste in all things, and his proud delight, to force nature, and this could not be checked by the pressure of war, or by religion.

Saint-Simon did not like the king; he had outgrown his former tastes, and poured bitter scorn on all the achievements of the *grand siècle*. But the works produced in the years between 1677 and 1684 in this marshy woodland valley, which later on fell a victim to the Revolution, leaving scarcely a trace behind, belong all the same to the most perfect achievements of the century. As Saint-Simon's statement that Marly cost more than Versailles may cause surprise, the following quotation may be made:

Marly est devenue ce qu'on le voit encore . . . cette prodigieuse machine, dont on vient de parler, avec ses immenses aqueducs, ses conduites et ses réservoirs monstrueux uniquement consacrée à Marly sans plus porter d'eau à Versailles; *c'est peu de dire que Versailles tel qu'on l'a vu n'a pas coûté Marly*. Si on ajoute les dépenses de ces continuels voyages . . . on ne dira point trop sur Marly seul en comptant par milliards.

The king had meant to build a hermitage, and he had in mind something like the one in the quiet solitude of the white house in the park at Gaillon; and even when Marly had grown to what it was in the end, he made a pretence of laying aside his royal majesty at the gates, or at any rate of replacing the etiquette left behind at Versailles with a new kind, specially made for this *Buen Retiro*. As a fact, the Duchess Lieselotte was horrified at the want of formality that prevailed at Marly. Hers was a nature that had never cared to take part in that mixture of grandeur and frivolity, severe etiquette and passionate desire for novelty, which was the accepted programme of court life, though it was one that could be changed easily on another stage and in another setting. "You no longer know where you are," she said. "If the king is going for a walk, everyone puts on his hat. If the Queen of Burgundy would walk, she takes the arm of one lady, and the others walk by the side, and nobody can see which she is. Here in the drawing-room we all sit in the presence of the Dauphin and the Duchess, and some actually lie at full length on the sofa. . . . I find great difficulty in getting used to this confusion. You have no idea how things go on; it does not seem in the least like a court."

It was a great honour to be taken out to Marly on these three-day expeditions, and the list of those invited was eagerly awaited. Marly was the exact opposite of Trianon, with its blue porcelain and scented flowers. Elaborate parterres of flowers were not wanted at Marly, for they would not have suited the idea of a hermitage; and even the modest parterre which lay to the south of the castle was made with a sunk lawn and only a narrow ribbon of flowers. The lay-out of the garden was not, in the main, unlike Versailles (Fig. 417); here too there was a large show-garden, but it was all around a great water axis, and it filled out that narrow woodland valley of which Saint-Simon speaks. As a fact, the king, with the purpose of getting a view, had had a hill that barred the end of the sloping garden completely removed, and had demanded a whole regiment to do it. The small castle is half-way up (Fig. 418), and behind it a great cascade passes grandly through the park, which is on rising ground to the south: this cascade ends in a pond of half-moon shape, whence there is a descent by the southern castle terrace (Fig. 419). From here you pass from terrace to terrace on wide convenient steps in front of the house. The middle of each terrace is marked by a pond and many beautiful fountains, *le font des quatre gerbes, le grand jet, la nappe,* and *l'abreuvoir;* and also a final one at the place where the hill was taken away.

On the great terrace stood the charming building of Mansart, rich in frescoes painted by Le Brun, Rousseau, and their pupils. It had a balustrade on the roof, and was only

FIG. 419. MARLY-LE-ROI—THE REAR GARDEN AND HEAD OF THE GREAT CASCADE

FIG. 420. MARLY-LE-ROI—VIEW OF THE GARDEN AND SIDE AVENUES

FIG. 421. MARLY-LE-ROI—LE BASSIN DES MUSES, WITH BOSKET

made big enough to accommodate the king's own family. All the middle part and the top were used for the octagonal dining-hall, and round this were four small rooms. The idea, first conceived at the Trianon de Porcelaine, was thoroughly carried out here, where we also get the first complete example of a central building for pleasure-houses which yet has the character of a hermitage; and this was a pattern repeated endlessly in Europe in the eighteenth century.

On both sides of the house there were originally four basins adorned with faience, which later were supplanted by the so-called green cabinets (Fig. 417)—rooms hedged in greenery where the ladies plied their embroidery, and took their midnight meal (*média-noche*) on festal days. The king's honoured guests were put up in twelve little pavilions attached to the castle, decorated with frescoes and balustrades. They were always arranged as for a small household, and were interconnected by *berceaux*, which formed a semi-circle round the castle as far as the foot of the cascade. There were three other avenues separating these outrunning pavilions from the open terraces of the water-basins; first there came clipped yew, then leafy trees cut into globular shapes, and lastly an artificially made portico (Figs. 417 and 418) of greenery. The eye passes over from the wide show-garden, designed for lively social gatherings, to small separate pavilions, out of which people could slip into quiet secluded groves (Fig. 420). One of these, the *Bosquet de Louveciennes* on the east side, had within it a *cascade champêtre*, an amphitheatre, Baths of Agrippina, and a Hall of the Muses (Figs. 417 and 421), all very attractive little places. On the west side the *Bosquet de Marly* included a large tennis-court, other places for games, such as a toboggan run, a belvedere, etc. All these were in the park on the hill. We must now imagine this charming picture enlivened with a great many statues; even as late as 1753, when the Abbé Delagrive sketched out his plan, there were over two hundred works of sculpture, some of them masterpieces, as for example "The Two Horse-Tamers" by Coustou, which to-day adorns the entrance to the Champs Elysées, but formerly stood by the horse-pond (Fig. 422), and other works of Coysevox, now to be found in the Tuileries Gardens.

At Marly, as elsewhere, Louis was perpetually altering something, first in one place, then in another, as long as he lived. He had scarcely achieved his first complete scheme, in 1684, when it was made the theatre of long-continued festivities of a splendid kind. In 1685 the marriage of the Duke of Bourbon-Condé with Madame de Nantes was the occasion of a great fair, where four stalls were set up to represent the four seasons. Two of these were reserved for the bride and bridegroom, the third for Madame de Montespan and the fourth for Madame de Maintenon. But Madame de Montespan probably saw for certain at this fête that her clever scheming rival had won the victory over her. She herself had persuaded Scarron's widow to come to her as governess and nurse to her son; and slowly, step by step, Madame de Maintenon had contrived to make herself indispensable to the king. His triumphant faith in his own good fortune had been shattered by the miscarriage of his foreign policy; he had grown weary, and in this mood he came under the influence of a woman who had in all walks of life shown great ability as a teacher and a ruler. She was always unnoticed, yet she was master where-ever she came. But in one matter she had never been successful with the king—she had never made him economical, hard as she tried at Marly to check his everlasting alterations and embellishments.

FIG. 422. MARLY-LE-ROI—THE HORSE-POND

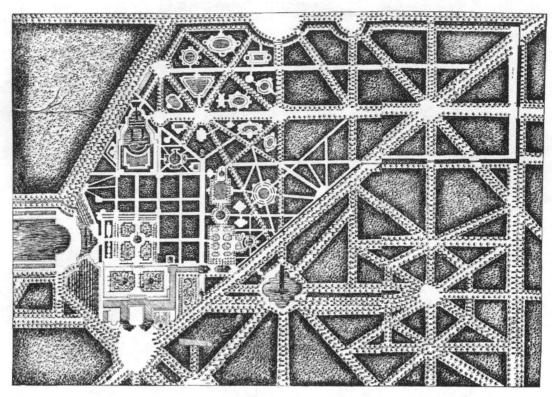

FIG. 423. THE GREAT TRIANON—GROUND-PLAN

Marly held its ground as long as the king was alive, but it fell into decay quickly during the Regency. The fact that it was not utterly destroyed—though the order was actually given—is due to the energetic protest of Saint-Simon, who represented to the Regent the incredible folly of such an intention. So Louis XV. more or less restored it to honour, and there were many fêtes held at the pretty country house; indeed when, on a day in October in the next reign, Louis XVI. was dragged from Versailles to Paris by a howling mob, preparations were actually being made at Marly for receiving the court at a fête. The storm of the Revolution raged over it and destroyed it so utterly that now in the quiet woodland valley nothing can be seen except here and there a straight avenue, a clump of trees, and in its last neglected state the horse-pond, the beautiful *abreuvoir*, to mark the spot where once stood a graceful work of art.

When one reads, in Dangeau's writings, that the king found a palace "more beautiful than ever," one must be prepared for great alterations. "Anything already there that cannot be improved must be destroyed, and something new must be made." In this mood Louis XIV. found himself before the wonderful little erection of the Trianon de Porcelaine. Marly was ready in essentials, and Versailles was practically finished. But this little blue and white tea-house, with its waves of sweet flower scents, what more was it than one of the groves which the master had already cast aside? So he summoned Mansart, who was architect enough never to say no if anything was to be pulled down and rebuilt, and who knew how to entangle the king in projects that were constantly changed. Also it was an argument that Madame de Maintenon, now the actual wife, must not be made inferior to the discarded mistress, and he wanted to give her a little castle as a wedding present. In short, Mansart was commissioned to pull down the porcelain house and to set another of marble in its place. Fifty-six sculptors at once set to work to prepare the necessary carvings. Trianon was to be a convenient garden-house: it had only one story, and its gallery on the roof was decorated with vases and statues. The garden was approached immediately from the house.

The novel feature was a certain irregularity about the ground-plan, which apparently was liked (Fig. 423). People seem to have begun to feel that a very strict axial line was rather oppressive. As at Clagny, you crossed two semicircular moats and passed into an inner court, but this one was formed by two wings, and divided the long main façade into three parts. The whole of the middle division adjoining the court was an open pillared hall, used by Louis in summer as a dining-room. It looks straight into the garden, which lies in the axial line with terrace parterres, boskets, and a fountain to mark the end. It is clear that in this place Mansart had no idea of repeating the plan of separate pavilions; but still the king was so used to living alone in his summer residences, that a guest wing had to be built apart, though connected with the main dwelling by a gallery. This part for guests joins the castle at right angles, and is on one side only, so that an open

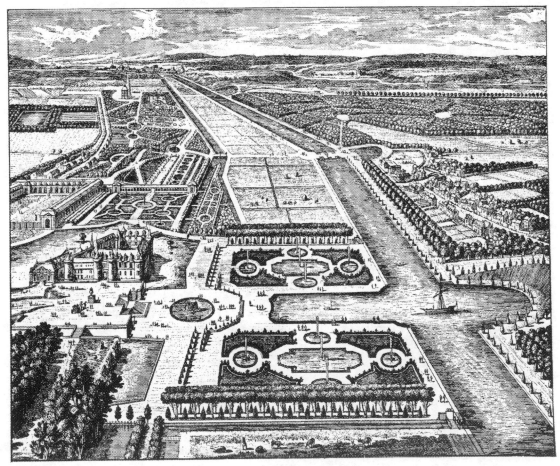

FIG. 424. CHANTILLY—GENERAL VIEW

II—H

FIG. 425. CHANTILLY—VIEW OF GARDEN AND CASTLE FROM THE CANAL

view can be left on the canal side; for here you had the second and most agreeable way
out into the garden, whereas on the other side you climbed up on steps, round a semicircular basin, to the second garden terrace. From the hill-slopes sweet scents were
wafted from the flowers that covered them; and this was perhaps the same in the days of the
old Trianon. The garden of the new Trianon was certainly rich in flowers, and there were
very unusual and magnificent ones grown in the *Jardin du Roi,* which was really a
giardino secreto, situated at the back of the right wing of the *corps de logis,* under the
king's windows.

The guest wing, known as Trianon-sous-bois, got its name from being in the middle
of the side boskets, one of which was greatly admired by reason of a certain novelty in it.
The Duchess Lieselotte, who at one time inhabited this wing, thus writes: "One
bosket is called Les Sources, and it is so dense that the sun cannot penetrate it even
at midday. There are fifty springs, and more, with little streams scarcely a foot wide which
you can easily step over. They make small grassy islands, just big enough to put tables
and chairs on, so that you can work in the shade. There are steps down on both sides,
for the whole place is slightly inclined. Water flows down these steps, and makes a waterfall on either side." On the front of this wing an avenue leads beside the bosket to the
cascade, which shuts off the path in the form of a so-called *buffet.* There are a great number
of other beautiful fountains in the park, which stretches a long way farther on this side,
and is also quite unsymmetrical. There can be no doubt that this garden was intentionally
less enclosed, and treated more simply, so as not to bring it into competition with Versailles, which was near at hand; they took pains to keep a more distinctly country style with
less conventionality.

If Le Nôtre busied himself generally with the plans for Trianon, this will have been
one of the last things he did. He had now grown old, and his prince had shown him
honour; he had ennobled him, and given him the Order of Saint-Michel. The old man
appears to have kept his childlike pleasures to the end. When the king invited him to
Marly shortly before his death, and let him ride beside him in one of the little wheeled

chairs, Le Nôtre suddenly exclaimed, "Oh, my poor father, if only you were alive, and could see how your own son, a mere gardener, is driving in a carriage by the side of the greatest king in the world, there would be nothing wanting to my happiness." One month after this he died. Saint-Simon writes about him: "Le Nôtre died in 1700, after a life of eighty-seven years passed in perfect health, intelligent and upright, with keen enjoyment of his own ability, and honoured because he designed the plans for those lovely gardens that are the glory of France, and that have extinguished the fame of Italian gardens, which are indeed nothing to compare with them; now all the most famous masters of this art come here from Italy, to learn and to admire. Le Nôtre was so candid, so trustworthy, and so upright that everyone respected and honoured him."

Le Nôtre had not, however, during his active career given his mind wholly to the works that he actually wrested from Nature, and which were the visible mirror of the *grand siècle*, for it is clear that the king did not want the gardens already in existence at his old castles to be left quite uncared-for. And the more the reputation of Le Nôtre increased, the more was his advice sought, and also his plans, by the great families of France. One of the earliest cases is the garden of Chantilly, which Le Nôtre always thought one of his finest works. The great Condé made use of the undesired leisure forced upon him by Louis' disfavour between 1660 and 1668, to remake the Renaissance garden at his castle, which, although broken up into many divisions, was really quite a small place. Le Nôtre, whose plans Condé was using, found here what was always wanting at Versailles—water in abundance, though divided into many small canals. He collected all these into the broad band of canal which he made to cut off the main garden crosswise as at Vaux-le-Vicomte (Fig. 424). Since the mediaeval plan of the castle prevented him from

FIG. 426. CHANTILLY—THE FLOWER-PARTERRE

FIG. 427. CHANTILLY—THE GREAT CASCADE

throwing all the gardens and buildings together to make one whole, as at Vaux and Versailles, he made a great stairway plan as architectural conclusion for the parterre, at the castle terrace leading southward. In the centre the canal cross-cuts into the parterre, and on the other side spreads into a semicircular bay, with avenues and meadows adjoining (Fig. 425).

In the parterre of this chief garden water reigns supreme; the open flat spaces are all laid out with groups of five round ponds surrounded by grass, box, and strips of flower-bed—an effect that Le Nôtre tried to get afterwards at the castle terrace at Versailles on a different scale. This water-garden, corresponding to the water-castle in its style, was bordered on one side by a great colonnade, which in its main lines is still standing. On the other side, to the east of the castle lake, there was a second parterre with wonderful flower arrangements (Fig. 426). Behind it many groves were made, of which the finest was the Great Cascade (Fig. 427). The engravings by Perelle, Rigaud, and others have preserved for us pictures of these fine garden scenes, of which nothing now remains, thanks to the Revolution and the change to the picturesque style. If one searches among all the little details to find some remains of the old time, as for example the ruins of the fountains below the tennis-court, one is almost startled at the traces left of the great age of art. On the far side of the castle terrace a piece of the formal park remains with its beautiful high hedges, though the winding paths tell of a somewhat later period. A charming little garden-house, the *Maison de la Silvie*, dating from 1684, with its quiet parterre still preserved, has given its name to this part.

Among the royal castles, Fontainebleau keeps its last state intact (Fig. 428). Le Nôtre finally got rid of all the smaller parts that Henry the Fourth's garden had kept to some extent. The great parterre shows still its simple lines; the canal, though begun by Henry IV., now first shows its full importance for the park. The view from the back of parterre and castle is kept, as at Vaux, with cascades and grottoes in the supporting wall of the great parterre (Fig. 429). No other garden lies so clearly before our eyes in its long stretch; and even to-day there is only a comparatively small bit of it, the old *Jardin des Pins*, that is changed into the picturesque style.

At Saint-Germain, where the fine building of Henry IV. had hardly been inhabited at all, Le Nôtre's chief work, apart from certain enlargement of parterres, was the main terrace, which extends in front of the upper garden as a walk across the river.

The gardens at Meudon were more important. In Louis the Fourteenth's time the castle had passed into the possession of his proud and ambitious minister Louvois, who eagerly completed the enlargements which his predecessor had undertaken, with a view to making an imposing castle, and then employed Le Nôtre to lay out the gardens. Le Nôtre began by extending the castle terrace (Fig. 430) into a large flower-garden commanding a lovely view over the Seine valley, and Paris by its side. The main axial scheme is carried out from the castle by two terraces, descending to the orangery, to rise again, marked out by fountains and basins, to a woody hill above. By the side of the castle there was still the old grotto site with its parterre. Only after the death of Louvois in 1691 did the castle pass into the possession of the Crown; and Louis had a new castle built by Mansart in the same place, to be appointed as the residence of the Dauphin. This little castle stood until 1870, and by that time the front parterres had long been remodelled in the English style. After the demolition of the place one part was rebuilt and utilised as an observatory. The main castle, greatly injured during the

FIG. 428. FONTAINEBLEAU IN THE TIME OF LOUIS XIV

FIG. 429. FONTAINEBLEAU—THE GREAT CANAL

Revolution, was razed to the ground in 1803 and 1804. Of the beautiful prospect that was the work of Le Nôtre there remains only the great terrace, now turned into a public walk which leads by steps into the badly-kept orangery, and to the lime-trees.

Another task, similar in many respects, was awaiting Le Nôtre at Saint-Cloud (Fig. 431). The house of the Gondi family had passed from hand to hand. About 1625 it was owned by an ambitious banker named Herward, who was aiming at court favour. He was a protégé of Mazarin, but the wily cardinal sacrificed him to the king, when his master gave out that he wanted to get the property for his brother, the Duke of Orleans.

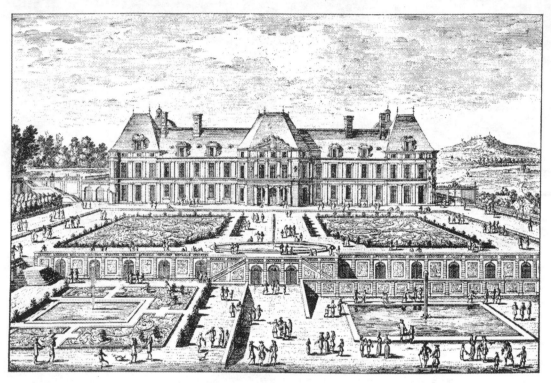

FIG. 430. MEUDON—THE PRINCIPAL PARTERRE

Mazarin managed to buy it for a very small sum, and in due course the duke took his young bride there. Henrietta was greatly adored. It was in her honour that Fouquet gave the first feast in his castle. Here at Saint-Cloud Molière, with his actors, passed the time before he went into the king's service. The sudden unexpected death of Henrietta in 1670 threw the country into the utmost consternation, and all hearts were sad when Bossuret began his funeral oration with the words, "*Madame se meurt, Madame est morte.*" A very different woman came to the palace as her successor, the Palatinate princess Lieselotte, who was blunt and had the straightforward character of her family. She could never feel sufficiently at home in the atmosphere of this court to conquer her homesickness and her longing for her paternal castle on the Neckar. But she loved Saint-Cloud, and in her letters often expresses her delight in the beautiful gardens. She chose it as her home when she was a widow, and always took refuge there when Versailles and Madame de Maintenon annoyed her too much.

FIG. 431. SAINT-CLOUD—GENERAL PLAN

Le Nôtre found beautiful gardens ready to his hand, and best of all the waterfall that was already famous in the time of the Gondi. Here he could do no more, nor did he wish to, than make enlargements and bring together all the little parts. The cascade, which was away from the Seine in a dip in the valley of this very irregular ground, was left where it was, yet he transformed it into an imposing triple waterfall, which flows over steps and into a huge semicircular basin (Fig. 432). But even in this work Italian

FIG. 432. SAINT-CLOUD—VIEW OF THE CASCADE WITH POOL IN FOREGROUND

FIG. 433. SAINT-CLOUD—THE GREAT CASCADE

influence was plain: the French had not forgotten the old Italian art in their treatment of waters falling from a height (Fig. 433). The king and his brother were always rivals in building, for Monsieur wished to do as much as the king did, and the cascade at Saint-Cloud was his pride for a long time, as being unsurpassed in France. It was only when

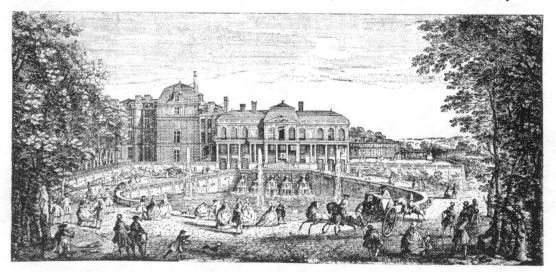

FIG. 434. SAINT-CLOUD—THE SMALL CASCADE

the beautiful water came rushing down at Marly that Louis considered that he himself had the best cascade. Now, when the king had made his first Little Trianon, which, as we saw, was so soon to attract imitators — Monsieur must needs have a hermitage hidden away in the park. There was a charming little *Pavillon de Breteuil* at the end of the lower avenue, and certain *berceaux* leading to the parterre produced a peculiarly homelike effect, while at the same time there was a magnificent view from the top which made the place a favourite with the owners of Saint-Cloud. This pavilion and the cascades

FIG. 435. THE LITTLE TRIANON—THE GARDEN SIDE

are now the only things that give us indications of the great beauties that were to be found on the banks of the Seine even until 1870. The show-garden proper was to the west of the castle, and here also the parterre was adorned with little cascades (Fig. 434) and its climbing axis defined with fountains.

The king lost in André Le Nôtre his oldest fellow-worker. All the same, the great gardener's spirit lived on, and there were always only too many willing hands to do the behests of a king who was unwearied in his desire for change, and whose interest in his garden never slackened to the day of his death. But when his eyes were closed, and a child of five was on the throne, all Louis' works were threatened with the gravest danger. True, the Regent had no such barbarous intention with Versailles as with Marly, but it was soon felt that the garments of a giant were too big for the new race. In the opposition of Saint-Simon to Versailles, to be sure, we chiefly feel his general dislike to the king.

But in the course of time there arose continually new voices that were depreciatory. "You can only get to cool shade when you have crossed a burning hot part, at the end of which you have no choice but either to climb up or to climb down, if you can walk farther at all." The violence done to nature everywhere put people off, and annoyed them. "The artificial waters which are too abundant on every side are getting green, thick, and muddy; they emit an unhealthy, enervating dampness, and a still worse stench. Their effect is incomparable, but it can only be enjoyed with caution; there is nothing left but to admire—and run away."

FIG. 436. VERSAILLES—THE APOLLO GROUPS AS THEY ARE TO-DAY

Louis XIV. made a somewhat dangerous, but philanthropic, innovation when he ordered in 1704 that all the gardens and all the fountains were to belong to the public, and he himself opened the boskets to the people. But after his death very little took place in the gardens, and this meant that they were soon half ruined. But in 1722 the young King Louis XV. moved his court to Versailles, and the place became once more the scene of fêtes. In 1740 the Fountain of Neptune was ornamented with the greater part of its sculpture, which as a fact Mansart had already designed. But the king so soon felt uncomfortable at Versailles, which was too big for him, and therefore seemed empty, that he went more and more often to Trianon, especially after the temporary relation with Madame de Pompadour had become a permanent one. And now the king's circle must always be inventing something new; this was the "martyrium" that his mistress had taken upon herself, for it was her task to subdue the ennui, the yawns, that overcame king and court. Already the longing for nature pure and simple had penetrated their life, and already Louis XV. had made himself a wing in the roof of the Trianon, and had had his *petite ménagerie* built, with a home farm, a cow-shed, a sheep-pen, and a dairy. In 1759 he had the Little Trianon (Fig. 435) built, in the vain hope of finding greater peace in a smaller place. At that time the Little Trianon still had a formal garden, but later on this house is associated with the figure of Marie Antoinette, and is noteworthy as the first example of the English garden on French soil.

But what became of Versailles? We possess two pictures of the year 1775, which show us the frightful devastation in the garden. All the tall trees have been cleared away, and between them branches are lying about, fountains drying up, and the white bodies of fallen statues sticking up, while here and there one fountain still flows. But the spirit of Le Nôtre was still powerful. When a certain desire for regeneration, prompted by new tastes, asserted itself, no one ventured to upset his ground-plan, and it was only here and there in particular groves that the new spirit gave free rein to its fancies. Thus the man who made these pictures, Hubert Robert, designed a "natural rock," to provide shelter for the poor white groups that were taken out of the grotto (Fig. 436). In 1817 the Grove of the King's Island, which had long before degenerated into an evil swamp, was converted into a small picturesque flower-garden. The new way of planting was not calculated to induce

FIG. 437. A VASE DECORATION IN A FRENCH GARDEN

Louis XVI. to make a long stay, especially as his own taste did not abjure formal regularity. But then came the Revolution, and the shrieking mob hustled the king and his family out of Versailles. The Assembly gave orders that castle and garden were to be destroyed. But there must have been something compelling in its size that protected the royal castle, in spite of the fact that hatred would naturally be chiefly directed upon it. To save the garden, the wise director suggested that the boskets should be used for the cultivation of vegetables, and as potato fields, and this proposal calmed men's minds, and there was no more talk of destruction. As a fact, however, the gardens gradually fell more and more into decay, and at best they only appeared in the nineteenth century as glorious remains. Then people gradually began to search in the groves for anything that might be preserved from the past. Late movements in taste, favouring the old style, helped towards a sort of resurrection of the ancient giant. There was some restoration of what it was possible to restore. The parterres once more glowed with a wonderful array of flowers, and thousands of people came each month, attracted by the spectacle of the playing waters.

Although to-day the picture is wanting in its old wealth of sculptures, in the many coloured accessories of royal processions, in the high green walls of the hedges, and in the great beauty of most of the groves, there is a lively consciousness in all French people of what a noble legacy from their *grand siècle* is left to them at Versailles.

CHAPTER XIII
THE FRENCH GARDEN IN EUROPEAN COUNTRIES

CHAPTER XIII

THE FRENCH GARDEN IN EUROPEAN COUNTRIES

IN the time of Louis XIV., the formal garden had reached a height that could never be surpassed. There were then associated an ingenious artist, an enthusiastic ruler with unlimited powers, a technical skill that nothing could baffle, and a host of practical fellow-artists to make the individual sections contribute to a successful whole. It followed that the art grew to its utmost height, and became an organic thing, essentially independent.

The northern garden style originated in France, and became the one shining example for Middle and Northern Europe. All eyes were fixed on the magic place Versailles; and to emulate this work of art was the aim of all ambitions. No imitator, however, could attain his object completely, because nowhere else did circumstances combine so favourably. The great importance of the style lay in its adaptability to the natural conditions of the North, and in the fact that it was easily taught and understood. Thus we have a remarkable spectacle: in spite of the fact that immediately after Louis' death the picturesque style appeared—that enemy destined to strike a mortal blow at a fashion which was at least a thousand years old—for some decades later there came into being many specimens of the finest formal gardens, and the art flourished, especially in countries like Germany, Russia, and Sweden.

France did not become mistress of Europe in garden art merely because of such of her examples as could be copied; of almost equal importance was the wide popularity of a book which first appeared anonymously in France in 1709 under the name of *Théorie et Pratique du Jardinage*. In the third edition this work was fathered by the architect Le Blond, who had distinguished himself in the construction of gardens. Some had thought D'Argenville Dezalliers to be the author. Never before did a book lay down the principles of any style so surely and so intelligibly in instructive precepts. It claimed to be the first work entirely devoted to the pleasure-garden, the kitchen-garden being dismissed with complete indifference. "In large gardens there are good vegetable plots worth looking at, but they are kept away from the house and do not contribute to its grandeur or beauty." The author will accept only Boyceau and Mollet as his predecessors, and then only in certain departments.

The great diversity in garden art, which gives a place to every other art, compels the garden student to receive a many-sided education. "He must be something of a geometrician, must understand architecture, must be able to draw well, must know the character and effect of every plant he makes use of for fine gardens, and must also know the art of ornament. He must be inventive, and above all intelligent; he must have a natural good taste cultivated by the sight of beautiful objects and the criticism of ugly ones, and must also have an all-round interest and insight in these matters."

Le Nôtre had brought up a generation of pupils who were educated in these qualities and could easily apply what they knew, and Le Blond, who was busied with the drawings, at any rate, for this work, was one of them. He explains the garden in a methodical way. After preliminary tests have been made, a site is to be preferred where the land is either flat or gently inclined, and not a steep hill. He objects to very high terraces, commanding stone steps, too much trellis, and too many figures. Here we clearly get the opposite of the Italian Renaissance style. In vain had the attempt been made on the French side of

FIG. 438. ORNAMENTAL VASE IN A FRENCH GARDEN

the Alps to imitate Italian gardens; it was labour in vain to do here what came so easily to an Italian. This is expressed in that classical sentence: *Le cose che si murano sono superiori a quei che si piantano* (The things that are walled in are better than the things that are only planted). The French garden produces a plant-architecture to which statues, fountains, and water must accommodate themselves. The house must, of course, be somewhat raised on a terrace overlooking the garden, and the site must be fixed in obedience to four main principles, (1) art must be subject to nature, (2) the garden must not be too shady, (3) it must not be too much exposed, (4) it must always look bigger than it really is. The first principle, soon to be put forward by the picturesque style as a destructive criticism, only emphasises the opposition between French plant-architecture and Italian wall-architecture. The other principles refer to the effort made by the French garden to combine the greatest possible variety with the strictly formal style. House and garden are so united by a single idea that their size is relatively and immovably fixed, and the open garden, the parterres, and their contrasting boskets must exactly correspond to them.

It is perhaps in the laying-out of the parterre that Le Blond has least gone beyond Boyceau. He was acquainted with all the kinds, including the *parterre de broderie*, with arabesque patterns marked out in box and combined in one large design—this was now the favourite kind—and the other sort that had geometrical shapes of flower-beds edged with box, now somewhat out of fashion, and generally used, in combination with the *broderie* style, to give greater variety. From England had come the fashion of laying out the parterre in great stretches of lawn, with a pattern in coloured clay, and a strip of flowers or dwarf trees round. The boskets were now made into novel and hitherto unheard-of forms, and these "contain all that is most beautiful in a garden." We have become familiar with such arrangements in Le Nôtre's great works. Every garden must needs have boskets of the kind as a necessary background for the open parterre, to conceal the secluded parts and the *variété* from spectators on the house terrace, whose view over the open parterre was to be checked here; in these places there was the desirable unbroken shade, the theatre for fêtes, protection from every rough wind, and solitude. The splendour and importance of a garden depended on its many-sidedness; but even the most simple and unadorned could show beauty and symmetry, with a background of thicket, and with pretty paths cut in the *massif* of the hornbeam with which these small woods were generally planted.

In spite of the love for variety, the book utters that cry for simplicity which inspired the last period of the creation of Le Nôtre. It warns people against dividing and subdividing,

FIG. 439. CLIPPED HEDGES IN A PRINCELY GARDEN

a habit in which the author thought—rightly as the future showed—that he saw the greatest enemy of the French garden. The porticoes of many kinds that were cut in greenery, the winding trellis which was overdone, the extravagant clipping of trees into the shapes of animals, men on horseback, men on foot, and many other things—all this was disliked by the writer. What the French garden needed, he said, for its main lines, was most of all simple tall hedges. Everything mean and shabby, even in garden sculpture, should be avoided: better no statues at all than bad ones.

Le Blond's treatment of water corresponds to this idea in the main. When avenues and squares are planned there should be a really useful surround of water, but he is contemptuous of petty detail in the way of shell-work and small basins— and calls them *colifichets* (gewgaws). All the important fountains ought to be visible from one central point. It is clear that the art of Le Nôtre could not have found a better or more lucid exponent. There must needs be powerful, if unseen, reasons at work, if so noble an art was to be brought to ruin. The success of the book was remarkable: edition after edition appeared, then pirated issues and translations. And it had significant results. To its influence was due the improvement in skill and the lightness of touch which came about in gardens at that time.

France was behind other countries in the matter of new works in the eighteenth century, especially in those districts in the north that were influenced by the Parisian court. They always harked back to Versailles, without which French taste could not have produced so manifold a progeny in the rest of Europe. But the court, as we have seen, changed its taste; the new century was not one of fêtes and displays, because for one thing money, exhausted by the Thirty Years' War, was scarce in the state treasury and was not forthcoming for new creations, which could only have compared unfavourably with those of the seventeenth century. After Louis the Fourteenth's death the spirit of the time expressed itself in places like the Little Trianon at the time of its first garden. In the ever-increasing artistry of the parterre there developed very markedly that transition state, of which we shall speak hereafter.

FIG. 440. JARDIN DE LA FONTAINE, NÎMES—
GROUND-PLAN

Before we turn our attention to the influence of France on other European nations, one more garden, standing outside the limiting circle of the court, must be considered—the so-called Jardin de la Fontaine at Nîmes (Fig. 440). This is perhaps the most important work that exhibits directly the newly awakened interest in antique art. When the foundations of mighty Roman remains were discovered in the thirties of the eighteenth century, the enthusiasm of the people was so great that they demanded restoration. The work was entrusted to Maréchal, a fortress-builder, in 1740; and he proceeded to design a most imposing scheme of terraces, steps, basins, statues and gardens, mostly on the old foundations. It was the best kind of baroque work, and translated the spirit of Roman life into the style of the great age.

At one time there had stood in this place temples, baths, corridors richly adorned with statues, and a theatre. The chief garden is in a straight line with the main street of the town, the Boulevard de la République, and old foundations of baths were utilised as canals, flowing round the different terraces. At a spot where there is now reposing, on a high pedestal, a nymph with children, at the top of the basin of the baths, there was in former days the statue of Augustus on a stylobate, with decorated columns at its four corners. The spring itself lies somewhat removed from the main axis, exactly at the foot of the hill; and on the top of the hill stands a Roman watch-tower, La Tour Magne, while farther towards the side is a Temple of Diana, where the nymph of the stream was worshipped. This enforced bending from the axial line, in which we discern a sure indication of the Roman spirit, is here only a special case of rhythm, for there is evident

FIG. 441. NÎMES—JARDIN DE LA FONTAINE

everywhere a strong feeling for unity, shown in the all-pervading balustrades, statues, vases at the corners, steps, and bridges (Fig. 441). The true feeling of the antique world, which restrained the architect, served as a protection to this late work (as also to the Villa Albani in Italy, whose date is much the same) from all the pettiness and prettiness of the court style in Northern France.

And now we must consider the period when all the countries of Europe directly or indirectly felt the influence of Versailles, that central sun of France, so long as it maintained its full and original splendour.

ENGLAND

England was probably affected least of all, at any rate permanently, by the direct influence of French art. When Charles II. came to the throne after years of severe Puritanical rule, he could not carry out his own plans so swiftly as Louis XIV. had done. A great part of the gardens of his forefathers had been destroyed, or at any rate neglected. Nor did his means put at his command those far-off artists whom his royal friend was able to employ on the other side of the Channel. Moreover the utilitarianism and sobriety of the Puritans, always controlled by reason, continued to influence in no small degree the gardens of England, even after the Restoration. In his Diary, that mouthpiece of contemporary fashions, Pepys tells of a conversation which he held with Hugh May, an architect in the king's service, about the ruling styles soon after the Restoration:

Then walked to Whitehall, where saw nobody almost, but walked up and down with Hugh May, who is a very ingenious man. Among other things, discoursing of the present fashion of gardens to make them plain, that we have the best walks of gravell in the world, France having none, nor Italy; and our green of our bowling allies is better than any they have. So our business here being ayre, this is the best way, only with a little mixture of statues, or pots, which may be handsome, and so filled with another pot of such or such a flower or greene as the season of the year will bear. And then for flowers, they are best seen in a little plat by themselves; besides, their borders spoil the walks of another garden: and then for fruit, the best way is to have walls built circularly one within another, to the South, on purpose for fruit, and leave the walking garden only for that use.

This increasing hostility to flowers was sometimes felt in France, but was not so evident because of the plentiful water, statues and other sculpture. The English garden, with its love for wide convenient paths, and very small provision of sculptures, must at that time have looked empty and dull. As a fact, good taste was turning the other way; and in the year 1665 a violent protest was raised. Rea writes in his garden book *Ceres, Flora, and Pomona*: "A choice collection of living Beauties, rare Plants, Flowers and Fruits, are indeed the wealth, glory, and delight of a Garden," and he goes on to say that the new plan of gravel walks and close-cut lawns is only suitable for town houses, though leading features at many a stately country seat whence garden flowers, "those wonders of Nature, the fairest ornaments ever discovered for making a place beautiful," are banished. He adds a hope that this "new-fangled ugly fashion" will disappear together with many other alterations.

This protest against the hostility to flowers was repeated twelve years later by the much-read author, John Worlidge. But those two men of understanding, Pepys and May, had happened on the very centre-point of the requirements of an English garden, when they spoke of wide paths, seats to rest on, and objects to make for on a walk, because here we have the essentially English delight in active exercise in the open country. And so Rea's objection to wide paths was not endorsed by Worlidge himself, who thought that it was by no means the least part of the pleasure given by a garden to go for a walk in it with friends or acquaintances; or to go alone and so get refreshed, free from the cares of the world and society which are often burdensome; then if one were tired, or if there were more great heat or rain, one could take a rest under a fine tree or in a covered arbour before again enjoying the open air.

Of this sort of garden to stroll in, no Southerner or even Frenchman had ever dreamed. For them the open parterre was to be looked at from above, or enjoyed at leisure when it was cool; in the sunny squares people preferred to be carried, or to drive in little carriages along the broad paths into the shady boskets, and even there they did not walk. Quite peculiar to England, both then and now, are the smooth garden walks of grass. The unrivalled beauty of the lawns, which England owes to a damp climate, produced a reaction in France in favour of the grass parterre *à l'anglaise*. In England such a parterre was of course often an unbroken lawn, with flowers only on the borders; or it might be, as was the fashion at the end of the century, adorned with vases, statues and small green trees. It was in England too that were first made with thick short grass the wide and sometimes very long alleys for playing bowls. It is evident how little the other countries were likely to have lawns of this kind when one sees the odd meaning of the word "bowling-green" as used in French garden language. It is possible that they did not know that "boulingrin" was an English word at all; it was understood at the date of the *Théorie et Pratique* to mean a sunk piece of grass, which formed the centre of a bosket and often had a fountain on it. So far the lawn had only been regarded as attractive to look at, and it was not connected in the least with the French *jeu de paume ou mail*—a game which, if not played under cover, had a course of earth stamped flat. Later on, the word "boulingrin" was derived from *boule* (= bowl) and *green,* so meaning a green hollow place.

It is not surprising that one innovation from France found a welcome in England, namely, the avenue. Not that these long straight lines, closely uniting house and garden, were an invention of the French gardeners, because we found them in Italian and Spanish gardens at the end of the sixteenth century; but their regularity is the first expression of the large all-embracing scheme of the French style. For there would be three to five walks, all starting from a single point, which was as a rule in the central axis of the garden, and leading through the park in different directions, often with some distant church or fountain as *point de vue,* but sometimes going up to the main entrance as a grand carriage-drive. Led by this fashion, Charles II. had wide avenues cut through both parks at Hampton Court, reaching from the east nearly to the front of the palace. The great canal also, of which Charles at once took possession after his return, seems at first to have led up close to the house-front on the east side. Thence started the star-shaped avenues of the park, out of which the great semicircular garden was afterwards made. Evelyn visited the place 9 June, 1662, saw these innovations, and mentioned them as adding to the beauty of the park. He found many pretty bits in the gardens to praise, but thought that they were very much too small as a whole; so he can have seen only the old Renaissance gardens on the south front. Even if the semicircular site already existed in Charles's plan, he certainly had not made much progress with it. In any case he felt a lively desire to learn all that was possible from France, and for this purpose sent his gardener Rose to Paris to be educated. Indeed, Charles went farther, and asked at the French court if Le Nôtre himself could come on a short visit to England; and Louis appears to have given Le Nôtre a somewhat hesitating permission, though nothing is heard of his coming to stay in England at that time.

The chief credit for bringing the gardens at Hampton Court to their present form is due to William and Mary. Both of them had a great liking for the palace, and made it their permanent residence. Christopher Wren, at that time beyond dispute the greatest living architect, was summoned to build a new, important palace on the east side, round the pretty old Tudor building, which still was encircled with a moat. The style of this palace shows that men's eyes were directed towards Versailles. London and Wise, both pupils of Rose, were commissioned to lay out the new gardens (Fig. 442). In front of the lately erected east wing they cut off a large semicircular piece of the park, and laid it out as a flower-garden; for this the walks which led up to the old castle had to be put back, and then other walks, and also a canal, were made round the semicircle on the outside. On the inside it was laid out with *parterres de broderie,* the paths being kept in the form of a star which led to the castle. There were thirteen fountains, some large, some small, and a great many statues. By the side of the house ran a gravel path, 2300 feet long, following the whole length of the house and its side wings, and this path had to serve instead of terraces.

Although William had large plans, he could not see his way to making those enormous earthworks which sunk gardens would have involved, and this was the only way of getting a terrace. His contemporaries, and writers of a later date, all bewail the want of a terrace

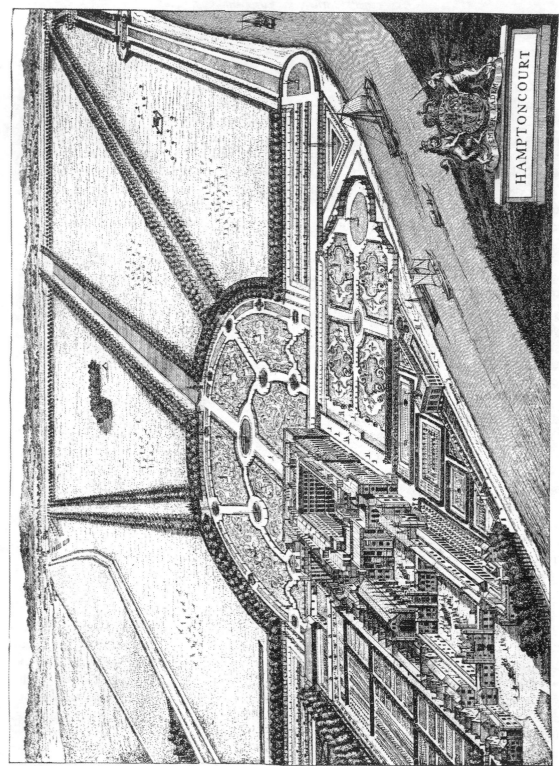

FIG. 442. HAMPTON COURT UNDER WILLIAM AND MARY—GENERAL PLAN

to give a general view. The gravel walk was furnished in summer with a row of fine orange-trees. These were thought very appropriate by the royal pair, who saw in them a half-political allusion to the House of Orange. The south gardens were also remade: the lesser ones were now turned into special gardens for flowers, and among these the so-called pond garden (Fig. 344, Vol. I.) is even now a charming piece of Renaissance work. The hill for a view rose behind, and looked towards the Thames, in the so-called private grounds: there is a summer-house on the top, which is seen in the view from Christopher Wren's new wing. It was levelled, and flower-beds were set out.

The back of the semicircle, where the "mount" had been, was shut off by an iron trellis containing twelve gates, among the most beautiful works of art of this kind. The designer was the Frenchman Jean Tijou, who produced many other works in iron for English people (Fig. 443). These particular gates were removed in 1865, at the time when garden art was most degraded, and placed in the newly founded South Kensington Museum; for the institution was in want of works of art, and there was a wish to accustom the public to their exhibition. Fortunately, it was recognised a few years later that things of this kind had the best effect in the places they were intended for, and the iron-work was restored to its old home. One of the most

FIG. 443. HAMPTON COURT—THE WROUGHT-IRON GATE OF TIJOU

private and secluded parts of the garden, the wonderful covered walk called Queen Mary's Bower, belongs to an earlier date, for it was seen and admired by Evelyn.

The last change made by William was the conversion of the old orchard on the north side into the so-called Wilderness. This is a significant indication of the conservative feeling in the English gardens of the period. The great idea in the French garden was the shaping out of the bosket or thicket as "relief," with a view both to variety and the provision of grand displays, but this notion seldom took hold in England. People were content in the royal garden with a wilderness plan, which was adopted at the beginning of the century, with winding paths cut in the thick growths, the greenery mostly held back by trellis at the side. The small importance of the grove was a consequence of its being ill-suited to the damp climate, and shade was preferably sought in long airy avenues and walks.

The gardens of Hampton Court belong to the very few which in their main lines have kept their original form. True, the fountains in the great parterre have disappeared to the last one, and out of the beds with box borders have grown large lawns of trapezium shape, while the park avenues have been continued into the garden as avenues of yew;

but the sameness of this part is enlivened by a marvellous show of flowers, and the surround and main lines of the garden are just as they were. This fact seems surprising, for in England the great revolution in taste raged high. Were we not aided by excellent engravings, it would be hard indeed to get a tolerably comprehensive view of England's gardens about the year 1700. But a flood of copper engravings, mostly Dutch, such as the work of the engravers and draughtsmen Knyff and Kip, who tramped up and down trying to get views for their drawings and plates, and more especially the views of important castles and gardens, had now reached the country. They are preserved in a great

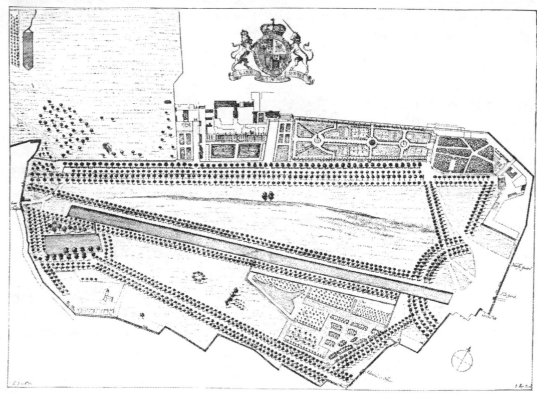

FIG. 444. ST. JAMES'S PARK, LONDON—GROUND-PLAN

series of pictures, which we often find repeated in the different books. The best of these, for the size and beauty of the drawings, is *Le Nouveau Théâtre de la Grande Bretagne*, which appeared in 1714.

The pictures show first and foremost that the French fashion for large lines had made an impression even in England. The actual size of the gardens was more imposing than it had been hitherto. The *bosquets* were thought less of, as we have said. The many kinds of water arrangements had lost their importance; even the great canal is not universally present, and when it does appear, is not situated so favourably as at Hampton Court, but lies at the side and does not connect with the garden. It is not unlikely that such a peculiar situation is due to Dutch influence, for in Holland, as we shall see later on, it seems to have come about from natural causes. St. James's Park, also made by William III., is a typical example (Fig. 444), and the canal is at the side of the park in

FIG. 445. BADMINTON, GLOUCESTERSHIRE—GENERAL VIEW OF THE MANSION AND GARDENS

one long strip, with straight avenues running alongside and enclosing wide-stretching meadows. In front of the palace there are two fine lawns, also bordered with trees. The side of the palace looks on a flower-garden with finely laid-out parterres, but the boskets are unimportant. In spite of the fact that it is all so very unlike the French style, there is a legend that Le Nôtre drew up the plans for this garden also.

We have already spoken of the English parterres, which almost always put the lawn itself in front. An entirely level situation of the garden is the first requisite for its beauty,

FIG. 446. BRAMHAM PARK, YORKSHIRE

even more here than in France, and on this account the hill to give a view over it has often been kept in a later garden. But here too a raised terrace with balustrades has been much liked, attached to the house. Where a hilly ground favoured terraces, as so often in Scotland, and also in the more important English gardens, they were used, but surprisingly seldom, for cascades.

One peculiarity of English country houses at the time of the Renaissance often conditioned the site of their front gardens: the old Tudor house was always entered through a front court; carriages drove up to it, and visitors had to go on foot through a second court on a paved path; there were lawns on both sides, with fountains or perhaps parterres. It was not before 1700 that people began to alter these inconvenient approaches. Then at Hatfield House and at Montacute the entrance was changed to the other side, close to the house.

One of the finest and also most interesting places of the period is Badminton in Gloucestershire (Fig. 445). Henry, Duke of Beaufort, built the house in 1682. He had a real passion for avenues, and his park grounds were traversed by numbers of walks, twenty of them starting from one point like the centre of a star. It is said that he infected his neighbours with his own enthusiasm, so that they let him extend the avenues into

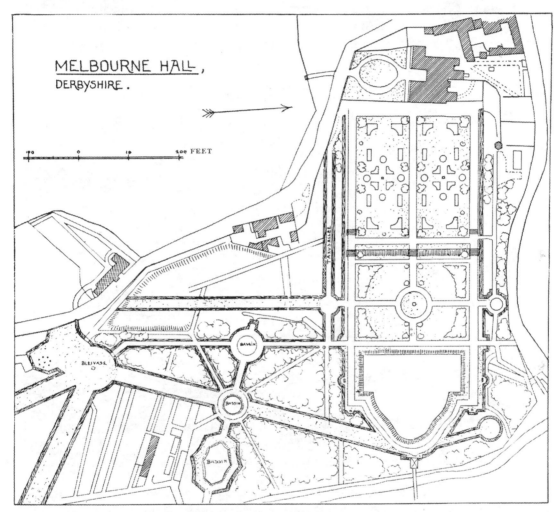

FIG. 447. MELBOURNE HALL, DERBYSHIRE—PLAN OF THE GARDEN

their territory, and in this way he obtained more distant and glorious views. But the gardens too cover a very large tract of land. They lie round a house in the middle of a great park, with the chief avenue two and a half miles long leading to the entrance. On the left of it there are parterres and a bowling-green; behind the parterres, and in a straight line with them, are boskets with fountains and finely designed paths; at the very end is a semicircular little room cut out of the hedge and containing two fountains.

The work of the duke at Badminton has perished, like nearly everything shown in the engravings. But here and there something has remained of the less famous garden of a less ambitious owner. A garden in the north of Yorkshire has preserved the beauty of its grand avenues bordered by straight-rimmed ponds, and its *points de vue*. This is Bramham Park, and the drawings show a scene of much beauty and character (Fig. 446). There is a place (Fig. 447) of that date at Melbourne in Derbyshire, laid out in the years 1704 to 1711 by the king's gardener, Henry Wise, for Thomas Coke, who was later on Vice-Chamberlain to George I. Here the parterres end in a wide pond, with a pretty summer-house above it, and farther on a park-like meadow, which is reached by a bridge. This parterre is enclosed on either side by shady avenues of yew, cedar and wellingtonia; the last was of course only planted in the nineteenth century, when it was

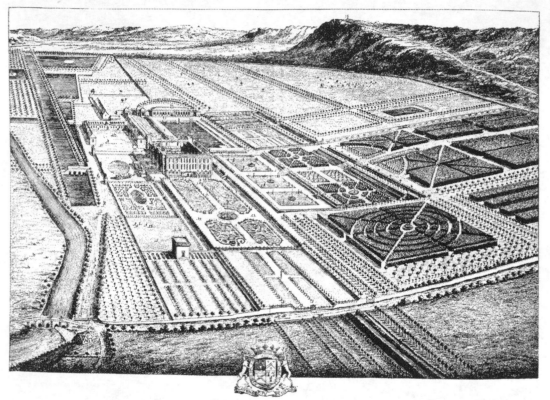

FIG. 448. CHATSWORTH, DERBYSHIRE—AN OLD VIEW CF THE MANSION AND GARDENS

introduced. A park adjoins the place at right angles, and paths, sometimes forming a star shape, are cut through it, and are ornamented with fountains and pretty leaden vases of French make.

The Duke of Devonshire's seat, Chatsworth (Fig. 448), which, like Melbourne, is in Derbyshire, has had a changeful history, for each century has completely altered its character. We know nothing of its Renaissance gardens, but in 1685 the duke had house and garden altered in the style of the period, and this design is shown in the above engraving by Kip. The gardens ascend the hill in several terraces, the upper ones made use of for groves; these, however, are not so important as they might be, for each one has its own particular axial lay-out. The house is on the second terrace, and the view is over a great *parterre de broderie*. One axis at the side climbs the hill, and ends in a cascade that cccupies thirty steps. The chief beauty here, which is by no means exhibited in the picture, is really in the great amount of water. From· the River Derwent, which flows past the garden, a canal of great length branched. The water in the garden itself is all alive with great playing fountains, adorned with dolphins, and sea-gods, and many small jets and water-devices. But in comparison with French art one misses at Chatsworth, with all its spaciousness and many-sidedness, the unity given by straight lines and distant views. Even the cascades can only be enjoyed when one goes outside the house, although the ponds on the lower terrace must have been all one picture with the glittering waters that came from the boskets. Le Nôtre must, it would be thought, have laid out these gardens also; but in reality it was another Frenchman named Grelly, who made at least the water-devices. He adorned one of the thickets with a fountain which he directly imitated from the Marais, at that time so much admired at Versailles. It is only made of tin, with willows painted in natural colours, pouring water over large stones, and weeping, as it were, out of the tips of the leaves: in this fashion an ancient idea was transplanted into a northern land. The fountain and many other objects were preserved till after the English style set in, or possibly they may have been reinstated by some intelligent person who was looking after the garden in the nineteenth century.

By this insistence on a garden for strolling about, which demands a less formal and more cheerful arrangement than the French garden could allow, England was perhaps already paving the way for that revolution in taste which was soon to occur.

GERMANY

Germany had to begin almost all over again from the middle of the seventeenth century, at the end of the Thirty Years' War. The cultivation of the garden is a peaceful art; and it was only exceptional men such as Wallenstein and Maurice of Nassau who tried to keep the country to its peaceful occupations while they were in the midst of war, weapons in hand. For the most part the war had left wasted lands bare of inhabitants, but there was more than this—the tradition that was never very strong in Germany was completely destroyed. It was just this state of things, however, that drove a generation hungry for peace to seek for teachers whose instruction it could follow with delight. One important factor in making garden art flourish in Germany was the increased power of the many princelings, great and small. The feeling of sovereignty showed itself in the second half of the seventeenth century, when prosperity was increasing, in the creation of splendid homes. For most of the princes, especially those in the north and west, Versailles served as a fascinating visible example. Only a few, who were interested in Italy, took their inspiration in these days from the old forms of art on the other side of the Alps. Le Nôtre's was the truly great name, and as soon as his reputation had once extended across the Rhine, it was considered good luck to secure a garden artist who had somehow or other got his education by actual study of the works of Le Nôtre.

Duke Ernst Johann Friedrich of Hanover reckoned himself one of the fortunate ones when he secured Charbonnier, who belonged to the school of Le Nôtre, to lay out his garden at Herrenhausen. The architect for the house was Quirini, a Venetian, and he gave it an Italian look with two wings of one story, which jutted forward and showed a flat roof with balustrades. At small German courts, we often find, as late as the middle of the eighteenth century, a partnership of Italian architect and French garden artist, for the French style in building arrived later in Germany than the garden style, and was never really naturalised. The duke loved magnificence, and he rejoiced in the stir and bustle that a tribe of foreign artists, French and Italian, brought to his place.

Although the keeping up of the pleasure-grounds at Herrenhausen cost nearly six thousand dollars in 1679, the year of Duke Ernst Johann's death; and although his successor, Prince Ernst August, was very angry about the extravagance, it was this very

FIG. 449. HERRENHAUSEN, HANOVER—GENERAL PLAN

successor who extended the garden to double its size, and gave it pretty much the appearance that it still has (Fig. 449). It is natural to think of the close relationship between the Hanoverian and French courts, which was kept up in the liveliest way in the correspondence of the gay Princess Sophia of Hanover with her niece Lieselotte, Duchess of Orleans; and it may easily be believed that as the two ladies took such an interest in gardens, they shared some direct advice and even plans by Le Nôtre. The plans were as formal as any we know, giving the impression of an example in a school-book. There seems to be a kind of anxiety not to omit any of the rules or injunctions: first there are the fine parterres with a central fountain, behind them four almost square ponds, then a simpler parterre with two little pavilions, which have now disappeared. They formed the connection with the boskets, which were traversed by regular star-arranged paths with tall hedges of box, and which all had a basin in the centre. There was a very large round pond at the end

FIG. 450. HERRENHAUSEN, HANOVER—GARDEN THEATRE IN ITS PRESENT STATE

of the middle walk, and the two side paths led to summer-houses built like temples. Avenues of limes encircled the whole garden, with canals running beside them, which formed a semicircular bay behind the round basin in the middle axis. The first half of the garden, which lies nearest to the house, shows clear traces of the earliest phase of Versailles. The grotto occurs at exactly the same point; but as complete regularity demanded a corresponding site on the opposite side, here were the so-called cascades and a wall with grotto and shells, enlivened by waterfalls and springs. Here also was the attractive orangery beside the castle, and corresponding to it on the other side a garden for flowers or vegetables.

The only part that was not formal was the theatre on the east of the great parterre. This stands on a made terrace, varying the monotony of the otherwise level ground. The back of it is occupied by the stage, from which steps lead to the garden beside a beautiful fountain at the supporting wall. The side scenes are trapezium-shaped, meeting together at the back, and cut out like small green dressing-rooms, with statues in front of them (Fig. 450). The stage is separated from the amphitheatre for spectators by a low wide gangway, on a level with the garden, and approached by steps from the stage. This must have been a great help to the performances, as it served as a sort of orchestra. The garden was quite finished by 1700, but the theatre was so placed in the body of it that one may perhaps assume that it was adopted into the ground-plan, and it thus would be one of the earliest of the kind. The garden at Herrenhausen had no particular park of its own; from the treatment of the canal surrounding the whole place, this would have been impossible. The omission may have been due to Dutch influence, for gardeners from Holland were working here later.

This first attempt at Herrenhausen to imitate the French style was carried out too stiffly, too academically. Fortunately the artistic interest of German princes was so many-sided, and their love of building so extreme, that the danger of a rigid style was averted. And in the North France did not reign alone. Close by, at Cassel, about the time when Herrenhausen's buildings were nearly finished, the young Landgrave Charles came back from his travels in Italy with a project which, though never completely carried out, fills the present generation with wonder and admiration for the force of will that speaks therefrom: this is the plan for Wilhelmshöhe on the Weissenstein near Cassel. On his return home Charles summoned as architect a Roman named Guernieri, and he made the great cascades that overtop the park. French influence is absent here; everything that theorists and the example of Versailles strictly enjoined had now gone to the winds. The landgrave was so full of the impressions formed in Italy that he allowed Guernieri to work entirely according to his native traditions, and thus on a northern soil there arose for the first time a work in which walled architecture in conjunction with water played the leading part.

If Guernieri's plan had been completely carried out, there would have arisen a work which for size, grandeur and completeness would have been almost unequalled in the whole of Europe (Fig. 451). The predecessors of Charles had erected a hunting-seat in the place where there now stands on a hill the castle of Wilhelmshöhe, built at the end of the eighteenth century; and above this towered the steep Habichtsberg. According to the plan of Charles and Guernieri, the whole of the great wooded hill was to be converted into an enormous terrace site, and the main lines, dominating all else, were to be formed by a

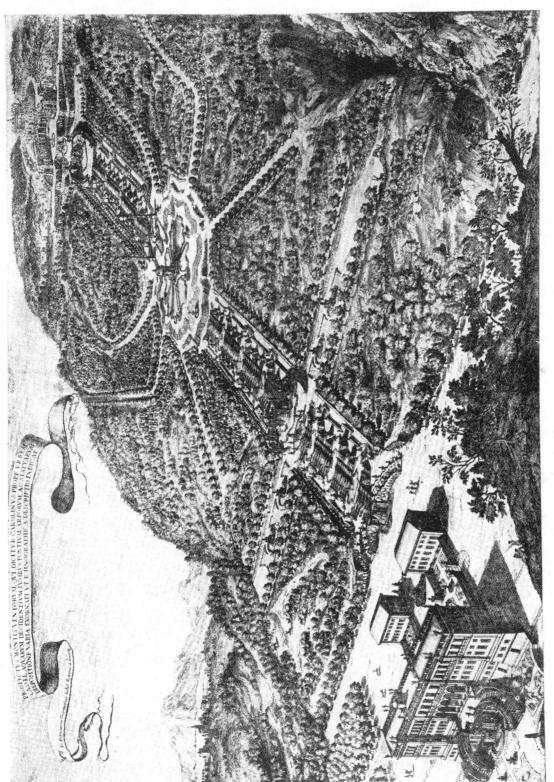

FIG. 451. WILHELMSHÖHE, CASSEL—VIEW OF THE KARLSBERG

FIG. 452. WILHELMSHÖHE, CASSEL—VIEW OF THE GIGANTOMACHIE

great series of cascades. But only the upper part was actually carried out (Fig. 452). This the Italian pictured in a great copper engraving in 1706, just as in earlier days Salomon de Caus had pictured the castle garden of Heidelberg. The top of it, a summer-house, which was to serve as a huge reservoir and also for a fountain, was begun first. It was an octagonal building of three stories, whose two lower floors were to look as though they were growing out of the rock, and were to contain alcoves and statues: one story was a little behind the other, and the uppermost was an airy hall, protected on the top by a balustrade which ran round the flat roof. On this roof stood a pyramid thirty metres in height, on the top of which was the colossal figure of the Farnese Hercules made of copper. The reclining hero looks down upon the third terrace below, where a giant's head is squirting out a jet of water more than twelve metres high. Between these two there is a terrace with a grotto of the god Pan and all sorts of water-plays and devices. The main stream descends to the terrace of the giant's head, and glides down on either side over grotto-work. From here it makes another plunge in the form of a cascade 250 metres in length and 11½ in breadth, falling over steps, which are interrupted by steep broad landings. Lastly it falls in one tremendous plunge over the grotto of Neptune with the figure of the god, so that anybody inside the grotto looks out under the stream of water, which finishes in a large basin.

This is the only part of Guernieri's plan that came to completion, and it was about a third of the intended length. The cascade was to have gone all the way down to the castle in two more great descents. At the foot of the first of these there was to have been a large round water-parterre with a fountain pavilion in the centre, and the second drop was to end in a great semicircular theatre on a terrace, cutting the whole park transversely in one broad strip and also adorned with various other fountains. The castle at the bottom of the third cascade was designed in a style purely Italian with *giardini secreti* behind it, and in front there were semicircular steps leading out of an open Florentine pillared hall to a fine ornamental parterre.

The park itself, except for the terraces, was only to be interrupted by straight avenues, but during the course of the eighteen years of preparation this plan was greatly altered and enlarged. The fact is that the landgrave could not entirely divorce himself from the influence of France. After 1715 he had eight fine views made by a painter of Haarlem, Johannes von Nichole, of the Weissenstein and the Karlsberg, as they were then called; and these preserve the plans intended to be carried out at that period. According to them, there were to be certain wide terraces furnished with parterres, fountains, and little waterfalls below the cascade round a great castle somewhat like Versailles, so that the Italian cascade would only cut into the park like a side-scene a long way back. But of all these great plans not one was ever completed. The death of Charles in 1730 interrupted the work; and when his successors turned their attention to his proposed plans, the English style had become so influential that the cascade dominated a greatly altered park as a self-contained independent thing (Fig. 453). The idea of a battle of giants was essential to it; and as the conqueror Hercules in a reclining attitude looks down upon his enemy, who rears up with no strength left, so does this great place gaze upon the many small new developments that are scattered about the park, owing their existence to a sentimental time and to various phases of architecture. Wilhelmshöhe, which derived its name from Prince Wilhelm, who built the castle now existing, stands alone, not merely because it is large and self-contained, but because it provides a visible and tangible proof of what Italian genius could create in a German garden, even at so late a date.

As we have said before, Italians were for long the ruling architects, but gardens were left for the most part to the care of French artists. Even at the court at Munich art was not wholly inspired by Italian feeling in the second half of the seventeenth century. In the sixties a whole tribe of Italian artists arrived at Munich in the train of Princess

FIG. 453. WILHELMSHÖHE—THE ENGLISH PARK

Adelaide of Savoy. This lady, with her great *joie de vivre*, contrived to drag her dull
husband into a whirl of gaiety; and when in 1662, after eight years of marriage, she
presented him with a son, not only did he fulfil his vow of building a Theatine church,
but, to please his wife very specially, he arranged to build a pleasure-castle near Munich,
to be called Nymphæum (Nymphenburg). For both these buildings an architect from
Bologna, Agostino Panelli, was employed, and after him Enrico Zuccali, who was also to
carry out the new work at Schleissheim.

The early death of the duke and duchess put an end to their work; so their son Max
Emanuel, who was proud and ambitious, must be regarded as the real founder of both
Schleissheim and Nymphenburg. In the first splendid years of his rule the warlike prince
was so much abroad that the building did not begin with full vigour before the eighteenth
century; but very soon, disaster befell Bavaria, and in 1704 the country was seized by
Austria. The duke was deprived of the crown and had to flee, and he lived the next eleven
years in Paris. Here in exile he had leisure to study French gardens thoroughly, and he
had scarcely returned home before he showed in the laying-out of his gardens at Nymphenburg and Schleissheim that he had learned to some purpose. Garden artists of the French
school were busy from the start. Before his exile Max had summoned Carbonet, a Belgian
by birth; but the chief merit for the complete work is due to François Girard, who was
responsible for the outside of these remarkable places, when he came to Munich after

the duke's return. In both gardens the whole effect depends on the position of the canals, which form the centre line of the castle.

At Schleissheim (Fig. 454) the parterre, rising by a few steps from the terrace, is particularly well designed, partly in its size and the variety of *broderie*, and partly in the number of its springs and fountains. Besides the two basins whose waters play in the parterre, there are two narrow canals alongside the middle walk, and twenty-six water-jets, which make a sort of balustrade—an idea first carried out by Le Nôtre at Vaux-le-Vicomte. The middle avenue leads to the cascade which faces towards the parterre at the beginning of the canal, where there are waterfalls, fountains and various figures. This canal is in the middle line, running from the new castle parterre, with hedges and boskets full of statues along its course, until it arrives at a small casino, which is older than the great house, and was built in the nineties. It was named Lustheim by Maximilian, who wanted it to be his Trianon, as is indicated by the two pavilions flanking it. Later on these were connected by a semicircular gallery with the castle between the two (Fig. 455); this was at the time when Lustheim was at the end of the large garden, and so marked out the middle axis for the buildings of the new castle. At the same time an ornamental parterre was made in front of the casino, while the canal was conducted as a small strip round the galleries, and then joined on to the park that lay behind by six separate paths. But by these arrangements the little house was sacrificing its peculiar feature as a Trianon, that is to say, as an independent place, removed from the stir of outside life; for it was now, on the contrary, an actual *point de vue* for the whole picture as seen from the castle—a very unusual plan in a French garden. It required an unimpeded view over the open country, for which some church tower in the far distance would generally give a good resting-point.

FIG. 454. SCHLEISSHEIM—PARTERRE AND PARK FROM THE CASTLE TERRACE

FIG. 455. SCHLEISSHEIM—LUSTHEIM AND PARTERRE

The second of Max Emanuel's two castles, Nymphenburg, was less restricted. In the grand style of its waters, and the variety in the park, it unquestionably holds the first place among the level gardens of Germany (Fig. 456). Just as at Schleissheim, one small branch of the Würm was converted into a canal, so that the approach to Nymphenburg from the town side is marked by this long piece of water, which ends in two broad ponds with fountains. This fine approach is carried out further in the garden; for there are narrow canals passing round the court of honour, the castle, and the great *parterre de broderie* with its ornamental fountains. These smaller canals come together again in a wide lake

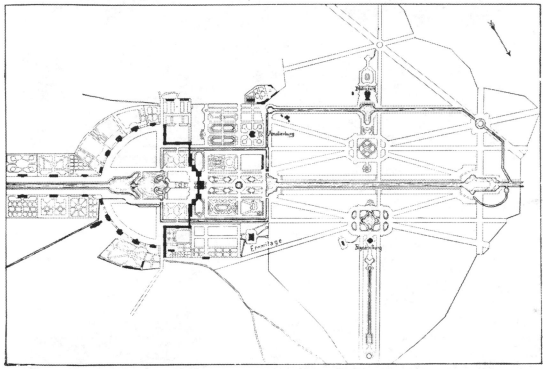

FIG. 456. NYMPHENBURG—GENERAL PLAN

that includes six springs, and forms the head of the large canal, which as a middle axis cuts through the raised boskets that lie on either side of it (Fig. 457). It ends in a wide pond, into which a fine cascade discharges its many waters towards the castle. This falls over marble and is decked with many statues, and loses connection with the canal, which then proceeds farther into the park, whence the eye can range as far as to the church tower of Pasing.

A traveller, a nobleman called von Rothenstein, visited Nymphenburg in 1781, and thus describes the splendour of the place, which up to the end of the eighteenth century remained undisturbed by the assaults of the new fashion (Figs. 458 and 459).

The garden has 19 fountains, which give out 285 jets; and such a number of water-devices, gilt vases and statues, meet the eye that they are better imagined than described. The great flower-parterre is 138 fathoms in length, and has one large fountain, four smaller ones, and a six-headed one. The parterre is laid out with box, and with vases, and beds between with many flowers, which each month present a different picture. . . . Right in the front stand six gilt urns, 3½ ells in height . . . next there are dragon fountains to right and left with ever so many dragons and snakes separately lying on hills of stone. . . . In the parterre there stand 28 gilt statues, groups, vases, and urns, and near the box-espaliers 17 statues made of white marble. After the dragon fountains come two of children, each child

FIG. 457. NYMPHENBURG—THE CANAL AND CASCADE

seated on a gilt whale. . . . Finally in the centre comes the great fountain of Flora, which is octagonal, made of white marble, and over 100 feet in circumference. In the middle is a great basket with flowers, from which there springs a jet 30 feet high, and as thick as a man's body. On the side of the basket you see the goddess Flora seated, 12 feet high. Beside her is a Zephyr, who holds a great wreath of flowers in one hand, while with the other he expresses his astonishment at a monkey who is working the water from the basket. On the hill stand a lion, a shaggy dog, 3 large swans, 2 storks, and a great deal of sea-

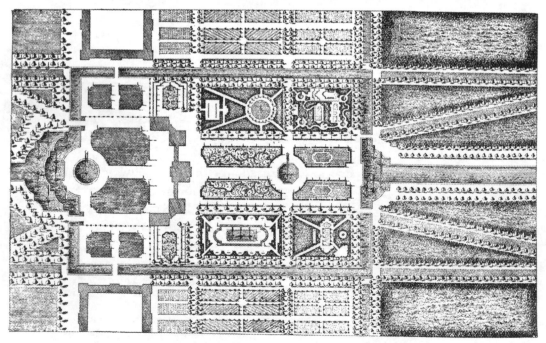

FIG. 458. NYMPHENBURG—PLAN OF THE PRINCIPAL PARTERRE

FIG. 459. NYMPHENBURG—THE PRINCIPAL PARTERRE

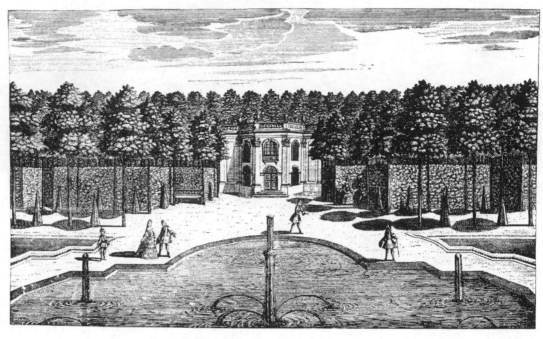

FIG. 460. NYMPHENBURG—THE PAGODA PLACE

weed. Further there are in the pond 8 tiny gilt mountains, with love-gods on four of them, and on the other four tritons, holding in their hands corals, pearls and the like: they are seated on whales. On the edge of the tank, on the border round it, there are 8 gilt frogs, spurting water upward in arches. This grand fountain cost 60,000 gulden, and used 250 cwt. of lead. Then you come to another great tank, which has 6 springs all in a row, and into this basin the canal runs right and left, passing onward to the great cascade.

FIG. 461. NYMPHENBURG—THE BATH PLACE

FIG. 462. KARLSRUHE—THE CASTLE GARDEN IN 1739

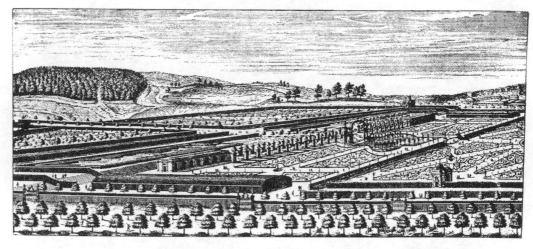

FIG. 463*a*. LUDWIGSBURG CASTLE, STUTTGART

When we read this, the fantastic ideas for "princely" fountains that Becker suggests in his *Fuerstlichen Baumeister* (Architects for Princes) do not seem so preposterous. The question is always insistently arising, as to where all these ornaments came from, for the number of new things of the smaller sort are peculiar to Nymphenburg, and they are actually in excess of those found in the French places we know. The separate pavilions, to live in, correspond to similar ones at Marly-le-Roi; but here they appear for the first time in a half-circle, surrounding the handsome court of honour, and with the town canal in the middle. This arrangement was very popular in Germany, and at Nymphenburg there are several extra pavilions at the side of the canal (Fig. 456).

The small scattered houses with their separate gardens, which remind us in their variety and number of the hermitages at Buen Retiro at Madrid, form an important feature of the park. Their erection was due to the differing requirements of the princes, but they divide the great park with a pleasing regularity. One cross-road through shrubberies was laid out as a tennis-court, and near this was the garden theatre, with green side-scenes. Max Emanuel set up a pavilion for actors and spectators to rest in and take refreshment, and had it made *à l'indien*, giving it the name of Pagodenburg (Fig. 460). The prince had certainly never seen the old Trianon de Porcelaine since he lived in Paris; still its name was in everybody's mouth, and the little blue and white central building at Nymphenburg is a direct imitation of it. In front of the Pagodenburg was a large pond adorned with fountains, and on the opposite side was the theatre, approached by several steps. Seats for spectators were not provided, and perhaps people sat round the pond to see the performances. On the other side of the little house there was a narrow canal in the park. Also there was the bath-house (Fig. 461), containing the bath itself and a number of rooms beside a large piece of water, and on the other side a pretty parterre.

In addition to these garden pavilions there was wanted a real hermitage, which was to unite religion and fashionable life in the Spanish-French style. The hermitage with the chapel of St. Magdalene was first put up in 1725–8, and its neo-Gothic architecture bears the stamp of the growing Romanticism of the period. Later on, the opposite erection, the shooting-box called Amalienburg, was built by Charles Albert, Max Emanuel's successor. It served as a resting-place, after the hunt, for the prince's wife, who loved the

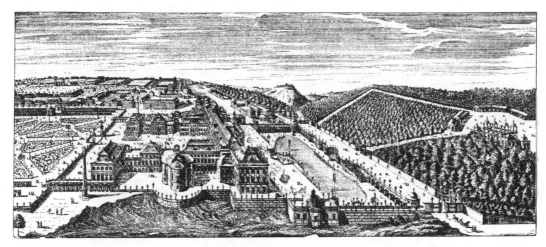

FIG. 463*b*. LUDWIGSBURG CASTLE, STUTTGART

chase. All these agreeable rococo houses, that so badly need their own surroundings, were absorbed into the English park. We can only recognise the old design in the lines of the canal, the basins, and the parterre, now poor and barren of ornament.

The princely seat of Karlsruhe (Fig. 462) shows how completely the idea of the garden dominated the eighteenth century—that century of princes, as one may call it. In 1709 the Margrave Charles William of Baden-Durlach built himself a little shooting-box in the middle of the Hardtwald to serve him as his Trianon. A hunting-tower which was in an isolated position in front of the house, looking towards the wood, was the middle point for thirty-two walks that were cut in the surrounding wood. Even the ground-plan of the castle had to be arranged to suit the prince's whim, and its side wings were set at an obtuse angle in a line with two avenues. The segment of the circle enclosed by these wings and the buildings that adjoined them farther along was laid out as a pleasure-garden; the front part was enclosed by groups of buildings for the court nobles, or for servants' use—one group between every two avenues—arranged in precisely the same way as at Nymphenburg. The oddness of the plan at the back was really grotesque; in the little circular bit round the tower there were twenty-four small houses, one at the starting-point of each avenue, varying in ground-plan, but all alike in size; each was provided with its own little garden, used for different purposes, as fountain-house, bath-house, pump-house, etc. The place round the tower was adorned with four fountains, and there were others put about in the thick of the wood where the avenues are cut.

All these separate pieces, taken from the idea of the French garden, are in this place stiffly designed and too bizarre. The margrave soon preferred Karlsruhe to the Residence at Durlach, which was growing slowly because of the opposition of the burghers, so he quickly made up his mind to establish himself firmly there; and it was soon taken as a permanent home to settle in. The front walks in the park, which led from the dwellings of the court servants, were made into streets for the new town, which now received the name of Karlsruhe. The town accommodated itself all the more readily to the symmetrical order of this park, because it embodied an ideal aimed at at the time—the uniformity of the burgher houses: these had the Residence for central point, threw it into relief and encircled it, and this was the first thing demanded.

The garden at Karlsruhe in itself was never of great importance, even when it was enlarged behind the house near the new castle building. In some respects the fine garden made by Charles William's neighbour, Duke Eberhard, Ludwig IV. of Würtemberg, at Ludwigsburg near Stuttgart, and at much the same time, was far superior (Figs. 463*a* and 463*b*). The arrangement of the park on the side opposite the pleasure-garden has a certain resemblance to Karlsruhe. But water, which is entirely wanting at Karlsruhe, is present here in a series of fine cascades connecting the main castle on the side of the park with the little casino called Favorite standing high on the hill. On the other side also the pleasure-garden, which is remarkably large and fine, is on slightly rising ground, so that the castle is the lowest of all. On one of the higher terraces stands the famous orangery, which had distinguished the Renaissance gardens of the lords of Würtemberg in earlier days. The architect who gave the final form to this garden was Giuseppe Frisoni, an Italian, who began his career as a worker in stucco, but afterwards gained experience by travelling in France.

The princes of the Church soon began to rival those of this world. It was with difficulty that they had been able to preserve their right of rule during the religious struggles which took place in Germany, but after the Peace of Westphalia had lastingly ensured their existence and their safety in the South and West, they felt it to be more and more necessary to express by outward signs, in the same way as their worldly friends, that feeling of sovereignty which was added to their spiritual dignity, since they were princes of the blood and gentlemen of standing. The war had been hard on their estates, which were always in dispute, and they came for the most part into possessions that were utterly spoiled; it was sheer necessity that forced them to build, for otherwise they had nowhere to live. The passion for building, which inspired them all, can only be compared with that of the Roman princes of the Church in the time of the Renaissance. They felt in just the same way that their works were only to endure for the short time of their own rule, and would in no way affect the future. This acted as a spur which made them strive to do their utmost for this limited space of time, and so connect their name with the pride and splendour of their buildings.

The archbishopric of Cologne had to endure the troubles of the Thirty Years' War long after it was over; and it remained a bone of contention even between Louis XVI. and the State. Clement Joseph of Bavaria, who after the Peace of Nymwegen in 1689 was able to enjoy his rule, stood in the nearest relationship to the Bavarian court; indeed for more than a hundred years after this date the archbishopric was a kind of right-of-the-second-son for Bavaria, and this comes out in the resemblance of ideas for building. When Clement Joseph was in power, his residential castles, Bonn and Brühl, were mere rubbish heaps. Still, although he betook himself almost at once to the rebuilding of his castle at Bonn, it was only his nephew Clement Augustus who was able to finish the place there, and to build a series of other castles, among which the pleasure-castle of Brühl takes the chief place, with its fine gardens. "The first wish of the owner, the first care of the architect, is to get a garden planted before he begins on the buildings": thus does Blondel define the position of the garden in relation to architecture in his *Cours d'Architecture*. In any case, house and garden must not be treated independently of one another.

Clement secured the very best of helpers in Girard, who during at least ten years proved in a splendid way his skill and experience, both at Schleissheim and at Nymphenburg. He made the plans for the park at Brühl, and was often there in person to see them carried out (Fig. 464). The water was very finely diverted, especially into canal and pond, but the way it was divided up was unusual at that time. Not only was the whole somewhat irregular garden plot encircled by canals, which also went round the individual thickets—

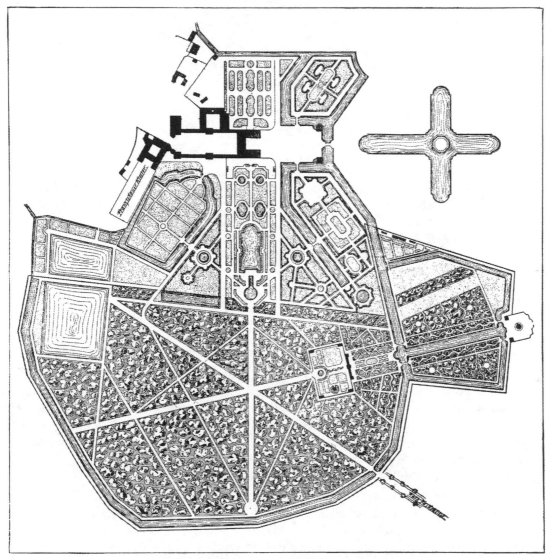

FIG. 464. THE CASTLE GARDEN, BRÜHL—GENERAL PLAN

a reminder of the French Renaissance garden—but the chief feature was a great canal in the form of a cross, in the middle of which there was a small island, not where the parterre is lengthened out, but in the axis of the court of honour; it was reached by a bridge over another narrow strip of water. The position of the parterre on the side of the south wing is also rather unusual; the garden front is built out in a wide terrace with two wings. The parterre itself is handsome, and on large lines, divided by waters and beds, which clearly show the hand of Girard: the only thing wanting is the wide view of the water that we get in both the Bavarian gardens.

II—L

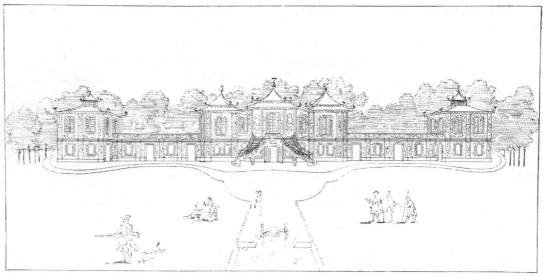

FIG. 465. CASTLE OF BRÜHL—THE CHINESE HOUSE

The numerous small buildings that enliven the park remind one of Nymphenburg, but everything here seems rather casual. The charming little casino called Falkenlust is at the end of a side walk, which issues from the great central star in the park. Nearer the castle is the amusing little Chinese house, very suitably dubbed the "Maison Sans Gêne" (Fig. 465). The wish for private life, and detachment from the more and more burdensome shows, made itself felt in things little and great, as the eighteenth century grew older. More was known about Chinese building than when the Trianon and the Pagodenburg were put up, and this little new house had curved roofs hung about with bells. The so-called Schneckenhaus (Snail-shell) is a thoroughly baroque affair (Fig. 466) set up in the middle of a circular pond: it is a kind of compromise between the old Schneckenberg and a Chinese tower. The park at Brühl shows all manner of indications of a new style, without abandoning the large straight lines required by the French garden, but yet more decidedly than most places that belong to the first half of the eighteenth century. Particular boskets, especially near the Chinese building, show wavy lines, though of course they are controlled by the feeling for symmetry.

The great importance that the smaller ecclesiastical princes attached to the land was not so much because of particular castles, but rather for the cultivation and improvement of their whole estates, which were not too big to be looked after personally. Like the old Romans, these spiritual lords required their accustomed luxury wherever they went: accordingly Clement Joseph of Cologne had a *corps de logis portatif* constructed for use on his travels, and this was made by a French architect and decorator called Oppenord; unfortunately no trace of it has survived. So these lords built themselves pleasure-houses and shooting-boxes in every pretty spot on their estates, so that they might go from one to another along the pleasant paths and streets and avenues, which broke up the country as though it were a park. Clement Augustus made an avenue of four rows of trees, from the Residence at Bonn to the castle at Poppelsdorf, which was a fine little erection with a round open interior court—a design we have often met with in Spain and in Italy. Both castle and garden were the work of Robert de Cotte, who built the castle at Bonn. The house was very greatly altered afterwards, and of the garden little is known; but in addition to the ordinary water arrangements it contained a cascade, a theatre, an arena for wild beasts to fight in, and butts. It was chiefly meant as a place for a park and for boskets, which could not be had at Bonn, where there was no room except for laying out a parterre.

How firmly and prominently the garden stood in the forefront of men's minds is indicated by Clement Augustus's institution of a " Confrérie des Fleuristes." Their sanctuary was the chapel of Poppelsdorf, which had to be decked with fresh flowers every day. A figure of Christ as a gardener, with Mary Magdalene, composed the altarpiece under an open *berceau*. From the castle of Poppelsdorf, which had the advantage of a view over the Kreuzberg and the Siebengebirge, a road led to the middle of the Kottenforst, and at the end of it stood the handsome castle of Herzogenlust, awaiting its master. A second road was planned to lead from Poppelsdorf to Brühl, while another led from there to Cologne. Clement put a charming little hunting-box on the top of the Humeling. This place, Clemenswert, consisted of one central building and eight detached pavilions placed round it in a circle. The direct influence of France is undeniable, but the task is accomplished in a rather original way at Clemenswert: eight paths start from the eight pavilions, making a sort of star; the three at the back are connected by a rectangular canal with three basins, and in front the middle path leads to the stables.

More important for garden history was the Schönborn family, a race of nobles occupying the greater number of ecclesiastical properties in mid-Germany during the first decades of the eighteenth century. After the famous and ambitious prince Johann

FIG. 466. BRUHL—THE SCHNECKENHAUS

Philip of Mainz had raised the family to honour and dignity, a great number of his nephews, who had been destined to the career from their youth up, attained the high position of ecclesiastical lords. This peculiar sort of nepotism appears once more at Cologne as a late flower of the Italian Renaissance growing on German soil. And these fundamental relationships come out in a similarity of ideas, a proud and masterful spirit, an unbounded love of building, and also a sense of responsibility directed less to politics than to art. Wherever the Schönborn family came, life was full of activity. Their castle (built by the non-clerical part of the family in Lower Austria on the River Enns) shows the artistic feeling in all of them, both by its situation and in the importance of the garden

FIG. 467. CASTLE OF SCHÖNBORN—GROUND-PLAN

at the new home. It is one of those works whose *tout ensemble* is one of great magnificence in strict adherence to French rules (Fig. 467).

Of the clergy the most conspicuous was Lothar Franz, who became Bishop of Bamberg in 1693, also Archbishop of Mainz and Elector in 1695. The castles he built have been engraved, gardens and all, by Salomon Kleiner, and the engravings were published in 1728. These pictures are the only abiding witness to the gardens, scarcely one of which has endured to the present day. The shooting-box, Seehof, one of the smaller places, near Bamberg, was received as a legacy by Lothar Franz from his predecessor Marquard von Stauffenberg, after whom it took another name, Marquardsburg (Fig. 468). Lothar Franz finished it, and liked to live there while his residence at Bamberg was being rebuilt. To suit its character as a shooting-box, the central building was quite open. The original plan may have been to make it easy for huntsmen from the district to effect an entrance on all sides by four approaches, and so assemble at the door. There is a large show-garden round the castle, and the whole ground is divided into three equal rectangles lying side by side. The house lies in the middle on a raised square terrace, which adds a slight contour to the level garden. The fine low-lying parterres with their cascades and fountains belong to the middle rectangle, the other two being laid out left and right in boskets, in one of which is the theatre, and in the other some fountains.

The most important garden made by Lothar Franz, the Favorite at Mainz, has the same threefold arrangement of axis (Fig. 469). The territory here was certainly quite different from that at Seehof. The land, again a regular rectangle, ascends gradually from

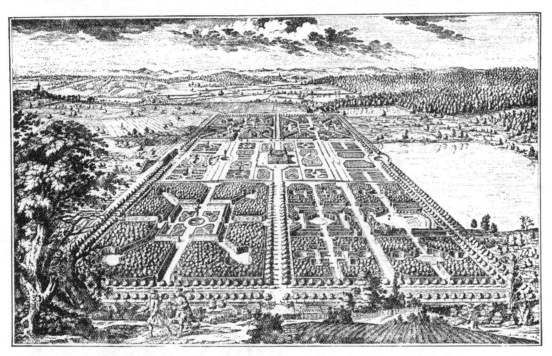

FIG. 468. MARQUARDSBURG, NEAR BAMBERG—GENERAL VIEW

the Rhine, only separated from the river by a carriage road. This garden is divided as it ascends from the Rhine into three parts, cut off from one another, but lying side by side. The first ornamental garden mounts upward from a parterre that has wonderful water-works, ending in a grotto of Thetis, to a still larger pond with statues and cascades. This main parterre (Fig. 470) is enclosed by six pavilions arranged with the chief building in the form of an amphitheatre. This originally was to be the actual pleasure-castle, but was afterwards converted into an orangery with a banqueting hall. The second garden, close by, was overlooked by a grotto terrace on the Rhine; again the axis is marked by the arrangement of the water, and a great pond comes to an end with a Neptune cascade, from which we mount farther to a Ring Cascade, and finally to the Grotto of Proserpine. The third and last garden has a hedge, which cuts off from the river a so-called *boulingrin*, the name now given to a sunk part with trees—in this case chestnuts—and a basin in the enclosure. From this spot we ascend to the great promenade between high hedges and on grassy steps: this walk has crossway avenues of chestnut, and water-works at

FIG. 469. FAVORITE, MAINZ—GENERAL VIEW

the end. The horse-chestnut was the favourite tree at the turn of the century, because it was new.

The Elector called this beloved creation of his "a little Marly," and certainly the first ornamental part, with its shining waters leading straight down to the Rhine from the main building, by way of pavilions, reminds us more of the much-admired French model than of any German garden. But the plan of the whole is most original, and to get the necessary balance between show parts and private parts these are placed side by side. But the river with its life and power binds the whole together, like a mighty canal across the end; and a most pleasant landscape frames the picture. When in 1726 the fourteen fine engravings of Salomon Kleiner were published, the editor added a pre-

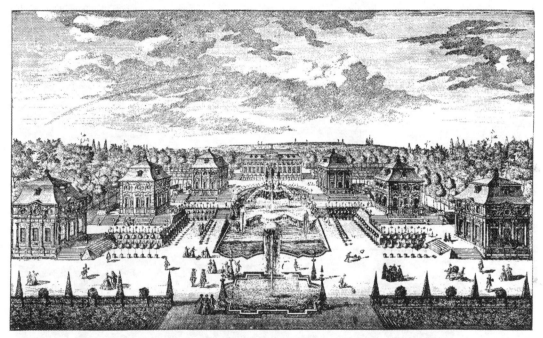

FIG. 470. FAVORITE, MAINZ—THE PRINCIPAL PARTERRE

face, which says that, "although the almighty and omniscient Creator of the Universe gave it the perfection of beauty, still Art has given certain aid to Nature herself, by providing noble buildings and beautiful gardens, always more and more gay and elegant." This work (Favorite), however, "that can never be enough admired for its exquisite architecture," as the title says, perished in the storm of revolution, leaving no trace behind. The last Elector of Mainz, the Coadjutor Dalberg, entertained six thousand *émigrés* here in 1792, and arranged a magnificent feast for them in the garden. Favorite opened its hospitable gates for the last time to the Congress of Princes held on 19 July, 1792. On 21 October the French arrived, and a few months later the whole place was razed to the ground.

At the time when Lothar Franz was still busy over his darling place, his small private castle at Gaibach (Fig. 471) was altered. It was still a water-castle in the old style; and the Elector retained the moat as an ornament to the place, and only added two wings to the garden façade. People walked over the bridge into the garden, where Lothar Franz succeeded in combining the most delicate products of modern feeling with the older character of the house. The fundamental plan of the garden shows a sentiment of the Renaissance, that union of the ornamental and useful which the age of Louis XIV. persistently challenged and rejected. After a fine *parterre de broderie* with a Triton fountain, there comes a place that again points to the Elector's preference for cross-roads, and perhaps also shows direct Italian influence: a plantation on the right is laid out as a round botanical garden for foreign plants, and answering to this on the other side there is a sunken round basin, with parterre beds and a high hedge round it, and across the end a

FIG. 471. GAIBACH CASTLE—VIEW FROM THE ORANGERY

grotto hill, ornamental waters, and a little house, which perhaps was a relic of the former garden. The next things we find are two large plots of greensward planted with fruit-trees, some tall, some dwarf. One of the two has a pergola on it like a great cross. From this part of the ground two gently sloping terraces rise, with a semicircular orangery and grand hall on the top (Fig. 472). To right and left we have *berceaux* and pavilions, and in front of the orangery there are parterres. All these separate parts might easily belong to a Renaissance garden; but in the place taken as a whole there is a severe regularity of plan; and the placing of the main and side avenues, the marking of the middle axis by fountains, all show the style of the eighteenth century.

A fourth castle that Lothar Franz inherited he altered in 1711 after Favorite was finished. This was Pommersfelden at Bamberg (Fig. 473). The peculiarity of the fine terrace gardens is that they end in two great ponds, which in most modern gardens would be superseded by a canal. But the canal is always kept in the background in the gardens of the Schönborn family, and thus they again have a character in common, such as we find in the juxtaposition of the different kinds of garden scenes, in spite of the great variety they show.

From among the seven nephews of Lothar Franz no less than four princes of the Church were appointed. Perhaps the most original figure among them is Damian Hugo Philip, who held the bishoprics of Speyer and Constance till the year 1743. His means were small, and he came to a neglected land, but still he knew how to make a flourishing and attractive place out of the little Residence of Brüchsal, which was (to use his own words) "a peasant's hole, so to speak"; and from this, with the greatest difficulties to surmount,

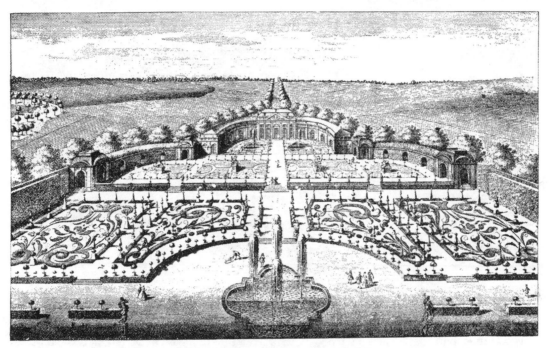

FIG. 472. GAIBACH CASTLE—THE ORANGERY

he grouped the various parts about his own house, where he began to build with great zeal and energy in 1720. His love of rule matched this energy. "I will be master," he said; "not till I am a cold corpse will I cease to be master." He always had it in his mind that he would get a claim to immortality in the world to come if he made his name famous through his buildings. Once when he returned from his travels, he found that certain burghers had dared to make some change in the stucco-work and the height of the windows, so he made a formal protest to posterity in the form of a protocol: "Thus do we herewith make protest, and no reasonable man will impute this to us for blame— seeing that outside and inside our land we have built and set up many fair buildings worth millions of money, under our own direction and ordering, to the approbation of all men—saying that in our old age we are foolish and base for devising arrangements so contemptible and worthy of laughter. We herewith once more protest most emphatically, and we disapprove of everything that has been done in this business, against our will, and against our own arrangement."

FIG. 473. POMMERSFELDEN CASTLE—GENERAL VIEW

The garden at this place was never very important, and its greatest charm was due to the wide terrace—the middle walk in its present altered state. This was edged by a pergola in a half-circle, and adorned with parterre-beds, which led by a few steps to the greater parterre. There is not the least doubt that there must have been an Indian house and other erections in the park beyond. A Chinese pavilion stood on the hill at a distance from the castle, and it is still preserved as part of the so-called Belvedere, which was in a small pleasure-garden, named Wasserburg because it contained the reservoir that fed the waters. When the work of restoration was completed, the court of honour was once more planted in the old charming fashion. The *parterre de broderie* round the two fountains has a very pleasing effect, especially as seen from the balcony of the castle. The enclosed court, bounded on the town side by gates and pavilions, was perhaps still more attractive to look at when shimmering waters flowed into the trenches that were cut between these buildings. It is a pity that an attempt was not made to restore the garden itself in the old style.

Brüchsal was not to remain as the only castle that Damian built and beautified with a garden. He raised the fortunes of the bishopric by the exercise of extreme personal economy, and there arose not far from Brüchsal another place called Kisslau, now quite abandoned, and a few miles farther on the peculiar hermitage of Waghäusel. The spiritual lords of Germany were well suited to maintain vigorously that union of worldly pleasure and unworldly piety in the eighteenth century which began with the Renaissance and was especially observable in Spain. They found an outward symbol in the hermitage. Damian's personality, his true piety, his passion for building and the chase, and his love of art, all made him a typical representative of this spirit of the age. So his hermitage was close to an important cloister, with beautiful woods suitable for hunting, and therefore to his liking.

He was quite in earnest when he set about making his stucco-work in the hall "in hermit fashion," giving his instructions from Rome in 1730. One piece of the ornament is still extant, showing in the painting of the dome what this hermit-decoration meant (Fig. 474). The picture has a double interest: it shows a hermit's hut built into an ancient ruin; supports are seen fixed two by two, holding up a defective roof of straw, which is made to cover wooden beams exposing through their arches all sorts of other bits of ruined material; and sacred utensils are hanging on the temple pillars. The idea was to depict a garden hall with a hermitage roof. Very far removed as such a painting seems from the ancient frescoes, it certainly belongs to the same stage of development. It is here that the sentimentality of the age first makes its appearance, for it was not very much later that men began to put up buildings in the parks outside, which came into existence from the very same feeling. The sickly fancy for ruins and hermitages, not unknown in the days of the Renaissance, but at that time quite overmastered by experiments in other

FIG. 474. WACHÄUSEL—THE ROOF

directions, has now grown more and more into a real passion, and the picture we give here is only a very early indication of what this movement will produce later on. From its structure Waghäusel ranks with open central buildings, and also clearly shows that it is a hunting place. Just as at Clemenswert, there are four pavilions at the openings of four avenues. Nothing of its garden surroundings is preserved, but a later plan gives fruit-trees in a concentric arrangement, and this may have been in the original design.

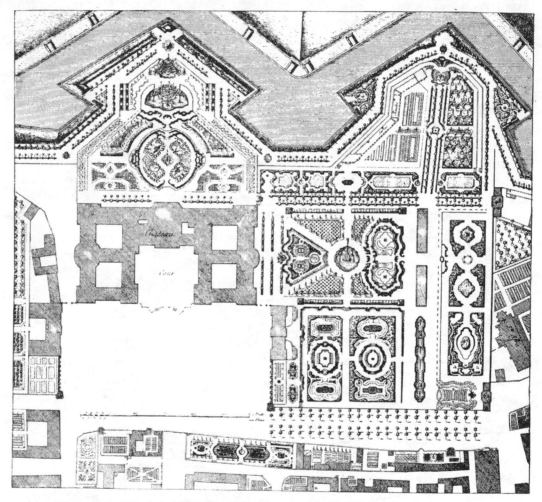

FIG. 475. WÜRZBURG—GENERAL PLAN OF THE CASTLE GARDEN

The two brothers of Damian Hugo held the bishopric of Würzburg one after the other, with a very short interval. The Residence was built for them by the most important architect of Middle Germany, Balthasar Neumann. The foundation of the beautiful castle, one of the most conspicuous of the eighteenth century, was laid in 1720, but building was postponed owing to the early death of Bishop Johann Philip Franz, and it was first finished in rough-cast in 1744 by his brother Friedrich Karl. Neumann certainly had the sketch for the garden in his hands in 1730, but it was only gradually made along with the castle, so that when Salomon Kleiner published his album of Würzburg in 1740, it only contained a page giving a bird's-eye view of the garden at its earliest stage. The

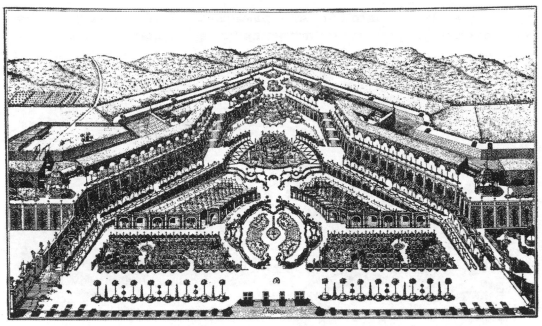

FIG. 476. WÜRZBURG—THE PRINCIPAL PARTERRE

plan is determined in connection with two corners of the town fortifications (Fig. 475)
The middle axis of the castle was intentionally directed towards one of these, and the
flower-garden had to conform, finding its appointed end behind the parterre at the raised
steps of the citadel terrace, which, according to the first sketch, had two summer-houses
at the top. The second corner was in the line of the vegetable-garden, where there were
also boskets, a labyrinth, and the orangery. There was another part on the side, according
to the favourite plan in gardens of Middle Germany.

As at Würzburg, the garden at Mannheim made use of the fortresses in its design.

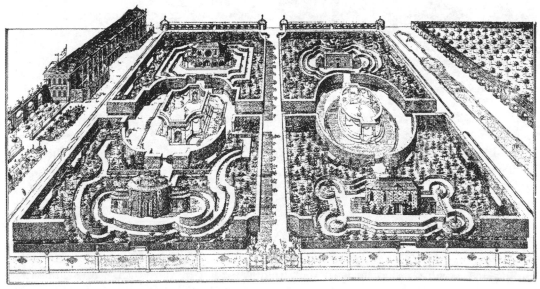

FIG. 477. WÜRZBURG—THE CASTLE GARDEN AND SO-CALLED LABYRINTH

The garden had to be laid out to agree with the extremely formal lines of the plan of the town, and then had to fit into the three corners of the fortifications in a shape that was practically the same in each case: the number of separate parterres was restricted by the raised "surround" which ran alongside the walls.

The Würzburg garden owes its completion and its historical importance to the second owner after the Schönborns, one Adam Friedrich von Seinsheim, who in 1770 summoned to his aid as inspector of gardens the famous Bohemian botanist Johann Procopius

Mayer. He enlarged the place, and gave it the form which it still keeps in the main (Fig. 475). Mayer, an artist of much taste and intelligence, published a book in 1776; though it was first and foremost botanical, it treated the art of the garden in a theoretical and pedagogic fashion.

At the time Mayer wrote, the English style of gardening had already made a victorious onslaught upon Germany: but he defended the formal style at Würzburg with the full sympathy of his employer, and made his defence in an essay at the beginning of the book. "Here we are to have no simple shepherdess, plucking meadow-flowers to adorn herself withal, but some proud court beauty must appear in all her paint and finery, one who is not debarred by dress or station from the free use of ornaments and gold, but must shine in array worthy of a palace—and of what a palace, one of the finest in Europe!" The real nature of the rococo in Germany could not be better described, and the garden was laid out in just this spirit. The ornamental part (Fig. 476) had no real axis, and the balustraded terrace beside the

FIG. 478. WÜRZBURG—GROUP OF CHILDREN IN THE GARDEN

parterre, and the plan of steps, which includes cascade and grotto and is continued to the garden by a semicircular trellis, produce a picture that for gaiety and splendour suggest Italian rather than French models. The natural sphere of interest for the lords spiritual was Rome, and they were attracted so strongly to Italy, and stayed there so often, that it is not surprising that the artistic bent of France in the garden was often interfered with by that of Italy.

At Würzburg the orangery was close to the charming flower-garden at the narrow side of the palace, and the kitchen-garden was beside the second corner of the fortification. Mayer himself was careful to draw attention to the gradation of his garden. After the orangery there follows what he calls the strolling garden or labyrinth, "of a kind that really comes nearer to the country." It is a curious place (Fig. 477): it has hedged paths

FIG. 479. WÜRZBURG—PERGOLA IN THE GARDEN

FIG. 480. VEITSHOCHHEIM—A CORNER OF THE GARDEN

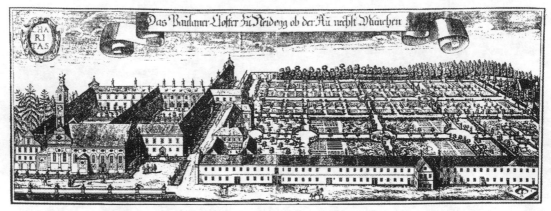

FIG. 481. PAULANERKLOSTER, NEAR MUNICH

that have nothing to do with the old sort of labyrinth, and within them a host of small
erections: temples, Gothic ruins, coal-sheds, barns, hermitages, all these being required
by the new style, though grouped in regular order; for this was a tribute that Mayer felt
he ought to pay to his own age. Farther along the plan shows us nothing but a sunk
bosket and a grotto with figures from Æsop's fables. Many statues are still preserved
(Figs. 478 and 479).

The Prince-Bishop Adam von Seinsheim was still very earnest about the old style,
and he had every garden kept thus in his castles at Bamberg and Würzburg, although
many of his artists, and among them the Inspector Jacob, "admired the English style in
their hearts." Veitshochheim also, the pleasure-castle of the bishops of Würzburg, was,
if not designed, at any rate finished by him, and furnished with fine statuary (Fig. 480),
which has survived till now, giving a picture of German rococo, wherein, in many colours
and combinations, we find mythical gods, shepherds and shepherdesses in fancy dresses,
and peasants in the costume of the day. This is a fashion that Italy probably introduced
in the seventeenth century, but which constantly lent a charm to German gardens, always
picturesque though often verging on the grotesque and even on caricature.

The castle of Veitshochheim was originally a centre building, a typical shooting-box.
It lies on a high balustraded terrace with groups of children and a small parterre. Here,
too, the garden lies in a side direction. Four avenues traverse this part lengthwise, dividing
it into three sections, more or less answering to the French boskets. From the side façade
there starts an avenue of fig-trees, first passing an out-of-door theatre, then various
boskets, round temples, and statues till it comes to an end, where the corner is occupied
by a pretty octagonal summer-house, which has its ground-floor treated as a grotto. The
second axis also passes by boskets and round spaces ornamented with temples, fountains,
and statues. In the third division there is a very large and somewhat elongated basin with
a wavy border, and hedges and boskets: in the middle of the basin stands a bold group
of Pegasus, which was originally painted, thereby showing something of the grotesque as
well as the picturesque. This axis is finished off with a small pond. Cross-paths are cut into
the long ones, so that you see temples, ponds, cascades, and summer-houses from every
point of intersection. It is a pity that the clipped hedges, which were still there in 1830,
have not remained in fashion, for with proper training and the use of the shears this
garden might very easily have kept its original character.

It was only to be expected that when clerical magnates were thus flourishing, the monasteries would not be left behind. If at that period the style and ornament of churches reached the highest point in art and splendour, it was natural and right that their gardens should match them. The owners had long abandoned the simple laws of the kitchen-garden, and now were laying out grand flower-gardens, whose high walls showed a special desire for seclusion as at the Paulanerkloster, near Munich (Fig. 481).

Important as the great church lands were, especially for the central parts of Germany, there was still something unexpected and therefore capricious and unstable about their aspect. In the history of garden art we come to a surprisingly important cross-track—a thing that often happens in Germany. If we direct our attention once more to the then capital of the empire, Vienna, we find that a really quiet process of evolution begins very late, and that is so for outside reasons. Long after the Thirty Years' War was

FIG. 482. BELVEDERE, VIENNA

II—M

over, the oppression of the Turks kept back the development of gardening. The final conquest of the enemy came about with the last inroad in 1683, and then the court and nobles ventured once more to establish themselves before the gates of the city. This was done cautiously at first, but then with an impetuous desire for building, and a strong feeling for peace. Under the auspices of Leopold the Holy Roman Emperor came an ever-increasing prosperity.

Among the first who secured land in front of the glacis south of the town was Prince

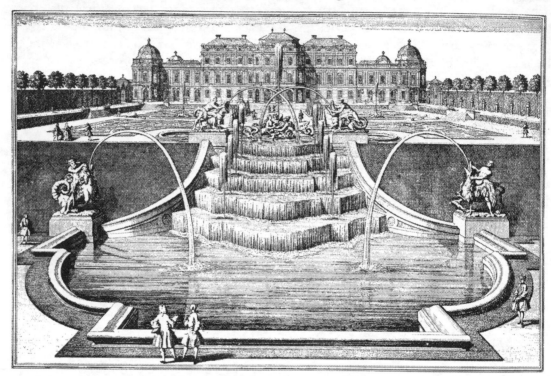

FIG. 483. BELVEDERE, VIENNA—THE GREAT CASCADE

Eugen of Savoy, the conqueror of the Turks and a much-admired hero. He bought a property where the vine-clad hills rise steeply from the racecourse as early as 1693, and a little while after he found a neighbour there, his old adversary in councils of war, Count Fondi-Mansfeld, who had been before him, in that he had had a palace built for himself at the foot of the hill, by Fischer von Erlach. Later on the empress established a convent for the nuns of the Order of the Visitation, whose garden was on the other side of the hill (Fig. 482). In all these gardens we have to deal with the type of a suburban villa, which means that they are somewhat limited in size. There could not be any park, and the view was necessarily over the city and its many towers. This explains the townish appearance of these places.

They are suitable for pomps and festivities, especially the Belvedere, as the prince's estate is called. The garden divides into two chief parts: the upper one has a very large parterre in front of the house and its fountain, and a second much lower and simpler parterre, with a magnificent cascade (Fig. 483) falling into the middle of it, and on both sides a flight of very shallow steps (Fig. 485). Round this parterre there runs a narrow

FIG. 484. BELVEDERE, VIENNA—THE UPPER ENTRANCE-GATES

FIG. 485. BELVEDERE, VIENNA—THE SIDE STAIRWAY AND CLIPPED HEDGES

terrace at the level of the upper entrance to the castle, and the whole place is entirely
without shade, with tall hedges and Hermæ against the wall. The entire breadth is closed
in by the castle, one of the lordliest of pleasure places, for which this high part of the
garden must be regarded as a gigantic open dining-hall. The prince never lived here,
and the great entrance-gates (Fig. 484) were only thrown open for festival occasions.
Hildebrand the architect built the dwelling-house for the family by the lower road,
cleverly adapting the slanting lines to the shape of the garden ground: this also took the
whole breadth, but was laid out in a simpler, homelier fashion. After the parterre with its
fountains there come boskets, four shady places with hedges, and lawns in the middle.

FIG. 486. THE BELVEDERE, VIENNA—THE ORANGERY

The two pairs are separated by imposing basins and fountain groups. An avenue of chestnuts leads along the supporting wall to the upper garden, whence there rushes out a second
powerful cascade decorated with statues. At the sides the two gardens are connected
with grand straight steps, and groups of children as ornaments (Fig. 485). Thus the upper
and lower gardens are, so to speak, interwoven, with a most harmonious result.

The pressing problem of how to pass from the showiness of publicity to the
comfort of privacy, from sunshine to shade, is admirably solved. If a spectator of
to-day is worried by the want of shade in the higher part, he must bear in mind this
fundamental requirement. There are little bits of garden attached to the large rectangular
part, which skilfully and intelligently reconcile the want of regularity in this estate.
Below on the right there is the charming orangery near the dwelling-house, with its
arched trellis and attractive pavilions on the second terrace (Fig. 486). And by the upper
villa there is the kitchen-garden, unobserved at the side of the grand approach for

FIG. 487. SCHWARZENBERG, VIENNA—THE GARDEN, WITH BELVEDERE ON THE RIGHT

FIG. 488. THE LIECHTENSTEIN PALACE, VIENNA

carriages but reached from the garden side, and there is also the interesting menagerie. The idea of a concentric arrangement was adopted from Versailles, but in this place it is all on a smaller scale and more consistently worked out. Instead of finding a small casino in the centre as at Versailles, the spectator stands in front of an iron grating, whence little fan-shaped parterres spread out towards the animals' winter quarters. In 1731, when the prince was still living, Salomon Kleiner published some very fine engravings of views of the garden, and in the peculiar title of the work paid a personal tribute to the warrior-hero, calling it " The Wonderful Home of the Incomparable Hero of our Time in Wars and Victories; or the actual Presentation and Copy of Garden, Court, and Pleasure Buildings, belonging to his most Serene Highness, Prince Eugenius Franciscus, Duke of Savoy, etc." After the death of the prince in 1736, the Belvedere passed into the hands of the imperial family. Both before and after, the garden witnessed those brilliant fêtes for which it was intended: in 1700 there was a masked fête on 17 April, such as Vienna had never yet seen; and to accommodate six thousand dancers a great dining-hall was built out in the garden, covered over the top with 15,000 ells of linen; on the walls and roof there was painted a *berceau* of gigantic size, ornamented with flowers and festoons. The eighteenth century had still to learn how to keep a great fête!

The neighbouring estate, laid out by Count Fondi-Mansfeld in 1694, passed into the possession of the Schwarzenberg family in 1715, and they completed both house and garden. A similar but somewhat simpler problem was presented here. The garden had only to consider, in the way of buildings, the castle below, whence it climbed upward in terraces from the fine parterre, growing ever denser and more shady with groves. Its beauty lay in the well-marked middle axis of water: this formed two cascades with many

FIG. 489. HOF CASTLE, VIENNA—VIEW OF THE GARDENS

sculptured figures at the dividing walls, and ended with a great reflecting pond, that occupied nearly all the space there was (Fig. 487).

Just as the Belvedere supplied the town with a central point for a garden quarter in the south, the house built by the wealthy Adam von Liechtenstein in the north made with its splendid gardens a focussing point (Fig. 488) for other grounds round about the Alserbach. The beautiful Belvedere with its open pillared hall, cutting off the Liechtenstein garden on the river side, and giving a grand view of Kahlenberg, has been absorbed, view and all, into the town.

If we turn over the engravings by Delsenbach and Kleiner, which keep alive for us these Viennese gardens that belong to the first half of the eighteenth century, we are necessarily struck by their formal, prescribed type. The canal plays no part here, but instead many cascades are found in the strictly marked middle axis, conditioned by terraces which are everywhere supreme. The boskets are simpler than those a French garden requires. All these peculiarities show that at Vienna the prevailing influence was not French, but more and more Italian, although there are many French details worked into an Italian background. This state of things suited the political situation—not only the violent animosity towards the French court, but also the long alliance with Italy, so sympathetic to the main interests of court life at Vienna. Under Maria Theresa all literature and art took their colour from Italy. The greatest architects of the period: Hildebrand, and Fischer von Erlach, who gave a new character to Vienna in the early decades of the eighteenth century, were full of Italian ideas. It is true that Prince Eugen had employed Girard, the French gardener at the Bavarian court, to lay out the grounds of the Belvedere, but it is noteworthy how the *genius loci* constrained this artist to work in an Italian style. One has only to compare his work at Nymphenburg with the Belvedere to realise the facts. In both, the water is the main feature of the design, yet what different pictures we get! In one there is the French canal garden, in the other, Italian terraces. Even the choice of ground shows the effect of the different styles.

All the gardens hitherto treated lie almost in the precincts of a town, and the absence of large parks may be put down to want of space, but one has only to look outside Vienna to discover the same thing. In the estuary between the March and the Danube stands the Schloss Hof (Fig. 489). This Prince Eugen inherited at the beginning of the century, and he made a garden there. Later it became an imperial property, and in the years 1758–60 Canaletto painted his fine pictures here, at the bidding of the Empress Maria Theresa. This garden, whose extension was really unlimited, was governed by the same spirit we have met with in the towns. Six terraces ascend to the palace, some wide, some narrow, and there are parterres on three sides of it (Fig. 490). We reach the actual ornamental garden by three steps, and a terrace projects from it enclosed by a balustrade, and forming three outstanding parts. Below there is a grotto in the middle, cut off by a wrought-iron gate. A fine flight of steps leads up to the next narrow terrace and to another parterre which has thick arbours of lattice at the side, and pavilions roofed with copper. There is a cascade in the middle, falling over a supporting wall which is well made architecturally. A simpler cascade plunges down from the fifth to the sixth terrace, with shrubberies to walk in at the sides. On the highest and lowest terraces there is a large fountain with groups of figures. The River March, which flows past the garden, brings this region to an end in the valley below. The peculiar southern character of this garden was kept

FIG. 490. HOF CASTLE, VIENNA—THE UPPER TERRACE

up by the use of evergreen hedges of juniper. French influence is seen in the great number of parterres, which are to be found on almost every terrace, but otherwise the Italian style is prominent. In the Austrian gardens of this date we can see a happy mixture of the two.

The greatest of the imperial castles, Schönbrunn, is not wholly wanting in the same spirit, in spite of its immense size, and in spite of the propelling power of its well-known rivalry with Versailles. Schönbrunn had belonged to the emperor ever since the sixteenth century, but in 1683 the shooting-box, built in the Italian manner, was destroyed by fire. Thirteen years later, with new noblemen's castles growing up everywhere round Vienna, we hear about fresh building at Schönbrunn. The bold first design of Fischer von Erlach shows the spirit that was then alive at Vienna in matters of art. The castle was to stand at the altitude of the present Gloriette, and the hill below was to be converted into great terraces, the supporting walls held up by architectural niches, with water-devices and parterres on the terraces. Above, in front of the castle, there was to be a circular pond on a terrace of a similar kind. A great court of honour, meant for games, was to form the end of the place in front, and it is like what we have to-day in the same part in front of the castle. A cascade in seven streams falls foaming over the rocks. This plan might truly rival the fairest villas of Italy—nay, it might eclipse them. An inscription under the picture says that there was a desire to get the grand view from the top, and that the park was to be laid out on the land that sloped gently towards Hetzendorf.

It was an architect's dream, which even the wealth of the Viennese court could not realise. Fischer had to content himself with a second plan, which was to build the house at the foot of the hill, and to lay out the garden right up to the top; and this scheme was

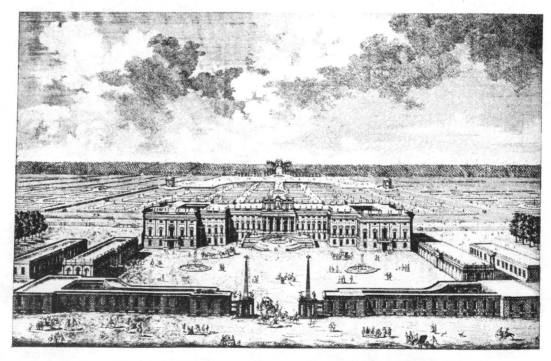

FIG. 491. SCHÖNBRUNN, VIENNA, WITH THE FIRST PARTERRE BEHIND

FIG. 492. SCHÖNBRUNN, VIENNA—THE GLORIETTE

in the end carried out. The absence of a view from the main house was to be compensated for by setting a little pleasure-castle on the top of the ridge. In the very large parterre shown in the sketch by Kraus (Fig. 491) Fischer makes other concessions to French taste. A canal flows all round it, and two small pavilions mark the corners. At the back the canal curves in a half-circle; and between it and the hill (shaped to make a semicircle as though for a theatre) there lies an open space; you go up past a fountain by a middle walk to the casino. The garden is formal, and differs very much from its last state; that, however, was due to Maria Theresa. Hesitation seems to have been felt chiefly over the way the hill part was laid out, for a pleasure-house was firmly established on the top. One design, which perhaps was never carried out at all, solves the difficulty very happily with a broad cascade foaming down into a pond. Another plan, which was carried out under Maria Theresa's rule, gives the hill divided up into different terraces with steps and grottoes and semicircular colonnades about the pleasure-house. It was not till 1775, when the garden received its present form, that the architect Hohenburg set up the pretty Gloriette (Fig. 492), an ornamental building with a room in the middle and open halls on both sides. Unfortunately its fine silhouette is only shown nowadays against a background of empty field, cut up with ugly zigzag paths. This leaves a sensible gap in the whole garden picture of the castle.

One of Canaletto's pictures gives a good view of the whole parterre, embracing all the level ground from the castle to the hill slope (Fig. 493). They gave up the canal plan, in spite of the fact that a Dutchman, Steckhoven, educated in France, was the gardener at that time; they had too little sympathy with the Austrian feeling for art. The parterre itself was of course often changed according to the caprices of fashion, and about the middle of the century it was laid out with patterns on the grass *à l'anglaise*, and with narrow ribbons of flowers. Then in the eighties it became the mode to decorate all gardens with baskets of flowers on the lawns, and unusual flowers in bloom were set in them (cf. Fig. 547). But the flower-gardens proper were partly sunk on both sides of the

FIG. 493: SCHÖNBRUNN, VIENNA—THE PRINCIPAL PARTERRE

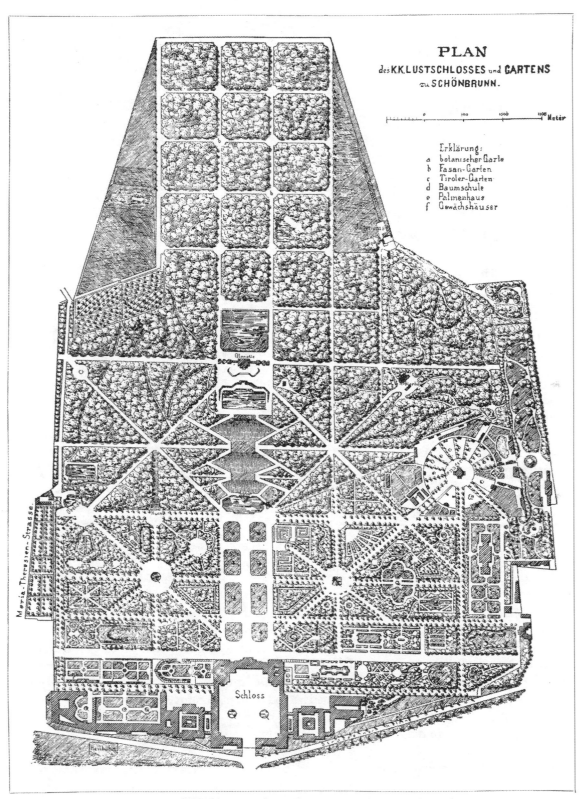

PLAN
des K.K. LUSTSCHLOSSES und GARTENS
zu SCHÖNBRUNN.

Erklärung:
a   botanischer Garte
b   Fasan-Garten
c   Tiroler-Garten
d   Baumschule
e   Palmenhaus
f   Gewächshäuser

Schloss

Reitbahn

Maria-Theresien-Strasse

FIG. 494. SCHÖNBRUNN, VIENNA—GROUND-PLAN OF CASTLE AND PARK

house like *giardini secreti* (Fig. 494); they were separated from the great gardens by a pergola, which protected them. It was in the seventies that the parterre had the fine end-piece added, the Neptune fountain, which decorated the foot of the Gloriette hill. From the beginning the great parterre was bordered on both sides by groves, which were originally laid out as a sort of labyrinth with so-called "apartments" (Fig. 493, on the left). At the crossing of the main paths the eye is arrested by fine fountains with sculptures.

In the park beyond there are certain sites which, though they show different tendencies in the course of the long period when Schönbrunn was building, may very easily be admitted into the grand movement of the style that was now predominant. The

FIG. 495. SCHÖNBRUNN, VIENNA—THE ARTIFICIAL RUINS

menagerie shows a most marked leaning towards Versailles (Fig. 494), and this was the cherished creation of Francis I., the husband of Maria Theresa. Its plan of concentric circles was more rigidly carried out than at Versailles, and in the centre there was a pleasure-house, from which the cages could be inspected, running out like the spokes of a wheel. Then Hohenburg put up a sham ruin at the foot of the hill in 1776, only a little later than the menagerie. It was on the left of the Fountain of Neptune, and fitting properly into the scene of the park (Fig. 495) it formed a good termination for one of the side avenues.

We have often encountered this fancy for ruins in the history of gardening, but until now it has never found a suitable home. In Italy ancient ruins were utilised in the same way as antique statues. The French gardens of the *grand siècle* were opposed to all fancies of the kind; they were too honest, too fond of their own proud forms, altogether too magnificent. But in the eighteenth century sentimentality had so grown into the German nature that ideas of this sort were greedily seized upon. We found forerunners in the paintings at Waghäusel and others like them, though the artists could not free themselves entirely from the formal style. But at Schönbrunn artificial ruins came into fashion as the result of another tendency, which arrived in the second half of the eighteenth century, side by side with the sentimentality of the northern countries. This tendency was classical; it came from the South, and exercised a strong influence on garden decoration. Archæological interests, as we said before, were particularly active in Italy in the middle of the century: the important work of Winckelmann is an eloquent sign of this. But, side by side with the desire for real scientific knowledge, there went (as so often at such times) a certain lack of discrimination between what was false and what was genuine. This is indicated by the number of forgeries that flooded first Italy and then other countries. Hand in hand went a delight in imitating antique building, though in this there was no doubt a real archæological interest, as we can see in many of the erections in Italian villas, above all in the ruins at Villa Albani. In 1786 Winckelmann spent some time at Vienna as an honoured guest. It is not known for certain whether the idea of building the Hohenburg ruins emanated from him; but here, as at Albani, there is no sentimental romanticism to be detected, only delight in successful imitation; it would have been contrary to the instincts of Maria Theresa, who was gay, determined, and self-willed, and she made Schönbrunn her usual residence.

All the other statuary was of the antique character. An artist educated in Rome might revel with his fellow-workers in Greek gods and heroes, but although Italy would sometimes go wrong in her models, here they went wrong always, for they had no models. Instead, they relied on archæological handbooks and lexicons to inspire their imagination, and the result was rococo, and the affectations that we find to-day. It was in the same spirit that they first undertook the delightful task of reconstructing Roman villas with the help of Pliny's letters. But the attempt of the court architect Krusius at Dresden, and later the similar effort of von Schinkel, never transcended the limits of their own age; and in both cases nothing more came of it than the gardens of a French castle.

The rivalry in garden culture was always very keen in the Austrian crown lands. All the gardens, such as Hellbrunn and Mirabell at Salzburg, were decked out with fashionable parterres and statues; and, most of all, men's eyes were directed towards the groves, where decorative effects were achieved (Fig. 496) that were quite foreign to what had gone before. At that time the inspector at the Salzburg gardens was Franz Anton Danreiter, who translated the French instruction book, *La Théorie et Pratique du Jardinage*. His activities were very important, even in the garden plans that he made himself, which, to be sure, were often as much over-valued as were those of Dekker.

Hungary had no special development to boast of as distinct from Austria. Till late in the seventeenth century the Renaissance flower-garden held its own with interests purely botanical, and these were also furthered by learned travellers. In 1631 a physician of Upper Hungary was granted a title of nobility with the affix *ab hortis*, as a recognition of his services in this field. It is also said of Protestant theological students that, on their return from Dutch universities, they often brought some cutting or rare bulb in their modest knapsacks, and in their heads the knowledge of all sorts of garden constructions that they made use of at home. When in 1664 the first garden book in the Hungarian language, *The Garden of Pressburg*, appeared, its author, George von Lippay, described exhaustively the garden belonging to his brother, the Prince and Archbishop of Gran. He owned the most famous Hungarian garden of the time, and took the utmost pride in the number of flowers which were raised there, and which made a fine many-coloured picture on their beds with the marvellous borders. But at the beginning of the eighteenth century French influence made its way without restraint into the gardens of Hungary. They vied with those of Austria, and the same nobles owned properties in both countries.

The Austro-Hungarian nobility at the time were most opulent, and felt that their first duty was to art. This was the time when Prince Esterhazy built a castle in the style of Versailles at his family seat in Hungary, and there erected a theatre and opera-house, where his *Kapellmeister* Joseph Haydn performed his works before a select audience that

FIG. 496. MIRABELL, SALZBURG—SMALL FOUNTAIN WITH BOSKET BEHIND

numbered four hundred persons. The theatre opened on the beautiful garden, where the fêtes were held. The chief garden was laid out entirely in the Austrian taste, with *giardini secreti* at the side of the castle, and a great lawn with innumerable baskets of flowers on it, with high espaliers enclosed by boskets. Under these princes, who liked to be in the height of the fashion, it was only in the pleasure-park that the new style crept in, with its "winding walks of the English sort"; for Austria had long been contending against its onslaughts. Although Vienna and the country round about took time to arrive at the undisturbed reign of peace that is demanded for garden art on traditional lines, still it had arrived at a style of its own, due to a position rather different from the rest of Germany, and one which had apparently a surer footing than any of the other local developments in the eighteenth century.

The gardens of Middle Germany were far more affected by the growing movement, and especially in the kingdom of Saxony. From an early date there had been a fine show here; for even in Renaissance days it had not been only the princes and the rich nobles who kept gardens well tended, but the towns had played an active part as well. And now Leipzig, which was wealthy, thanks to its central position for trade, had as citizens men who ventured in their self-confidence to live their lives on an equal footing with the great lords, even in a century when the power of court and nobles seemed overwhelming. The town gardens of Leipzig in the eighteenth century enjoyed an international reputation. Two brothers named Bose laid out gardens of marvellous beauty. The place belonging to the elder, Caspar Bose (Fig. 497), was at the Grimma gate, and not only did it excite the observation of Italy, but (we are told) the Pope inquired into the nature of its arrangements. The most interesting part is the site of the sunk orange-parterre, with radiating beds in a great semicircle, in the middle of which orange-trees were planted.

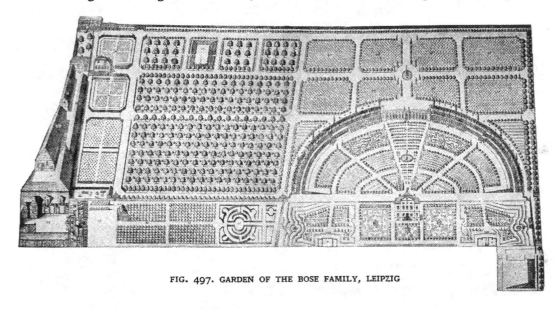

FIG. 497. GARDEN OF THE BOSE FAMILY, LEIPZIG

The ground that shuts off the orangery in the middle axis of the villa rises in the form of an amphitheatre; and fountains and statues enliven the scene. At the top there are flower- and tree-gardens, grottoes and more fountains. We see here the Renaissance ideas firmly maintained in the town garden, in spite of all the ornament in particular parts; and this is especially noticeable in the immediate connection of use with pleasure. Moreover the orange-parterre recalls the arrangement of botanic gardens, which, ever since the one at Padua was founded, have been apt to show the trapezium shape in the beds.

The creator of this place was the architect and copper-engraver David Schatz, who also laid out the next garden we speak of, which is peculiarly interesting for having attracted and delighted the eyes of Goethe when he was a young student. He writes to his sister in December 1765: "The gardens are so beautiful, I have never seen anything like them in my life. I may send you a view some time of the entrance to Apel's garden. It is glorious; the first time I saw it, I thought I was in the Elysian Fields." And Goethe was right, for it far excelled the ordinary town garden in size as well as arrangement. True, the idea of the parterre, like an amphitheatre, was also present in the Bose garden; but here there were fan-shaped avenues stretching out behind, and coming to an end

II—N

in a great circumference, richly stored with statues and with all sorts of thickets. Although the picture (Fig. 498) does not do it justice, it is evident if one looks at this garden that it was meant for great fêtes as at a court. King Augustus the Strong was a welcome guest, and the older writers assert that he had a close relationship with the rich merchant's wife. One fête on the great canal at the side, the so-called Fish-sticking, for which Apel engaged fishermen from Naples, was celebrated here for the first time in 1714 in the presence of the king, and has been kept up till the present day as a people's festival.

The garden must have been grand at that time, to please the ruler of a court recognised as the most brilliant in Germany. Augustus the Strong liked to think he had about him a kind of copy of the court of Versailles, and his own love of building in no way fell

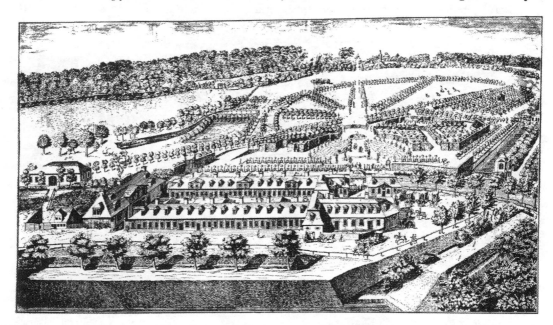

FIG. 498. GARDEN OF THE APEL FAMILY AT LEIPZIG

behind that of the French king. In his own country he found garden art in a flourishing condition, and water-castles in particular are still more numerous in Saxony than in any other German territory. The greatest number, if not actually made in the seventeenth century, were altered in the proper style, and a garden suited to the castle was laid out according to the fashion of the period. The character of the Renaissance is shown markedly in the way that canals are turned into ponds or moats. But in the eighteenth century also they knew how to suit their gardens to water-castles of this kind. We may compare the plans of old and new lake-houses, or still better, the form given by Augustus to the old water-castle of his ancestors, the Moritzburg at Dresden. This house was originally on a tongue of land in a pond. The king had it connected with the bank by means of a bridge and a dam, widening the pond on one side so that the castle was on a little island, much in the same way as at Chenonceaux. Outside the flower-garden, which encircled the high-lying island castle as a second terrace, there was a great semicircular region beside the bank, in the middle of a network of forest paths in straight lines, which cut the woody district into regular divisions.

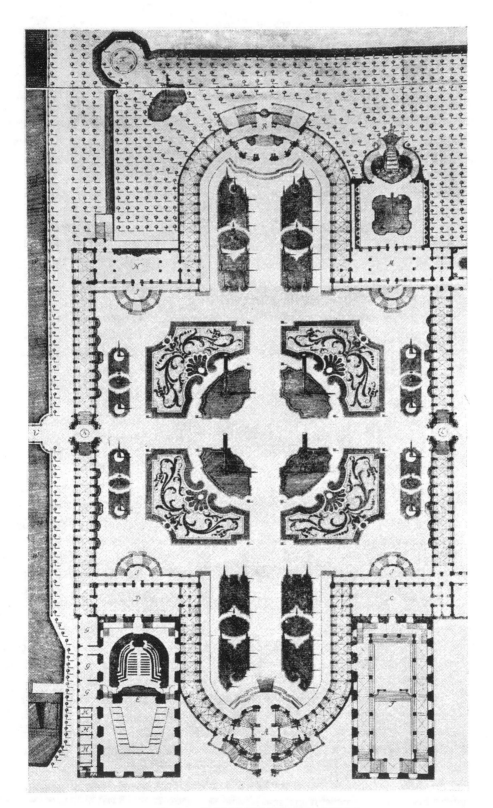

FIG. 499. PLAN OF THE ZWINGER, DRESDEN

The French feeling for lines and wide spaces came into Saxony with Augustus; and first and foremost in his mind was the thought of building a new castle as his residence. According to the plans made by his architect Poppelmann, it ought to have been one of the grandest of the kind. The area can only be computed now by the original orangery, which, because of its enclosed situation, was named Zwinger (cage) (Fig. 499). It was characteristic of the time that they should have begun with the orangery; and only when one considers the Zwinger as that can one understand the plan of the building. In the seventeenth century orangeries were made in more and more lordly fashion. People specially liked to make use of them in summer, when the trees were standing in the garden, as comfortable cool rooms for guests; and later on they mostly connected a grand dining-hall with the conservatories where the trees were kept, and made greenhouses in a semi-circle on both sides, so that they could be approached that way, and fêtes could be held there in the winter as well. A good example is seen in the end orangery at Castle Gaibach (Fig. 472), where the semicircular greenhouses have beyond them semicircular *berceaux*.

In the decoration of these buildings they felt that they could give free play to their fancy. Before now Salomon de Caus in his plan for the first stone orange-houses at Heidelberg hung his pillars with flowers and foliage. Taken all in all, we may say that in the Dresden Zwinger we find the plan most highly developed. The design shows a repeat of the semicircular nursery galleries, widened on the sides by straight wings. Thus an enormous garden court is shut in by corridors, which are interrupted by four corresponding monumental gate-pavilions (Fig. 499, Plan A, R, K, L). In the four corners fête-rooms were built, these also in accordance with the ground-plan, unsurpassable in size and splendour: a great dining-hall with an ante-chamber (F), a theatre also with an ante-chamber (E), a grotto-room (N), and a nymphæum with a bath-room (M, Q) filled out these corners. The architect, to judge by his plan, had intended the whole court to be treated as a garden parterre, with basins in the middle and water-devices at the sides. It is clear from the ornamentation of the south side as it then was, i.e. fountains and statues, that a plan for a gigantic nymphæum was hovering in his thoughts. In the summer this parterre was further decorated with the treasures from the greenhouses. But as a real nymphæum such as they had in the Renaissance days there was made a small court for a bathing-room—a place lying deep and cool, with alcoves fitted out with fountains, and statues between its pillars. Opposite this room a cascade fell over some steps between the statues into a semicircular basin. The middle was occupied by a large tank and fountain, and even in its neglected state at the present day it makes a fine picture (Fig. 500). When the plans for the castle were more and more advanced, the galleries also were made use of for fêtes. The orangery was at a greater distance, and the garden court was used also as a place for show processions, tournaments, and the like.

While the king was busying himself with these far-reaching building plans, he spurred on his nobles to other extravagant works. They built away cheerfully, and the more it cost the better, for they knew the king would be delighted to buy their estates at a high price and give them to one or other of his mistresses. The Dutch, or Japanese palace (Fig. 501), so called because of its famous porcelain collections, was built by the minister, Count von Flemming, apparently with the idea that, if the king's buildings were completed, there should be a fine garden scheme leading from the royal castle to the Elbe, and on to his own grounds, which lay upon its banks. In 1717 the king bought the place,

FIG. 500. THE NYMPHENBAD AT THE ZWINGER, DRESDEN

and began by laying out the parterre along the Elbe in gently sloping terraces. Later on the garden was enlarged at the side of the palace, and behind the great semicircle of the fortification walls, which served as point of view for the parterre, there was a great canal with ornamental waters, and also pretty boskets, a theatre, and other places.

On the other side of the town, the south-east, Johann Georg II. had made himself a shooting-box in the middle of the wood, in the eighties of the seventeenth century, and this now bears the name of the "Great Garden" (Fig. 502). The house, in the middle of a pheasantry, exhibits the familiar characteristics of a hunting-seat. It is one of the earliest on German soil, for with its eight pavilions it had been built by 1698. The garden developed gradually out of a little hunting-place and a mere pheasantry, where "hedges and underwood made a pleasant wilderness," into a wide expanse. The French spirit is shown in the complete mastery over all the materials, so that a mere imitation discovered, so to speak, a way to an original scheme. When the garden was finished, at about the year 1720, the castle looked out on a narrow parterre like a ribbon, which passed round it. To right and left you saw parterres divided two by two in the boskets—a plan much favoured in Saxony. The eight pavilions mark out the broad middle tract of garden. They are united in pairs with terraces, that are designed to exhibit orange-trees in the summer. The four front ones enclose a part that is planted with clipped trees, and later on with parterre-beds, and in those at the back there is the great basin like a canal, which ends in a pretty open pavilion. From here semicircular avenues pass round the water, and meet in a central one, which leads through the park to the gate on the north. Round these inside show-gardens there are ornamental thickets, showing a variety of arrangements to a bird's-eye view. The out-of-door theatre also belongs to the beginning

FIG. 501. THE JAPANESE PALACE, DRESDEN, BEFORE THE ALTERATIONS

FIG. 502. THE GREAT GARDEN AT DRESDEN IN 1719

of the century: it is set out very finely in the northern grove on the right. The side scenes are made of trellis-work, with a background that is shortened by perspective and encircled by round paths. Just as at Herrenhausen, the stage is separated from the auditorium, which is built like an amphitheatre reached by a sunk path. The great boskets surround this ornamental garden in the form of a Greek cross.

According to a plan of the Bavarian court architect Cuivillié, drawn in his own magnificent style, there was to have been a still more important feature shown in this garden about the middle of the eighteenth century. The canal was to have been lengthened from the wide basin on to the end, and finished off there with a great water scheme. But as the little castle was to be discontinued and rebuilt nearer to the town on a much larger scale, it is a good thing that the plan never arrived at completion.

FIG. 503. GROSS-SEDLITZ—THE ROYAL GARDEN, "STILLE MUSIK"

If the king and his architect had had the means of Louis XIV., they would certainly have gone beyond their French model. But there was a personal limit to what was possible for Augustus the Strong, and in this he differed from "Le Roi Soleil," who carried out all his plans. The works of the Saxon king which are still in existence are on so large a scale that it is surprisingly difficult to grasp the fundamental plans, until we remember that these are only those portions that were actually completed. This is particularly the case with the still imposing Gross-Sedlitz. There are two rows of terraces beside one another, leading down to a cleft in the valley (Fig. 503). The park mounts again to the other side. The one set of terraces forms a semicircular orangery with gardens going up from it; and on the top of the other stands the castle. The water axis of this garden forms on the opposite side a fine cascade, pouring its waters into the dip of the valley. Originally the orangery design was to have been repeated on the other side, so that the castle with the water axis in front of it would have been in the middle, and thus would have made a truly regal pleasure-house.

In the same fashion Poppelmann, helped by Longuelune, who worked with him as garden architect, made a series of equally great plans for the royal place of Pillnitz on the Elbe. What actually was worked out was one side of the original garden scheme with two pavilions (Fig. 504). It was a huge toy, which might have enchanted some giant's daughter, it was so charmingly carried out. But at that time people were more and more absorbed in an endless variety of games, and in this respect Pillnitz is quite a typical place. There were two little palaces, exactly alike, one on the river and the other by the

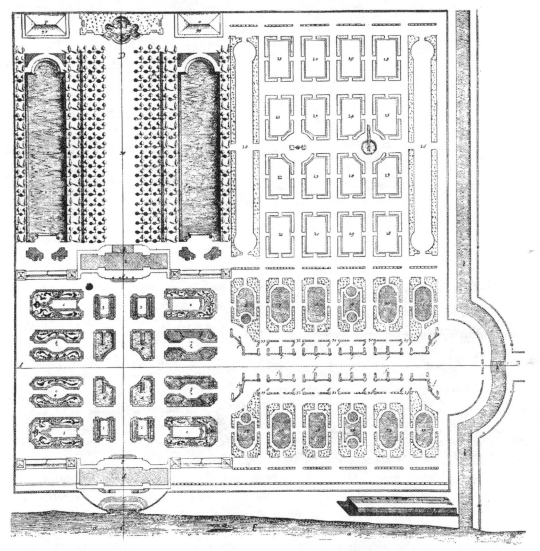

FIG. 504. PILLNITZ, DRESDEN—GROUND-PLAN

hill, and these are clearly shown by their style to have been originally mere summer-houses: they are so-called "Chinese" pavilions, with a baroque underground part, on stumpy columns, an open front hall, wavy "Chinese" roofs, and cornices with all sorts of Chinese decoration. The whole garden, which extended as a parterre between the two houses and round them on three sides between river and hill, was laid out with a view to games of all kinds. There were forty-four little square plots for playing on, mostly with hedges round them, and also a shooting-ground behind the house by the hill (Plan D) with the butts in a kind of grotto in the hill-side. Over against the garden was the tennis-court (29), a little billiard-house, and another for wrestling (O, P). It is a sign of the great skill and the certainty of plan in garden-making at this period that the designers knew how to give not only symmetry but also a certain grandeur to this region of games, by making an imposing approach from the Elbe, and by laying out large courts in a clever way with narrow water-terraces and fountains (called in the plan "water-lights") (*f*), and thereby emphasising a crossway axis.

With King Augustus I. the great time of Saxony, even in the art of garden-making, quickly came to an end. Nothing new was made except certain attractive minor features, and there were no ideas for great estates. Many of these lesser features have been preserved at Dresden. In the matter of fountains a high degree of virtuosity was arrived at, the erection of statues being combined there with water. One masterpiece is the fountain in the Marcolini garden (Fig. 505), with its playing waters. The charming picture of this place recalls what would otherwise be lost. And also at this time the pretty trellis-work of the Renaissance, as well as fountains, was revived, after being little used in the later days of Louis XIII. The interest in the garden that Count Brühl laid out on the narrow Elbe terrace lies only in the ground-plan, which shows how he managed to get over the difficulties of the space he had to deal with. Its chief beauty was in the covered walks made of trellis, which enclose the large garden in the front, and lead down from the belvedere in the garden at the back in a half-circle. This part, united with the terrace of trees by several avenues, cleverly joins on to the round part in the informal corner that had the Belvedere above it, a building now replaced by a new one. By the side of the trellised walks were the famous fountains that are still working, and have always been a marked feature. This late time can only claim originality when it gives itself up to play, either entirely or in part. And what Pillnitz did on a large scale in part only, the smaller places did altogether.

FIG. 505. MARCOLINI GARDEN, DRESDEN—THE NEPTUNE FOUNTAIN

A really amusing place for games, and consistently carried out, is the little pheasantry called Falkenlust at Moritzburg. This is a very neat round building on an extremely

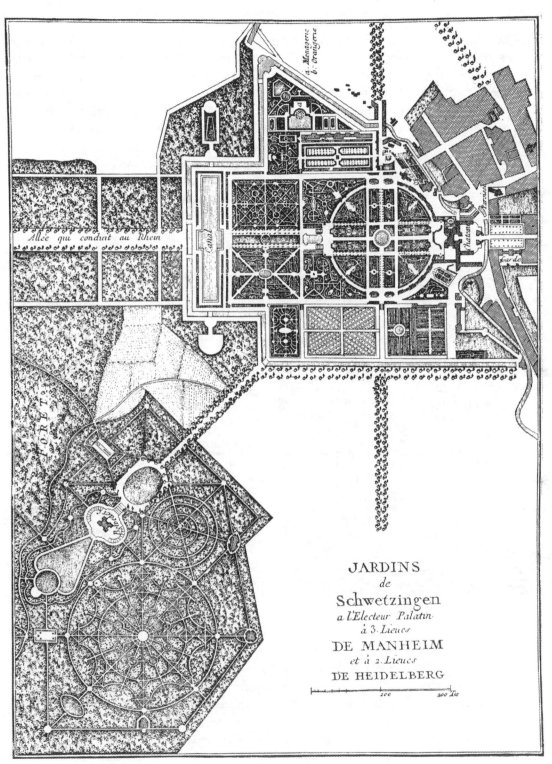

FIG. 506. SCHWETZINGEN—THE CASTLE GARDEN

small scale, with a Chinese roof and figures, standing on an artificial rock decorated with creatures of all sorts. Here to begin with were the pheasantry buildings, enclosed with trellis overgrown with greenery, and ornamented with various fountains which are more or less well preserved. The little castle is at the junction of the great avenues of the park. In the main walk towards the Moritzburg there is a great canal with artificial rockery, and by the side of it an elaborate garden piece. Hedges of pine are trained to an immense size, so that they slant obliquely from a height of one metre to eight or ten, and then "show clearly and distinctly the initials A.F.A., of their exalted Majesties." This main walk leads down from the castle to the lake, and there we find a well-fortified harbour, with pier, lighthouse, bastions, forts, and a little frigate, all very pretty and dainty. Saxony may boast of having reached a point, not too high, of the prevalent Chinese fashion.

FIG. 507. SCHWETZINGEN—THE APOLLO THEATRE

The formal style of garden was now threatened from every side. There were attempts, in adopting features of the new fashion, to combine them with the old severe restriction of the straight line and symmetry, but in the end there was complete capitulation. It even came about that the same owner and architect worked the changes in the same places. One good example of a transition garden is found at Schwetzingen. It was laid out for Charles Theodore of Bavaria by the architect Pigage and the head gardener Petri, in the formal style. It came into existence in the years 1753 to 1770 in a fashion that was entirely French (Fig. 506). The *Théorie et Pratique du Jardinage* had a great influence in the garden-plan.

The parterre ends in a longish basin. On the other side the *tapis vert* forms a middle axis between the boskets; it is a wide strip of grass adorned with statues, reaching to the great canal that crosses it, and makes a slight widening inside the park. The French garden book also recommends a semicircular end of the parterres in covered trellis. At Schwetzingen these are finished off with galleries that are also in semicircles, which join the two sides of the castle into a whole circle: this gives a peculiar stamp to the garden, so much the more because the parterre, always quite open in France, is traversed by avenues both lengthwise and crosswise, and these are of course kept clipped very low. This was a concession to the taste of the day, which preferred seclusion and shade to the open show-garden. Still from the *pian nobile* of the castle one could get a view of the whole:

the round parterre with its narrow beds in the middle stocked with flowers, and the side plots of trapezium form. High trees now conceal this effect, beautiful in their masses of greenery.

In its great lines the garden belongs to the best traditions of France. On the other hand, it shows some wavering in the laying-out of the boskets, for there is plainly an effort to accommodate itself to the gospel of the wavy line which is foreign to its own style. Of course the patterning of the parterre had already been working towards the change of line. Still, however elaborate the *broderie* in Louis the Fourteenth's patterns, they were never simply wavy and niggling. This was checked by the feeling for bold masses and clear symmetrical designs; but under Louis XV. and Madame de Pompadour the little broken-up patterns became ever more numerous, and were mixed up with a completely separate arabesque. In groves more and more licence was allowed, and *Les Sources* at the Grand Trianon actually showed the wavy line. But what was at first tolerated as variety now began to spread, because even avenues were laid out *à l'anglaise*, although they were still kept in check by the formal style.

The boskets in the Schwetzingen gardens were free from the peculiar sentimentalities that marked the first period. Pigage, the architect of the little buildings in the park, was no genius, but he showed a delicate feeling for grace and beauty in such places as the out-of-door theatre with the Temple of Apollo as background (Fig. 507), and the Bath Pavilion with its charming aviary. To show to perfection, both these places need their old surroundings of severely clipped hedges to serve as side scenes of the natural stage, and the trellised walks and aviary for outside bordering. Also the rock fountain with its cascade has nothing of the picturesque colouring of the English park, set as it is in clipped hedgerows. Scarcely, however, had Charles Theodore finished here than he was unable any longer to resist the current of modern thought, and he entrusted to Ludwig Skell, the rising star in the garden world, the task of laying-out his grounds according to the new taste. All the same, Charles Theodore felt so great a respect and reverence for what he had made that he set this new garden only as a girdle round the old French part. The canal, which formed the boundary, was left there by the artist fronting on the old garden, and forming a clever transition in such a way that it lent itself to the picturesque style by dipping into the opposite bank with all manner of deep-cut inlets and curves.

Every court, great or small, had given some flowers to the wreath of lovely gardens that now covered the provinces of the German Empire. Even the young and aspiring court at Berlin had its peculiar, many-sided picture to show. We have already seen how the Great Elector (Frederick William) had worked in the furtherance of the art. At his Residence he had too much to do in establishing his house; and the pleasure-garden of the castle at Berlin was only a smaller garden, though a flourishing one. It was Frederick, the first King of Prussia, who first found in castles and gardens an ample field for his love of magnificence; and the gardens of Oranienburg and Charlottenburg were laid out in a style purely French. They were among those that boasted—though it was but a legend —of having been laid out from the plans of Le Nôtre.

Charlottenburg was remarkable because of its unusual supply of water (Fig. 508). The Spree ran along the whole length of the garden, and beside it there was a large so-called carp pond, with six rows of trees shaped like a hippodrome round it. Outside, a number of tributaries of the Spree were conducted through the back part of the park, all

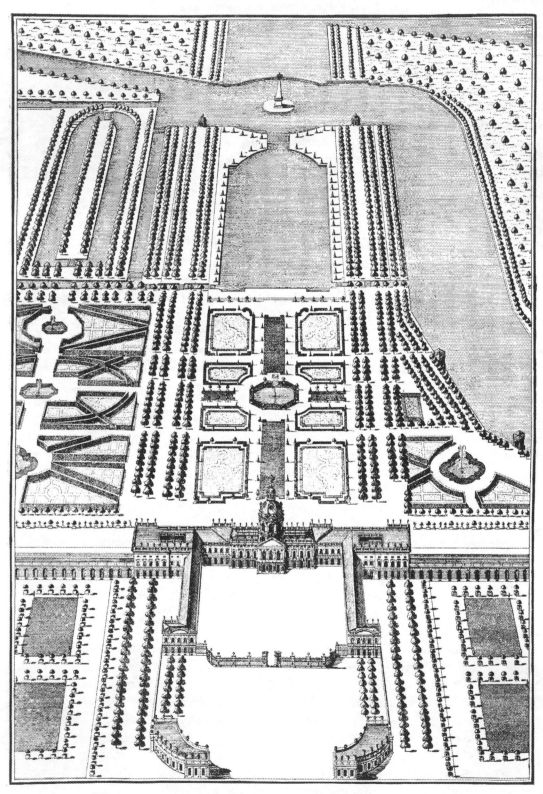

FIG. 508. CHARLOTTENBURG—GROUND-PLAN SHOWING CASTLE AND GARDENS

of them used for gondolas or to rear fish of various kinds. Next to the castle was the tennis-court, and also the orangery. Many statues stood in the very large parterre which was known as the pleasure-ground, and twelve busts of emperors were especially famous; these stand to-day in a pergola at the left of the entrance.

In the time of Frederick William I., the minimum of ornament was insisted upon, and the maximum of use and profit. He converted the pleasure-gardens in front of the castles at Berlin and Potsdam into exercise-grounds, and the place he chiefly used for

his own recreation was a fruit- and vegetable-garden, which he laid out on the north-west of Potsdam round a little pleasure-house, on which he conferred the pompous title of the Marly Garden. "I cannot tell why," says the Margravine Wilhelmine, who shared her father's recreations very unwillingly. It is the best proof we can get that Marly was the actual name for a small pleasure-house removed from the Residence; the king might just as well have called it a hermitage or some other name of that sort.

His son, the future Frederick the Great, had when he was crown prince shown the strongest inclination towards the art of gardening. Lovingly and with great care he had ornamented his castle of Rheinsberg. This love of the garden he inherited from his mother, who converted her little summer place, a very charming house near Berlin on the Spree, into a real treasure which deserved its name of Mon Bijou

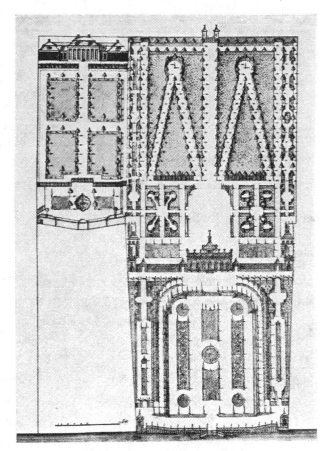

FIG. 509. MON BIJOU, BERLIN—GROUND-PLAN

(Fig. 509). Until 1725 it was only a small house in the middle of a garden, a real Trianon, made with the same object and in the same spirit, with porcelain and small choice works of art, and on both sides trellised walks and pavilions. A pretty parterre with flowers and fountains and attractive side scenes led to the river from the house, which stood rather high. Near the road the garden was laid out with boskets, leaving a large empty space in front of the house, presumably for games. At the side was the orangery with its own parterre, and opposite this a little menagerie for tame animals. In 1717 this retreat was endangered by a visitor, the Tsar Peter, who spent a couple of days quartered here with his retinue. It was known beforehand, of course, what sort of guests to expect, and the queen had had everything fragile removed. But in spite of this, after their departure house and garden ·looked "like Jerusalem after the sack." So says the Margravine of Bayreuth, and as a fact the queen had to have the whole place restored from top to bottom. The mother of Frederick the Great lived here for forty-six years; she was a lover of art, and she had castle and garden greatly ornamented and enlarged, but never altered the fundamental plans.

When Frederick the Great came to the throne, he wanted to build a little hermitage near Berlin according to his own fancy. For this he chose the hill opposite his father's Marly Garden, and in 1744 laid the foundation-stone of his castle Sans-Souci. This place would in any case deserve special consideration as the creation of Frederick himself, but there is more than that, for in garden history it fills a page of its own. His nature, a happy blend of the characters of his parents, is very clearly brought out in this his darling home. Even while he was crown prince, he was busy pursuing his love for fruit culture,

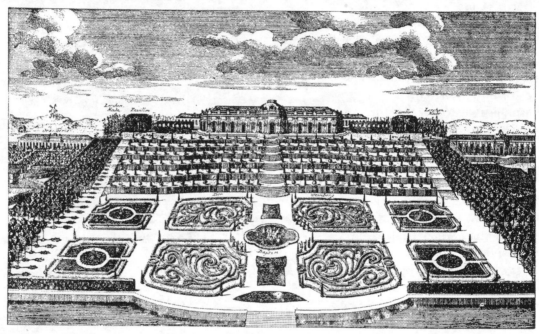

FIG. 510. SANS-SOUCI, POTSDAM—THE TERRACE GARDEN

and in his garden at Neuruppin he had made himself a vineyard and an orchard, which he had called Amalthea, proving that he read his Cicero with love and understanding. At Rheinsberg he tried yet another experiment, for he had his vineyard made in the form of a labyrinth with a temple of Bacchus in the centre.

Now that he was king he desired to make nothing more nor less at Sans-Souci than a great pleasure-garden with a hill planted with vines in the middle. Whether at first his sole intention was to make the hill and the little house—at first called Vigne, which suggests this idea—is not of moment now, the less so as in the first year the garden already showed an important extension in proportion to the house that was actually built. The main thing for us is that in the eighteenth century, which had created gardens purely for pleasure and banished all and every special kind of cultivation of particular things, when every minor prince and every wealthy private person regarded these luxurious pleasure-grounds as theatres for displaying their riches and their power, this king unconcernedly harked back to a stage in garden history that seemed long since superseded and abandoned. And just as that primitive garden was made, which we hear of in the ancient days of the Egyptians, so Frederick devised a scheme whereby everything was grouped about the vines, which served as a centre. For the vineyard terrace (Fig. 510), on whose high ground the king erected his pleasant summer-house, the most southerly aspect was carefully selected. The grapes must have friendly sunshine, and so the supporting walls were sloped, and made a sort of para-bola, so as to get for the more precious kinds every ray of sun that the northern clime could yield. In winter they were protected from frosts by glass. On this centre for the cultivation of vines Frederick the Great con-centrated all the skill that the science of his own age could supply. In the middle of his projecting terraces, whose very object made their lines beautiful, there rose broad stairs, and at the sides were gently ascending paths made secure by stone-work. The narrow terraces were cut off in front by low hedges of yew clipped into pyramids, and at the back there were in summer orange-trees and pomegranates in flower.

The castle at the top is reminiscent of the first state of Mon Bijou with its low windows, and in spite of its grandeur it never loses its character of a country place. On each side of the upper terrace was a bosket adorned with statues, and covered walks with pavilions flanking the house: in one of these was the beautiful Praying Boy now in the museum at Berlin. There were many busts and vases, but in this upper part there was no parterre. The first one was laid out at the foot of the terraces round a large basin, with a group of gilded statues in the middle. The end of the parterre was made by a canal that crossed one of the terraces and went on farther round the whole park. On the far side of the canal was an avenue with two sphinxes at the upper end, leading to two summer-houses: rows of trees bordered this garden. It had been the king's first idea to make a little vintage-house here, "*une espèce de vide bouteille,*" writes the friend of his young days at Rheins-berg, Bielefeld, in 1794, "*mais ce vide bouteille commença par être une retraite de Roi et finit par former un palais d'été digne de Frédéric. S.M. en traça Elle même le premier dessein.*" Two pen-and-ink drawings confirm these statements.

The king's architect Knobelsdorf carried out the

II—O

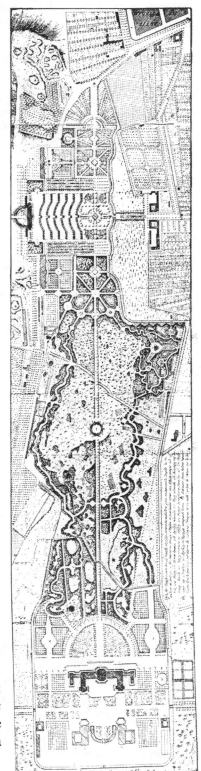

FIG. 511. SANS-SOUCI—GENERAL PLAN

completed plan, but Frederick insisted on making the final decisions about every detail, hence no doubt amateurish features are to be seen in places. Sans-Souci really became little by little a Marly to the king. It was not in the main axis of the castle that the garden was widened, but at both sides on the slopes of the hill (Fig. 511). On the top the king made on each side of his house another one-storied building, on the west side a small shelter for horsemen (at first intended for an orangery, and looking out on a cherry-garden), and on the east the picture-gallery. From this a terrace-garden descended. The dividing walls, which were covered with shells, led down by steps and grottoes into the so-called Dutch garden, a *parterre d'émail*, laid out with glass beads and Dutch vases for ornament. Semicircular *berceaux* led to a lower balustraded terrace. Another small

FIG. 512. SANS-SOUCI, POTSDAM—THE COLONNADE IN THE PARK

secluded garden, which still breathes the spirit of Frederick, lies farther to the east of this one; on the top is a grotto of Neptune with the god, and nymphs and tritons escorting him. A sort of nymphæum was in the mind's eye of the king in this part of the garden. He had planned handsome water-works, which were to feed this grotto and others; also cascades, but he was badly misled—a most unusual thing at that time—and much to his chagrin, he could never get enough water for them.

By the side of the nymphæum, at the foot of the hill, was the chief entrance to the garden, a semicircular gateway with beautiful doors in the middle, and made with double columns, which at one time also had convenient low doors. They were unfortunately replaced by the immense, ostentatious gates fit for an exhibition, which now quite spoil the delicate beauty of the entrance. From this part there is a wide, very long avenue leading first through a front garden, and then right through the main parterre and across the whole of the park (Fig. 511), which extends on the west of the castle, and was originally meant for deer and pheasants. Behind the parterre this main axis again passes through boskets ornamentally laid out with statues and basins, and goes on to the wall at the end, which separates the pleasure-garden proper on the west side from the park, meeting near the middle of the park the Marble Colonnade (Fig. 512), a charming building made by Knobelsdorf. This was an imitation of the colonnade of the bosket at Versailles, but unfortunately it was broken up later, so that its columns could be given to the marble palace built in the reign of Frederick William II. As an end and *point de vue* a great grotto construction was contemplated, but after the Seven Years' War the New Palace was built a little behind, and the grotto had to give place to this.

FIG. 513. SANS-SOUCI, POTSDAM—THE CHINESE TEA-HOUSE

In the park itself the king had several small buildings set up, so as to bring the northern hill into connection with it, and among them the beautiful Belvedere, a two-storied rotunda with Corinthian columns round it on the north-western slope. Below was a Chinese pagoda; and farther to the east, immediately above the castle, Roman ruins were built on the hill. Round them there was to be a reservoir, and they were to give a *point de vue* for the rather long colonnade, which is in front of the back façade of the castle. In the south-east corner of the park there was still standing, a little while before the Seven Years' War broke out, the Chinese Tea-House (Fig. 513), one of the most freakish attempts in this foreign style. The cheerful little rotunda has an overhanging roof with a China-man on the top, supported on gilded palm-trunks; round these columns are grouped Chinese figures taking tea. The whole place is merely a fancy, and it lies prettily in the middle of a space bordered with hedges, which opens on three of its sides into small grass paths, narrowing as they advance. Vases made of Misnian porcelain stand at the end.

Before the Seven Years' War the park itself was crossed by straight avenues. But in the meanwhile the new style had penetrated into Northern Germany, which was nearest to England. There were actual examples to be found here and there, and people were beginning in a cautious way—sometimes painfully and unintelligently—to obey the new gospel. The change in the park at Sans-Souci, which was effected after 1763 when the New Palace was put up, is a good example of the somewhat uncomfortable kind of compromise. What chiefly inspired the designers of these transition gardens was the dogma of the wavy "line of beauty." In various other gardens we have found the same idea, especially in the treatment of the small boskets, still more or less under the control of the laws of symmetry. But here the walks were allowed to run about, right and left of the middle axis—itself straight and happily not tampered with—in the oddest curves and windings. On both sides these paths were hedged, and would have given the impression of an endlessly long labyrinth to anyone strolling there, had they not been interrupted now and again by an opening with a view of some meadow landscape, which adjoined the narrow thicket at the side of the path. Especially unrestrained were the serpentine paths round about the Chinese tea-house; so possibly the notion of a labyrinth was intentional there. When we come to study the development of the fundamental principles of the English style, we will consider again the complete misunderstanding shown in this park.

It is certainly not to be regretted that the next state of this garden was not merely in part, but entirely modern. Would that instead Sans-Souci's pleasure-gardens had been kept in their old form, and above all the great parterre! But here everything is upside-down. Because the new taste was partially adopted by way of compromise, the finest effect of Frederick's estate was lost through introducing woodland at the foot of the terraces. All the same, we can still enjoy the sight of the king's work, one of the greatest achievements of his days of peace, and easily disentangled from foreign trimmings. On the threshold of a new ideal of art, Frederick produced a work which was so clearly his own that it is difficult to speak of any decided influence, and to separate what is due to France and what to Italy. We must accept it as an expression of his personality.

The children of Queen Sophie Dorothea were all great garden lovers. One of the sisters of the King of Prussia, Louisa Ulrica, we shall meet again in connection with her work in Sweden. The favourite sister, the Margravine of Bayreuth, most like her brother in many ways, says herself that she has so beautified her hermitage that it is now "one of the loveliest places in Germany." Wilhelmine married into a family which loved building. From the beginning of the eighteenth century her husband's forbears had built one pleasure-castle after another round the Residence at Bayreuth. At the attractive place called St. Georgen (Fig. 514) there was a little lake with a round island in it, gay with garden-beds and fountains. Here there was a mock-fortification afloat, much the same as at Moritzburg. The margrave could watch water-fights and competitions from his pleasant castle. At first this was in separate buildings, each divided into three parts, and with parterres on both sides; later, when a larger castle was built, a small town was added at the back of the Residence, and was laid out in regular lines. The founder of St. Georgen, the Margrave George William, also had a hermitage set up. This was certainly not intended to imitate Marly; but from outside the unhewn stone was to create the impression that the rooms were really cells; and if they did not succeed in looking so, at any rate the margrave, his wife, and the court, lived there dressed as hermits, and had a little bell that summoned them to prayer.

FIG. 514. ST. GEORGEN, BAYREUTH—CASTLE AND GARDEN

Wilhelmine altered the castle when she had it as a personal gift from her husband, but without quite taking away its original character. In her graphic description of her "Tusculum," she says:

The house has no architectural adornment outside, and it might be taken for a ruin among rocks. But in the front there is a parterre with flowers, and there is a cascade in the background, which seems to start from a fissure in the rock; it falls towards the hill, and empties into a great basin. Every path through the wood leads to some hermitage, and they are all quite apart. The view from mine is the ruin of a temple, made on the model of those that have remained from old Roman times. I have dedicated it to the Muses; and inside there are portraits of all the famous learned men of the century: Descartes, Leibnitz, Locke,

FIG. 515. THE HERMITAGE AND ORANGERY, BAYREUTH

Newton, Voltaire, Maupertuis, etc. There is a round hall at the side; there are also two smaller rooms, and a little kitchen which I have had decorated with old Raphael porcelain. Going out of the small rooms you come to a little garden, at whose entrance stands a ruined portico. . . . Higher up you meet with a surprise in the way of novelties. You come upon a theatre of freestone, with arches at intervals, where operas can be performed in the open air.

Below this truly sentimental place the same princess had a new castle built, the so-called orangery, which was to be an imitation (of a kind) of the Great Trianon, and gives a surprisingly good and full picture of the great style of the seventeenth century, by its striking arrangement of arcades (Fig. 515) and wonderful water-works, all within a severely regular garden parterre. Bayreuth offers us a good example of the inquiring, ever-restless eclecticism of those days. There are ideas and to spare everywhere, and old and new are alike seized upon, and sometimes subjected to the former criterion of style. The inner part of the orangery includes a curious room, which no doubt came into existence after Wilhelmine's time. On the walls is depicted a garden, with a fountain in the middle and a low grating, with climbing plants reaching up to tall trees; in the foliage there appear ripe oranges and exotic birds. All this is carved and painted in many colours, and mounts up to the ceiling, which is like a blue sky with white clouds (Fig. 516). A long chain of ancestors unites this fantastic place with ancient garden-rooms, but still the chain seems to have reached its end in the clumsy naturalistic hermitage at Bayreuth. What most attracts us there is the atmosphere of the lively sister of Frederick the Great.

FIG. 516. THE HERMITAGE, BAYREUTH—A GARDEN-ROOM

The hermitage, however, was no exception at that time. If we read the descriptions of gardens in travellers' books about the middle of the century, we shall understand that the pleasure-castles, with the precious collections, and the gardens, were the main stages or goals for travellers of the period. And this was not so in Germany alone, for we meet the same thing in countries which were now first open to the research of eager travellers, such as the Scandinavian peninsula and Russia.

SWEDEN

Queen Christina first introduced the ideas of France into her country in the middle of the seventeenth century. In the first bright days of her actual reign she hoped to gather round her a splendid court of the Muses, and she summoned to her side André Mollet, the son of that Mollet whom Henry IV. had employed in France. He had left England some time before because of the Civil War and he now came to Sweden. In 1651 Mollet published at Stockholm his book called *Le Jardin de Plaisir*, which he dedicated to the queen, and which influenced the theory of gardening. Unfortunately he does not say for which garden he designed his *parterres en broderie, compartiments de gazon, bosquets,* but the style seems to indicate one of the best Swedish seats, Jakobsdal, which after the year 1684 was known as Ulrichsdal, after Prince Ulrich. The castle came into being between 1642 and 1644 (Fig. 517). A set of fine engravings by Perelle show the state of the garden, as it was laid out about the middle of the century by Magnus Gabriel de la Gardie. It was improved by Queen Hedwig Eleonora, widow of Charles X., who possessed the place after 1669.

The castles of Sweden owed their chief charm and beauty to their great stretches of water. The house as a rule either has the sea close to it, or has one of the great lakes, as here, on two sides. The main approach is by water, the others are by the garden. Jakobsdal

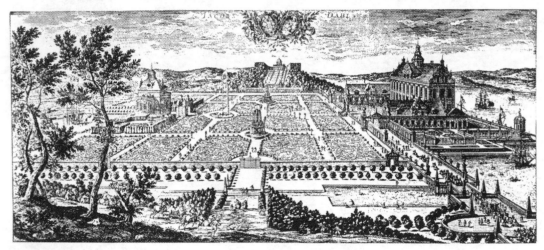

FIG. 517. JAKOBSDAL, SWEDEN

stands on a terrace, which is washed on three sides by the waters of the gulf. A wonderful balustraded approach leads from the landing-place to the fine Renaissance building, which on the garden side has two wide wings standing about a paved court. A carriage road with balustrades serves as the approach on one side to this terrace. There is an ascent by very wide steps to the large parterre. There are sixteen *compartiments* laid out in the style of Mollet, connected by dwarf trees with statues in the corners and three fountains. In the main axis and in front of the castle is a domed building, in a grove which contains a grotto. Farther along on both sides of the middle walk there are two vivariums, the one on the left ornamented with a dragon fountain, at the foot of a wild grotto hill: here Andromeda is fixed to the rock, and Perseus appears leaping down from the heights. To the right of the castle the park climbs up the hill, and broad steps lead to the summer-house called Marienberg, after Magnus Gabriel's wife. Lying along at the foot of the hill is the orange-garden. On the other side a canal widens out into two basins and joins the sea. This garden is one which ranks Mollet, if we may assume that the design is really his, among the outstanding artists of his time, just as his father was. The style of the French Renaissance, visible everywhere, has adapted itself most happily to the magnificent situation of this castle.

In Dahlberg's *Views of the Swedish Castles of the Year 1735* we find several smaller gardens clearly showing the Renaissance character, and of these Jakobsdal is the finest. The paths are covered with trellis arbour-work running crosswise, as in the pretty little garden at Eckholm, or the parterre at Mirby engraved by Perelle. But most of them show the influence of Le Nôtre, which was quite irresistible here after the end of the seventeenth century. Hedwig Eleonora in particular was devoted to the laying-out of gardens and their beautification. Her favourite place was the Castle of Drottningholm (Fig. 518). She found a mediaeval castle on one of the islands in Lake Mälaren, and she began to make alterations in 1661. This castle also stood on a raised terrace, and its approach was by water from an oval-shaped harbour. The garden lay towards the south, and showed the influence of Versailles more than any other in Sweden; its fine parterre was laid out in patterns of box, with a border of clipped trees and flowers. In the centre, steps in the open mount up by a wide middle walk to a basin with a Hercules fountain. After this comes the water-parterre, some steps lower, with eight round or oval basins and

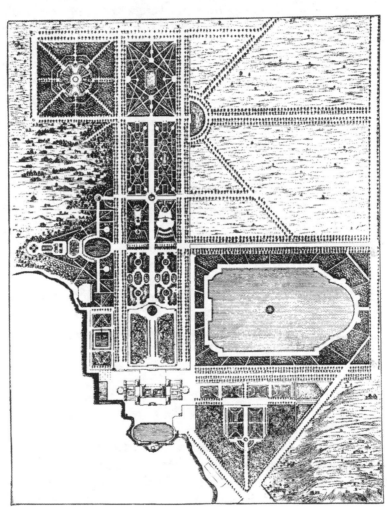

FIG. 518. DROTTNINGHOLM, SWEDEN—GROUND-PLAN

fountains. At the end of it there is one of those water-roads that French gardens adopted from Italy. This part ends on the left with an oval basin, with a cascade blocking the view on the other side. Beyond the water-parterre are elaborate boskets. On the west of the large parterre is an immense pond with an island on it, and adjoining it a park for deer with straight walks cutting through it. The fine scheme is completed with a menagerie site made in concentric circles on the south-east.

In the second half of the eighteenth century this garden was destined to flourish once again, loved and tended by another queen, Frederick the Great's sister, Louisa Ulrica. She, like her brother and her sister at Bayreuth, found the greatest joy of her life

in the study of literature and art, and she also had inherited from her mother a delight
in collecting things, and in the splendour of her surroundings. "All the queen's rooms
are most beautiful," says a traveller of that day. She, like Hedwig Eleonora, chose Drottningholm for her favourite dwelling-place and furnished the inside of the castle with
collections of porcelain, Chinese and Japanese carpets, pieces of furniture, and pictures.
The garden, and still more the park, were now equipped with all kinds of fashionable nooks
and corners in accordance with the feeling of the time. In the park, a gun-shot distance
from the garden, Louisa Ulrica set up an entire little colony, which she called China.
Round a plot of ground there were pretty little houses in the Chinese style, the chief
one adorned with tables of lacquer and Chinese figures. There was a Chinese pagoda
with a bell-tower, also Chinese vases in porcelain, and gilt statues. The whole place was

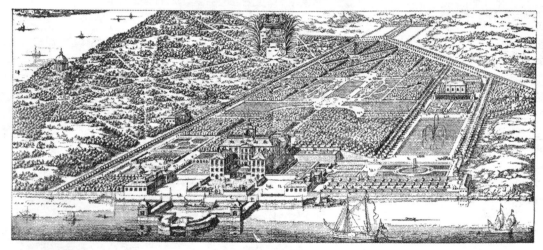

FIG. 519. CARLSBERG, SWEDEN

enclosed with a dense border of firs, that made it dark. Hirschfeld speaks of another place
with little buildings which shows the queen's affection for China: "It lies at the end of
the French garden and is called Canton."

The largest of the Swedish castles is Carlsberg (Fig. 519), but the garden has not
the individual character of Drottningholm, though the approach is similarly by way of
the lake. Behind the house is a semicircular piece with a narrow border of flowers, and
behind that are dense thickets, with immediately at the back, parterres round a large pond
in a wide open space. As this garden lies towards the north, the tall trees round the house
are to keep off the cold winds. Near the house itself this was made up for by handsome
side parterres, which keep a decidedly Renaissance aspect, being treated partly as hanging
gardens. Round about them there is a great park, penetrated by avenues that run out in
the form of a star, and this park has kept its French character to a great extent even
to the present day.

It follows from the situation of castles like these, that the real importance which the
canal acquired for French gardens is here superseded, for the outlook over the lake or
the sea fulfils the object of the canal, though in quite another fashion.

DENMARK

The gardens of Denmark developed under similar conditions to those of Sweden. Very little has remained of those that lay around the proud castles of the Renaissance, and the ones that show most traces of the period are the gardens of Frederiksborg at Copenhagen (Fig. 520). The front courts and castle buildings are on three islands con-

nected by bridges, at the side of which is a small parterre, also on an island. The chief garden reaches to the north bank of the lake, and because of this situation has renounced all allegiance to a connected ground-plan, though the house does stand in the central axis-line. The arrangement of the parterre, which has a canal going out from it and cutting through the boskets, certainly belongs to the seventeenth century. The plan of the castle of Hirschholm is like this, and stands on islands which are joined by bridges upon a lake. Both house and parterres follow an axial line upon the islands, and are shut in by groves round the bank (Fig. 521).

A typical castle of the French kind is Fredensborg (Fig. 522), which was built to commemorate the Peace between Sweden and Denmark. The garden and park are now the best in Denmark. The castle lies high above the Essommer Lake, and its grounds reach down on the left side of the building, on the east. The garden is in many respects like Hampton Court, but its semicircular parts are united with its buildings in a more original way. There is an octagonal court, and in front of the garden there are seven avenues passing like rays into the park, but first leaving room for a semicircular parterre, and to the right and left for two boskets enclosed by hedges. A semicircular avenue of limes passes round this decorative place, and makes an

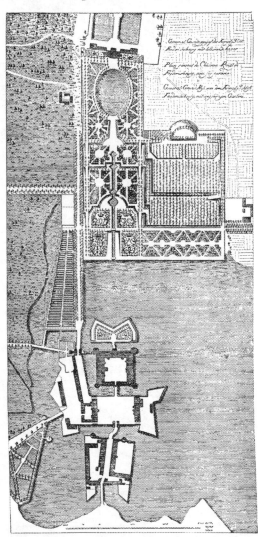

FIG. 520. FREDERIKSBORG, COPENHAGEN—GROUND-PLAN

outside border for the avenues. The middle walk has four rows of trees like Versailles, and a wide view over lawns with statues on them. Instead of the canal cutting across here, there is the shining Essommer Lake to help the view. It forms the end and *point de vue* of the western avenues, which all wend their way to the water.

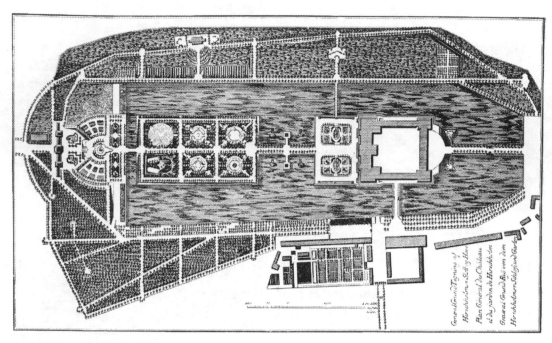

FIG. 521. HIRSCHHOLM, DENMARK—GROUND-PLAN

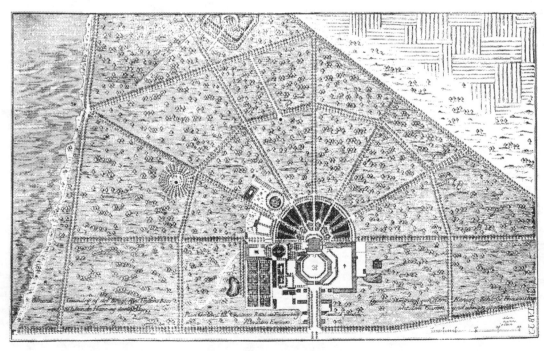

FIG. 522. FREDENSBORG, DENMARK—GROUND-PLAN OF THE GARDENS

RUSSIA

Russia enters into the history of gardening much later than any of the countries we have written about, and it is impossible to speak of any individual cultivation before the days of Peter the Great, in the first third of the eighteenth century.

The information is very scanty about the summer residences of the emperors at Moscow. Up to the end of the seventeenth century we find nothing but wooden buildings, very subject to danger from fire, and these first had gardens made round them in the reign of Peter the Great. But after Petersburg was founded, there arose not only the palaces of emperor and nobles, but gardens as well; the Tsar had learned to recognise the earlier examples, having studied them on his travels in Holland, England, and Germany. He knew far better how to protect the gardens at his own home against his barbarous troops, than how to protect those foreign gardens and homes which he had ravaged on his travels. In 1714 he made a great garden at the summer palace on the so-called Admiralty Island, which was made by the River Neva with her canalised arms, and has now disappeared. Here he and his followers adopted all the ideas that the style of the time had to give. The parterres with their grand waters, cascades and playing streams, the plantations with their tall espaliers,

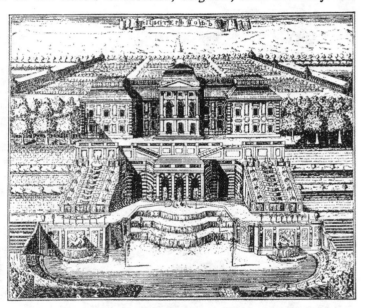

FIG. 523. PETERHOF, RUSSIA—THE CASCADE AND UPPER GARDEN

were all adorned with works by famous Italian sculptors; also antiques were brought, for they were always to be had, while in the park there were various summer-houses, and a bosket with fountains illustrating Æsop's *Fables*, as at Versailles, as well as a menagerie with valuable beasts.

For the grotto which Peter made, as also for the water-works, he engaged the great architect Schlüter of Berlin, who had left his own home, in a state of discontent, to find a new field of activity in the Tsar's service, but the very next year died in Petersburg without having done anything. The French artist Le Blond fared better, for he was at once entrusted by the Tsar with a most important piece of work. Opposite the town, at the south of the gulf, the Tsar had built a little house on the shore, before he was attracted in 1715 by a beautiful spot where he built a pleasure-castle, which he named Peterhof. This was on a natural terrace twelve metres in height, where the hilly part of it falls away somewhat towards the land (Fig. 523). Of course it was intended to rival the French Residence, and so French artists were called in. The plans came straight from Paris, and there was nothing to hold Le Blond back from getting on with the castle and garden.

The great advantage here was that they did not have to concern themselves much with underground operations; but the planting of the vegetation was no trifling matter, and whole shiploads of trees and plants were procured. The interior of Russia supplied elm and maple, and we are told that 40,000 trees were brought. Then came beeches, limes, and fruit-trees from Western Europe. Foreign specimens were brought from the ends of the earth, and in spite of the long winters flourished and still flourish. With these the whole of the lower part, from the sea to the lofty terrace, was planted and laid out as a park, with a great variety of fountains, which marked the crossways of the main avenues.

There is one cross-road which starts from a little house called Monplaisir, built

FIG. 524. PETERHOF, RUSSIA—CASCADE AND CANAL

on the strand by Peter in a pretty little garden in the Dutch style. It leads to a second small building, and this is named Marly—another reminiscence of France. Behind the Marly pond falls a cascade, glittering on gilded steps. The boskets contain, among many other water-devices, some weeping trees; little did Madame de Montespan know what an effect she would work with that boscage which she designed. But Peter also was very fond of fairy-tales and fantasies; in his little hermitage there was a real "table-be-covered," which at the sound of a bell rose out of the ground and vanished again. This park was divided in half by a sort of large waterway in the middle axis of the castle (Fig. 524). A double cascade falls from a terrace in front of the castle down into a wide basin. There is a grotto beside it, with sets of seven steps in coloured marble, and on them a series of gilt statues. In the basin is Samson on a rock, tearing open the lion's mouth, from which a great column of water goes up. From here a quiet canal flows seaward, and buildings at the harbour help the disembarking and landing from the royal ships. On both sides of the canal there is a walk with fountains which throw silver showers up and down upon the dark tall firs, and various masks spurt their waters into the canal. There is a cheerful open garden beside the cascade, and the terrace steps on either side are decorated with dwarf trees, while on the flat there is always a basin with beds of flowers. Above, in front of the castle, there is an incomparable view, for right over the lower garden and the water's edge, which so soon was covered with fine country houses and gardens following the king's example, the eye sweeps right over the sea to the town with its golden domes, while far away on the right the Finnish coast appears. Behind the castle lies the upper garden (Fig. 523) with its fountains and the Neptune in the middle; here all travellers praise the lovely clear waters that the hills of Peterhof pour out in profusion. Hence proceeded wide star-shaped avenues, passing through the park above, and meeting at one point on the hill, whence it is possible to see all the views skilfully and pleasingly combined.

The French artists, using the nature of the ground, cleverly created a wonderful picture. This garden is clearly a symbol of all Petersburg culture, which at that time was the scion of a French stock. For western eyes there was too much gold and glitter and too many colours used in other castles as well as in this one, in accordance with the Russian taste and feeling. There is a story that the French ambassador, when he first saw Karskoje-Selo, Queen Catherine's castle, exclaimed that there was nothing wanting but a case to protect this jewel in bad weather. The short reign of the French garden came to an end with this castle. Catherine was so modern a ruler that she laid out her garden in the new English style, the first one in Russia, and greatly admired.

In all these countries of Northern Europe the art of gardening reached its highest point in the eighteenth century. We have not found any fine display of new ideas, but French art absorbed and embraced within its wide boundaries much of the individual peculiarities of the different countries, and their various changes in taste during these hundred years. Variety was the charmed word that led them to their pinnacle. All the same, this variety had to be united with a definite and abiding form in the main lines. We have seen as we went along how economic and political conditions prepared the soil for garden development in the course of this century.

ITALY

We do not find the southern countries, especially Italy, so much inclined to adopt a thoroughgoing change as Russia was. Italy had no doubt given up the leadership to France in the seventeenth and eighteenth centuries, and was fully conscious of that country's superiority. So the art of Le Nôtre was supposed to have done the impossible, and villas like Ludovisi and Albani were quoted as works of his, whereas one was much before his time and the other was made after his death. But this want of precise knowledge at Rome was the best possible proof that the spirit of Le Nôtre and his northern feeling for style had not vanished from Roman minds. Only thus could it have been possible for such a really Roman villa to be created as the Villa Albani so late as the year 1740. Certain concessions made by the cardinal to the taste that was then the fashion penetrated (somewhat timidly and, so to speak, unintelligently) into nooks and corners. The small garden, sunk below the so-called coffee-house, tries to be a sort of bosket in the park. At the side of the coffee-house there are steps leading down to an artificial ruin, such as was talked about so much, and this one was built as a bird-house. The part at the back of the coffee-house had a country-like appearance, and it looked out on a cascade, which plunged down into an elongated basin like a canal. But the canal passed between walls to the back entrance of the garden. This part, which was quite small, would have found its proper place in a northern park; but here, shut in by walls on every side, it looks most odd when one walks down to it from the thoroughly Roman parts on the upper terraces.

FIG. 525. PALAZZO VENIER ON THE BRENTA

All the classical alterations made in the larger villas had been at least hinted at before; and there was seldom any progress made in this direction, even though there was nothing at present so fatal as the English landscape style, whose approach was imminent.

For Northern Italy the French style was necessarily important, because the level ground of the plains of Lombardy was suited to it. The nobility of Milan and Turin always had French sympathies, and, like the towns on the other side of the Alps, these looked to France for the patterns of their own villas. Moreover, country houses sprang up on what there was of terra firma in Venice, and these were clothed with French gardens and parks. A collection of Da Costa's engravings, called *Delizie del Fiume di Brenta* (*The Beauties of the River Brenta*) show the villas and palaces (Fig. 525), as they were when Goethe saw them, when leaving behind "many a lordly garden, many a lordly palace," he proceeded by the beautiful Brenta to Venice, his longed-for goal, on 28 September, 1786.

Nowadays the gardens of these villas of the Venetian terra firma have either completely vanished or have been badly kept. The largest and most important of them is Palazzo Pisani, built in grand style for the noble family of the Pisani about 1740. The garden is on large lines, but has no importance for the history of art. In the same way the neighbourhood of Milan felt the influence of France; and of the gardens there, we have engravings on copper in *Ville e Delizie di Milano* by Dal Res, published in 1773. Almost all are of much the same design: parterres and groves in the French style, and a small summer-house here and there in the park.

The Villa Castellazzo has in some measure preserved its old state. The great parterre,

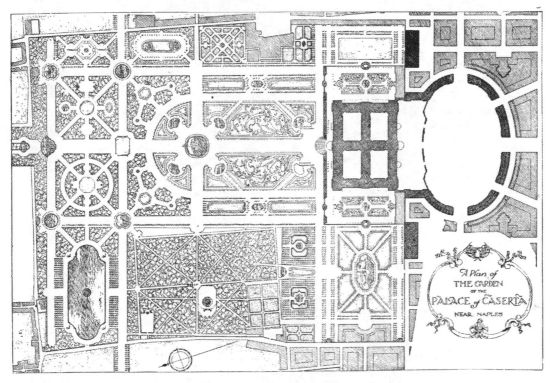

FIG. 526. CASERTA, ITALY—GROUND-PLAN OF THE PARTERRE

with boskets all round and mostly on the north, lies in front of a side wing in which is the principal room of the house. The park begins behind at a closed door, near a semicircular plot with a Hercules on it. By the side of the boskets, which enclose fountains, labyrinths, and statues, a wide avenue leads past a green theatre to the Diana fountain, a very charming but quite Italian spot, which is the more important through having to the north of it a thicket with bird-decoy and aviary, forming the end of the garden.

Wherever the Bourbons came with their changing fortunes we perceive French influence. When the son of Philip V. of Spain, Charles III. (who was sole heir through his mother of the Farnese family), came to Parma in 1731, he laid out the great public garden that is still there by the side of the little Palazzo del Giardino, built in the sixteenth century. In its present condition this garden suggests something foreign that has pushed in from the North, especially as the high box hedges bordering the paths that run through the boskets are stripped of their statuary, and the parterres are not kept up with their

II—P

FIG. 527. CASERTA, ITALY—PRINCIPAL VIEW OF THE CASTLE AND GARDENS

old patterns. Charles proceeded to Naples as king in 1734, and after he had at length secured recognition from the Powers, he desired to give proper expression to his might and
dignity by building a gigantic castle in the fashion of his northern ancestors. In 1752 the
architect Vanvitelli began to make a castle at the little town of Caserta, and this was to be
the greatest palace in the world (Fig. 527).

There was no doubt in the mind of the founder that this place would rival Versailles,

Photograph by Mrs. Aubrey Le Blond.
FIG. 528. CASERTA, ITALY—THE POOL OF THE GREAT CASCADE

garden and all. The building, with its four great courts, was entered through a gigantic
fore-court, oval in shape. By the side were parterres of flowers, and an orangery was on
lower ground on the east, as at Versailles. On the other side there was a riding-course
bordered by avenues in the Roman fashion. A series of steps leads from the castle terrace
to the great parterre, which was embedded in boskets with basins and fountains (Figs. 526,
527). Neither the parterre nor the surrounding groves show any original features. The effect
of the whole must have depended on the cascade, which fell from the hill opposite into
the main axis of the castle. Above, it is cut off by a terrace which gives a fine view. The
water falls sheer fifty feet down into a pond, which is decked out with groups in Carrara
marble, representing the story of Diana and Actæon (Fig. 528). Thence it falls from step
to step, so superabundantly beset with statues that their white curving bodies are mingled

with the foaming, plunging waters, looking now like a solid stream, now like a moving shape, till they end at last in a great basin with statues representing Neptune's seat, close to the great parterre. There were tall hedges the whole length of the canal, and at the different stages there were marble steps on the bank. It made a majestic view, most astonishing for anyone who saw it suddenly from the frame of the gate as he emerged from the great corridor of the castle. But like the house with all its vastness, the cascade is wanting in good proportions, and one is struck by the absence of a great connected coherent scheme. We have only to compare with Caserta places of the good period, such as Villa Aldobrandini at Frascati, which show intelligent consideration of every detail, and therefore are never wearisome, but keep the whole picture uniform and noble.

SPAIN

The Bourbon Charles III. did not take any Italian villa nor even Versailles as the immediate model for Caserta, but the garden where he had spent his time as a boy, La Granja at San Ildefonso, north-west of Madrid. We saw that Buen Retiro was the last great creation of the Renaissance in Spain; and after Philip the Fourth's death, the whole of Spain slumbered while his son, a weak, lazy man, ruled, and things got worse and worse. In 1665 Aranjuez was burnt to the ground, and the king allowed it to lie in ruins during the whole of his reign—one thing typical of many.

After the War of Succession was over, the influence of France was entirely victorious with the House of Bourbon, both in architecture and in gardening. It is significant that there was now no question about the park at San Ildefonso being laid out by French artists, Carlier and Boutelet, in spite of the fact that Philip V. was melancholy, quite unfit to rule, had no will of his own, and was completely governed by his Italian wife, Elizabeth, a princess of Parma, of the Farnese family. She would naturally have preferred Italians as advisers and if possible as artists. We learn, however, from the whole history of the summer residence how much this French grandson of Louis XIV., yielding and easily influenced as he was, succumbed to the peculiar spiritual atmosphere of Spain. After the end of the War of Succession the king lived in an old, beautifully situated castle that had belonged to the monarchs of Castile, on the western slope of the Sierra de Guadarama, and from here he used to visit the hermitage of San Ildefonso, which was higher up the mountain. Henry IV. had before this, in 1477, laid out a farm there; it was near the hermitage, and he had presented it to the monks of Porral as a summer retreat. King Philip was very melancholy, and was only affected in a serious way by the beauties of nature and by music; so he was greatly attracted towards this lovely spot, which stood at a height of 1191 metres, and was overhung by the powerful crest of the Pico de Peñalara. He at once determined to put up a shelter here that should be the kernel and centre of a mighty castle and important large gardens. This reminds us of the wish of Louis XIV., who also intended to make himself a hermitage at Trianon, and at Marly, both of them being turned into castles and parks. Here in Spain, however, his grandson was directed to other paths; for added to his personal recollections of Versailles there was the thought of the Escorial, which was so sympathetic to his temperament; and the actual grouping of the castle buildings was very strongly influenced thereby.

FIG. 529. LA GRANJA, SAN ILDEFONSO—THE PRINCIPAL PARTERRE

It has rightly been conjectured that the kernel round which Louis XIV. framed his enormous place at Versailles was his own bed-chamber. At any rate, his private rooms were in the old castle of Louis XIII., and they were the actual centre for all later extensions. At San Ildefonso the centre was the old farm with its cloister court, the Patio de la Fuente, with the large collegiate church adjoining on the north-west, completely dominating the palace court. It was the first thing built and was consecrated as early as 1724. La Granja (the farm) was the name by which the whole royal estate was known. There can be no doubt at all that the park, which was the result of diligent anxious work during twenty years, really was meant to copy and rival world-famous Versailles. But in the struggle to overcome nature by art, La Granja is rather to be compared with Marly. Monstrous subterranean works were needed, on quite another scale from those at Versailles, so as to get level terraces and plots for gardens in the steep mountain cleft. Of one thing, however, they had enough and to spare, and that was water; it was used so lavishly that we are full of admiration for the artist's ideas. Some guardian angel seems to have protected this mountain-girt garden from such storms as worked destruction on the French castles, so that the greater part of it has been preserved in its original condition. Nature allowed herself to be mastered, and then added to art her beauties of marvellous plants, grand trees, and the giant frame of her everlasting hills; and this beauty Louis XIV., for all his powerful will, could never attain in the pleasant vale of Marly, to say nothing of the flat marshes round Versailles.

The artists at San Ildefonso never even remotely approached the noble schemes of Le Nôtre, and his masterly grasp of a whole. To the south-east and south-west the garden rises steeply, so that the part on the north-east is considerably lower. The chief parterre (Fig. 529) is treated in a comparatively simple way, and does not produce the same striking effect in relation to the waters as one gets in French places with the view from the middle windows of the castle. There is a semicircular basin, cutting off the two parterre beds, and a fountain of Amphitrite, at the foot of a marble cascade, which shows at the top a pretty two-shelled fountain of the Graces. The view at the back is cut off by a charming octagonal garden-house. The small garden here is quite independent and almost cut off from all the others, with statues at the side and hedges at the corners, and tall trees overhead which were once more closely cut back. The court of Philip V. had abandoned the idea of a show-garden, which as a middle axis should subordinate to itself every detail of the endless number of separate parts. The eighteenth century with its tendency to seclusion always growing stronger, shows feeling conspicuously here.

On the east of the palace, great steps lead to a parterre on a lower level, which is bounded by the canal on one side, and runs in the same direction as the chief parterre, ending in a far more splendid water arrangement. There is a basin approached by steps, with in front a Neptune fountain (Fig. 531), and behind it love-gods riding on sea-horses, called *Carreras de Caballos* (Fig. 530); this idea was developed by Charles in a far larger way later on at Caserta, and here the fountain formed the *point de vue* from the side windows of the castle. A further parterre, also in the same axial direction, is on the east, and ends in the Andromeda Fountain, a work of incredible boldness in the grouping of its figures, showing the utmost flexibility that statuary is capable of. A labyrinth of no great importance adds an attraction to the bosket on the east side.

FIG. 530. LA GRANJA, SAN ILDEFONSO, CARRERAS DE CABALLOS

All the western part of the park is absorbed in a region which has an importance of its own, in that it unites the tradition of Spain with a more unfettered French spirit. In the centre is a large octagonal place from which go out eight avenues, *ocho calles*. It is decorated with a single group, of Apollo and Pandora. The avenues are held together in pairs by a fountain with a large marble arch overhead. At the end, as the crowning point for these walks (arranged in two groups of four), there is set up another fountain of some importance, which independently acts as a central point for a round plot. The wide path which runs alongside the façade of the castle ends at the so-called Bath of Diana

FIG. 531. LA GRANJA, SAN ILDEFONSO—NEPTUNE GROUP

(Fig. 532), which is a place of waters similar to the French water-buffet, over-ornamented with marble architecture and bronze statues. From here the water descends in falls at the side, by fountains and in light cascades, into a large semicircular basin. The south-west façade forms the wall of the great court of honour, the Patio de la Herredura, which is joined by a parterre on the side (Fig. 533), which again ends in a colossal fountain-site, the Fuente de le Fama. This is a rock with figures attached to it in the most unnatural positions, and is crowned by a statue of Fame on a winged horse. It is a Mount Parnassus, which no Spanish garden from the earliest Renaissance days cared to be without. ·

Quite on the top, apart from the castle, there is another great reservoir, called El Mar, which is not utilised for the general view of the garden, although it is in the main axis, and can be approached by park avenues. Standing alone in the solemn surroundings

FIG. 532. LA GRANJA, SAN ILDEFONSO—THE BATH OF DIANA

FIG. 533. LA GRANJA, SAN ILDEFONSO—THE SIDE TERRACE

of the dark mountain forest, it makes a happy contrast with the cheerful exuberance of the garden. It is clear that the artists were at this point thinking of particular views, which in the tradition of the Spanish Renaissance are always desired for the surroundings of fountains (Fig. 534). But they are without connection, and so lack the rhythmical feeling needed in the picture of the whole; and that last achievement of the French garden, the subordination of the parts to the whole, is absent here.

The real moving spirit and head of all was Elizabeth, the queen. When Philip abdicated in 1742, feeling himself absolutely unequal to the cares of state, he went back

FIG. 534. LA GRANJA, SAN ILDEFONSO—LEFT, FOUNTAIN; RIGHT, VASE

to La Granja, and reserved for its rebuilding an enormous sum which the exhausted country could barely supply. But after nine months the death of the king's son caused the weary Philip to take the reins once more. The queen hoped to spur on her husband to take an interest in the work at the gardens of Granja by eagerly pressing them forward, and in the course of his long absence in 1727 a great deal was effected, among other things the Bath of Diana being completed. But the king could only give a weary smile, and, it is said, utter these words in front of the Diana: "It has cost me three millions of money to get three minutes' entertainment."

At the same time Philip gave orders for Aranjuez to be rebuilt in the same style as far as possible, but larger. Towards the east the great parterre was added, with a double fountain in front of the new wing. A canal was made round the parterre and a second stone bridge to the Jardin de la Isla. Later the present Jardin del Principe beside the great walk, Calle de la Reina, was made, very like the island garden with a long series of fountains

FIG. 535. ARANJUEZ, MADRID—JARDIN DEL PRINCIPE AND CERES FOUNTAIN

FIG .536. ARANJUEZ, MADRID—FOUNTAIN IN THE JARDIN DEL PRINCIPE

(Figs. 535 and 536). The little summer-house at the end of the garden, the Casa del Labrador, conceals under a modest name a luxurious park house such as we have often met with in the North. About the middle of the century a very important change was taking place for the whole town attached to the court. A new plan of building was now to create a regularly-designed little town out of the small wretched village, whose hovels hitherto had leaned against the castle. This was so linked on to the east parterre that two of the star-ray avenues of the park were now extended to form streets in the town, and the third was the old Calle de la Reina; on the other side the walks approached till they were quite close to the castle. We have seen in the North this tendency of the period to build towns on rational principles, and orientated to the centre point of the residence, and it was from the North that this idea came into Spain. The actual originator was the Marquis Grimaldi. He had been ambassador at The Hague, and thus this last form of royal residence came from Holland.

HOLLAND

No country has had to suffer so much from the hostility of supporters of the picturesque style as Holland. Everything freakish, petty, in bad taste, or contrary to nature, was stigmatised as "A Dutch garden," in the eighteenth and nineteenth centuries. Indeed the country had every fault of the old style laid at her door. This happened more and more, and in England, where these hostilities began, they may be explicable by local conditions. But if we look at the part played by Holland in garden history—and this will take us back to an earlier period when she was flourishing—we shall be enlightened as to her peculiar position. People were misled by the term "Dutch garden," as it came to be used derisively in the eighteenth century, and so they were always trying to discover some special definite Dutch style.

The very important part played by the canal in Dutch landscapes and therefore in Dutch gardens has caused people, and especially later writers, to think that the Dutch garden was a leading factor in the development of the French canal plan. Although at the first glance this seems an illuminating idea, considering the close relations of the two countries, the whole course of our story shows that the merit must be accorded quite simply to the natural development of the gardens of France. Although, however, as Hirschfeld perceived, Holland does not play so important a part in garden history as this, she has evolved a peculiar style of her own which is due to geographical conditions. In garden history, just as in the general history of art and culture, Holland appears late. The gardens of the Netherlands, with which we dealt in our account of the Renaissance as a flourishing branch of German gardening, were developed in the southern provinces of Belgium. In those fine engravings that de Vries has left us, the canal plays no part, and water altogether plays a much smaller part than in many another country, including Germany. Later on, Belgium was here, as everywhere else in the world of art, only a branch of French development. We can understand such a garden as the Dukes of Enghien (Fig. 537) laid out in the middle of the seventeenth century, only if we bear in mind that the chief idea of Louis the Fourteenth's grand style was still apprehended in only a vague way, and was confused with the works of the Renaissance. Separate gardens

FIG. 537. ENGHIEN, BELGIUM—THE CASTLE GARDEN

are still small and numerous, and are not controlled by one great plan. The canal, too, is made use of as an important central axis for a side garden. The bosket arranged in great concentric circles is peculiar, and gives a fortress-like character. Long avenues proceed from it all through the park. But in Holland the nobility and the inhabitants of towns were so completely occupied with the weary war for their deliverance from the Spanish yoke during the later part of the sixteenth century (which was so important in the history of gardening) that they had neither the time nor the security needed for adorning their open country, in its perilous condition, with the gentle, peaceful arts of the garden.

At the close of the war came the great boom in trade, whereby Holland became one of the wealthiest and most flourishing countries in Europe. In this favourable time many things contributed to help the culture of all sorts of gardens. First and foremost stood the battle between sea and land, and the Dutch owe the security of their homes to the work they had done consistently ever since the twelfth century, in setting up barriers against the sea. When trade, which was constantly improving, had provided enough capital to make them feel strong and inclined for new undertakings, they began to consider how their growing population could get more room, and how they could win tracts of territory from the sea and reclaim the inland marshes. One part of the country lay below the level of the sea and was enclosed by high dams, but it needed wide canals to carry off the excess of water and to serve as conduits; for when the land is quite flat, with no descents, canals are absolutely necessary for regulating the flow. By a water-loving people the canal, as well as the rivers, had always been used for convenience in traffic. A great deal of particularly fertile land was acquired at the beginning of the seventeenth century by the reclamation of a number of these inland water-spaces.

A formal style was natural for Holland, a country which had from the first conceived the idea of conducting its farming operations by rule and reason; and not only were kitchen-gardens and fields subjected to such regulation, but so also were country houses and villas and their gardens, which often were happily established on the ground reclaimed in the neighbourhood of a large town. One example may here stand for many. In the years 1624 to 1658 the so-called Diemermeer was drained by the town of Amsterdam and a great part of the reclaimed land was given up to suburban houses. Each of these, none large, had a small piece of garden, reaching as far as the canal. The canal formed the end of the garden, and served as a connection, while together with a path on the opposite side it made a road for general traffic. Then when people got accustomed to thinking of water as a connection and not merely as a division, they were separated from their neighbours only by a narrow canal, so that most gardens were divided off on three sides in this way.

For the most part such gardens were no bigger than a simple parterre, generally with a fountain in the middle, and clung to the Dutch feeling for all that was thoroughly clean, neat, and pretty, being laid out in the fashion of that day with box and coloured clays, which made neat patterns round the flower-beds. The owner being rich and his space small, it often happened that there was an excessive amount of ornament, as much being crammed in as the place could possibly hold. Statues, small pavilions, much clipping of hedges, all made an attempt to satisfy the pressing requirement for variety in these little gardens. No one man or place could fulfil the demand for greater stretches of park and so common land was made use of at the side of the houses. Thus in a drawing of the Diemermeer there is a large tennis-court behind the house with several avenues, and this is meant for the people who live there to use in common. This type of building for suburban houses and gardens is everywhere typical of Holland; it extended for miles, and even now is often to be met with.

The kind of flowers that were cultivated in Holland had another considerable effect on the extension of gardening. As already remarked, soon after the end of the sixteenth century the growing of bulbs was the first interest for botanists, and the trade of Holland was the centre of the flower market. Speculation grew to a mad frenzy and seized upon this trade in bulbs, especially tulips, but it scarcely deserves mention as a curious matter of history, and did not affect the world of botany, to say nothing of the art of gardening. All the same, the passion for flowering bulbs made a great difference in Dutch gardens. It certainly was a protection against the hostility to flowers that was prevalent for a time in England and in France, though it also caused a certain stiffness in the beds, and perhaps more than anything else an inclination towards excessive variety in colour. When the eye grew accustomed to the many brilliant colours of these perishable flowers, it also came to pass that Dutchmen demanded a substitute to serve in those long periods when they had to wait for their beloved flowers to bloom. To be sure, the many-coloured globes which mirrored the garden in all sorts of hues were not a Dutch invention, for in the Renaissance-gardens of every country this form of decoration was popular, and one has only to call to mind such a description as Bacon's; but it was first in Holland that this kind of ornament was conspicuous, together with little bells, coloured clays, and coloured statues—all in such a small space that taste became satiated, and somewhat later a great change was inevitable.

In the Holland of the seventeenth century statuary found a place in the garden, where it could make its way to extreme realism, as though it were one branch of the great art of painting. We saw in other countries how at the first beginning of a baroque style a naturalistic statuary of a *genre* kind found its way into gardens, and how even Alberti did not stand aloof from it. In Italy, later on, naturalistic modern statues took the place of the Muses among the great sculptures of the gardens. But what they achieved was always too little to hold its own against other objects of art; and in the lands on the other side of the Alps these carvings had to take their place with the purely ideal forms of gods and nymphs. In Holland, however, the naturalistic style proved more and more victorious, and became increasingly noticeable because of its variety in colouring, which was often conjoined with sound and movement produced mechanically. In order to get as much ornament as possible into the little gardens, these scenes were frequently expressed in miniature; and a conglomeration of grotesque figures in clipped trees was the last characteristic feature of the Dutch garden in the seventeenth and eighteenth centuries. The effect was naturally most impressive as it appeared to the eyes of a traveller who passed along comfortably by the water-way, repeated as it was a hundred times, for "the love of a garden prevails everywhere, and much money is spent on it. Everybody who can possibly manage it, owns a garden nearer or farther from the town, where he lives with his family from Saturday to Monday."

Johanna Schopenhauer visited Holland in 1812, and looking at the gardens with very friendly eyes, saw to her sorrow that the English style was encroaching rapidly. But she gives a description of the gardens in the village of Broek which show, even in an exag-

FIG. 538. A SMALL GARDEN AT BROEK, HOLLAND

gerated way, complete specimens of Dutch taste pure and simple. "The gardens in front
of their houses," she says, "are just as wonderful to look at. You find everything there
except nature. There are trees, which no longer look like trees, so clipped are their tops,
and whose very trunks are painted with white oil paint to make them ornamental. There
are all kinds of possible and impossible animals from the known or unknown world cut
out of box, and columns, pyramids, and grand gates, all carved out of yew-trees. In the
middle of the garden stands the choicest decoration, perhaps a Dutchman sitting on a
tub, and very highly coloured, or perhaps the figure of a Turk smoking his pipe, or an
enormous flower-basket with the figure of a gardener looking out of it roguishly, painted
white with gilt extremities. The ground is covered with countless scrolls and flourishes,
as neat as though they were drawn with a pen; the spaces are filled in with coloured
glass beads, shells, stones, and pots in all manner of colours; and in their perfect symmetry they resemble embroideries of colossal size and the very worst taste." And yet
in this little remote village the prevailing modern style has had its destructive effect.

A specimen of this sort of little garden has been preserved here and there (Fig. 538),
but a colourless photograph gives an incomplete idea, and makes the pretty but overladen
erections look like cardboard boxes. The little temple at the side, clipped to an antique
style and lacquered white, has in front of it a small musical box, painted in many colours
and standing on a support like a table; there can be no doubt that it used to produce
music by some mechanism. On similar supports there are little wooden figures variously
painted. The centre piece is the shell fountain; there is a heart made out of different-
coloured stones bearing the emblems of Faith, Hope, and Charity, with the figure of

Fame on the top, and all round it a parterre laid out with box and filled with flowers in many colours. As usual in the decoration of eighteenth-century gardens, the Indian or Chinese pagoda is to be found here.

Such were most of the private Dutch gardens of that period, but the latest efforts have only remained as curious survivals even in Holland—just as in the case of the garden called by Riat the *Jardin minéralogique*. That means that there was no trace of vegetation kept, the walls were decorated with coloured shells, and, in order to get still more decoration, with vases of faience, gilt birds, statues made in shells, men and animals, and waterfalls made of glass which fell into basins of tortoiseshell. Only a few beds with tulips were left.

It goes without saying that important and beautiful gardens, as well as these small ones, were made in Holland and the Netherlands generally in the seventeenth century, for the country was growing rich. But it must be clearly understood that France exercised the chief influence in the grand style; and that every effort to prove that Holland took a leading part in the garden history of Northern Europe, either before or after the time of Le Nôtre, is labour in vain. In 1668 there appeared the most important and the most widely read of Dutch works on gardening, Van der Groen's *Den nederlandischen Howenier*. This book was translated into French and German, and was very widely studied, because of its useful hints to gardeners. Its two hundred plans of parterres show plainly—compared with the earlier French ones by Mollet and Boyceau—how very strongly inclined the Dutch taste is towards what is simple and small, even petty. The debt to France is frankly

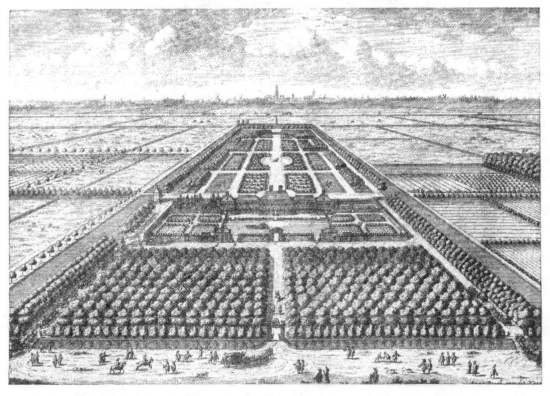

FIG. 539. THE HAUS NEUBERG, NEAR RYSWICK, HOLLAND—GROUND-PLAN

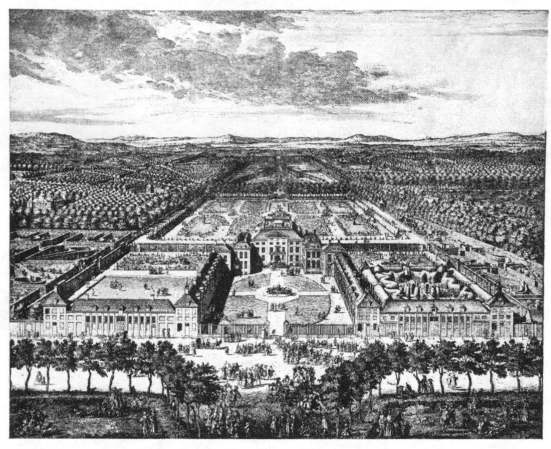

FIG. 540. HET LOO, HOLLAND—GENERAL VIEW

acknowledged. Groen was gardener to Prince William of Orange. The house called
Neuberg (Fig. 539), which William II. built in the neighbourhood of Ryswick, may
serve as an example of the famous gardens of the middle of the seventeenth century.
The completely level ground made it far from easy to get the rhythm and variety which
were so rigidly required by French taste. The whole middle axis was partitioned into
four squares, which were parterres held together by fountains or statues of some kind.
A semicircle made by *berceaux* formed the end of these. In France itself at that time the
parterres in any typical garden would scarcely have been separated even by very low
fences or corner pavilions. Here there were pairs of rectangular basins to right and left,
placed quite symmetrically, and separated by a bosket, those in front adorned with pretty
fountain groups, with narrow ones at the side bordering the whole garden in a charming
fashion. The outside limits were defined by a canal and avenues of trees, which were
carried round the whole estate. The special park groves were on the other side of the
house, and on both sides there were *giardini secreti*.

William III., whose love of gardens we have heard about in England, had produced
fine examples, fit for a prince, in Holland also, during the last third of the seventeenth
century. At the command of Queen Mary, and inspired by her, Harris, the king's physician,
helped her to give the good impression of Holland which she desired by writing a close

description of the country, ending with the statement that "the Dutch deserve from us much respect and kindly feeling." Through him we obtain a very distinct picture of the gardens of that day.

The chief home made by the royal pair, and the one they loved best in their old country, was Het Loo (Fig. 540). Mary laid the foundation stone herself, and they never forgot the place, even after they had settled down at Hampton Court. In Holland, Het Loo naturally got the name of a second Versailles. Some liveliness of contour was cleverly contrived, for the chief garden with its eight parterres was sunk, and it had a terrace surround, and behind was brought to an end by an avenue of oaks. The middle part was open. The view was not impeded over the upper garden, which was somewhat raised, and which ended in a semicircular gallery and a series of fine water-works. The idea of the sunk parterre, which originated in the Italian Renaissance, had been already taken up and frequently carried out with success in the French garden. Two large gardens, one on either side of the house, still keep the Renaissance character at Het Loo, and the whole place shows the same feeling in many other ways. There are groves on either side of the middle part, and in their form, which is partly baroque, they differ, far more than the show-gardens do, from the grand style at Versailles. The canal divides the halves of the garden with two arms, but has not here the important task of uniting garden and

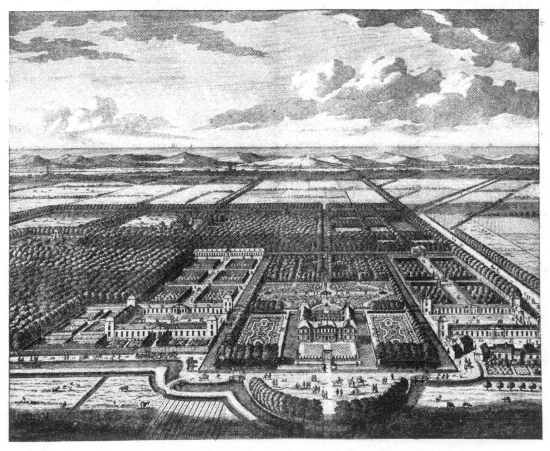

FIG. 541. HONSLAERDYK, HOLLAND—A BIRD'S-EYE VIEW

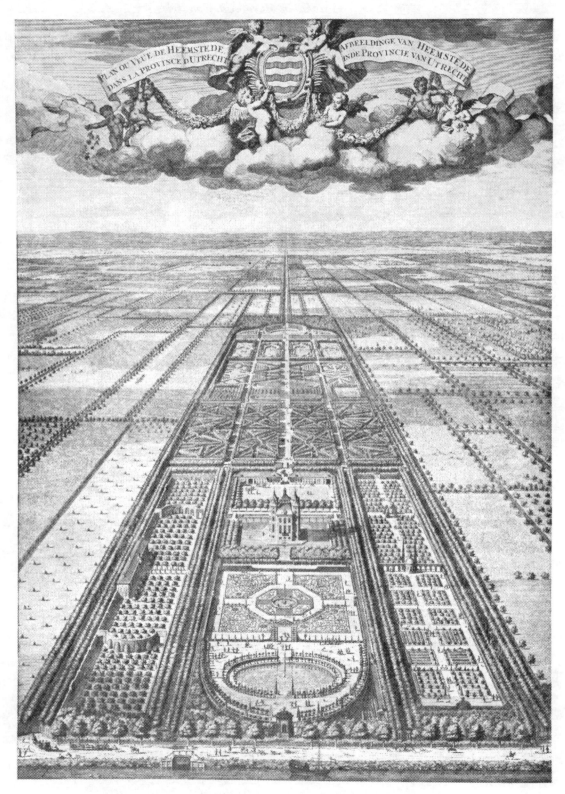

FIG. 542. HEEMSTEDE, UTRECHT PROVINCE—GENERAL VIEW

park. It need scarcely be said, considering the small measurements of the house, that the total area is nothing like that of Versailles.

Very similar in plan to Het Loo is another pleasure-castle called Haus Honslaerdyk (Fig. 541), which William built not far from The Hague. The canal not only goes round the garden, which adjoins three sides of the house, but also round the house itself, thus giving it the character of a water-castle. The boskets, in accordance with their smaller dimensions, are treated far more simply than at Het Loo. A good many other royal castles

FIG. 543. THE HAUS PETERSBURG, HOLLAND—VIEW FROM THE GREAT BASIN

of the period are preserved for us in writings and engravings, but none of them are so large or so important in their decoration as Het Loo.

The beautiful garden of Heemstede in the province of Utrecht has a country road between it and the cross-canal used for traffic (Fig. 542). The drive up to the dwelling-house, a small one in the middle of a large garden and surrounded by a canal, is as usual on the land side, whence the house is reached by way of numerous shrubberies; beside it there is an open grass plot. The real flower-garden is alongside the canal; and next to the house there is a very fine parterre, and on it one of the oval basins so much beloved in Holland. About the beginning of the eighteenth century the passion for gardening, and also real skill, attained their highest point in Holland; indeed the form of the garden had a peculiar quality because of its unusual grouping, separate gardens being arranged within one encircling band of canal or river. This gave a local self-contained unity, so that each individual garden acted as one factor in the gigantic scheme. It is noticeable that the

FIG. 544. THE HAUS PETERSBURG, HOLLAND—THE ARTIFICIAL MOUND

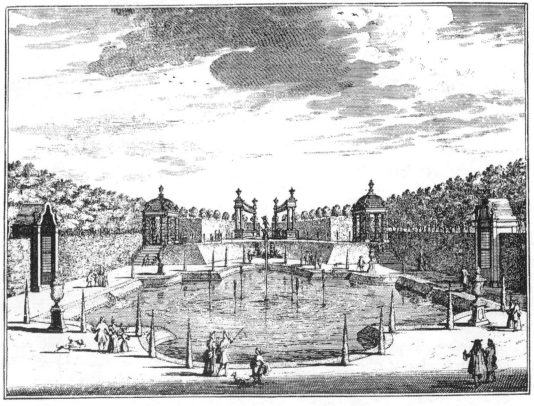

FIG. 545. A LARGE POOL IN A DUTCH GARDEN

issues of engravings recognise this. In the eighteenth century Dutch engravers were incontestably at the head of their art, and they not only supplied Holland with numerous pictures of its own gardens and houses, but also satisfied a great part of the demand in the rest of Europe, as we have seen, and especially in England. In Holland the chief complete sets are *Watergreefs of Diemermeer bij de Stadt Amsterdam, Het zegenprahlende Kennemerland,* and the *Zegenprahlende Vecht;* there were many others of the same sort.

FIG. 546. THE HAUS PETERSBURG—CANAL AND ARBOUR

All the villas have an open parterre by the canal or river, and the travelling on water-roads like these cannot be too much praised, especially on the Vecht, which is one of the most beautiful. Hirschfeld has a description of what it was in his own time (1785).

The country houses and gardens [he says], on both sides, make travelling on this river and in such surroundings one of the greatest pleasures a human being can imagine. Every moment the view changes, first to a labyrinth in a garden, showing thousands of shapes cleverly cut out of lime-hedges, elms, or yews, then again avenues of lime-trees and chestnuts. Sometimes a canal cuts through, sometimes two gardens are separated by a little meadow; or another garden has thickly woven arbours and covered walks. Sometimes hard by the bank there is a pretty house built of brick, another time the gardens are enclosed with iron trellis-work. The traveller can see into gardens and paths, which are ornamented with statues, and along the banks there are long beds with flowers, the tulips making a splendid border. They are especially pleasing to the eye; and a traveller passes by so quickly that he does not feel wearied by the uniformity and regularity everywhere, because he sees such a constant change and succession.

Thus circumspectly speaks the traveller who is kindly disposed to the picturesque style.

At the greater estates the main garden is extended behind the house to the side away from the water-road, according to the style familiar to us in French gardens and in the instruction books. The houses, which are small according to French and German ideas of the period, are hardly ever raised on a terrace. It is only the real old castles that have still mostly kept their character as water-castles, and still present a more important appearance with their moats and drawbridges than the many villas of a later date that are mostly like pavilions (Fig. 543). In the gardens the actual ground is seldom tampered with, and if it is, it is only to get an artificial hill with a summer-house on the top (Fig. 544) or a sunk parterre—a *boulingrin*.

The demand for variety had to be satisfied, first and foremost, by water. But fine and multiform as its devices were—sometimes the great mirror-like basin (Fig. 545), which was often at the end of a large garden, sometimes the fountains in countless forms with shells and all sorts of figures, sometimes a canal enclosing either the whole garden or separate parts of it (Fig. 546)—there was always wanting what was naturally one of the greatest attractions, the cascade.

There can be no doubt that at the turn of the seventeenth century the standard in Dutch garden art was astonishingly high. We very often find Dutch gardeners in foreign service, where they were sought for to cultivate flowers. It must not be forgotten, too, that Peter the Great first studied in Holland the gardens that he afterwards copied in his own land. But in this connection the Dutch garden must be reckoned as of the French school, and must find its place as one member in the whole body of French development.

FIG. 547. A BASKET OF FLOWERS—EIGHTEENTH CENTURY (*See page* 169)

CHAPTER XIV
CHINA AND JAPAN

CHAPTER XIV

CHINA AND JAPAN

WHEN we consider the far-reaching influence of the French garden style, we find that its highest point was reached in the middle of the eighteenth century. The new picturesque garden had done little in the way of propaganda outside England.

At one moment it seemed as if the great art which Le Nôtre had raised to its utmost glory was to get a strong foothold in the Far East. Letters came to France from French missionaries in China, saying that the emperor had commissioned the Jesuit Father Benoît to lay out in the French style a portion of his park at the Summer Palace, Yuen-ming-yuen. At the same time, Father Castiglione built the first European houses in the Chinese capital.

The first thing made by Father Benoît was a highly artistic cascade. There were two pyramids of water beside a semicircular place where the battles of fishes, birds, and wild beasts were represented in the mirror of the waters. As a special compliment to the Celestials, Benoît set up a skilfully-made clock; and the twelve hours of the day were represented in Chinese fashion by twelve animals, each spurting out water for one hour (Fig. 548). This work of art was erected in the second *pavillon à l'italien* like a large water-buffet. The writers show great pride in this site and similar ones in the Chinese Versailles.

The emperor liked to adorn gardens and buildings with treasures from Europe, of which he had a great number. He had copper-engravings made of his new "European" grounds, and sent them over to Europe. They could be compared with the boskets of Versailles. They are a mixture of Chinese ideas and exuberant Baroque (Fig. 549), showing a lively fancy among the Jesuits. The Chinese were amazed at the wonderful hydraulic machinery which the foreign missionaries, magician-like, had at their command. The natives learned to manage it quickly; but when they had mastered the supposed mysterious arts they lost interest in them. After Benoît's death the machines were never repaired, for the emperor, who was now an old man, seldom came into that part of the park: if those responsible heard that he was coming, men were sent to fill up the basins, so that the waters should play while he was passing by. Soon, however, nothing more was heard of the European works of art, and China was just as far as ever from being affected by the influence of the West.

As regards the effect of the East on the West, while the many tales sent to their homes in Europe by missionaries in the Far East, and the accounts brought by travellers on their return, had their influence, the Western mind could only grasp slowly the garden ideas of the East. It was the first time in history that an entirely new style had appeared with nothing in the least foreshadowing it—the style of the picturesque, opposed in fundamental principles to that garden art which all the rest of the world had adopted.

233

海晏堂西。一

FIG. 548. EUROPEAN GARDEN AT YUEN-MING-YUEN, PEKING

西正觀外方

FIG. 549. EUROPEAN GARDEN AT YUEN-MING-YUEN, PEKING

The first ample accounts of the gardens of China are given in the history of his travels by Marco Polo the Venetian, undertaken in the years 1272–93. When Marco arrived at the court of Kublai Khan, the great Mongol emperor, he saw the deer-park at the summer Residence at Xanadu, which one could only reach, as is the case in all Chinese gardens, by going through the palace, in the middle of a small wood with a summer-house supported on pillars. He also saw the palace of the Great Khan at Cambalu (Kambalu), and described its double row of encircling walls, between which there are animal parks. The footpaths were paved and somewhat raised, so that the rain ran off them; thus they were never dirty, and the vegetation through which they ran was always well watered. At the corner of the circumvallation, on the north-west,

FIG. 550. PEKING—VIEW OF THE GREEN HILL

there was a fish-pond. A river ran through it, with gratings at each end to prevent the fish swimming through. In the garden proper Marco Polo especially admired an artificial mound, fully a hundred feet high, standing on a base of "about a mile" (perhaps a thousand feet?). This mound was made of earth dug out for the lake, and on it there were handsome evergreen trees. As soon as the emperor heard of a beautiful tree anywhere, he had it dug up with all its roots and a great deal

of earth, and conveyed to this mound by elephants, however heavy it might be: thus he acquired the finest collection of trees in the world. He had the mound itself covered with green earth, so that not only the trees but the place on which they stood were green; therefore the mound was called the Green Hill (Fig. 550). On the top a palace was set up, which was also entirely green inside and out. The whole conception, hill, trees, and building, produced an impression of wealth and splendour; moreover, the blending of shades of colour was wonderful. Between the Great Khan's own palace and that which he built for his son, lay a wide lake, over which a bridge passed, connecting the two.

The description of the capital of the conquered dynasty, Kin-sai, now Hang-chu, makes it even more wonderful and grand than Peking, which the Mongol princes first used as their Residence. To Marco Polo, and to all travellers of his time, it seemed the greatest and most beautiful town in the world. It was bounded on one side by the river Tsien-tang, and on the other by the Hsi-hu or Western Lake. This landscape has always been, and still is, considered by the Chinese as their very loveliest. It has been praised by artists and poets alike.

FIG. 551. PEKING—ISLAND PALACE IN THE PURPLE FORBIDDEN TOWN

On the borders of the lake [says Marco Polo] are many handsome and spacious edifices belonging to great magistrates and men of rank. There are likewise many idol temples. . . . Near the central part are two islands, upon each of which stands a superb building (Fig. 551). . . . When the inhabitants of the city have occasion . . . to give a sumptuous entertainment, they resort to one of these islands, where they find ready for their purpose every article that can be required, such as vessels, napkins, table-linen, and the like. . . . There are upon the lake a great number of pleasure vessels or barges . . . and truly the gratification afforded in this manner upon the water exceeds any that can be derived from the amuse-ments of the land; for as the lake extends the whole length of the city on one side, you have a view, as you stand in the boat, at a certain distance from the shore, of all its grandeur and beauty—its palaces, temples, convents, and gardens, with trees of the largest size growing down to the water's edge.

Marco Polo's account was confirmed, and even added to, by later travellers, who admired the pavilions built on pillars. They stand either on paved terraces or artificial mounds, and are connected by arched and balustraded bridges (Fig. 552). In this town is the palace of the deposed emperor. Although it was half in ruins when Marco Polo saw it, he was able to give a good description of it from the accounts of his guide and from Chinese writings. He saw wonderful gardens between the walls, orchards, hunting-parks, groves, and lakes of fish. The women's quarters, now entirely gone, were on a lake, and the emperor laid aside the cares of his exalted station and held splendid fêtes on its lovely banks with his wives. He was to lose all these pleasures eventually at the hands of the usurper.

It is clear from later writers that Marco Polo's descriptions were accurate. The fact that they did not produce the effect they should have done in the West was partly due to their being taken for fairy-tales, partly to the absence of any standard of measurement or comparison; the great difference between Eastern and Western gardens was not under-stood. The other travellers in the thirteenth and fourteenth centuries, such as the monk Odoric, who attempted to convert the heathen in China in the years 1325 to 1327, bore out what Marco Polo had said. But Mandeville's fanciful stories of other lands at about the same date reacted on Marco Polo and brought him into discredit; thus, China was again lost so far as the West was concerned, and remained so until the Portuguese arrived in the country, or at any rate at its boundaries, in the middle of the sixteenth century. For more than a hundred years repeated attempts, for the most part in vain, were made to force a way into the interior of the country. One finds a few accounts, but they are very confused and inconsistent. At last, in 1644, the old Ming Dynasty was succeeded by the

FIG. 552. BRIDGE OVER A LAKE IN A CHINESE GARDEN

Manchu Dynasty and the position of the Jesuit missionaries was immediately changed. Ever since the death of Xavier, they had waited expectantly for this moment, and now proceeded as victors to the capital, where they were destined very soon to exercise their political skill and to play a great part at the court. For it was not only as men who taught the truths of salvation, but as engineers, mathematicians, and astronomers, that they gained influence and won the position which they sought. The Jesuit Adam Schall, an accomplished man from Cologne, became president of the Imperial Astronomical Tribunal in 1643 and filled the post till 1645. Other Jesuits held it, with only short interruptions, to the worst days of the persecution, and into the nineteenth century. In 1655 appeared the *Novus Atlas Sinensis*, by the Jesuit Martinus Martini. It was published at Amsterdam, and in 1663 was included in the magnificent work of Blaeus, the *Welt-Atlas*. It was compiled from Chinese authorities and maps, and supported by the author's own observations. The merchants followed the lead of the Fathers. The accounts of the first legations of the Dutch in the sixties received the authority of the Jesuit settlers, and were endorsed by them.

After the arrival of the five French Jesuits, when from 1688 regular news and letters came to Paris from the Fathers, it was supposed that the doors of China were to remain open for ever to the Christian mission; and the desire to press on into the intimate secrets of this marvellous civilisation grew in foreign countries beyond all bounds. (We shall see how the reaction in Europe affected garden art.) The original hopes very soon proved deceptive, for in the first quarter of the eighteenth century there began the persecutions of the Christians, which were carried on with deliberate intention by the Emperor Kien-lung, whose long reign extended into the last two-thirds of the century. He kept at Peking the Jesuits who were so useful to him, got all the good he could out of their learning, gave them tasks of a lofty scientific character, paid great honour to the Italian painter Castiglione, and made European houses and gardens; but in respect to their own calling he left them severely alone. Thus it came about that China was completely closed to foreigners in the eighteenth century, for even the famous embassy of Lord Macartney was economically fruitless. All the same, there arrived an uninterrupted stream of news from the interior of China. Indeed, it was the eighteenth century that first brought about some real understanding and some sort of comparison with the West.

This short sketch indicates the sources from which the nature of garden art in the countries of Eastern Asia came to be known by Europeans. This art, with its own technique and its own sentiment, is perhaps easier for us moderns to understand, despite its marvellous foreign nature, because we are able to place it among other kinds of art in Eastern Asia. We have passed out of a period of the picturesque style in Europe, and we are able to compare it with a similar effort made in Asia, taking into account the great difference in the means at the disposal of the two, and in their aims. For Europeans in the seventeenth century both the name and the nature of "Landscape Gardens," about which travellers spoke, were something quite foreign and unfamiliar.

Sir William Temple writes of it with some surprise in his Essay in 1685, in which he collects with much discrimination the various tendencies of gardening art in his time.

> What I have said of the best forms of gardens [he writes] is meant only of such as are in some sort regular; for there may be other forms wholly irregular, that may, for aught I know, have more beauty than any of the others. . . . Something of this I have seen in some places, but heard more of it from others who have lived much among the Chinese, a people whose way of thinking seems to lie as wide of ours in Europe, as their country does. Among us, the beauty of building and planting is placed chiefly in some certain proportions, symmetries, or uniformities; our walks and our trees ranged so as to answer one another, and at exact distances. The Chinese scorn this way of planting. . . . Their greatest reach of imagination is employed in contriving figures where the beauty shall be great, and strike the eye, but without any order or disposition of parts that shall be commonly or easily observed. . . . And whoever observes the work upon the best India gowns, or the painting upon their best skreens or purcellans, will find their beauty is all of this kind, (that is) without order.

A comprehension, however slight and imperfect, of the true Chinese feeling was a great help to the next generation, which brought the picturesque style to the front. People laughed at Sir William Temple's concern lest the difficult job should be attempted of imitating a Chinese garden, and nobody suspected how right he really was. His idea that people should study Chinese gardening by way of their paintings ought to have been a fruitful one, for the two are connected very closely.

This new and peculiar kind of art came into our circle of vision as a complete, perfected thing—a style built upon a distant tradition, which is disengaged from what we call garden architecture, and from any of those useful purposes that we found to be fundamental elsewhere. From one point of view Chinese art is the purest of all, and the questions of origin and history are most enticing. But there are difficulties which we cannot overcome, since there is no country that shows fewer traces of old historic gardens. The curious etiquette of Chinese emperors, and indeed in a great degree of other important men, forbids them to live in the home of their predecessor; moreover, each new dynasty is apt

II—R

to make a new capital. Then, since in China more than in any country the imperial court was the centre of art and culture, the earlier Residences fell into decay. Even till quite recent times there were continual complaints of the bad condition of the homes of the emperor and of the great men, when they were not being occupied as Residences. It was really a help to the country when the emperors were fond of travelling, as were most of the members of the Manchu Dynasty, who after 1644 took a great many journeys into their Tartar lands. For all the palaces which were visited had to be kept up and well cared for. But this can only have been for a short time, since complaints about decay always arose as soon as any emperor had grown old and given up travelling. Sums of money were often embezzled that had been earmarked for keeping up the estates. It was useless to hand over one or other of the imperial palaces to a rich man on condition that it should be kept up. The Emperor Kienlung entrusted the beautiful gardens at his palace at Ou-yen to a wealthy salt merchant, but this man was certain that as the emperor was old he would never come back, and so he did nothing for the place. Travellers from Europe found it in such a state of decay a few years later that they scarcely ventured to step on any bridge or wooden veranda.

On the other hand, there is an unexampled continuity about Chinese culture in every department, and not least in gardening, for the centralisation at the emperor's court was a great help. Etiquette at the Chinese court has always been an asset to the historian. To the Chinese nation the love of what is old is truly a passion. They were not wanting in historical research of every sort, though the unbroken development of centuries offered so little in the way of contrasts that the origins of an art which grew slowly were lost in the darkness of antiquity.

It is clear from the information that comes through about the doings of the Han Dynasty, members of which occupied the throne from 206 B.C. to A.D. 201, that it was under these emperors that the fancy first started of making great mounds and building palaces on them, and then linking them up by bridges. It is expressly said that the capital city was not built on formal lines like the earlier ones, but made a kind of "star-picture." The Chinese certainly tell of far older gardens; but they do not conceal the fact that there is something legendary about the miraculous gardens of Kuen-luen, though they also tell of grand show-gardens of the Emperor Chou, which go back to two thousand years B.C. Parks, which were laid out chiefly with a view to the chase, were always disliked by the people, who thought they wasted good land. And it is obvious that historians regard gardens and parks as an error and a snare for princes, whose life of pleasure in their gardens and consequent neglect of rule made it so easy for the greedy heir to deprive them of throne and life. But there is nothing to show how these gardens were laid out.

Wu-ti, an emperor of the Han Dynasty, appears to have been conspicuous in his love for extensive grounds. The historians say that the gardens were fifty miles in area, and that every valley between the mountains had palaces, pavilions (Fig. 554) and grottoes scattered over it. This emperor also made the state treasury and its surroundings. He built gardens and palaces for his beloved Fey-yen, about whose beauty writers such as the great lyrical poet Li-Tai-pe composed their inspired verses eight hundred years later. The summer palace of Chao-Yang is described by the poet as a sumptuous paradise of spring, where the emperor spent nights of love with Fey-yen.

All these descriptions are vague. Although the love felt by the Chinese for artificial

FIG. 553. OVERHANGING ROCKS—A CHINESE GARDEN SCENE

hills became famous in the Han Dynasty, we know that in the gardens of Western Asia the Assyrians and Babylonians also showed a liking for them. At the beginning of the great period of Chinese art, under the Tang Dynasty, we find in the lyrics, with their depth of tenderness and feeling, certain hints and allusions which prove without question that the art of gardening, together with that of painting, attained a very high standard. But just as in painting no original work of the period has come down to us, so there is no really intelligible picture of a garden earlier than the one which we get in a poem of the eleventh century, the noteworthy description of the garden of Hsi-Ma-Kuang in the year 1026. Hsi-Ma-Kuang was a great statesman, and (as a later emperor says of him) "he was a benefactor to his own age because of his wisdom, his philanthropy, and the mildness of his rule."

He has described his own estate in graphic language:

Other palaces may be built, wherein to escape from grief or to subdue the vanities of life. But I have built a hermitage, where at my leisure I may find repose and hold converse with my friends. Twenty acres are all the space I need. In the middle is a large summer-house [the word in the French translation is *salon*, but apparently it is one of the separate buildings] where I have brought together my five thousand books, so as to consult the wisdom therein, and to hold converse with antiquity. On the south side there is a pavilion in the middle of the water, by whose side runs a stream that flows down from the hill on the east. The waters make a deep pond, whence they part in five branches like a leopard's claws; numbers of swans swim there and are always playing about. At the border of the first stream, which falls in cascades, there stands a steep rock with overhanging top like an elephant's trunk. At the summit stands a pleasant, open pavilion [Fig. 553] where people can rest, and where they can enjoy any morning the red sunrise. The second arm is divided after a few feet into two canals, which twist and turn about a gallery, bordered by a double terrace. The eastern arm turns backward toward the north, beside the arch of a pillared hall, which stands in an isolated position, and is thus made into an island. The shores of this island are covered with sand, shells, and pebbles of different colours. One part of it is planted with evergreen trees. There is also a hut made of straw and rush, just like a fisherman's hut. The two other arms seem alternately fleeing and pursuing, for they follow the turns of a flowery meadow, and keep it fresh. They often over-flow their bed, and make little pools, which are edged with soft grass; then they escape into the meadow, and flow on in narrow canals, which disperse in a labyrinth of rocks that hinder their course, confine them, and break them. Hence they burst forth in foaming silver waves, and so pursue their proper course. There are several pavilions on the north of the large summer-house, scattered about here and there; some of them are on hills, one above the other, standing like a mother among her children, while others are built on the slope; several of them are in little gaps made by the hills, and only half of them can be seen. The whole region is overshadowed by a forest of bamboos, intersected by sandy footpaths, where the sun never penetrates. Towards the east there is a little level of irregular shape, protected from the cold north wind by a cedar wood. All the valleys are full of sweet-smelling plants, medicinal herbs, bushes and flowers. In this lovely place there is always spring. At the edge of the horizon there is a copse of pomegranates, citrons and oranges, always in flower and in fruit. In the middle there is a green pavilion to which one mounts by an imperceptible slope along several spiral paths, which become narrower as they get near the top. The paths on this hill are bordered by grass, and tempt one to sit down from time to time, so as to enjoy the view from every side. On the west one walk of weeping willows leads to the bank of the river, which comes down from the top of a rock covered with ivy and wild flowers of all kinds and colours. All round there are rocks piled anyhow, with an odd effect rather like that of an amphitheatre. Right on the ground there is a deep grotto, which gets wider the farther one goes, and makes a kind of irregularly-shaped room with an arched ceiling. The light comes in through a somewhat large opening hung round with wild vine and honeysuckle. Rocks serve as seats, and one gets protection in the blazing dog-days by going into the alcoves and sitting there. A small stream comes out on one side and fills the hollow of a great stone, and then drops out in little trickles to the floor, winding about in the cracks and fissures till it falls into a reservoir bath. This basin has more depth when it reaches an arch, where it makes a little turn and flows into a pond, which is down at the bottom of the grotto. This pond leaves only a little footpath between the shapeless rocks, which are oddly heaped together in piles all round. A whole family of rabbits is established among them, terrifying the fishes in the lake, and in turn terrified by them. What an enchanting spot this hermitage is! The second pond has little sedgy islands on it, the larger ones full of birds of every kind, and bird-houses. The way to get from one to the other is by the big stones that stick out of the water, or by the small wooden bridges that are scattered about, some of them arched, some in straight lines or zigzags, according to the space that has to be filled up. When the water-lilies near the bank are in full flower, the pond seems to be wreathed in purple and scarlet, like the edge of the southern sea when the sun rises. Pedestrians must make up their minds either to go back the same way they came, or to climb up the rocks that close in the place on every side. Nature intended that these rocks should be approachable from one end only of the pond. They seem to be fastened together where the waters have opened up a thoroughfare among the willows that stand between them, breaking through on the other side, and forcing their way with a roar. Old fir-trees conceal the dip, and nothing can be seen among their top branches but stones that have become imbedded in a groove or in some broken tree-trunk. Leading up to the summit of this rocky wall there is a steep, narrow stairway; and this has been chipped out with a hatchet; the mark of blows is still visible. The pavilion which is set up here as a resting-place is quite simple, but is remarkable for its view of a wide plain, where the River Kiang follows a serpentine course in the rice-fields.

The prose poem here quoted ends with a description of the writer's occupations in the country, the visits of friends, a laudation of solitude, and a farewell to his beloved garden, because his life was devoted to his fatherland, which summoned him into the town. Hsi-Ma-Kuang belongs to the Sing Dynasty, that is, to the period of Chinese history when every art reached its highest stage of development—to the classical age, in fact, which is comparable with the epoch of the Renaissance in Western Europe. Poetry and painting were both at their best. "A picture is a painted poem"; and paintings show the poetic temperament in the same degree as poetry exhibits the spiritual elevation of the artist. Gardening, however, is only one branch of landscape painting. Nothing that concerns our art of gardening so surprises the Asiatic mind as our eagerness to hang landscapes on our walls, while yet we never arrive at making round our houses such pictures as are composed by the actual works of Nature.

Perhaps we ought to detect in the peculiar reverence which is shown by the Chinese for stones and mountains an early state of religion, and also the fundamental reason for the changes they wrought in the garden. Martini, in his *Chinese Atlas*, speaks of the peculiar Chinese superstition with regard to mountains: "They investigate the psychology of a mountain, its formation, its actual veins, just as astrologers examine the heavens,

FIG. 554. PAVILION IN A CHINESE IMPERIAL GARDEN

or chiromancers the hand of a man." If the Dragon (which means the flowing water, the bringer of all good fortune) is seeking the mountain for its dwelling-place, then it is indeed the bearer of good things, and there will they set the graves of their ancestors and their sacred shrines. In Eastern Asia mountain, rock, and stone enjoy the reverence which is given to trees in Western Asia and many parts of Europe. The first visible and practical sign of this reverence which we find in gardens is the creation of artificial mounds. Since certain natural hills were honoured as particularly propitious, it is easy to understand that people would make artificial hills where they lived, on their own lands. The garden might be small, and in that case the hill must have proportional dimensions. In the pictures there are examples of mountain shapes, in sixteen formal designs, and these portray the possibilities which tradition has accepted, in marks that to us are hardly intelligible when we look at mere pictures. It is probable that plans of a like kind were worked out when hills were artificially made in gardens. How strange some of these were is shown by the elephant's-trunk shape in an overhanging rock (Fig. 553) in the garden of Hsi-

FIG. 555. CHINESE GARDEN WITH SESSEL BLUE STONE

Ma-Kuang, so frequently copied in Chinese gardens and also in Chinese pictures. Beside the mountains are stones, which form one important characteristic of Chinese gardens.

Delatour, who attempted to explore the very foundations of Chinese garden art at the beginning of the nineteenth century, tells us that he has in his cabinet a collection of several hundreds of drawings of stones, of all kinds and colours (Fig. 555). He says that they are mostly in one piece; but some of them are like pyramids and pieced together; some are like obelisks; and several are broken up in irregular shapes. When they are like rocks, water flows through the crevices, and they are all of different sizes. Frequently they are set on a wooden base, and then are usually close to some building or in the foreground of a small garden, where their peculiar shape may serve as table or stool for the great ladies whose whole life is spent in the garden (Fig. 556). They are very often of light blue, made of a stone which is hewn in Southern China. The important towns have large shops where these stones are for sale, and the best of them fetch a high price.

Water acts as the veins and arteries of the mountain, and is essentially important, partly in itself, and partly in connection with the course it takes, the hollows it scoops out, and the lakes in which its streams are collected. When the Chinaman makes artificial hills and mounds in his garden, he is also making the bed of an artificial lake. Hsi-Ma-Kuang lets the overflow (in his chief garden) run down into the lake from the hills on the east, for the course from east to west is lucky; and probably the five streams "like a leopard's claws" have the same sort of significance. Martini says that any lake which is

FIG. 556. CHINESE GARDEN WITH BLUE STONE TABLE

FIG. 557. A GARDEN WITH A LOTUS-POND, HONG-TSHU

fed from nine sources is considered to be lucky, and thus we have the lake of Kiu-Lung, the Lake of the Nine Dragons. The art of the Chinese, like their feeling for Nature, is at bottom a profound sentiment beyond anything which shows on the surface. Symbolism is essential to it; and however accurate the observation of natural features in individual objects may be, this symbolism is its sole inner significance. The cultivation among the Chinese of a religious tradition in matters of art is often hard for us to understand. Among a host of pictures by a famous Chinese artist, there is a drawing of flowers painted by

FIG. 558. THE GARDEN WITH THE THOUSAND SNOW TRACKS

the Emperor Kienlung, and on it the following words are written: "In a happy hour of summer there came into my hands the picture of Ku-Kaichih, and inspired by him I have sketched in black ink the spray of tree-orchid to express my admiration for the deep feeling and inner meaning of the picture."

This power of suggestion in Chinese art made it possible for garden artists to gratify that desire which was the main thing for the owner, i.e. to possess some copy of a famous landscape of his own country. Great admiration was felt for the Hsi-hu (Western Lake). To enjoy its beauty, nothing more was needed than a lake and one or two islands, with pavilions on them connected by bridges, and mountains by the shore with buildings on the top. Some kind of stone, native to the place; some kind of plant which would especially stir up memories (and before all others water-lilies, which are very plentiful at the Western Lake)

(Fig. 557)—these things are quite enough. Through them the spectator, on the veranda at his house or in some special pavilion, can summon to his sight the whole beloved view of the beautiful landscape. For the rest, the artist is allowed an entirely free hand, and he can fill in objects, great and small, according to his own fancy, with such means as he has.

Besides the imitations of well-known landscapes, original pictures were painted, in which was an attempt to express some mystical or imaginative fancy that was in the mind of the artist. The French engraver found on his Chinese original, which he carefully copied, the words "Garden of the Thousand Snow Tracks" (Fig. 558), and the grouping of artificial mound, numerous pavilions, zigzag steps, stream with waterfall and

bridge, trees and stones, really expresses this idea in a definite landscape. For traditional feeling acted as a help, or even as compulsion, prescribing the three main elements, viz. hill and stones for the first, lake and stream for the second, pavilions and other erections for the third. The overhanging rock, with its pavilion at the top from which Hsi-Ma-Kuang admires the red hues of the rising sun, has its appointed task to perform in a Chinese garden. We shall see later how in Japan too, where literature on the subject is instructive and convincing, tradition and original art were closely connected.

FIG. 559. ROSE LATTICE AND PINES BEFORE A HOUSE

The Chinaman is not familiar with the idea of going for a walk. Well-born ladies were incapable of walking very far, because of their compressed feet, which were never allowed to grow. They spent their lives largely in the garden. Chinese women are so fond of flowers that they wear them in their hair even after they have grown old and grey, and this applies to the lowest class; but as they never walk in the fields, interest in wild flowers does not exist. In the gardens the flowers are not set all together in beds, but arranged in groups like flowering shrubs, especially lilies of various kinds, and peonies. The little picture (Fig. 559) shows a rose climbing on a trellis, but we ought perhaps to see European influence in this, though the pine-tree on one side is really Chinese, and so are the foliage plants on the other side and the finely decorated vessel with a stone standing up in it shaped like a flower. All this, combined with trees in flower, which are very highly thought of in Eastern Asia, gives to their gardens an appearance of wonderful colour; and when the season no longer allows of it outdoors, the effect is helped by plants in pots.

The Chinese enjoy their gardens sitting down, and hence the many pavilions—a

FIG. 560. PAVILION AND BRIDGE IN THE GARDEN OF LI-CHING-MAI, PEKING

feature which always strikes the European first. They approach these places of rest by winding paths paved with coloured mosaic tiles. Every pavilion has its fixed purpose of enlivening a particular scene, or providing a rest at some special view-point, or showing the garden in the varying light of different times of day, like the one turned to the morning sun at Hsi-Ma-Kuang's garden. Pavilions play a prominent part in lyrical poetry as meeting-places for all social parties (Fig. 560). To feel the whole charm of Chinese garden life we should read the little poem *The Porcelain Pavilion*, by Li-Tai-pe:

> In the middle of the lake we made
> Is a porcelain house all green and white.
> Thither we walk on a bridge of jade
> Arched like the back of a tiger.
> In this pavilion sit our friends,
> In garments light, and drink their wine,
> With merry talk, or writing verse.
> Their heavy headgear they discard,
> And turn their sleeves a little up.
>
> And in the lake the little bridge appears,
> A crescent moon of jade, turned upside-down;
> So standing on their heads our friends are seen
> In lightest garments clad, and drink their wine,
>   In a porcelain house.

The writing of verses is an essential part of Chinese hospitality. They have a favourite game, which was afterwards transported to Japan, in which a wine goblet is made to float on a stream, and anybody whom it passes must either compose a verse or drink a great draught of wine. There are two pictures dating from the Ming Dynasty which show garden fêtes of this sort in the second half of the fourteenth century. The first depicts the garden of Villa Kin-Kou at Honan, founded by Shi-Tsung between the third and fourth centuries in the Tsin Dynasty. Shi-Tsung was a high official in the family of the emperor. He is seen in the picture at the side of his beloved lady, for whose sake he ruined his career, in the company of pleasure-seekers and wine-bibbers. The second picture shows a garden fête given by the poet Li-Tai-pe himself, who is holding a festival, together with his three brothers, in the garden of Villa Tau-li, which was famous at the Tang capital, Tsi-Nan-Fu. This work was an illustration of a poem by Li-Tai-pe: "Any night," he says, "when lovely flowers smell sweet and a light breeze blows cool, is a gift from Heaven for our delight. Nothing better can we do than be happy, light the candles, lift the wine-glass, and write poems. But he who is no poet must drink three glasses, as once of old was done in the garden feast at Kin-Kou." Both pictures give only the foreground of the garden beside the house; we have also bridges, and flowering trees, and pine-trees with tables set out below them.

Hsi-Ma-Kuang also mentions the number of pavilions that were scattered about his grounds. All the ambassadors who later on gained admittance into the imperial gardens relate that they were entertained first in one pavilion, then in another, as the emperor might happen to be dining in this or that. Every large garden in China comprised the most diverse arrangements, and each place was designed for the enjoyment of its own peculiarities as a place of rest. The first view was as a rule to be seen from the house, or might open out to the visitor as he approached, like that described by Hsi-Ma-Kuang: a larger or smaller lake, according to the size of the estate, always provided the centre-point of the open, smiling valley. If large enough, it had one or more islands, all decked with small summer-houses. It was approached either by a path of flat stones or by a bridge. According to its length, the bridge is either constructed of flat rectangular stones, or is of wood with a balustrade to it, and in that case it takes a zigzag course. If the lake is large and deep enough for the owner to indulge his love of boating, the bridges have arches, made of all sorts of different materials, under which the boats can pass (Fig. 560). But, however varied lake or bridge might be, the Chinese were always mindful of the original pattern for every Chinese lake, the Western Lake (Hsi-hu). There they found bridges, between the roads that led across the lake, which house-boats could get under. These boats were hired by the day for whole families. Very often there is a pavilion in the middle of a bridge, or a "Gate of Triumph" at each end. There were gates of this sort at both ends of the bridge leading across to the imperial palace at Cambalu, told of by Marco Polo.

The garden of Hsi-Ma-Kuang, which in its simplicity seemed to mirror the taste of that quiet man of learning, had its library as the central feature. The chief part of the garden was treated very simply. There is nothing said about an island in the lake, but there was a stream flowing down from the eastern hills to form the cascade, which in the east part of the chief garden was quite indispensable. Also, to north and west, hills encompassed the banks, falling back a little in the middle and so making the view more open. The eye was attracted onward to half-hidden ravines and bamboo groves, and farther still to the horizon and the little woods of flowering pomegranate trees, citrons, and oranges. From the middle of these rose the spiral hill with its green pavilion.

In Chinese scenery smiling landscape must contrast with something terrible, such as overhanging threatening rocks, deformed trees apparently broken by the force of a storm, dark hollows, foaming waterfalls, or buildings which look partly like ruins and partly as though they have been destroyed by fire. With these accessories the effect of the terrible is produced. The hermitage of Hsi-Ma-Kuang has a gentler look, due to its quiet lake with encircling rocks. In some of its features it partakes of the romantic and idyllic, which comes out strongly in the little island with the fisherman's hut. The poet conducts us to the hermitage by way of dark cool caves. In the artificial hills and rocks of all gardens of any size, there were hollows and even actual rooms, often made at great expense. Martini also speaks of this: "In the beautiful gardens of China I have seen artificial hills in which have been cut most skilfully hollowed recesses, rooms, and stairs, even ponds, trees and other objects, where art really was a rival to nature. This is done

FIG. 561. CHINESE GARDEN WITH PINES

to get rid of the heat of summer in the cool of such caves, when men want a place for study or for a fête. Still more beautiful is the place where the labyrinth is made, for although the area is not extensive, you can walk about there for two or three hours." The chief aim, however, of Chinese architects was to find some central point where all these minor pictures, which were enjoyed separately, could be seen in one comprehensive view. Hsi-Ma-Kuang does not emphasise this point particularly, but probably for him the view on the top of the hill, whence he could also look down on the level of his own river, the Kiang, was the chief attraction.

The garden of a minister was far surpassed by that of an emperor, both in size and in splendour. In all cases, whether the estate was large or small, the chief effort in a Chinese garden was to make the picture in the right proportion for the given space. Although it was practically never possible to repeat in its own dimensions the natural scene that was imitated, there was a constant endeavour to have the place appear more important than it was by a clever management of the perspective. It is said by Staunton, who was in the retinue of the English Embassy, that in one garden which he noticed there was a slight wall, which, looked at from a certain distance through the branches of a thicket, gave the impression of a magnificent house. The Chinaman, even in his own large garden, had to make a miniature copy of his model, and Staunton writes of the imperial palace at Peking:

It stands as the middle point of the Tartar town, which lies unregarded in the dusty plain; yet the walls of the palace coincide with every winding or contour which nature in her most capricious mood had imposed on the surface of the ground, but always in less degree. Hills, valleys, lakes, rivers, the bold precipice and the gentle slope, all are here, and not where nature placed them; but in their relative sizes they are so exact and in such perfect harmony, that (if the whole aspect of the surrounding country did not give the lie to this deception) any spectator would feel a doubt whether this was a natural site, or only a felicitous imitation of the beauty of nature. This world in little has been brought into being at the command of man and at his pleasure, but through the bitter toil of thousands.

Seeing that even a place of importance, like an imperial garden, must have its proportions reduced, the artists had to accommodate their work to really tiny places; and this was the reason why they imposed a limit on the growth of trees, but even dwarf trees, however small they were, had to keep every peculiarity of those that had grown to their full natural size. The native of Eastern Asia particularly admired the diversity of form, the queer irregularity, of trees in the open landscape, where he likes old willows, pines and firs, also cherry-trees; and these are commonly shown in the pictures (Fig. 561). It almost looks as though the trees, if left to themselves, have a more varied form and significance in China than with us, just as his mountains are full of weird and curiously formed overhanging rocks. These landscape pictures, taken as a whole, are often scarcely to be distinguished from some particular large garden, for which they have doubtless served frequently as models (Fig. 562).

The trees are of all sizes, from fully - grown specimens in the large gardens to dwarfish ones found with the little hills, valleys, and rocks of a tiny plot of ground; and to produce the miniature trees was the main occupation of Chinese gardeners. After the art of regulating the growth of

FIG. 562. A CHINESE LANDSCAPE GARDEN

trees was learned they went so far as to use dwarf trees in vases to decorate indoor rooms (Fig. 563). Father Cibot says in his Essay that he saw trees, such as pines and cedars, only a few inches high; and to match the little trees there would be a miniature landscape in a vase, with everything set out in the right proportions. For a Chinaman a tiny landscape of this kind presents all the beauty of nature. Lord Macartney,

FIG. 563. VERANDA WITH PLANTS IN POTS AND DWARF TREES

the English Ambassador, says that the understanding of dwarf tree culture was a secret, and was very highly esteemed. One of the Chinese poets sings the praises of the art thus: "It makes our nature cheerful, and fills the heart with love; it destroys ennui and evil passions; it teaches us how to change flowers and trees, and brings distant landscapes to our view: we need no journeying to behold the wave-beaten shores, and mountains, caves, and cool grottoes; we behold the course of the ages, but not their decay." All this experience the Eastern Asiatic gets out of a little landscape which is only a few feet in area.

The peculiarities of trees are also brought out in the kind of avenues used, both as a grand approach to some temple and to form a straight line which guides the eye to some particular point, as does the avenue of willows in the garden of Hsi-Ma-Kuang. Houses in China, as elsewhere in Eastern countries, are adjacent to greater or smaller courts; and these, even in the humblest homes, give character to the garden by containing flowering trees and shrubs, or pot plants, which are liked still more. One passes through these courts before coming to the real garden, which is seen from the veranda at the house. Close to the house the pool often adopts the form of a regular basin; and here also the decoration is prettier and better arranged, with perhaps a lattice border and climbing roses, which remind us of the gardens of small European houses, or again with covered leafy walks, concealing the garden walls; these things are known and beloved by the Chinese as well as by us; but informality still persists, and groups of trees and stones accompany the road till we are close up to the terrace of the house.

In the immediate neighbourhood of the house the ever-present symbolism plays a great part in determining the decorations. In every plant and in every tree there is a meaning applicable to human life, and so everyone chooses the flowers for his own house to suit his private wishes. As in the building of a house, the laying out of a garden is subject to fixed laws, so as to propitiate the kindly deities, and keep away the bad ones. These laws are called Feng-shui by the Chinese, and by their guidance they attribute valuable powers to certain trees and plants according to their kinds. All men like to find a peach-tree before the door, because it is the symbol of immortality. Since the stork was said to promise a great age, this bird was naturally set there as a companion picture; and human forms, to represent genii, were made of the green wood, exactly as they were made in European gardens.

Of all the imperial gardens the best known to Europeans are those of the eighteenth century. At that period news was coming in from every side, and there were continual fresh accounts either of complete gardens or of particular beautiful things in them. The Manchu emperors had laid them out to the west of the capital, but in the Ming Dynasty, which began in 1368, and to which the Chinese ascribe the highest development of their garden art, the summer palaces and gardens were on the south side. Yuen-ming-yuen, the Round Bright Garden or Garden of Heaven, the summer palace of the emperors, is on a large, rather steeply ascending plain, which would make the construction of a garden less difficult. Father Attiret, who gives a specially detailed description of this garden, says that the artificial hills put up there reach a height of twenty to sixty feet, and that there are an immense number of little valleys between the hills. The bottom is filled with clear water in canals which make pools and lakes in all directions. Highly decorated boats float on the water, and some of them have little houses on the top. Paths paved with pebbles wind about the place; they lead close up to the water, and then turn off again and go still farther. Every valley has its own house, of one story only, and with small surroundings, but considerable enough for the accommodation of the emperor and his suite. They seem to stand on the rocks like fairy dwellings; and fairy steps lead up to them, looking quite natural. Some of the houses are made of cedar-wood, others stand on pillars, and are finely painted. "Astonishment increases when one is told that within this magnificent territory there are actually two hundred such palaces."

The different buildings are invariably separated by water or by artificial mounds. Many sorts of bridges, frequently winding, and often supplied with white marble railings, beautifully divided and carved, lead across the waters. On those bridges which command a specially fine view there are pavilions; others have triumphal arches. One of the largest ponds measures almost half a mile across, and in the middle lies a real treasure, an island that might be called a rock, rough and very steep. A palace is built on it, and small though it is, it actually contains no fewer than a hundred rooms. The architect selected this particular spot so that the eye might cover at a glance all the beauties of the park, which on a walk could only be revealed one after the other. From this point there were visible all the mountains that closed in here, all the canals that poured their waters here, all the bridges near and far, all the pavilions, and the arches of triumph; all the little groves that show green either between the palaces or in front of them. The banks of canals and lakes were most varied in kind; some had stone quays, and galleries, in some cases the paths were covered with shells. Here and there were pretty terraces with steps leading up to the palaces. Above there were higher terraces with buildings like an amphitheatre, surrounded by groves of flowering trees. Between the rocks flowers bloomed everywhere; and beyond there were thickets of wild trees which will only grow on quite uncultivated mountains, very tall, and the abiding-place of shadows. At Yuen-ming-yuen there was also the European part, with water-devices hitherto unknown in China, and living but a short life there. Round about the walls of this princely seat there were a great number of dwellings made by the artist and architect, and work was going on continually for the garden and the palace. And near this imperial summer-house there were a great many others.

The monarch had a great affection for "the Hill of the Wide View," which was quite near; he would have liked to make his residence there, but etiquette forbade him to live in the home of his predecessor. According to the descriptions we have, this garden was laid out in a simpler style, and deserved its name; for the buildings were set up in terrace form on a high hill, whose base ended in a great lake scattered over with innumerable small structures. In the case of other places, their very names stir our imagination, as, for example, Shang-chuen-yuen, "Garden of Everlasting Spring," and Tsing-ming-yuen, "Garden of Lordly Peace." As we have said before, the Manchu emperor was fond of travelling; and every year he went on a seven days' journey to the fine hunting-seat—Jehol. At each stage there was a palace, provided with everything necessary, and surrounded by a garden. Lord Macartney visited him at Jehol, and found its gardens incomparable. He praises a green valley full of huge ancient willow-trees, which reached down to the bank of the great lake. They crossed over it, between the water-lilies, to the small palaces on the shore, until their way was checked by a bridge that their boat could not pass under; behind the bridge the lake appeared to lose itself in the blue distance. The palaces each contained a large room with a throne, and were adorned with European works of art. The ambassador especially observed that the valleys could be planted with northern oaks and with the most tender plants of the south, although this palace was in a wild, inhospitable part of the country.

The Chinese had public gardens also. We have already spoken of the house for entertainment on the island of Hsi-hu. The rich salt merchants had presented the emperor with an exceedingly handsome Summer Palace at the town of Yang-Shou, between the rivers of Kiang and Hoang-ho, and this was a highly decorative addition to the town, and made a public promenade. Although it took three-quarters of an hour to walk through it, everything was open to the public. A person going in stood on the bank of the great lake at the particular central point which gave him a view of the whole. He could sit down in one of the little tea-pavilions which were on the shore in great numbers, and gaze at the spectacle of pleasure-boats on the lake, or look right across over the heads of merry groups to the hills covered with buildings. The top was brightened by the emperor's palace, and from here the best view of all was obtained.

The gardens at the temples were also open to the public, but their importance for garden art in Eastern Asia was first clearly recognised not in China but in Japan. Still, in the surroundings of Chinese temples we sometimes find to-day very old and very beautiful trees, which in many cases were planted at the time of the foundation of the temples. The fantastic style of architecture and the gay colouring in the temples fit well into garden surroundings. The pagodas with their irregular wavy roofs almost look like trees frozen rigid. The important pagoda of Lung-hua at Shanghai (Fig. 564) is in

FIG. 564. THE LUNG-HUA PAGODA, SHANGHAI

FIG. 565. PAGODA AT FATSHAN, SOUTHERN CHINA

the midst of ancient trees, and the wonderful building of the pagoda at Fatshan (Fig. 565) shows the corner of the temple behind which it stands. The trees by the temples are just as inviolable as those which the Chinaman plants about the graves of his fathers, for the Oriental trait of reverence is particularly widespread and very deeply rooted. Every Chinese family, except the very poorest, has its family grave, which is sacred to it; and the surrounding of trees has often grown to be a park. In a country so thickly built over, where every foot of open land is used for cultivation, these sacred groves are almost the only preserves for trees.

JAPAN

The civilisation of Japan is modern compared with that of China, but it is of the Chinese school, as is unreservedly recognised. The relationship is made very clear by the fact that the Japanese first obtained a notion of literature when they accepted Buddhism, which happened with astonishing quickness in the fifth century A.D., and it was from China that this religion overflowed into Japan. The Japanese authors, who had to invent a written speech, adopted the single-syllable language of the Chinese, by means of which they wrote their own language phonetically. The Japanese are a surprisingly receptive as well as a very zealous and diligent people. They absorbed the results of Western civilisation in a single generation. In the domain of art the Japanese have originality, but in all else they appear only as an offshoot from the great Chinese tree, and only began to exist when that tree was independent and full-grown. This is especially true in the case of painting, and in the garden art which is so intimately connected therewith. China offers riddle after riddle to an inquiring student, in spite of the fact that in her life and art there is such remarkable continuity and coherence, and problems occur all through her history; Japan, on the contrary, has all her cards on the table and allows a greedy inquirer to make investigations into the innermost recesses of her spiritual history.

At the present time the garden art of Japan is known and admired in Europe. It is examined and imitated as few others are. This is not simply the result of unrestricted trade between the nations and the incomparable advantage that the Japanese have themselves seen European gardens, but is also and perhaps still more due to their philosophical type of mind. In life, thought, and action the Japanese have always been in complete harmony with their environment and history, and the greater part of their literature consists of works that are meant to be instructive. Many of the national characteristics that incline the Japanese towards gardening are shared by the Chinese, or indeed by most of the people of Asia; but we are particularly well informed as to the love felt by the Japanese for nature and the garden. It is not a mere æsthetic exhibition made by the aristocracy, although we are dealing with a people so aristocratic that all culture so called was for a long time confined exclusively to the one class. Delight in the beauty of nature is expressed in great festivals shared by all classes, and in this particularly festive nation flower fêtes take the first place.

It was a fortunate circumstance in ancient Japanese civilisation that its calendar was a lunar one and made New Year's Day fall in February, so that the day and the month of the new year combined to create a feast of welcome. On this happy day the first plum-blossom, *prunus mume*, so dear to the Japanese heart, was greeted with great rejoicings.

The tree itself, dark and gnarled, and in old age often quaint-looking, is a great contrast to its delicate flowers, with their snow-white or blood-red bloom, appearing before the leaves. The people come in crowds, and though some are noisy, others gaze with silent rapture. Poets of all times have taken this tree for their theme, and it seems to embody hopes for the coming year in the endless allegories and legends that are bound up with it. Royal princes and great men alike planted parts of their parks with the mume, after-

FIG. 566. UYENO PARK, TOKIO—CHERRY-TREES IN BLOOM

wards admitting the common people, and receiving the nobility on certain special days marked for ceremonial fêtes. Rejoicings are still more vehement when the Japanese cherry, *Prunus pseudo-cerasus*, is in flower in April. Both these trees are grown for the sake of their flowers, not for their fruit, which is of little use. The year has come to its chief beauty, and now even the very poorest person brings out a gay dress, to share with friends and children a humble picnic under the flowering trees (Fig. 566).

The double cherry, with its large white and pale pink flowers, is exceedingly beautiful. In the old garden, the Fukiage, which once belonged to the Shogun, and is now the property of the Mikado, are held the annual ceremonial feasts of the Cherry-blossoms. There is a saying, " If anyone asks where is the heart of a true Japanese, point to the wild cherry blossom, where it glows upon the tree." Thus the Japanese people, like the Chinese, have a complete flower-calendar according to their favourite blossoms; and every month is indicated by its own most beautiful flower. Flowering trees are cultivated in both countries entirely for the sake of the beauty of the blossom (Fig. 567). Between plum and cherry comes peach, in the month of March. In May appear the flowers of the fuji, the wistaria, after which the Japanese have named the most beautiful of their hills, the Fujiyama. The most distinguished of Japanese families, which, like the Mikado himself, traces back its origin to the gods, bears the name of Fuji-wara, i.e. wistaria-field. All the

FIG. 567. FLOWERING PEAR-TREES IN A PEKING GARDEN

arbours are covered with this tree, in fact the verandas are only there at all so that they may show from below the great beauty of the bunches of flowers (Fig. 568).

In June the iris follows; and in the fields of iris herons are to be seen, slender birds, tame, and fond of company. In July there is the lotus, changing every pond as with a magician's wand into a charming garden of flowers. In the days of Li-Tai-pe the Chinese also used to go out to the rivers and lakes which were so famous for water-lilies. They went in hosts, to admire and to pluck the flowers, as is told in the tender poem, *On the Banks of the Yo-Yeh*. June brings peonies, which are also much delighted in by poets. August and September are the months when the many kinds of hibiscus are in flower. In October the chrysanthemums bloom, and again people come together, to see the exhibitions of these flowers. Once more the nobility hold festivals, which are really one mass of tradition, in the imperial gardens. The chrysanthemum appears in the emperor's coat of arms. Now also the fine foliage of the maple takes colour, and is the pride of many places. November and December produce a particular sort of camellia, and also the flowers of the

Chinese tea-plant. Last of all, in January, we have gardens full of the more common kinds of camellias.

Japan has not a very large number of different kinds of flowers, and the attempt to cultivate a greater variety only began with the influence of the West. But we can add to the list of the popular flowers already named both azaleas and orchids. Such Western methods of training as they could not actually see were quite unknown to the Japanese. Even in their imperial flower, the chrysanthemum, they were surpassed by gardeners of the West. Yet Europe has had no notion of the innate, profound, and highly developed love for flowers that is felt by the Japanese, and which is remarkably exhibited in one branch of art, viz. the way that flowers are arranged. This they call *Ike-bana*, or living flowers (Fig. 569). The notion is to find for a flower, which is taken out of the earth,

FIG. 568. THE WISTARIA ARBOUR

some place wherein it is as far as possible in its natural atmosphere. Because it has been moved and placed in a vessel it has to be under special protection and care. The stalk or the branch must keep its own particular form. The vase must above all be suitable for the flower, and must, so to speak, counterbalance it as did its native earth before, while at the same time it must add something to its beauty, and set it in the right light. It is easily understood that the trained eye of the Japanese detects the idiosyncrasy of any plant; and as Nature does not often give these forms in their perfection, art must be summoned to her aid.

The art of arranging flowers is closely connected with the underlying symbolism that we shall find in Japanese gardening. There is faithful observation at work, together with delicate æsthetic sensibility, both for line and for symbolic meanings. This is always present with the Japanese; and out of it there arises some form which is sanctified by their tradition, and has been elaborated in many volumes of theoretical treatises. Thus the very curves which the plant is to take are prescribed, and also the little branches with three, five, or seven blossoms or leaves. All the details are systematically worked out, and each has its definite symbolic importance. Moreover, care has to be taken about the surroundings in the room, and where the vase is to be placed. Seeing that the room looks out upon a garden, it really belongs in a sense to that; and so this garden must not be neglected in the scheme.

It follows from the relation of the Japanese both to Nature as a whole and to trees and flowers in particular, that theirs are not flower-gardens in the same sense as ours are. The fundamental principle of Chinese art is so to bring Nature before the eyes that only selected beauties are seen, and this is the end and sum of Japanese art also. This art, as indeed were all the arts, was at its height in the military epoch of Japanese history. But the Japanese received their style as such ready-made, for it was delivered over to them from China at the same time that the country was penetrated by Buddhism. At any rate, it was then that attention was first directed towards the building of temples, and Buddhist monks have from the beginning been wont to take the utmost care of the surroundings of their temples and monasteries. Japanese writers do indeed talk of a primitive style, existing before the Chinese influence came in, and we read of a lake with island, bridge, and plum- and orange-trees; but as there was no written literature at the time they would have been made and planted, this tradition is too vague.

It was not till the warlike period of Heian, about 800 to 1150, that the court nobility began to build palaces in Kyoto, the capital, and that gardens were laid out in the front of the strange collection of houses, which were joined to each other by corridors. At that

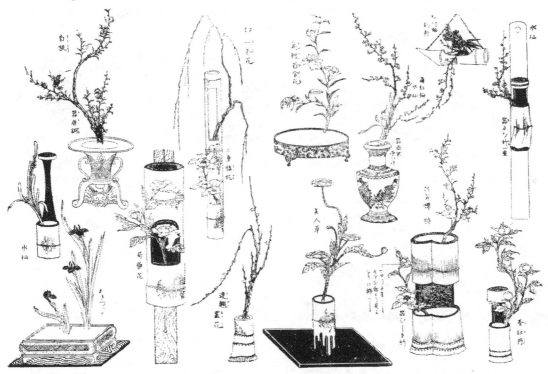

FIG. 569. ARRANGEMENT OF FLOWERS IN JAPAN

period there existed only one typical style, imported from China, which brought into a single picture the view of lake, island, bridge, water falling from an artificial hill in the background, and different sorts of trees and stone-work oddly shaped and twisted. There would be a grotto at one corner, and some sort of hermitage or hut to enliven the scene. People were wont to come to a place like this on fine summer evenings. There were covered walks leading from the house to the garden, which also served as walls for the whole enclosure. In the ninth century the famous artist Kanaoka was busy, drawing such houses and gardens. His death was followed by the closing years of the Heian epoch, a period of much revelling and licence, when the nobles took to all kinds of extravagances; but famous gardens were laid out at Kyoto.

The greatest admiration was excited by the house of a chieftain of the Minamota clan. This house was encased in brick—a style which at that time was positively unique. The owner made a park round his house which copied in miniature a landscape that was

famous in Japan—the salt coast (*shiwo-mama*) of the province of Mutsu. It is characteristic of the nobility of that day that they had hundreds of tons of salt water evaporated
there, so as to give the fresh-water lake the proper taste of the sea. In the same spirit of
exaggerated æstheticism, they would cover tall trees with artificial cherry- and plum-
blossom, in order to recall the spring, or would hang garlands of wistaria on pine-trees
in the autumn, or would pile up great masses of snow, so that they might still see traces
of it under the sunny skies of spring. The mixture of snow and flowers is, of course, one
of the things that the Japanese have delighted in at all times.

The art of gardening shows what seems to be an entirely new feature in the next,
the so-called Kamakura, epoch, lasting from about 1150 to 1310. The Buddhist monks,
as we have said, were the chief cultivators of the garden and also the chief teachers. Even

FIG. 570*a*. SUIZENJI PARK, KUMANOTO, JAPAN—
ARTIFICIAL FUJI MA IN THE BACKGROUND

FIG. 570*b*. FUJIYAMA, JAPAN

to this day we find, as in China, the best old gardens round the temples and monasteries;
and these look very fine against the dark-green background with the complementary red
colour so universally used in religious buildings.

We have already noticed the importance of stones in the Chinese garden. It is probable
that the monks took from China into Japan the custom of naming the most important
stones, which had special places assigned to them, after certain Buddhist divinities. In
the garden of the Abbot of Tokuwamonu Lafcadio Hearn saw represented the legend
of the Buddha, before whom the stones bow down. And even in the latest times stones
bearing the names of gods are to be seen in monastic gardens: they are mostly nine in
number, five standing, four lying down. Each had to have a fixed place, and they were
to serve as protectors against evil. Probably these temple gardens had predecessors in
China that were equally or more important. Hills are sacred to the Japanese as well as
to the Chinese. Nearly all are dedicated to a particular deity, and have temples to which
pilgrimages are made. The highest hill, and the most perfect the gods have ever made, is the
Fujiyama; and to have a copy of it in one's own garden is the best thing possible (Fig. 570*a*).

A legend says that Fujiyama (Fig. 570*b*) rose in one night, together with the Biwa Lake, for the gods piled it up with the earth obtained from the lake basin. This hill forms a lofty example for the enduring work of a gardener; lake and hill serve one another.

The first Japanese treatise on the art of gardening appeared at the beginning of the thirteenth century, and was a kind of moral or philosophical essay. The art is founded, according to Yoshitsune Gokyogoku, on the first principle of the prevailing philosophy of Laotse, viz. that everything in nature is split up into two corresponding parts, the male and the female, the active and the passive, the ruling and the ruled. In accordance with this principle, we must proceed in gardening as in every other form of art. Thus in a picture there is produced the effect of contrast and proportion; if there is a stone that has in it the male element, and is of a tall upright shape, there has to be a female stone as a complement, flat and lying down. Again, if on an island there is a tall spreading tree, there should be a slender lamp-post set up near it; and beside a hill of some particular shape there should be another hill of different shape. But this is not all: answering to the suggestive fancy that was adopted from China, and to the Chinese delight in symbolism, the Japanese supplied some positive meaning and definition for every single thing, all held together and controlled by regular fixed rules. In this way there came about a kind of grammar, a formal instruction in garden art, which was eagerly seized upon by the followers of Gokyogoku, and carried out in as accurate and detailed a way as possible.

The monk Soseki, a leading member of the Zen sect, and the artist-priest Soami, rank as the most important teachers and supporters of garden art in the fifteenth century. Soami makes a division of landscapes and sea-scapes into twelve principal kinds. When a garden is to be laid out, the designer must first have certain main points fixed in his mind, and these are generally marked by stones, which are therefore spoken of as the skeleton of the garden. And here we must concede to the dwellers in Eastern Asia that they have a fellow-feeling with Nature, and especially with inanimate Nature, which is unknown to Europeans. Lafcadio Hearn says that if we cannot feel with conviction that stones have a character of their own, as well as colours and values, the artistic idea of a Japanese garden will never be revealed to us. Soami classifies separately lake-stones and river-stones, stones that glide along, stones that float on the waves, stones that interrupt a river and stones that the river avoids, stones on which the water breaks and stones that stand apart from it, stones that stand up and stones that lie down, stones where water-fowl dry their feathers, stones for mandarin ducks, and Sutra stones, to enumerate only some of the most important. But later on 138 kinds, and even more, were classified as necessary for a perfect garden, though in smaller places they could be reduced to five, which must be regulated as to size and position by the area of the ground. By the scheme of male and female—and some of the stones and other natural objects seemed to partake of both—a happy proportion was kept. There were, moreover, as many helpful stones as were needed to characterise a landscape of river, lake, or rock. Stones were often brought from a distance when they were required to elucidate some particular landscape. If the blocks of rock intended for a large place were too heavy for transport, they were broken up on the spot, and set up again with cement in the park, of exactly the same size and shape.

The garden artist had to take into account the suitability of the ground and the inclination of the owner before he chose between the two chief kinds of garden, the mountainous and the flat. In large places, belonging to one of the Daimios, or terri-

FIG. 571. DESIGN FOR A HILL-GARDEN IN JAPAN

torial nobles, the first of these styles was as a rule used for the principal view, which could be seen from the chief reception-rooms of the castle. But there was plenty of choice as to whether the landscape should be a combination of land or of lake features. (In the latter case the Seiko—the Chinese Hsi-hu or Western Lake—was always the ideal in Japan.) A number of subdivisions were ready for selection. The style of sea and rock demands a great waterfall, firs beaten by the storm, rocks, and the stones of a beach. The style of the wide river, on the other hand, demands a smaller waterfall, stones such as are found in rivers, and an actual river widening out into a lake. The style of the natural stream demands a mountain brook and a small pool; whereas the style of waves must have no island, but many water-plants and only a few stones. The style of the marshy grass calls for hills that are round, like dunes, with flat stones. On one side of the sea there must be heath-land or moor, and beside the water either willows or gnarled plum-trees.

It must be borne in mind that in Japan all forms of art, including that of gardening, show three grades in execution. There is the perfect and completed style, the intermediate, and that which is merely sketchy. For every sort of garden a particular plan has to be made. Thus for the completed style of the mountain garden there must be at least five hills (Fig. 571). The middle part of it is occupied by a broad hill declining on both sides (1), of which the original model is Fujiyama; 2 is its complementary picture, lower and quite near to it; 3 is on the side that faces 1, and stands in the foreground with a valley between, thickly planted, and either actually containing, or at least suggesting a hidden stream; 4 is entirely in front, and emphasises the hill-landscape; and with the same object there is a fifth hill in the background, so as to give the necessary impression of depth, between the two chief hills.

In a garden of this kind there have to be ten principal stones with their proper functions and names. The largest, 1, is the Watchman Stone; 2, its feminine counterpart, is on the other side of the waterfall. The Stone of Adoration is generally situated on an island, and the Stone of the Complete View is either in front or on one side, and shows by its position where the finest outlook can be obtained over the garden. In the same way all the rest of the stones have their places and names, which are thoroughly understood.

Correspondingly, the chief trees are easily recognised by Japanese eyes; 1 is a stately oak or some other leafy tree set in the middle, and has to be perfect in growth and beauty, so as to attract men's gaze to it first; 2 is the favourite pine-tree, which stands on an island; 3 is the Tree of Loneliness. There is also the Tree of the Cascade, and so on. A maple is generally selected for the Tree of the Setting Sun (5), and is placed so far to the west that as the sun sinks it shines through the red leaves. Every perfect garden must have seven special trees such as these.

There must always be water also, whatever kind it may happen to be. A lake must take the form of a tortoise, or of a crane, or some other definite shape. The main picture must always be brightened by a waterfall. Running water must take the direction of east to west and should flow swiftly. The water-basins and the waterfalls are mostly placed near the house, so that they can be used for people to wash their hands. When water does not

FIG. 572. A JAPANESE DESIGN FOR A LEVEL GARDEN

arise from the lake, there must be a spring; but if that is not forthcoming, there has to be a path which loses itself, and so suggests that the source is a long way off. Where no water at all can be got, it can at any rate be suggested symbolically by overhanging trees, and by river stones, or by a bed of firm sand. Lafcadio Hearn describes a certain garden which was almost entirely sand and stones. He says that the effect the artist intended was an approach to the sea over a chain of sand dunes.

The other inevitable ornament of every garden of this kind is the bridge, which is no less common in Japan than it is in China, but here there is more restriction about its

FIG. 573. A WASHING-BASIN BESIDE THE VERANDA

treatment and situation. The bridge is always to be found, when there is an island, with trees and lamps. As to the lamp, it appears in its special garden form because of tradition. It really seems to have been of purely Japanese origin, and is not found in Chinese gardens. It has been imported into the gardens of the laity from Buddhist temple-gardens, where lamps are often arranged in rows as votive offerings. These lamps are not used for lighting as a rule; indeed, they are very seldom lighted. By their ornamentation and symbolic signs they are apparently meant to stand for something sacred, for some sort of religious feeling.

In the older Japanese gardens there are no grass lawns, and the ground in front up to the lake is only firmly trodden earth, kept damp, or else strewn with fine white sand, often marked out in ornamental shapes. In neither case may it be trodden on, so there are stepping-stones made of irregular blocks in winding patterns, which lead from the veranda to the bridge or to some other particular spot. The footpaths themselves are made either of stones or firm sand. Ornamental bars and gates are of more importance in the two other grades of gardens, which are content with fewer cardinal points in their lay-out (Fig. 572). Charming bamboo screens are nearly always used as an ornament close to the veranda, and here there have to be washing-basins, which are decorated and placed among trees, stones, and lamps (Fig. 573). Another ornament, also found in China in a somewhat different form, is the so-called Torii. These are wooden posts with two cross-beams, of which the upper one is curved. Like the lamps, they serve no useful purpose, and are often set up in long rows. At one time they were probably intended to ornament the ends of bridges. They came into gardens out of the Shinto temples, but their religious meaning has never been explained, and they have never been used as gates. In the gardens they may perhaps serve as a reminder of something that had a religious nature, such as a temple.

The gardens of the less exalted grades, those of the middle and the sketchy styles, differ in that they have, as mentioned above, a smaller number of fixed cardinal points to serve the artist when he has to make a garden of smaller size, or of a simpler kind, to suit the rooms in the house. Aiming at this change, he has at his disposal the second group of the so-called level gardens, for which, moreover, a threefold expression is ready to his hand. However simple the place might be, every garden had to reckon with the meanings of hills, stones, and trees; and with the necessity of exciting the imagination of the owner in a language of suggestive symbolism. For "it is the most striking gift of the Japanese that they can seize upon the fundamental characteristic objects in Nature and reproduce them in those smaller writings and pictures which serve the ends of decorative art." It is the artist's business also, bound as he is by the limitations of an obviously rigid rule,

not only to realise all the possibilities of imitating actual landscape, but always to create some special meaning that suits the taste, the station, or the profession of the owner, and to impress this meaning on his whole picture. It is stated of the gardens which are famous in history that they have represented Calm Retirement, Happiness, Age, Wedded Love, The Book of Change, and many other things.

The Japanese liked to look on purely poetic or historical scenes with imaginative eyes, even when he was in his own garden. To behold Elysian isles, for example, he had only to make a lake, with water-lilies on an island to which his own bridge would take him; and with the addition perhaps of an old fountain the scene would be complete.

FIG. 574. JAPANESE DWARF TREES

A garden monument, shaded by a group of firs, would suggest to his mind some specially sacred temple. Thus his garden is not a mere picture to him, but he is also able to converse with it in the language of the poets. In every larger garden there is a series of pictures, a series of views such as there are in China, and they have to express their general unity by way of contrasts. It was a favourite custom to connect eight different scenes, answering to the Hak-kei or eight views, which made particular points in Japan so famous. Other gardens would often show far more than eight. In one famous park at Tokio, which was destroyed in 1867, there were the thirty-six views which took the travellers so greatly by surprise on the way there from Kyoto. So faithfully were these copies made that "a tour through the park was just like a journey from one capital city to another."

The miniature gardens, which had come about, as in China, through the cultivation of dwarf plants, became by degrees objects of extreme luxury and value. Like dwarf trees in pots (Fig. 574), the tiny landscapes became almost priceless, according to our ideas. Early visitors to Japan spoke of them: Kämpfer saw growing together in one box four inches long, one and a half inches wide, and six inches high, a bamboo cane, a pine-tree, and a

flowering plum-tree, which together were valued at two thousand marks, and this was not
an exceptional price. In 1910 the Japanese Government presented to the City of London
two transportable miniature gardens as a most precious gift, one of them a hill-garden,
the other a tea-garden. According to an inscription the trees varied in age from thirty to
a hundred and fifty years; the rocks and stones had been brought from different parts of
Japan, and so had all the distinctive marks of a place remarkable for its beauty. The palaces
and shrines had been modelled in exact conformity to the ancient style, which went back
two hundred or even five hundred years. This was read by the London public on the
inscription, and they gazed with astonishment on these apparent playthings.

A perfectly new style made its appearance in the fifteenth century. This was the
so-called Tea-Cult, which was a reaction against the luxury then spreading in the
palaces of the rich and great, and was connected with a ceremonial partly æsthetic and
partly moral. The word *Cha-no-yu* means literally "tea-with-water," and is meant to
express in a veiled way the fundamental idea of this ceremonial, that is, an education in
simplicity of life. To attain this end there was a ceremonial that went beyond all European
ideas, a Tabulature, the study of which might be the task of a lifetime. Its full significa-
tion is for ever the secret of the Eastern-Asiatic mind. In this case also the germ, the
essential element in the ceremonial, was imported from China, together with the tea.
Not that tea-plants and the use of them were altogether unknown to the Japanese, for as
early as the twelfth century the choice kinds of tea had been brought over by a Buddhist
monk belonging to the Zen sect. But cultivation went on so slowly in Japan at first that the
finest present one could give to great warriors was a chest of Chinese tea.

To bring the ceremonial into relation with European ideas, it has been compared
with freemasonry, and in both cases the point is to educate people in certain virtues by
wrapping these up in a covering of quaint ceremonial. The Japanese consider the cardinal
virtues to be politeness, kindness, purity, and equanimity. It was towards the end of the
fifteenth century that the ways of the feudal nobility in the country began to approach
the more refined manners of the court nobility, which had been centuries ahead of them.
These ceremonial tea-meetings, to which, as in freemasonry, only men were admitted,
supplied a ground on which the two classes could meet. The warrior class, who, as following
the severe and simple customs of the preaching sect of the Zen, had so far only respected
such virtues as ascetic practices, were readily inclined to exercise them in connection with
a ceremonial which was on the lines of extreme refinement. The court nobility, who on
their side were accustomed to ceremonial, were prepared to bring within their sphere of
influence something that was lofty and moral, having up to this time worn ceremonial
only as a cloak for extravagance and dissolute living. Allied with all this, and arising from
the aristocratic side, there was an æsthetic sort of education coming from the numerous
objects required for ceremonial purposes.

The tea-meetings were originally very simple. Special and rare kinds of tea were put
before a friendly group, who were expected to use their delicate, aristocratic sense of
taste to classify the different kinds. By the end of the fifteenth century, however, this
had grown into a highly complicated ritual. The combination of severe simplicity with
æsthetic refinement of taste, of an outward conformity to ceremony (which ruled every
smallest movement) with an education in moral freedom and complete submersion in
Buddhism—all this must have made a special appeal to the Japanese nature.

Like the Chinese, the Japanese is fond of setting some particular task at his parties. At one time it was the fashion to write verses, and this became almost a mania. Before tea-tasting came in there was a fancy for distinguishing between different kinds of agreeable scents. But there had been nothing so famous as the tea ceremonial became. In its long history there appear the names of all the well-known artists, who were also the instructors; and of the heroes of war, who were pupils and furthered the movement. Even at the present day, though much has been simplified and reorganised, the cult is a great force. The house or pavilion, where the host received his guests, was strictly regulated. The size of the two rooms included in the pavilion varied according to the importance of the

FIG. 575. A TEA-HOUSE GARDEN FOR THE CEREMONIES

different teachers, who were really the rivals of the great artists and warriors; and everything was fixed and arranged, from the size of the rooms down to the thickness of the window-bars and the number of nails in the doors. Then the garden had to correspond to the house, which in its simplicity was only meant to be a kind of symbol of a dwelling-house (Fig. 575).

We have observed already how the garden had to accommodate itself to the places from which it was overlooked, the picture being made to suit the sort of rooms and their arrangement, according as this was simple or grand. The garden at the tea-house was generally divided into an outer and an inner garden. The outer one contained a little hall, where the guests had to wait and to change their clothes. In this place there were only the most necessary objects, such as washing-basins. There was a path of stepping-stones leading past a few bushes into the inner garden. Here the Japanese could indulge themselves with their signs and symbols. Here they might express such ideas as Feeling

FIG. 576. A JAPANESE TEA-HOUSE GARDEN

for the Country, Humility, Simplicity, even Gloom—and all these must be conjoined with an immaculate purity. The fencing of the garden must be of a very slight and delicate kind, unless it was desired to produce the impression of melancholy and sadness, and in that case walls of earth were thrown up.

The plan of imitating famous landscapes was not given up in these places. The Tamagawa tea-garden is so called after one of the six great rivers of Japan, but it sufficed to have a clear stream that was winding, and was spanned by several bridges. Provided the chief stone, the Watchman Stone, was there, also a good lamp, a vessel for water, and a few trees and bushes, the landscape was complete. To represent a mountain moor, only a bank of stones, some grass and a few moorland plants were required; for the mind of the spectator at once grasped the scene that the artist intended to convey. Conder gives an account of a garden at a tea-house (Fig. 576), which presents a scene near Fujisan; its river is the Fujikawa, and the pines on the far side of the hedge indicate the Miopine forest. Another river is hinted at by flowers that grow near it; and so on.

The first person who offered a site for the tea ceremonial at his own house and garden was Yoshimasa, one of the Shoguns of the Ashikaga, under whose patronage the arts of Japan (again influenced by China) attained to their highest glory. He was the patron of Soami, the painter and garden artist, whom he commissioned to design the famous Silver Pavilion after he had himself retired from the cares of state in 1472. This place was the wonder of its age, and traces of its beauty remain to the present day. Soami made a whole series of different scenes, which he called by names that give a good idea of the aims of his art. Some of these are: The Law of the Waters, The Roar of the Storm, The Soul of Scent, The Gate of the Dragon, The Bridge of the Mountain Spirits, The Valley of Golden Sand, and The Hill that beholds the Moon. The last is the name of a view-point that is still excellent for observing the effect of the moon upon the landscape.

Soami drew the attention of his patron to the priest of the Zen sect, Shukô, who was the most distinguished teacher and professor of Cha-no-yu, and who saw in this cult an occasion and opportunity that would lead to complete absorption in religion. Yoshimasa seized on the new teaching with all eagerness, and had the first tea-house and its proper garden set up in his park under the direction of the master. He decided on the name Shukô-an, and wrote it with his own hand on the shield fixed above the pavilion. At one stroke he hereby brought to his side the whole of the aristocratic class, and it is obvious from Japanese history that these curious ceremonies had the power of inspiring and stirring to action men of all conditions. It is very easy to see how favourable to garden art the feudal system would be, and this was continually growing stronger in the centuries which followed. It is also very clear that the surprising adaptability of Japanese gardening to the size of the place and the means of its owner was greatly assisted by an organisation so strict and so regulated.

In the epoch of Tugugawa the peaceful art of gardening had been able to extend

FIG. 577. A DAIMIO GARDEN IN JAPAN

widely, ever since at the beginning of the seventeenth century the important Shogun Iyeyasu had completed the feudal constitution of the Japanese kingdom, a matter of great moment, after a long period of warfare bringing in a time of peace. It was not that any special ideas came to the front, but the general regulation of home life that was established under the military feudalism offered a fine opportunity for further development. Every Japanese baron, or Daimio, had his Joka, his residence, surrounded by moats. The Shiro, his palace, was generally on a hill, with a large garden adjoining it. Even now a great number of these Daimio gardens are in existence (Fig. 577), although unfortunately such of the land as was left after the earthquake in 1867 is constantly subsiding. There was a high wall round the castle estate, and the houses of the Samurai (knights) were next to it, each of them with a well-kept garden of its own. The whole of the Yashiki (upper town) adjoined the Machī (lower town) with the homes of hand-workers and tradesmen. In this part the poor people had to be content with one court, in which there was probably nothing but a couple of dwarfed fan-palms; but the richer persons even there had houses with gardens, and in a merchant's home the so-called flat garden style was preferred as a rule. The same arrangements were repeated at the capital, Yeddo (Tokio).

In making his plans Iyeyasu had had the ingenious idea of obliging the Daimios to stay for a while every other year at Yeddo, attending upon the Shogun at the same time. Politically this meant no less than the informal supervision of the restless country magnates, and the suppression of every personal desire for independence. When a Daimio lived at Yeddo, his relationship to the Shogun was exactly the same as that of the Samurai to himself when he was in the country; and this was apparent in the actual method of living. For the residence of the Shogun throned it in the middle. Within its walls there was no want of park lands. Indeed, even the huge earth embankments with their deep trenches, which excite the surprise and amazement of modern spectators, were planted with dark pine-trees by the Japanese, who cannot forgo their pleasure in Nature; and these trees make a good contrast with the light green of the Korea grass that covers the slopes. Round this castle the Daimios set up their temporary residences, which often embraced wide stretches of land in parks and buildings, and changed the capital into a beautiful city of gardens; the owners vied with one another in adorning their houses with choice works of art.

In Lafcadio Hearn's sympathetic fashion he has given a description of a garden of the Samurai, where he lived for a time at Tokio, and we are thus shown the private home life in its natural beauty. As always in dwellings of this kind, a simple garden gate leads in from the street, and the house is only separated from the castle by a wall. It is airy, and has only one story. At the entrance to almost every Samurai house stands a tree, called Tegastriva, having a small trunk but large leaves, and this has its own symbolic meaning and legend, bringing the house security and blessing. "Gardens are on three sides of the house, and wide verandas overshadow them; from one particular corner of a veranda two gardens can be seen at the same time. Bamboo hedges and interwoven bushes form the boundary of the three different parts; the wide openings have no gates." One of these gardens, which has a great many curious stones in it, hollowed on the top so as to hold water, contains miniature hills and old dwarfed trees. There must of course be water in a landscape of this sort, but it is only suggested by overhanging green trees and a winding river of fine, pale-yellow sand. On the sand one may not tread, but must make use of the stepping-stones provided. There are trees to protect the garden from anything destructive, and among them are five pines, which not only form the fixed centre of all this greenery, but with their inner virtue drive off evil spirits.

On the northern side is the favourite garden of Lafcadio Hearn, where there is a miniature pond round a miniature island full of curious plants, such as dwarf peach-trees, maples, and azaleas, many about a hundred years old, but none more than a foot high. We are told that from a certain part of the guest-room this garden does not look like a miniature, but like a real landscape—the bank of a lake with an island, only a very long way off. The third garden has now been allowed to grow wild. It was originally a grove of bamboos, with a spring that provided the household with ice-cold water, a small sanctuary in front, and close by a plantation of chrysanthemums supported on bamboo canes; these are still to be found there. This quiet little property is in the heart of the capital; and, like everything else, it will soon be swallowed up.

The appearance of the towns, with these seats of the mighty and their gardens, was further improved by temple gardens and the open places round the public tea-houses. Round all the towns there was an endless number of both temples and tea-houses, mostly on the hilly parts just outside. Even at the present day the chief ornaments of Tokio are really the temple gardens of Uyeno, Shiba, and Nikkô, all three the burial-places of the great Shoguns of the Tujagawa period. The remains of their founder Iyeyasu lie at Nikkô, where stone steps amid gigantic ancient trees lead up to the handsome tomb. His son founded Uyeno to the north of the town, and Shiba to the south; and six of his successors were laid to rest in each of the temples. People liked to have avenues to make a dignified approach to the tombs and the temple gates, and gradually these have grown up to be magnificent trees. In the same way they liked to put up long rows of Torii, votive lamps, and pictures of Buddha—rows so long that the eye could not reach to the end; it was one of the religious duties of the devout to count them. Belonging to the Japanese temples, and outside the shrines, there were a great many other houses, with gardens round them; these differed in no way from the gardens of the laity, and were a great ornament to the town near which they lay.

The public tea-houses rivalled the temples in their garden surroundings, and as a fact there is scarcely any difference in Japan between a temple and a tea-house. People fly to either, if they want a pleasant refuge from the noise and bustle of the town. In both are to be enjoyed the loveliest arbours, the choicest dishes, and the sweetest music. The neighbourhood of Nagasaki was described by one of Lord Elgin's suite, who says: " It has been computed that on the hills around there are 62 temples, great and small, and 750 tea-houses, all of which provide good tea and a fine view to a man who wants a rest: moss-grown steps lead up to the hill, and going by wide stairs and through grand gates you come to a place like fairyland, on a projecting point, with gardens and shady groves behind, leading to the grottoes, where the gleaming waters fall down from the hill." The enthusiastic Englishman who wrote about this place left it open to doubt whether in his account he was describing a temple or a tea-house.

It is in these public gardens that at the present day most of the beauty of the past is preserved. It is to be hoped that the Japanese will not only preserve the old glories, so far as they exist, but will guard their art from the influence of Western lands.

CHAPTER XV

THE ENGLISH LANDSCAPE GARDEN

CHAPTER XV

THE ENGLISH LANDSCAPE GARDEN

IN considering the rise of the English landscape garden, we cannot ignore Chinese influences on the art of gardening in England and in Europe generally.

This influence has often been met with before, but at the beginning it did not affect essentials in the least, only ornamental details.

After Louis XIV. had set up his Porcelain Trianon, the signal was given to erect Chinese buildings of every sort and kind in gardens. As knowledge of Chinese architecture increased, certain pleasing features were adopted, but these were carelessly mingled with native work. In the early porcelain pavilions people had in mind that wonder of the world, the porcelain tower at Nanking, even if they were only putting up a baroque pavilion with Dutch tiles, and setting in the garden vases of a colour to match.

Later on we even find slated roofs, as at Pillnitz, laid without a qualm of conscience over a baroque house! Then there came into fashion the Chinese parasols, later on Chinese bridges, and though such things had very little to do with genuine Chinese art, they served to remind Europeans of the eighteenth century of the wondrous land in the East. The pleasure taken in the Chinese parts of the park was so universal that it is impossible to mention any one of the larger French or German gardens of the period that had not at least a Chinese pavilion. But often, as in the case of the Swedish country estate of Drottningholm, and at Wilhelmshöhe, we find complete little Chinese villages, or (as at the Rheinsberg of Frederick the Great) a Chinese fisherman's hut near the Chinese pleasure-house, and also a Chinese winged court.

It is difficult to say whether the number of buildings in Chinese gardens had a direct influence on the increasing number in parks, but at any rate it is a fact that the desire felt at all times to put up in a garden buildings of an unusual character did grow in the eighteenth century to an excessive height. The chief reason must be looked for in the spirit of the time, which was opposed to the pompous publicity of Louis XIV. and his age. People now only cared to find peace, and relief from their boredom, and they sought this in circles that were small and intimate; they showed more and more markedly their longing for solitude. In spite of all this, boredom was lurking in the background; indeed it is evident that the wish for variety had grown up before now, but that it was no longer sought for in showy plantations, but rather in the restful seclusion of a pavilion or some other small place. Thus Chinese art was snatched at as one sort of variety; for every type of art was welcomed, and the more the better.

This accumulation of buildings in a park had intrinsically nothing to do with the great change of style in gardening on whose threshold we now stand. We find these buildings in both styles, side by side, in the eighteenth century; and we cannot even say that the garden of the painter adopted them from the garden of the architect: there

277

is a parallel development equally strong in both, although, as we shall see, it arose from different causes.

The straight-lined, architectural or "formal" garden style which, except in England, held undisputed sway over the whole of Europe in the first half of the eighteenth century, contributed no essentially new ideas (in earlier chapters this theme has been sufficiently laboured) although it was by no means fixed and rigid. It was able to adapt itself to the needs of the time, and especially to the conditions of German life. It was even flexible enough to adopt, after a time, all sorts of influences from the artist's garden, as for example the serpentine curve, provided that these innovations did not disturb the firm unalterable outlines of the main design, and also satisfied the desire for variety.

FIG. 578. SUPPOSED ENGLISH PARTERRE IN TRANSITION STYLE

Curious errors and misunderstandings are made by French architects, for instance, when a sketch shows a supposed English parterre (Fig. 578) laid out in front of a castle in order that the plantations may be drawn within straight lines. And so we are able to speak of a transition style (Fig. 579). Much the same happened as in the great battle between Romantic and Gothic; for after the former had reached the height of its charm in the transition period, it had to yield in the end to the great Gothic revolution. In the gardening art the victorious revolution destroyed so utterly the foundations and pillars of the old style that there was never any question of gaining the final victory by way of transition or concessions made by either party: there was nothing for it but the complete destruction of the old.

And now how do matters stand as to the influence of China on this new style? It is clear enough that there is a connection between the two when we consider their common enemy, the architectural garden. We have seen that information had reached Europe in 1685 about a new style of garden without straight lines, and Sir William Temple's tentative glance had been directed towards China, and not without sympathy; but it had been at once diverted, as from a task which, though not lacking in charm, was far too difficult. And Sir William was perfectly right; for never could a really living style have developed by way of vague imitation of the art of some utterly foreign race: it must have remained a changeling here, as indeed it was with many other Chinese imitations, architectural and artistic. True, the revolt against the old style did at first get helped by the news from

China; but the revolt actually started entirely from within, and that is why it is so very important. It is also a proud boast for England, much as people are nowadays disposed to look down on the whole movement. For in its true beginnings it was an intellectual movement; around its cradle stood poets, painters, philosophers, and critics. It was the first child of that new love for Nature which was setting itself in conscious opposition to barriers and forms.

The strange fact (overwhelming at a superficial glance) is that this movement occurred in the very heart of English classicism—that its leaders were the very same men who upheld and carried out the classical ideals in literature. To understand it properly one must grasp the essential character of English classicism, and realise how, for this nation, one influence never quite supplanted the other.

The rationalistic spirit of English society in that day had certainly come in contact with those rigid laws of form that ruled supreme in France; but Addison's admiration for Boileau by no means prevented him from being the first man to gaze with enthusiasm upon the wild lawless beauty of the folk-song. And Shaftesbury was able to pass on from his optimistic theism to the deification of untouched Nature, which, he said, is in itself good, so long as no outside hand checks or destroys it. This "extravagant love" of nature he graced with the fundamental axiom, that all healthy love and admiration is Enthusiasm; and this thought leads by a logical process to his admiration

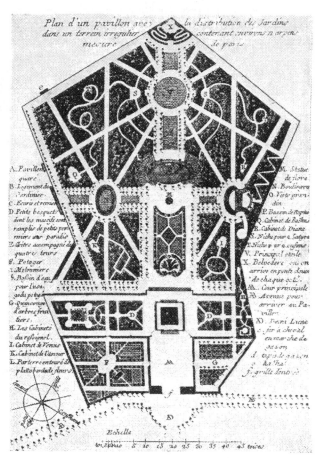

FIG. 579. PLAN OF A GARDEN IN THE TRANSITION STYLE

of open landscape as opposed to the formal gardens of his day. No more will he resist the love of Nature,

where neither Art, nor the Conceit or Caprice of man has spoiled their genuine order, by breaking in upon that primitive state. Even the rude Rocks, the mossy Caverns, the irregular unwrought Grottoes, and broken Falls of Waters, with all the horrid Graces of the Wilderness itself, as representing Nature more, will be the more engaging, and appear with a Magnificence beyond the formal Mockery of princely Gardens.

Here for the first time we find untouched Nature set in opposition to the formal garden. It is true that people in the seventeenth century were very well acquainted with the so-called open landscape; but it was the scenery of pastoral poetry, bound by conventions, and by tradition attached to the theatre of love-romances handed down from antiquity and the Middle Ages. This romantic landscape was best pictured as a garden, and could often take its place there without any change in sentiment; for what the hand of man did in a garden was the same as that which "Nature with her wondrous art" had accomplished in open country. Aye, and we shall learn to prize still more the striking merit of Shaftesbury and his like-minded English friends when we see how, decades later, everywhere outside England the feeling for Nature was inspired in the main by the artistic beauties of the garden. As we have said before, one pet idea is the identification of Nature or even God with the gardener. In the same year that Shaftesbury confesses his faith, a German poetaster sings:

> Willst du die Gartenlust des grossen Schöpfers schauen,
> So sieh den grünen Strich, der schönsten Bäume Pracht.
> Betrachte die Alleen, die bunt bemalten Auen,
> Die Grotten, die Er selbst mit eigener Hand gemacht.

[Wouldst thou behold the great Creator's joy in His garden, then look upon the stretch of green, and pride of fairest trees; gaze on the avenues, the many-coloured meadows; and look upon the grottoes that His own hand has made.]

Even travellers in foreign lands are not without their garden thoughts and fancies. In Guinea a traveller comes upon woods which, he says, "are just as flat at the top as if they had been levelled down and cut with shears." Another traveller standing on the Brocken says, "At the top of the hill are set trees in a circle, just as if they had been carefully planted, not one of them out of the proper order." And yet another in the Black Forest in 1760 sees everywhere in Nature a pattern for the ideal garden of that day. "The wood," he says, "is incomparable; in it one can see how a forest looks if left to itself for hundreds of years. There are no imaginable beautifications, fancy clumps, avenues, arbours, that one cannot find here in their untouched beauty. I was immensely delighted with the groups of firs that grew here and there in endless variety, each tree a magnificent pyramid from its base to its lofty top."

Shaftesbury's words, however, were not to die away unheard in England. A few years later Addison published in *The Spectator* of 25 June, 1712, an essay which—following Shaftesbury—purported to explain the differences between open wild nature (the landscape) and art (the garden) in their effect on the mind. In the whole essay Addison is less extreme than Shaftesbury. Though he sees in Nature more grandeur and sublimity than art can ever attain, still he says that "we find the works of Nature all the more agreeable, the nearer they approximate to the works of art," and we may feel just as sure that works of art derive their chief merit from a similarity to Nature. Addison in this his first thesis is quite on the side of Nature in combination with art, and he is able so to draw together his art and his nature that he can take the combination of the two as a ground for his campaign against the British garden, which, instead of helping Nature, has done everything it can, he says, to banish her altogether. He complains that the "trees are made to look like ninepins, like spheres, or like pyramids: we find the trace of the scissors on every bush and on every plant."

Addison travelled a great deal on the Continent. Italian gardens were often greatly neglected at that time, but this very fact gave the tall southern vegetation such a luxuriance that an eye wearied with the stiff formalities of well-kept northern gardens could only rejoice therein. Moreover, England had never seen much of the working of French art, and the Dutch soberness, prettiness, and actual limitations (partly perhaps because of the alliance between the two nations) were very much in favour. Addison set Italian and French gardens against English; in them he found something more worthy. Travellers' tales declared that the Chinese laugh at the plantations of our Europeans, which are laid out by the rule and line; because, they say, anyone may place trees in equal rows and uniform figures. They choose rather to discover the genius in trees and in Nature, and therefore always conceal their Art.

In spite of this notion, however, Addison was far from understanding the real nature of Chinese gardening art, for in a second essay, published a couple of months later in *The Spectator*, he shows what his ideal really is. A foreigner, he says, who had lost his way here, would fancy he was in a natural wilderness, a medley of kitchen-garden and parterre, of fruit-trees and flowers—the flowers growing in all sorts of places. So little does the owner value these flowers on account of their rarity, that he often likes to bring wild flowers home and plant them in his garden; and he feels it is delightful not to know, when he is taking a walk, whether the next tree he comes to will be an apple, an oak, an elm, or a pear; and he has carefully guided the little wandering stream so that it runs exactly as it would in a field between "banks of violets and primroses, plats of willow or other plant."

Addison had exclaimed at the end of his first essay:

I do not know whether I am singular in my opinion, but for my own part, I would rather look upon a tree in all its abundance and diffusions of boughs and branches, than when it is cut and trimmed into a mathematical figure; and cannot but fancy that an orchard in flower looks infinitely more delightful than all the little labyrinths of the most finished parterre.

Addison well knew that he would not long be alone in his ideas, though he might be the earliest to venture to stir public opinion by his writings, and later to dominate it.

The first man who came to his aid was Pope. He too adopted Shaftesbury's thoughts, and Addison's also in a modified form: "I believe it to be a true observation, that men of Genius, the most gifted artists, love Nature the most; for they feel very strongly, that all Art is really nothing else than an imitation of Nature."

Here again we have a sort of solution of the problem of Romance, Genius, and Nature, attempted by a classicist. Indeed the import of these words must have been something new, as they gave a fresh impulse even to poetry. Pope, however, went in search of Nature chiefly because of his opposition to the tricks of art. He pours out his witty sarcasms upon the "garden tailor" who compels tree and bush to take the shapes of beasts and men. He tells of a certain cook, who decorated his country place with a coronation feast spread over the grass; and adds a most amusing catalogue of his stock of wares: among others he finds recommended the following:

Adam and Eve in yew, Adam a little shattered by the fall of the tree of Knowledge in the great storm; Eve and the serpent very flourishing; St. George in box, his arm scarce long enough, but will be in condition to stick the Dragon by next April; a green Dragon for the same, with a tail of ground ivy for the present. (N.B. These two not to be sold separately.) Divers eminent poets in bays, somewhat blighted, to be disposed of a pennyworth. A quick-set sow, shot up into a porcupine, by its being forgot in rainy weather. . . .

Pope was determined, however, to effect more than mere jesting. He wanted to give an example; and when a few years later (1719) he moved into his villa at Twickenham on the Thames, he resolved to copy "Nature unadorned" in his own garden. Now although the Muses' seat is often to the fore in his letters and his poems, the whole place was much too small to exhibit more than a negation of the old style: no more clipping of trees, no more symmetry, that was the first command, and Pope proudly declared that two weeping willows beside the house, with a lawn in front leading down to the river, were the finest in the kingdom. His favourite of all was the grotto so often mentioned, a kind of sub-terranean tunnel leading from the front garden under the main road to the back garden. With touching childlike enjoyment the poet toils during the summer months at this darling work; his friends send him specimens of rare minerals for the walls and ceiling; he cannot find words enough to praise the many light-effects and river-views. The descriptions he gives his friends sound like the scenery in a fairy-tale, and yet it was only a grotto such as you could see in endless variety in all the earlier gardens.

Pope's gardener kept a picture of the place, grotto and all, just as it was when the master died. At the garden entrance to the grotto there are stones heaped up to imitate an ancient ruin. Very likely Pope with his sense of humour would have laughed at this some other time, just as he made merry over the minuteness of his garden, saying that "when Nebuchadnezzar was turned into an ox he could have finished the grass in one day." Certainly Pope was no practical genius, and Twickenham is only interesting as a first experiment, begun at a time when people, in spite of their appreciation of the ideas of the protagonists of the new style, could not for some decades venture to desert the old, because of the lack of examples to follow.

England also had her transition period; but whereas other countries clung to the ground-plan of the old gardens and only yielded concessions within its borders, in England the new ideas made a curious but more important attack upon the ground-plan, while the details remained much the same for a long period, again because there were no patterns to copy. In a description of the garden at Stowe in 1724, which was much admired by Pope, we are told: "Nothing is more irregular in the whole, nothing more regular in the parts, which totally differ the one from the other," and finally: "What adds to the beauty of this garden is, that it is not bounded by walls, but by a Ha-ha, which leaves you the sight of a beautiful woody country, and makes you ignorant how far the high-planted walks extend."

This new method of enclosure was due to the landscape gardener Bridgeman, who designed Stowe: it was simply a ditch which one could not see till one was quite close to it. The popular but erroneous explanation is that the name Ha-ha was given owing to an exclamation of surprise by a visitor. Sometimes there was a hedge sunk in a trench, which answered the same purpose of concealing the boundary between garden and open country.

Horace Walpole, who half a century later wrote the history of the great revolution in the garden, saw rightly in this invention a leading factor towards the victory of the new movement. The walls of an architectural garden had been its peculiar characteristic support; they shut it out from the surrounding country, so that it was alone and free in a world of its own; in French gardens also this was a leading principle, so highly esteemed was a fine prospect seen from the end of an avenue. But now it came about that there was no border-line before the eye, and the garden was just a foreground for the wide landscape beyond.

England had discovered by way of poetry, so to speak, the northern lands that were her home. By the second half of the seventeenth century there was already a species of poetry which later on Dr. Johnson called "local." This was descriptive verse, which aimed at inspiring the reader with love for an individual landscape locally limited: Waller, Denham, and Cowley had made such attempts before Pope wrote his *Windsor Forest* in 1712. Although they show very little of the "Romantic" feeling for nature, the writers know how to observe, and to depict, the characteristic features of a given landscape. In Thomson's *Seasons* we have a giant's stride in the same direction.

These poems were received with enthusiasm, which proves that the Scotchman Thomson was not the only man who went for walks, and with open eyes enjoyed the beauties of nature. He is sensitive to the peculiar northern character, and it is no doubt typical that he begins with Winter, a season which only in the North can express its whole strength and beauty. Englishmen had before now been lovers of long walks, and this was why at the end of the seventeenth century the garden paths were laid specially with a view to this fancy for walking about, and it is worth noticing that the garden itself was intended and designed to be a place for walks: now, however, when the boundaries had gone that cut it off from the surrounding country, people felt with increasing impatience the contrast between the artificial "artistry" inside and the natural landscape. There they saw meadows enclosed by bushes following the natural curves of river and brook, or the magnificently grown single trees, or again a picturesque group, and on these they gazed with happy eyes and loving hearts. Pope, who was always ready to do battle for the new creed, gave expression to the feeling of oppressiveness associated with the enclosed garden in his epistle to Lord Burlington (1730), in which he lashes at "Timon's Villa," the showy, tasteless, conventionally regular garden, in lines that soon, like winged words, were in the mouths of all men:

> On every side you look, behold the wall.
> Grove nods on grove, each alley has a brother,
> And half the platform just reflects the other.
> The suffering eye inverted nature sees,
> Trees cut to statues, statues thick as trees. . . .
> Here Amphitrite strays through myrtle bowers,
> There gladiators fight or die in flowers.

And elsewhere he says that all the rules of the art may be reduced to three, viz. Contrast (including picturesque effects of light and shade), Surprise, and Concealment of Boundary.

Lord Burlington, perhaps moved by the sarcasms of Pope, was one of the first to make all sorts of experiments in his own garden; and these are evidence of the unsettled state of things. The plan (Fig. 580) shows, as Stowe showed, no regularity of outlines, but it fails to avoid in its details the crazy serpentine twists and turns. It was out of the enemy camp that auxiliary troops had to come, to straighten the wavy lines and to furnish a model.

A little later (1732) Joseph Spence began his *Polymetis*, an attempt to remove the barriers between the arts, especially emphasising the close connection between painting and poetry. Spence was an enthusiastic gardener himself, spending all his spare time in the garden, where he was actually working at the time of his sudden death. To him horticulture seemed a natural branch of painting, and thus the garden artist was a garden poet as well.

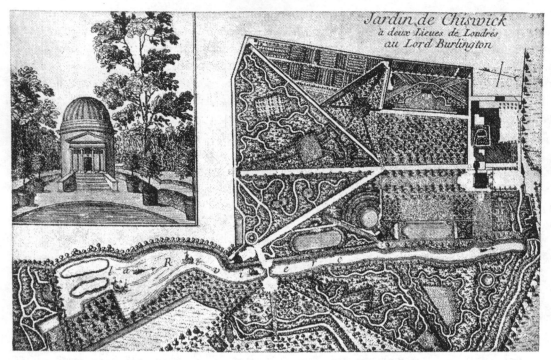

FIG. 580. AN OLD GARDEN AT CHISWICK, NEAR LONDON

The ever-increasing dislike which was felt by men of that time for regular forms, and still more for symmetry, is singularly marked in art criticism, which makes use of the dogmas that were fashionable, and declares war on regularity of every kind. In 1745 Hogarth drew (on the palette he is holding in a portrait of himself) an undulating line, which later in his theoretical work, *The Analysis of Beauty*, became the famous "Line of Beauty." Hogarth tried to prove that beauty is not a *je ne sais quoi*, but a clear and positive quality in things. The formula for the highest beauty is a waving line, which at no two points is the same; it shows the utmost variety, and has an advantage over the line of the circle, in that it stimulates the imagination by disappearing from sight and again reappearing.

Edmund Burke soon took up this idea of Hogarth's, and whereas Hogarth had only refused to accept the earlier view that regularity and symmetry were in the essence of beauty, Burke declared war on symmetry altogether, which no more than regularity had any relation to beauty as such. They are not to be found in nature; it is only man who has had the unlucky inclination to confine his view within them. The best example Burke can find is the old style of horticulture.

Having observed that their dwellings were most commodious and firm when they were thrown into regular figures, with parts answerable to each other, they transferred these ideas to their gardens; they turned their trees into pillars, pyramids, and obelisks; they formed their hedges into so many green walls, and fashioned their walks into squares, triangles, and other mathematical figures, with exactness and symmetry; and they thought if they were not imitating, they were at least improving nature, and teaching her to know her business. But nature has at last escaped from their discipline and their fetters; and our gardens, if nothing else, declare, we begin to feel, that mathematical ideas are not the true measures of beauty.

In place of incongruities like these, Burke holds that smoothness is the determining character of beauty—so much so that he knows of nothing beautiful that is not also smooth. Burke did not appeal in vain to horticulture, which from this school of thought was to encounter an influence by which it was stimulated at first, but alas, too soon devitalised.

In consequence of all these theories and tendencies of the age, the Landscape Garden came into being. The first man who in practice attempted to "pluck the ripe fruit" was William Kent; and little as he was able to accomplish in this form of art, he saw how to imitate those models which were required by the new style, if it was to be freed from its fetters by landscape gardening. England, to be sure, was by no means opposed to the imitation in pictures of the kind of landscape that the poets had revealed; but the Continent was well ahead with its great landscape painters: Claude Lorraine, Salvator Rosa and Poussin for the South; Everdingen, Ruysdael, and other Netherlanders for the North, though the English were beginning to study the work of these men with enthusiasm. If at first Claude and Poussin were preferred to the Northerners, it was because their fine, well-kept, formal landscapes were more suited to the sentiment of an age which approved Addison's dictum, that Nature was at her happiest when she came nearest to Art.

Kent himself, who was only a coach-painter before his patron, the Earl of Burlington, noticed him and sent him to Italy, learned in the South, if not actually to paint pictures, at any rate to see things, and compare them. It occurred to him, as it had done before to Addison, that the style of Italian gardens was not nearly so out of character with the landscape as was that of the North. He saw how the painters of the South had often taken an actual garden as subject for a picture because they admired it; and the merit of Kent is that he did not, full of this idea, rush on to a further imitation of the same things; but when, after his return, he had acquired an authoritative position in matters of taste, he was the first man to lay out a garden in an unfettered artistic manner, taking his ideas from the surrounding landscape. His motto was, "Nature abhors straight lines." It is easy to see how hereby he was making war on the old style, not only in regard to the ground-plan but in every detail. Straight paths were to be carefully avoided, all water-works, even fountains, were tabooed; only a lake with irregular banks, or a river that flowed like a snake through the grounds, might remain. The artist's scheme of light and shade was expressed in trees and bushes: they were allowed to grow freely but were planted according to Pope's rule of contrast, and gardens were like pictures in this respect also, that perspective had to be taken into account, so that one way led on to a second view, or even to a third. The lawn, which always played an important part in the English garden, often spreading out close in front of the windows of the house in a square of dark green with no path through it, is required by Bacon and even by the theorists of the early eighteenth century to serve as a carpet contrasting in colour with groups of bushes and trees.

Contemporary with Kent and living after him was another garden artist, Lancelot Brown—a great name in the middle years of the century. Brown was inspired with a veritable passion for rooting out the "unnatural bad taste" of the old style, which in the previous forty years had captivated a people so very conservative and sedate. The old pretty gardens, whose features are preserved in the collections of Kip, Atkyns, and others, were in the next thirty or forty years completely transformed; and to-day it is all but impossible to find any trace of them, especially in the Midlands and the South. Brown was the original advocate of Hogarth's wavy "line of beauty," which he must have in every part of the garden. Even the ground must have its gently-waving contours; terraces are abolished as unnatural. The paths leading to special views are real examples of the line of beauty. In particular, the path which went right round the whole park and was spoken of as "The Belt," was meant to give to the estate an effect of greater size because of its countless windings.

The laying-out of the later gardens at Stowe (Fig. 581), chiefly the work of Brown, is a masterpiece of this kind. But his chief strength was in the water plans: he was the first to give movement to a lake (later this was greatly exaggerated) by cutting up its banks into creeks and curves; rivers he treated in the same way. Once Brown was so enchanted with his river banks, which he thought surpassed the Thames in beauty, that he is said to have exclaimed, "Oh Thames, Thames, never will you forgive me." Certain scoffers nicknamed him "Capability Brown," because he was for ever talking about the "capabilities" of his garden grounds; but his vanity seems to have taken kindly to the epithet.

Flowers found no place to speak of in this type of pleasure-ground, especially in the earlier days. They took their place in the background, as they had done before in England, i.e. in the kitchen-gardens, which, in spite of the mania for destruction, had been allowed to keep to their own enclosures, with high brick walls round the regular beds. And so we still find them.

In such a state of things the boundary-lines between the pleasure-grounds and the open park-lands tended to disappear. The garden writers and critics have, it is true, spoken of an ornamental plot close beside the house; but as there were no barriers, and a change was bound to come, though gradually, gardens of this type were felt to be less permanent and therefore less attractive.

On the other hand, in this style of garden the individual tree found for the first time its proper place and full development. It is well worth noticing how, simultaneously with the advance of the artistic school, there was a sudden and steady influx of different sorts of American wood. Botanical science did not as yet exercise a directing influence, and it was not till the nineteenth century that it became a real guide. But the acclimatisation of these foreign shrubs and trees, with their beautiful foliage, their noble growth, their elaborate ramification, was an immediate consequence of the change in fashion, as is shown by their universal popularity. This acclimatisation was confined to England until the middle of the century, and then, when the picturesque style reached the Continent, acclimatisation suddenly made its appearance there also.

In the course of the eighteenth century the English garden took on still more the character which we now call park-like. The so-called "enclosures," fenced-in fields and meadow-land, which had been little by little reclaimed from common land and had become private property, were for the most part separated from their neighbours by hedges and ditches: instead of very small enclosed bits of woodland, trees were planted as landmarks, standing separately, but growing well in the moist climate. True, during the whole of the eighteenth century, the English were a farming folk, but the individual farms were mostly small ones, with a few fields. Because of the closer approach of garden to park, and of park to open nature, there arose the idea of beautifying

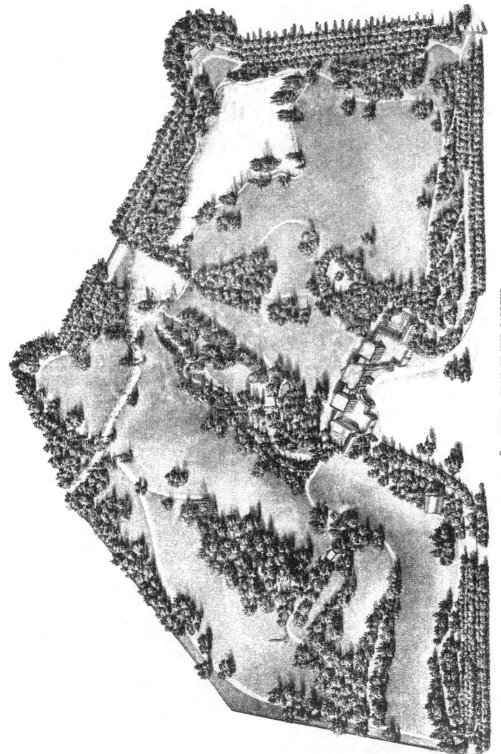

FIG. 581. STOWE—PLAN OF THE GARDENS

a whole property, an estate, and of subjecting it to the rules of a regular garden scheme, yet without making any sacrifice of its usefulness.

At this stage of English landscape gardening the poet William Shenstone played an important part. In the year 1745 he laid out his paternal estate, The Leasowes—the name means pasture-land—in the new style. Dr. Johnson calls this the life-work of Shenstone, a poet who was a forerunner of the Romantic movement. Shenstone at once began, says Dr. Johnson, "to point his prospects, to diversify his surface, to entangle his walks, and to wind his waters; which he did with such judgment and such fancy as made his little domain the envy of the great and the admiration of the skilful; a place to be visited by travellers, and copied by designers."

There is hardly any garden tract (for so we have to call the place) of which we have such full descriptions as of The Leasowes. The poet, who invested the greater part of his fortune in this place, discusses the principles of his art in an essay called *Random Thoughts on the Art of Gardening,* only published after his death in 1764. Shenstone seems to have been the first person to use the term "landscape-gardener," and he certainly expressed hereby a close relationship with the artist. "I have used the term 'landscape-gardener,'" he says, "because in accordance with our present-day taste, every good landscape painter is the proper designer of gardens." If we put this beside Shenstone's other dictum, "Man must never venture to tread in Nature's domain," and add to that the words of the German writer Hirschfeld, "The garden artist works at his best when he discards all that the architect reveres," we shall see how utterly opposed are the ideals of the picturesque garden and the architectural.

The importance of The Leasowes lies in the fact that here Shenstone has materialised the idea that "the garden is no longer limited to the place from which it takes its name, but makes subject to its own laws both the laying-out of a place and the embellishment of the park, the farm, and the cart-track." The descriptions take us by dark woody valleys to open heights with lovely views, over pastures and cornfields to a hidden boat-house and rushing waterfall, or by the winding banks of lake or river. Close beside the dwelling-house are these scenes with contrast and change as their leading motives.

Thomas Whately, one of the chief critics, blames the farm for venturing so close to the house, because the landlords have to be too near to the tenants. But he admires greatly the invention of the so-called "ornamented farm," which beautifies even the fields devoted to kitchen produce with the gay attire of the garden. He gives a description of Wobury Farm in Surrey: everywhere there are seats to be found on the walks, and buildings of many kinds; round the cornfields there are rose hedges, with little clumps of all sorts of flowers at the corners. And though the different parts are so gay and garden-like, they are all open for farm use: the cattle are grazing, the sheep are bleating, the fields are tilled and the crops harvested. With some hesitation Whately has to admit that with all this beauty and charm the simple country-like character of the farm is lost.

The English of those days were well aware that for the first time they had done something original in the province of gardening, and had also advanced in the art of painting.

The writer Henry Home, Lord Kames, says that his countrymen "are still far from perfection in the fine arts, but are on the right way, though making very slow progress except in gardening." Whately ascribes a wonderful importance to gardening among the pictorial arts, and even sets it above painting in so far as he rates all reality higher than imitation.

The poet Gray writes:

> The only proof of our original talent in matter of pleasure is our skill in gardening and the laying-out of grounds. And this is no small honour to us, since neither Italy nor France has ever had the least notion of it, nor yet do the least comprehend it when they see it. It is very certain, we copied nothing from them, nor had anything but nature for our model. [It is not forty years since] the art was born among us; and it is sure that there was nothing in Europe like it, and as sure, we then had no information on this head from China at all.

The history of the movement, as we have already seen, fully justifies these words of Gray. It was the offspring of the sentiment of the earlier half of the eighteenth century, and England was fully alive to its existence before Père Attiret sent his report to France in 1747, with the description of Yuen-ming-yuen, the Chinese emperor's marvellous pleasure-castle. The fashion for China, with its various effects on the art of the eighteenth century, was perhaps less marked in England than in other countries, but in garden art it is clearly traceable. From Sir William Temple we discover that the news about the unsymmetrical gardens of China had not escaped notice; but when Addison enlists this vague information as an auxiliary force in his campaign against the old style, it is not with the desire to imitate China. So also in the descriptions of the transition style, as seen in the early gardens of Kent and Brown, there is no mention of Chinese pavilions, though there is a great deal said about other kinds of summer-houses. If therefore after the middle of the century the cry is everywhere heard, "the English-Chinese garden," this must needs be explained by what happened in France.

France had so far been quietly adapting the baroque gardens of her country to the requirements of a rococo style, and meanwhile was exercising an effective influence upon Europe generally. The few voices that were raised against Versailles, Saint-Simon's for example, were really only objections to the king, Louis XIV. The first who showed a decidedly hostile feeling to the conventional style was Langier in his *Essay on Architecture*, which was published in 1753. Père Attiret's account, which appeared in France in 1747, was first translated into English about 1752. It had had time to work its effects in France, and to be compared with the ideas that came flooding in from England. French people were astonished by the similarity of the leading thoughts, and it was no wonder that they (who were in the very centre of the Chinese fashion) assumed that England had taken the whole novel idea about horticulture from this much-belauded China. So France began to adopt the new style under the name of Anglo-Chinese. This term was not altogether without justification in the second half of the century, for one important current of the picturesque style began to make a delicate approach to the Chinese, and no wonder, for Père Attiret's verbal descriptions were now reinforced by pictorial evidence. The Emperor of China had engravings made of his gardens, and sent the pictures to the French court. Moreover, travellers brought other pictures to Europe, and these were engraved by skilled Belgians and Frenchmen who had a marvellous feeling for style, and then were sent out into the world as models.

A French publication, *Le Jardin Anglo-Chinois*, which was published 1770–87, gives (among illustrations of many European gardens) over a hundred striking drawings of the gardens of the Chinese emperor. Some of the originals were the property of the French king, but for the most part they were brought over by the Swedish Ambassador Cheffer (Fig. 582). The work is explicitly intended for the furtherance of horticultural art, because "everyone knows that English gardens are only imitations of the Chinese."

The engravers had at any rate a more sympathetic feeling for their own subject-matter than the editors of the whole series had; for in the sixteen parts one finds a miscellaneous collection of every style, bearing witness to the instability of French art and its experimental nature. In one thing all seem alike, that is, in their love for piling up buildings

FIG. 582. A CHINESE GARDEN IN EUROPE

in the garden. There is a little book, called *Livre des Trophées Chinoises*, with illustrations of the newest fashion in gardens, full of scrolls and flourishes that people were pleased to call Chinese. Similar publications appeared in England, where the two brothers Half-penny, who called themselves "architectural joiners," became well known. If one looks at these pages, one clearly perceives that the two styles were combined because there was a very poor understanding of either. The fancy for Gothic architecture, which for hundreds of years had been the last word in bad taste, appeared in England rather early, and in 1747 an anonymous book came out called *Gothic Architecture*, with sixty-two illustrations but wanting text. But the "shelters, porches, and pavilions, which complete the view" are hopelessly like Halfpenny's Chinese objects (Fig. 583). In both we get the twirls and flourishes that were used for decorating roofs, balconies, and window-frames. Goethe, moreover, in his *Triumph der Empfindsamkeit* gives the name of "Chinese-Gothic" to the picturesque grottoes of the English garden.

The real influence on English landscape gardening that came from China is connected with the name of Sir William Chambers. In his early life he was in the service of the Swedish East India Company in China. There he made a series of sketches of Chinese buildings, costumes, etc., which he published in 1757, with the intention of showing the real thing as opposed to the pseudo-Chinese. After his return to Europe he travelled a long time in Italy to prepare for an architectural career, and like Addison and Kent, he could not resist the charm of the Italian gardens.

As an Englishman, Chambers was too much captivated by the new ideas not to be fundamentally opposed to the old style; but he compared the wealth shown on the one hand by the southern continental gardens, and on the other hand by the real or only pictured beauties of the gardens in China, with the ever-increasing uniformity in the new style at home. Much had been done in various ways to make the new gardens look empty, one difficulty being that the thickly planted groups of trees were still very young, and the places appeared bare, since the chief effect of the whole picture depended on the beauty of these plantations; again there were not nearly enough men trained to the difficult work of laying out a garden picturesquely. The helpful architectural and plastic ornamentation was now confined to summer-houses, bridges, and the like.

FIG. 583. A GARDEN DECORATION IN THE GOTHIC-CHINESE STYLE

At that time the garden in England was so far opposed to the Chinese habit of overcrowding its buildings, that Chambers wrote his famous essay, *On Oriental Gardening*, as a direct protest against the devastation and emptiness in the gardens of the Homeland. He urges that artists and connoisseurs both lay too much stress on Nature and Simplicity; and that this is the cry of half-educated chatterers, a sort of refrain lulling them into a lazy condition and complete want of taste. He goes so far as to say that if likeness to Nature is to be adopted as the measure of perfection, we must confess that the wax figures in Fleet Street are superior to the divine works of Buonarotti.

In spite of this Chambers would not say a good word for the old style, and especially jeers at Dutch gardens, calling them "cities of verdure." What is needed, he thinks, is the example of a land like China. The Chinese also take Nature as their model, and copy her lovely irregularity, but the main reason of their success lies in the fact that they demand a long training for their garden artists, so that the effect of their taste is

FIG. 584. THE PAGODA IN THE ROYAL GARDENS, KEW

visible in the whole scheme; whereas on the continent of Europe horticulture is a secondary affair for the architect, and in England is even relegated to the kitchen-gardener. But the Chinese have collected every charming thing for their gardens, lavishly bestowing the beauties that love or money can buy, though they still remain the humble servants of Nature. This Chambers demands for the English gardens, especially emphasising the need for contrast. He declares that the spectator must be amused all the time, and his powers of observation kept awake; his curiosity must be aroused, and his whole soul stirred by a great variety of conflicting emotions.

Chambers had used Attiret's letters as the chief authority for his descriptions, but he now enriches them with a lively and purposeful fancy. It was due to him (though the movement was transitory) that in England the garden landscape was enlivened with buildings. He was himself, in 1758–9, the leading architect for Kew Gardens. The Crown Princess Augusta, George the Third's mother, had built a villa at Kew, and looked about for an architect to provide the proper accessories for her garden, which she wished to have laid out in the picturesque style; and Chambers wanted to show what he could do. Hirschfeld criticises Kew Gardens as being laid out in too narrow a space, for being concentrated upon the lake in the middle, and for making no use of the lovely surrounding country of the Thames. But Chambers relies, as excuse, on his lofty eight-storied pagoda, which affords a very wide view (Fig. 584). In this "wilderness" you see at a glance the Pagoda, a Mosque, and a Moorish building called Alhambra. There was a row of Greek temples (and their particular style gave great joy to Chambers, who was a scholar), there was a Roman ruin (Fig. 585), and there were other lesser monuments; these objects prevented the place from being "boring," which Chambers so greatly dreaded. Kew has often had to bear the reproach of being too full of buildings, yet the excess is not nearly so great as we find it in many a continental garden. The importance of this place, which from the first was richly provided with foreign plants, especially American climbers and conifers, became extraordinarily conspicuous in the nineteenth century, when it was the leading botanical garden in Europe.

Chambers always overlooked one unbridgeable difference between English and Chinese requirements. The Englishman likes to stroll about; he likes to be tempted to seek view after view, by paths that are as winding as possible; and the little, necessarily restricted views which the Chinaman can enjoy sitting down, would of course be unattractive to him. What wonder that a lively protest greeted the writings of Chambers! It happened, moreover, that the *Dissertation on Oriental Gardening* (1772) appeared after the fashion for China had declined on the Continent.

In England there arose at once contradictory voices, and in the forefront of the battle stood the poet William Mason. He had begun a poem in 1772 about the garden, and this he published in several volumes during the next decade. He felt himself to be the herald and poet of the new style, a warrior fighting for the rights of Nature, who laughs at her fetters, and allows no beauties that are foreign to the soil which she bestows on us.

It is true, says Mason, that the gardener must learn from the painter; and he adds that design is indeed a wide province, but gardening is only one of its districts. With this thought he begins an essay (of the same date) on *Design in Gardening*. The garden artist, he maintains, has robbed Nature of her fairest and best, and we must learn to restore into flowing curves all that is now straight, angular, or parallel; or, otherwise put, the serpentine line uniting grace and beauty is the true line of Nature. He also

FIG. 585. RUINS IN THE ROYAL GARDENS, KEW

demands variety, which was the everlasting cry of the French garden of the old style; but the crowding in of images and ornaments, as advocated in Chambers's essay, seemed to him an affectation, and a danger to the ideal of Nature.

In full sympathy with Walpole, Mason answered Chambers in a sarcastic epistle worthy of Pope, and his work went through numerous editions. Thus did the first quarrel break out. It took different forms in England, and lasted till the beginning of the new century. The principles underlying the movement were many-sided, and it might be that the differences were only differences of direction in that imitation of Nature to which all the parties aspired.

The situation was acute when in 1794 an essay by Sir Uvedale Price appeared, *On the Picturesque as compared with the Sublime and Beautiful*, followed by a *Dialogue on the distinct Characters of the Picturesque and the Beautiful*. The former was specially directed against Burke to whom in chief is ascribed the folly which is known as "improvement"; for it was he who so seductively put forward smoothness and perpetual change as the attributes of beauty. Brown was no better with his monotonous cry of clumps, belts, and artificial lakes. So Price now turns seriously to the idea of copying painters like Claude and Salvator Rosa. Every feature of Claude's work he would have repeated in the garden, especially his vague and partial concealment of the chief objects, buildings, and the varied arrangement of water, saying that for the sake of contrast not only may wild things but even ugly things be brought into the picture. His friend Richard Payne Knight, to whom he confided his plans, resolved to help him with a poem.

Practical men like Repton, then a busy garden architect, answered these attacks. Without extolling or defending the poverty of many of the new pleasure-grounds, he was the first man to free himself from the exaggerated idea of a similarity between painting and landscape gardening. He laid his finger on the difference between them, caused by the constant alteration in the spectator's point of view, and by the changes of light in a garden. To his mind what is much more to the point is concord between the architect and the gardener, for a house is presupposed for every garden.

There is no doubt that in the last years of the eighteenth century Repton exercised a most important influence in England. He was unwearied in his study of the essential nature of a given landscape, so that he might suit his improvements to the particular place. In his numerous books he appealed to the eye, schoolmaster fashion, for he first gave a picture of nature unadorned or of an old garden, and then another picture of his improvements drawn on the same scale. He used what he called his "red books" for all his creations, and later on published a collection of them.

While these quarrels were agitating people in England, the whole movement had overflowed its banks. In France the "Anglo-Chinese Garden" was only one of the causes why the new style won the day. Far mightier, because its workings lay deeper, was the influence of Rousseau, who by his gospel of the sacredness of Nature in her purest, most abstract form, had expressed the feeling of England. The famous garden, to which Julie in the *Nouvelle Héloïse* takes her future lover Saint-Preux, is a wilderness, wherein, by means of the highest art, all human work and every trace of it is concealed. The flowers grow as though in their natural home in the meadow, or beside the edge of a brook that meanders along or falls foaming over the stones. The winding unconfined paths are shaded with climbing plants; the birds are not prisoners, but are tempted thither by food and by bird-houses; and there they build their nests, singing their songs to the delight of every visitor.

Addison had already asked for these things in his "wilderness," and Rousseau's picture is no doubt partly indebted to him. Rousseau was forced, out of consistency, to turn away from the Chinese garden; for if one were endeavouring with every art to conceal art—a condition which he thinks essential to good taste—there would be far too many costly objects in a Chinese garden to produce a natural effect. Rousseau will tolerate no building at all in his nature garden, nothing shall betray the hand of man. But though his attitude aroused deep interest, his demand could not be complied with in practice.

Far more widespread was the feeling mentioned by Shenstone, that no country

FIG. 586. ERMENONVILLE—LA TOUR DE GABRIELLE

scene can be thought of without some building. Moreover, the park which Rousseau's friend and last patron, the Marquis de Girardin, laid out at Ermenonville, though it was to have been made exactly after Rousseau's ideas—one of the scenes in the park shows Julie's famous Elysium at Clarens—was not without buildings. There was even a temple sacred to the philosophers, and in the wilderness which professes to be a copy of Saint-Preux's place at Meillerie, there is a little hut perched on the top of a high rock with a lovely lake at its foot. Inscribed on the rocks every here and there one finds the names of the lovers. Farther on is the tower of the fair Gabrielle (Fig. 586), the beloved of Henry IV., in the depths of the wood, "quite in the old taste, a little winding stairway leading to different rooms. . . . The walks in this garden are not only charming to the eye but also to the ear. For the marquis entertained a number of clever musicians, who might be heard sometimes on the banks of the lake, sometimes actually on the water, either performing alone or in concert." But the crowning glory of the place was the tomb of Rousseau, a sarcophagus enclosed with poplars, rising on an island of the lake. And the thought, "Here lies Rousseau," is all we need to complete the touching loveliness of the scene (Fig. 587). Hirschfeld, who repeats Girardin's own description of this park, gives in "this perfect example of improved Nature" the picture of a real masterpiece of the sentimental park of the period.

Another French park, which won an even greater reputation, was that laid out by Louis Philippe's father, Philippe of Orleans, at that time Duke of Chartres, about the year 1780. The "Parc Monceau" (Fig. 588) gets its name from a little village south of Paris. Carmontelle, who was an artist, designed the plan. The chief pavilion, where the duke

held gay festivities and also open-air assemblies, is surrounded by a tract of land with parterre and plantation. The ground itself was cleverly made undulating, and the picturesque park exemplified all the variety that the age demanded: close beside country-like meadows, vineyards and brooks, stood kiosks and spiral hills, and side by side with Gothic ruins was the marvellous affair in the north-west corner, the "Naumachia." This was a large oval marble tank with Greek ruins round it in artistic confusion (Fig. 589), and it was dominated by a lofty column. The most noteworthy feature is the colour-garden, a round space enclosed by small regularly laid-out flower-gardens—blue, red, and yellow patches right in the middle of the park.

FIG. 587. ERMENONVILLE—THE ISLAND OF POPLARS AND
ROUSSEAU'S GRAVE

It need not be repeated here that it was not the sentimental park which created the love of separate little erections, nor did it particularly encourage this love, since an outspoken hostile opinion was constantly reasserting itself; but the meaning of this particular adornment, which had been adopted from the old style, became a completely different thing under the new regime. In the old style the erections were no more nor less than rendezvous, or shelters against sun or bad weather, but now it was not the visitor, it was the spectator, whose interests had to be considered. The building had become an accessory to the landscape, a principal factor for bringing about a desired Mood, which the picture had to express. The Mood, or frame of mind, gave to life at that time its interest and charm; it was the leading feature of sentimentalism, a peculiar mixture of reason and feeling.

From the union of Rationalism and Emotionalism there had naturally sprung Senti-mentality. Every impression was to be clothed in feeling, but man must always have an explanation and a sort of justification of the feeling, and this was most easily to be found in the so-called "animated nature" form. It was understood by people of the late eighteenth century that they must feel melancholy at the sight of a ruin, that a hermit's retreat incited

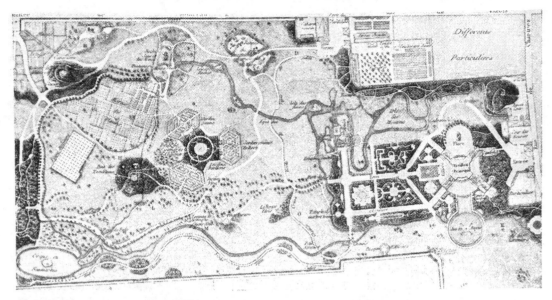

FIG. 588. THE PARC MONCEAU, PARIS—GENERAL PLAN

to silence and solitude, that a Greek temple stimulated the *joie de vivre*. And if this was not enough, surely the Mood would be helped by a suitable inscription! The oddest part about it was that they never needed to be alone, but were sure to be seized upon by this Mood (which was produced by external things) when they were in a large company of people.

The accessories are found also in very small rococo gardens, especially ruins: these were intended to "cheat the reason," e.g. the beautiful antique temple at Schönbrunn.

FIG. 589. THE PARC MONCEAU, PARIS—THE RUINS

But true Sentimentality will have nothing to do with effects such as this, and it is interesting to see the two parallel streams that never meet and blend.

Home is perhaps the first writer to accentuate Feeling strongly. The garden as a work of art excites in him sentiments of greatness, charm, mirth, melancholy, wildness, even surprise and wonder. In order to keep each of these feelings clear and strong, he desires that scenes which are next to one another shall be different in kind; and it is a good plan to mix rough uncultivated places with wide open views: these are not attractive in themselves, but in the long run they increase the sentimental charm. And so he excuses

FIG. 590. THE LITTLE TRIANON—THE HAMLET

Kent for introducing here and there into a landscape withered trees or broken trunks—a plan that the other side repudiated with scorn.

But Home also requires Simplicity as the leading principle of garden art, saying that only an artist who had no genius at all could be inclined to put up triumphal arches in his garden, or Chinese houses, temples, obelisks, cascades, and endless fountains.

One cannot urge against those who were endeavouring to guide with their pens the art of gardening, that they encouraged the fast-growing inclination to overload the park with buildings. They almost all fought against it; and so did Horace Walpole, who, when approaching his seventieth year, in his elegant *Essay on Modern Gardening*, treated the subject historically, and opposed the excessive decoration of Chinese gardens, which to him seemed quite as unnatural as the decorated formality of the old style. Walpole had long postponed the publication of his essay, and it appeared first in 1785 both in French and in English, dated from his house, Strawberry Hill. Walpole was very proud of the first Gothic erection which he built at Strawberry Hill—in the "pure style," as he hoped; we must say, however, that it was only in connection with pseudo-Gothic, and not pure Gothic, that he exercised his real influence on the development of the new style in England. In the matter of taste there is no doubt that Walpole was one of the most influential men of his age. His essay, which had remained for more than ten years in manuscript, had a great effect, and especially in France, for it showed much enthusiasm and learning.

FIG. 591. THE LITTLE TRIANON—THE DAIRY

We saw before how France was drawn to the picturesque by Rousseau's introduction of English ideas. And yet she remained half-hearted; for there was always the inner voice, the romantic, deep-rooted love of formality, which restrained the French from that complete destruction of the old state of things which took place in England. It is true that they turned away from the larger designs, and made parks where they could enjoy senti-mental pleasures. But Versailles was still there, and they could restrict themselves to the Little Trianon. Here with the *hameau* (Fig. 591), the mill, and the dairy (Fig. 591) they had a background for their games and their fashionable dresses. Here Marie Antoinette trifled away the last years of her glory with her ladies and cavaliers. No threatening voice of the coming revolution penetrated to that rippling lake, where they played blind-man's-buff on the banks; or to the beautiful round temple, whence the little god of love looked down on their happy games. The noble groups of trees that were set round the lake were calmly

growing higher and higher; men who were used to this pretty little place shuddered to
think of the long broad avenues at Versailles, where one was lost and felt so small. Versailles slept the sleep of the giant, slept through every danger that threatened, until the
day of its awakening came.

There are similar grounds in the park at Chantilly and at many castles on the Loire.
Hirschfeld tells of a whole array of gardens in and near Paris, all laid out in the new
fashion. But seldom did they venture to put the dwelling-house right in the middle of
the landscape garden, as in England, where a lawn, the "lawn-carpet," with side-walks
of regularly planted trees was usually found until the beginning of the nineteenth century;

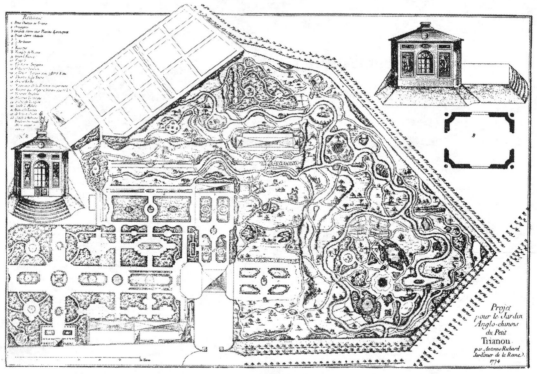

FIG. 592. THE LITTLE TRIANON—THE FIRST PLAN FOR THE ENGLISH PARK

this was the first view to be seen in a picturesque garden, and the lawn reached close up
to the house. Even at the Little Trianon there were still formal gardens about the tiny castle
(Fig. 592), when in 1774 the plan for an English garden was designed by Richard; at
any rate, important alterations in these formal parts had been intended in the final
execution of the plan.

As in England, so in France, a singer was found to extol the new movement;
and in France it was the Abbé Delille. In many respects he sides with Mason, although
when he is trying to get the better of Rapin, he himself derives his authority from Virgil.
But whereas Mason, though he is poet enough to see with pain and regret the fall of aged
trees, yet pulls himself together with a determined "It must be," the whole soul of
Delille trembles at the thought that Versailles, "the masterpiece of a great king, of Le
Nôtre, of the age," may fall. Nothing of Rapin's could have described more pathetically
and more movingly the beauties of the old garden than the verses wherein Delille tells

of his fears for their ruin. But in spite of this he sees in their monotony no subject worthy of a poem, for the old gardens were the offspring of architects, the new of philosophers, painters, and poets. The programme he sets out has no single original feature; its leading thought is for picturesque Contrast. "Imitate Poussin," he cries, "for he paints the merry dance of shepherds, and beside it sets a grave with the inscription, 'I too was a shepherd in Arcady.'"

Somewhat later, Germany also experienced a change, no doubt under the influence of England, and it brought about her own characteristic flowering. In her attitude towards the new style she filled a far more important rôle than France. The circle of Swiss poets around Bodmer had eagerly seized upon Spence's first principle, *ut pictura poesis*, but had also accepted Shaftesbury's and Thomson's gospel of the greatness and ennobling effect of untouched Nature, and had developed the idea at the same time as their compatriot at Geneva, if not perhaps so violently. And so we find in the poets of this school, Kleist and Gessner, the first expression of ill-will and actual revolt against the "cleverly laid-out gardens with their green walls, and labyrinths, and obelisks of yew rising in stiff ranks, and gravel paths, so laid that no plant may annoy the foot of the stroller." "Too bold man," Gessner cried, "why strive to adorn Nature by using imitative arts? What I love is the country meadow and the wild hedgerow." But in his pictures (which this poet-artist paints in the utmost sympathy with his idylls) he seems quite happy with lattices and bowers; and in the graceful idyll, *My Wish*, he depicts the garden behind the house, "where simple Art assists the lovely fantasies of Nature with helpful obedience, not endeavouring to make her the material for its own grotesque transformations"—though all the same this garden is enclosed by walls of nut-bushes, and in each corner stands a little bower of wild-currant.

Kleist, who borrows from English sources the inimical feeling towards the old style, hails the tulip, afterwards so much despised as stiff and formal, as the "Princess among Flowers."

This group, which is like the coterie around Gleim in North Germany, is still at the stage of the first plan of Stowe, the pre-Kent period. Gleim gave to his "Hüttchen" with its little garden a touch of the antique, by arranging that he was to be buried there, surrounded by memorial stones put up to his friends, who had so often gathered round him in this garden. Thus in Germany too we see that poets started the movement; and there they reappear, though some decades later, as critics and theorists. Sulzer in his *Theory of the Fine Arts*, dated at the beginning of his seventieth year, adheres closely to Home's first principle, that Nature is the supreme gardener, but that horticulture, like every other art, is an imitation of Nature, to be classed with the arts of drawing and design. Mason, with his insistence on design, was immediately ahead of him. And then, as Chambers had meanwhile come forward with his glorification of Chinese gardens, Sulzer cut himself adrift, especially from the Chinese style, which he disliked and opposed.

A little later began the activity of a man who, because of the popularity of his style as a painter, was of immense importance in Germany, the Professor of Philosophy at Kiel, Christian Hirschfeld. As early as 1773 he wrote his first paper, *Observations on Garden Art*, which was followed two years later by a short *Theory of Horticulture*; and from 1779 his great work in five volumes, *History and Theory of Horticulture*, began to come out. Hirschfeld wrote at a favourable moment for Germany. Before him in

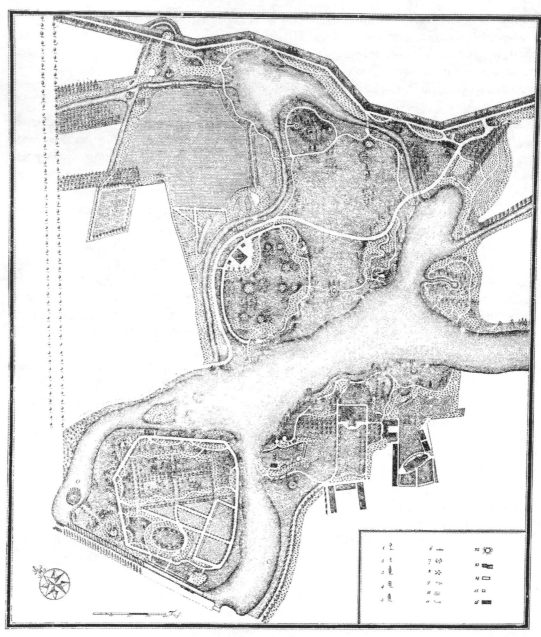

FIG. 593. WÖRLITZ—GROUND-PLAN

England there was not only a great literature of gardening, but a goodly number of examples. He had steeped himself in the new ideas, and felt that he was the representative of good taste, and its champion in the fatherland. So, as Goethe says, he lighted with his own fire the emulation and enthusiasm of the rest. In his five volumes he embraces all the essentials of the garden: theoretical, æsthetic, and historical. He tries to inspire the artist, the amateur, the gardener, by giving many examples, by his treatment of individual cases, by criticism and instruction. His most valuable gifts are a series of pictures illustrative of English, French and German gardens.

Walpole had in his high and mighty way indicated that he did not believe the new style of garden would meet with much approval on the Continent, and especially that "the little German princes, who set out their Palaces and Country-Houses so extravagantly, will not be able to imitate us." When Walpole published his essay, this prophecy had been put to the test of truth; and the German princes who had been making their gardens with ever fresh and capricious fancy in the French fashion, were prompt enough to procure for themselves, one after another in quick succession, anything new that came from England.

In 1769–73 Duke Francis of Dessau built a summer home close to his residence at Wörlitz (Fig. 593). Goethe (in his *Dichtung und Wahrheit*) dates it farther back, to before Winckelmann's death in 1768; but that was because he wanted to extol the greatness of the prince, as shown in his friendship and admiration for Winckelmann as well as in his park, which at that time was unique. But, however this may be, Wörlitz is one of the first remarkable examples of the new art. The prince found a beautiful lake of a good shape; he cut fresh creeks and inlets, and united these with small streams passing through canals, so that each island thus formed was a complete picture, with one or more buildings as accessories. In this way the prince attained the first essentials: variety and contrast.

The Prince de Ligne, who composed one of his witty and eloquent descriptive pieces about Wörlitz, attempts to divide the whole place into five acts with seven scenes each. The first, which he calls *Champs Elysées*, is an island garden on the north-west of the castle, laid out as a private winter-garden, and enclosed by an evergreen hedge. In

FIG. 594. WÖRLITZ—VIEW OF THE GOTHIC HOUSE

the middle of it there is a maze adorned with busts of Lavater and Gellert, which you approach by complicated paths, sometimes underground. On one side this garden opens on the wide lake with two islands in it, one of them copied from the Rousseau island at Ermenonville and bearing his name. On the other side of the broad basin of the lake is the garden of the Gothic house (Fig. 594), which is the most prominent feature of the view from the castle.

From a small gardener's cottage this house has been enlarged gradually, and is to-day the Art Museum. Very wisely the prince had always provided a surrounding for his numerous erections—generally a small straight-clipped hedge—and in this way the houses, temples, and grottoes are isolated and seem more important. The subterranean paths offered an opportunity—welcome in that sentimental age—for an awesome sense of solitude. But one cannot deny that an impression of grandeur is given by the Louisa Rock at the east end of the lake, rising as it does from an underground labyrinth of rocks in steep, bold outline—the more so because of the calm smooth surface of the lake, which adjoins this scene and most of the others. The many bridges are considered to give the finest points of view, and from them one can always get a new picture with some building as its centre. Variety is still farther increased by the inclusion, in the wider parts, of large meadows, even cornfields, and so the park is gradually converted into a farm estate.

With still more surprise, in the so-called new part, one walks straight into the most marvellous building in the whole park, one that pays its tribute to the freakish spirit of the age, which only too easily degenerated into childish folly. This construction is the notorious fire-spitting mountain, called Vulcan by the Prince de Ligne: on the outside it is uncommonly like a baker's oven, but inside there is a Temple of Night with light-effects made by coloured glass, which to-day would not produce the desired illusion even for children. Men of that day, however, had a great fancy for such toys, and nothing in the garden was more popular with visitors.

The prince describes a similar grotto in the castle of Schönau, with a waterfall tumbling over it, inside of which one could, with the aid of torches, decipher the profound meanings of inscriptions and emblems, until one arrived at the throne of the veiled Goddess of Night, seated on a chariot by a triangular table, on which the "bird of Minerva" presents—the Visitor's Book! The picture of this park, imposing even now, must not be prejudiced by these little tricks of the moment; and the happy use made of the lake contributed fresh beauty even to those new sites which were visited and studied by Goethe in his early years at Weimar.

Goethe confesses that his interest in horticulture was due to Hirschfeld's work and the Wörlitz park. The park at Weimar and the book, *Elective Affinities*, are the culled fruits that came to maturity in his all-embracing mind. In a charming little essay, *Das Luisenfest*, which was destined to find a place in the autobiography he had already planned, the poet describes the origin of the park. At an improvised fête on the princess's name-day a hermitage had been set up beside an alder clump on the bank of the Ilm. Friends wearing monks' cowls received the court, and prepared a successful surprise for the company. To this little idyll all the other parts of the park were adapted under the immediate supervision of the poet. It had before consisted of gardens in the old style, which at the beginning of the eighteenth century were not exactly poor, but rather broken up, and these adjoined the old castle, which still kept its mediaeval style. The pleasure-garden proper was in great part destroyed through the burning of the castle in 1774.

There was no more talk in Goethe's time of the park and the fertile canal gardens alongside; but in the park there was now the so-called "Stern (star)," then a public walk, a well-known spot, a space full of trees and shrubs—ancient trees planted in straight lines, trees which rose high into the air; also many avenues and broad plots for meetings and entertainments. Besides all this, there was a high place, the Schneckenberg (Fig. 595), with winding paths up to a castle—still standing even in the nineteenth century—in whose green-clad walls were windows and little turrets; this was always the special sign of an old park. Quite near here Charles Augustus in 1776 had built a summer-house to give to his friend,

FIG. 595. THE PARK AT WEIMAR—THE SCHNECKENBERG

with a terrace-garden adjoining. The fire at the castle and the destruction of the old gardens had not only made a bare space, but "the lordliest persons, robbed of a home suited to their comfort and their station, betook themselves to the open."

Directly after Goethe had composed his little idyll about the Luisenkloster, as the hermitage (Fig. 596) was called, "people loved to go back to the place. The young prince liked to spend the night there, and for his pleasure they erected the ruin and a sham campanile." What looked like a ruin was an old shooting-stand, built out of the stones of the burnt castle. And now that the paths were also transformed to suit a romantic necessity; "they wound about (Fig. 597), now over rocks, now under arches, now passing out into the light; with their empty, wild aspect, and here and there a hollow place or a seat, they gave some idea of the famous rock-paths of Chinese gardens."

Here we find the first picture of the kind of garden which Goethe so happily called "æsthetic," to indicate the sentiment of his time. Thus out of a romantic necessity came into existence one picture after another, mostly started as a setting for some merry fête: there was the house of the Knights Templars (originally a Gothic tea-house), the Roman house, another ruin, a temple, monuments with inscriptions, and so on. All the separate scenes were united into one by great field views ringed with trees and bushes, brooks with bridges, paths that lost themselves in the distance, vistas with far-away church towers. The complete want of connection with a house—to which men gave no thought because there was now no house near by—matches with the want of an original plan for the whole: they were led on by one single desire, "to beautify the landscape while winning from it its own peculiar charm" (Fig. 597). Thus the park at Weimar taken as a whole gives perhaps better than any other a clear exposition of the feeling of the time

FIG. 596. WEIMAR PARK IN GOETHE'S TIME—THE HERMITAGE

on this subject, and is doubly important from the fact that Goethe clothed the fancy in a fair form. What he learned here, with all the limitations of practical execution, he gave out in a thoughtful tale in *Elective Affinities*. Human nature in this story is shown at its purest in the activity of a garden life, and a peculiar harmony in the book is brought out, especially when Charlotte and the Captain are working together: they are complementary to each other—on the one hand the clear-headed, careful woman, attending to details, on the other the man trained in a military school, and keeping in view only the things of final importance. Gradually, as at Weimar, the separate parts of Charlotte's park appear; by uniting three ponds the great central feature of the lake is made. The pleasure-house on the hill overlooks it, narrow footpaths and steps in the rocks lead to imposing points of view.

Goethe had zealously studied English engravings, and it is not unlikely that he had Repton's work in his mind when he tells of opening the books, "wherein one always found a picture of the ground-plan of the place and also a view of it as a landscape, and then on another page a picture of what had been made of it so as to use the good points it already had, and to enhance them." Meanwhile, side by side with this creation of new landscape-gardens, the old castle-garden was still green with its lofty avenues of limes and its even plots, which were the work of the last generation, intact and unaltered; but all the time secretly waiting for a new, distant resurrection which Goethe possibly foresaw even then. And so this business of beautifying the land, which permeates the whole tale, combined with a constraint that limits it to what is objective, creates a

FIG. 597. WEIMAR PARK—A WINDING WALK

happy state of equilibrium between the violence of passion and the peaceful background that frames the tragic fate of its hero.

We see what competition there was in Germany, from the middle of the eighteenth century, in this matter of laying out gardens according to the new fashion, if we look at the number of descriptions in the appendices to the different volumes of Hirschfeld's work. As we have seen, it is often the same princes and sometimes the same artists who supplant the old style—though it is still beloved—with the new. In the early days strange mongrels appear; indeed it almost looks as though the childish complaints incidental to the new style had broken out with peculiar virulence in Germany. Charles Augustus had hardly given the last touch to his fine gate-house at Schwetzingen when Skell came home from England, and began to put the new belts round the old parks (Fig. 598). From the very first he seemed unconvinced and hesitant. In 1784–5 Hirschfeld came to Schwetzingen, and he felt that bad taste ruled supreme. Work was going on at the Turkish Mosque:

Look at the Mecca scene, for example . . . this Mecca is in the middle of the French part . . . from the mosque one looks straight into the Egyptian part, where work is still going on, and this, like the Turkish, seems to have fallen from the skies. It is a hill on which a monument of King Sesostris is being put up. This monument ought not, if the illusion is to be preserved, to be very different from

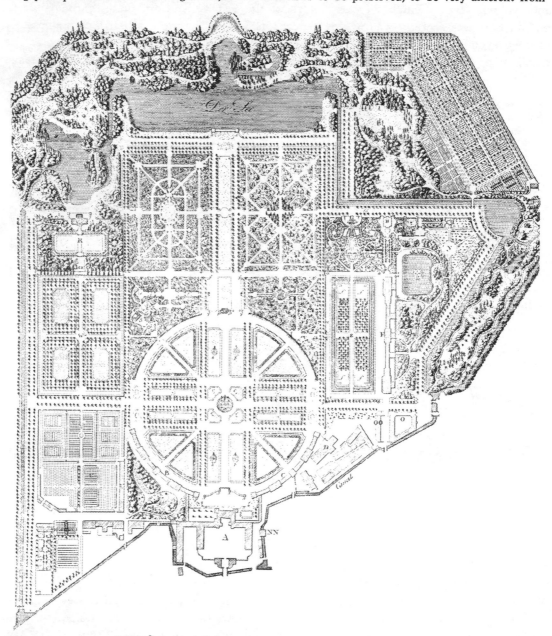

FIG. 598. SCHWETZINGEN—ALTERATION INTO A PICTURESQUE PARK

ruins that are nearly worn away by the hand of time; but here everything is new, perfect, ornate, and time has altered nothing. In the caves of this hill there are to be mummies and graves . . . round the hill the Lake Mœris is to be dug out.

Hirschfeld came at an unpropitious moment, and Skell abandoned the whole Egyptian plan, contenting himself with the ruins on the hill, which was made of earth taken from the bed of the lake. Even to-day this corner of the English part does not look so harmonious as the illustration (Fig. 599) represents it, though the treatment of the great lake is much more successful. In the English garden at Munich, which is Skell's chief work, he has been almost too sparing of his buildings, and Uvedale Price would certainly have reckoned it among the wearisome creations of which he talks.

Like nearly all the theorists of the new style, Hirschfeld fought to the utmost against the overcrowding of buildings and especially the mixing up of different styles; but so long as the same theorists clamoured after variety and contrast, the worried practical people had to catch at the help that garden-buildings gave. It was unbridled licence in taste,

FIG. 599. SCHWETZINGEN—THE MOSQUE

and the extravagant desire to bring every fancy to completion, that brought to birth such a monstrosity as the garden of Rosswald near Troppau in Silesia, which was laid out by that queer being, Count Hoditz, who in his last days of poverty had to be supported by his friend and patron, Frederick the Great. Everything was heaped up there that people had thought of for hundreds of years. Beside a Chinese garden and temple there was the Holy Grave; after Christian hermitages came Indian pagodas; here a picturesque hill, there a little town for dwarfs, with a royal palace, church, etc. And from want of dwarfs the count for a time had children to live there. Next came Druid caves, with altars; then an antique mausoleum, to which sacrifices for the dead were brought.

It was a great joy to the count to have fêtes corresponding to all the various parts of his garden—to suit the Chinese garden, or the wilds of America. He introduced the gambols of naiads and mermen in the lake, but best of all he loved his Arcadian fêtes, when he dressed his peasants as shepherds. None the less, people took him seriously. Frederick the Great was inspired to write him an admiring letter in verse. The fancy of the time for masquerading both in outward and inward ways—which we must always remember had nothing whatever to do with the new style of horticulture or its real principles—seemed at times to tend to such an exuberant growth that one could not recognise its original intention.

While Count Hoditz sinned in his senseless conglomeration of disconnected scenes, a famous garden at Hohenheim, a couple of hours from Stuttgart, was going astray in quite another fashion. Here one idea only dominated the whole place. The designer's intention was to represent a colony settling on the ruins of a Roman town (Fig. 600) and the effect was hailed with admiration by his contemporaries.

> We take a lively interest in these dwelling-places, and believe them inhabited; we are astonished at the remains of temples and strong walls, which stand exactly as though they had been rescued from destruction hundreds of years ago. . . . Anyone just passing through the garden and looking at it can get no clear impression, because of the number of buildings; but it is quite different for a person who enjoys choosing some particular part, and staying awhile where he finds the right nourishment for his mood . . . he soon comes to a spot for the *dolce far niente*, then to another which, because it bears the stamp of simple benevolence, has the power of pouring blessed peace into his soul.

But even now we have not had enough deception! We go into an apparently simple hut, and meet the last surprise; for inside it are wonderful rooms furnished in princely fashion, bath-rooms, silk tapestries, paintings, and so forth. The Prince de Ligne counts sixty different views in the comparatively small space of sixty to seventy acres, which can be strolled round in four or five hours.

It is not surprising perhaps that the lively, fashion-loving prince—a man for whom the *dernier cri* was the topmost peak of civilisation—took delight in this garden. In the park of his family place at Hennegau in Belgium, he laid out a Tartar village, where all the aboriginal character of the shepherd's life was staged, with young bulls as well as young students at the dairy. When work was over, the shepherds played on instruments which the prince had brought away from the Alps with his cows, and wore a uniform worthy of the beauty and simplicity of Nature, whose high priests they were. In this Tartar village the dairies are concealed in the mosque, whose minarets serve well as dovecots.

And other men of more weight than the Prince de Ligne yielded to the charm of the "idea of uniformity" as shown at the Hohenheim Park farm. It even finds favour in the eyes of Schiller, who gives his opinions on the garden in the *Pocket Calendar* for 1795. Though he severely condemns the overcrowding of scenes into the gardens of the day, "where the whole number of her [Nature's] charms are displayed as in a book of patterns," and though he considers it a mistake when horticulture takes painting as a model, because it has "no reduced scale," he still hails with joy the idea of this kind of garden. Although he thinks it an affected sentimentality to hang little tablets with mottoes on the trees, he gains a point by urging that the Nature which we find in the English garden is no longer the same Nature as the one we have left outside. It is a Nature enlivened with a soul, a Nature exalted by art, which delights not only the simple man, but the man of education and culture: the one she teaches to think, the other to feel.

Schiller had always had a certain local interest in the garden of his own father's house; but all this talk points to a general mistrust of the new art. There was a strong feeling that in this particular department, where Nature must and ought to provide the material to be copied, it was impossible to attain an artistic style from a simple imitation of her.

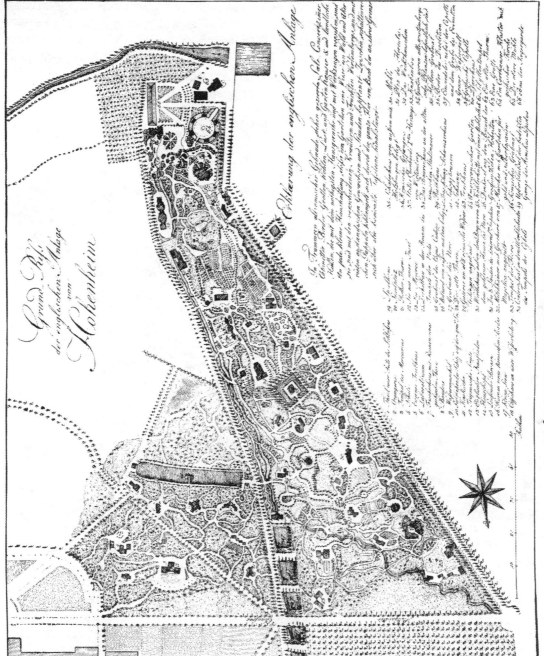

FIG. 600. HOHENHEIM—GROUND-PLAN OF THE PARK

Certainly Goethe found himself very uncomfortable on his visit to this garden during the tour in Switzerland in 1797. What he said was, "Many little things put together do not, alas, make one great thing." He wanted to take notes of this garden with a view to a later treatise on the subject, for which he had already collected material.

Without any consistency the critics wavered this way and that. They could not apparently make up their minds to give to horticulture a definite place among the arts, as Lafontaine had done, and yet no art critic ventured to ignore it, so strong was its

FIG. 601. WILHELMSHÖHE, CASSEL—THE FELSENBURG

position in the foreground of their interest. "To separate harmony and discord, to know and to make use of the individual character of each locality, to cherish an active desire to exalt the beauties of nature—to collect them, if this is not a fine art, then there is none." Thus Herder speaks in the second part of his *Kalligone*. He wishes Horticulture to be joined with her twin sister Architecture, but not to be subject to her laws. It will be seen that he is in opposition to Sulzer, Hirschfeld, and others, who would rank her with painting. Immanuel Kant and his school also see in horticulture a branch of painting; and if this is only an outside opinion, it arose from the need of fixing the direction for future developments.

Schiller with his phrase, "Nature exalted by Art," had involuntarily provided a very effective watchword for the phase of horticulture described above. Only such an exalted Nature could serve as a changing background for transitory moods. In the last quarter of the eighteenth century Germany stood at the head of the new fashion of chivalry with its romantic, sentimental love of ruins. When the Landgrave William of Hesse, immediately after his return to power in 1785, took up the task of completing the mighty work of his forefather, Weissenstein, the plan for building the castle at the foot of the hill, so long delayed, had to be carried out. So entirely must this castle correspond with the new sentimental parks, that the landgrave decided to add a wing to it in the form of a ruin; but this grotesque plan was fortunately never carried out, and only one of the series of

FIG. 602. WILHELMSHÖHE—AQUEDUCT AND ARTIFICIAL RUINS

view-plans has been kept, which the prince commissioned the elder Tischbein to paint for him. But the landgrave did not give up his idea about ruins, and some years later had them erected in a better place in the corner of the great park: they were called the ruins of a knightly castle, Felsenburg, afterwards Löwenburg, and their foundations were laid in 1793 (Fig. 601).

The prince was in deadly earnest about these freaks, and the place had drawbridges, bulwarks, entrance-towers, moats, all in the proper style. It did not matter if the bulwark reached no higher than a man's knees! Many churches in the district had to give up their ancient glass to adorn the chapel windows at the castle. Every kind of ornament was lavished on the other parts of the estate. The landgrave gave orders for a laurel or box hedge to be set round one of the gardens, clipped in the Dutch manner, but the gardeners of that time had forgotten how it was done, and he had to content himself with red-fir hedges, which grew up quicker. In the castle was installed a bailiff, and also a garrison, who wore a very choice uniform with bearskin caps. Later, when after seven years of exile the old lord came back to his estate, where King Jerome had been living, as the Elector William I., the time of absence was by his command counted as nothing: he strode across the drawbridge at Löwenburg, and the castle warder stepped forward and announced, "Nothing new has happened."

Knightly castles of this sort (Fig. 602), where in all solemnity chivalric scenes were acted, now appeared in the larger parks. Close to the Löwenburg was a tilting-ground with wooden barriers, but this was soon replanted, for nobody would make use of it. But at Laxenburg, near Vienna, where the Emperor Francis had a lovely park, a tilting-ground still exists near the castle (which itself stands in the middle of the lake as the *chef-d'œuvre* among many fine views), and shows the old arrangements with the proper approaches and lists. "There," says a guide-book, earnestly and naïvely, "in 1810 a wonderfully brilliant tournament took place on the name-day of the Kaiser's third wife, when he and all the archdukes took part in it. . . ." "In 1841," the book goes on, "there was another lively occasion at the tilting-ground . . . a company of equestrian performers gave an exhibition of their skill in the presence of the Kaiser and his court."

We see how seriously this knightly business was taken when we consider the Rosicrucian League (Brothers of the Rosy Cross), who lived as a sort of mystical, masonic sect at the Prussian court of Frederick William II. The new garden of the marble palace in Potsdam was their proper theatre. There they held their meetings in the small grove of conifers, where stood a statue of Cybele with many breasts as the Mother of Nature. This garden, which for a time supplanted the interest in Sans-Souci so entirely that the ornaments of the latter were actually removed, must be regarded as a masterpiece of the masquerade craze. For a long time no building in a park was allowed to appear what it really was. Near the castle, which was built with the marble pillars taken from the Rotunda at Sans-Souci, one noticed a half-buried temple highly decorated with columns, capitals, and the like; but when one looked closer, a kitchen was revealed. An Egyptian structure with sphinxes on guard is an orangery. Under a pyramid an ice-cellar hides. Far away in the park one comes upon a hermitage: one steps in, and there is a luxuriously appointed bath. Farther on there is another kitchen disguised as a little house made of bark, with an iron tree-trunk for a chimney.

In England too there were some extravagances of this kind, but they never went so far. In Windsor Park there was a hay-wagon, with a room inside it—an idea which proved attractive, especially in South Germany. In the now non-existent park of Ludwigsburg at Saarbrücken a hay-cart stood in the middle of a meadow, and concealed in its interior a dining-room. The servants must have had to be content with queer housing, seeing that the court marshal, for whom there was no room in the somewhat small pavilion of the princes, lived in a place in the park which was disguised as a pile of wood; and similar crazy tricks were to be seen in the park of Klärlich at Trierschen.

It was not the slightest good that theorists and artists, whatever point of view they had once taken, were continually abusing these extravagances, or that poets poured their scorn upon them. Although in the early days at Weimar Wörlitz had attracted Goethe to the point of imitation, he now, in his *Triumph of Sentiment*, rebukes these follies sharply. In the park in Hell Ascalaphus is made to give his orders thus:

"The home for Cerberus' dogs," he says, "is to be turned into a chapel, for, mark well, in a park everything must be of an ideal nature, and—saving your grace—each bit of rubbish we must wrap round with some lovely covering; for example, behind a temple we will put a pig-sty, and instantly that pig-sty will turn into a Pantheon."

Goethe's *Triumph of Sentiment* was the gauntlet which he threw down as a challenge to that epoch of sentimentality which he had outgrown; but just as the despised gardens continued to flourish, so did the offshoots of sentimentality.

In 1784 Jung Stilling describes a scene where he and Selma are walking in the garden of Herr Schmerz. He wonders how its creator "could have made every little hill, dale, tree, bush, individually beautiful." He finds first strength, then terrible beauty, then dreamy melancholy, then again riotous luxuriance. Many inscriptions draw the two friends into the mood they desire. They turn into a rock chamber for refreshment.

> When it was dark Schmerz says, "Come, friends, it is very lovely outside." I took Selma with one hand and another lady with the other, and Schmerz walked quietly beside us. We wandered on: forward in the path. . . . Good God! a bright green light shone on the wood, and a hundred lamps lighted up the urn! Ah! what a sight! . . . The skies shed mild lightnings above us, and now this spectacle! . . . Selma sobbed and swooned. . . . I tore myself away. . . . Tears rolled down my cheeks . . . a soft-sounding music was heard, behind the urn a clear green light . . . there floated towards us an Adagio out of *Zemire and Azore,* and I cried, "Schmerz, illuminate Christina's urn, for the lightnings pour their wealth on me and Selma, and the wood breathes out a gentle peace." . . . Then Selma and I swore eternal love, but we also swore to love God and mankind, to the utmost limit of human power.

Before Goethe, Justus Möser had poured scorn upon the new fashion in his *Phantasien,* which Goethe greatly admired. It seems almost incredible that the little satire, *English Gardens,* was written as early as 1773. And yet it is significant that before the English style had entrenched itself in its greater works, the frivolous fashion of Anglo-Chinese taste had made an appearance, and probably was at its height in the Hanoverian states, which felt English influence first. Möser loved what was old and native to Germany; he was pained by foreign taste, by the childish and transitory. In one of his works a girl tells her grandmother in a letter how the grandmother's bleaching-ground, fruit-garden, and cabbage-patch have been converted by the writer's husband into an English garden with little hills and dales: "but now it is called a shrubbery, or as other people say, an English bosket." On the hill, she says, they sit under a Chinese canopy, which is a sun-shade with gilt metal lining. Of course there are Chinese bridges, and a Gothic dome as a summer-house. Still more plainly than in the *Triumph* does Goethe follow the lead of Möser in his *House Park,* written in 1797. The daughter is complaining to her mother that her playmates laugh at her:

"I ought to feel what Nature does in the open . . . and they cannot bear stiff green walls, for they can see right through them from one end to the other. Our leaves are cut down by shears, and the flowers too; what a shame! Our dear cousin Asmus calls it just a tailor's game." (Asmus—Mathias Claudius—in the Serenata in *Wandsbecker Boten* scoffed at the old park-gardens, in which "nothing can be seen any longer of the great full heart of harmonious Nature," as a mere "tailor's game.")

In the midst of this contest which raged to and fro in their own camp, there very soon arose other voices which condemned the whole fashion of landscape gardening uncompromisingly. It was from English Classicism that the style, so often and justly called romantic, had originated; the Romantics themselves were only in part admirers of it,

FIG. 603*a*. MUSKAU—VIEW OF THE PÜCKLER PARK

indeed the first important attack was to come from their side. Among the leaders of the English Romantic school Wordsworth stands high, with his so-called æsthetic style. He feels proud that painters and poets are the creators of English horticulture, and that now they will win for themselves the high praise that they are the fathers of better taste. In this sense he writes to his artist friend Sir George Beaumont, at whose house he stayed. "As to the grounds," he adds, "they are in good hands—the hands of Nature."

He adopts Coleridge's principle that the house and garden must belong to the land-scape, the landscape must not be subsidiary to the house. One ought to lose oneself in the beauty of the actual countryside; and this is what business people will never do, but only painters and poets. Thus did Wordsworth express his innermost feeling; and it is from his personal idiosyncrasy that there arose in him an aversion for places that by their emptiness have cut out all life, and so make one think of the legend of the upas-tree, which breathes death and devastation around. The poet of the garden, he says, should weave into one sympathetic whole the joy of every living being, of men and children, birds and beasts, hills, rivers, trees, and flowers. We are warned, however, against "dressing the whole of a landscape in the livery of man"; Nature must take entire charge, so that everything we do is suited to her beauty. This is the fundamental law that Words-worth always obeys, and it is a law of nature purified from sentimentality.

A hundred years had passed since the birth of the picturesque style, when Sir Walter Scott, in 1824, describes in *Waverley* the old garden of Tully Veolan, and throws down the gauntlet to the formal style. The park, he tells us, comprises several square fields with walls round them and a short straight avenue of horse-chestnuts and sycamores, which is the approach from the lower to the upper gates. Three steps lead down from the main terrace, which has a balustrade in stone, and animals as ornament; in the middle is a bear holding a sundial. From the garden with fruit-trees, flowers, and evergreens cut in

FIG. 603*b*. MUSKAU—ANOTHER VIEW OF THE PÜCKLER PARK

grotesque shapes one passes along wide terraces to a canal, which has a waterfall at the end of it, and an octagonal summer-house with a bear on the roof.

Scott admits that he had in his mind a real Scotch Highland garden, and many specimens like it can still be found. He defended his love for this garden, which he had described as a poet, writing of it later in a little paper in the *Quarterly Review*. The garden was, he confesses, in the highest degree artificial, but it was a lovely sight, a triumph of art over the elements . . . nothing being so distinctly a work of art as a garden. He even defends the beauty of walls, with the warm tones of English bricks, in contrast to the green, and also the tabooed water-works; and although clipped hedges and trees are rejected for these new gardens that he would like to have, he still wants them in the old places where they give him such a feeling of quiet and seclusion.

Earlier, and much more decidedly, the German Romantics spoke out for the old style. George Jacobi in a series of letters had written disapprovingly of the English garden, but in favour of the kitchen-garden, for every deviation from the original purpose appears to him a concession to luxury. He mocks at the notion of attempting large landscape effects, and thinks the formal garden is the proper contrast to the country that surrounds it.

Tieck, in *Phantasus*, takes up this idea, and in particular thinks that in a hilly country the formal garden is not only the most suitable but the most attractive. In majestic surroundings any imitation of the landscape would be silly. "This garden lies at the feet of a giant with his forests and waterfalls, and quietly and humbly plays among its own flowers, arbours and fountains as a child plays with its innocent fancies." This scene appears to him "a bright miniature taken from the parchment manuscripts of the olden time"; and he confesses that he loves above all others such gardens as were dear to our forefathers, "which were merely a roomy, green extension of the house. . . . There they were encompassed by enchanting Nature, governed by the same laws as men of understanding and reason, the laws of the inward unseen mathematics of life."

Tieck does not quite exclude landscape gardening, but whereas the French style seems to him, as it did to Sir William Temple, to have scarcely any fault, the English garden, he thinks, should never be copied and repeated, for each is unique. There should be no failures or confusions through losing the real personal feeling for nature. An English garden ought to be a true and perfect poem, a lovely individual thing, sprung from one mind only. This thought he expresses elsewhere, using words that Wordsworth would have heartily endorsed.

But Goethe also towards the end of his life had lost his affection for the English style. On a walk with the Chancellor von Müller to the Belvedere, he commended the way French gardens were laid out, at any rate at the great castles. "The spacious arbours and bowers, the Quincunx, allow of a large party coming and going in a decorous way, whereas in our English places (which I might call Nature's little jokes) we keep knocking against one another, and either get boxed in or quite lost." In such words as these Goethe seems to express almost a repulsion from his early enthusiasm and the silly sentimentality on which it was grounded.

It must also be remembered that Goethe, who created the park at Weimar, never contemplated treating the garden at his own town house in the picturesque style; and to this day it bears traces of the older fashion. The walled-in square has two summer-houses in the corners at the back, and is divided into regular straight-lined sections. At one time the poet put masses of flowers in it, when the trees—now overgrown and casting broad shadows—only marked the edges of the beds, and did not exclude sun. In the background is a pergola, leading from one summer-house to the other, and this makes a quiet walk for anyone strolling there from the house.

Thus were voices raised in opposition, especially in Germany, and even the leaders and promoters of the art detected faults and bad taste in this or that particular example, though not in the style itself. The whole of the nineteenth century must complete its tale of sins before the foundations are shattered. In its first decade there appeared in Germany a new inspired prophet of landscape gardening in the young, good-looking, enthusiastic Prince Pückler. His personality was his chief asset; and through that he brought to realisation his ideas and his wishes; and the performance in which he took most pride, the park of Muskau, shows his handiwork at every glance. He came into his family estate, Muskau, when it was in a rather neglected state, only the old castle with not much park-land, an insignificant part with warm springs, a few fir-planted fields, and a great deal of marsh. He at once formed the resolution that he would create a place that should surpass the much-admired English masterpieces (Figs. 603*a* and 603*b*). In his *Hints on Landscape Gardening*, which was published in 1834, he sets forth his ideal, which is to convert a whole estate without any particular demarcation into an improved landscape, not paying too much attention to economy, yet effecting the purpose at no great cost—perhaps less, indeed, than people generally incur in such cases.

In his own scheme at Muskau Pückler did not set a good example so far as economy was concerned, for he put all his great wealth into it and went bankrupt. But the work itself was a masterpiece, and having been cared for by pious hands ever since, is preserved intact. In a few decades he changed the whole vast region of the

Neisse valley, with its border of mountains that enclose the pleasant bath-buildings, into a great park.

Quite at the beginning the prince commissioned the artist Schirmer to paint him some landscapes of the park as he saw it in his own mind, and these pictures he used afterwards as patterns. He had a predecessor here, for Count Girardin, Rousseau's friend, had had Ermenonville laid out from pictures which he had ordered. Pückler, with a view to getting rid of the wretchedly bare look of a young plantation, transplanted large trees in their own earth, and this turned out a great success. And yet the gardeners tell you that the park has only to-day achieved the beauty of the inspired picture which

FIG. 604. HOUSE IN A MEADOW AT KEW

the prince put before the painter's mind. The importance of the scenes which one sees in a long series in ever-changing groups in an hour's walk, lies mainly in the arrangement of individual trees, which Pückler especially loved, groups of beeches and a border of forest trees, with meadow ground and water. It is most surprising how variety can be gained by the help of colour and light, "which ever leaves something for Fancy to guess."

Pückler does not despise the aid that buildings lend; besides the castle, and the towers of the little town, he has two temples, a vaulted church, a ruined tower, and some country houses farther off; but these are only meant to enliven the picture, not to force on any particular mood. In the whole scheme, sentimentality is banished; even the inscriptions from Goethe in the park at Weimar Pückler prefers to read from the master's books.

The passion for ruins now took on an historical character; it was no longer excited by the thought of the transitoriness of life, but by the recollection of some actual or imagined incident. "A garden on a grand scale is a picture gallery, and a picture must have a frame."

II—Y

FIG. 605. MUSKAU—THE FLOWER-GARDEN NEAR THE HOUSE

One secret of art is, so to contrive that each path leads on to some new picture, on which the interest may be freshly concentrated. And this great end the spirit of the prince did attain when he was working upon things that were large and distant. But his noblest triumphs, such as his wise plan of extending the park beside the castle into the open wild park of the Kur-region, we seek in vain in his private garden, the English pleasure-ground adjoining the castle. "Round the house one has to be satisfied with a charming garden within a small compass, as far as possible in contrast with its surroundings; and in this narrow space it is not the variety of landscape, but only convenience, grace and elegance that we desire." Pückler applied this principle in deliberate opposition to English parks, where the house generally stands cold and bare in a monotonous green meadow, at best only enlivened by cattle (Fig. 604). "Among the English it is almost an obsession that one can never have a cheerful landscape without animals in it."

In gardens near the house one may exercise one's own taste freely, and alternate at will formality and the opposite. But, successful though Pückler had been in his larger efforts, his idea miscarried here, and his taste became childish to the point of caricature. Men seemed from now on to forget what true formality meant; and although the prince in his travels all over the world had seen and studied the gardens of very many different lands, and although in words he was for ever setting up as models the Italian gardens of the Renaissance, yet the garden round his own house is frightful, and disastrous in its effect. The lake that bordered one side of the castle bore no trace of the grandeur everywhere remarkable in the park at Neissefluss, and the other sides of the building are surrounded by so-called flower-gardens. Here for the first time in Germany we meet with Carpet Bedding. Beds which are called "artistic," that is beds without any order or plan, are strewn over the lawn—here a cornucopia, here a star, there a flower-basket or pyramid of flowers, which have nothing to do with their surroundings, and make the actual flowers look ugly and mean, mixed together so badly and packed so close (Fig. 605).

The whole of the nineteenth century suffered from this lamentable invention of the carpet-garden. It was one of the most mistaken attempts ever made to keep something of the old brightness of the parterre, and to bring back again the flowers which had been drawing away more and more shyly from the neighbourhood of the house so that they could be seen from the windows. The result is only a sign of barbarous taste. Thus Pückler appears on the one hand as the man who proved the grandeur and importance of landscape gardening and made it live, and on the other as the man who proved the utter impotence of this style for the garden in its narrower sense, i.e. in the immediate neighbourhood of the house. But landscape gardens for long decades to come were to exercise an almost unlimited influence—an influence, however, which on the side of art was entirely unproductive.

CHAPTER XVI

TENDENCIES OF GARDEN ART IN THE
NINETEENTH CENTURY

CHAPTER XVI

TENDENCIES OF GARDEN ART IN THE NINETEENTH CENTURY

AT no period was the art of gardening thrust forward so strongly into the sphere of literary interests as when the style was revolutionised in the eighteenth century. This movement went far beyond the circle of the parties immediately concerned, i.e. garden artists and owners of gardens; and one of the chief objects aimed at in the second half of the century seemed to be to study the nature of art in general through the medium of gardening. It was inevitable that a reaction should set in. The principles of picturesque art had never won a decisive mastery, indeed, we see counter-influences rising up, first in one place and then in another, sometimes on the theoretical side and sometimes on the practical. They were so frequently and so openly discussed, that they penetrated into the consciousness of the general public.

The subject took the tighter hold (in Germany perhaps even more than in England, and certainly more than in France) because men's sight became somewhat blurred when the picturesque style approached nearer and nearer to nature. There was no more interest taken in the great contrast of style in the old formal gardens, nor even in the romantic gardens which laid stress on buildings and sentiment. Little by little, people forgot to look for art at all. The unavoidable consequence was that as far as the generality of mankind was concerned there was an increasing lack of interest in the garden. Goethe observed this apathy as early as 1825, when he expressed to Varnhagen von Ense his surprise at the change of sentiment. "Park-sites, once the ambition of all Germany, especially after Hirschfeld's book was widely circulated, are now quite out of fashion. People neither hear nor read, as they used to, that somebody or other is still making crooked paths, or planting weeping-willows, and it looks as though the fine gardens we have will soon be broken up to make potato patches." Certainly there was no fear of this, for destruction on a large scale is only the result of active revolution. But the development hitherto helped forward by the general interest in art, had now become affected by two other powerful influences: science and democracy, which modelled and even controlled civilised life in the nineteenth century, and had a very marked effect on the art of gardening.

When in its earliest stages, the picturesque style found an ally, helping to a final victory, in the powerful impetus towards the knowledge of plants which occurred in the eighteenth century. The cultivation of individual trees could not amount to much in the stiff formality of the French style; and it was impossible to use groups of shrubs of different kinds and colouring. Plants were not wanted unless they could be used for architectural purposes, and foreign ones were only acceptable in so far as they accommodated themselves to that kind of art. This applied also to flowers in the parterre. Attempts to acclimatise exotic plants were crippled by the masterful influence of the formal style, and such plants were relegated to the botanic garden. When, however, people began to admire trees and shrubs for their individual beauty and natural growth, they were attracted more and more towards places abroad, whence travellers and explorers brought back novelties, first singly and then in large numbers. Things moved slowly at the beginning; for the scientific and geographical interest taken in any special tree was not very strong. It had to express some feeling—must have, so to speak, something to say to men.

The judicious Kasimir Medicus, in his *Materials for the Art of Beautiful Gardening*, published in 1782, utters a warning against the destruction of the old gardens, and also regards with some anxiety what he considers the over-strenuous efforts made in the interests of botany. He thinks that the introduction of foreign trees, though in itself praiseworthy, has nothing to do with the laying out of an "English wood," but is the affair of the botanist. The garden artist ought to study a tree with the eyes of a landscape painter and the spirit of a poet. To him the plane is the tree of reflection, and the annual shedding of the bark typifies the shedding of prejudices by the wise man. The maple is the tree of joyful companionship, and the Babylonian willow the tree of sorrow and mourning.

The whole botanical movement was only a slow and gradual prelude to the great concert that was to be performed in the nineteenth century, after the extensive acclimatisation of foreign plants had been accomplished, in which England took the first place. In the Physic Garden at Chelsea, maintained by the Society of Apothecaries, an attempt was first made in 1683 to acclimatise the cedar of Lebanon. The next year, to the great surprise of Sir Hans Sloane, young trees were growing happily on English soil, with no help from a forcing-house. Yet nobody could have guessed that in the nineteenth century cedars were to be almost the leading feature of the English garden.

A great deal went on henceforth in the way of importing American trees. At first it was only individual travellers and learned persons who explored temperate, hilly countries, especially America, in search of trees and shrubs. The American oak, many species of firs, poplars, magnolias, and the so-called acacias, only to mention the most familiar names, were introduced. The foundation of the Horticultural Society in London in 1804 was an important step forward. A few years later George Johnson wrote a history of garden architecture in England. He dedicated this work to Thomas Andrew Knight, the president of the Society, and a younger brother of Richard Payne Knight, author of the poem *The Landscape Garden*. Thomas Knight, who was quite different from his dilettante brother, advanced the gardening of his time because of his energy, his profound knowledge of botany, and his work as a hybridiser.

It was one of the functions of the Horticultural Society to send plant collectors into all quarters of the world. The plants which they found were tried in England, and subsequently an account of them was published in the Proceedings of the Society. One of the men sent out was David Douglas, who had good fortune in discovering conifers in America. He wrote to Sir William Hooker, the Director of Kew Gardens, when he was sending some new things: "You will begin to think that I am raising fir-trees just as I feel inclined." Early in the century the chrysanthemum and the wistaria were introduced. One of the most successful collectors in the East was Robert Fortune. In the year 1842 he sent from China and Japan *Anemone japonica*, *Dicentra* (*Dielytra*), and other plants which were destined to be great favourites. Dahlias, which had been introduced in the last years of the previous century, became very popular fifty years later. Fuchsias were brought in during the first half of the century; and about the middle came many orchids and innumerable quantities of hot-house plants.

Kew Gardens, founded by the Dowager Princess of Wales in 1759, soon acquired a European reputation. As early as 1789 between five and six thousand kinds of plants were growing there; and when a second list came out in 1810–13, the number had increased to eleven thousand. In 1839 the Botanic Society was founded in London, and did everything that could be done to help the study of scientific botany. The effect of this enormous

FIG. 606. BRANITZ—VIEW IN THE PARK WITH SOLITARY TREES

increase of plant material on the one side, and the growth of scientific botany on the other, with the accompanying knowledge of the geographical distribution of plants, soon become apparent.

The sentimental period had passed away, having lived itself out. The fashion of composing poetry and talking philosophy in such garden scenery as was supposed to express a definite thought or feeling—a fashion which appealed strongly to German theorists at the turn of the century—had dwindled away to nothing. Moreover, people grew weary of tricks and playthings in gardens. "*La nature*," says Alphand in the seventies, "*finissait par triompher de tout cet ameublement baroque.*" The garden artist was drawn more and more towards the actual plants, and since he wanted to get a real acquaintance with the bewildering number of things offered to his choice, his chief study must needs be botany. Thus the garden fell entirely into the hands of the gardener and the botanist, seeming to elude the architect altogether. People plunged into a study of the conditions necessary to plants, especially trees and shrubs, and were proud to think that "every tree and every plant had been assigned the place which Nature had intended for it, some on the mountain tops, some in the valleys, some in the shade, some in the sunny meadows, and others on the borders of the forest."

It was thus that the tree standing alone came to occupy the centre of the picture, as we saw so clearly in the grounds laid out by Pückler at Branitz (Fig. 606). Later on arose a delight in planting a pinetum or plantation of conifers, and fortunately a great many new conifers were to be had. In England especially the pinetum became a thing of great beauty, for people learned that they must control thickness of planting, which at first was excessive. At the beginning it was thought that the soil of England could not supply nourishment enough for the cone-bearing giants when they had reached their full stature, as they were tropical trees. The first pinetum was established in Kew Gardens in 1843. Plantations of this kind were in a sense the sign of a new influence which hailed from Italy, but these plantings were very unlike those of the Italian Renaissance, when dark pines stood in even rows, like pillars with a green roof. Italy herself had, however, changed; the new pinetum with its picturesque groups at the Villa Doria Pamfili was planted near the gates of the gymnasium, the old park itself being made as far as possible like the English.

Although botanical interest was so strong and active, it was inevitable that English parks, with their exclusive care for trees and shrubs, should become in the long run uniform and dull. The means of expression were limited. The cry for variety which had kept artists and owners on the move for hundreds of years was subsiding more and more. Fortunately, however, the plants brought over by explorers in foreign countries were not limited to trees and shrubs; on the contrary, there were many different kinds of flowers. Only a few of these could endure northern winters, and the first result was the erection of new forcing-houses. The orangeries of the old style had fallen into the background since the picturesque fashion came in, and new ones were not made. Thus the number of hot-houses in private gardens increased at the expense of orangeries.

In 1833 the Englishman Ward invented an air-proof glass case, on the principle of the circulation of water through earth and air at an even temperature, and the transportation of tender plants to Europe with comparatively little trouble was thus possible. People who cultivated flowers grew more and more skilful in hybridising the original plants that were brought over. We ought to remember, for example, that innumerable kinds of roses were raised in the nineteenth century. And very soon there arose great trading firms, which sent out their own explorers to every part of the world in search of new plants; indeed some firms, such as Veitch of Chelsea, Bull of Chelsea, and later Sander of St. Albans, and Vilmorin of Paris, had collectors scattered about in every country.

It was natural that people who had nursed tender plants through the winter and spring should want to enjoy them in summer in their own gardens. The most imposing effect was produced when they were planted in great masses, but in the North such an effect was unnatural, and it was not easy to produce. The method first hit upon was neither systematic nor artistic. Carpet-gardening, introduced into Europe by Pückler (Fig. 605), must be regarded as the first stage of a new alliance between flower-growing and the picturesque garden. The beds were filled with different flowers according to the time of year, and were mixed with plants that had various-coloured leaves. Unfortunately this ugly and stupid style is to be found in certain public gardens even in our own day.

The first attempts to reform the garden were made by two artists, and were started from different directions; but every attempt aimed at the same thing: to make a more worthy home in the garden for all the plants that now arrived in such numbers. In the forties and fifties of the nineteenth century the eyes of architects, especially Englishmen, had been turned once more to Italy. Sir Charles Barry, the English architect, travelled in the South, and especially in Italy when a young man. He took back with him to England a knowledge of Italian art as treated more or less from the historical point of view, and

FIG. 607. TRENTHAM CASTLE, STAFFORDSHIRE—GENERAL VIEW OF THE DUCAL GARDEN

applied it in a series of country places, sometimes entirely new, sometimes only altered and restored. The buildings which he designed show a strong likeness to Roman suburban villas, such as Villa Borghese and Villa Doria Pamfili. We feel the resemblance to the parterre of the Doria Pamfili when we walk through an "Italian garden" at an English country seat. Almost all Barry's work was done between 1840 and 1860. A piece was cut out of the picturesque garden, generally close to the house and as a rule only on one side of it, and was then laid out as a sunk parterre. The beds were edged with box, and here the treasures of the greenhouse were "bedded out," to be changed several times in the course of the year. There were fuchsias, lobelias, heliotrope, shrubby calceolarias, and in particular different kinds of zonal pelargoniums. Somewhat later there were begonias. These, with many others, formed a brilliantly coloured carpet of flowers. The corners of the

FIG. 608. DRUMMOND CASTLE, PERTHSHIRE—TERRACE GARDEN

beds were marked by dwarf trees, but as the whole parterre was to be allowed no shade, tall trees were banished. Wherever the ground allowed, this part of the garden was laid out in level terraces adorned with Italian balustrades, which contributed the chief or even the only architectural feature. Where terraces were not possible, the parterre was sunk, with the just belief that the view could be best seen from above. One famous example of a house of the period was Trentham Castle (Fig. 607) which Barry altered for the Duke of Sutherland. The open arcade and the balustrade on the roof made the chief decoration. The idea was carried out in very wide and rather low terraces, all with balustrades. Another ornament was found in small open summer-houses. There were also a few Italian statues, for example the Perseus of Benvenuto Cellini. The great lake at the end had its straight side next to the parterre, and its curving banks stretched out into a picturesque pleasure-ground. The garden was designed by Nesfield, the artist, who often worked with Sir Charles Barry.

All these Italian gardens showed the same character as a whole, though in particular parts and in particular circumstances they might differ. Colour effect, as produced by bedded-out plants, was more desired than general design. Most important English landscape gardens of the period had a semi-formal character. If one turns over to-day the pages of beautiful illustrations of English gardens which are given in the periodical *Country Life*, one is surprised to see how many of the older ones are faithful to the formal idea. Some examples are Harewood House in Yorkshire, Holland House in London, and Longford Castle in Wiltshire, where Italian parterres have been wrested from picturesque parks. In Scotland more especially we find a revival of the earlier terrace structures. To this class belongs Drummond Castle (Fig. 608), where the old plan of terraces received a much altered stamp from the new style of planting. Another English place, Shrubland Park, near Ipswich, which was laid out in the first instance by Sir Charles Barry, in conjunction with Nesfield, for Baron de Saumarez, could not have acquired the perfection of its terrace-building without the help of old examples of the same kind. Between an upper and lower Italian parterre there runs an elegant flight of steps at the head of five steep terraces, all bordered by balustrades, and emphasised at top and bottom by an open Italian summer-house (Fig. 609).

When once the historical interest in particular styles had become active, whether through imitation or revival, people attempted to accommodate what they already had to other types. When the Duke of Westminster cut out a flower-parterre from his picturesque garden at the neo-Gothic castle, Eaton Hall, Cheshire, he set up statues of knights and

FIG. 609. SHRUBLAND PARK, IPSWICH—VIEW OF THE TERRACES

ladies instead of Renaissance works of art. A dragon fountain was placed in the beautiful pond, and Gothic pointed arches were put on the balustrades. The time was gone by when the life and feeling of a nation expressed itself in one particular style, which had both educated and limited its own artists. Every style recorded in history lay open to choose from, and architects, from 1850 onwards, made it their pride to be able to build in any style they pleased. Sir Charles Barry, who introduced the Italian style into his

FIG. 610. NEWSTEAD ABBEY—THE MONKS' POND

own land, was the architect of the Gothic Houses of Parliament, and had also attempted purely classic buildings, even adopting the Palladian style of architecture.

This feeling for history, this eclecticism, which so characterises the art of the nineteenth century, tried to find a place in the garden parterre, but there it was never more than a name. People liked to put a French parterre by the side of an Italian, or even a Spanish; but it would have been a difficult thing, even after a close study, to say where the difference between them came in. Newstead Abbey, that old Augustinian monastery which has gained an imperishable renown from the name of Byron, had not seen good days under the ownership of the poet's family. His immediate predecessor ruined the park, but he set up a little fort and a tiny flotilla in the lake, which was all in accordance with the taste of the moment, and is described by Horace Walpole. The old estate showed signs of its monkish origin in having a fountain at the crossways, and also a fish-pond. After Byron's death the place passed into other hands, and the gardens were laid out in the prevailing fashion about the middle of the nineteenth century. There was now an Italian parterre, a French, and a Spanish. These had to endure each other's company amid the far from modern surroundings of the monks' fish-pond, called the "Eagle Pond" (Fig. 610). The beds of all the parterres were edged with box, and had Gothic railings.

The largest garden of the formal type which was made at that time in England was at Chatsworth, Derbyshire. The Duke of Devonshire, who was President of the Horticultural Society, wanted to make new gardens, and found a highly gifted man in young Joseph Paxton, who was on the point of going abroad, as he was in want of work. In 1826 the duke made him his head gardener, and commissioned him to lay out the grounds at Chatsworth. It is not easy to determine what was the state of the place when Paxton took it over, or how much was in existence, but certainly a good part of it had been laid out previously in the picturesque style. However that may be, Paxton took note of the main paths belonging to the old formal style and retained them. He and Barry were among the very first who set out parterres in formal lines.

As at Newstead, we find one Italian and one French garden. The latter at least has a certain claim to its name, because it had a *parterre de broderie* in front of the former orangery, and also statues standing on tall pillars at the end of it. These certainly did produce a somewhat French effect. But Paxton went beyond the parterre, and the great ideas and scope of the time of Le Nôtre were recalled by the axial line from the south front of the house leading to a canal-like pond at the very end of the garden. All the same, the wide gravel walks and lawns marked a great difference between this place and anything French. Paxton allowed the old cascade, which had been much esteemed, to remain in its place, but improved it. He restored water-works as far as possible. Where there were any weeping-willows he kept them. The style of the place did not differ widely from that of the English garden as it existed, under a certain amount of French influence, after the Restoration; but the way it was planted, especially in the western parterre, certainly gave a modern appearance.

Paxton was a person of clear understanding who reached what he aimed at quickly and in a practical manner. His own personality comes out in his works in the garden. His ingenuity was exhibited the most markedly in the great palm-house, which he began in 1836 and finished in 1840, and which became the horticultural wonder of the world. This iron and glass construction made it possible to cultivate great palms, tree-ferns, and other tall tropical plants in northern lands; and to create a garden of the Torrid Zone even in the depth of winter. It was 300 feet long, 123 feet wide, and 67 feet high. With its wonderful system of heating it created as much enthusiasm as the beauty of the tropical gardens which it enclosed. Not only was Paxton's building the model for many others, but it also served as a model for himself, when in 1850 his plan was accepted in preference to 233 others for the palace of the first great exhibition. The Crystal Palace at Sydenham, by which he made a great name, is one more in the long series of his buildings and gardens. A knighthood was conferred on Paxton for all he had done. He became a personal friend of the duke, whose service he had originally entered as an assistant gardener, and died full of honours.

Following Paxton's example, glass houses with their tropical gardens sprang up everywhere, and the Temperate House at Kew gained an international renown, which it still enjoys. It was now so easy to study foreign plants as they grew, and their conditions of life, that gardeners were emboldened, and not only bedded out exotics for decorative purposes in summer, so contributing a new touch to the carpet-garden as well as to the parterre, but also endeavoured to grow them outdoors throughout the year. The climate of some parts of England was favourable to such attempts, as for instance Cornwall, where camellias grew into large trees. The rhododendron was also successful, and about the middle of the century there were many kinds to be seen in gardens as well as in conservatories. In the south of England, especially in the New Forest, the rhododendron serves as a sort

FIG. 611. CHANDON DE BRIAILLES, EPERNAY

of underwood; and in gardens and parks it has often been made into great boskets. It was a favourite plan to set large beds at the sides of the long carriage-drives, which generally described a wide curve and were very ornamental.

Although in the nineteenth century the English were easily first in their zeal for seeking out and rearing new plants, development was not confined to them. The French carried out similar ideas logically and completely. The Revolution had cleared away nearly all the old gardens, and those that were left had been neglected. The works of the eighteenth century had passed away, and the new style of gardens fitted in with the new notions about the rights of man and the return to nature. All sentimental adjuncts had been cast aside, and nothing received attention except the natural qualities of soil and plant. Thus under the First Empire the picturesque attained almost entire mastery, and when the old nobility returned to their estates, so often found in a ruinous condition, they did not feel much inclination to restore their gardens in the old style. For one thing, they had no money, and for another they had learned from England new ideas. Nevertheless, there remained in some quarters a certain undercurrent of feeling for form as seen in the old gardens of France, and therefore it was possible for the garden at Chandon de Briailles at Epernay to be laid out as it was in the days of the First Empire (Fig. 611). Isabey, the miniature painter, the "darling of the *Incroyables*," made a picture of it. It was kept in the traditional, formal style at a time when nobody on the other side of the Channel would have thought of such a thing. The whole garden was laid out as a sunk parterre in three divisions, with a raised surround furnished with clipped trees. The middle part was formed by a large basin, the slopes at the edge being of grass and flower-beds. There was an orangery at the back with a semicircular parterre in front of it. To the right and left were the other two parts of the parterre, also with beds of flowers.

A few gardens of this type continued to exist in France until, towards the middle of the century, landscape gardens won the day, and became practically universal. In 1835 appeared Vergnaud's book, *L'Art de créer les Jardins*, in which the success of this style appears to reach its highest point. Vergnaud wrote as an architect, and warned people against letting their gardens fall into the hands of gardeners; yet he did not think of making any connection between the house and the garden; and even close to the house he could not tolerate anything formal. When he enumerated and described beautiful English parks, Chatsworth was a thorn in the flesh to him on account of its formality. Very soon, the only mistress of the garden in France as well as in England was the plant itself. Tree and shrub gained every advantage from the picturesque style. Flowers were permitted to encroach on certain fixed parts and display their brilliant colours. Carpet-bedding seems to have been carried out very early in France. Hirschfeld reproached the French for liking this wretched kind of decoration, but they kept to it, even up to our own day, more obstinately than the English.

In the time of the Second Empire there was a fashion for formal flower-parterres both in France and in England, and the French gave them the name of *jardins fleuristes*. Edouard André, a pupil of Alphand, who in the middle of the century was the unquestioned leader and chief of garden art in France, thus defines the *jardin fleuriste*: "Ground reserved expressly for the cultivation and arrangement in an ornamental way of plants that have beautiful leaves." We find attempts at the Gothic and Italian in a pseudo-historical manner, but difference of style was only apparent in the balustrades and statues. A formal bed at the Gothic castle of Bois Cornille was supposed to be laid out in the Gothic style.

We have already considered what was done by Prince Pückler in Germany in the first half of the century. His importance lay in the lofty enthusiasm with which he maintained the interests of the garden, though he largely shared the weakness of his own day, in that he had no notion of how to lay out the parts near the house. It is difficult to see how the colour-gardens of which he speaks, the blue and the yellow, were managed. He may have taken the idea from Paris, possibly from Parc Monceau. He certainly had carpet-beds with flowers of many colours. Peter Lenné made a great name in North Germany as director of the garden at Potsdam. It was perhaps only the respect for the past felt by the master of the house that prevented the whole garden of Sans-Souci from being turned into a landscape park, with even its axial lines obliterated, for we know that Lenné had already executed designs for such a change. Happily he was content with having boskets all round the place, but even then much was lost of the original form. Lenné mellowed as time went on, and when he had Schinkel as a fellow-worker, and as patron a man of so much artistic taste as King Frederick William IV., he helped to create

II—Z

at Sans-Souci a work typical of the time of the Italian Renaissance. Under this king Potsdam with its surroundings grew to be one of the finest princely seats in existence. As early as 1825, while crown prince, he received a small piece of land to the south-west of Sans-Souci as a gift from his father, and the next year he began to build the house he called Charlottenhof, in the style that was then considered to be classical Italian. Schinkel infused a strong flavour of his own masterful personality into this type of building when he was working as architect in Berlin, and by his treatment had given it much character and individuality. He had a great plan for Charlottenhof, where there were to be gardens with pillared corridors round them, and ornamental parterres in the Italian style. The limited means of the crown prince would not permit of a large house. He seems to have had the Villa Albani in mind, and made a terrace (Fig. 612) which ended in an arbour of vines

FIG. 612. CHARLOTTENHOF, SANS-SOUCI

with a Roman seat and a shell fountain. Certain parts of the place must have reminded him of happy days spent in Italy, for here were the so-called Roman baths, which properly represented an Italian *villa rustica*, with its impluvium and its terrace.

When he became king, as Frederick William IV., he conceived far-reaching plans for embracing not only Sans-Souci but also Potsdam itself in a general scheme. His father had already greatly enlarged the circle of princely castles which lay around Potsdam, and had given a special seat to each of his sons. Babelsberg was assigned to Prince William, whose taste differed from that of his brother, and he had this house, which was in the neo-Gothic style, set in the middle of a real landscape garden. Its beauty, apart from the old trees, mainly lay in its fine view over the Sacred Lake. Prince Charles, whose taste resembled that of his elder brother, decorated Glienicke, which fell to his lot, with a great many small sites imitating an antique style, but also with some very fine genuine antiques. Frederick William now wished to close in the circle of stately gardens on the north-west, and had an idea of connecting the Pfingstberg, where his father had had a little tea-house, with the new garden at the Marble Palace by means of a handsome terrace construction. Only the castle and the upper portion of these terraces were ever finished; and the town, as a fact, made its way in between the various parts of the circle of gardens. What was actually made was not really a series of garden terraces, which would have had to be continued farther down. The building was a curious and unmethodical mixture, reminiscent of Italian and Moorish styles, and it produced a bizarre effect.

The king had some plans for developments at Sans-Souci, but these were only partially carried out. His chief innovation was the new orangery, which very clearly showed the Italian ideal that then dominated Northern Europe. Towers, pierced and furnished with loggias, were connected by corridors on pillar supports, and these were the most striking feature; then three terraces, of which the highest one had a flower-parterre on it, descended to the main road. The most conspicuous of the smaller parts was the pretty Sicilian garden. The king had it put in the place of the old orangeries of Frederick's day, and laid out as a *jardin fleuriste* in front of a fine balustraded wall furnished with alcoves containing many statues. Unfortunately the effect of this graceful work was utterly spoiled by the introduction in 1902 of the colossal bronze figure of the Archer into the foreground. What Frederick William accomplished at Sans-Souci was superior architecturally to anything done in the way of Italianised gardens in other countries, but even in Germany it remained a personal and unique fancy of the king. Lenné was dragged in on the artistic side by the king's strong influence and by the powerful personality of Schinkel. Whenever Lenné designed gardens outside Potsdam, he adhered strictly to the forms of the *jardin fleuriste*. In the Instruction Book, which was published in 1860 by his best-known pupil, Gustav Meyer, the ideas of the time, and principally those of Lenné, were adopted for the treatment of the formal parts. "The amount of space to be used for laying out formally near the house must be dependent on the size and style of that house. At a country place it is necessary to have formal paths and a couple of rose-trees on the lawn. Palaces need more, but Nature must never be altered by clipping or the like, and Gothic buildings need very little regularity." Such teaching in a German treatise, which enjoyed unbounded popularity right on to the end of the century, showed how little account was taken of places of this kind. Lenné's own heart was entirely in favour of the free development of a natural style, and he found great opportunities at Potsdam, at the park at Babelsberg, at Glienicke, and in the garden of the Marble Palace; for at all these there was much that he could do.

Even more peculiar than the great work at Potsdam—indeed one might say absolutely unique—were the works of King Ludwig II. of Bavaria in the second half of the century. The Bavarian royal house had early shown an inclination for the formal style. Ludwig I. took pains that the long-neglected garden of Schleissheim should have its parterres made once more after the old drawings, and there were similar plans for Nymphenburg. King Max wanted to build himself a castle in the unfinished park of Feldafing, after Lenné's plan, and with formal surroundings, but this remained no more than a vision. His chief gardening activities were restricted to the hanging winter gardens at the Residence in Munich. There he wanted to have scenery of the kind found in Upper Italy, so the tiresome greenhouse was to be concealed as much as possible on one side by evergreen oaks and fig-trees, and a pergola on pillars with a fountain at the end of it must entice one's steps forward, and give the illusion of a garden in a southern land (Fig. 613).

His son, yielding to a love for what was disproportionate and fantastic, played a higher trump card than this winter garden, and combined the flora of India with the architecture

FIG. 613. WINTER GARDEN OF LUDWIG II. AT MUNICH

of the Moors. However, the lake which he put on the roof of the house soon proved to be insecurely supported from below. Soon after his death, the garden perished, like so much else of his work; and, as he had strictly excluded the public, very few people can ever have seen it except himself and the gardeners. The gardens which he laid out round the castles were much larger and still more fanciful. Ludwig was like Philip V. of Spain in that he was fond of lonely valleys in the mountains. As Philip felt about San Ildefonso, so Ludwig felt on seeing a little hunting-box at Graswangthal near Oberammergau, and he wanted to make a Versailles in the solitude of the hills. This scheme was never carried out at Linderhof, the place originally proposed, for the king changed his mind and selected another, the island of Herrenchiemsee, which was still less suitable. At Linderhof, however, he had a great terrace-garden made in the later baroque style. He had no perfectly clear pattern before his eyes, and the place reminds one most of certain castles built by Augustus the Strong, and in particular of Gross-Sedlitz, for in the same way the main axis passes right across the valley, and has to ascend the slope on both sides. At Linderhof (Fig. 614), so called from a lime-tree which was kept standing in the middle of the architectural design, the castle stands on almost the lowest possible level. Close behind was a cascade. On one side stood three imposing terraces in succession to a sunk parterre; they mounted up by elliptical stairs to a temple, and each terrace carried flower-beds graduated in size. The most successful parts of the arrangement are found in the small *giardini secreti*, which are attractively laid out at the side of the house, with parterres bordered by trellis paths.

The whole formal picture was set in a landscape which was treated rather carelessly. On either side of the water stairway, parts of the hill-slope were brought into the formal portion of the grounds by the aid of semicircular leafy paths. Unfortunately much of this is unfinished, and all of it very badly preserved.

People regarded the whole idea of an architectural garden at Linderhof as a mere freak or hobby of the king's, but even less defensible was the disastrous imitation at Herrenchiemsee of the mighty model of Versailles. In the first place, the site was a little island, about two kilometres in length, and, small as it was, it could not all be used; moreover, it had nothing in common with Versailles. To give the appearance of a canal, dams were put in the lake, and planted round with hedges. It was very characteristic of the time that the parterre from the Latona to the Apollo fountain was copied exactly, whereas the boskets were reduced to a minimum, for there was no intelligent grasp of the scheme. Ludwig was perhaps told that the two large fountains in front of the castle had central groups originally; in any case he adorned them with copies of groups in the park at San Ildefonso. Ludwig's gardens have remained the monument of a too excitable personality.

In the course of the nineteenth century it came about that princely gardens lost the special interest which attaches to the best models. Public gardens, which grew ever more important, and being every man's property alike, captured all hearts and all eyes, took their place. In the course of our history we have over and over again met with public gardens in towns, where all the people could go. When the Greek *polis* became democratic, the first real city park found its way into the gymnasium. Cimon embellished places like the agora with shady trees. In towns like Antioch and Alexandria we have seen how

FIG. 614. LINDERHOF

Hellenistic cities developed in a modern spirit; they took over their inheritance from Greece and then extended it to a size and magnificence truly oriental. In Rome the emperors were careful that round the narrow crowded dwellings of the townsfolk free spaces for recreation should be provided in a belt of gardens and beautiful grounds. The development of the public garden took a different direction in the Middle Ages. No doubt the burghers found open walks in the gardens of the guilds before the town gates, but they did not need them much, as the pasture-lands were so near at hand. Afterwards, in the days of the Italian Renaissance, the fine private gardens of the gentlefolk came into existence, and it became a point of honour to open them to the public. Travellers from northern countries, where the feeling of the Renaissance was not so fully active among the townsfolk as it was in Italy, and the love of private possession was much stronger, recognised this with surprise, noting in Rome, above all other places, what they considered the liberality and magnanimity of the rich. It is seldom indeed that we hear of hospitality being extended to all comers in the patrician gardens of the great towns of the North, as it was in the Roman gardens of Montaigne's day. However, northern princes became more and more imbued with the new spirit in the seventeenth and eighteenth centuries. Whenever possible the smaller princes made their residence in garden cities, and most of the parks were thrown open to all their subjects. The people were not, however, at home there—not the real masters. True, there were not many places where an inscription was put on the front gate threatening common people with a cudgelling if they presumed to sit down on a seat where some noble visitor wished to sit; but this was actually done at the entrance to the Herrenhausen garden; and particular gardens were often closed at the owner's pleasure; in Paris, for example, places which the people had supposed to be theirs by right were suddenly closed to them by some caprice of the actual owner. Thus, in 1781, the Duc de Chartres closed the Palais Royal Garden, which had been open ever since it was founded by Richelieu. In 1650 Sauval writes of the Luxembourg Garden, "It is often open and often closed, just as it may please the prince who is living at the castle." The Duchesse de Berri had all the doors but one blocked up, so that she might be undisturbed at her gay parties.

In England Queen Caroline, intelligent though she was, held the views of the despotic little court which she came from, and had the fancy to shut up Kensington Gardens. She inquired of Walpole, who was at that time her Minister, what it would cost, to which he gave the significant reply, "Only three crowns." In London the great parks were the property of the Crown, though in the eighteenth century they were completely given up to the use of the people. Queen Caroline took a lively interest in these parks, which crossed the interior of London like a broad green belt, and particularly in Kensington Gardens; the fine avenues and the great basin in the middle are due to her. This garden never quite lost its formal character, whereas in the reign of George II. Hyde Park was converted into a picturesque garden, with the artificial long lake that is known as the Serpentine. Both in London and in Paris parks of this kind were an indispensable theatre for the world of fashion and wit, as it existed in the eighteenth century. But the people also gained from the parks, particularly in England, for they served as large club-rooms, and provided open spaces suitable for public meetings. During the nineteenth century it was the citizen class which really carried out the traditions of the parks.

In France the intellectuals were delighted to meet one another in the garden of the

Palais Royal. It is the scene of Diderot's work, *Le Neveu de Rameau,* and there many of those ideas were hatched which matured in the great Revolution. It is curious to notice how the rendezvous of fashionable society changed according to the caprice of the owner of the place or the leading coterie of the moment. The Champs Elysées and the Cours de la Reine were among the most famous meeting-places for people of fashion in the earlier days of Louis the Fourteenth's reign, but when he became less attached to Paris they were neglected. In Louis the Fifteenth's time Madame de Pompadour lived at the palace of the Champs Elysées, and considered making a view right through to the dome of the Invalides, but this she only carried out in part. The Champs Elysées remained in high favour until the end of the First Empire, and Napoleon set up the Arc de Triomphe to complete the view. After the Restoration people turned away from this part of the capital, which was associated with so much bloodshed and misery, and allowed what the invaders had spared to fall into decay. The promenades were gradually made what they are now, and came to form a prominent feature in modern Paris.

The parks, all open to the public, cannot be regarded as of much importance from the point of view of art in the earlier half of the nineteenth century. The special interest of the princes was cooling down, and as yet there was no active superintendence, for no one had a right to undertake the care of them. Most of what we hear about the parks in Paris during the first three decades is a tale of decay, and the ornamental parts disappeared. Dead trees were replaced in the long straight walks, but little more was done. There was hardly a trace of the picturesque style in these parks. Even as late as 1835 Vergnaud complained that the French would never get rid of the notion that a public park had to be a formal one. He proposed to include the whole of the Tuileries Gardens as far as the Bois de Boulogne, and to convert the place into one connected park such as would be made in England.

At the time when several of the chief streets of London were being altered under the auspices of Nash the architect, and so came to look very different, especially Regent Street, there was an attempt to beautify the belt of park-land. St. James's Park was remodelled, and its long canal received the new form of a natural lake with a wavy outline: everything else was in a style to match: the single trees, the meadow-like surroundings of the woodland parts, and all the rest. As early as 1811, when an insignificant and neglected park at Marylebone became Crown property, an Act of Parliament was passed by which a public garden was to be provided for the north of London, called Regent's Park after the Regent who was later on to become King George IV. In the design for Regent's Park there appeared all those fixed schemes which were now being carried out in every public park: there was a large lake of the picturesque kind for rowing-boats, and by the side of its winding banks there were shady walks, with views continually changing; there were wide meadow-grounds, suitable for various games and for large gatherings of people, and these various arrangements filled nearly the whole space. In fact, here we have the type of park which is so familiar to everyone who knows London. The architect had nothing to do with it; indeed, great efforts were made to conceal the neighbouring houses by planting trees in front of them. The refreshment-houses inside the grounds were small, and had to hide among the trees as well as they could. By the middle of the century London seemed to set an unrivalled example in the size and beauty of these parks. A total area of 1200 acres was covered by Kensington Gardens and Hyde

Park, with the Green Park and St. James's adjoining, and Regent's Park on the north. But Paris also was very well off with the Tuileries Gardens, the Champs Elysées, the Palais Royal, and the Parc Monceau on the right bank of the Seine, and the Jardin des Plantes and the Luxembourg on the left. In both towns the parks were Crown property.

In the forties, consequent on the increasingly flourishing condition of the towns, there suddenly came a new advance in the matter of public gardens. Now, for the first time in the history of the garden, America took an important place. It was not that she had not progressed, but that she demanded here, as for her art in general, a certain independence and originality. After the American Civil War town-dwellers began to build houses in the country, which served them for a rest and holiday during a part of

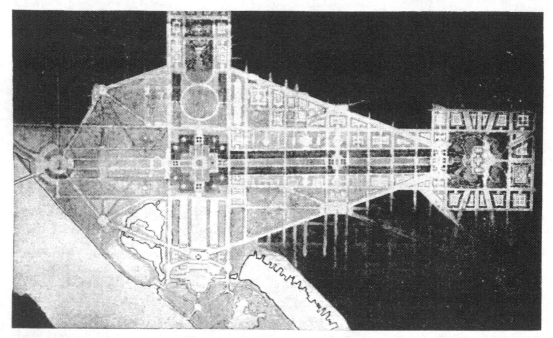

FIG. 615. WASHINGTON—L'ENFANT'S PLAN

the year only, or for week-ends. In architecture and in garden planning America made progress side by side with Europe, or more accurately, followed its leader England. As early as 1682 William Penn had laid out Philadelphia according to a regular plan, with square ornamental plots; and at the end of the eighteenth century the French architect, L'Enfant, had made, at General Washington's desire, a complete plan for the town which bears his name. It resembled one of the great residential towns of Europe, transported, so to speak, *en masse*. The central point was the dome of the Capitol; broad avenues were to form an approach, with boskets and parterres at the side (Fig. 615). But the plan was only on paper, and a hundred years passed before a fresh movement towards laying out park-lands carried it through, and on a grander scale. When, in consequence of the amazing growth of her population, due to the influx of emigrants, America had to deal with New York, provision was made by laying out a great park in the very middle of the city, about 850 acres in area. This was a municipal act which deserves all praise. Frederick Law Olmstead, the most important landscape artist that America had produced,

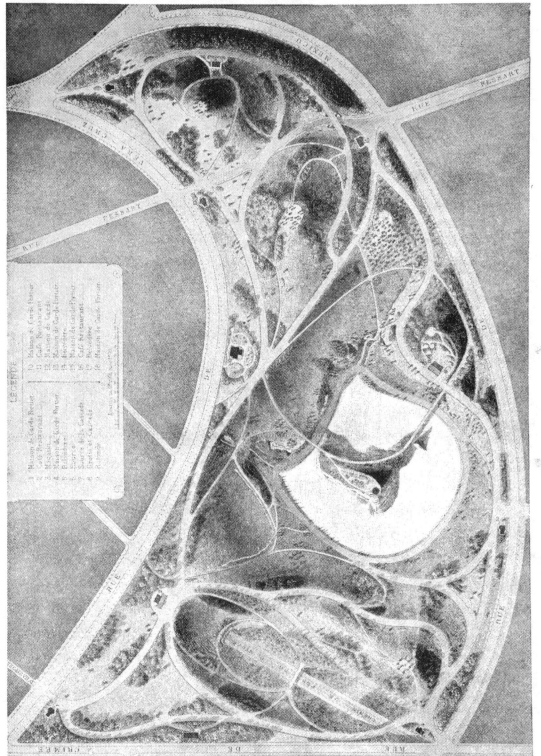

FIG. 616. BUTTES-CHAUMONT, PARIS—GROUND-PLAN

laid out the park in 1854, in the picturesque style universal in his day. He wanted to effect a strong contrast with the distressing surroundings of the great city. The park was bounded by a wall, and had some extremely beautiful parts.

The first American parks owed their existence and their character to the desire for some refuge for citizens from the nerve-racking life of a great town. They were meant to be a real incursion by Nature into the ever-increasing sea of houses. General wheel traffic was not allowed, and vehicles could only pass through by four sunken roads spanned by bridges. This creation of people's parks was among the first acts of importance

FIG. 617. BUTTES-CHAUMONT, PARIS—THE ROCKS

done by citizens for the good of citizens, of a democratic power that had now grown to maturity.

At the same period, but with less to contend with, all Europe was making ready for similar efforts. In 1852 Paris took over the Bois de Boulogne from the Crown, on the understanding that in the next four years two million francs were to be expended on improvements. Many changes had the Bois seen since in 1528 Francis I. erected the grand Château de Boulogne in the middle of the wood, since Henry II. had put a wall round it to make a hunting-park, and since Margaret of Navarre had built the charming little castle of Muette. Louis XIV. had had the wood pierced by wide avenues, with crosses at the intersections. The Revolution made its home there in terrible fashion. Later on Napoleon had seen in the site a great opportunity for associating his name with a place whose size and grandeur could compare with what Louis himself had created. So after the little King of Rome was born, Napoleon thought he would build a marble palace where the Trocadero is now, the gardens of which should extend as far as the Bois. On the other side of the palace the Champ de Mars was to form a great promenade, flanked with public buildings. Two skilful architects, Percier and Fontaine, were commissioned to carry out the project, and several months of work had been devoted to the levelling of the hill and the making of terraces when the Battle of Leipzig put an end to it all. The Bois was often the field of operations for troops in 1814, and was nearly given over to destruction. Its preservation meant for Paris the retention of her finest park-site.

The place was laid out eventually by Hittorf, the architect, and Varé, the landscape gardener, in confessed imitation of the parks in London, and especially of the lake features of Hyde Park. We must look upon Alphand as the real author, for his name is bound up with the creation of new and the restoration of old parks, in the same way as Haussmann's name is connected with the altered appearance of the town itself after the great new streets had been formed. The Buttes-Chaumont (Fig. 616), which was considered at the time a triumph of garden art, was made a little later. A very short time before it had been a wretched, squalid place, with a gibbet and a skinner's yard—lurking-place for criminals. This unpromising site was converted into a fine landscape park in the early sixties. The steep walls of a chalk cliff were partly pulled down and partly raised. The whole had a romantic hilly appearance (Fig. 617).

Paris intended to have parks at all four corners of the town. On the south Monsouris was added. The Bois de Vincennes was restored, for the park attached to the ancient royal castle had come into the possession of the town.

In London the royal parks, great as they were, soon had to be added to. More and more thought and care were given to the subject, especially with a view to providing parks, great or small, for the districts east and south of the Thames, which hitherto had been much neglected. Victoria Park and Battersea Park were the largest, but a considerable number of smaller ones were added to these. A few figures will show how London was feeling the need of parks and was trying to meet it: in 1889 the whole area of town gardens and ornamental grounds, with the exception of royal parks, was 2656 acres, whereas in 1898 it was 3665 acres.

A peculiar system prevailed in the case of certain ornamental plots. Originally the so-called London squares were semi-private gardens, to which nobody had the right of entrance except the people who lived round them. For this reason they were generally enclosed in some way. There were hedges or at least bushes cutting them off from the streets; and within there was a garden of the picturesque type laid out with paths, groups of trees and shrubs, and even a small lake: There might also be a couple of carpet-beds on the lawn as the chief decoration. In many cases, however, the private character of the places was lost, passers-by could go in, and the places were treated as ornamental appendages to the street. Paris adopted the name "square" as well as the style of laying out, and Alphand in particular planned a great many of these (Fig. 618).

Germany did not hold aloof from the movement for making parks in the towns, and one of the smaller towns, Magdeburg, had the honour of being the first to come forward. As early as 1824 Lenné was commissioned to lay out a public park there. He replied in a letter that plainly shows how unusual this was: "It is nothing new to me that princes and wealthy private persons should spend large sums on the beautiful art of the garden. But an undertaking of this kind, which from a rough computation will cost, exclusive of buildings, no less a sum than 18,000 dollars, undertaken by the town authorities, is the first example I have ever encountered in the whole of my life as an artist." Berlin could

FIG. 618. SQUARE DES BATIGNOLLES, PARIS

not lag behind. At the centenary festival of Frederick the Great's accession, the town
council at Berlin resolved to lay out a People's Park, to be called Friedrichshain, on the
eastern side of the city. No happier occasion for inaugurating a town garden could have
been found. It was a way of showing homage and gratitude to the creator of Sans-Souci,
who loved gardens, and also a proud demonstration of the growing spirit of citizenship,
which now felt prepared to take out of the hands of royalty the ornamentation and the
hygiene of the towns. But the Crown made a presentation to the people at the same time,
handing over the garden for wild animals as a public park. Ever since the sixteenth century
this woodland tract of about 600 acres had been in the possession of the Princes of Brandenburg, and its conversion into a pleasure-park was due chiefly to Frederick the Great,
who had the avenues made in the shape of a star with basins and statues, small green
rooms and labyrinthine walks. Later on, one part of the grounds, to the south of the
great star, had been changed into the English style; the large lake and the Rousseau island
are really picturesque in the best sense of the word. On the north of the great star there
is also a landscape garden at the Castle of Bellevue. In 1840 Lenné was entrusted with the
modernisation of the Friedrichshain and also of the animal garden. But whereas the
first was laid out entirely after the pattern of an English park, in the second he kept the
avenues and walks of the old place, and confined his renovations to what was already in
progress on the southern side, where he made a second picturesque lake (Fig. 619). It
seems as though the spirit of Frederick William IV. must have had some share in this,
at least there is one sketch in existence, which he made while crown prince, of a
garden round a great hippodrome, laid out in a purely classical style, to which he wished
to annex a portion of the animal garden. The great avenues which still cross the park were
in one way a drawback; for traffic did not cease, as in the American and English parks,

at the outer boundary, but was everywhere carried through, and consequently the roads were cemented or paved, which is out of harmony with the park spirit. In 1837 Gustav Meyer, a pupil of Lenné, and author of the Instruction Book, was appointed the first director of the town garden at Berlin, and he was very active there until his death. He laid out all the parks at Berlin including the later Schiller park, and his advice was asked for by many other towns.

In this first period the idea was before all else to make the towns beautiful. People were quite satisfied to make one park after another according to the English pattern described above, wherein certain æsthetic principles prevailed, but which grew more commonplace and spiritless as time went on. England differed from other countries in one respect: her fields and lawns were thrown open to the people, whereas on the Continent there was always a peremptory order, "The public must not walk on the grass." Wherever possible, the uniform level of a park was varied by an imitation of a hilly landscape, as at the Buttes-Chaumont in Paris and the Victoria Park in Berlin, the latter being an imitation of the fall of the Zacken River in the Riesengebirge.

For a fairly long time public parks were unaffected by botanical interest. It is true that foreign trees and shrubs were planted, so far as climate permitted, and so far as they were needed for picturesque grouping; but the rarer and more expensive kinds, especially those conifers which gave a distinctive character to private gardens, were generally absent. Uncommon exotic plants, especially flowers, were relegated to the Botanic Gardens, where

FIG. 619. THE NEW LAKE IN THE TIERGARTEN, BERLIN

artistic arrangement was superseded by purely scientific division and classification. There were scarcely any flowers in the parks in these early days, only stiff carpet-beds in the squares, on the promenades or on the ramparts of fortifications.

In due course civic authorities were faced with far wider demands. The democratic feeling of the masses grew until it exercised a powerful and irresistible influence in all domains of art, and it now turned the treatment of public gardens into new directions. Once more America pressed to the front in the development of People's Parks. The enormous growth of the population in many of her towns was felt to be a menace. If the people were to be saved from asphyxiation there must needs be more open parks.

Chicago made a successful experiment by converting a sea of houses into a place that has earned the honourable title of "Urbs in Horto." It was said, no doubt with truth, that before the change was made there were fewer green trees in the whole town than there were rooms in one of the gigantic business houses. Chicago's plan, which succeeded, was to separate the blocks of houses at fixed intervals by interposing People's Parks, larger or smaller as circumstances would allow, and varying from two and a half acres to sixty-three. In a comparatively short time about twenty-four playing-grounds were made, at a cost of 42,0c0,000 dollars, and one or other could be reached from every house in the town in a few minutes. The features which are found in all these parks, even the smallest, are a football ground with walks round it, gymnasium, a playground for children with a shallow pond in the middle—and finally a swimming-pool with baths attached. The larger parks have facilities for rowing. There are club-rooms, after the pattern of private clubs, with a large central hall and private rooms for meetings. Round the inner park Chicago has set a belt of outer park that takes in more and more ground as time goes on (Fig. 620). The Grand Park in the south is one of the finest. In addition to all this, Chicago planned a Lagoon-garden, a thing never heard of before. The Grand Park had been made on a mound artificially constructed from the town rubbish-heaps. For the Lagoon-garden the refuse was to be used in a systematic manner, so as to form tongues of land in the lake, 100 to 300 metres in breadth, each strip to be laid out as a garden, with the quiet waters of the lagoon between, providing a shore for bathing.

What Chicago, that great centre of trade, did, other towns accomplished in their several ways. Boston made park-like streets radiating into the interior of the city from a great belt of park outside. Washington, St. Louis, and Philadelphia all laid out magnificent streets inside the towns. The administrative bodies of the giant cities of America consider it one of their chief duties to provide parks and garden grounds, and there are Park Societies whose aim is to support the municipal authorities in their work.

In the Old World, Paris made not very long ago a circle of parks where the belt of fortifications used to be. This was effected by means of a grant that extended to milliards of francs, and even America was amazed. At Vienna plans were considered for laying out a belt of meadow- and wood-land, with the same objects in view. In Germany societies were formed to extend parks. The efforts of the towns to develop gardens and parks seem to extend to private property in Germany. The Schreber gardens, as certain small plots are called, have acquired an importance which may increase. About the middle of the last century a certain physician of Leipzig, Dr. Schreber, made over to the town a considerable sum of money on condition that land was bought and leased out to the citizens in small garden plots, about two hundred square metres in area. Private societies,

FIG. 620. CHICAGO—PLAN OF THE GREAT PARK AND HARBOUR

called "Schreber-Vereine," carried the plan farther, and subsequently enlarged the grounds by the addition of gymnasiums and halls. The playgrounds were put in the middle, so as to connect the tiny plots of garden, which were left to each individual to cultivate. The intention was to give poor people pleasure in a little bit of land. The plan was copied in many towns, and in 1901 Schreber gardens were laid out at Breslau.

The Garden City movement owes its origin to the development of parks. The idea arose in England of checking the ever-growing congestion of large towns, by making little towns outside. These recall the Residence towns of the eighteenth century, but in place of the castle garden which as a rule was in the centre, there is generally an open space for games, and streets radiating from it with small separate houses, each with its own garden. The leading idea was to provide light and air for the working classes, and to lead them back to Nature, which they had ceased to know. It is not a part of our subject to discuss the economical and social bearings of this movement.

Space for games became increasingly in demand in the public parks. America had imbibed, chiefly from England, the love of sport, of games in the open air, and had cherished it in a more democratic fashion. There were two demands to fulfil: large level grounds must be acquired, with a view to people's fêtes and meetings, and also to provide sites for sport and games of every kind. The older parks were adapted for walking in— "the ideal place for taking a walk," as Meyer says; and people had not yet quite broken away from the delight in views. A solitary stroller, or a small party of friends, found pleasure in a constant variety of pictures, and it was a real triumph of beauty when some lake, some meadow, was so placed that, owing to ingenious turns in the path, it was impossible to see where it ended. But it was only the quiet, observant strollers who noticed these things; the majority wanted to meet for games in smaller or larger parties, and to mingle with their fellows so that each individual or each group should feel like a member of a corporate body without, however, being overwhelmed by numbers. The games ground, which is best in a regular shape, must be open to the view of players and spectators alike. The places designed for people's fêtes and shows must not be lost in woodland surroundings. There must not be deceitful effects of distance, with twists and turns. The public wishes to see and to be seen. The water, which was not, as in early days, to serve only for people in rowing-boats, who might prefer a winding course, must now follow a new plan, so as to serve for swimming-baths and for skating. All these requirements led more and more towards a formal design; yet the public park, so strongly bound by tradition, could hardly have found the way to an essentially new type had it not been helped by a great movement which came from another direction.

A fresh victory was gained by the architectural garden, but it did not spring from the public park, nor from the private gardens of princes: it was democratic, starting from the small town house and its garden, and extending far beyond. It was a movement from outside, like the old invasion of the landscape-garden style which came from the artists, but this time it was the architects who took action. Long in the background, they now became conscious of their rights, and seeing that the time was ripe, entered the lists. In 1892 there appeared a little book called *The Formal Garden in England,* by the architect Reginald Blomfield. For the first time the view was advanced, in outspoken language, that landscape gardens are in bad taste and an absurdity. Blomfield put the question whether the garden is to be considered in relation to the house or whether the house is to be ruled out when the garden is arranged. Only ill-wishers, he thought, would call the old style formal: it ought to be called architectural, for its object is to bring house and garden into harmony with each other, and to let the house grow out of its surroundings; otherwise one has to transgress the laws of architecture, the inner order of firmly fixed lines of symmetry, or at the very least of proportion.

Landscape gardeners have intentionally avoided this question, and with a half-concealed affectation are wont to speak of some sort of connection between house and garden, following a "method" which is the systematic avoidance of all method. Blomfield attacked the chief maxims of this unsystematic method, and especially the cardinal doctrine of the imitation of Nature. He asked what nature really is, and what is "natural" in connection with a garden. In his opinion the landscape garden is just as artificial as any other. Nature in herself has nothing to do with either curved or straight lines, and it is an open question whether the natural man would prefer a straight or a crooked line. As to the realities of nature and the forces we work with, a clipped tree may be quite as natural as a woodland tree, and it is no more unnatural to clip a tree than to cut grass. The landscape gardener turns his back on architecture, that is, the house, so as to unfold natural scenes; his chief aim is to create a deception as to the size and surroundings of the garden and of its different views. Writers on the subject, Blomfield said, concern themselves as little as possible with the question of art in the garden as a whole, but hurry on to their own special interest—horticulture and hot-houses. This province he would gladly leave to gardeners. Horticulture ought to be subject to the architectural plan just as building is; and it was the intention of the book to restore to the architect the province which had been stolen from him. All Italian and French ideas were as far as possible discarded, for the style that Blomfield was anxious to get rid of had only too often attempted to make an unmethodical and worthless compromise with these.

The only feature approved of as English was the little Renaissance garden with a protecting border, which in a direct way, and with no deception about it, cuts off this plot from its unsuitable surroundings, and connects it with the house most harmoniously, by the help of brightly coloured beds and patches of green shade. But in this department also there was to be a reform, and the small town and country gardens were to be laid out in another fashion. Blomfield held that the unnaturalness of the "natural" style is shown more tiresomely and more stupidly in small gardens than elsewhere, because there the connection with the house is most obvious. The smaller the place, the more the bad taste of the gardener was apparent, with his kidney-shaped patches of grass, his paths twisted and tortured, his artificial mounds packed with trees and shrubs, and his carpet-beds with a group of foreign foliage plants in the middle. There were thousands of such designs in the smaller gardens which proved only too clearly Blomfield's contention that architectural sense had been lost. The little book made a great impression in England; and its effect was all the surer, because the author took no separate steps to arouse men to action, but let the result follow as it might.

We have seen how interest in gardening proper had been diverted to botany, and this fact comes out very clearly in literature. Books about botany and horticulture appeared in great numbers, and the place taken in them by real garden art soon reached the vanishing point, so much so, indeed, that one of the later writers on æsthetics, Heinrich von Stein, speaks of it with the greatest surprise, saying that in former days people actually regarded gardening as one of the fine arts. The garden had so ceased to be a work of art that it had become merely a place where trees, shrubs, and flowers could develop in their individual ways and in as favourable circumstances as possible. In the eighteenth century the architectural feeling for proportion and harmony, for the sense of space, as one might put it, had gone, but in place of it there was at least sentiment, and people consciously drank in the beauty of the picturesque scenes which were unfolded to their view; but even this was lost in the course of the nineteenth century; indeed it was bound to vanish entirely, for all thoughts and all efforts were now directed to the nurture of plants.

At the end of the seventies, when a strong interest sprang up in England for arts and crafts, people turned their eyes to the treatment of houses, both outside and in, and an impetus was given to gardening also. William Morris, who did so much to revive various art industries, said that a garden, great or small, ought always to be entirely planned and be good to look at. A garden, he thought, ought to be shut off from the outside world, and should on no account imitate the caprices and wild conditions of nature, but should be a something not to be found anywhere except close to a house. Others had expressed opinions, sometimes whimsically, to the same effect. As early as 1839, the architect T. James said that if he must have a system, let it be the good old system of terraces and right-angled walks, with clipped yew-hedges and splendid old-fashioned flowers shining in the sun against their dark-green background. He liked topiary work because of its confessed artificiality, and because it discarded the deceitful cowardly maxim *Celare artem*. The carving of trees and shrubs was a natural transition from the architecture of a house to the untouched beauty of meadow and grove. Even Blomfield could not have expressed himself more clearly than this. Again, Sedding, one of the leading architects in the sixties, made much the same protests as Blomfield.

Practical results had come about from these teachings, and probably the best was Penshurst, which Blomfield praised as one of the most beautiful gardens in England, not large, but very tastefully arranged. At the side of the sunk flower-parterres, where an attempt was made to cultivate old-fashioned flowers within box borders and in simple beds, there are orchards shut in by tall hedges treated in the style of boskets, and joining on to the garden. But the most attractive feature is a long path between hedges leading to the pond-garden, a sunk basin with sloping sides of mown grass.

Landscape gardeners advanced in serried ranks against Blomfield, and in the nineties there arose a feud which was carried on with extreme bitterness. In the second edition, which was required in the same year, of Blomfield's book, he added an argumentative preface, explaining his views on art, and even going beyond his earlier standpoint. What he started from was not a question of fashion, but of principle. It was simply the eternal question of art, how far is man the slave of Nature? Or more accurately, how far ought man to subject the expression of his ideas to an actual imitation of what one may for this purpose call (unscientifically) Nature in the rough? The answer which Blomfield would give to such questions was sharply opposed to all that the eighteenth century had won after so many battles, and all that the nineteenth century had been content for a time to accept unchallenged. It was soon evident that Blomfield and the architects, with the other artists who speedily joined them, had made a great impression. The most eminent supporter of the other side, the important landscape gardener William Robinson, sprang forward in determined opposition. He wrote two essays on garden-planning and architectural gardens, to show that the clipping and shaping of trees, with the object of bringing them into harmony with architecture, is a barbarous practice. Many people agreed with Robinson when he preferred a noble tree grown freely to any tree clipped and cut about. Addison had said much the same before him, but everything has its place. Although the landscape gardener still controlled the great parks, it looked for a little while as though in private gardens there would be a complete change.

When, eight years later, Blomfield was preparing the third edition of his book, he left out the violent preface to the second edition as no longer necessary, warning people, moreover, not to set up one artificial plan instead of another, which seemed to be the danger of the moment. In their admiration for certain old gardens, they must not attempt to repeat them in circumstances where success would be impossible. What made him most uneasy was the danger of a growing dilettantism. The interest of English people, so long leaning to the botanical side, was turned to the real art of gardening. Perhaps, however, they tried to combine two ideals. Old traditions, once broken, are not easily restored.

One result of the new movement was a study of such of the old gardens as had survived. Old garden-houses were visited, and pictures were made of them, and were taken as models. Antique sundials and leaden statues were accorded places of honour, and were imitated and recast. But most of all was the fashion revived of clipping trees and shrubs. People once more wanted hedges, which were indispensable for creating a sense of seclusion, of being at home. There was a general hesitation and shyness about cutting trees, but many people thought that this kind of ornamentation gave an old-fashioned

FIG. 621. TOPIARY WORK AT LEVENS HALL, WESTMORELAND

charm to a garden, and pilgrimages were made to such places as Levens Hall in Westmoreland (Fig. 621) and Elvaston Castle in Derbyshire. At Levens a great deal of topiary work which dated from the beginning of the eighteenth century had been preserved in an old garden. Elvaston Castle is a notable example (Fig. 622) of what can be done in a short time by diligence and with the help of modern technique to adorn an entirely new garden with the most audacious kinds of topiary. The whole of this garden, with the exception of the great conservatories and kitchen-gardens, was devoted to clipped trees. The yews, in particular, were cut into all imaginable shapes, and the owner had old specimens brought from other lands, in the wish to give his figures an appearance of antiquity. The only colour in the place was supplied by the astonishing variety of shades in the foliage of the yew-trees, ranging from the darkest green to golden yellow.

The literature of those still-recent years was a sign of the ever-growing interest in garden art, and showed us how far it was identified with the formal style. At the end of the nineteenth century the journal *Country Life* published in three volumes the book *Gardens Old and New*, which gives a wonderful array of formal gardens with only here and there, for politeness' sake, an illustration of a landscape garden. The tendency is perhaps most clearly to be seen in the fact that in the many pictures given of the gardens of the middle of the nineteenth century, the relatively small piece known as the Italian garden appears, whereas the landscape park is not shown. No doubt England was herself surprised to find how many formal sites she possessed, and especially what rich fruits the new predilection for the formal garden had to choose from.

FIG. 622. TOPIARY WORK AT ELVASTON CASTLE, DERBYSHIRE

A prominent feature at this time is the return to separate small gardens; and this is easy to understand, for as the whole movement arose from the small garden, it connects itself with the English Renaissance custom of laying out large places in little divisions, and, in particular, with making use of hedges. The little pond-garden at Hampton Court became a very popular model. The pergola also, which at the time of the Italian parterre had not been much favoured, no doubt because there was a tendency for it to be nothing more than a support for plants, came again into the foreground. But with all this passionate interest in the art and the design of the garden in the architectural sense, botany was not forgotten. On the contrary, the two interests were combined, and herein lay the attraction which brought both gardeners and botanists, so long hostile, into one camp. For a long time the eye had been wearied with the handsomely coloured but monotonous and very expensive products of the greenhouse; the meaningless carpet-beds were too stiff to suit the light, rhythmical beauty of the new architectural garden, and indeed were as inimical to this as to the "natural" style.

Attention was turned to groups of plants long banished from gardens, which could stand the winter in their native soil, and could live for years as hardy herbaceous plants. Poor country people had always grown them, and as a rule their gardens had been kept in a simple way and had followed straight lines, both in beds and paths. These plants were speedily converted by gardeners from the condition, so to speak, of country children into creatures of wonderful beauty and colour, suitable for every class. They could be used in all sorts of ways. With their variety of colour they decorated the square beds in their box edgings; and when carefully chosen and set side by side showed a river of bloom against the dark green of the hedge from the earliest spring days till the end of autumn. These beds, or borders, are the chief feature of the flower-garden of to-day, and in their interest for the botanist they have saved the garden from becoming a mere soulless repetition of former days. There was some danger to this new style in an ever-present eclecticism, which made people constantly seek for something new, only to be disappointed when the novelty turned out to be a copy of something old. "In some of the best English gardens," writes Rose Standish Nichols, "there is a combination of classical statuary, Renaissance fountains, French perspectives, Dutch topiary work, and the flowers from all over the world."

This many-sidedness was not found only in the formal garden, but the natural style also made an effort to win new forms for itself at any cost, or at least to revive some of the old ones. Mr. William Robinson was really returning to the ideas of Addison and Rousseau—even Bacon had a similar fancy—when he recommended a wild garden. The plants, especially flowers and shrubs, must be so placed that they are as far as possible in their natural habitats, perhaps in marshy ground, perhaps in a rock-garden where the flora of the Alps can be grown, perhaps in a water-garden where aquatic plants of many kinds will offer a charming sight. To such places as these a gardener cannot be followed by a "crazy architect," for only the botanist will pursue. But since it is only possible for botanic gardens and large parks to give to such parts of a garden proper professional attention, this sort of place is often a sad caricature in a small place. The fundamental ideas that were worked out by way of experiment in small plantations were next transferred to the huge natural parks, where, especially in America, primeval forests might be found in their original beauty, untouched by the hand of man. To complete the many-sidedness of the modern picturesque style, the imitations of Japanese gardens may be mentioned, since in a perfect garden one part is generally given to them. It is true that great difficulties stand in the way of a good imitation, which is often confined to the provision of certain decorations, such as lanterns, bridges and stones, with possibly a dwarf tree here and there.

The reason why the average garden is so good in England is that a surprising number of people are educated in gardening and botanical lore, although they are not professionals. It has been said that whereas ten years before the end of the nineteenth century the care of a garden was the favourite occupation of the few, it is now the passion of the many. To this we may add the important fact that women in England, especially of the educated class, take an important part in gardening, either because they have a place of their own to cultivate, or else because they adopt it as a profession. A decision to admit women to the Horticultural College at Swanley was arrived at in 1891; and there are now more than seventy studying there.

While England must be esteemed the leader in the modern gardening movement, other countries have followed at a certain distance. Out of the countless currents of thought in the modern world which cannot yet be historically treated, we will consider here only what has come about in France and in Germany, not because other countries, especially America and Russia, have failed to play their part conspicuously and actively, but because France and Germany have sounded a distinctive note in the general harmony. When France joined the new movement, she began by restoring some of her historical gardens.

M. Duchesne, father of the garden artist of our own days, spent his life in an intensive study of old gardens and especially parterres, with their marked style of severe formal lines and regular plantation. He supplemented his study by forming a fine collection of old pictures. In our history of French gardens we noticed that the centre of interest before the time of Le Nôtre was the parterre and its gradual development. But in the modern approach to the formal style there was fortunately never the slightest wish to restore the bosket as it was at the time of Le Nôtre. Here we see the true mark of the people's garden, which is more noticeable in France than in England. It was recognised that *parterres sans futaies* (and without boskets), arranged with taste and in proper proportions, were exactly suitable for small estates. The disagreement in France was less vehement than in England, but there was a strong opposition to the *jardin fleuriste*, because the severe lines in such plots made the plants entirely subordinate to the design. This is shown in

FIG. 623. CHENONCEAUX—VIEW OF THE PARTERRE

M. Duchesne's restorations of old gardens, e.g. those at certain castles on the Loire, such as Laugais, the parterre of the castle of Condé-sur-Iton (Fig. 624) and a great many others. Chenonceaux (Fig. 623) was restored at this time. But in his own garden, planned by himself and his son, Duchesne served the interests of botany by laying more stress on the beauty of flowers, and by including plants that were not known in the days of his models. These served at the same time as "the frame and the picture." Before all other considerations the gardens had to be brought into close connection with the house, far more than the *parterre fleuriste* had been. For the garden is not only an ornamental plot, as it used to be; but when it belongs to a small house, to a villa, it is a living-room in the open air.

Opposed to this scheme, which is very nearly that of the old formal parterre, is a different one, which is frankly an enemy to tradition. According to this, there will indeed be a garden near the house, but it must be an expression of the modern man and of modern art. It demands something personal, something intimate, quite other than the grandiose impersonal style of the eighteenth century. "Just as the modern alexandrine differs from that of the eighteenth century, so must our garden differ from the garden of Le Nôtre." So says André Vera, the author of *The New Garden*, and he even goes so far as to urge that the garden should imitate a modern woman's dress in its colour scheme. Gardening, like architecture, must show the sign of human thought. The fruit-garden, moreover, must be regarded as a real part of the whole, and as having the same claim as the other parts: the flower-garden or the rosery. Everything must form one whole, consistent with the particular house to which it belongs. But Vera stops when he comes to the park. "I have nothing to say against a landscape park," he says at the beginning of his preface; "for that may be left to the landscape gardener, and it will form an easy transition to the open scenery of Nature." We shall shortly see how the landscape gardener is threatened from another quarter, and is in danger of being driven out of the park as well.

FIG. 624. CONDÉ-SUR-ITON—THE PARTERRE

In Germany we find the first signs of a real reversal of taste in the last years of the nineteenth century. On the whole the dispute is on the same lines as in England, and is just as violent, though in Germany it is more pedagogic, and of a more decidedly democratic character. Two men made it the business of their lives to awaken the artistic sensibilities of the *bourgeois*, and to inspire the masses, who are so hard to move, with fresh life and thought; these were Lichtwark and Avenarius, and they may be said to stand at the very apex of the new movement. They wrote at much the same time. Avenarius makes fun of a "Piepenbrinkgarten" in his journal *Der Kunstwart*, apropos of two kidney-shaped beds of grass and other beds like a child's "brezel" (twisted) cake. Then Lichtwark in 1892, in his essay *Makartbukett und Blumenstrauss*, directs attention to the peasants' gardens at Hamburg, which he prefers to the English landscape parks. From now onward *Der Kunstwart* never ceased to admonish artists and architects, urging that it was their business to bring "fresh air" into the enclosures of the professional landscape gardeners.

The century had to come to an end before there was any real response. And now the architects themselves joined in the fray, both as theorists and as practical men. The most prominent among them was Muthesius, who had been drawn towards the English movement very strongly during a stay in England, when he made an intensive study of non-ecclesiastical architecture. In a course of lectures delivered by him in 1904 at Dresden and Breslau, on the subject of English houses, he attacked landscape gardening on similar and often identical grounds with Blomfield, finding it still the leading fashion in Germany. In these lectures Muthesius enjoyed the great advantage of having in his mind a style that had already made some progress in England, and which he could present to his audience. His definition of the English garden is very characteristic of the trend of the German movement: "The garden of the present day shows a mutual interdependence of particular formal parts, which may be compared to the ground-plan of a house, only that rooms (terrace, flower-garden, lawn, and kitchen-garden with greenhouses) are open on the top. Together with great variety in particulars, there is in the whole a form essentially regular and closed in; and all the separate parts lie horizontally, with their limits and boundary lines clear to be seen." Very soon "Rooms in the Open" was to serve as the fanciful name for these modern gardens. Thus Muthesius feels the need of harmony between house and garden, not so much in regard to what was thought important in Renaissance days, viz. that the lines of the house which are visible from outside, both vertical and horizontal, the carvings and other ornament, should be repeated in a show-garden, and so in a sense should introduce architecture into the open; but rather he desires to have the inner part, the living-rooms of the house, reproduced as far as possible in any garden which he would consider ideally planned. Also in what he called the indoor furniture of the garden, the seats, the borders of hedge or pergola, the paths—all should show some likeness to the inside arrangements of a house. Similar demands were fulfilled after their own fashion in the Greco-Roman court-gardens.

About the same time there appeared Schultze-Naumburg's book *Gärten*, which was the second part of his *Kultur-Arbeiten*. This book, which belongs to the school of Avenarius and Lichtwark, with whom Schultze-Naumburg had worked from the beginning, shows clearly that the eyes of these men were really directed towards the gardens of small houses. It was a recognised aim of theirs to dignify the formal treatment of the "garden room" by making it conformable to rules of æsthetic art. This object the author tries to attain in a pedagogic fashion by drawing a distinction between the Beautiful-Suitable and the Ugly-Unsuitable.

The whole tribe of landscape gardeners closed their ranks against these intruders into their province. If we turn over the pages of the *Zeitschrift für bildende Gartenkunst*, we cannot fail to notice a remarkable energy among the chief contributors, the landscape gardeners themselves. In 1887 there took place the first meeting of the "Garten-Künstler der Lenné-Meyerschen Schule," which expressed an active dislike towards their enemies the architects, "with their stony hearts, with minds that harp on mathematical formulas, and ideas that can never transcend fixed rules." The next year one of their champions shows more and more anxiety, because "the fatal shears, without the least consideration for what their work is destroying, are for ever widening their path, and without a glance at the principles of art." In these circles England was feared most, as it was seen that more and more of the English gardens were suffering disfigurement from the clipping-shears. "It looks as though," says Fintelmann, "gardeners were on the way to return to the days of Le Nôtre." In 1904, angry and perturbed, they were still combating the assaults of Muthesius, but all the same the course of events proceeded surely. The longer the gardeners stood waiting and grumbling, the more energetically did the architects acquire possession of the gardens.

General attention was directed towards the works of architects when they began to hold exhibitions, in which they were of course the sole masters. In 1897 the first exhibition of garden buildings was held under the influence of Lichtwark. This was at Hamburg, the headquarters of the formal party; and here the exhibitors had to bring their own

FIG. 625. GARDEN SHOWN AT MANNHEIM EXHIBITION, 1907

materials, and set them up, often with great trouble and difficulty. But it was at Düsseldorf in 1904, at Darmstadt in 1905, and at Mannheim in 1907 that proper models of architectural gardens were first publicly exhibited (Fig. 625). Much was experimental, and there were many wild ideas at these early shows. Some men, like Peter Behrens, saw salvation in the new lattice-work, and his garden was mostly pergola and bordering. Läuger at Mannheim let the plant world retreat into the background, with his basins and statuary. Olbrich at Darmstadt on the other hand borrowed from landscape gardens the idea of coloured plots, and laid out miniature gardens.

At the beginning only effects suitable for an exhibition were produced; but the professionals soon saw that they would be thrust back, little by little, unless they admitted the enemy into their hitherto restricted circle, and unless they joined hands with the architects, both as learners and teachers, complying with their demands and working loyally with them. This *volte-face* was described in the journal *Die Gartenkunst*, when in

FIG. 626. GARDEN OF A LEIPZIG HOUSE

1906 the editorship passed into new hands, and under the guidance of an open-minded man the controversy was carried on with much struggling and many discussions from both camps, but with less bitter feeling.

A new generation of young German architects is now busy designing gardens, which will grow to be more and more like the people's gardens in England. Their style is astonishingly fresh and original. They are mostly from the north and middle parts of Germany, from Hamburg and Bremen, Cologne and Leipzig. They are garden architects. Their knowledge of plants is very thorough, and of quite another sort from that of the architect proper, for they work according to the outspoken maxims of Schultze-Naumburg, Muthesius, and Olbrich. Their gardens differ from the English in that there is less clipping of hedges. The hedge is not unknown in the work of Gildemeister at Bremen, Leberecht Migge at Hamburg (Fig. 632, p. 364), Eneke at Cologne and Grossmann at Leipzig, but it is restrained and is less marked as an architectural feature. But all these gardens show that the hints of Muthesius are followed, that before all things they are living-rooms in the open, and that they deserve the name they are known by, "Open-air Houses" (*Freilufthäuser*) (Fig. 626).

The public garden could not long remain uninfluenced by this movement, judging by its own history and the tasks which it had to perform. It was the smaller pleasure-grounds in France that first returned to the formal style, when the brief fashion for English squares had gone by. These were most adaptable because of the architectural style of their surroundings, and also because of their *raison d'être*, as ornamental places away from traffic, with their flower-beds like parterres, or as recreation grounds with places for games or meetings, as we see them in American towns; in either case a formal site with plenty of architectural ornament was most required. Nowadays practically none but formal places are laid out in the greater towns; yet we have to remember how recently battles were fought—such as the one over the Friedrichsplatz at Mannheim (Fig. 627), which was designed by the architect Bruno Schmitz—if we are really to grasp the fact that this movement is a new one.

It was rather late when decorative flowers found a place in the real People's Park; and even in England, always to the fore in such matters, it was only at the end of the nineteenth century that the experiment was made (in Regent's Park) of introducing a flower-garden. In Waterlow Park, in Ravenscourt Park, and in Hyde Park, we have "the old English garden."

In other countries, under the leadership of America, a different development appears. In the latest form of the park there hides a germ whence the old grove might come to life again, and possibly has done here and there. In the manifold forms of playing-fields there is a possibility of making separate gardens which might be, with a difference, what the boskets of Le Nôtre were in the old royal park. But since a People's Park in a large town requires an open show-garden in the centre, which will have to cover far more space than even the great courts of Louis XIV., it is a necessity due to the spectacular demands of modern days to have also one great central view. The club house with its ornamental adjuncts in front now takes the place of the royal castle of former days. Some of the latest American parks on the gigantic scale, such as the one at Chicago, have completely adopted the formal style. In America the large grounds given up to ornament and parterres are

FIG. 627. ROSE-GARDEN IN THE FRIEDRICHSPLATZ, MANNHEIM

introduced to help the view of the whole picture, but not at all in the same fashion as in England.

France also has adopted the idea of enlivening the old perspectives of her parks in a

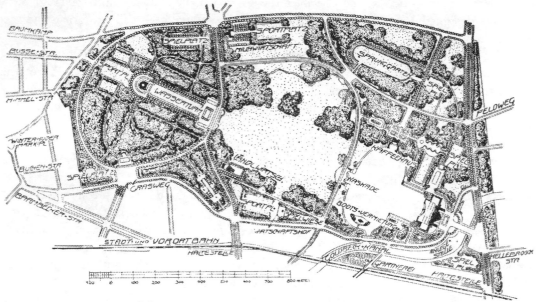

FIG. 628. PLAN FOR THE TOWN PARK OF HAMBURG-WINTERHUDE

new way. The new part in Paris which leads from the Trocadero over the Champ de Mars to the Ecole Militaire—thus partly carrying out a grandiose plan dreamed of by the first emperor—is regular and uniform, which perhaps is one of the signs of its newness. All the same, however, it shows some of the former grandeur of a French design.

FIG. 629. CLUB HOUSE IN THE TOWN PARK, HAMBURG-WINTERHUDE

The latest parks in Germany are making efforts towards future developments of a novel nature. In the parks at Cologne, which have developed under the auspices of Encke, a garden artist who early escaped from the fetters of the narrow landscape tradition, there is practically nothing to be found but the formal style. The Schiller Park at Berlin and

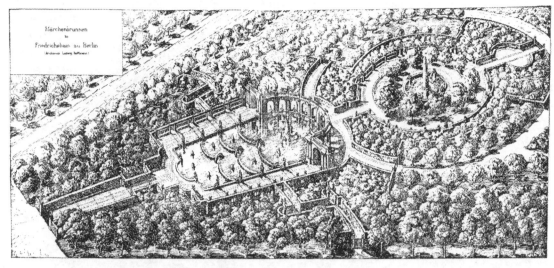

FIG. 630. THE MÄRCHENBRUNNEN, FRIEDRICHSHAIN, BERLIN—A BIRD'S-EYE VIEW

the town park at Hamburg, both modern, are working out these great perspectives (Fig. 628). Schumacher in particular, the designer of the Hamburg park, attempted to get an imposing effect by combining the large lines of the perspective (Fig. 629) and the ornamental grounds and playing-fields with picturesque plantations between them, especially by making edges

FIG. 631. THE MÄRCHENBRUNNEN, FRIEDRICHSHAIN, BERLIN—VIEW AT THE ENTRANCE

of woodland to the club field. In details we may recall Italian models once more, and adapt these to modern feeling. Ludwig Hoffman made a monumental entrance to the Friedrichshain, the old landscape park at Berlin, and this undoubtedly inclines towards the ideas of the Frascati Villas. The favourite triangle, formed here by two diverging streets, may perhaps have attracted the architect (Fig. 630). The style of a pillared hall, rounded like a theatre, cutting off the wide water stairway (Fig. 631); reminds us in its distinctly classical form of such erections as the coffee-house at Villa Albani, the Gloriette at Schönbrunn or the colonnaded bosket at Versailles. Perhaps one would like it better if this beautiful place stood at the end rather than at the entrance of a public park, for in that case the crowds would be somewhat dispersed on the broad walks after they left streets crammed

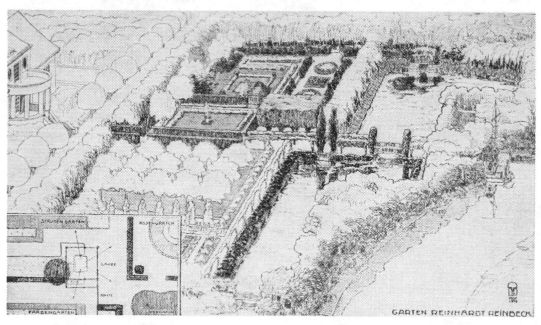

FIG. 632. HERR REINHARDT'S GARDEN AT REINBECK, HAMBURG (*see page* 360)

with traffic and before they arrived at this retired spot, with its feeling of privacy pervading it in spite of all its grandeur. But the whole place has been thought out with a noble feeling for art both in the general plan and in the details; and it certainly marks an important step in the progress of modern garden development.

In the new garden movement the burial-ground also finds a place. Goethe dealt with this problem in his *Elective Affinities*. He thought that the ugliness of churchyards, which was shown both in actual monuments and in the separate treatment of the graves, might be remedied by the laying out of a Campo Santo. The practice of cremation had been growing, and this fact was an argument for setting up certain places where urns could be housed according to the rules laid down at the crematorium. But before that idea was carried out, an experiment was tried in America with a view to avoiding the ugliness of separate graves, and to place them without any regular arrangement in a landscape park, where they were scattered about and concealed from sight. In Europe similarly there were a few burial-grounds made in landscape surroundings; but it soon became apparent that this was only a way of making things worse by concealment. The woodland cemeteries that are to be seen nowadays are probably an experiment of the same nature. Most people's taste has reverted to the piece of ground that is laid out formally, and attempts are made to give it an appearance of size by an imposing perspective through the middle, while particular graves are grouped together to form separate gardens. But the problem has not yet found a solution, and architects and garden artists are actively concerned with it.

If the future is not clear, at all events life is everywhere full of movement, not yet ripe for historical treatment. All garden lovers and artists may rejoice in the consciousness that in our own time a new development has come about, and one that is full of promise. There are important tasks ahead, in small and great matters alike, and we must hope for strength and energy to carry them forward.

CHAPTER XVII
MODERN ENGLISH GARDENING

CHAPTER XVII

MODERN ENGLISH GARDENING

By Walter P. Wright

SOME hard things are said in the present work about a style of gardening which under the description of "English" was supposed to have laid a blight on many Continental gardens of the eighteenth century. But in those years England had no style so firm and distinctive as to deserve to be called national.

It is true that she did not adhere permanently to the formal style which was introduced from Italy in the time of the Renaissance and from France during the spacious days of Louis XIV.; and there was a reason for this which was entirely dissociated from Art, although eminently sensible and practical. The fact that her two great university towns, Oxford and Cambridge, lay in relaxing situations, supplemented the ordinary effects of a damp climate in creating the necessity for much more physical exercise than was needed in drier countries, and consequently young members of wealthy families acquired habits of activity which they carried into their homes. The result was that grounds were extended and walks lengthened. But parterres and fountains could not be provided to accompany every winding path. Groups and belts of trees and shrubs could, however, be planted.

This, perhaps more than anything else, even including Addison's essays and Pope's satires, explained the development of a supposed English "natural" style. As a matter of fact, the actual old English garden always was more or less formal. Grass thriving naturally in the humid climate, it was, and is, brought close to the house and most of it mown and trimmed into "lawns," often broken with groups of beds. Right and proper as this procedure may be, one can hardly describe it as "natural."

THE FORMAL GARDEN STILL LIVES

And some formality will continue. The lawn will probably remain what it has always been, one of the chief glories of the English garden. It is a part of us—of our native climate, of our native character. Small gardens and large will have their area of shaven grass. Our robust outdoor games also demand turf. And beds will be put on a good many lawns, being none the less beds because they contain salvias and snapdragons instead of scarlet "geraniums" and yellow calceolarias.

There does not seem to be anything to worry about, anything to apologise for, in all this. It fits in with the prevailing circumstances. It is a part of the home, and one might pardonably describe it as domestic. But it does not stand alone. Supplementary to it there has come into being a system of Garden Art which may truly be described as English, inasmuch as it is the work of English reformers, who have borrowed their ideas only from earlier generations of Englishmen. Under this system hardy plants play a predominant part. Harmony of colour has become more important than design of parterre. Alpine plants are utilised extensively and with great effect. Plants are given more individuality.

It is a little curious that while foreign critics should have condemned the so-called English style for its informality, home critics should have complained of its formality. Until comparatively recent years, English flower-gardening was undoubtedly formal. Even within the recollection of many living people, carpet-bedding was practised in private as well as in public places. It has to be confessed that this style is still pursued in the public gardens of industrial districts, and enjoys much favour with the populace. But it is dead in private gardens. It would no more be tolerated there than the still older monstrosity of coloured earths.

The worst in the way of formalism that is to be met with now in the gardens of the nobility is encountered in those places where reasons of economy born of high taxation and a dwindling agriculture dictate that the garden staff shall give most of its attention to growing fruit and vegetables for market, so that nothing special can be attempted in flower-gardening. There are, unfortunately, a good many such cases. The gardeners consider that if they plant a few flower-beds with bulbs or wallflowers in autumn and antirrhinums in summer, and mow at least a part of the lawn, if only a ten-feet belt, they have done all that can reasonably be expected of them.

Happily a wave of garden-love has spread over the powerful commercial and professional classes of the country—classes with higher standards of education and culture than they are generally credited with possessing. We are all familiar with the tradesman or merchant (he is apt to be a soap-boiler) of the novel and the stage, who misplaces his aspirates. He lingers in fiction and at the play because those forms of "entertainment" are hopelessly archaic. In reality, he is to-day a well-educated man with a *flair* for public life and with cultured womenfolk who have a passion for flowers, which they study closely.

This wave has amply compensated for the loss — if indeed there is a loss —in the gardens of the aristocracy. It has swollen immensely the membership of the Royal Horticultural Society. It takes greater crowds than ever to the larger shows. It has raised the standard of gardening in all the more important public resorts. It has added pretty flower-gardens to thousands of good new middle-class houses and villas. We have, indeed, become a flower-growing people.

HAS ENGLAND A NATIONAL STYLE OF GARDENING?

Have we, however, evolved a "style"? Has gardening in England become so distinctive and at the same time so homogeneous that we can fairly claim to have developed a national system? This is a debatable question. Some things we have certainly done: we have got rid of elaborate "parterres," furnished with plants most of which are too tender to endure our winters. We have got rid of or reduced elaborate water-devices, statuary, and labyrinths. We have practically got rid of carpet-bedding. We have got rid of coloured earths. We have got rid of ribbon borders. If to have made a clean sweep of the principal components of the formal style of the past is *pari passu* to have formed a style of our own, then our position is impregnable.

FIG. 633. DELPHINIUMS IN THE HERBACEOUS BORDER IN MR. H. BEVIS'S GARDEN, WINFIELD, HAMPSHIRE

Before, however, we attempt a decision based only on a series of negatives, let us see what we have put in the place of the discarded things.

We have in the first place made a great stride in the direction of simplicity by substituting to a considerable extent (not wholly) hardy plants for tender. With certain qualifications, which honesty imposes in consequence of a partial revival in Dutch clipped and Japanese dwarf trees, we may claim to have made a farther step in the direction of good taste by growing our trees, shrubs and plants in their natural forms. We have

FIG. 634. ROCK BORDER BESIDE A LAWN (RIGHT) AND WALK WITH HERBACEOUS BORDER (LEFT) IN MR. MARK FENWICK'S GARDEN, ABBOTSWOOD, GLOUCESTERSHIRE.

unquestionably impressed contemporary nations with the vast progress which we have made in the creation and use of hardy herbaceous plants, such as delphiniums (Fig. 633), phloxes, peonies and Michaelmas daisies. We have corrected the stiffness of our lawns by providing rock borders and herbaceous walks beside them (Fig. 634). We have softened the severity of terraces by planting them informally with hardy things (Fig. 635). And we have done one other really great and significant thing: we have shown that the contours of our gardens can be improved better by the formation of rock gardens and the cultivation thereon of alpine plants than by elaborate, costly and unnatural terrace-building, construction of artificial ruins, and so forth.

These are not small achievements, indeed, when they reach fruition, when it can fairly be claimed that the English flower-garden is firmly based on naturally grown trees and shrubs, on well-furnished herbaceous borders, and on rock gardens, the whole mellowed with the good lawns of our past, then indeed we shall have an "English style" again, a style very different from the old—more artistic, more free, more simple—yet equally robust and coherent. And that time is drawing nearer with rapid strides.

THE GREAT CHANGE IN ENGLISH GARDENING—A TRIBUTE TO RAISERS

It is interesting to speculate on the principal causes for the change in English gardening which began about the middle of the nineteenth century. One of the most influential was undoubtedly the appearance of raisers of great genius, who had the foresight to turn

FIG. 635. HERBACEOUS PLANTS ON THE TERRACES AT ABBOTSWOOD, MR. MARK FENWICK'S RESIDENCE IN GLOUCESTERSHIRE

their attention to hardy plants. The present writer can remember when the introduction to commerce of a new variety of zonal "geranium"—one of the principal components of the old "ribbon border"—still created a furore. But that memory of childhood was fleeting. Superimposed upon it, and growing yearly stronger and stronger, was the recollection of displays of new roses, new gladioli, new peonies, new carnations, new phloxes, new sweet peas, new Michaelmas daisies, new delphiniums, new irises. Such names as Paul, Bennett, Kelway, Burrell, Martin-Smith, Douglas, Eckford and Michael Foster became increasingly prominent, as did those of Barr and Engleheart with daffodils.

We might take the gladiolus as an example of a garden plant which sustained great development. As a bedding plant far superior to any "geranium," as a flower for cutting beyond comparison with any old bedder, as a border plant almost perfect, the rise of the gladiolus is one of the romances of modern gardening. There were probably hybrid gladioli in the seventeenth century, for there are several varieties illustrated in Besler's *Hortus eystettensis,* which was published in 1613, but it is improbable that they played a prominent part in the famous gardens of that epoch. We have to take a leap of nearly two centuries to find gladioli exciting the least attention. Then in 1810 we find Dean Herbert raising new varieties, and not only so, but twenty-four years later telling of seedlings raised by one Bidwill, an Englishman, from parents that were destined to become more famous through their offspring than they had ever been of themselves. One of these parents was certainly a species called psittacinus, a native of South Africa; the other has been variously stated as cardinalis, floribundus, and oppositiflorus. Under the name of gandavensis, so called because the hybrids, obtained from the garden at Enghien in Belgium, were distributed by a Ghent (Gand) nurseryman, the progeny of this cross, in the hands of Kelway, Standish, Burrell and other English raisers, as well as in those of the Frenchman Souchet, made the gladiolus one of the greatest of garden flowers.

This is not the place to pursue the subject of flower-creation, important and fascinating though it is; it must suffice to say that what happened with gladioli happened also with other great garden flowers. The English banker Martin-Smith and the Scots gardener Douglas did for the carnation, Eckford did for the sweet pea, and Michael Foster, Dykes and Bliss did for the iris, what the florists already named had done for the gladiolus. James Kelway improved the peony and the delphinium as much as he had improved the gladiolus; and these two splendid flowers in themselves had an immense influence on the extension of herbaceous borders. It is one thing to have a conception, it is another to have the material with which to develop it. Had there been hundreds of magnificent gladioli, peonies, delphiniums and irises in 1650 there might have been no Versailles, for "Le Roi Soleil" might have become enamoured of herbaceous borders.

A TRIBUTE TO WRITERS

Having paid one tribute to raisers, let us pay another to writers. At least three authors of force in William Robinson, Gertrude Jekyll and Reginald Farrer played a great part in changing the style of English gardening. As a writer of pure genius not less than as an intrepid collector, the ill-fated Farrer stands supreme. But Robinson was the earlier and probably on the whole the more influential writer. In such books as *The English Flower Garden, Alpine Flowers for English Gardens* and *The Wild Garden*, he attacked unsparingly the parterre, bedding-out and ribbon-border systems, with their stiffness, formality and garish colours; and he argued with great force the superiority of a more natural system, in which hardy plants (both shrubby and herbaceous), alpine flowers, and trees and shrubs naturally grown, should play the most prominent parts.

The first edition of *The English Flower Garden* appeared in November 1883. Forty-three years later, and at the age of eighty-seven, the author produced a fourteenth edition. This is a remarkable record, testifying at once to the virility and stamina of both book and writer. There can be few, if any, such cases in the history of literature. The strength of the book lay no more in the principles of natural beauty which the author so forcibly propounded, appealing though they were, than in the thousands of cultural details which he gave relating to the many thousands of species described.

Robinson's theories were not in all cases new; for example, in attacking "absurd 'knots' and fashions from old books" and "attempts to . . . get colour by the use of broken brick, white sand and painted stone," he was very near to Bacon with his "As for the making of knots, or figures, with divers Coloured Earths . . . they be but toys; you may see as good Sights many Times in Tarts." But the modern writer had even more justification than the ancient one for his onslaughts.

Miss Gertrude Jekyll, whose eigh y-third birthday in 1926 made her no less than sixty-four years older than the first edition of her finest book, that on garden colour, was probably scarcely less influential than Robinson in advocating the charms of natural gardens. There are two great groups of gardeners; one consisting of people whose main interest lies in plants as plants, with no particular regard to their place in the garden, the other of persons who think of plants in terms of gardens. Miss Jekyll is whole-heartedly a member of the second group. Observe with what cogency she supports her views:

> Merely having plants or having them planted unassorted in garden spaces, is only like having a box of paints from the best colourman; or, to go one step farther, it is like having portions of these paints set out upon a palette. This does not constitute a picture; and it seems to me that the duty we owe to our gardens and to our own bettering of our gardens is so to use the plants that they shall form beautiful pictures; and that, while delighting our eyes, they should be always training those eyes to a more exalted criticism; to a state of mind and artistic conscience that will not tolerate bad or careless combination or any sort of misuse of plants, but in which it becomes a point of honour to be always striving for the best.
>
> It is just in the way it is done that lies the whole difference between commonplace gardening and gardening that may rightly claim to rank as a fine art. Given the same space of ground and the same material, they may either be fashioned into a dream of beauty, a place of perfect rest and refreshment of mind and body—a series of soul-satisfying pictures—a treasure of well-set jewels; or they may be so misused that everything is jarring and unpleasing.

All this, of course, is not to condemn every type of specialist. The raiser, for example, must consider the flower first. He must grow it in colonies in his nurseries, partly in order to provide it with the particular conditions which suit it best, partly for convenience of comparison and crossing; but the amateur, the garden-maker, is under no such compulsion, and it is to him that Miss Jekyll addresses herself.

Farrer was no less gifted and scarcely less a power than the two great horticultural writers already discussed. If a somewhat wayward genius, he was still a genius. If his literary style was sometimes flamboyant, it yet remained arresting and persuasive. And it must be remembered in his honour that he made an intensive study of Alpine plants and sought them all over the world, never shrinking from hardship and danger, and penetrating fearlessly the most distant recesses of unknown China and savage Tibet. It can scarcely be doubted that Farrer's books had an immense influence in spreading an interest in and love for alpine flowers, or that he was the means of inducing thousands of people to take up rock gardening, just as *On the Eaves of the World* and other of his travel books must have inspired many a bold spirit to go out into the wild in search of new plants.

Having paid tribute to the influence of raisers and writers in bringing about a more artistic system of flower-gardening, we, the people, may perhaps permit ourselves to believe that neither class could have been completely successful had there not been something responsive in ourselves—some desire, one might even say some yearning, for guidance in an art that a higher standard of education, a more widespread love of beauty, taught us had not been done justice to. It was because the seed which had been sown had fallen on fertile, receptive ground that germination was so swift and growth so strong. The new gospel spread with amazing rapidity, so that England became a land of gardens in which true ideas of Art were conspicuous.

COLOUR-PLANNING—A BLUE BORDER

Naturally there have been disappointments and failures. Many people have taken up colour-planning with herbaceous plants, for example, under the impression that it is a comparatively simple matter to get continuous colour-harmonies from spring to autumn in borders. It is, however, complex and difficult. Considerable forethought, much care in choice of material, and no small amount of skill in cultivation, are called for. Weather vagaries, and attacks by insects and fungi, are often overlooked, yet they have a vital

FIG. 636. BOLD FLOWER GROUPS NEAR A MAIN PATH IN SIR ARTHUR LEVY'S GARDEN, THE MOUNT, COOKHAM DENE

bearing on results. It is perhaps wiser on the whole to be satisfied with bold groups, particularly in selected places near the principal paths, as shown in Fig. 636, where the formality of bridge, steps and paths is broken by noble border groups.

Colour-blends have, however, a peculiar fascination. The writer was deeply impressed with a "blue border" under a terrace wall at Chilham Castle, near Canterbury, in June, and may mention some of the plants which composed it. As might be expected, there were delphiniums (perennial larkspurs), and one recognises with gratitude the good work of raisers in providing such a host of beautiful varieties, single and double, in all shades. There were anchusas, splendid in colour, but the larger kinds a little gross in habit. There were lupins, and here again one is deeply sensible of advancement; the habit of the plants is perfect for the border, and the range of colour has been widened to a remarkable degree.

Irises, both of the German and Siberian types, were a host. Campanulas were charming. One of the deepest blues available in a border plant was there in veronica True Blue, which, as a June bloomer, is perhaps of the Gentianoides class. The early-flowering flax (Linum perenne) was a valuable plant, albeit a little loose-habited. A variety of pentstemon with lilac, purple-shaded flowers, not often met with, named Scouleri—an effective plant —was present. Lastly there were violas old and new, the best of all low-growing blue flowers.

It is not easy to get a blue wall-covering for a blue herbaceous border, and here a point may be strained, for there is a splendid evergreen blue-flowered wall shrub, used at Chilham Castle, which has every merit: ease of propagation, vigour of growth and remarkable profusion of bloom in June, and that is Ceanothus veitchianus, one of several good evergreen ceanothuses.

SHRUBS AND HERBACEOUS PLANTS

It is only the stickler over terms who will refuse to use a shrub, deciduous or ever-green, in or in association with an herbaceous border. Many irises are not truly herbaceous, nor are pentstemons in all circumstances, nor all phloxes. Some spiræas are herbaceous and some shrubby, but why exclude any spiræa from an herbaceous border except on such grounds as coarseness? If it comes to that, why exclude all roses? It is really a question whether "herbaceous border" is not an unfortunate term, for it may cause some people to shrink from using any shrubby plant, while other people not only mix the types without hesitation—perhaps, indeed, without knowing the difference—but exhibit flowers of shrubs in classes for "herbaceous" plants at the shows, thereby running the risk of disqualification.

A flower border should be made up of such plants, capable of passing most of the year out of doors, as are suited to the circumstances, whether they lose their stems in autumn or continue to hold them throughout the winter. And again, the background, whether wall, or fence, or open lattice-work, may consist largely of shrubs, for otherwise roses and clematises would be excluded. Nor, with a strict adherence to the herbaceous character, could beautiful tubs or vases of hydrangeas be used for special sites: low walls (Fig. 637), pillars at the head of steps, terraces, etc., noble objects though they are.

BEAUTY OF BACKGROUND

The matter of background does not receive sufficient consideration. It may be admitted that there are cases, as for example hardy flower borders on the margins of lawns, where existing trees and shrubs provide as suitable a background as could be devised. But there are others the boundary of which is an ugly wall, fence or building, and it becomes exposed in autumn and remains so all the winter where the contents of the borders are strictly confined to herbaceous subjects. In such cases a background becomes important, and it can be provided with espaliers or lattice-work. A large bed would be transformed into two borders if such a background was fixed along the middle, and thus there might be a "border" down the centre of a flower-garden in place of a group of beds.

It is particularly in small gardens that the plan of a central division is helpful, because it increases the area available for plants. The support, whatever it may be, can be planted

FIG. 637. VASES OF HYDRANGEAS ON A LOW WALL IN SIR ARTHUR LEVY'S GARDEN, THE MOUNT, COOKHAM DENE

on both faces, so that room is provided for an increased number of plants, while at the same time those planted in the border itself, on the level, are not robbed. Supports may be of various kinds and materials, including rustic work; but those who are prepared to go to the first expense of iron espaliers must reap their reward in time, for metal espaliers are practically everlasting. Such woodwork as is attached to them—and some is almost indispensable—need not come into contact with the ground, and consequently it also has a long life. Admittedly the erection looks somewhat crude at the outset, but it is covered in a year or two, and thereafter is an object of great beauty. A similar framework may be used at will for the back of a border on the outskirts of a lawn or elsewhere.

THE BEST HERBACEOUS PLANTS

One of the difficulties in the way of colour-harmonies in herbaceous borders is the unexpected extension of coarse-growing kinds, which often affects less rampant subjects injuriously. Thus there are certain plants, by no means without ornamental value—moon and ox-eye daisies, anchusas, Japanese anemones, even Michaelmas daisies—which are apt to become a nuisance. The Japanese anemones are very beautiful, but they are terrible rovers in soils which they like, and their roots penetrate so deeply that if once they get out of hand it is almost impossible to bring them under control again.

As plants which do not encroach, which are easily kept under control, and which yet are exceedingly beautiful, hollyhocks, phloxes, pyrethrums, peonies, lupins and delphiniums must be considered best for the herbaceous border. Hollyhocks are never more beautiful than when used as the back of an informal border beside a walk (Fig. 638), and it is a mistake to suppose that they must necessarily become unsightly through disease, which can be kept in check by spraying with Bordeaux mixture. Phloxes are almost ideal, and it is very gratifying to know that there are now large numbers of splendid varieties. As a matter of fact, phloxes are well worth growing in beds to themselves, so far as summer effect is concerned; earlier blooms can be got by planting bulbs with them. At Gravetye Manor, Mr. William Robinson's place in Sussex, one sees a bed of summer-flowering phloxes which in June is brilliant with the little-grown but exceedingly attractive Tropæolum polyphyllum, with its curious, twisted, creeping growth, grey leaves, and masses of canary-coloured flowers; the bed is edged with lavender.

Pyrethrums are good or bad according as they are tended and staked or left uncared for. The stems are not strong enough to sustain the flowers, consequently neglected plants are ugly, but if some light, semi-natural support is used, such as the upper twiggy parts of hazel pea-sticks, both foliage and flower-stems receive welcome support and the plants show their full beauty; they are then capable of giving charming colour-effects in the border.

The one drawback of herbaceous peonies as border plants, and particularly in colour-blending, is their spreading, rather floppy habit, which prevents them from fitting in well with more slender and upright growers. A better plan where space permits is to bed them, using orange-coloured tulips to harmonise with the bronze of the young peony stems in spring. It is easy to blunder with the staking and tying of peonies, and as a rule the less the better. Nor should root interference go farther with peonies than is absolutely necessary.

In light, rather poor soils it is far better to place dressings of manure or fresh soil round the plants than to take them up, divide them, and replant them, for they are apt to sulk after disturbance.

With increased experience of delphiniums, one is disposed to believe that they also resent division, and are best invigorated by the same means as peonies. One has seen cases in which, with a praiseworthy desire to give clumps of delphiniums a new lease of life,

FIG. 638. HOLLYHOCKS IN A BORDER WITH BOX EDGING IN FRONT IN THE LATE SIR HENRY WHITEHEAD'S GARDEN, STAGENHOE PARK, HERTS

they have been lifted to permit of the soil being deeply trenched and liberally manured; yet the result has not been good for a long time.

The truth is that the great quartet of hardy herbaceous plants may be divided into two pairs so far as toleration of frequent division is concerned: peonies and delphiniums resenting it, Michaelmas daisies and phloxes appreciating it.

In well-managed herbaceous borders, large plants will never be allowed to predominate, however ample the area may be. The best species and varieties of smaller things, such as geums, campanulas, irises, veronicas, gladioli, pentstemons, columbines, and lilies, will have their place.

Whatever may or may not be done with herbaceous borders, there is always a strong case for flower-beds on any level spaces near the house, with grass or paving as the case may be. Crazy paving has deservedly a powerful vogue, but unfortunately one often sees the interstices filled with ugly weeds instead of with dainty low plants like campanulas, alpine pinks, saxifrages, and portulacas; and then one sighs for plain, wholesome grass. The fact is that crazy paving, like most other things in gardens, needs cultivation. It cannot be left to itself.

MONTHS OF ROSE-BLOOM

With flower-beds the line of least resistance in the old days was to cram them with zonal pelargoniums. Now the "geranium" was, and is, a very brilliant and useful plant, which no wise person will despise. But it is garish and it is tender. Being garish it may easily become tedious and even offensive; and being tender it needs winter protection. Again, it is not a valuable cut flower. Here are three reasons why zonal pelargoniums are inferior to, for example, roses. In the old days it was held against roses as bedders that being leafless in winter, and being purely short-period summer bloomers, they left the beds without interest for the greater part of the year, and consequently the only people who grew roses in beds were exhibitors, who cared little what their beds looked like nine months in the year, provided there were fine examples of bloom on a few particular exhibition days in July.

Times have changed in several respects. In the first place the flowering season of roses has been lengthened. In the second, a greater demand has arisen for cut flowers. In the third, there has been a development of material suitable for forming a groundwork to roses in beds, and flowering before the roses make much growth. The first and second points may be connected in a system of feeding and pruning under which roses in beds without a groundwork of dwarf plants give a long succession of cut flowers from June to October. By feeding, mostly with superphosphate or other chemical fertilisers, the plants are induced to push long flower-stems and these are cut low in order to get as great a length of stem as possible for the vases and bowls. The low cutting becomes a kind of summer pruning, which, combined with the feeding, induces the plant to push up quickly a fresh set of long flower-stems. So the process goes on, but low pruning is not done after mid-August, because late growth would not ripen. This system of what might be called "pruning-for-cutting" is carried out very successfully at Lympne Castle in Kent, among other places. For many weeks there is always bloom in the beds, because all the plants are not pruned at the same time; and there is always bloom in the vases, because there are always some flowers ready for cutting.

It is not likely that this system would answer without abundant moisture and liberal feeding, and those who are unable to provide both would probably do better to carpet their rose-beds with some close-growing plant.

There never was an absolute dearth of carpeting material, if people had only thought to look for it. At Gravetye, for example, one sees Viola gracilis, Phacelia campanularia and Verbena chamædrifolia used in different beds, and all these are old plants. Violas are almost ideal carpeters, and the number of beautiful varieties is legion. Nor is there any valid reason why pinks and carnations should not be associated with roses, as at Gravetye; holding their leaves as they do throughout the winter, the ground is never bare.

In the reaction from the old-style bedding with zonal pelargoniums, many people have discarded bedding altogether, satisfying themselves with rock-gardens and herbaceous borders; but beds are often useful, and it is no more necessary to plant them with tender

FIG. 639. ROCK JUDICIOUSLY PLACED AND SKILFULLY PLANTED IN MAJOR J. F. HARRISON'S GARDEN,
KING'S WALDENBURY, HERTS

things than it is to associate them in elaborate and intricate designs, as in the formal
"parterres" of the past. There is literally no end of material amongst hardy plants from
which to choose; it is merely a matter of studying books and catalogues, and of picking
up ideas in good gardens and nurseries. At Gravetye one sees beds of hardy ferns, to which
life and outline are given by pillars of clematis. There is no brilliance in such beds, but
there is interest, there is beauty, there is character. One should beware of sameness:
antirrhinums may easily become as tiresome as "geraniums."

ROCK PLANTS AND ALPINE GARDENS

Rock-gardening and the cultivation of alpine flowers (these are often, but not as often
as they ought to be, permanent parts of one whole) form between them the most remarkable and distinctive development in modern English gardening. Probably the progress
that has been made in this branch of ornamental gardening has been due at least as much
to displays by trade growers at the larger shows as to the efforts of writers and artists,
influential though the latter have undoubtedly been. Small but beautiful gardens, ingenious
assemblages of rock planted with alpines, generally having a cascade and pool, appearing
in the grounds of large shows, struck the public fancy, were repeated on a larger scale
and in increased numbers, and so became firmly established as features of the more
important exhibitions.

While it might be possible to exaggerate the effect of miniature alpine scenes in a
show-ground in London, one can believe that by bringing the charm of alpine flowers
before the eyes of large numbers of cultured people the displays in question must have
exercised a very real influence. Certainly the demand for suitable stone as well as the sale
of alpine plants increased rapidly. Flower-lovers came to realise that rock-gardens were

capable of adding a new and very pleasing feature to their places without making excessive demands on space. Perhaps all did not realise as clearly as they should have done that a rock-garden needs a good deal of attention, because one meets with cases in which gross plants, or even weeds, have been allowed to overgrow the more delicate kinds. There should be no misunderstanding on this matter. Hardy plants, including alpines, need as much attention in their way as the old-fashioned tender bedders; some even need glass, although in the form of protecting squares to throw off winter rains rather than in the form of frames and greenhouses.

One might add to this hint a reminder that a rock-garden should have several aspects; or conversely, that if there is but one aspect the choice of plants should be such that the aspect suits the whole of them. One sees cases in which this point has been overlooked. Similarly, the presence or absence of lime in the natural soil should be considered in connection with the selection of kinds.

Rock-building has been developed into an art by experts, and where alpine gardens on the grand scale have been established inquiry generally reveals the fact that experts have been employed. It was so at Wisley, the famous garden of the Royal Horticultural Society. It is so at almost every considerable undertaking of the kind in public or semi-public gardens. Alas that such cases are so few and that bedding-out still reigns so strongly in these places! And the rule prevails in private gardens where extensive rock-gardening

FIG. 640. ROCKWORK EFFECTIVELY ARRANGED WITH STEPS IN MAJOR J. F. HARRISON'S GARDEN, KING'S WALDENBURY HERTS

is done, such as Tongswood, near Hawkhurst, in Kent, and other places. He would be a genius indeed who, untutored and unsophisticated, proved capable of handling hundreds of tons of stone with good effect and at the same time with due economy of labour. An error in choosing a place for a shrub, even a mistake in selecting a site for a flower-bed, can be rectified at no great cost; but a blunder in placing a pile of rock is a serious matter.

At a period when the cost of transport as well as raw material is high, it would be pardonable for a lover of alpine flowers of limited means to govern his operations in rock-building by the supply of local stone. If ample both in quantity and quality, then he might certainly build up to the full area available, for a rock-garden is not an ephemeral thing, but on the contrary improves year by year. If, however, local supply is deficient or even wholly wanting, then may the alpine home be restricted in extent. One makes these perhaps rather trite suggestions with the less hesitation because there are many rock-gardens in which stone is palpably out of proportion to the plants grown in them. When, however, they are liberally and skilfully planted, as in the frontispiece to Volume I., and in Figs. 639 and 640, there is no preponderance of stone, no forbidding bareness, but on the contrary charming natural pictures.

ALPINE TREASURES

There can be no successful rock-garden with an inadequate supply of plants. And while, alike in the interests of economy and of the garden itself, well-known, easily increased plants may occupy the greatest amount of space, there should yet as far as possible be the interest of rarities. In this connection let us glance at some of the plants on a first-class rockery, such as that at Tongswood. On a cool, steep rock-face, imitating the conditions under which the plant thrives in its Pyrenean home, is a colony of ramondia, obviously very much at home. That exquisitely beautiful grass-leaved gromwell which is supposed to be unsuited to our climate, Lithospermum graminifolium, covers a broad ledge, and is not less delightful than the better-known prostratum and the form Heavenly Blue; among profusely bloomed blue-flowered plants there are few greater June treasures than these gromwells. Overhanging the rockery from a lofty summit are the white masses of the Tasmanian daisy tree (Olearia stellulata, once called Eurybia gunniana), which is not supposed to be hardy, but is here sound and healthy after a bitter spring. On lower sites, spreading into dense mats and sprinkled respectively with blue and with rose flowers, are two botanically related but very different plants in Rhododendron fastigiatum and Azalea rosæflora.

The old garden antirrhinum, often home-sown on ancient walls, is used for many purposes in these days, but few people, perhaps, recognise at sight its sister the sulphur-coloured species asarina, for the heart-shaped toothed leaves bear little resemblance to the better-known type, and it is only on a closer inspection that the characteristic snapdragon shape of the flower stimulates recognition. Drooping from a sheltered, dry crevice to a depth of two feet or more, it makes an instant impression.

A flowering strawberry which does not set fruit, but has leaves which should perhaps be more shiny than they are, considering that the species bears the name of [Fragaria] lucida, is an interesting novelty. It covers a broad ledge with a mat of typical strawberry leaves, runners and white flowers.

FIG. 641. THE WELL-HEAD IN MR. J. R. UPSON'S GARDEN, SARACENS, SURREY

Among floral treasures unknown to earlier generations of rock gardeners is Celmisia spectabilis, an evergreen of low growth which bears a profusion of large flowers, the ray florets white, the disc yellow, a plant which is not of the hardiest, and yet which thrives on a sunny rockery in well-drained friable soil. Rock roses (cistuses), however, and sun roses (helianthemums), are among the oldest and also the best. The cistuses are shrubs, and what plants of this class are more beautiful? The large blossoms are fleeting, but flower follows flower in such rapid succession and in such profusion that for several weeks the plants are covered with bloom; and as, when suited by the conditions, they grow to a large size, they become glorious objects. Although less vigorous, the sun roses spread widely and form broad, low, dense masses smothered in beautiful flowers. It is not for spacious rock-gardens alone that sun roses are suitable. Wherever there is a low wall or sunny ledge to cover they come into their own.

The orchid family is represented by a good plant in Orchis foliosa, which forms handsome masses of oblong leaves and shortish spikes of purple flowers. It likes a cool sheltered spot. The same may be said of many beautiful old and new primulas, such as japonica, cortusoides, bulleyana, littoniana, pulverulenta, cockburniana, and sikkimensis; also the little low-growing species rosea, which is never so happy as when growing in a cool, moist, shady spot.

What for the rock-garden is a rarity indeed, being generally grown under glass, is an evergreen with lance-shaped leaves and scarlet, lily-like, curiously toothed flowers of great beauty, borne on stems about two feet long—a plant that likes a sandy mixture of peat and loam and will only thrive outdoors, if at all, in a warm, sheltered spot; this is Tricuspidaria lanceolata, otherwise Crinodendron Hookeri. And yet another rarity for outdoors is an exquisitely beautiful amaryllis-like plant called Habranthus fulgens, which few gardeners dare grow in the open air. The secret of its success at Tongswood is that the bulbs are protected by a mat of heaths.

Special plants like these give interest and distinction to the rock-garden, but the unsophisticated amateur feels more at home with familiar things—the many beautiful alpine pinks, low-creeping phloxes, saxifrages in almost multitudinous variety, and lovely gentians like verna, bavarica, Andrewsii, freyniana, and the splendid but rather capricious acaulis. Nor will he despise the true geraniums, which perhaps receive less attention than they deserve because their name has been usurped by the zonal pelargonium—that gorgeous, glittering king of the old-time flower-garden. In large rock-gardens and in borders alike, true geraniums such as pratense, sanguineum, and Endressi are equally at home; while the smaller cinereum and argenteum are good alpines. The beginner finds it helpful and encouraging to handle plants which are responsive, and he will certainly find responsiveness in geraniums, just as he will in sun roses, in rock cresses (aubrietias and arabises), in most of the bellflowers (campanulas), in alpine pinks (dianthus), in geums (equally good for the border and the rockery), in alpine candytufts (Iberis gibraltarica, sempervirens, garrexiana, etc.), in gromwells (lithospermums), in certain evening primroses (Œnothera) suitable for the rock-garden, such as cæspitosa and fruticosa, and in the charming little blue Omphalodes verna, provided it is given a cool, moist, shady spot with other shade-lovers, such as many anemones (blanda, hepatica, etc.), hardy cyclamens (Coum, Europæum, etc.), American cowslips (dodecatheons, especially integrifolium and Meadia), the fumitories (Corydalis), some of which are responsive in shade almost to the extent of weediness, the epimediums with beautiful foliage, Orobus vernus, the mossy saxifrages, the graceful Tiarella cordifolia, the white wood lily (Trillium grandiflorum), Sisyrinchium grandiflorum, and other shade-lovers. Under the stimulus of success with accommodating plants one can go on hopefully to the more difficult gems.

WATER IN THE MODERN GARDEN

Readers of gardening history cannot fail to observe how prominent and yet how false a part water as the medium for an elaborate fountain-system played in the great formal gardens of the past. They can even see it to-day, for the fountains at Versailles, as at some other famous old-time gardens, still exist and play on special occasions. As a holiday entertainment fountain-playing on the grand scale is equally as legitimate as municipal bands, or lawn-tennis, or horse-racing, or county cricket; but it is assuredly not gardening, for it does not, subordinating itself to plant-life, unite with Nature in making places more beautiful, more fragrant, more peaceful. Rather is it a holiday spectacle, and as such, and only as such, to be tolerated, being by no means unpleasing of its kind, refreshing the weary eyes of jaded townsmen, and even perhaps stirring within them a certain national pride.

In the modern English garden the rôle of ornamental water is essentially different. It is not forced. It is not crammed into distorted tubes and thence hurled startling distances in jets and sprays. It is not turned into something forceful and spectacular. On the contrary,

FIG. 642. A CHARMING POOL IN MR. A. MILLWARD'S GARDEN NEAR READING

it is treated as a gentle, soothing spirit, that harmonises with the peace-inspiring influences of gardens, and sustains beautiful and fragrant flowers. Such pictures as Figs. 641, 642, 643 and 644 tell their own story. Runnels and pools find their places in the rock-gardens large and small which have sprung up by hundreds and even by thousands in English gardens during recent years, and ferns and moisture-loving flowers have peaceful homes by waters that if not still are only faintly murmurous.

FIG. 643. DELIGHTFUL TREATMENT OF PAVING, PATH AND POOL IN SIR ARTHUR LEVY'S GARDEN, THE MOUNT, COOKHAM DENE

In some old gardens attached to stately homes the parterres of the past are no more, but the pond with its water-lilies lives on, reflecting in its calm waters the ripe wisdom of the centuries, and giving even in the most hectic times a reassurance of stability. Thus, at Penshurst Place, in Kent, one sees in Diana's Pool a piece of placid water, stone-bordered, with flights of steps softly lapped, and breadths of giant white nymphæas which lie restfully on a surface so still that it seems to sleep. And perhaps the sense of repose that creeps up from the pool into the spirit of the watcher above is increased by the grey walls of the castle and the green squares of the ancient yew hedges.

At Chilham Castle, at Gravetye Manor, and in many other famous gardens of England, water is treated in the same reposeful spirit. One approaches the lakes by mown paths

FIG. 644. THE WATER GARDEN OF SIR JEREMIAH COLMAN, BT., GATTON PARK, SURREY

beside long grass, with banks of rhododendrons, magnolias, spiræas, mock oranges, heaths and roses. Colonies of yellow irises at the edge of the water, and breadths of nymphæas on its surface, give life and beauty without violence, without even stir, with no more than the gentlest imposition of the hand of man. The rush and roar of fretted water, symbolical of the fevered life of the camp, appealed, one may suppose, to the warrior kings of the middle centuries, but their works have passed with their lives.

ROSE-GARDENS

To speak of rose-beds amongst beds of other plants in connection with the modern system of bedding, in which hardy flowers of delicate colour take the place of the tender and garish bedders of days gone by, is not to exhaust the special value of roses in the gardens of to-day; for there is such a thing as a rose-garden proper: a garden in which all the beds are beds of roses, plus perhaps a carpet plant, the beds forming a design, with or without rose pillars, arches, arbours, or pergolas. With such gardens it is general to restrict the beds to one variety, each chosen rather for what is often called its "decorative" or "garden" quality than for perfection of bloom; this quality lying in vigour of growth, profusion of flower, and relative freedom from mildew and other pests. Happy the modern rose-lover who has space enough to form such a garden—a garden, indeed, within a garden—for there is in modern varieties a range of colour, a vigour, and a toughness of foliage which did not exist a few generations ago.

It is not usual to have intricacy of design: that is a thing of the past in every type of bedding. Modern flower-lovers, rosarians among them, realise that the more intricate the design the greater the departure from nature, the greater the cost, the greater the temptation to force the plants out of their native shapes. Thus, the grouping is on the simplest of plans.

The old-time rose-growers can live again in the beautiful modern varieties, many having the blood of the yellow Austrian brier in their veins, and being in colour shades of yellow, orange, salmon, flame, and apricot. These "pernetiana" roses have glossy foliage in varying degrees, the leaves large and laciniated, so that they give a clear reminder of holly.

One can suppose an imaginative rose-grower having a vision of a long wide border of roses in which there shall be a subtle blending of tender colours: cream, canary, buttercup-yellow, salmon, salmon-pink, orange, orange-cerise; the whole based on a preliminary study of the habits of the varieties chosen, so that they shall correspond as closely as may be. Consider such a border under the light of a June dawn. There would be such a charm of colour-blending as few enterprises in herbaceous borders would be likely to provide.

WALL-BEAUTY

Wall-beauty was not forgotten by former generations of gardeners. They had not so wide a choice of material as we of the present day, who have had a wealth of new plants provided for us by modern raisers and collectors; yet when we see an old Banksian rose covering something like six hundred square feet of wall, as we may at Chilham Castle; or a wistaria at its best, with drooping clusters ("terminal racemes") two feet long, as at Springfield, Maidstone; or in a cool sheltered place, with sun exposure only on the eastern

FIG. 645. PINKS AND ROCK ROSES ON A DRY WALL WITH IRISES ABOVE, AT WISLEY IN JUNE

face, an exquisite breadth of the flame nasturtium (Tropæolum speciosum), which some people mistakenly suppose can only be grown successfully in Scotland and in the Lake District; or on a southern terrace wall the brilliant honeysuckle (Lonicera sempervirens) as at Gravetye; or even a villa wall with so homely a covering as that provided by Clematis montana, we realise that our forbears were not stinted. There are, however, many new climbers and creepers from which we can choose.

But there are low walls to consider—walls where the most suitable creeper would be a selected variety of Japanese pear (Pyrus japonica), which does not extend rapidly, but is very beautiful in spring; walls, however, that can be treated without creepers, by sowing in their crevices seeds of small campanulas, aubrietias, encrusted saxifrages, alpine pinks, wallflowers, snapdragons, sedums and yellow perennial alyssums ("gold dust"). At Gravetye one may see a wall which has been thus sown clouded over with clematis trails rambling on a trellis; and at Wisley one sees a dry wall clothed with beautiful things, amid which pinks and rock roses are conspicuous (Fig. 645). In cool spots ferns would be at home.

SOME BEAUTIFUL SEED-RAISED HARDY FLOWERS

It is not the least gratifying feature of the modern English garden that while it is more artistic and more varied than the old, it is potentially and in most cases actually less costly. It is true that extensive rock-gardens, carried out on the grand scale with large blocks that have to be transported a considerable distance, are expensive; but these are the exceptions. In the vast majority of gardens, where the bulk of the flowers are hardy kinds raised from seed or by division out of doors, the cost is comparatively small. Flower-lovers have learned that in seeds they have a means of obtaining very large numbers of beautiful plants, annual, biennial and perennial alike, by sowing in the open air at three seasons: spring, early summer, and autumn. And seedsmen have met them at least half-way by raising improved strains, not only larger and more freely bloomed than the older types, but with a greater range of colours.

In this connection one may doubt, however, whether beginners do best for themselves by sowing where the plants are to bloom; and whether the old gardener's way of sowing in prepared beds common greens for subsequent transplantation, is not worthy of imitation with flowers, with the possible exception of spring-sown annuals. It is in such beds, the soil brought to a fine tilth, adequate moisture provided, the drills drawn far enough apart to admit of regular hoeing, the kinds neatly labelled, the seedlings thinned betimes, and in as many cases as possible transplanted to nursery beds in summer where they can strengthen for the autumn planting—it is in such beds that the amateur can provide himself at small cost with large stocks of beautiful flowers.

Let us see what kinds, among others, he can thus provide: long-spurred columbines (aquilegias); tall, bushy anchusas of the most intense gentian hue; mauve, lavender, and rose aubrietias for the rock-garden; single and semi-double peach-leaved campanulas (persicifolia); white and coloured perennial candytufts (iberis) which form beautiful breadths of white spires on the rockery; Canterbury bells, of both the ordinary and the cup-and-saucer type—plants which one must remember are as good for pots as they are for the garden; orange Siberian wallflower (cheiranthus Allionii), which lasts so much longer than the ordinary wallflowers; also cheiranthus (or erysimum) linifol um, the mauve alpine wallflower; the large-flowered yellow coreopsis grandiflora, one of the gayest of medium-height border plants; giant white, pink and crimson daisies for spring beds; foxgloves, especially the pure white; mixed hybrid delphiniums, and also the old light-blue favourite Belladonna, which comes true from seed, likewise the dwarf scarlet delphinium nudicaule; gold and crimson-banded gaillardias, and others of pure yellow; those wonderful geums, the orange-scarlet Mrs. Bradshaw, and the yellow Lady Stratheden; apparently quite double and yet producing seeds which reproduce them true to type; the chalk plant or gauze flower (gypsophila paniculata), which some people find difficult to grow, mainly perhaps from want of lime; the coral-red heuchera sanguinea, with its low masses of rounded, wrinkled leaves and slender stems crowned with charming flowers, one of the very best of June plants for the semi-shaded border and rockery; and single and double hollyhocks, those old-time favourites which so often disappoint because (in the absence of spraying with Bordeaux mixture) the leaves get badly diseased and the plants become unsightly; even so, they often hold their flower-stems in full beauty an appreciable time.

The list, strong as it is, by no means exhausts the supply. Beautiful pictures are made in English gardens with the modern improved herbaceous lupins, pink, yellow, mauve, and other colours, apart from the white and the yellow shrubby kinds; with myosotis (forget-me-nots), giving spring carpets of their precious blue in bulb and other beds; with giant pansies, white, primrose, mauve, purple, and other shades; with primroses and Munstead and other polyanthuses, most beautiful of front-place border-plants for spring blooming (note that it is even better, but not always so convenient, to sow these in frames in February); with perennial poppies of the orientale and bracteatum types, having giant flower-stems in June, also Iceland poppies and the modern sunbeam poppies; with double and single giant pyrethrums, having in June (given proper staking) masses of beautiful flowers; with scarlet, white, pink, purple and rose Brompton stocks, flowering in May, and having large, long, sweet spikes; with sweet williams, in white, pink, scarlet and other colours, purchasable separately or mixed; with tall white and yellow biennial mulleins (verbascums); with bedding violas (tufted pansies) in white, yellow, bronze, and violet (the florists' named varieties are best raised from cuttings); and finally with wallflowers, a host in themselves for beds and borders, in yellow, crimson, chestnut, maroon, palest cream and—perhaps best of all—in the beautiful modern shade of orange.

If one omits antirrhinums (snapdragons), it is in no way from want of admiration, but rather from the belief that for outdoor sowing early autumn is soon enough, when at the same time one may sow many beautiful hardy annuals, such as candytufts, chrysanthemums, clarkias (the doubles under glass in September for early bloom in pots), collinsia bicolor, annual coreopsis (or calliopsis), cornflowers, godetias (the doubles under glass in September for early bloom in pots, like clarkias), larkspurs, the little limnanthes Douglassi, linarias (toadflax), the pretty blue nemophila insignis, the long-lasting blue phacelia campanularia, various poppies, the dwarf pink saponaria calabrica and the slightly taller but still dwarf silene pendula, sweet peas, sweet sultans and viscarias.

Of course antirrhinums, together with the beautiful yellow, pink, orange, scarlet and crimson nemesias, also with China asters, ten-week stocks, marigolds, zinnias, phlox Drummondii, nicotianas (white and coloured tobaccos), salpiglossis, etc., are often sown in gentle heat in winter and hardened in frames in order to provide material for planting out in June, when late bulbs, such as Darwin and cottage tulips, have been carefully lifted from beds and borders and laid-in to ripen. It is by these and other means that modern English gardens are made beautiful at low cost.

BEAUTIFUL ENGLISH GARDENS

One would repeat with emphasis the advice to garden-lovers to seize all opportunities of visiting good gardens. In this connection one would say that the movement organised in 1927 by the Women's Committee of the Fund for the National Memorial to Queen Alexandra, to obtain the opening to the public, by payment of a small sum, on selected dates, of some of the most beautiful of the private gardens of England, was in every way an admirable, one might almost say an inspired, step. The results were highly gratifying. In the first place, a substantial addition was made to the fund. In the second, many thousands of people were provided with opportunities of seeing garden art in its highest phases, and thereby of receiving a stimulus at once pleasant and instructive.

Where the gardens were so numerous, and of so high a standard of beauty, it would be invidious to particularise, and one must be content with naming a few, such as the beautiful gardens of his Majesty the King, Sandringham, Norfolk; of the Duke of Devonshire, Chatsworth, Derbyshire; of Mr. C. E. Keyser, Aldermaston Court, Berkshire; of Viscountess Hambleden, Greenlands, Berkshire; of Mr. L. de Rothschild, Ascott, Buckinghamshire; of the Duke of Westminster, Eaton Hall, Cheshire; of Mrs. Tremayne, Carclew, Cornwall; of Major Dorrien-Smith, Tresco Abbey, Isles of Scilly; of the Earl of Carlisle, Naworth, Cumberland; of Lord Walter Kerr, Melbourne Hall, Derbyshire; of Sir Randolf Baker, Ranston, Dorsetshire; of Lady Barnard, Raby Castle, Durham; of Miss Ellen Willmott, Warley Place, Essex; of the Duke of Beaufort, Badminton, Gloucestershire; of the Marchioness Curzon, Hackwood Park, Hampshire; of the Earl of Carnarvon, Highclere, Hants; of the Hon. Vicary Gibbs, Aldenham House, Hertfordshire; of the Marquess of Salisbury, Hatfield House, Hertfordshire; of Sir Charles Nall-Cain, Brocket Hall, Hertfordshire; of Sir Otto Beit, Tewin Water, Hertfordshire; of Sir Edmund Davis, Chilham Castle, Kent; of the Lord Sackville, Knole, Kent; of the Lord de Lisle and Dudley, Penshurst Place, Kent; of Mr. A. C. Leney, The Garden House, near Hythe, Kent; of the Women's Horticultural College, Swanley, Kent; of Mr. C. E. Gunther, Tongswood, Kent; of the Marquis of Exeter, Burghley House, Lincolnshire; of the Lady Battersea, The Pleasaunce, Norfolk; of the Duke of Portland, Welbeck, Nottinghamshire; of the Duke of Marlborough, Blenheim Palace, Oxfordshire; of Viscount Ullswater, Campsea Ashe, Suffolk; of the Lord de Saumarez, Shrubland Park, Suffolk; of the Duke of Northumberland, Albury Park, Surrey; of the Earl of Dysart, Ham House, Surrey; of the Lord Dewar, East Grinstead, Sussex; of Mr. William Robinson, Gravetye Manor, Sussex; of the Lady Loder, Leonardslee, Sussex; of Mr. J. G. Millais, Compton's Brow, Sussex; of the Marquis of Lansdowne, Bowood, Wiltshire; of Earl Beauchamp, Madresfield, Worcestershire; and of the Lord Bolton, Bolton Hall, Yorkshire. These and other lovely gardens were open to the people.

Visitors to great private gardens find remarkable differences in treatment. Such gardens as, for instance, Chatsworth and Aldenham, both nobly beautiful, have practically nothing in common. One gladly takes the opportunity of referring briefly to the features of each in turn.

CHATSWORTH

The appeal of old buildings which have been sanctified by time and history is irresistible, and not less so is that of old gardens, such as those of Chatsworth. Over the meads of our pleasant land of England there lie spread the stately seats of her nobles, set in age-old gardens. In many cases the names of the founders may exist, while those of the architects and the landscape gardeners have been lost, and then there is an unavoidable sense of incompleteness. In other instances, however, the names of all concerned are enshrined in irrefutable records, so that the history of the place is complete; and it is precisely in these, given the necessary distinction in building and garden, that the interest of cultured people is keenest.

In seeking for an example of a place with which distinguished workers of all classes have been identified for several centuries, one need look no farther than Chatsworth. Many generations of noble owners, all eminent in public life; a series of stately buildings (for the original structure has entirely gone, and more recent ones have been altered and extended); landscape gardeners of unequalled reputation; famous architects, decorators, sculptors and painters—all these are associated in a work of almost unique force and attraction.

The building, its gardens and its galleries having long been thrown open to public inspection on approved occasions, Chatsworth is well known to countless thousands of English-speaking peoples, by whom it is revered. It is, indeed, one of the national treasures —a private possession, yet a possession which is freely shared with others. It is a symbol of that tradition of faith, fidelity and substance which is so precious, so vital to an old nation, sometimes shaken but never overthrown, standing steadfast and four-square in a world of turmoil and alarms: a nation whose fibres sink like the roots of its ancient oaks into the depths of an unconquerable soil, into the hallowed memories of a past which, if not wholly unstained by passing human frailties, is yet in sum noble, pure and magnanimous.

It would be impossible to the most expectant garden-lover, as it would be unwise for the artist, the architect, or even the simple holiday-maker, to separate Chatsworth from its environment. It is not an agglomeration of different things which can be detached and examined as objects of art-value, horticultural interest, or mere curiosity; it is a great unit of inseparable elements. Fully to appreciate Chatsworth, one must take in with one comprehensive sweep all the impressions and implications which it is capable of conveying —natural objects such as the swelling curves of the surrounding hills and the course of the winding river; works of art like the building itself, with its lakes, fountains and gardens; the steep eastern slopes with their massive boulders, down whose stained faces unending streams of water pour; the whole informed with that mysterious yet intimate appeal, that absorbing human interest, which binds past and present, tradition and reality. It is not a brief nor an easy task, yet it is one worth the making, because the Chatsworths of England are not merely of her Yesterday nor even of her To-day; they are of her To-morrow. For in them, in the very ground on which they stand, in their walls, in the pleasure which they convey and the education which they impart to visitors, in the impression of substance and security which they make on people from overseas, and not least in the lesson they teach of the sense of public duty which inspires the owners—in these things there is instinct an assurance that faith, liberty and prosperity will remain secure in England in the future as in the past.

One passes on to the famous gardens. In Chapter XIII. of the present work will be found one of those fine old engravings by the Dutchman, Jan Kip, from a drawing by Leonard Knyff, which have so great an interest for garden-lovers. The engraving, one of many which were published in that rare French book of 1714–16, *Le Nouveau Théâtre de la Grande Bretagne* (in which Kip, Knyff and other gifted artists collaborated and which is now available in *English Houses and Gardens* (B. T. Batsford)), shows the Chatsworth of other days. But there have been many changes. One must assume that Knyff saw a wide canal, with bridge connecting the western terrace, between Chatsworth House and the Derwent, since it appears in his drawing; but it no longer exists, nor is it known what purpose it could have served. Gone the range of low buildings to the north-west of the house, gone the *parterre de broderie* to the south of them. The garden on the south front lives, but with less elaborate adornment. Gone the vast series of intricate bedding on the east front.

There were great gardeners about in the early days of Chatsworth. Whether or no the famous Le Nôtre played any direct part there—and probably he did not—he had able disciples. Our author mentions one Grelly, a Frenchman, who was particularly clever in water-devices. Among the records in the Chatsworth library is a large volume, beautifully kept in a scholarly hand, showing payments made to various artists and workers late in the seventeenth century, and one of them named Grillet was perhaps the "Grelly" of our author.

Grillet (or Grelly) may have anticipated Paxton in the first garden on the west front, now called the Italian garden, also with the *parterre de broderie* on the southern portion. The Italian garden exists to-day, and very beautiful it is, although there is no trace of the elaborate bedding shown by Knyff and Kip. Instead, there are wide walks and broad areas of grass, broken by vases and clipped yews, with stone-framed mounds carrying golden yews amid which are drifts of yellow barberry. There are roses on the terrace walls, and here and there belts of tawny snapdragons, but of bedding so called there is none whatever. Nor, standing at the front of the terrace, and looking down to where there may once have been a large canal, but where indubitably there is to-day the river with its picturesque bridge on the right, can one feel that it would be in tune with the surroundings. But there is at the middle of the terrace garden a round pool with what is known as the Duke's Fountain, and that is more in keeping than the gayest of flower-beds.

No more garish than the Italian garden is the garden on the south front. The same note of cool spacious lawns, wide walks, and ample water is struck. Flowers there are, admittedly, but not in the form of wide borders and large beds. When one says that the brightest floral objects are the hedges of monthly roses, one has perhaps paid the best tribute that could be paid to the standard of taste which governs the planting.

Where, then, are the flower-beds of Chatsworth? Of formally grown flowers there are few anywhere. Perhaps the nearest approach to bedding is in the French garden (Fig. 646), which is close to the buildings on the east side. It fronts what was once the orangery (readers of this work will have grown familiar with the orangeries in the great gardens of the past), but which is now a camellia house. Here there is really bedding, albeit of no gaudy kind—simply a group of beds of bright old-fashioned flowers, flanked by rows of

FIG. 646. THE FRENCH GARDEN, CHATSWORTH

tall pillars bearing statuary that was once within the building. A charming place, this French garden. One lingers by it.

The hand of Sir Joseph Paxton is not apparent everywhere in the Chatsworth garden of to-day, although it might almost be said that he belonged to the place, since he went as a young man and stayed all his life. One may believe that when he found there the gardens and the fountains of Grillet (or Grelly) he was not ill-content to leave them, while dispensing with most of the parterres. One can conceive that he widened the lawns and walks, in order to impart that air of dignity which is now so obvious and so satisfying. It is well known that one of his greatest achievements was the building of the vast conservatory described in Chapter XVI. *Sic transit gloria mundi!* The conservatory has followed the canal and the parterres into the limbo of past things. A vestige remains, no more. The Great War brought about its destruction. And perhaps, if Paxton could emerge from the shades to revisit the scenes that must once have been so dear to him, he would not repine. For after all the conservatory was not his greatest work, and there remain, ever becoming more and more beautiful under skilful hands, imperishable in their setting of stone, the gardens which he made on the hillside to the east. Gardens they are, despite the absence of shaven lawn and trim walk. There must be several miles of paths winding in and out over the declivities, every yard skirted by cunningly placed rocks and shrubs, for countless tons of stone were brought down to the lower slopes and there used to form an immense variety of erections and homes for plants innumerable.

The treatment of this hillside, carried out by Paxton under the sixth duke, was a great achievement, and just as the abundant supplies of water from the higher elevations were used by Grillet (or Grelly) for the great cascade and fountains around the house, so they were utilised by Paxton for his gullies and ravines. The great cascade is avowedly artificial. It has pleased many eminent persons and displeased others. But Paxton's smaller cascades on the slopes of the hill, amid masses of rhododendrons and much other semi-wild growth, are so close to Nature as to have all her own native charm.

It is there that the chief gardening work at Chatsworth is now going on. The Italian garden remains, and will remain, the Italian garden. The French garden needs, and will receive, little renovation. The lawns, the great walks, the ponds, the pools, the fountains, will not be tampered with. The Chatsworth of to-day will remain, to become the Chatsworth of posterity. But up there, beyond the confines of the formal garden, where Paxton's great work was getting to be more and more overgrown with every passing year, where much that he had accomplished was actually hidden by ever-encroaching masses of vegetation—there active renovation is being pursued. Choked ravines are being opened out, new vistas are being cut, fresh plantings are being made. There is not the remotest fear, however, of any violation, however slight, of the spirit of the past. The traditions of Chatsworth will be maintained. It will remain an abiding monument of much that is best in English life, and a beacon to art-lovers in the years to come.

ALDENHAM HOUSE, HERTFORDSHIRE

Among the famous gardens which are described or named in various chapters of the present work, some are renowned as picturesque, some as formal; in some the interest is horticultural, in some botanical. There are few gardens which command attention from every point of view: gardens which contain numerous exquisite pictures in the natural style, yet in parts are formal; gardens in which trees and shrubs are used lavishly to produce fine landscape effects, and yet are treated as individuals of botanical interest, forming collections which embrace the newest and rarest species; gardens which have the area of great public parks and yet have the distinction and refinement of the best private places; such a garden exists, however, a few miles from London.

In the whole history of garden art, long and remarkable as it is, there has been no achievement more admirable, more satisfying, than that which has been accomplished at Aldenham House, in Hertfordshire.

Admittedly there have been vast undertakings in the past, as the earlier chapters of this book show, and at Aldenham there is nothing comparable with the grandiose and spectacular works, carried out at fabulous cost by men who were half landscape gardeners, half engineers, which amazed the peoples of past centuries. In those stupendous operations pure gardening was not the first consideration. Ambitious kings vied with each other, and stimulated for political ends the rivalries of their peoples. At Aldenham, however, neither landscape gardener nor engineer has ever been employed. There has been nothing theatrical, nothing sensational. From first to last garden art, pure and undefiled, has been the object in view.

The knowledge that art is paramount at Aldenham must inevitably have its effect on people of culture. They will go, as they do go in their thousands on the days appointed for visitors, with the reassuring conviction that what they are to see is gardening and nothing but gardening, that it stands or falls by the extent to which art, and art alone, has gained the ends it sought.

The fact that the great work of developing the garden at Aldenham is comparatively recent, having been started in 1898, has led many to consider the whole place modern. That is not the case. There was probably an Aldenham House in being before Shakespeare was born, for the first structure was reputedly built in or about 1550. Naturally there have been changes in the mansion, as there have been in the owners. The point is that Aldenham has just the same claim to antiquity as Chatsworth or Hatfield, although the sequence of owning families has not been so continuous. In turn Thomas Sutton, Henry Coghill, Robert Hucks, Miss Noyes, and Lord Aldenham have been owners of Aldenham. If the first house was really erected in 1550, the Thomas Sutton who founded the Charterhouse could not have been the first owner, as he was born in 1552, but he may have been a son of the founder. And not only has Aldenham House the hallowing charm of age, but it is richly decorated in the grand manner and plenteously stored with art treasures.

The modern garden at Aldenham is the work of a man of genuine horticultural genius, the Hon. Vicary Gibbs, a London banker, devotedly seconded by a gardener of exceptional parts, Edwin Beckett. Neither had the special training of an architect or a landscape gardener. The horticultural education of both has been based—and this is significant—on a deep love for plants. Out of that all the rest has sprung. Superimposed upon it there has grown a garden (one might say a whole series of gardens) not only of vast extent but of almost bewildering diversity and overwhelming beauty. As to area, it is only necessary to compare Aldenham with, say, Hampton Court. The flower-garden at Wolsey's master-piece extended to four acres, that at Aldenham approaches two hundred acres. But area is after all a minor thing, the real core is treatment; and here one realises how inadequate

II—2 D

is a single visit in any year, even if one's steps have the privilege of guidance by both the great workers. For Aldenham is garden within garden, repeated (with variations) a score, even a hundred, times. It is a garden of all seasons. Standing out above the majority of gardens in its wealth of shrubs, it shows colour of leaf, or stem, or twig alike in spring, summer, autumn, and winter. There is apparently no week, perhaps no day, in which colour cannot be found.

It is particularly in connection with shrubs that one realises the remarkable versatility of Aldenham. A garden most beautiful, it is also virtually a botanic garden, one might

FIG. 647. THE DIPPING POOL IN THE HON. VICARY GIBBS'S WATER-GARDEN AT ALDENHAM HOUSE, HERTS, WITH HOLY-WATER STOUP FROM A CHURCH AT VENICE

almost say a nursery. For there are acres upon acres of shrub-beds, planted, not with commonplace aucubas, but with all the rarest introductions from the Far East. Famous modern collectors in China and Tibet like Wilson, Farrer and Kingdon Ward have had no more liberal supporter than Mr. Gibbs. Hundreds of thousands of seedlings, many wholly unknown to cultivation, are raised on the place every year, each labelled, the pedigree of each docketed.

As with shrubs, so with trees. There are vast collections of species, some exotic, some native, of the kinds best known in British gardens, parks and woodlands. That fine old English tree, the yew, is a case in point. The best species and forms may be found in the grounds. In this connection one may refer to the valuable article by the Hon. Vicary Gibbs on Taxaceæ (yew family published in the *Journal of the Royal Horticultural Society*, vol. li. People interested in the yew (and who is not?) will find in it a mine of information on this venerable old tree, so long associated with British churchyards. They will learn that although normally diœcious (male and female forms in separate trees as with the aucuba, the hop, and several others), a female branch will be found occasionally on a male tree. They will learn of at least one variety of which no male form is known, but a female only. They will learn of noble specimens here and there on the countryside, such as those on Mickleham Downs in the neighbourhood of Leatherhead, Surrey. And they may suffer the shattering of a delusion held by many—that yews were planted in the churchyards in order that our bowmen could have the best of wood for their bows, whereas most of the

FIG. 648. THE FORMAL GARDEN AT ALDENHAM HOUSE, HERTS, WITH SPRING BEDDING

wood used for bows by English archers was imported from the Pyrenees. The trees of Aldenham are indeed a wonderful source of interest and pleasure.

Cross-fertilising is carried on extensively with many kinds of plants, from choice hot-house things to plain kitchen-garden vegetables. The interest of, the joy in, Aldenham will, however, lie for most visitors, as would naturally be expected, in its lovely ranges of flower-garden. One uses the word "ranges" advisedly, for there are numerous pieces which are gardens in themselves, and yet which form but a part of the wonderful whole. Here, for example, is a piece of water enclosed by banks of shrubs, its calm surface enlivened with a host of water-lilies; there, at the turn of a walk, is a belt of rockwork clothed with exquisite bloom. Then there is the Wrestler's Pond, having the shape of a Maltese cross, with fountain in the centre and beautiful colonies of water-lilies. From a bridge one looks down in another place on water which plashes from dripping stone, and passes on over a rocky bed to a ravine whose sides are planted with beautiful things. And there is human interest too, for in one water-garden one sees a dipping-pool (Fig. 647) with a holy-water stoup brought from a church at Venice.

One passes from garden to garden, from avenue to avenue, from vista to v.sta. While the note of nature is truly and firmly struck in almost every part; while, far away from the house, one feels in solitude among ranks of trees or groups of shrubs, there yet comes a momen when on the terrace behind the house one sees a formal garden (Fig. 648) which

FIG. 649. THE GREAT BORDER OF MICHAELMAS DAISIES (PERENNIAL ASTERS) AT ALDENHAM HOUSE, HERTS

in spring is furnished with tulips, primroses, polyanthuses, and other gay things, and in summer is alive with equally beautiful flowers.

Landscape gardening in its purest, its most beautiful form is pursued at Aldenham on a scale rarely equalled, yet its garden interest is no more exhausted there than in the shrubberies and greenhouses, the orchards and the kitchen-gardens. For many great hardy herbaceous plants, particularly including Michaelmas daisies and delphiniums, are given special treatment, all the best varieties being grown and new ones raised annually in large numbers. It is because of this that on an autumn day the visitor may come upon a broad belt of perennial asters, giving light and fire to a long border (Fig. 649). Pictures such as this—and it is but one of thousands—live in the memory.

And memory, recalling the Aldenham of twenty years ago not less vividly than the Aldenham of the present; recalling, too, hundreds of gardens seen in our own and other countries, can find no parallel to the achievement which it represents, having regard to the configuration of the ground, the soil, the vast extent, and the short period in which everything has been accomplished. Nature gave a flat surface and stubborn London clay; art has produced range and elevation in infinite variety and an amenable earth abounding in fertility. Moreover, this wonder-garden equals in area the combined parks of many an important town. Unexcelled as a work of pure art, a storehouse for thousands of ornamental plants which are now unknown but are likely to possess great artistic and commercial value in the future, a birthplace and testing-station for numerous utilitarian members of the kitchen-garden, almost equally important artistically and educationally, Aldenham stands for English gardening in its highest, its greatest phase.

THE PLEASAUNCE, OVERSTRAND

Among the gardens on the eastern seaboard of England, where the great East Anglian shoulder thrusts out into the North Sea, there are few which can compare in beauty with The Pleasaunce, that exquisite gem of horticulture founded by the late Lord Battersea a few miles from Cromer.

One recalls one's first visit, when the man who called it into being, famous alike as politician, sportsman and artist, himself acted as guide, and afterwards quietly asked a guest still struggling with his impressions and emotions to suggest improvements. In reply one could speak only of learning, not of teaching.

The Pleasaunce was, and remains, an artist's garden. It was the original, the finished work of a man on whose walls hung some of the best paintings of Botticelli, Leonardo da Vinci, Moroni, Burne-Jones, Bassano, Rubens, and Whistler; a man to whom the importance of line, form and colour was a law. Cyril Flower had brought both training and imagination to bear on the task which he had set himself. Despite this, he was not troubled by horticultural tradition. It was nothing to him that seventeenth-century architects and landscape gardeners had tied themselves to severe axial lines, terraces, elaborate water-devices, fountains and statuary, for these things were not art as he understood it. He had no sympathy, indeed, with the formal system as such. His respect for form did not blind him to the demands of Nature. He could not visualise the garden as a mere appanage of the house, although he was quite prepared to associate the two in harmonious ways. Above all things he set before himself the task of making a garden which should be beautiful in all its parts—a garden that conformed to the laws of art in line and colour and yet was entirely informal, creative, stimulating and original. He achieved success in a very remarkable degree—so much so, indeed, that The Pleasaunce became one of the distinctive gardens of modern England.

It remains a private possession, but just as, in mediaeval times, the great nobles of Italy threw open their grounds to the public, so, in these days, do many liberal-minded proprietors of English gardens give the people access to them on stated occasions. Garden-lovers may, therefore, visit The Pleasaunce at particular times, as they may the royal gardens at Sandringham a few miles away; and one can hardly imagine a more pleasant and inspiring pilgrimage than that which is made to embrace both these beautiful places.

Visitors to The Pleasaunce will find roses, hardy herbaceous plants, alpines, shrubs and aquatics used with equal taste and skill. They will see delightful pergolas, loggias and

FIG. 650. WATER-GARDEN WITH HYDRANGEAS ON THE LEFT AND SUMMER-HOUSE AT THE BACK IN LADY BATTERSEA'S
GARDEN, THE PLEASAUNCE, NORFOLK

summer-houses. They will find enchanting ponds and pools, the banks of one of them
planted with beautiful shrubs, beyond which a summer-house looms (Fig. 650), the water
carpeted with nymphæas. The many beautiful walks will particularly arrest attention.
It was in his treatment of walks that the creator of the garden displayed his greatest skill
and originality. He was one of the first, if not the first, to edge walks with small borders
of rock, planted with attractive alpines; and these stone-lined paths remain one of the most
pleasing features. But there is much of the now familiar (and in many other places gravely
overdone) crazy-paving, and some walks are wholly flagged. Of the wider walks, some are
bordered with bright-hued shrubs, such as golden yew and golden box, clipped to a neat
yet not excessively formal shape; and these give a note of both colour and distinction.

The rose-garden, with the neighbouring summer-house whose pillars are clothed at
the base with flowering evergreens and wreathed above with ivy, is another beauty-spot.
The approved plan of growing only one variety in each bed is adopted in most cases,
for this facilitates securing that general effect of colour-harmony which is so desirable, and
yet so difficult to obtain when several varieties are mingled in a bed. The sundial round which
some of the beds are grouped gives its sedate, mellow and soothing note to the scene.

There can be few visitors who will fail to note the striking effect of flowering plants
and shrubs grown in large tubs and vases. Particularly are these conspicuous in the Italian
garden, with its flagged courts that are interlaid with mosaics (Fig. 651). The low walls
of the bays, with their time-stained stones, the counterbalancing promontories of gay

flowers, the massed shrubs on the boundary walls, and the bold groups of colour beyond, combine in an entrancing picture. No shrubs are used more effectively in tubs and tall vases for selected places than fuchsias, which in some instances are lofty bushes, bearing myriads of graceful and beautiful flowers.

Pergola and herbaceous border combine to form another charming section. The herbaceous plants are of the approved kinds; and here, as elsewhere—in the rock-edged paths, in the Italian garden, in almost every part of the place—attention is devoted to the finish, a foreground of suitable dwarf plants being provided for the purpose. The pillars of the pergola, mostly brick or flint, are clothed with roses; but there are pergolas in other parts of the grounds which bear different burdens, in one part fruit-trees, in another laburnums, which in their abundance and their grace, though not in their colour, recall Mr. William Robinson's wistaria-covered pergolas at Gravetye Manor, in Sussex.

The Pleasaunce is not a garden which description can portray, nor one which can be understood and appreciated by the casual looker-on; but it is one of those works of art which study confirms as great, and in its finished beauty creates ineffaceable impressions.

WISLEY GARDENS, SURREY

In days that the present writer recalls vividly, it was held as a grave reproach against the Royal Horticultural Society that its appeal was less to horticulturists as such than to members of the nobility whether or not they had a real interest in plants and garden art.

FIG. 651. LADY BATTERSEA'S ITALIAN GARDEN AT THE PLEASAUNCE, NORFOLK

The society's headquarters were then at South Kensington, where it held periodical meetings, much as it does now at Vincent Square, Westminster. Its membership was small, it was poor, and it strove to make ends meet by letting its grounds for functions remote from horticulture. It had a garden at Chiswick, small, but well conducted on plain lines by a Scottish gardener.

Gardeners were not ill-satisfied with Chiswick. A minority, neither small nor uninfluential, urged long and persistently that the South Kensington centre should be given up, and Chiswick Garden made the headquarters of the society. Perhaps at one time this minority was nearer success than its own leaders ever knew. Be that as it may, the society's fortunes took a change for the better with the great revival of gardening interest which set in about the end of the last century, and with the gift to it by the late Sir Thomas Hanbury of Wisley Gardens.

It may be a moot point whether the acquirement at Westminster of a large and well-equipped hall of its own, or the acquisition of Wisley, was the stronger influence in the further great increase in the society's strength which followed. Certain it is that Wisley swiftly became, and still is, an asset of immense value and importance. Its history is not without interest. Great horticulturist though Sir Thomas Hanbury was, and famous as were his achievements in gardening, he was not the founder of Wisley Gardens. They came into being through the wanderings of a nature-loving, flower-loving Battersea candle-manufacturer, one G. F. Wilson, whose name lives only in its attachment to a few plants. This nature-seeker found Wisley during a characteristic ramble, acquired it, and made the plantings which were destined to grow into so rich a heritage.

One must ever remember to the society's credit that it has dealt generously and spaciously with Wisley. It has enlarged the garden enormously, built glass-houses and laboratories on a handsome scale, and made extensive plantings of fruit-trees and ornamental trees and shrubs.

Founded mainly as a home for hardy plants, growing in a more or less natural environment, Wisley retains its charm for nature-lovers, while assuming wider interests. It was embellished with a large rock-garden (see frontispiece to Volume One) which alone in the season of its chief beauty attracts visitors in large numbers. Constructed by specialists, it was made on a scale which precludes complete repetition except by people who have both wealth and ample space; nevertheless, it is capable of conveying lessons to alpinists of all classes.

The great rock-garden is not, however, the only attraction for lovers of hardy flowers, who will learn valuable lessons of the possibilities of plant-culture in cool, moist, shady places by allowing themselves time for slow, tranquil rambles in its dells and coppices and by its hedgerows. Conversely, they will learn what delightful pictures can be made under drier, hotter conditions by observing how much at home are pinks, rock roses and other plants growing on the face and summit of dry walls (Fig. 645, page 391).

The ever-growing legion of shrub-lovers find rich pabulum at Wisley. And when, as in many cases, the love of aquatic plants goes hand in hand with that of shrubs, there is a still greater reward. The pond in the shrub-garden (Fig. 652) unites two interests, and enviable is the lot of the horticulturist who can seek this place not only in spring but also in summer. The greatest feast of shrub-beauty is obtained in May, of nymphæas and other water-flowers in July; but at all seasons Wisley has its rewards and its lessons.

FIG. 652. "SEVEN ACRES," THE POND IN THE SHRUB-GARDEN AT WISLEY

Wisley Gardens possess a source of interest and beauty which ordinary gardens, large or small, lack—a feature, in fact, which is not to be found elsewhere except in the grounds of a few of the great trading firms—namely, extensive trials of the different species and varieties of popular plants. On July days one may see, for example, a great array of modern roses, or of sweet peas, or of irises, or of poppies, or of dahlias, or of phloxes, or of delphiniums. It is not suggested that all these plants will be on trial in the same season; but it is safe to say that in most years July will show extensive trials of several important kinds, each a beautiful display in itself, and with the additional interest of educational value. The Wisley trials, indeed, form one of the most useful items in the work of the Royal Horticultural Society. It is obvious that when a selected number of important plants are selected for cultivation in a particular season, and thereupon a large number of the best varieties, old and new, of each, are sown or planted side by side, the whole forming a considerable area—it is obvious, one repeats, that something may be expected which will not only be very beautiful as a spectacle, but will also be intrinsically educational.

Lying near the heart of the Surrey pine-woods, catching something of their piquant odours when the heat-haze quivers over the surface of the adjacent ponds, Wisley Gardens draw with irresistible force nature-lover and gardener alike. There is no branch of horticulture which is not touched there with distinction. Considering the poor sandy soil, hardy fruit is grown with a success which surprises; and the stock vegetables of the kitchen-garden are also made to flourish. Homely and uninspiring, they are nevertheless important, like the fruit, in the economy of the garden and the household. Far from negligible, therefore, are those phases of the society's operations which concern themselves with food plants.

Still, having regard to the circumstances in which the garden came into being, it is natural and right that its chief interest should lie in the skill with which the woof of its natural amenities has been interwoven with the warp of modern flower-gardening in its highest aspects. It is in that triumph—for triumph it most truly is—that Wisley Gardens will live and grow, a joy to garden-lovers of the present, and certain to be a still greater joy to those of future generations.

PUBLIC PARKS AND GARDENS

In an industrialised and densely populated country there are few matters more important than the provision of public places for games and recreations. Modern nations realise that a "C3" standard in a large proportion of their manhood is incompatible with their obligations and aspirations.

This view has gained ground rapidly in England since the Great War, with the result that there has been a strong movement in the direction of providing more parks and open spaces in or near existing towns, and also in the direction of regional town-planning in areas, such as parts of East Kent, where new towns are springing up.

The increase of playing-fields as such does not come within the scope of the present work, but in so far as public spaces embrace gardening (as the majority do in some shape or form) they have an undeniable claim to attention, and it will therefore be relevant to take note, even if briefly, of developments, alike as to area and manner of treatment.

LONDON

We may begin with London. References are made to some of its principal parks in Chapter XVI., and figures are quoted showing that in 1889 the area other than royal parks was 2656 acres, whereas in 1898 it had grown to 3665 acres. These figures showed great expansion, as might be expected, considering that in 1889 the chairman of the Parks and Open Spaces Committee of the London County Council was so pronounced a believer in progress as the Earl of Meath. The expansion was continued, with the result that in October 1927 the spaces under the control of the Council comprised an acreage of 5659½. Large as it is, this area would probably have been greater but for the effects of the war of 1914–18, which checked development considerably. That great efforts were made by the Council to regain lost ground is shown by the fact that in the ten years 1918–27 further new parks and open spaces, with a total acreage of 570, were added. Most of these are situated in such densely populated industrial districts as Greenwich, Woolwich, Plumstead, Stepney, Poplar, Wandsworth, Southwark, Lewisham, Beckenham, and St. Pancras. Most of them, too, are in the Eastern and South-Eastern districts.

So much for area, what of treatment? The first point to recognise is the inevitability of games and recreation grounds, as distinct from gardens, occupying by far the largest area of public spaces under the control of the London County Council. But public opinion asks, and rightly asks, that reasonable space shall be devoted to ornamental gardens. It asks more: it asks that as far as possible the gardening shall be modern. It has long been a reproach against flower-gardening in public places that it is conducted on the old-fashioned formal bedding pattern of mid-Victorian days, even embracing carpet-bedding. One must recognise that to a certain extent public flower-gardening will always be formal. To keep the cost of public parks and gardens in industrial districts within such limits as shall avoid unnecessarily oppressive local rates, entails the cultivation in bulk of a restricted number of amenable plants, and precludes elaborate schemes in which a considerable number of different kinds are associated. But that is not to imply an unchanging fare of tiresome bedding-plants, still less carpet-bedding.

To do the Parks and Open Spaces Committee of the London County Council justice, it has pursued an enlightened policy in this not unimportant matter. In the various properties which it has purchased that embrace informal grounds surrounding mansions, it has in many cases preserved the established amenities, as at Ravenscourt Park, near Hammersmith. At Ken Wood, comprising 121 acres in the borough of St. Pancras, it has retained cover for bird-life. At Golder's Hill, near Hampstead, an old English ornamental garden has been laid out in what was once the kitchen-garden. At The Rookery, once a private property, in Streatham, there may be seen a genuine example of typical landscape gardening. At Avery Park, Eltham, one finds a beautiful winter garden. Even in Battersea Park, with its plebeian surroundings, the visitor may see near the north-western entrance a delightful garden, with lily-pond, pergola and sundial. Standing in this garden from time to time both in spring and summer, the present writer has become familiar with its informal old-world charm, and earnestly impresses on other flower-lovers its undoubted claims to attention and admiration.

These remarks, brief out of necessity, may serve to show that flower-gardening under the London County Council is not hopelessly archaic and stereotyped, but on the contrary is imbued with the modern spirit of artistic treatment. Nor in those large areas in London which are not controlled by this body does the cultured visitor find himself constantly bored, and even affronted, by floral platitudes and monstrosities. Many important London parks are controlled by the Government itself, through the Office of Works. Such familiar places as Hyde Park, St. James's Park, the Green Park, Regent's Park, Greenwich Park, and Kensington Gardens, are under State management, equally with Kew Gardens, Hampton Court Park and Gardens, Bushey Park, Richmond Park, Woolwich Common, and other large areas which, with smaller places, comprise an area of about 6000 acres.

During recent years the flower-gardening at Kew, in Hampton Court Gardens and in Hyde Park, always notable, has become of increasing beauty and importance. There was a time when the value of Kew lay almost solely in its work as a botanical station. Not less eminent in that respect to-day, it now enjoys the distinction of being a true and highly precious national garden, where plants are grown for their beauty and interest as plants, not exclusively as morphological objects. In fine, Kew is a place of outstanding interest to lovers of flowers as well as to botanists. The rock-garden, the azalea-garden, the bamboo-garden, the rhododendron-garden, the rose-garden, the herbaceous garden, the water-garden, not less than the vast collection of ornamental trees and shrubs, the greenhouses, the lakes, the lawns and the flower-beds (Fig. 653) provide between them beauty and interest throughout the whole of the year. One would be only too glad if space permitted of that detailed description of the principal features of Kew which they so well deserve.

The gardens of Hampton Court, whose history is so well described in an earlier chapter, claim increasing public attention every year. The flower-gardening is conducted on an ambitious and enlightened scale, and few garden-loving people visit this historic place, particularly in spring, about the time when the famous horse-chestnuts in Bushey Park are in full glory, without receiving agreeable impressions.

Similarly in Hyde Park, and to a smaller but not negligible extent in Regent's Park, the flower-gardening is conducted on bold, impressive lines, under the influence of which old-fashioned bedding plants have given place to modern plants, more free, more striking, and yet equally full of colour.

The parks and open spaces of London do not end with those controlled respectively by the Government and the London County Council. Areas, most of which are much smaller, but collectively constitute a considerable acreage, are governed by the City Corporation and the Metropolitan Borough Councils. Thus, the former controls, amongst other areas, St. Paul's Churchyard, Finsbury Circus, and the famous burial-ground of Bunhill Fields, in addition to larger areas beyond the confines of London, such as Epping Forest and Burnham Beeches; while the latter administer a large number of small recreation-grounds, playgrounds, churchyards, greens, squares, triangles, commons, gardens, and parks of a few acres. These enclosures are spread over the whole of the principal London boroughs. The largest has an area of only forty acres, and many have the dimensions only of a small allotment-garden. All, however, serve a real purpose.

FIG. 653. A BEAUTIFUL BED OF TORCH LILIES (KNIPHOFIAS) IN THE ROYAL GARDENS, KEW

GLASGOW

When one turns to provincial cities, one is almost embarrassed in presence of the large
number which take a genuine pride in their public parks and open spaces, so impossible

is it to do justice to them in
the space available. It may be
truly said that there is scarcely
one town of importance in the
United Kingdom which is not
imbued with a healthy spirit
of emulation in this matter,
and seeks to increase its
amenities to at least an equality
with other large towns. We
may take a typical Scottish
and a typical English city as
examples.

With a rise in its population from 329,096 in 1851 to
761,712 in 1901 and 1,034,174
in 1921; with an increase in
its area from the original 1768
acres to the 19,183 of 1927;
with a growth in its rateable
value from £5,840,256 in
1906–7 to £10,480,454 in
1926–7, Glasgow attained to a
strength and wealth which are
a legitimate source of pride,
while entailing momentous
responsibilities.

The city has not failed to
fulfil its obligations, and in
addition to providing the
various services which are
most vital to public health, has
greatly extended its parks and
open spaces—a task in which

FIG. 654. THE BEECH WALK IN DAWSHOLM PARK, GLASGOW

its corporation has been greatly aided by public-spirited citizens and neighbours. Thus, Mr.
A. Cameron Corbett (later Lord Rowallan) not only gave the beautiful Rouken Glen Park
of about 228 acres, but also the great Ardgoil estate, of nearly 15,000 acres, which lies
about forty miles from Glasgow. Ardgoil, with its mountain ranges and its glens, presents
perhaps the nearest approach which is to be found in Britain to the national parks of
America, so graphically described by Professor Frank A. Waugh in Chapter XVIII.

Glasgow Green, the first of the great parks of the famous Clydeside city, was founded as far back as 1662, and with its area of 136 acres is still a source of enjoyment to the citizens. Several of the more modern parks are larger, notably Queen's Park, Bellahouston Park, the Linn Park, Knightswood Park, and Ruchozie and Frankfield Park. On the other hand, many are much smaller, including Springburn Park, with its beautiful winter garden; and Dawsholm Park, with its delightful beech walk (Fig. 654). In connection with the great attraction of Springburn Park just mentioned, Glasgow justly prides itself on its great strength in winter gardens. The London County Council has its Avery Hill Winter Garden, but in point of numbers has to yield precedence to Glasgow. It is gratifying to know that the corporation has a keen sense of natural beauty, and preserves the amenities of many of its parks with jealous care, while providing no stint of flower-beds in other places.

The thirty-two parks administered by the Parks Department of the City Corporation of Glasgow by no means exhaust its responsibilities in the form of open spaces, for it also has charge of small areas in various parts of the city to the number of over ninety.

MANCHESTER

In turning one's eyes from the greatest of Scotland's industrial cities in search of a thickly-populated town in the north of England one naturally looks towards Manchester. This great city was an early mover in the provision of public parks; indeed, it claims to have been the very first town to provide parks maintained solely out of the rates. The year was 1846, the names of the parks were respectively Philips and Queen's. Twenty-two years later it acquired Ardwick Green and also Alexandra Park. Thereafter it rested almost supine for some twenty years, adding only forty-two acres in two decades. Then, perhaps in emulation of the progressive work done in London from 1889 onwards under Lord Meath and other public men, it set out in real earnest to increase its resources, with the result that in 1927 it was able to boast the possession of seventy-four parks and other open spaces with an area of nearly two thousand acres.

As in London and many other large towns, the greater part of the acreage at Manchester is devoted to courts, greens and courses for various games, the fees for which make a substantial contribution (in 1927 alone upwards of £14,000) to the cost of upkeep. But ornamental gardening is not forgotten. Thus, in Alexandra Park there are charming alpine, rose and herbaceous gardens in addition to a botanical garden; in Heaton Park there is a beautiful old English garden (not to speak of extensive spring flower-gardening); and at Platt Fields there is a Shakespeare garden.

An item in the gardening operations pursued by the Parks Department of Manchester which is not common, and might be imitated with advantage in other cities, is that of placing several hundreds of tubs containing handsome trees and shrubs in the principal squares and around the most important public buildings. Window-gardening and allotment-cultivation are also encouraged and given concrete assistance.

The sister-town of Salford is also active. Thus in 1927 it sought permission from the Ministry of Health to raise a loan of no less than £28,076 for acquiring eighty-seven acres of land with which to provide playing-fields.

GARDEN CITIES, SATELLITE TOWNS, AND TOWN-PLANNING REGIONAL SURVEY

Proof accumulates year by year that those who look with apprehension on the extension of the great industrial cities, fearing that the nation will soon begin to suffer from the reduction of open country, have good cause for their disquietude. Especially is this the case in the south-east of England, where the influence of London is so powerful, and where that of the East Kent coalfields must exercise increasing pressure.

It is but a few years ago that a radius of a dozen miles covered the railway lines which served the metropolis and its suburbs. Already (1928) the radius has spread to over thirty miles. A few years hence it will have extended to sixty miles or more: in other words, the metropolitan suburbs will have spread to the vicinity of Reading, Guildford, Brighton, Folkestone, Canterbury and Whitstable. In case this should appear exaggerated, one may point to the national scheme for extending electrical power, and the provision which is being made for linking it up with the new mining towns of East Kent. The widening of old and the formation of new roads will supplement these influences.

It requires but a very brief spell of reflection to realise that the amenities of the countryside and the interests of garden art must be affected by such developments. And the influences might easily be adverse. Unless the establishment of new towns is conducted with due regard to the provision of adequate gardens and the reservation of ample open spaces, the interests of the nation must suffer. Garden cities and garden suburb trusts must be multiplied; the National Trust must be strengthened in every possible way. There are many who would go farther, even to the extent of forming a department of the Government vested with the particular duty of scheduling and preserving large tracts of country, such as downs, woodlands and commons, as national amenities, safe alike from builder, manufacturer and farmer. There might be many worse objects, but the times are not propitious for the multiplication of State bureaucracies, and it will be well if an enlightened public opinion can be left to work out its own ways and means of attaining the end in view.

In this connection it is satisfactory to realise that much has been done during the past quarter-century. The mere fact that the present book is printed in a garden city carries its own significance. It is unnecessary to devote much space either to the objects or achievements of Letchworth. It has long been an accomplished fact. As it presented itself before the vision of Sir Ebenezer Howard and his supporters, so it stands to-day: a garden city in being, a town of decent dwellings, airy factories, flourishing gardens and reasonably large open spaces; a town where indoor work is carried on under conditions vastly superior to those which prevail in London; a town where life is lived on a higher plane than in the average provincial town with its petty social distinctions and narrow interests.

It is unnecessary, too, to describe at length such foundations as the Hampstead Garden Suburb Trust, of which Dame Henrietta Barnett became chairman at the age of seventy-six, twenty years after she had cut the first sod. A wider social outlook, a better understanding between the different classes, is implicit in the spirit which animates the trust. In Dame Henrietta's own words:

The houses are for persons of every class of income, from houses costing £10,000 at a ground rent of over £100, to single-room tenements with a patch of border at 4s. 6d. a week. I think a great many social troubles are caused by the ignorance of one class of another. The rich and cultivated have always considered the poor as an object of pity. The ignorance of the disinherited concerning the wealthy has caused grave and dynamic consequences. In St. Jude's Church, the Free Church, and the Institute, at the Hampstead Garden Suburb people of all classes of thought mix together, thus enriching their lives.

Equally typical in their way are the garden towns, such as Bournville and Port Sunlight, which owe their foundation to broad-minded, far-seeing commercial magnates like the late George Cadbury and the late Lord Leverhulme. These places, with their roomy squares, wide streets, convenient and well-built houses, large gardens and spacious playing-fields, are also well known, and need only be mentioned as reminders that there are examples in the immediate past for the work which calls for accomplishment in the near future. Would that Bournvilles and Port Sunlights could arise near every great town, with concomitant destruction of slum areas, but short of that there are amenities well worth aiming at: as, for example, houses only in pairs and never in terraces, each pair with a good slice of ground. This must be claimed as the "irreducible minimum." Such new London County Council estates as Downham, at Bromley, in Kent, and St. Helier, south of Mitcham, in Surrey, adding as they will within a few years of the time of writing something like 80,000 to the population of Outer London, have immense possibilities both for good and evil.

The fact is often overlooked that the housing problem is two-sided: there is not only a war-shortage to be made up; there is also a past heritage to be replaced. It is a question if the latter—so often entirely forgotten—is not the greater part of the problem. These lines are written in an age-old city which has attracted thousands of visitors (pilgrims, art-lovers and the merely curious) annually for many centuries, and attracts them still. It is doubtful if one per cent of these people take the trouble to observe that the immediate surroundings of the cathedral and other historic buildings consist of mean streets, where humanity is herded together in small, gloomy dwellings utterly subversive of the simplest standards of comfort.

Garden art can penetrate but slowly, if at all, to places where humanity lives under conditions so debasing; and when one wanders reflectively through the great industrial towns, and sees the miles of dingy streets in which factories, small shops, smaller dwellings, and taverns jostle each other in their thousands, one can but marvel that culture is able to rear, however feebly, an aspiring head. That it does so is a tribute to the good instincts of large masses of the people, who seek, mainly through music and flowers, to quench the thirst of parched souls not wholly subdued by squalid conditions and gloomy surroundings.

Art gets its opportunities slowly in these places through the provision of garden suburbs and still more slowly through the elimination of slums. The building errors of the past were too gross and deep-seated for swift redemption. But the Town Planning Act of 1909, with the amending Acts of 1919 and 1923, all ultimately merging in the Town Planning Act of 1925, gave local authorities great opportunities. They now have such power over the land within and near the areas under their control which is not built on as can effectually prevent a repetition of past mistakes, by adopting under expert guidance schemes which make ample provision, not only for home gardens, but also for allotments, parks and recreation-grounds; and when equal powers are given to them of dealing with areas already built on, they can pursue schemes of replanning that are equally beneficent, although necessarily more difficult of accomplishment. Since the passing of the Act of 1925, local authorities have expedited the important work of town improvement, and hundreds of thousands of houses, each with decent provision of rooms and a useful piece of garden, have been built in or near towns.

There is, however, still room for organisations such as the Garden Cities and Town Planning Association and the National Housing and Town Planning Council, not to speak of the younger Rural Community Council, which has made so promising a start in Kent. It must not be forgotten that Letchworth was launched under the auspices of the Garden Cities and Town Planning Association, a limited liability company being formed to raise capital, purchase a site, and found a city. Welwyn Garden City was founded in a similar manner. Such bodies have the initial advantage over the local authorities with their existing towns that they have no commitments—no slums, no cottages on potential factory sites, no factories on potential cottage sites. They can start with a clean slate. And with vision and courage they can proceed on bold, comprehensive lines. Thus, at Letchworth, an area of 4500 acres was purchased, although a large town was not aimed at, but rather:

> A town planned for industry and healthy living; of a size that makes possible a full measure of social life, but not larger; surrounded by a permanent belt of rural land; the whole of the land being in public ownership or held in trust for the community.

As a matter of fact, the population was not estimated to exceed 30,000. Nor was it proposed that the area to be used for building, including roads and open spaces, should exceed one-third. Thus, out of the total acreage of 4500, less than 1500 acres was to be thus used, the balance of approximately 3000 acres being land to be devoted to farms and gardens.

With land unbuilt on, experts can choose the best sites for each particular purpose: public buildings, factories, shops, private residences, parks, recreation grounds, allotment fields, schools, and so forth; at the same time making new or adapting old roads. Proper provision can be made for water-supply, sewerage, heating, and lighting. In this connection attention may legitimately be drawn to Mr. C. B. Purdom's valuable book, *The Building of Satellite Towns*, published by Messrs. J. M. Dent and Sons Ltd.

Town-planning regional survey goes beyond the work both of local authorities with town-planning schemes and of bodies such as the Garden Cities and Town Planning Association with schemes for new garden cities. It goes beyond the abolition of slums, the erection of council houses, the building of garden cities, and the establishment of suburban trusts, although in a sense it embraces all of them. It takes in not only the great town, with its surroundings and its manufacturing and residential requirements, but also the adjacent countryside. In special cases, as in East Kent, where a whole series of new towns seemed likely to grow up in the modern coalfield areas without regard to one another, or to the rural amenities, a large section of a large county was embraced in the purview. The present writer, knowing intimately as he did every village and hamlet in East Kent; knowing Nonington, Tilmanstone, Eastry, Woodnesborough, Staple, Wingham, Ash, Goodnestone, Northbourne and Betteshanger equally with Sandwich, Deal and Dover, was one of many who viewed with anxiety the future of East Kent under such a development as that which threatened. But the wisdom of her public men, happily shared by the mine-owners, averted what would have been almost a national calamity. Professor Patrick Abercrombie of the University of Liverpool was called in, and a regional survey prepared by him with the assistance of Mr. John Archibald. This survey took in the whole countryside. It was not satisfied with considering the relation of its streets in one town, but it embraced the relation of each town to every other, and not only so but in relation to the whole of the surrounding district with its various amenities. There is now every hope that the foundation of a series of entirely new towns, housing in the aggregate more than a quarter of a million people, will be accomplished without outrage and with due regard to the interests of all classes.

In *East Kent Regional Planning Scheme* (University Press of Liverpool and Messrs. Hodder and Stoughton, Ltd.) the reader anxious that more and more opportunities should be provided for increasing rural amenities, including gardens, finds a masterly Report which arose out of a meeting at which no fewer than seventeen local authorities decided to join in a regional town-planning scheme. Broadstairs, Deal, Canterbury, Dover, Folkestone, Herne Bay, Margate, Ramsgate, Sandwich, Walmer and Whitstable were among the number of adherents.

In its comprehensive survey the Report embraces:

Natural Features, including Rainfall, Topography, Surface, Underground and Economic Geology.
Agriculture and Vegetation.
Archæological Factors.
Administration.
Population, Health and Housing.
Coal, Ironstone and other Economic Minerals.
Communications.
Open Spaces and Natural Reservations.
The Old Cities, Towns and Villages.
The Principal Resorts.

In its outlook for the future the Report includes:

Zoning Outlines.
Ports.
Population.
New Towns.
Communications.
Coal-working Considerations.
Water Supply and Drainage.
Electric Power.
Small Holdings and Allotments.
Social Life and Education.

In its broadminded survey of methods of realisation, it asks for:

A Regional Plan and a Regional Committee.
A Development Company or Public Utility Society for the new towns.
An Advisory or Civic Society.
An Art Advisory Committee.

In short, the Report aims at a great scheme embracing more than 186,980 acres of land, a rateable value exceeding £1,797,644, an existing population of 300,000, an additional population consisting of miners, steel-workers, and employees in ancillary trades with their families amounting in the near future to 278,000, and a thirty years' increase of 100,000. A scheme so vast merits unbounded respect and the most careful consideration, more particularly in view of the fact that the minimum standard of open space within accessible distance, and not counting natural tracts farther away, proposed by Mr. G. L. Pepler, F.S.I., in the *Town Planning Review*, namely, five acres per thousand persons, is accepted.

The remarkable illustrations included in the Report are themselves a feature of inestimable interest and value.

Although East Kent is one of the greatest it is by no means the only comprehensive scheme of regional survey combined with town planning, and there is ground for hope that the next few years will see still more. The fear that there is insufficient land to supply largely extended schemes without robbing agriculture is ill-founded. In his very useful *Town Planning Handbook* (Messrs. P. S. King and Son, Ltd.), Mr. Richard Reiss points out that while the acreage of England and Wales is rather more than 37,000,000, the population is between 37,000,000 and 38,000,000, or an approximate average of one person per acre; therefore, with average families of 4½ persons per house, little more than 1,000,000 acres would be required even if the houses numbered no more than eight per acre. As a matter of fact, town housing schemes allow fifty per cent more.

No purpose would be served by exaggerating the potential value to garden art of the multiplication of garden cities, satellite towns, town-planning schemes, and schemes of regional survey. Let it suffice to say that there would inevitably be a gain, and that it would probably operate at increased speed with every passing year.

CHAPTER XVIII

LANDSCAPE ARCHITECTURE IN NORTH AMERICA
(UNITED STATES AND CANADA)

A HISTORICAL AND CRITICAL SURVEY BY FRANK A. WAUGH

Professor of Landscape Gardening, Massachusetts Agricultural College; Member American Society of Landscape Architects; Author of "Textbook of Landscape Gardening," "The Natural Style in Landscape Gardening," "Formal Design in Landscape Architecture," "The Landscape Beautiful," Etc.

CHAPTER XVIII

LANDSCAPE ARCHITECTURE IN NORTH AMERICA
(UNITED STATES AND CANADA)

IN studying the progress of garden art in America, especially whenever any comparison with Europe is implied, one fundamental difference should always be taken into account—namely, that by comparison with Europe, America has never had a large number of great private estates. A certain number were indeed created, but many of them have already been abandoned, and none has ever had a permanent leadership or influence. At most they represent a transitory phase of American culture. On the other hand the American taste in small home grounds represents something permanent, general and significant; and this may be said to be a natural corollary of the earliest traditions.

Civilisation in America began, as it were, full-fledged. The early colonists came direct from the settled civilisations of Europe, particularly from England. Many of them were persons of education and refinement; some were men of substance. Under such circumstances one might expect that evidences of culture, including the making of gardens, would be shown very early, and that some of the slow and painful stages of progress as witnessed in the Old World might be altogether elided. This is in fact what happened. Other circumstances contributed to the same end. Every colony was compelled under threat of imminent starvation to gain an immediate living from the soil. Practical gardening and simple agriculture began at once and in great earnestness. The colonists were forced to strain every nerve, not alone to make a living, but to make homes. These they conceived inevitably in English terms—a house surrounded by a garden, and in the garden always plants both for food and for delight. There were flowers for colour and for perfume.

Even the first-comers brought seeds and cuttings and with these began at once the experiment of growing the English favourites: apples, plums, cherries; beetroots, turnips and carrots; catnip, marjoram and thyme; gillyflowers, poppies and roses. While some of these failed, others happily succeeded. Then there were the native plants of the New World, which were not to be neglected. Here were fruits and shrubs and gay flowers ready to be pressed into cultivation. Their enlistment moved more slowly than one might have expected, but it went on. There is to be considered the further fact that the first colonies were planted in regions of propitious soil and climate. Gardening came easily.

The first colonists were practically all English. They came from a country of gardens. They had been bred in the tradition of gardens and some of them were skilled in garden practice. And the great preponderance of English blood and of English culture, so marked in the beginning, has continued to rule American life even to the present day, in no realm —not excepting even literature and common law—more strongly than in gardening. In later years America received large levies of immigrants from other nations, notably from

Germany and the Scandinavian countries, and quite recently from Italy, Greece and their neighbours. These immigrants, especially the Germans and the Scandinavians, contributed substantially to some departments of American thought and culture—to education, science and technology, for example—but not appreciably to gardening. To conclude in a sentence this very brief account of foreign influence in American landscape architecture, it may be noted that French contributions have been nil: only two French settlements survived on the continent, a small one at New Orleans and a larger, more proliferous and more permanent one in Quebec. Neither has affected American culture, least of all American gardening.

In quite recent times, however, a certain amount of Latin influence, mainly Italian, has been manifest. This has flowed in through two openings. First has been the stream of wealthy (and largely parvenu) Americans who have travelled and lived abroad. They have found Paris a place convenient for the spending of money by persons of limited imagination. If they have returned to America at all, they have returned measurably Europeanised and in a temper to imitate the customs of France and Italy, even in the making of gardens. Since the Latin garden forms offered special opportunities for extravagance, it was natural that some of them should adopt this way of showing their wealth.

But the old garden forms, especially the Renaissance gardens of Italy, have strong attractions for more cultured minds also. Thus it happened, in the second place, that Americans of refinement began to be moved by Italian traditions. Here entered the new profession of landscape architecture, with a group of ambitious young men eager to learn all that Europe had to teach. The architecture and gardening of the Italian villas were studied intensively, sympathetically, and with some regard to their acclimatisation in America. These two groups—persons of wealth and persons of education—both helped to introduce French and Italian ideas, especially the latter, into American landscape architecture. Later an attempt will be made to estimate more exactly the results of this impact.

Before this topic is dismissed mention should be made of the truly remarkable cultural unity of the North American people. Though they are derived from many races and nationalities, there is an astonishing uniformity of speech, thought and feeling. There are of course appreciable differences of dialect, but not more over the whole continent than may be found in two adjoining counties in England or than can be discovered between the German spoken in Hanover and that of Bavaria. The newspapers, inordinately read, are highly standardised, printing the same news and the same "features" from Maine to California and from Texas to Canada. Everybody on the continent sees precisely the same "movies." Everybody listens at the same instant by means of the universal radio to the same lectures, the same songs, the same ball games. Schools are graded exactly alike from the kindergarten through to the college. Every article of daily use is "nationally advertised" and continentally sold. One buys precisely the same toothpaste, collars, canned foods or cigarettes in Montreal, New Orleans, San Francisco and Boston.

These conditions obviously affect every phase of life in America deeply, distinguishing it from Europe, with its multitudinous races and tongues. The strong national tendencies extend even to gardening. Nationally known brands of oranges, apples and bananas are eaten everywhere. Strawberries, onions, celery, early potatoes, peaches and watermelons are shipped in heavy carloads across the continent. And the nurseryman who introduces a new rose or a new philadelphus advertises it impartially to Canada, California and Virginia. Every book on landscape architecture is written for sale over the whole breadth of the land.

GEOGRAPHIC AND PHYSICAL FACTORS

America is a large country, and no one can gain any comprehension of the garden-making problem there without due consideration of factors of geography, topography and climate. In latitude and longitude the inhabited portions of North America cover a territory equal to the British Isles, all of Western Europe, all Eastern Russia, one-half of Siberia, and the whole Mediterranean Basin, including Turkey, Persia and Northern Africa. If it is necessary, in writing of European gardening, to discriminate carefully such areas as Italy, Germany, Russia and Great Britain, it is equally necessary to examine the peculiarities of California, Florida, the Mississippi Basin, New England and Canada in speaking of gardening in North America.

Aside from its mere physical vastness, this North American continent has a highly varied topography. Beginning at the eastern seaboard there is found a narrow coastal plain marked by low hills, often rocky. Back of this lies the geologically old Appalachian mountain range, heavily wooded and watered, and in its northern reaches strongly glaciated. Next comes the Mississippi valley, very wide, generally level, considerably varied in its soil but largely of limestone derivation, exceedingly fertile and mostly well cultivated. The eastern two-thirds of this basin has an ample rainfall, ranging roughly from twenty-five to thirty-five inches annually. The western third verges toward arid conditions, the rainfall diminishing westward to the Rocky Mountains. In this system of high mountains is found a remarkable range of physical conditions, varying from narrow, sunny, fertile, well-watered valleys to arid steppes and mountain peaks capped with eternal snow. West of the Rocky Mountains lies the great interior plateau, about the size of France and comprising several states. The elevation ranges from 2000 to 6000 feet above sea-level, with many local mountains running considerably higher, a few up to 10,000 feet. Rainfall is deficient, but a few small areas under irrigation are highly fruitful. This brings us to the Sierra Nevada range, almost as high as the Rockies and perhaps more picturesque. These mountains are heavily wooded on their western slopes but nearly arid on their eastern side. Between them and the Pacific Ocean lie the rich, varied and mild areas of the Pacific slope in British Columbia, Washington, Oregon and California. Here the rainfall is generally heavy, especially northward, forests are made up of enormous trees and crowding undergrowth, and the climate is much milder than in corresponding latitudes eastward. This amelioration of the Pacific Coast climate by the warm ocean currents from Japan is a factor of commanding importance.

This glance from the Atlantic to the Pacific Coasts necessarily ignores many local conditions of great importance. And it leaves the necessity of retracing steps to speak of Canada at the north and the Gulf States at the south. It is true that, in general terms, the topographic features just sketched extend northward across Canada; true also that, lying farther north, each Canadian zone has a slightly shorter growing season and a lower summer temperature than the corresponding zone in the United States. Yet Canada is a highly fertile arable area, and has a large population of cultivated citizens who have made great progress in horticulture and landscape architecture. The areas bordering on the Gulf of Mexico constitute another zone of quite individual qualities. This zone includes the whole of Florida, with portions of Alabama, Mississippi, Louisiana and Texas. Altitudes are low, usually hardly above tide-level, the surface is flat, and there is much swamp land. There is naturally much heavy forest in which southern species of pine are conspicuous. Rainfall is ample and the temperature is warm and equable.

Emphasis must be placed upon the fact that these large areas represent major physical subdivisions of the continent, characterised by substantial differences of soil, rainfall, altitude or temperature, such as exert a determining influence upon plant culture. Nor may the complementary fact be overlooked that within these areas lie many smaller sections with very diverse conditions. The full development of local possibilities under these peculiarities has not, generally speaking, been accomplished in America, perhaps from lack of time; and this lack of intensive local refinement is one of the distinguishing characteristics of American horticulture as compared with that of Europe. In America, where everyone from coast to coast buys the same manufactured articles, reads the same garden magazines, and patronises the same nurseries, and where they even buy standardised ready-made houses from mail-order merchants, the tendency toward uniformity is very strong and the development of local specialities is correspondingly impeded.

NATIVE FLORA

In every land and in every time the art of gardening must have shown some impress from the native flora. In North America this impress has been very considerable. The following reasons may be alleged for this influence, though their exact measurement is obviously impossible: First, the severity of the climate has made the introduction of exotic plants always difficult. Second, the inhabitants have always shown a keen delight in the natural landscape and the native plants. Third, the natural style of landscape gardening for which a distinct preference has been manifest would tend to favour native scenery. Fourth, there has been working at sundry times a strong propaganda for native plants. Fifth and last, the native flora is extensive, varied, and exceedingly interesting, commending itself to the skill of every garden lover. How cogent is this appeal may be read in hundreds of volumes written by early explorers on these shores—by Michaux, Rafinesque and scores of others. For upwards of two centuries the importation and acclimatisation of American plants in Britain and on the continent of Europe was the vocation and delight of all botanists and gardeners.

The index of American plants is a very long one, owing to obvious physical and climatic conditions. There are many notable species of trees well suited to planting for forest and landscape use—dozens of species of pine, fir, hemlock, maple, elm, and oak, not to mention such particularly interesting sorts as the tulip-tree, the live oak, the catalpa, and the magnolia. The species of shrubs suitable for ornamental planting probably exceed a thousand, many of them of signal beauty. The rhododendrons, azaleas and kalmias supply a suggestive illustration. Likewise the great number of desirable herbaceous species should be emphasised. The asters, solidagoes, pentstemons and aquilegias may be cited merely by way of example. In spite, however, of this abundance of native flora it is quite certain that the ultimate effect upon American gardens would have been less had it not been for urgent preaching in a country where waves of propaganda have a powerful influence. A good many nurseries have been established which specialise in the collection, propagation, improvement and sale of indigenous plants; and of necessity their advertising has supported the doctrine that native plants are to be favoured. Yet it is a curious fact that some of the very best garden varieties of American plants have come from the nurseries of Europe, where they have been raised and large quantities sent to America. The selected varieties of asters grown in England, and the delightful coreopsis from Erfurt, Germany, exemplify this point.

It may be said, by way of summary, that at the present time a catholic taste prevails. Landscape architects and home gardeners use freely all kinds of plants with little respect to their nativity. Japanese species show a rather peculiar adaptability to the Atlantic seaboard region, yet the unquestioned merits of American species, especially trees and hardy shrubs, give them a conspicuous ascendancy in nearly all American landscape gardening.

EARLY AMERICAN GARDENS

The first permanent settlements in America were made in Virginia in 1607 and in Massachusetts in 1620. Other colonies were planted soon after, notably the one at New Amsterdam (now New York), the one in Maryland, Penn's settlement at Philadelphia, and the Carolinas. The early colonists found some crude gardening already practised amongst the Indians; they found many useful native fruits and herbs (they were, for example, greatly impressed by the abundance of native grapes); and they were all under the stern necessity of making the utmost efforts towards supplying their own wants. Thus they were gardeners by example and by compulsion. They immediately began the cultivation of all economic plants. They formed small enclosures about their homes, and in what were literally gardens, they soon brought to blossoming, urged by a higher spiritual need, the favourite flowers of their old English homes.

Some of these early gardens were reasonably commodious and notably fruitful. Abundant records remain of Governor Endicott's garden in Salem, Governor Winthrop's garden in Plymouth, and of the gardens of Charleston dating back to 1682. Yet for the first hundred years there were no great gardens of princely scope, nor indeed anything more than cottage gardens, properly speaking. A few were larger and better furnished than the others; but the typical picture is that of a small garden plot next the humble dwelling, in which cabbages, beans and corn were grown for food, and hollyhocks, rosemary, pennyroyal, coriander and sweetbrier were cultivated about the windows and in the front yard. No particularly fine or famous gardens have come down to us from those colonial days. Yet there are remembered Mount Airy, built in 1650; Tuckahoe, from 1700; Stratford Hall, in 1725; and Westover (Fig. 655), the home of Colonel Byrd, built in 1726. Magnolia-on-the-Ashley (South Carolina) dated from 1671, and John Bartram's famous botanical garden in Philadelphia from 1728.[1]

Mount Vernon, the home of George Washington, is the only one of these colonial gardens which has ever appealed warmly to the popular imagination. This was not formed on its present lines until the colonial period had closed with the revolutionary war.

---

[1] References to these items are to be found in Earle, *Old Time Gardens*, New York, 1901; Tabor, *Old Fashioned Gardening*, New York, 1913; *Historic Gardens of Virginia*, Richmond, 1923.

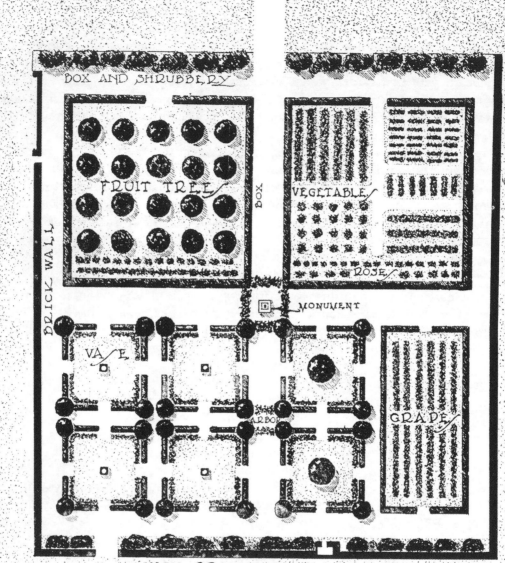

FIG. 655. THE ORIGINAL BOX GARDEN AT WESTOVER, BUILT IN 1726

It was not in any sense an elaborate "estate," but was a very simple country home. Tens of thousands of citizens to-day own places larger, more elaborate or artistically better. Yet it was one of the best of its time; it was well planned; above all it was the home and the handiwork of the "Father of his Country," and it has been carefully preserved to the present time. These circumstances have contributed to its celebrity. Yet even Mount Vernon has had no perceptible influence on American gardening, though the house has been copied many times in many forms. It should be added that the "colonial period" in general has been held in high repute in later years and has had a considerable influence in many fields of art. Colonial architecture, colonial furniture and colonial gardens belong to this category; and there has been a good deal of conscious effort to reproduce the atmosphere of those simple, dignified old homes of pre-revolutionary times.

From the War of Independence to the Civil War (1776–1861) the country changed little except for its westward expansion, both in Canada and the United States. Such building and gardening as were done followed the colonial models, but with decreasing fidelity. Towards the close of this period certain new movements were observable, but it is more convenient to treat of them under the period of their fruition, which came after the Civil War, than here. The great Civil War began in 1861 and gripped the nation wholly till 1865. At the end the country was exhausted and nearly bankrupt. A new era then opened, but unfortunately one of very bad art. In architecture, sculpture, poetry and gardening, inspiration was denied and taste fell to the lowest levels. Yet beneath this ruck of stupidity good beginnings were being made. For there had already arisen that great luminary of landscape gardening, Andrew Jackson Downing, who had flamed a moment in the sky, and then gone down to his untimely death. All this was before the Civil War. Downing was born in 1815 at Newburgh on the Hudson, New York, and died near the same spot in 1852. He represented very clearly the Reptonian tradition in America. He preached the doctrine of the English or natural style (the two terms were interchangeably current) of landscape gardening, and his preaching found a ready response in the best American thought. In his editorials in the *Horticulturist* (1846–52) and in his classic treatise on *Landscape Gardening* (first edition 1841) he completely captured the popular taste.

This noble leadership, though implanted before the Civil War, flowered and bore its fruit after that interregnum. And in this later period it was reinforced and ably continued in the leadership of Frederick Law Olmsted, Senior. Olmsted was born at Hartford, Conn., in 1822 and died in Brookline, Mass., 1903. He came to public notice during the war, but his real work as landscape architect had begun in 1857, when he was appointed superintendent of the new Central Park, New York City, then under construction. This project was resumed under his direction after the war; and at about the same time he began to make plans for other important parks in Brooklyn, N.Y., New Britain, Conn., San Francisco, Calif., Chicago, Ills., and other cities. With him was associated for a time Calvert Vaux, a capable English-trained architect, who had previously been the professional partner of Downing.

Olmsted continued the traditions of Downing. He strongly favoured the English or natural style of landscape architecture (this term has to be used rather inexactly, since a strict analysis will show that every "natural style" is more or less conventionalised, and by each worker in his own way). He was the first man in America to organise and practise the profession of the landscape architect on a large scale. For a time he had associated with him Charles Eliot (lamented for his early death); also his stepson John C. Olmsted and his son Frederick Law Olmsted, Junior. The firm is still very active. During all these years, mainly the time from 1870 to 1890, a number of young men worked with the firm, afterward setting up for themselves, thus propagating in wider circles the Olmsted influence.

Olmsted, with sore misgivings, took the style of landscape architect, discarding the earlier nomenclature; and his example, more than anything else, fixed the use of "landscape architect" and "landscape architecture" on America in place of the older English terms, "landscape gardener" and "landscape gardening."

He wrote little for the public. His great influence was exerted through his personal disciples and through his works. These works were very many, of large proportions, widely placed throughout the United States and Canada, and lay in the main trend of the developments of the period. This period might fairly be called the park era. The construction of Central Park in New York (already advocated by Downing before his death) advertised widely both the park idea and the landscape architect in charge. American cities were multiplying and growing under the impetus of heavy immigration and the first burst of modern industrialism, and the park idea matched the times. Olmsted and his associates designed many parks besides those enumerated above, insomuch that his ideas were easily dominant throughout this distinct chapter in American landscape architecture.

Briefly, the Olmstedian principles may be described as follows: (1) preserve the natural scenery and if necessary restore and emphasise it; (2) avoid all formal design except in very limited areas about buildings; (3) keep open lawns and meadows in large central areas; (4) use native trees and shrubs, especially in heavy border plantings; (5) provide circulation by means of paths and roads laid in wide-sweeping curves; (6) place the principal road so that it will approximately circumscribe the whole area. These principles may still be seen exemplified in several of his parks, perhaps best of all in Mount Royal Park, Montreal, and in Franklin Park, Boston.

Along with the park movement, and as an integral feature of it, came the park cemetery. This idea is a distinctively American contribution to landscape architecture, for, while park cemeteries have been made in other countries, the first and most numerous successes were those in the United States and Canada. The first park cemetery to attract wide attention was Mount Auburn, near Boston, founded in 1831. Spring Grove Cemetery at Cincinnati came about twenty years later and was generally admired. But perhaps the most influential example of all has been Graceland Cemetery in Chicago (Fig. 656), of still later date. Graceland was designed and constructed by Mr. O. C. Simonds, who soon became famous as a designer of park cemeteries. In following years he designed some hundreds of these. Various factors contributed to the vogue of the park cemetery in America. The comparative cheapness of land, the popular taste for naturalistic garden design, and the coincident rise of the park movement, may be enumerated. Beyond all these, however, lies the fact that the idea is inherently sound and appealing. At the present time park cemeteries are the rule, not the exception. They are made by corporations, religious societies, municipalities, and even by the Federal Government.

In order to complete the discussion of American park design a few observations on later work may be added. Two new conditions began to change the park problem soon after Olmsted's death. The first of these was the further growth and industrialisation

FIG. 656. THE POND, GRACELAND CEMETERY, CHICAGO—A TYPICAL EXAMPLE OF PARK CEMETERY DESIGN IN AMERICA

of American cities, requiring "neighbourhood" playgrounds of a new type. The second was the introduction of new means of passenger transit, at the outset the electric tram and later the much more influential automobile. The local playgrounds had to be relatively small and had to bear very intensive use. Neither requirement was compatible with the open scenic park idea. The playgrounds were therefore a new development in American landscape architecture. Thousands of them were built and furnished, some of the most successful being designed by Olmsted's business successors.

In contrast to these small playgrounds, scattered thickly through the residential sections of cities, stand the large exterior parks made possible by the improvement of transport. The first fruit was seen in the Metropolitan Park System of Boston, founded under the leadership of Charles Eliot. Many other cities have adopted the same principle, such as New York; Minneapolis, where an excellent zone of outer parks has been acquired through the energy of Mr. Theo. Wirth; and Chicago, where the Cook County Forest Preserves have been built up under the leadership of Mr. Jens Jensen. A considerable portion of the energy in this park movement, however, was later diverted into the demand for state and national parks, forests and other rural playgrounds—a movement so important as to require extended treatment under a separate head.

Reference has already been made to the fact that in America the fundamental taste for the natural style of landscape gardening has developed in two different aspects. On the one hand has been the tendency to lay out private estates and city parks in a naturalistic, informal manner; on the other has been the movement to preserve considerable areas of native landscape for purposes of education, health, and recreation. Natural scenery is reserved for use. These reservations have been extensive; and this aspect of American landscape architecture is perhaps the most significant of all. While many of these reservations have been made by private purchase, or by private clubs holding the land for hunting and fishing, or merely as "country clubs" for general recreation, by far the largest and most important areas are dedicated to public ownership and use. The principal types of reservations are (1) the national parks, (2) the national forests, (3) the national monuments, (4) the state parks, (5) the state forests, and (6) sundry historic localities.

THE NATIONAL PARKS

The beginnings of the national park system in the United States were merely fortuitous. In 1832 the Hot Springs region in central Arkansas was reserved and has since been kept as a public park. The wonderful geyser basin in north-western Wyoming was set aside in 1872 and placed under the care of the War Department. In 1890 Congress created the Yosemite National Park, taking over the property from the state of California; and in the same year also the Sequoia and the General Grant National Parks for the preservation of the giant sequoias—the "Big Trees" of California. About the same time Mount Rainier in Washington became a national park.

As a system, however, the national parks came into existence in 1916 with the organisation of the National Park Service as a branch of the Department of the Interior to take charge of the entire group. At the time of writing (1927) the system consists of nineteen parks as shown in the following tabulation, taken from official sources:

Hot Springs National Park; middle Arkansas; 1½ sq. miles. 46 hot springs possessing curative properties. Many hotels and boarding-houses. 19 bath-houses under Government supervision.

Yellowstone National Park; north-western Wyoming; 3348 sq. miles. More geysers than in all the rest of the world together. Boiling springs. Mud volcanoes. Petrified forests. Grand Canyon of the Yellowstone, remarkable for gorgeous colouring. Large lakes. Many large streams and waterfalls. Vast wilderness, greatest wild bird and animal preserve in the world. Exceptional trout fishing.

Sequoia National Park; middle-eastern California; 604 sq. miles. The Big Tree National Park. Several hundred sequoia trees over 10 feet in diameter, some 25 to 36 feet in diameter. Towering mountain ranges. Startling precipices. Includes Mount Whitney and Kern River Canyon.

Yosemite National Park; middle-eastern California; 1125 sq. miles. Valley of world-famed beauty. Lofty cliffs. Romantic vistas. Many waterfalls of extraordinary height. 3 groves of big trees. High Sierra. Waterwheel Falls. Good trout fishing.

General Grant National Park; middle-eastern California; 4 sq. miles. Created to preserve the celebrated General Grant Tree, 35 feet in diameter, 6 miles from Sequoia National Park.

Mount Rainier National Park; west-central Washington; 325 sq. miles. Largest accessible single-peak glacier system. 28 glaciers, some of large size. 48 sq. miles of glacier, 50 to 500 feet thick. Wonderful sub-alpine wild-flower fields.

Crater Lake National Park; south-western Oregon; 249 sq. miles. Lake of extraordinary blue in crater of extinct volcano. Sides 1000 feet high. Interesting lava formations. Fine fishing.

Platt National Park; southern Oklahoma; 1⅓ sq. miles. Many sulphur and other springs possessing medicinal value.

Wind Cave National Park; South Dakota; 17 sq. miles. Cavern having several miles of galleries and numerous chambers containing peculiar formations.

Sullys Hill National Park; North Dakota; 1⅕ sq. miles. Small park with woods, streams, and a lake; is an important wild-animal preserve.

Mesa Verde National Park; south-western Colorado; 77 sq. miles. Most notable and best preserved prehistoric cliff dwellings in United States, if not in the world.

Glacier National Park; north-western Montana; 1534 sq. miles. Rugged mountain region of unsurpassed Alpine character. 250 glacier-fed lakes of romantic beauty. 60 small glaciers. Precipices thousands of feet deep. Almost sensational scenery of marked individuality. Fine trout fishing.

Rocky Mountain National Park; north-middle Colorado; 378 sq. miles. Heart of the Rockies. Snowy range, peaks 11,000 to 14,255 feet altitude. Remarkable records of glacial period.

Hawaii National Park; Hawaii; 186 sq. miles. Three separate areas—Kilauea and Mauna Loa on Hawaii; Haleakala on Maui.

Lassen Volcanic National Park; northern California; 124 sq. miles. Only active volcano in United States proper. Lassen Peak 10,460 feet. Cinder Cone 6907 feet. Hot springs. Mud geysers.

Mount McKinley National Park; south-central Alaska, 2645 sq. miles. Highest mountain in North America. Rises higher above surrounding country than any other mountain in the world.

Grand Canyon National Park; north-central Arizona; 958 sq. miles. The greatest example of erosion and the most sublime spectacle in the world (Fig. 657).

Lafayette National Park; Maine coast; 12 sq. miles. The group of granite mountains upon Mount Desert Island.

Zion National Park; south-western Utah; 120 sq. miles. Magnificent gorge (Zion Canyon), depth from 800 to 2000 feet, with precipitous walls. Of great beauty and scenic interest.

FIG. 657. A SIDE CANYON INTO THE GRAND CANYON, NATIONAL PARK, ARIZONA

In this list are several areas of great importance. Perhaps it will not be invidious to name the Grand Canyon first. By many judicious persons it is regarded as the most thrilling spectacle to be found amongst all the landscapes of the world. Its enormous scale, its stupendous mass effects and its glorious colourings combine to overwhelm the spectator. At the point where the canyon is most frequently visited it is thirteen miles wide and slightly more than one mile deep. The pictorial effect of this extraordinary landscape gains greatly from the dry, clear atmosphere of Arizona and from the altitude, which is from 7000 to 8000 feet. The Grand Canyon is notably versatile, and to be even partially appreciated, must be examined and explored at leisure, and must be seen at all hours of the day and night and at all seasons of the year. It is indeed an inexhaustible feast of its peculiar kind.

Just here a digression may offer a suggestion of some value. It would not be the whole story, but it would be a fair comparison to place the Grand Canyon beside Versailles, allowing one to represent the landscape architecture of America and the other to represent the landscape architecture of Europe. Each one is truly representative of a characteristic part of its country.

The Yellowstone Park is one of the oldest and most interesting. While it has many delightful landscape features—a noble lake, high mountains, a beautiful waterfall, much interesting wild life—the prime attraction is found in the wonderful exhibition of volcanic forces in action. There are great hot springs, mud volcanoes, and especially the famous collection of geysers. These spectacular sights attract hundreds of thousands of visitors annually.

Yosemite National Park in California preserves a very remarkable granite valley with several waterfalls of striking beauty, and many large trees. Glacier National Park is characterised by superb alpine scenery, including many glaciers, glacial lakes and streams. Zion National Park presents another deep and very beautiful canyon to the admiration of all lovers of the primitive landscape.

The Dominion of Canada has also set aside a number of national parks comparable in all respects with those of the United States. A brief summary of these parks is all that space will permit.

Rocky Mountain Park, in Alberta, was established in 1885 and has an area of 2751 sq. miles. Here is found some of the most famous alpine scenery on the continent, including the two noted alpine resorts, Banff and Lake Louise.

Yoho Park, in British Columbia, was established in 1886 with an area of 476 sq. miles. This park is characterised also by rugged mountain scenery.

Glacier Park, on the summit of the Selkirk Mountains in British Columbia, dates from 1886 and has an area of 468 sq. miles.

Revelstoke Park, also in British Columbia, has an area of 95 sq. miles and includes much accessible mountain country.

Kootenay Park, British Columbia, area 587 sq. miles, is notable for very wild mountains and unexplored country.

Jasper Park, in northern Alberta, established in 1907, is one of the largest and one of the most significant reservations on the North American continent. It has an area of 4400 sq. miles, or more than one-fourth the size of Switzerland. This region includes great ranges of mountains and much unexplored territory, and is called "the largest big-game sanctuary in the world."

Waterton Lakes Park lies in southern Alberta, adjoining the Glacier National Park of the United States, and has an area of 423 sq. miles.

St. Lawrence Island, in the Thousand Islands region of the St. Lawrence River, is a popular tourist section.

Buffalo Park, in Alberta, was dedicated in 1907 especially to assist in the preservation of the American bison, though large herds of moose, elk and other animals are maintained.

Other Canadian national parks, to a total of fifteen, are smaller and serve mainly to preserve points of historic interest.

FIG. 658. MOUNT HOOD, LOOKING ACROSS LOST LAKE, NATIONAL FOREST, OREGON

NATIONAL FORESTS

The picture of national scenic reservations would be seriously incomplete without the national forests. These are of vast extent, covering an area of nearly 250,000 sq. miles, or about double that of the British Isles. They include important mountain and forest lands in the eastern states, especially New Hampshire, Pennsylvania, Virginia, North Carolina, Georgia and Tennessee, and a few important though relatively small areas in the middle states; but the great bulk are to be found in the Rocky Mountain regions, and thence westward to the Pacific Coast. Quite naturally they include much of the wildest and most inaccessible lands on the continent; and just as naturally do they include much beautiful scenery and picturesque country highly eligible for hunting, fishing, exploration and other hardy forms of recreation. In fact the number of persons entering upon these national forests for recreation, attracted thither by scenery, by hunting, by fishing and by other sports, already runs into many millions annually.

Some of these areas are nationally famous for health or recreation, as the White Mountains in New Hampshire, the whole of western North Carolina about Asheville, the Rocky Mountains, including Pike's Peak, near Denver, and Colorado Springs, Colo., and the high sierras from Mount Baker near the Canadian border through Washington and Oregon (including Mount Hood [Fig. 658]) and California to the Mexican boundary. Popular scenic objectives are found in many places, e.g. in Lake Chelan, Chelan National Forest, the Columbia River Highway and Mount Hood Loop, Oregon National Forest, and the Mount of the Holy Cross in Holy Cross National Forest. While these wide areas are administered primarily with reference to timber production, water conservation and grazing, some supervision is also given to the highly valuable recreation uses. In particular is great care exercised not to impair the beauty of natural scenery, since this is generally held to be of great æsthetic, educational and social value to the nation.

As with parks, so with forests, Canada has placed herself in laudable competition with the United States. Correspondingly large areas of the notable forest lands of the Dominion have been reserved and are used for such immediate human needs as health, recreation and spiritual inspiration.

NATIONAL MONUMENTS

Another form of public holding which has considerable interest in this connection is found in the national monuments of the United States. These are relatively small areas, sometimes only a few acres, though two in Alaska extend to over 1000 sq. miles. They are set aside by Presidential order (as distinguished from the Congressional enactment required for the creation of national parks) and are administered by sundry federal officers. In general, each one preserves some historic monument, some prehistoric relic, or some feature of scientific interest. This may be illustrated by a few representative examples, selected from the catalogue of fifty-six monuments existing in 1927.

FIG. 659. CLIFF-DWELLING RUINS, NOW A NATIONAL MONUMENT, KNOWN AS MONTEZUMA CASTLE, ON BEAVER CREEK, BETWEEN THE COCONINO AND THE PRESCOTT NATIONAL FORESTS, ARIZONA

Montezuma Castle National Monument; Arizona; 160 acres. Prehistoric cliff-dweller ruin of unusual size situated in a niche in face of a vertical cliff. Of scenic and ethnological interest (Fig. 659).

Petrified Forest National Monument; Arizona; 25,625 acres. Abundance of petrified coniferous trees, one of which forms a small natural bridge. Is of great scientific interest.

Chaco Canyon National Monument; New Mexico; 20,629 acres. Many large pueblos

FIG. 660. AN OLD COMMUNITY DWELLING, EXCAVATED AND PRESERVED, BANDELIER NATIONAL MONUMENT, NEAR SANTA FÉ, NEW MEXICO

or communal houses, in good condition and of great interest. Considerable excavation done on several of the ruins.

Muir Woods National Monument; California; 426 acres. One of the most noted red-wood groves in California; donated by Hon. William Kent, ex-Member of Congress. Located 7 miles from San Francisco.

Tumacacori National Monument; Arizona; 10 acres. Ruin of Franciscan mission dating from seventeenth century. Being restored by National Park Service.

Gran Quivira National Monument; New Mexico; 560 acres. One of the most important of earliest Spanish mission ruins in the south-west. Monument also contains pueblo ruins.

Dinosaur National Monument; Utah; 80 acres. Deposits of fossil remains of prehistoric animal life of great scientific interest.

Casa Grande National Monument; Arizona; 472 acres. These ruins are one of the most noteworthy relics of a prehistoric age and people within the limits of the United States. Discovered in ruined condition in 1694.

Katmai National Monument; Alaska; 1,087,990 acres. Wonderland of great scientific interest as example of volcanism on a scale of great magnitude. Includes "Valley of Ten Thousand Smokes."

Mount Olympus National Monument; Washington; 299,370 acres. Contains many objects of great and unusual scientific interest, including many glaciers. Is summer range and breeding-ground of the Olympic elk.

Bandelier National Monument; New Mexico; 22,075 acres. Vast number of cliff-dweller ruins, with artificial caves, stone sculpture, and other relics of prehistoric life (Fig. 660).

Bryce Canyon National Monument; Utah; 7440 acres. Box canyon filled with countless array of fantastically eroded pinnacles. Best exhibit of vivid colouring of earth's materials.

Some of these national monuments are of extraordinary interest or beauty. Bryce Canyon, for instance, is one of the most remarkable examples of erosion known, with brilliant colourings and modellings even surpassing those of the Grand Canyon, though on a smaller scale. The Petrified Forest in Arizona has great scientific interest. Yet these are named only as examples to illustrate the character and quality of the national monuments as a series.

STATE LANDS

All the lands mentioned in the foregoing paragraphs are held and administered by the Federal Governments of the United States and Canada; but practically all of the forty-eight states of the United States and most of the provinces of Canada also own and administer public lands for similar purposes. In the aggregate, these lands mount up to many millions of acres, and their utility is greatly enhanced by their wide distribution. As an example, the Palisades Interstate Park, a mountain-forest area near New York City, receives over ten million visitors a year. Many thousands of these are children who remain in the forest camps for considerable periods, and very substantial contributions are thus made to their health and education. These state lands are held mainly in three forms, as (1) state parks, (2) state forests, and (3) historic reservations. No clear distinction is evident between these three classes of reservations and they may be fairly grouped together for social study. There is, of course, a theoretical distinction to the effect that parks are reserved for scenery, forests for timber, and historic areas for their educational value; but in actual service these distinctions become so blurred as to have little authority.

Finally, it may be repeated that so much attention is here given to these public landscape reservations for several carefully considered reasons, viz.: (1) they appear to illustrate a characteristically American form of organisation in answer to (2) a vital American taste for wild landscape; and (3) this seems to be one of the very strongest tendencies visible in the field of American landscape architecture.

HOME GROUNDS

Quite obviously it is the American social ideal that each family should have an independent home; that this home should consist of a detached house in a plot of ground; and that this plot of ground should be suitably planted with trees, shrubs, flowers and grass. This was the ideal from the days of the first settlements, and it is even now hardly obscured by the fact that increasing percentages of the population are going over to live in flats and hotels. Such makeshifts are still regarded as temporary and as tolerable only

FIG. 661. HOME GROUNDS PLANTED IN THE DOWNING MANNER, WITH EMPHASIS ON INDIVIDUAL SPECIMENS

under compulsion of circumstances. In the extensive literature of American landscape gardening a strikingly large proportion of attention is given to the discussion of the problems of home-grounds design and planting. The subject, furthermore, has been presented nearly always from the standpoint of the small home (cottage garden), it being felt apparently that practically all the home grounds in the land were reducible to this one type.

At the outset, and for many years thereafter, the majority of home gardens were enclosed. There were first rough stockades; but soon the neat fence of sawn wooden pickets became the recognised mode. This fashion persisted for many years. Wood being plentiful and woodworking a universal industry, much ingenuity was shown in elaboration. Posts were elaborately turned, sawn or built up of wood, and surmounted by turned or carved capitals, often of quite artistic design. The pickets themselves were shaped and spaced in various ways to gain effects pleasing to the eye; and the fences were nearly always neatly painted, white being the traditional colour. Naturally, also, the swinging gates in these picket fences received special attention, sometimes being real works of art.

The early colonial gardens enclosed by these fences were very simple. Nearly always they were made up of fruit-trees, kitchen vegetables and medicinal herbs, interspersed with flowering plants. Next the house and in the front yard, flowers and ornamental shrubs were grown. The lilac was an early favourite, as were roses, sweetbriers, hollyhocks, lemon lilies, and "flags" (iris). These front yards were narrow, seldcm more than six to ten feet wide, though the larger houses were scmetimes set farther back.

FIG. 662. FOUNDATION PLANTING ABOUT A TYPICAL WOODEN DWELLING ON A VILLAGE STREET—MODERN

Roughly speaking, the modern American taste for a wide set-back did not develop till after the Civil War (1865). Primarily these enclosures were made for protection against live stock running at large. As soon as pioneer conditions began to wane this necessity disappeared, and after a time the picket fences also disappeared. For although they were retained for a time on custom, there presently arose the counter style of having front yards all open to the street—a style which has ruled ever since. Correlated with this change was the movement of the dwelling-houses back farther from the street. The front yards thus became considerably larger at the same time that they became more open.

These front yards now began to be regarded as a major feature of the home grounds. They were large, open, democratic, and if they could be so dressed up as to be a bit showy, that quality was also in character. At the same time the English and German habit of living much in the garden was lost—indeed, seems never to have survived in America —and the demand for privacy, either in the front yard or in any other part of the garden, diminished, or vanished altogether. Indeed, American landscape architects and laymen of taste have long lamented this lack of privacy in home gardens.

The best of these front yards, as treated by Downing and his disciples (1850 to the present time), have one or two large shade trees, possibly more. In the northern states elms and maples were preferred; in the southern states, live oaks and magnolias; though many other species were used here and there. It also became the custom, less praiseworthy, to plant one or two showy exotic "ornamental" trees on the front lawn. Copper beech, weeping birch and Camperdown elm were old favourites: in recent times blue

FIG. 663. WALK, GATEWAY AND BORDER OF PERENNIALS, WITH BACKGROUND OF TREES—TYPICAL OF HOME GARDENING IN THE NORTH-EASTERN STATES

spruce has outdistanced all competitors. Fine shrubs, often as single specimens, sometimes in beds or groups, were also employed (Fig. 661). Of these lilacs were common and pleasing. Spiræas, syringas (philadelphus), weigelas, and deutzias were also used. In later times Hydrangea paniculata grandiflora became very popular—perhaps even too common. In extensive grounds rhododendrons, azaleas and kalmias made their appearance. On small places of the poorer sort flower-beds were often cut into the front lawn and filled with geraniums, zinnias, coleus, cannas or what not. The canon demanded always some clipped grass in this front-yard area. With the house well set back from the street and published so blankly to the world, and especially with the concomitant custom of using high foundations, it became very desirable to develop the planting of shrubs and vines immediately against the dwelling and its porches. These "foundation plantings" have come to be quite the style—a recognised necessity of the present mode. They soften

FIG. 664. A SMALL MODERN GARDEN OF OLD-FASHIONED FLOWERS ON LONG ISLAND, NEW YORK

the break between house and lawn, cover bare foundations and greatly ameliorate the bareness of a design otherwise somewhat meagre. (See Fig. 662).

During the middle period (1850–90) there was a strong movement, powerfully influenced by the teachings of Downing, towards the so-called natural style of landscape gardening, accompanied by some prejudice against the formal style. This preference showed itself most clearly in the park design of the time, but also in the design of the larger home grounds. Walks and drives were curved, sometimes without good reason, while trees and shrubs were scattered in asymmetrical groups. Where the land surface was altered, pains were taken to secure rolling and blending conformations. Although at first odd and exotic trees and shrubs were introduced into these "natural" plantings, even by Downing him-

FIG. 665. MODERN GARDENS OF AN ARTIST'S STUDIO AT GLENDALE, MASSACHUSETTS

self, they latterly were largely excluded on the theory that only indigenous species were proper to the natural style.

Towards the close of this period, and towards the end of Olmsted's time (1890–1910), there were built a number of new country estates showing the emergence of other influences, largely European. They were built, of course, for wealthy families; and as these persons had travelled much in Europe they were naturally hospitable to French and Italian ideas. Some of the smaller French and Italian works were openly copied; but more generally there were enclosed parks and formal gardens in the Latin manner. Much of this work was tentative, some of it ill adapted to American conditions, some of it vulgar and bad. Yet the net result has been excellent, especially in that it has broken down old prejudices and established a catholicity of taste highly advantageous to modern landscape architecture.

Having rid itself of prejudices and preconceptions, the American garden-loving public has quite recently made substantial progress toward a better domestic garden art. Privacy is again considered a desirable quality. Simplicity, snugness and intimacy are sought. And with an increased tendency to live in the garden, a new mode of design has been clearly developed. This is a genuinely native style of domestic design, although it bears a strong resemblance to the type of design used on small home grounds in England and Germany, where people of similar tastes have met the same needs in much the same way. In these modern home gardens the front yard is made small, severe, and simple. Clear separation is made between this public area and the service area and the private grounds—for now the desirability of strictly private gardens is generally recognised. All areas are kept small, not merely for the sake of economy, but also for the sake of intimacy. And since these areas are small, since they are necessarily rectangular, and since they are closely tied to the dwelling-house, quite the simplest thing is to give them formal treatment. This formality, however, is not elaborate. A single simple axis, suggested rather than defined, and terminated by such unpretentious figures as a bird-bath or a garden bench, gives the popular measure of formality. (See Figs. 664 and 665.)

Clipped trees or shrubs are not much used, and statuary of any sort is rare indeed. Enclosures are rarely made by masonry walls; they are usually formed by hedges, by vines on lattice screens, or by masses of informal plantings. Climbing vines on porches and on brick walls are popular. Flowers are grown in "old-fashioned gardens," simply formal, and in borders, or in reserve gardens. (See Fig. 663.)

CALIFORNIA

The fact has already been remarked that North America is too large and too diverse to be brought wholly under one-point of view. While all regions have much in common, there are some important particulars for which exception should be made.

California is in many ways an empire to itself. Quite significant for the present study is the different character of the Californian climate, determining a horticulture very unlike that of the central states or the eastern seaboard. There is also to be considered the history of the state, for it was at first a Spanish province, a part of Mexico, and its early traditions were Spanish instead of English. As to climate, California shares with the entire Pacific coast the warming influence of the Japanese current, which so ameliorates the temperatures as to make Oregon, Washington and British Columbia much warmer than regions in the same latitudes on the Atlantic coast. The effects of this warming are most important in the winter season, making it possible to grow most species of plants far northward of their natural range. Thus palms, araucarias, eucalyptus and pepper-trees are characteristic of California horticulture, being grown freely as far north as Sacramento; while in Portland, Seattle and Victoria there are luxuriant gardens of hybrid tea-roses and other half-hardy plants which can be grown in the east only under special methods of protection. The climate of British Columbia, Washington and Oregon, in fact, may fairly be likened to that of England. And for a similar reason, for England, too, is warmed by the mild Gulf Stream which flows to her shores from the western ocean.

These similarities, however, seem to run deeper than the thermometer would indicate. There are doubtless other factors involved. Long ago the famous botanist, Asa Gray, pointed out the interesting fact that the flora of the Pacific Coast resembled that of western Europe more than the flora of eastern North America; while the flora of the eastern states is more like that of Japan and eastern Asia. This likeness, visible in the indigenous flora,

FIG. 666. THE INTERIOR GARDEN, SANTA BARBARA MISSION, SANTA BARBARA, CALIFORNIA

extends to the artificial flora of horticulture. For example, the European grape, which thrives greatly in California, can hardly be grown at all in the eastern states. Thus it has come about that the horticulture and landscape gardening of the west coast have always had a more European cast than those of the east coast.

As to early history, it may be remarked briefly that the Spanish civilisation made a sufficient stand on Californian soil to leave a palpable mark and a considerable influence. But as these settlements did not extend northward beyond the boundaries of California, Oregon, Washington and British Columbia were wholly unaffected. The early Spanish settlements mentioned were those of the Franciscan missions. The first was established at San Diego by Father Junipero Serra in 1769. From this point the missionaries moved steadily northward, building missions every day's journey until a chain of these institutions, twenty-one in number, reached to the region about the Golden Gate where San Francisco now stands. Each one of these missions was an establishment of considerable proportions; some of them approached the dimensions of imperial colonies. Extensive buildings were erected, and large areas of land brought under cultivation. The work as a whole was prodigious and of a sort to fire the human imagination. Its romantic quality is certainly not dimmed for the present generation by being seen through the purple haze of one hundred and fifty years (Fig. 666).

The mission buildings were of substantial construction, built necessarily from materials found at hand, mainly brick (adobe or native sun-dried brick) and stucco. The mission fathers also invented a kind of tile for roofing. In design these buildings naturally followed Spanish models, though they were modified towards a greater simplicity and a certain crude though pleasing ruggedness hardly characteristic of their prototypes. The total result was indeed æsthetically quite satisfactory; and as many of these structures have been preserved up to the present time, they have become an authentic source of architectural inspiration. "The mission style of architecture" is generally admired, and has been successfully employed in many modern works, both public and private, in California and neighbouring states. In more recent times this "mission style" has been subjected to many dilutions, some of them acceptable, others less praiseworthy (Fig. 667).

Various strains of Spanish influence are blended in this modern architecture; mention is made of "Mexican," "Mediterranean" and "Argentine" styles, though they are not fairly recognisable as types. Under the general head of "Mediterranean" style there are demonstrable references to Italian and Moorish prototypes. Occasionally some fairly pure Italian Renaissance architecture is seen. Another type, still more curious though of much nearer origin, is taken from the Indian pueblos of south-western America. There is also found, endlessly multiplied, the so-called bungalow, though these small dwelling-houses hardly carry any reminiscence of India, where their name originated. They are snug, one-story houses, usually built of wood, with varied roof-lines.

The total effect of this modern California architecture is a kaleidoscope of oddities. The quieter examples of "mission style," and of Spanish and Italian inspiration, are altogether agreeable and are widely accepted as characteristic of California. This consideration of architecture is essential to an understanding of California landscape architecture, since the gardening receives its art impress primarily from the buildings. There has been much thought given, too, to the problem of making gardens which would give a proper atmosphere to the local types of architecture. One of the most fascinating problems in this

FIG. 667. MODERN ARCHITECTURE AND LANDSCAPE ARCHITECTURE IN CALIFORNIA, CONSIDERABLY INFLUENCED BY SPANISH IDEAS

field has been found in working out a "desert style" of gardening. For it must be remembered that large sections of California and of neighbouring states are arid owing to lack of rainfall. Yet many of these arid and semi-arid areas are inhabited by prosperous and home-loving Americans, who must have comfortable houses and who want gardens too. Now the native desert has in it many species of plants, some very interesting, some incontestably beautiful. These are impressed into the local gardening. To them are added xerophilous types (cacti, opuntias, cordylines, yuccas, etc.) from all over the world. The results are highly interesting from a horticultural point of view, and are sometimes artistically effective.

But the coming of the Franciscan missionaries brought more to California than Spanish forms of architecture. To the other side of landscape architecture they brought an equal contribution, that of horticulture. Father Junipero Serra himself planted seeds of the date palm as early as 1770, some of which grew and made fruitful trees. He and his followers also brought other palms, and all the fruits of southern Europe, the olive, the pomegranate, the fig, the lemon, the orange, the apple, the pear, the peach, and above all the wine grape. Because of the climatic affinity between California and Europe already remarked, these importations throve. Most of them were soon acclimatised, and were widely propagated by the industrious missionaries. It is recorded that the mission of San Gabriel near Los Angeles had over 2000 fruit-trees at the beginning of the nineteenth century, and even to-day there is exhibited at this mission an ancient grape-vine dating back to those early times.

Each one of the twenty-one missions on Californian soil was an active centre of civilisation. Schools were maintained, industries and crafts were promoted, and the native Indians were educated and to a notable extent taught the European handicrafts and the practices of agriculture and horticulture. There were also a good many immigrants from Spain and Mexico, Spaniards of both high and low degree, who received grants of land upon which they developed ranches for the cultivation of fruits, vegetables and live stock. This was the condition of affairs in California in 1848, when the cession of that whole empire from Mexico to the United States was almost simultaneous with the discovery of gold. The gold rush of 1849 filled the country with Americans—men mainly of English descent—and changed California abruptly from a Spanish colony into an American state. The Spanish influence lingered faintly in speech, in law and in customs; but its principal contribution to modern California is seen in architecture and gardening, as already stated.

FLORIDA

Like California, Florida was at first a Spanish colony. In 1565 Menendez built a fort and established St. Augustine, the oldest city built by white men on the western continent. Yet the Spanish settlements in Florida never flourished, their fragile military posts soon decayed, and the permanent occupation of this subtropical peninsula was accomplished by English colonists of the same general stock as those who settled in Virginia and the Carolinas.

Florida presents physically somewhat the same picture as California, yet with important differences. It lies well southward, and it is warmed by tropical ocean currents. Palms and other subtropical plants flourish and give their character to the landscape.

Yet the two floras are not the same. The divergence between the two sides of the continent already described, though less noticeable than in the temperate zone, extends into the subtropics. European and Mediterranean plants generally prove more at home in California. The many species of eucalyptus, so freely acclimatised in California, are less often seen in Florida. And while California is characteristically rugged and mountainous, Florida is characteristically flat and swampy. In terms of modern horticulture, however, the two states are similar. Citrus fruits are grown in large commercial plantations, and early fruits and vegetables for northern markets are produced in large quantities. Amongst ornamental plants the same favourites are noticeable—abelias, Cape jasmine, coprosmas, escallonias, hibiscus, myrtles, and bougainvilleas.

Southern California and Florida have other striking similarities of a more superficial origin. Both are popular winter resorts and are much patronised by tourists. Both have been "developed" by real estate "booms." In both there have been enormous areas of land sold in small lots under more or less artistic subdivision. This has led to the construction of vast, and usually rather flimsy, residence colonies of persons having their homes and businesses elsewhere. But the subdivision of land, the erection of multitudinous new houses with considerable effort at garden embellishment, and the provision of public works, buildings, parks, etc., for the new colonies, have offered extraordinary opportunities to architects, engineers and landscape architects. There has been a great deal of experimenting, and some of the results are highly pleasing. At the time this record is made, however (1927), no Floridan style of gardening has emerged, and it can only be surmised that the future has much to reveal in this fascinating land.

LOUISIANA

One side of Florida fronts upon the Atlantic Ocean, the other on the Gulf of Mexico. The shore-line of this Mexican gulf is thousands of miles in extent and marks a region interesting throughout its whole length. In the American vernacular "the Gulf states" include Alabama, Mississippi, Louisiana and Texas, it being customary to disregard Florida in this grouping. This Gulf region then centres upon New Orleans, an old city with a foreign history and a foreign atmosphere tenaciously held. This region of New Orleans and the Gulf coast has much of the character of Florida. There is much low, marshy land, wide, slow rivers, heavy native jungle-like growth, and a subtropical flora.

The old gardens of New Orleans have been much admired, mainly for their magnolias, their gardenias and similar plants, and not for their strength of design or sumptuous furnishings. Audubon Park, a large public playground, has been famous especially for its fine live oaks hung with grey curtains of the epiphytic Spanish moss. This whole region, though less exploited than other southern countries, has great charm and contains much of interest.

THE CENTRAL PLAINS

While exceptions are being made, and a few special localities more critically defined than the general picture of American conditions, a word should be said of the great central plains region, known at home as "the Middle West." It might be known as the upper Mississippi valley except for the fact that it includes Chicago, and all the region lying about

Lake Michigan, Lake Huron and Lake Erie, which waters of course belong to the St. Lawrence and not to the "Father of Waters." This Middle West is a vast empire of fertile level land, now highly farmed with staple crops, especially cereals, and is also a large producer of meat, especially beef and pork. Several large cities and many of moderate size have grown up. The region is wealthy, and the people are strongly devoted to education and to all forms of culture. Especially are they given to the making of comfortable homes, so that here we have the fundamental conditions for the development of architecture and landscape architecture. Rather unfortunately, perhaps, no indigenous architecture has emerged. Homes are mainly built of wood in forms common to the states farther east. In quite recent times the California bungalow type has appeared in considerable numbers. Brick and stucco have also begun to take the place of wood for buildings of the better class. A few public buildings of merit have been designed, e.g. the state capitols at Lincoln and St. Paul.

In the field of landscape architecture there is much promise, with, as yet, somewhat meagre realisation. The native flora is varied and exceedingly beautiful. This statement holds true whether we regard the open, grassy plains approaching the Rocky Mountains or the more rolling prairies of Illinois and Indiana, where the grassy meadowlands are richly interspersed with woodland. Throughout the whole region the watercourses are generally marked with ribbons of tree and shrub growth. Along these borders one finds wild apples, plums, hawthorns, dogwoods, redbuds and dozens of other strikingly beautiful species. Thus the natural topography and the native flora are full of grateful suggestion to the landscape architect.

A few able men have been quick to seize upon these opportunities. Mr. Jens Jensen, landscape architect in Chicago, may perhaps be specially mentioned, without forgetting others. Mr. Jensen has been an uncompromising advocate of the native landscape and the native materials, and insistent upon the duty of developing here a native style of gardening. Many others, also, moved by a love for the vast level plains, have sought to preserve the spirit of that landscape in their park lands and in their home gardens. A "prairie style" of landscape architecture is indeed sometimes discussed. While this is still incompletely formed and by no means widely accepted, some of the principles which govern such a garden form may already be observed. For example, the very level character of the topography makes the straight horizon line especially conspicuous, and suggests that it be adopted in the design. The whole composition, architecture and planting, may be given a general horizontality, with just enough of vertical lines to supply needed contrast. The use of native plants is strongly recommended. The use of wind-breaks supplies a very practical desideratum, since the plains are swept by strong winds for considerable periods both winter and summer.

Shade is more in demand here than in the eastern states, and accordingly more deciduous trees are planted. The proportion of deciduous plantings is increased because in the middle states fewer evergreen species can be grown with full success. As lawn grasses are hard to maintain, especially in the drier zone westward, lawn areas are made smaller and are less emphasised in the design. The obvious need for shelters, like the English garden-houses and the German *gartenlaube*, has not yet been met. All conditions so strongly invite the populations of these middle states to more and better gardening that every lover of this gentle art may confidently look forward to great advances in the near future.

CANADA

Frequent references are made in this chapter, both in foregoing and in following pages, to conditions in Canada. General statements, unless explicitly qualified, apply to the North American continent, including Canada and the United States. Special mention was made in an earlier section of the national parks and forests of the Dominion. A brief special reference in this place may be permitted, therefore, to cover an important area.

Canada's northern position on the map has fixed the belief in many minds that it has an arctic climate, a limited horticulture and little opportunity for gardening. Nothing could be farther from the truth. The Canadian climate, especially in the more thickly settled portions of Nova Scotia, New Brunswick, Quebec, Ontario and the Western Provinces, is agreeable, and wholly suited both to horticultural pursuits and to the development of the highest type of civilisation. Since garden operations are determined by the summer season rather than by the winter, it may be pointed out that the Canadian summer, though shorter than the summer of Florida, is warm and highly adapted to the growing of all popular kinds of plants, including such fruits as the apple, pear, plum, and even the peach; all kinds of forest and ornamental trees; all hardy shrubs, roses, etc., and of course every popular genus of hardy perennials, such as peonies, irises, delphiniums, etc.

In the English sections of Canada the gardening tradition is strong, being derived direct from the mother country. In these sections good gardening has been promoted to a marked degree by local and provincial horticultural societies. These organisations have been much more active and effective, generally speaking, than in the United States. With respect to kinds of plants grown, however, or to types of design, whether in cottage gardens, large private estates or public parks, there is hardly an appreciable difference anywhere between Canadian practice and that of neighbouring states across the border.

Mention has already been made of Mount Royal Park at Montreal, designed by Frederick Law Olmsted, Senior, as illustrating the development of the park idea in America in the years from 1865 to 1900. Victoria Park at Niagara Falls belongs to the same era. It is a fine public park of 1600 acres on the Canadian side of the great falls and is under the control of the government of Ontario. The superlative scenic importance of the Niagara Falls makes this park unique. Though it was first conceived under the earlier theory as a reservation of purely natural scenery, it has latterly been developed in the modern manner with ample refectories and other facilities for the entertainment of tourists and recreationists.

Mount Randle and Echo Lake in Banff National Park (Fig. 668) make this noble park famous.

Perhaps the most effective example of municipal park-making in Canada is found quite fittingly at the capital city, where the Ottawa Improvement Commission (established 1899) has created a comprehensive modern park system of the best sort. As at present constituted, this system comprises Rockcliffe Park, Central Park, Strathcona Park, Nepean Point Park, Macdonald Gardens, National Park, Somerset Street Park, Russell House Park, and Bronson Park; also certain beautiful islands in the Rideau and the Ottawa Rivers; also a system of park driveways through the city. In addition to these areas, formally dedicated as parks, the controlling commission also holds and administers various smaller tracts, more or less completely developed.

This brief inventory of the parks of Ottawa gives a fair picture of modern tendencies

FIG. 668. MOUNT RANDLE AND ECHO LAKE IN BANFF NATIONAL PARK, CANADA

as seen in most modern cities, both in Canada and the United States. But before dismissing finally the Canadian parks, mention should be made of a few others. Spring Gardens, in Halifax, Nova Scotia, have considerable historic interest and are highly cherished by the Nova Scotians. At Victoria, British Columbia, are the Butchart Gardens, especially famous as a rock-garden on a grand scale. Stanley Park, near the same city, is notable for its very large trees of Pacific Coast species. Further, the Central Experimental Farm at Ottawa deserves mention on account of its very influential position and the extensive work done there in testing and disseminating valuable trees, shrubs and flowers.

MUNICIPAL PLANNING

One of the most characteristic phases of American landscape architecture is found in municipal planning, i.e., in city planning, regional and country planning. Here is a field of professional activity which frankly requires highly technical work in architecture, engineering, traffic regulation, sanitation, social service and other lines, yet the field has been aggressively occupied by the American landscape architects. A large number of men in the profession, perhaps a majority, style themselves "landscape architects and city planners." Obviously this position could not be taken, and certainly could not be held, unless the profession were well organised and sure of its ground—unless, in short, it were made up of well-trained and competent men.

City planning doubtless received its first effective impulse in America from landscape architecture. Though L'Enfant was an engineer, and though his plan for Washington was always held in high esteem, the real urge towards general planning came as a part of the park movement already described. Downing, Olmsted and the other men associated with them all saw beyond the parks which they were then planning and spoke cogently of the need of applying better æsthetics to the entire physical community.

When this park movement was at its zenith came the World's Fair in 1893. This proved to be the beginning of a far-reaching revolution in American affairs. For the first time engineers, architects and landscape architects co-operated on a large scale to produce a unified result. The Fair Grounds were a revelation to America. The cities and towns of that day were mere aggregations of poor architecture in which individualism had run wild. The effort of each builder was often given to producing something totally different from his neighbours. Architectural freaks and monstrosities flamed on every horizon. Yet here at the World's Fair were unity, harmony, order, beauty, and withal practical convenience. The buildings were placed in connected groupings, they were embellished with significant sculpture, there were graceful bridges, and there was good landscape architecture shown in broad sheets of water, stretches of lawn, and plantings of trees. Every honest citizen gazed in wonder and immediately asked himself why his own home town could not adopt the same principle of methodical planning. The country was ready for the lesson, and the influence of this one demonstration was incalculable.

The first efforts to realise in local communities the dream of the World's Fair were not altogether successful. They were conceived on too narrow a basis. Attempts were for the most part directed to a rather superficial "beautification" of cities. More than one ambitious community launched a hopeful campaign for a "city beautiful," only to find that early enthusiasm presently faded before a paucity of permanent results. It was soon seen, therefore, that many other factors, especially convenience of business and health of citizens, must be combined with æsthetic improvement. It was seen, further, that this necessity greatly complicated all problems, so that long and thorough study would be indispensable if plans were to prosper. From another quarter also came a strong impetus towards city planning. This lay in the strong altruistic forces promoting housing reform. Tenement house life in the rapidly-growing cities was found to be unsanitary and immoral in a marked degree. The public-spirited men and women working for better conditions soon saw that the remedy must include far-sighted planning, not alone for better tenement houses, but for better streets, parks, transport and public service. Thus their objective was soon merged with that of the city planners.

Meanwhile, the architects, engineers and landscape architects, freshly inspired by the success of their work at Chicago, went forward with redoubled efforts. Architecture, in particular, which had suffered a severe lapse of taste in the years immediately preceding this period, began to show signs of a great revival. Public buildings of better and better design came into being and domestic architecture improved. On their part the city engineers, the builders of streets and of bridges, strove to do better work, both in quality of construction and in external appearance. Called forth by the public demand, the city planners, recruited largely from the ranks of the best-trained landscape architects, now entered upon the scene. The plans of Washington were re-studied (1900) by such eminent talent as Daniel H. Burnham, architect and leading spirit of the World's Fair; Augustus Saint Gaudens, sculptor; Charles F. McKim, noted architect; and Frederick Law Olmsted, Jun., landscape architect. Studies were instituted for a new plan for Chicago. Mr. George E. Kessler, landscape architect, made notable improvements in Kansas City. Many other cities developed plans, some thoroughgoing and useful, others superficial and transitory. Necessarily first essays were tentative and less fundamental than later plans. But the inexorable necessity for planning was now generally recognised.

Conditions in America at this time (1900–17) greatly favoured the city-planning movement. Many cities were growing rapidly both in population and prosperity. Large areas of farm lands were being converted into city property. The processes of industrialisation were in full swing. There was a strong spirit of emulation between cities. Then came the World War. There were immediate and serious dislocations of industry. New industrial centres had to be developed overnight. The city planners were mobilised, the United States Housing Corporation was organised, a large number of new towns and new suburbs were planned, and a few of them were constructed before the armistice of 11 November, 1918, put an end to war activities. But the demonstration of what could be done by intensive planning was impressive. The work of this period was characterised especially by the fruitful co-operation of architects, engineers, landscape architects and "realtors" (professional real estate dealers).

Since the war closed some American cities at least have continued their phenomenal growth, and new problems in city planning have been met with fresh knowledge and zeal. Some noteworthy industrial residence suburbs have been constructed. A good example is seen in Mariemont, a residence suburb of Cincinnati, planned by Dr. John Nolen. Another striking example may be seen in Palos Verdes, a residence and resort suburb on quite broad and rural lines, planned by Olmsted Brothers and lying on the hills next the Pacific Ocean near Los Angeles. This post-war period has also witnessed an extraordinary sweep of land speculation, touching many localities, but most fervent in southern California and Florida. The customary practice has been to subdivide large tracts of open country and sell the parcels as town lots, as small farms, or in any size between. Here the profession of municipal planning has had ample exercise. One brief comment will cover the entire case, viz. as might have been foreseen, some of this planning was admirably done, more of it was of very moderate merit, and some of it was jerry-planning of the shabbiest sort.

The war emergency also brought to the fore the idea of "regional planning." It was soon found that no municipality exists apart from its neighbours. Both communities and industries overlap. Any main thoroughfare, for example, runs from one city to another and must be planned with reference to mutual demands. Some cities are satellites, while others are pivotal. Fundamental planning problems are regional because populations and industries and commerce are regional. The study of these broader relationships, involving several municipalities at once, has come to be known as regional planning. It is manifestly more difficult to accomplish practical results on this large scale than on smaller areas; but the American genius for organisation and co-operation promises to overcome such obstacles and eventually to justify the serious attention given by the experts to regional problems.

Meanwhile a few persons had begun to talk of country planning, and the American Civic Association in its annual conference of 1915 had specialised on this topic. Again in 1921 the annual conference was devoted exclusively to country planning. It was conceived that the rural districts have a physical form of their own, quite different from the physical frame of the cities; that these differences of form and function imply different lines of growth, different planning—an understanding of a different category of causes and effects. And as rural interests in the United States and Canada are well organised under strong leadership, this idea has not remained fallow. There has been active work in the improvement of rural highways, much study given to rural schools, an extensive dedication of country parks and forests, and renewed study of the economics of farm-land subdivision.

Ambitious proposals have been made from time to time for carrying this planning conquest still farther, that is, for planning whole states and even the whole nation. Mr. Cyrus Kehr's recent book on *Nation Planning* expounds the idea on its theoretical side. But while there has been much real planning of national highways, national railways, national waterways, national parks, national forests, etc., it can hardly be claimed that as yet much has been done in unified study of national problems. Still less has been done by individual states. The emergence of such ideas, however, and their eager public discussion, give evidence of the extent to which the planning idea has captured the American mind.

LITERATURE

Landscape architecture in North America has a large, varied and worthy literature. As in any other extensive bibliography, there is, of course, much rubbish and much of only passing interest. However, there is also much of indubitable and permanent value.

In literature, as in practice, American landscape architecture grew out of horticulture. Any search for beginnings, therefore, must be made in the horticultural field. The English colonists brought over the English books of the time (middle of the seventeenth century), and though there never were many of these works in circulation they were accepted as authoritative. This fact is best measured in the notable extent to which they were quoted in the early books written in the New World.

The earliest instructions on gardening printed in America were those found in the more or less periodical (often annual) calendars or almanacs. These were numerous and widely circulated. (*Vide* L. H. Bailey, *Cyclopædia of American Horticulture*, 3: 1521. 1915.) The first garden book was probably Robert Squibb's *The Gardener's Kalender for South Carolina and North Carolina*, published in Charleston in 1787. Next in order was an American edition of an English work, Marshall's *Introduction to the Knowledge and Practice of Gardening*, published in Boston in 1799. In 1804 appeared another purely American work, though with many quotations from English sources, *The American Gardener*, by John Gardiner and David Hepburn, and printed in Washington. Bernard M'Mahon's *American Gardener's Calendar* was published in Philadelphia in 1806 and marked a notable advance. It was the work of an able man, was well written, and had a large circulation.

Passing over many interesting works, and ignoring those devoted primarily to fruit-growing or kitchen-gardening, a pause should be made to note the works of Peter Henderson. Henderson was a Scotchman, born near Edinburgh in 1822, who came to America in 1843 and who set up in business for himself in Jersey City in 1847 on a capital of $500, savings from his own small wages. He founded a famous seed and nursery firm, wrote several valuable books, and generally made a notable contribution to American horticulture. His first book, *Gardening for Profit* (1867), wellnigh pointed a revolution in American horticultural literature; for though Henderson was born and trained in Scotland, his book was intensely American, being derived strictly from experience under American conditions. It marked the final triumph of indigenous experience over foreign authority. It had a very wide sale. It was followed by other books on various branches of gardening, all of which proved popular.

In many of these early books on gardening were observations on the æsthetic aspects of the subject, or more specifically on garden design. But the American literature of this subject must be dated always from 1841, when there appeared the notable work on *Landscape Gardening* by Andrew Jackson Downing. This had an enormous vogue, and easily did more than any other printed work to form American ideals. It is much read and admired even to the present day, being still in the hands of the original publishers and now on sale in a tenth edition. Since 1841 there have been published hundreds of books in this field, but it would be invidious to mention particular titles, and a complete bibliography is obviously out of the question.

Periodicals of many sorts have had their place also in the development of American gardening. According to Bailey (*Cyclopædia of American Horticulture*), the first journal to devote any appreciable space to horticultural matters was the *New England Farmer*, published in Boston from 1822. The *Floral Magazine* was established in Philadelphia in 1832 and ran for some years. The *Horticulturist* was established in Albany in 1832 and had an influential career. The first seven volumes appeared under the editorship of Andrew Jackson Downing, and as a matter of course contained much important matter on gardening and on what is now called landscape architecture. Indeed some of the most notable portions of Downing's great work appeared in this magazine. From 1888 to 1897 there appeared weekly from New York the numbers of *Garden and Forest,* Dr. Charles Sprague Sargent, managing editor. This magazine devoted a large part of its space to landscape architecture during a critical and formative period; its work was of very high quality and its influence incalculable. In 1910 was established the quarterly magazine *Landscape Architecture,* the official organ of the American Society of Landscape Architects, and devoted exclusively to this special field. Several other excellent journals now current present various aspects of gardening, nursery practice, fruit-growing, forestry and general horticulture.

PROFESSIONAL EDUCATION

Landscape architecture in the United States shows one unique feature in the highly organised system of professional education now developed. The point which may be most conveniently marked as a beginning for this movement was the opening of a professional course at Harvard University in 1900. Other courses of a similar character were offered soon after in Massachusetts Agricultural College, Cornell University, and Illinois University, with still other institutions making a beginning from year to year. It would be an impossible task, and a needless one, to fix an exact chronology of these events. In many instances the initial work was quite unpretentious. Instruction in gardening, into which some elementary ideas of landscape gardening were introduced, was characteristic of the times and was to be found in almost every agricultural college (of which there are approximately fifty) in the country. Quite naturally some of these institutions found the demand stronger and their facilities better than others, and in such circumstances the courses grew. In brief, the teaching of landscape architecture nearly always began (under the name of landscape gardening) as a very modest development in horticulture. Wherever it received support and encouragement it developed, sometimes rapidly, by the steady addition of course after course, progress being towards higher academic standards, towards more professional quality and towards a broader foundation of engineering, architecture and general art training.

As a result of these developments there is found at the present time (1927) a well-developed system of instruction in landscape architecture, notably standardised, after the American manner. This system is distinctively of a university character. With only two or three exceptions, the courses are organised in colleges or universities, including many of the largest and most famous. In those exceptional cases where the work is given in a separate special school, it still follows closely the same pattern and conforms to the same standards. With considerable uniformity, therefore, this professional instruction in landscape architecture is standardised upon the regular college curriculum, which is almost invariably a course of four years beyond the completion of high school.

These college courses lead always to the degrees of Bachelor of Arts or of Bachelor of Science, though graduates in landscape architecture nearly always (quite paradoxically) receive the latter. During his four years in college the budding landscape architect is expected to receive, in addition to his professional training, a rather substantial education in general subjects such as languages, literature, history, and science, including economics and social sciences. By custom, the specialised professional courses are pressed mainly towards the end of this four-year curriculum. The professional studies generally include extended problems in the design of private grounds and public parks, in city planning and land subdivision, and in the construction of landscape work; also other courses, as extended as time will permit, in plants, mathematics, surveying and engineering, and architecture.

Instruction in these professional courses is nearly always given by means of lectures combined with laboratory and drafting-room study upon problems, with a strong tendency to emphasise the latter. Field studies and visits to public and private places of interest are also considered important. Many schools further lay stress upon the need for a considerable period of apprenticeship with established landscape architects after graduation from college. No very elaborate equipment is required for this instruction, and most of the colleges seriously attempting professional courses are well supplied with the needful libraries, class-rooms, drafting-rooms and gardens.

As in every other branch of formal education, however, more depends on the teacher than upon all other things combined. In those colleges where professional landscape architecture is emphasised the staff for strictly technical instruction usually numbers from three to six men (or men and women). Some of these are known to be teachers of marked ability; a number have gained national repute in this field. American university practice, however, strongly suggests still higher standards than those fixed by the four-year college course and the baccalaureate degree. The learned professions and many of the technical industries are taught by preference in graduate schools. Such schools require the baccalaureate degree for admission, their curricula extending for two, three or four years beyond the general college course. Throughout the fraternity of landscape architects, and especially in academic circles, there is a manifest desire to place instruction in landscape architecture upon this higher plane. The one outstanding example of such a programme is to be found in the School of Landscape Architecture at Harvard University, which is a graduate school exclusively. Here the course normally covers a period of three years. The work is highly specialised and professionalised, and an option is offered between landscape architecture and city planning.

Several other universities offer graduate courses in extension of their undergraduate curricula, but, generally speaking, these have not as yet been largely developed. The completion of such graduate courses, at Harvard University or elsewhere, usually earns for the successful student the degree of Master of Landscape Architecture.

PROFESSIONAL ORGANISATIONS

Commentators on American life have often pointed out the boundless tendency of Americans towards organisation. Societies for every purpose are formed everywhere. Many of these, of course, come to nothing; others enrol large memberships and exercise vast influences on a national scale. Americans are accustomed to operating through such societies and readily assign to them important functions.

The field of landscape architecture is no exception. As soon as there came to be any considerable number of landscape architects gaining their livelihood from the exercise of this profession, the American habit of organisation, operating in view of their common interests, brought them together, and the American Society of Landscape Architects was the result. The initial organisation was made in the year 1899 and the society was incorporated and made really national in its scope in 1916. Since that time it has gradually enrolled a large proportion of the men and women landscape architects of America. Naturally the efforts of this society have been turned first to the practical business interests of its members. Business methods, policies and ethics have been discussed, and to some extent standardised. But the society has also exercised a substantial public influence in such matters as national parks, state parks, war memorials, city-planning law and practice, and the promotion of art education in general.

From time to time there have been sundry horticultural organisations which have exercised some influence in gardening matters. The American Pomological Society (organised in 1848) deserves first mention. From the beginning this society gave most of its attention to fruit-growing, and with the increasing specialisation of later years passed over entirely into that field. Then came the Society of American Florists and Ornamental Horticulturists (organised 1883) which has had a lively interest in gardening, especially on its commercial side, and which has always included in its membership many eminent leaders in horticulture.

The more recent Garden Club of America (organised 1913) is made up of a large number of local women's garden clubs. The members are mainly women who own gardens and who are intensely interested as amateurs. This federation issues a monthly bulletin and other publications, and is a highly significant organisation. There have been, and still are, many other horticultural societies in the field, some of them of long record and honourable achievement, such as the Massachusetts Horticultural Society (organised 1829) and the Pennsylvania Horticultural Society (organised 1838); but on the whole this form of effort has been less effective in directing taste in landscape architecture than the other agencies mentioned. In the field of horticulture, more narrowly defined, their service has been of great value.

SELECT BIBLIOGRAPHY

(The following list comprises the principal books mentioned by the German author of the present work, together with certain others which have a definite bearing on garden art. It is not intended to be a selection of books on general gardening. In each case an endeavour has been made to state the year when the first edition was published, but complete accuracy throughout cannot be assured.)

ABERCROMBIE, PATRICK (with JOHN ARCHIBALD), *East Kent Regional Planning Scheme* (1925).

ADAMS, T., *Rural Planning* (1917).

ADDISON, JOSEPH, "On the Cause of the Pleasures of the Imagination," essay in *The Spectator* (1711-12); "Description of a Garden in the Natural Style," essay in *The Spectator*, No. 477 (1711-12).

AITON, WILLIAM, *Hortus Kewensis* (1789).

ALPHAND, *Les Promenades de Paris* (1867-73).

AMHERST, HON. ALICIA (HON. LADY CECIL), *A History of Gardening in England* (1895); *London Parks and Gardens* (1907).

ANDRÉ, *L'Art des Jardins* (1879).

ARCHIBALD, JOHN. See under ABERCROMBIE.

BACON, FRANCIS, essay *On Gardens* (1625).

BAIKIE, JAMES, *The Glamour of Near East Excavation* (1927).

BAILEY, L. H., *Cyclopædia of American Horticulture* (1900).

BETIN, P., *Le fidelle Jardinier* (1636).

BEYLIÉ, *L'Habitation Byzantine* (1902-3).

BLOMFIELD, SIR REGINALD, *The Formal Garden in England* (1892).

BLONDEL, *Cours d'Architecture* (1771).

BOLTON, A. R., *Gardens of Italy* (1905).

BOTTOMLEY, M. E., *Design of Small Properties* (1926).

BOYCEAU, *Traité du Jardinage* (1638).

CARMONTELLE, *Le Parc Monceau* (about 1775).

CAUS, ISAAC DE, *Hortus Pembrochianus* (1615).

CAUS, SALOMON DE, *Hortus Palatinus* (1618).

CAVE, HENRY, *The Ruined Cities of Ceylon* (1900).

CECIL, HON. LADY. See under AMHERST.

CHAMBERS, SIR WILLIAM, *A Dissertation on Oriental Gardening* (1772).

CHARDIN, *Voyages en Perse* (1811).

CHILD, STEPHEN, *Landscape Architecture* (1927).

CLEVELAND, H. W. S., *Landscape Architecture as applied to the Needs of the West* (1873).

COMPARETTI, *Le Pitture di Ercolano* (about 1885).

CONDER, *Landscape Gardening in Japan* (1886).

CRAWFORD, A. W., *The Development of Park Systems in American Cities* (1905).

CRIDLAND, R. B., *Practical Landscape Gardening* (1926).

CURTIS, WILLIAM, *The Botanical Magazine* (1787).

DIESEL, *Erlustierende Augenweide* (1899).

DIEULAFOY, MARCEL, *L'Art Antique de la Perse* (1884).

DOHME, ROBERT, *Barock- und Rokoko-Architektur* (about 1887).

DOWNING, A. J., *Landscape Gardening* (1841); *Rural Essays* (1869).

DU CERCEAU, *Les plus beaux Bâtiments de France* (1576).

EARLE, MRS. C. W., *Pot-pourri from a Surrey Garden* (11th edition, 1898).

ELLACOMBE, HENRY N., *The Plant-lore and Garden-craft of Shakespeare* (1878).

ELWOOD, P. H., Jr., *American Landscape Architecture* (1924).

ERMAN, *Aegypten* (1823).

EVELYN, JOHN, *The French Gardener* (1658); *Kalendarium Hortense* (1664); *Sylva* (fol., 1664).

FALDA, G. B., *Giardini di Roma* (about 1676); *Le Fontane delle Ville di Frascati* (about 1676).

FARRER, REGINALD, *The English Rock Garden* (1919); *On the Eaves of the World* (1917).

FARWELL, E. T., *Village Improvement* (1913).

FIRMINGER, T. A. C., *Manual of Gardening for Bengal and Upper India* (1864).

FONTAINE (with PERCIER), *Les plus beaux Bâtiments de l'Italie* (1809); *Choix des plus célèbres Maisons de Plaisance de Rome* (1824).

French publication, *Le Jardin Anglo-Chinois* (1770-87).

FURTTENBACH, *Architectura Civilis* (1628); *Architectura Privata* (1641).

GERARDE, JOHN, *The Herbal, or General History of Plants* (1597).

GRENIER, CH., *Habitations Gauloises* (1906).

GRIFFITH, F. LL. (with J. J. TYLOR), *The Tomb of Renni* (1895); *The Tomb of Paheri at El Kab* (1894).

GRINDON, LEO, *Shakespeare Flora* (1883).

GUERNIERI, *Delineatio Montis* (1706).

GUSMAN, *La Villa Imperiale di Tivoli* (1904).

HAARLEM MSS. 4425; *Roman de la Rose* (British Museum).

HERZFELD, *Vorbericht von Samarrâ* (1912).

HILL, JOHN, *The British Herbal* (Illus. fol., 1756).

HILL, THOMAS, *The Arte of Gardening* (probably 1563); *The Gardener's Labyrinth* (1577).

HOME, HENRY, essay on "Gardening" in the *Elements of Criticism* (1770).

HUBBARD and KIMBALL, *Landscape Design* (1917).

HUCHISON, MRS. MARTHA B., *Spirit of the Garden* (1923).

HUMPHREYS, MRS. PHEBE W., *Practical Book of Garden Architecture* (1914).

HYPNEROTOMACHIA, *Il Sogno di Polifilo* (1499).

JAMES, H., *Land Planning in the United States for the City, State, and Nation* (1926).

JAMES RIVER GARDEN CLUB, *Historic Gardens of Virginia* (1923).

JEKYLL, GERTRUDE, *Colour in the Flower Garden* (1908).

JELLICOE, G. A. See under SHEPHERD.

JOHNSON, G. W., *History of English Gardening* (1829).

KEHR, CYRUS, *A Nation Plan* (1926).

KELLAWAY, H. J., *How to Lay Out Suburban and Home Grounds* (1915).

KELSEY, F. W., *The First County Park System* (1905).

KENT, *Flora Domestica* (1823).

KERN, G. M., *Practical Landscape Gardening* (1855).

KIMBALL (with OLMSTED), *Frederick Law Olmsted, Landscape Architect* (1922).

KING, MRS. F., *The Well-considered Garden* (1915).

KIP, JAN (with L. KNYFF), *Le Nouveau Théâtre de la Grande Bretagne* (early edition, 1714-16).

KLEINER, SALOMON, *Wahrhafte Abbildung* (1726).

KNIGHT, RICHARD PAYNE, *The Landscape*, A Poem (1794).

KNIGHT, THOMAS ANDREW. Numerous publications (1797-1840).

KNYFF, L. See under KIP.

KRAUS, G., *Augsburger Gärten* (about 1894).

LATHAM, CHARLES, illustrations in *Gardens Old and New* (1905); *English Houses and Gardens* (1903); and *The Gardens of Italy* (1905).

LAWSON, WILLIAM, *A New Orchard and Garden* (1597).

LAYARD, A. H., *Niniveh*, and other works (1850–90).

LE MOYNE, *Country Residences in Europe and America* (1908).

LENOIR, ALBERT, *Architecture Monastique* (1852–6).

LEPSIUS, *Monuments of Egypt* (1897–1900).

LETAROUILLY, PAUL, *Les Edifices de Rome Moderne* (1860).

LEYLAND (with H. AVRAY TIPPING), *Gardens Old and New* (vol. ii.) (1905).

LONDON, G., and WISE, H., *The Complete Gardener*, from the French of J. de la Quintinge, translated by John Evelyn (1699).

LORRIS, GUILLAUME DE, *Roman de la Rose*. See also under HAARLEM.

LOUDON, JOHN CLAUDIUS, *An Encyclopædia of Gardening* (1822); *An Encyclopædia of Trees and Shrubs* (abridged edition, 1842).

MARCO POLO, *The Travels of Marco Polo*.

MARKHAM, GERVASE, *The English Husbandman* (1613); *The Country Housewife's Garden* (1617).

MASPERO, GASTON, *Histoire Ancienne des Peuples de l'Orient* (1893).

MATTHEWS, *Mazes and Labyrinths* (1922).

MAWSON, T. H., *The Art and Craft of Garden-making* (1901).

MAYER, J. PROCOP, *Pomona Franconica* (about 1776).

MAZOIS, *Les Ruines de Pompéi* (1824–38).

MENDEL, GREGOR, *Versuche über Pflanzen-Hybriden* (1865).

MEUNIER, L., *Diversas Vistas de las Casas y Jardines de placer da Rei d'España*.

MILLER, PHILIP, *The Gardener's Dictionary* (Illus. fol., 1731).

MILNER, H. E., *The Art and Practice of Landscape Gardening* (1890).

MITCHELL, D. G., *Out-of-Town Places* (1884).

MONTAIGNE, *Essays and Journals* (about 1627).

NASON, *Rural Planning* (1923).

NAVILLE, *Deir-el-Bakhari* (1835).

NEWBERRY, P. E., *El-Bersheh* (1894).

NICCOLINI, F., *Le Case ed i Monumenti di Pompeii* (1854).

NICHOLS, ROSE STANDISH, *The English Pleasure Garden* (1902).

NICHOLSON, G., *Illustrated Dictionary of Gardening* (1885).

OLMSTED. See under KIMBALL.

PAGET (with PIRRIE), *The Tomb of Ptah-hotep*.

PARKINSON, JOHN, *Paradisi in Sole Paradisus Terrestris* (1629).

PARSONS, SAMUEL, *The Art of Landscape Architecture* (1915).

PARSONS, SAMUEL, *Landscape Gardening* (1891).

PERCIER. See under FONTAINE.

PERELLE, engravings in *Bibliothèque de l'Art et d'Archéologie* (about 1680).

PETRIE, SIR W. M. FLINDERS, *Tell-el-Amarna* (1895).

PHILLIPS, H., *Flora Historica* (1824).

PIRRIE. See under PAGET.

PLACE, *Ninive et l'Assyrie*.

PRICE, SIR UVEDALE, *An Essay on the Picturesque as compared with the Sublime and the Beautiful* (1794).

PÜCKLER, PRINCE, *Andeutungen über der Landschaftsgärtnerei* (1834).

PURDOM, C. B., *The Building of Satellite Towns* (1925).

RAWLINSON, *Five Great Monarchies* (1862).

RAY, JOHN, *Catalogus Plantarum Angliæ* (1670).

REA, JOHN, *Flora* (1665); *Ceres, Flora, and Pomona* (1676).

REHMANN, E., *The Small Place* (1918).

REISS, RICHARD, *Town-planning Handbook* (1924).

REPTON, HUMPHREY, *Sketches and Hints on Landscape Gardening* (1795).

ROBINSON, WILLIAM, *The English Flower Garden* (1883).

ROHDE, E. S., *A Garden of Herbs* (1920); *Old English Herbals* (1922); *Garden-craft in the Bible* (1927).

ROOT and KELLY, *Design in Landscape Gardening* (1914).

ROSELLINI, *Monumenti Civilis* (1832–44).

ROSTOVTZEFF, *Pompejanische Villen* (1904).

SAMBON, *Fresques de Boscoreale* (1903).

SARRE, *Denkmäler persischer Baukunst* (1910).

SCHREIBER, *Hellenistische Reliefs* (1887–90).

SERRES, OLIVIER DE, *Le Théâtre d'Agriculture* (1651).

SHENSTONE, WILLIAM, *Unconnected Thoughts on Gardening* (1764).

SHEPHERD, J. C. (with G. A. JELLICOE), *Gardens and Design* (1928); *Italian Gardens of the Renaissance* (1927).

SIMONDS, O. C., *Landscape Gardening* (1920).

SIMSON, ALFRED, *Garden Mosaics* (1903).

SLADE, D.D., *The Evolution of Horticulture in New England* (1895).

TABOR, GRACE, *Old-fashioned Gardening* (1913).

TAYLOR, A. D., *The Complete Garden* (1921).

TEMPLE, SIR WILLIAM, "Upon the Gardens of Epicurus," in *Miscellanea*, vol. i. (1680).

TIJOU, JEAN, *A New Book of Drawings* (1693).

TIPPING, H. AVRAY, *Gardens Old and New* (1905).

TRIGGS, H. INIGO, *Garden-craft in Europe* (1913); *Formal Gardens in England and Scotland* (1902); *The Art of Garden Design in Italy* (1906).

TYLOR, J. J. See under GRIFFITH.

UNDERWOOD, L., *The Garden and its Accessories* (1907).

VAN RENSSELLAER, MARTHA, *Art Out of Doors* (1893).

VENTURINI, *Le Fontane del Giardino Estense*.

VERGNAUD, *L'Art de créer les Jardins* (1835).

VRIES, VREDEMANN DE, *Hortorum Viridariorumque Formæ* (about 1600).

WALPOLE, HORACE, *On Modern Gardening* (1785).

WARD, WILLIAM H., *The Architecture of the Renaissance in France* (1911).

WATERFIELD, MARGARET, *Flower Grouping* (1907); *Corners of Grey Old Gardens* (1914).

WATKINS, M. G., *Natural History of the Ancients* (1896).

WAUGH, FRANK A., *Country Planning* (1925); *Textbook of Landscape Gardening* (1922).

WEED, H. E., *Modern Park Cemeteries* (1912).

WHATELY, THOMAS, *Observations on Modern Gardening* (1770).

WILKINSON, SIR J. G., *Manners and Customs of the Ancient Egyptians* (1878).

WILSON, F. C., *Bôrô-Boudour* (1874).

WORLIDGE, JOHN, *Systema Horticulturæ* or *The Art of Gardening* (1677).

WRIGHT, WALTER P., *An Illustrated Encyclopædia of Gardening* (Everyman, 1911); *Roses and Rose Gardens* (1911).

YARD, R. S., *The Book of the National Parks* (1919).

YULE, A. F., *The Travels of Ser Marco Polo* (1903).

INDEX

A

Geoponica, the, ancient work in vegetable-gardens, I. 70, 78, 201
George II. of England, II. 340
George IV. of England, II. 341
George William, Margrave of Bayreuth, II. 194
Geranium, the, in England, II. 369, 373, 381
Gerard, John, botanist, I. 446
Germain-en-Laye, St. *See* St. Germain-en-Laye
Germany:
  Renaissance period: botanists, II. 3–4; Garden of the Holy Roman Empire, 4; Scholz, 6, 8; Gesner (q.v.); Nuremberg and Frankfort gardens, 8–9; a prince's garden, 11; Hapsburg gardens, 12; Neugebäude of Maximilian II., 13–15; Belvedere, Hradschin, 17; De Vries, 19–20; Rubens, 22; the first German garden book, 23; zenith of garden interest in, 23; theories, 23; Furttenbach's influence, 26–27; Italian influences, 27; Sittich, the work of, 27–9; Philip Hainhofer, 30; Munich, Residence gardens in, 31–5; Haimhausen Castle, 35; Stuttgart Castle garden, 35; Heidelberg Castle, 36–43; gardens during period of Thirty Years' War, 36; unity of style wanting, 36; Hessem Castle, Brunswick, 44; Wallenstein as town-planner, 44; Wallenstein's gardens at Prague and Gitschin, 45; Cleves rebuilt, 46; influence of Louis XIV.'s tastes, 47. *See also* Vienna
  After Thirty Years' War: Versailles as exemplar, II. 125; Le Nôtre's influence, 125; Herrenhausen, 125; Italian architects, 126; Wilhelmshöhe, Cassel, 128 (*see also under* Cassel); French influence in the Nymphenburg and Schleissheim, 132–3; Karlsruhe, 141; Church princes' estates, 142; Brühl, 143; Schönborn family's love of building, 146; Bamberg, 147; Favorite, Mainz, 147–9; spiritual lords' love of pleasure, 153; Würzburg, 154–7; English style, influence of, 156; old style preserved by Adam von Seinsheim, 156; Belvedere, Vienna, 159; Schönbrunn, *see under* Schönbrunn; artificial ruins, 172; Middle Germany, gardens of, 174; water-castles, 176; orangeries, 178; Bavaria and Saxony, royal gardens in French style of, 179–82; French formal style decadent, 186; transition period, 194; reaction against the formal garden, 301; Hirschfeld (q.v.); Wörlitz, 302; Weimar, 304–7; new style theorists, 308–9; horticulture as an art, 312; Goethe and Möser, challenge of, to sentimental epoch, 315; Pückler's ideas of landscape gardening, 318–19; carpet-bedding, 321
  Nineteenth century: landscape gardening, 335–8; Potsdam, 337–8; scenic illusions, 337; public parks, 345; architects attack the landscape garden, 358; Mannheim Exhibition models of architectural gardens, 359; approximation to people's gardens in England, 360; perspectives of modern public parks, 362–3
Gesner, Conrad, II. 4, 6; his classification, 11; "back to Nature" cry, 301
Geymülle, I. 402
Giardini secreti, I. 252, 261–2, 280, 288, 357, 381; of Fontainebleau, 427; Privy Garden, 439; of the Trianon, II. 96; Wilhelmshöhe, Cassel, 131; Dutch, 224
Gibbs, Hon. Vicary, II. 399, 400
Gildemeister, garden architect in Bremen, II. 360
Gilgamesh, hero-epic of Babylonia, I. 29
Giocondo, Fra, Charles VIII.'s garden artist, I. 394
Girard, François, II. 132; work of, for Church princes, 142–3; Brühl Park, 143
Girardin, Marquis, friend of Rousseau, II. 295, 319
Gitschin, Wallenstein's estate at, II. 45
Giustiniani a Bassano, Villa, near Sutri, I. 279, 329
Glacier National Park, Montana, II. 431, 433
Glacier Park, British Columbia, II. 433

Glasgow, public parks of, II. 412–13
Glass-houses, II. 333–4
Gleim, poet, II. 301
Glienicke, Potsdam, II. 336
Goethe, on waste of land for ornamental parks, I. 315; on the Bose family garden at Leipzig, II. 175; on Northern Italian villas, 206; on English grottoes, 290; on Wörlitz, 303; Schönau, Weimar, his interest in, 304; on Hohenheim, 312; his town-garden, 318
Golden trees of Bagdad, I. 149; in mediaeval poems, 191
Golder's Hill, near Hampstead, II. 409
Gomboust, II. 60
Gondi, Italian banking family at the court of Maria de' Medici, I. 431; II. 102, 103
Gonzaga's pleasure-house in Mantua, I. 234, 236
Gothic tradition, English architecture and, I. 436; II. 290; French, I. 436; and Romantic, II. 278; Gothic house, Wörlitz, 303; neo-Gothic, 331–2
Goujon, sculptor, I. 407
Graceland Cemetery, Chicago, II. 428, 429
Granada. *See under* Alhambra, Generalife
Grand Canyon National Park, Arizona, II. 432
Granja, La. *See* Ildefonso
Gran Quivira National Monument, New Mexico, II. 436
Gravetye Manor, Sussex, II. 379, 388, 392
Gray, Asa, American botanist, II. 443
——, Thomas, on English gardening, II. 289
"Great Garden," the, Dresden, II. 179, 181
Greece, ancient, no garden culture in, I. 53; evidences from Cretan and Mycenæan art, 53; garden culture in the Odyssey of Homer, 54–8; only the simplest cultivation known in Homer's time, 55; rose-culture of, owed to the Macedonians, 59; Adonis gardens, 59; pot-gardening in, 60; migration of Attic estate-owners to the towns, 61; development of sacred groves, 63; nymph sanctuaries, 63; hero sanctuaries and games festivals, 64–5; the gymnasium, 65; the Academy, 65; development of town-gardens, 65, 72–3; public gardens, 70; private gymnasiums and baths, 70; philosophers' gardens, 70; fruit from Asia Minor, 72; Alexander the Great's care for parks, 72; court-gardens, 74, 76; dwellings and the peristyle, 74; grottoes, 76; farm-garden descriptions in romances of a late period, 77; description by Longus, 78
Greenlands, Berks, garden of Viscountess Hambleden at, II. 394
Green Park, II. 410
Greenwich Park, II. 410
Gregory XIII., I. 295, 302
Grelly, French garden artist, II. 125, 396, 398
Grillet, French garden artist, II. 396, 398
Grimaldi, Marquis, II. 218
Grimaldus, Abbot of St. Gall, I. 180
Grimmer, Abel, painter, II. 19
Groen, Van der, II. 223–4
Grossmann, garden architect of Leipzig, II. 360
Gross-Sedlitz, II. 182; French model taken for, 182
Grottoes: ancient Greece, I. 76; Republican Rome, 86–8; the Rain Grotto, Villa di Castello, Florence, 247–8; Pope Julius III.'s nymphæum, 268, 269; Pratolino, water-tricks of, 285; tufa, of Pratolino, 286; Boboli Gardens, 293; Quirinal, 297; grotto-rooms, 317; Alcazar, Seville, 357; Tuileries, 413; *Grotte des Pins*, Fontainebleau, 414; Meudon, 414; natural animal decorations, 416; enamelled, 416–17; Versailles, 169; Schönbrunn, 169; Chinese, 242; "Chinese-Gothic," in English gardens, 291; freak, at Wörlitz and Schönau, 304
Groves: Versailles, II. 70; Clagny, 79; decorative, of Viennese eighteenth-century gardens, 173. *See also* Sacred Groves
Gudea, park of the Sumerian kings, I. 33
Guernieri, Roman master-builder, II. 128, 131

Printed in the United States
By Bookmasters

CAMBRIDGE LIBRARY COLLECTION

Books of enduring scholarly value

Botany and Horticulture

Until the nineteenth century, the investigation of natural phenomena, plants and animals was considered either the preserve of elite scholars or a pastime for the leisured upper classes. As increasing academic rigour and systematisation was brought to the study of 'natural history', its subdisciplines were adopted into university curricula, and learned societies (such as the Royal Horticultural Society, founded in 1804) were established to support research in these areas. A related development was strong enthusiasm for exotic garden plants, which resulted in plant collecting expeditions to every corner of the globe, sometimes with tragic consequences. This series includes accounts of some of those expeditions, detailed reference works on the flora of different regions, and practical advice for amateur and professional gardeners.

The Trees of Great Britain and Ireland

Although without formal scientific training, Henry John Elwes (1846–1922) devoted his life to natural history. He had studied birds, butterflies and moths, but later turned his attention to collecting and growing plants. Embarking on his most ambitious project in 1903, he recruited the Irish dendrologist Augustine Henry (1857–1930) to collaborate with him on this well-illustrated work. Privately printed in seven volumes between 1906 and 1913, it covers the varieties, distribution, history and cultivation of tree species in the British Isles. The strictly botanical parts were written by Henry, while Elwes drew on his extensive knowledge of native and non-native species to give details of where remarkable examples could be found. Each volume contains photographic plates as well as drawings of leaves and buds to aid identification. The species covered in Volume 2 (1907) include horse chestnut, buckeye, hemlock, walnut and larch.

The Trees
of Great Britain
and Ireland

VOLUME 2

HENRY JOHN ELWES
AUGUSTINE HENRY

CAMBRIDGE
UNIVERSITY PRESS

CAMBRIDGE
UNIVERSITY PRESS

University Printing House, Cambridge, CB2 8BS, United Kingdom

Published in the United States of America by Cambridge University Press, New York

Cambridge University Press is part of the University of Cambridge.
It furthers the University's mission by disseminating knowledge in the pursuit of
education, learning and research at the highest international levels of excellence.

www.cambridge.org
Information on this title: www.cambridge.org/9781108069335

© in this compilation Cambridge University Press 2014

This edition first published 1907
This digitally printed version 2014

ISBN 978-1-108-06933-5 Paperback

THE TREES OF GREAT BRITAIN AND IRELAND

PINUS LARICIO, FOREST OF BAVELLA, CORSICA

From an Original Sketch by the late Robert Elwes.

The Trees
of
Great Britain
& Ireland

BY

Henry John Elwes, F.R.S.

AND

Augustine Henry, M.A.

VOLUME II

Edinburgh: Privately Printed

MCMVII

CONTENTS

ILLUSTRATIONS

THUJOPSIS

Thujopsis, Siebold et Zuccarini, *Fl. Jap.* ii. 32 (1842).
Thuya, Bentham et Hooker, *Gen. Pl.* iii. 427 (1880).
Cupressus, Masters, *Journ. Linn. Soc.* (*Bot.*), xxx. 19 (1893) and xxxi. 363 (1896).

THIS genus is considered by many authorities to be merely a section of Cupressus or of Thuya. The foliage and cones, however, are remarkably distinct, and justify its retention as a separate genus.

Evergreen trees, belonging to the tribe Cupressineæ of the order Coniferæ, with reddish bark scaling off in longitudinal shreds. Branches in false whorls or scattered, giving off secondary branches, which terminate in very flattened branch-systems, disposed in horizontal planes. These resemble in their general arrangement those of Thuya and Chamæcyparis, and are mostly tripinnate, all the axes being covered with small coriaceous leaves, adnate in part of their length, and arranged in four ranks in decussate pairs. The leaves on the main and ultimate axes differ only in size.

The ventral and dorsal leaves are flattened and ovate or spathulate, with rounded apices; the lateral leaves are carinate, more or less spreading, with a slightly acute apex, which is bent inwards. The dorsal flat leaves are shining green, and marked with a central ridge, which is often hollowed in the middle line. The ventral flat leaves have a central green ridge, with a concavity white with stomata on each side. The lateral leaves, green on the dorsal side, exhibit a single stomatic concavity on their ventral side.

Flowers monœcious, solitary, and terminal, the male and female flowers borne on separate lateral branchlets as in Thuya. Male flowers cylindric, $\frac{1}{4}$ inch long, with six decussate pairs of stamens. Female flowers with five ovules on each scale. Cones globular, almost erect, with eight clavate, woody scales, in decussate pairs from a central axis, the upper pair abortive. Seeds three to five on a scale, laterally winged, the wing not notched at the summit.

The seedling[1] resembles that of *Thuya plicata*, but has broader and very blunt cotyledons, with shorter and broader primary leaves.

---

[1] See Tubeuf, *Samen, Früchte, u. Keimlinge*, 103, fig. 143 (1891).

THUJOPSIS DOLABRATA

Thujopsis dolabrata, Siebold et Zuccarini, *Fl. Jap.* ii. 34, tt. 119, 120 (1842); Franchet et Savatier,
 Enum. Pl. Jap. i. 469 (1875); Shirasawa, *Icon. Essences Forest. Jap.*, text 27, t. xi. 18-34
 (1900).

Thuya dolabrata, Linnæus, *Suppl. Pl. System*, 420 (1781); Masters, *Jour. Linn. Soc. (Bot.)*, xviii.
 486 (1881), and *Gard. Chron.* xviii. 556, fig. 95 (1882); Kent, in Veitch's *Man. Conif.* 236
 (1900).

The species has been described in detail above.

Two well-marked geographical forms occur, both confined to the main island
of Japan :—

1. Var. *australis* (*var. nova*). A small tree 40 to 50 feet in height, or a
shrub growing as underwood in the dense shade of forests. As a tree it has a
slender trunk, with drooping branches and a narrow pyramidal top. Branchlets
very flat and only slightly overlapping, the lateral leaves ending in acute points
bent inwards. Cones broadly ovoid, with scales thickened at the apex, which is
prolonged externally into a blunt triangular process. This is the form which is
known in cultivation in Europe, and described and figured in the works cited
above.

2. Var. *Hondai*, Makino.[1] A larger tree, attaining 100 feet in height, with a
stem of over 3 feet in diameter. The branch-systems are more densely ramified,
the branchlets being placed close together and overlapping one another by their
edges more than is the case in the preceding variety. The leaves also are smaller,
whiter underneath, and crowded more closely on the shoots; those of the lateral
ranks being usually blunt and not curved inwards at the apex. The cones are
globular, with scales not thickened at the apex, which is devoid of the process so
conspicuous in the other form, or merely shows it as an obsolete transverse
minute mucro. The seeds appear to be more broadly winged, the wings being
more scarious in texture.

This form has not yet been introduced. Elwes has brought home excellent
specimens of it in fruit from the Uchimappe Forest, near Aomori, in the extreme
north of Hondo. These differ in the characters given above from specimens of
the ordinary form obtained by him in the forest of Atera, Kisogawa, and Yumoto
(4000 to 5000 feet altitude) in Central Hondo. The smaller leaves, set more
closely on densely ramified branchlets in this variety, may be due to the influence
of dense shade. The difference in the cone is paralleled by what occurs in the
fruit of the different geographical forms of *Cryptomera japonica*. I am inclined to
think that var. *Hondai* is not a distinct species; but as it is very different, from
the point of view of cultivators, it may conveniently bear the name *Thujopsis
Hondai*.

[1] *Tokyo Botanical Magazine*, 1901, xv. 104.

Several horticultural varieties have been introduced, viz. :—

3. Var. *lætevirens*, Masters, *Jour. Linn. Soc. (Bot.)*, xviii. 486.

  Thujopsis lætevirens, Lindley, *Gard. Chron.* 1862, p. 428.

  Thoujopsis dolabrata nana, Gordon, *Pinetum*, ed. 2, p. 399.

A dwarf shrub having no definite leader, with slender and much-ramified branchlets, and very small and bright green leaves. This variety often shows acicular leaves, spreading all round the shoot, and is apparently a fixed seedling form. It was introduced in 1861 from Japan by J. Gould Veitch.

4. Var. *variegata*. This only differs from the ordinary cultivated form in having the tips of many of the branchlets pale yellow or cream colour. It was introduced by Fortune in 1861.

DISTRIBUTION

Thujopsis dolabrata was discovered by Kaempfer,[1] who mentions it in his *Amœnitates Exoticæ*, p. 884, as " a kind of Finoki." His specimen is still preserved in the Natural History Museum at South Kensington, and was figured by Lambert[2] in his account of the species. Thunberg long afterwards (about 1776) sent specimens to Linnæus, who first gave a scientific description of the tree. Thunberg[3] cites the locality as follows :—" *Crescit in regionibus Oygawæ et Fakoniæ, inter Miaco et Iedo.*" (A. H.)

Thujopsis dolabrata in Japan is known under the name of *Hiba*, and is found in a wild state north of about lat. 35°, and in the southern part of this area is a mountain tree only, occurring in the forest of the Kisogawa district from about 3000 to 5000 feet. In the vicinity of Nikko it is common between about 4000 and 6000 feet according to Sargent, but I only saw it here near Lake Yumoto where it did not appear to attain such large dimensions as farther north. The variety found in the forests of Atera is distinct in its fruit from the northern form. The excellent figure on Plate xi. in Shirasawa's *Essences Forestières* appears to be taken from the southern variety.

The northern form has been described by Makino as var. *Hondai*, but the latter is not mentioned either by Goto or Shirasawa, nor is it recognised as specifically distinct in any of the Japanese collections which I saw. Though the tree usually occurs in mixture with Tsuga at Nikko, and with Sciadopitys at Atera, yet in the extreme north of Japan, on the hills north of Aomori, it is found in pure forest on hills of volcanic formation from near sea-level up to about 3000 feet. An excellent account of the forest of Uchimappe is given in *Forestry and Forest-Products of Japan*, where it is stated that the mountains are of Tertiary formation, and the under-lying rock composed of tufa, sandstone, and slate. Pieces of this rock which I brought home have been examined by Mr. Prior of the British Museum of Natural History, who considers that in all probability they represent a rather basic andesite or basalt, but owing to the weathered and decomposed state of the specimens, satisfactory sections could not be made. I visited this forest in the

---

[1] See Salisbury, *Jour. Science and Arts*, 1817, ii. 313. [2] *Genus Pinus*, ed. 2, ii. tab. 68 (1842).
[3] *Flora Japonica*, 266 (1784), sub *Thuya dolabrata*, Linn

company of Mr. Shirasawa in June, and after passing through the flat rice-fields which extend from the sea to the foot of the hills, entered the forest, which consists mainly of Thujopsis naturally reproduced, though here and there, trees of *Quercus glandulifera*, *Magnolia hypoleuca*, and other species occur, whilst Cryptomeria and *Cupressus obtusa* are planted in the valleys, and *Larix leptolepis* on those parts of the hills where the natural forest has been destroyed by fire.   From observations taken at the meteorological observatory of Aomori, it appears that the climate of this part of Japan is cold in winter and the snowfall heavy, the thermometer falling in February to – 15° Centigrade, and rising in September to 32·5° Centigrade; the average temperature for the whole year being 9°, and the average moisture 78 per cent.   The average height of the trees here is about 70 to 80 feet, attaining in deep shady valleys 100 feet or perhaps more, and about 2 feet in diameter when closely grown, at the age of 150 to 180 years when it is considered ripe for felling.

The stems are often much curved at the butt from the pressure of the snow on the young seedlings, which require eight to ten years to get above its surface in winter, and these butts are usually cut separately and used for special purposes. The tree does not seem to have the power of reproducing itself from the stool, but produces abundant seed, which in dense shade germinates freely, though the growth of the seedlings is very slow at first.

The undergrowth of the forest is very different from what I saw in other parts of Japan, bamboo-grass (*Arundinaria Veitchii*) being much less prevalent, but in the damp places tall herbaceous plants were numerous, with Aucuba, Skimmia, and Ilex, and other evergreen shrubs on the drier ground, and many pretty liliaceous plants and orchids in places.

Goto says of this tree,[1] that it formed under the old regime, together with *Cupressus pisifera*, *C. obtusa*, *Thuya japonica*, and *Sciadopitys*, the so-called " Goboku " or Five Trees, which enjoyed careful protection at the hands of the feudal authorities; he also says that it is rarely planted, being regenerated naturally by seed, and that it forms extensive forests in a mixture with other conifers such as *Thuya japonica* and *Pinus parviflora*, in the mountains on the northern frontier of the province of Rikuchu, in Goyosan, and in the mountains of the Tone districts, Kozuké.   It has lately come to be in great demand for railway sleepers.

Plate 60 (in Vol. I.) represents a dense growth of trees of this species in the forest of Uchimappe very similar to what I saw in the Kisogawa district at about 3000 feet.   I am indebted to the Japanese Forest Department for the negative from which it was made.

The wood of Thujopsis is highly valued in those parts of Japan where it grows, on account of its great durability.   This is proved by specimens shown at the St. Louis Exhibition, one of which had been used as a gate-post for eighty-three years, another as a plank in a fishing-boat for eighty-four years, others as railway sleepers in use for fourteen years.   The wood has an aromatic smell, takes a fine lustrous polish when planed, and is yellowish white in colour, showing a fine grain, which makes selected planks from the butt length very ornamental.   Exceptional

[1] *Forestry of Japan*, 18 (1900).

cases occur in which the wood is curiously mottled and freckled. A ceiling and a screen made of such wood, which I saw in the Forestry Bureau at Aomori, were very beautiful.

The wood weighs about 30 lbs. per cubic foot, and is worth at Aomori from 40 to 50 yen per 100 cubic feet, or about 1s. per cubic foot. It is much valued not only for joinery and building purposes, but for foundations, ship and boat building, as it is stronger and more resinous than other woods of the same character.

The bark also, which is thin, tough, and durable, is much used for roofing and for partitions and walls of out-houses, fences, etc.

CULTIVATION

T. Lobb sent a plant from the Botanical Garden at Buitenzorg in Java, to Exeter in 1853, which died; and soon after, Capt. Fortescue, a cousin of Earl Fortescue, brought a plant from Japan which was planted at Castlehill in 1859. But this tree, as I learn from Mr. Pearson, the head gardener, has been dead for some time, though plants raised from its cuttings are still growing at Castlehill and elsewhere.

In 1861 Mr. J. G. Veitch and R. Fortune sent seeds from Japan to the Chelsea and Ascot Nurseries, from which plants were raised and generally distributed, so that the tree is now common in England.

From what I have said of its habitat in Japan it is clear that though this tree is hardy as regards frost in winter, it requires conditions which are rarely found in England to bring it to any size, and, as a matter of fact, it has not yet become a tree anywhere except in Devonshire and Cornwall, though perhaps if seeds from North Japan are obtained the results might be better.

Though no doubt it has ripened seeds elsewhere, I have never obtained any which germinated, except from a tree planted about 1881 by Queen Alexandra in the Earl of Northbrook's grounds at Stratton Park, Hants, which I gathered in October 1900. One of these grew, and is now a healthy plant about 9 inches high. It seems to suffer less from spring frost than many Japanese and Himalayan conifers.

The finest tree that I have seen in England is at Killerton, which in 1902 measured 35 feet 6 inches in height and 2 feet 4 inches in girth. It is growing on a slope facing south-west in a peculiar soil, which Sir C. T. D. Acland describes as "Trap, soft below the surface, but hard after exposure. This trap overlies red sandstone, but is rather darker and more porous." This soil evidently suits most conifers admirably, as I have seen no other collection which contains so many fine specimens as this.

At Boconnoc, at Carclew, and at other places in Cornwall there are trees approaching this in height, but we have not seen any specimen above 15 to 20 feet in other parts of England, though as a bushy shrub 12 feet high it exists in most modern gardens. In Scotland it seems hardy in the west and in Perthshire, whilst at Castlewellan in Ireland it has attained 30 feet in height. At Powerscourt and Kilmacurragh, Wicklow, there are trees with the lower branches layering and forming numerous independent stems. (H. J. E.)

ÆSCULUS

Æsculus, Linnæus, *Gen. Pl.* 109 (1737); Bentham et Hooker, *Gen. Pl.* i. 398 (1862).
Pavia, Boerhave, *ex* Miller, *Gard. Dict.* ed. 6 (1752).

DECIDUOUS trees and shrubs, belonging to the natural order Sapindaceæ, some authorities, however, making the genus the type of a distinct order Hippocastaneæ. Leaves in opposite decussate pairs, without stipules, stalked, digitately compound; leaflets five to nine, serrate in margin, pinnately veined. Branchlets stout, terete, with large triangular leaf-scars. Buds large, of numerous decussately opposite scales which are homologous with leaf-bases, the outer deciduous, dry or resinous, the inner accrescent and often brightly coloured.

Flowers in large terminal racemes or panicles, appearing later than the leaves, of two kinds, hermaphrodite and staminate, on the same plant; placed in the axils of minute caducous bracts on stout jointed pedicels. Calyx imbricate in bud, five- or two-lobed, the lobes unequal, united with an hypogynous annular disc in the hermaphrodite flowers. Petals four to five, imbricate in bud, alternate with the calyx lobes and inserted on the disc. Stamens five to eight, usually seven, inserted on the inner margin of the disc, unequal in length; filaments filiform; anthers two-celled, sometimes glandular at the apex. Ovary three-celled, rudimentary in the staminate flowers, each cell containing two ovules. Style slender, elongated, generally curved. Fruit a capsule; prickly, roughened, or smooth; coriaceous; three-celled, three-seeded, and three-valved, or by abortion one- to two-celled and one- to two-seeded, the remains of the abortive cells and seeds usually remaining visible. Seeds without albumen, rounded or flattened by mutual pressure; seed-coat brown and coriaceous, marked by a large whitish hilum. Cotyledons thick and fleshy, unequal, cohering together by their contiguous faces, remaining in the seed-coat during germination.

About twelve species of Æsculus[1] are known to occur in the wild state. They are natives of North America, Europe, and Asia. The genus was formerly divided into two sections, *Pavia*, with smooth fruit, and *Hippocastanum*, with spiny fruit; but this division is not a natural one. The following synopsis groups the species under sections, which are more natural, being dependent on the characters of the flowers and buds :—

I. HIPPOCASTANUM. Buds viscid. Calyx irregularly campanulate, four- to five-

---

[1] The two Mexican species, which have tri-foliolate leaves, are now separated as a distinct genus, *Billia*.

206

lobed. Petals four or five, claws not longer than the calyx; stamens exserted. This section includes all the old-world species.

1. *Æsculus Hippocastanum*, Linnæus. Greece.

2. *Æsculus indica*, Colebrooke. Afghanistan, north-western Himalaya.

3. *Æsculus punduana*, Wallich, *List* 1189 (1828). Sikkim, western Duars, Khasia Hills, Upper Burma, Tenasserim, Siam, Tonking. Large tree. Leaflets six to seven, very large, thinly coriaceous, stalked, acuminate, serrate. Panicles 12 to 15 inches or more, flowers white or yellow. Fruit brown, smooth.

Not introduced and not likely to be hardy.

4. *Æsculus chinensis*, Bunge, *Enum. Pl. Chin. Bor.* 10 (1835). Northern and Central China. A tree, 40 to 50 feet high. Leaflets five to seven, large, stalked, obovate-oblong, rounded at the base, abruptly acuminate at the apex, finely serrate, shining above, glabrescent below except for pubescence along the nerves, petioles pubescent. Panicles, 8 inches long, pubescent. Flowers small, white; sepals shortly and unequally five-lobed, pubescent. Petals four, minute. Filaments glabrous. Fruit[1] pear-shaped or globular, small (¾ inch diameter), one-celled, three-valved, brown, covered with warts, not spiny.

This species has been much confused with the next, from which it differs in every way. The flowers, though small, are numerous in the large panicle, and the foliage is very handsome. It is common enough in the mountains of central China, in Shansi, and in the hills to the west of Peking; and when introduced is likely to prove hardy in England.

5. *Æsculus turbinata*, Blume. Japan.

II. PAVIA. Buds not resinous. Calyx tubular, five-toothed. Petals four, yellow or scarlet.

6. *Æsculus glabra*, Willdenow. North America.

7. *Æsculus octandra*, Marshall. North America.

8. *Æsculus Pavia*, Linnæus, *Sp. Pl.* 344 (1753); *Bot. Reg.* t. 993 (1826). Middle United States. A shrub. Leaves with slender grooved petioles, the edges of the grooves jagged. Leaflets five, obovate, acute at the base, acuminate at the apex, finely serrate without cilia, slightly pubescent beneath. Flowers in loose panicles, 4 to 7 inches long. Petals red, meeting at the tips; upper pair longer, with claws about three times as long as the small spathulate limb; lateral pair shorter, with claws as long as the calyx, and rounded limb equalling the claw in length; margin of petals beset with minute dark glands. Stamens as long as the upper pair of petals. Fruit brown, without spines.

This species, though only a shrub, is mentioned here at some length, as it closely resembles *Æsculus octandra*, and moreover enters into such important hybrids as *Æsculus carnea, versicolor*, etc. All its hybrids may be recognised by the red colour of the flowers and the glandular margin of the petals. It is readily distinguished from *Æsculus octandra* by its smaller leaves and peculiar petioles. In winter it shows the following characters:—Twigs slender, glabrous, shining, with numerous lenticels.

[1] Cf. Hance in *Journ. Bot.* viii. 312 (1870).

Leaf-scars obovate or crescentic on slightly prominent cushions, with three groups of bundle-dots; opposite scars joined by a linear ridge. Terminal buds long oval or fusiform, pointed; scales numerous, the upper rounded, the lower pointed at the apex and keeled on the back, minutely ciliate in margin. Pith wide, circular, green.

9. *Æsculus austrina*, Small, *Bull. Torrey Bot. Club*, 1901, xxviii. 359; Sargent, *Man. Trees N. America*, 647 (1905); *Æsculus Pavia*, β *discolor*, Torrey and Gray, *Fl. N. Amer.* i. 252 (1838), in part. A small tree, attaining 30 feet in height, occurring in Tennessee, S. Missouri, E. Texas, and north-western Alabama. This resembles the last species. The leaflets, however, are usually more irregularly but finely serrate, and pale tomentose beneath. Panicles pubescent, 6 to 8 inches long. Petals bright red, meeting at the tips, unequal, oblong-obovate, rounded at the apex, glandular, those of the upper pair about half as wide as those of the lateral pair, with claws much longer than the calyx. Stamens longer than the petals. Fruit brown, slightly pitted. Not introduced.

III. Macrothyrsus. Buds not viscid. Calyx five-toothed. Petals four to five, white, claws longer than the calyx. Stamens exserted, very long.

10. *Æsculus parviflora*, Walter, *Flora Caroliniana*, 128 (1788). South-eastern North America. A shrub. Leaflets five to seven, elliptical or oblong-ovate, densely grey-tomentose beneath, finely serrate. Panicles erect, 8 to 10 inches long, slender, narrow. Flowers white, faintly tinged with pink. The long and thread-like stamens are pinkish white and very conspicuous.

This is a valuable shrub, as it flowers late, in July or August, some five or six weeks later than any of the other species except *californica*. Occasionally it forms a short single trunk, but generally it sends up a crowd of stems from the ground. It is figured in *Gard. Chron.* 1877, viii. fig. 129; and is often known in gardens as *Pavia macrostachya*, Loiseleur, or *Æsculus macrostachya*, Michaux. See *Bot. Mag.* t. 2118 (1820), where it is stated that the species was introduced by Mr. John Fraser in 1785. Canon Ellacombe reported in 1877[1] that he had at Bitton a specimen, which was at least forty years old, but that it remained a bush, not exceeding 8 or 10 feet in height.

IV. Calothyrsus. Buds viscid. Calyx two-lipped or five-lobed. Petals four, pink or white, claws not longer than the calyx. Stamens exserted.

11. *Æsculus californica*, Nuttall. California.

12. *Æsculus Parryi*, A. Gray, *Proc. Amer. Acad.* xvii. 200 (1881); Sargent, *Garden and Forest*, 1890, p. 356, fig. 47. Lower California. A small shrub, resembling the preceding species; but differing in the five-lobed calyx, and in the leaflets, which are small, obovate and hoary pubescent beneath. It has not been introduced.

V. Hybrids. The most important is *Æsculus carnea*, Hayne, which is a cross between the common horse-chestnut and *A. Pavia*. This is described fully below.

[1] *Gard. Chron.* 1877, viii. 691.

Æsculus plantierensis, André, a supposed hybrid between *Æsculus carnea* and *Æsculus Hippocastanum*, will be mentioned under the former species. *Æsculus versicolor*, Dippel, a hybrid between *Æsculus Pavia* and *Æsculus octandra*, will be treated under the latter species.

The following key to the species in cultivation is based on the characters of the leaves and buds. In Plate 61 the leaves of all these species are shown; and in Plate 62 are represented the twigs and buds of six species, *viz., Hippocastanum, carnea, indica, glabra, octandra,* and *californica* :—

A. *Leaflets sessile or nearly so; buds very viscid.*
  1. *Æsculus Hippocastanum.*
      Petioles glabrescent. Leaflets obtusely and irregularly serrate.
  2. *Æsculus turbinata.*
      Petioles pubescent, especially towards their tips. Leaflets regularly and crenately serrate.
B. *Leaflets stalked.*
      **Buds viscid.*
  3. *Æsculus indica.*
      Leaflets finely and sharply serrate, pale beneath. Buds very viscid.
  4. *Æsculus carnea.*
      Leaflets obtusely and irregularly serrate. Buds only slightly viscid, the brown scales having a dark-coloured margin.
  5. *Æsculus californica.*
      Leaflets shallowly and crenately serrate, pale beneath. Buds viscid, glistening with white resin.
      ** *Buds not viscid.*
  6. *Æsculus parviflora.*
      Leaflets densely grey-tomentose beneath, finely serrate in margin. Buds minutely pubescent.
  7. *Æsculus octandra.*
      Leaflets pubescent beneath, broadly lanceolate, shortly acuminate, with twenty or more pairs of nerves in the terminal leaflet; margin finely serrate but not usually ciliate. Petioles without jagged marginal ridges.
  8. *Æsculus glabra.*
      Leaflets glabrous beneath, except for a slight pubescence along the midrib and tufts in the axils, long-acuminate, with about fifteen pairs of nerves in the terminal leaflet, finely serrate with ciliate tufts in the bases of the serrations. Petioles with smooth marginal ridges.
  9. *Æsculus Pavia.*
      Leaflets slightly pubescent beneath, narrowly lanceolate, finely serrate but not ciliate in margin. Petioles flattened on the upper side, with marginal sharp ridges, usually jagged.

ÆSCULUS HIPPOCASTANUM, Common Horse-Chestnut

Æsculus Hippocastanum, Linnæus, *Sp. Pl.* 344 (1753); Loudon, *Arb. et Frut. Brit.* i. 462, iv. 2543 (1838); *Gard. Chron.* 1881, xvi. 556, figs. 103, 104.

A large tree, attaining in England a height of over 100 feet and a girth of 15 or even 20 feet. Bark smooth and dark brown in young trees, becoming greyish and fissured longitudinally in old trees, at the same time scaling off in thin plates. Leaves palmately compound, digitate, on a long stalk widened at its insertion. Leaflets five to seven, sessile, obovate, cuneate at the base, abruptly acuminate at the apex, unequally and coarsely serrate; green above; beneath pale, tomentose at first, but ultimately glabrous, except for small tufts of hairs in the axils of the veins and a few scattered hairs over the surface; middle leaflet the largest, with twenty-four or more pairs of nerves, lower pair smallest; venation pinnate; petiole glabrous. The leaflets as they emerge from the bud are at first erect, but soon bend downwards on their stalks. When nearly full grown they rise up and become horizontal. In autumn they turn yellow or brownish and fall early, each leaflet disarticulating separately from the petiole.

Flowers in large upright pyramidal panicles, the primary branches of which are racemose, the lateral branches cymose. Upper flowers staminate and opening first; lower flowers hermaphrodite. Calyx greenish, five-toothed. Petals four to five, crumpled at the edge, white, with yellow spots at the base, which ultimately become pink. Stamens seven, longer than the petals, the filaments bent down when the flower opens and the stigma protrudes, later moving up on a level with the style. Fruits few on each panicle, large, globular, green, with stout, thick conical spines, three-valved, usually one-seeded, occasionally two- to three-seeded. Seed large, shining-brown, with a broad whitish hilum. Cotyledons two, large, fleshy, distinct below, blended into one mass above.

Seedling [1]

The cotyledons are large and fleshy and remain in the seed, which frequently germinates on the surface of the soil or only slightly buried beneath it. The cotyledons have long petioles ($\frac{3}{4}$-1 inch), which are broad and flattened, with a concavity on their inner surface. The caulicle, very variable in length (1 to 4 inches), is stout, brownish, pubescent, and ends in a stout tap-root, which gives off numerous branching fibres. The young stem is stout, terete, brownish, striated and marked with numerous lenticels, puberulent or glabrous; it has no scale-leaves, differing in this respect from the young stem of the oak. In other respects the germination of the oak and of the horse-chestnut are almost identical. At a varying height

---

[1] Cf. Lubbock, *Seedlings*, i. 356 (1892), where it is stated that the seed is carried a considerable height above ground during germination owing to the great length of the caulicle. So far as I have observed, the seed does not change its position during germination.

above the cotyledons the first pair of true leaves are produced, which are opposite, compound, digitately five-foliolate, and closely resemble the adult foliage except that they are smaller in size. Successive pairs of similar leaves follow on the stem, each pair being placed decussately with reference to the pair immediately below it.

Abnormal Flowering

The horse-chestnut sometimes produces a second crop of flowers in autumn, which appear in much smaller panicles than those of spring. This is due to the premature fall of the leaves in July or August, usually following an excessively dry season. The buds are stimulated to premature energy and put forth young leaf-shoots, which are terminated by flowers. This phenomenon, which is equivalent to an anticipation of the opening of the buds by several months, as they would normally open in the following spring, is frequently observed in the trees planted in the boulevards of Paris.[1] In the dry season of 1884, a single tree at Kew produced small panicles of flowers in September, after previously shedding nearly all its leaves. In the following year it produced a few panicles of the ordinary size. At Hythe,[2] near Southampton, a horse-chestnut is reported to have bloomed and fruited three times in 1868, once in spring, again after the rain which succeeded the long drought, and a third time in September.

Identification

In summer the common horse-chestnut is unmistakable. The only other species with large sessile leaflets, *Æsculus turbinata*, is easily distinguished by their regular crenate serration. In winter the twigs and buds show the following characters :—Twigs stout, brown, glabrous or minutely pubescent towards the tip ; lenticels numerous. The large opposite leaf scars, flat on the twigs with no prominent cushion, are joined by a linear ridge, and vary in shape, the larger being obovate with seven bundle-dots, the smaller semicircular or crescentic with usually only five dots. Buds very viscid, larger than in the other cultivated species ; the terminal much exceeding the lateral buds in size, occasionally absent, and replaced by the saddle-shaped scar of the previous year's inflorescence ; scales imbricate, the external ones in four vertical ranks, rounded at the apex, glabrous, not ciliate, dark red-brown. The buds contain the next year's shoot in an advanced state of development, flowers being visible in them in October. The scales are morphologically equivalent to leaf-bases. In the interior of the bud, scales are observable with traces of leaf-blades, which gradually pass into the true leaves, visible in the upper part of the bud.

Varieties

1. Var. *flore pleno*, Lemaire, *Illust. Horticole*, 1855, ii. t. 50. A variety with double flowers, the pistil even in some cases becoming petaloid. Mr. A. M.

[1] See article by Roze, translated in *Gard. Chron.* 1898, xxiii. 228.   [2] *Gard. Chron.* 1868, p. 1116.

Baumann discovered in 1822, near Geneva, a horse-chestnut tree, of which a single branch bore double flowers ; and from this branch the variety was propagated at the Bollweiler nursery in Alsace.[1]   The flowers last longer than those of the single kind,[2] and no fruits are formed, which renders it useful as a tree in streets, where the fall of fruits is an inconvenience.   This variety is very hardy, and resisted well the severe winter of 1879-80 in France.[3]

2. Var. *laciniata* (var. *asplenifolia*, var. *incisa*).   Leaflets cut up into narrow lobes.   According to Beissner[4] this variety has been in cultivation for over forty years ; and a form of it was found by Herr Henkel of Darmstadt, which keeps its foliage much longer than the typical form ; but this is not the case in some localities.

3. Var. *crispa*.   Leaves short-stalked, with broad leaflets.   Tree compact in habit.

4. Var. *pyramidalis*.   Upright in habit.

5. Var. *umbraculifera*.   Crown densely branched, and globular in outline.

6. Var. *tortuosa*.   Branches bent and twisted.

7. Var. *Memmingeri*.   Leaves yellowish in colour, looking as if powdered with sulphur.

8. Var. *aureo-variegata*.   Leaves variegated with yellow.

Several other varieties of slight interest, which do not seem to be in cultivation in this country, are mentioned by Schelle.[5]

DISTRIBUTION AND HISTORY

The horse-chestnut occurs wild in the mountains of northern Greece.   Halácsy,[6] the latest authority, gives many localities in Phthiotis, Eurytania, Thessaly, and Epirus ; but states that it is not found wild on Mount Pelion or in Crete.   Baldacci,[7] in 1897, found the tree growing wild on almost inaccessible precipices below the lower limit of the coniferous belt near Syrakou in the district of Janina in Albania.

The native country of the tree was long a matter of doubt ; but the whole question was satisfactorily elucidated by Heldreich[8] in a paper, from which we extract most of the following account.   Linnæus considered the habitat of the tree to be northern Asia, and De Candolle thought that it came from northern India.   The tree is, however, not known wild in India, where it is replaced by *Æsculus indica*.   Boissier[9] states that it is recorded from Greece by Sibthorp, from Imeritia (Caucasus) by Eichwald, and from Persia by various authors.   It is, however, unknown in the wild state in Persia ; and Radde[10] mentions it only as a planted tree

---

[1] *Rev. Belgique Horticole*, 1854, iv. 216.

[2] See *Garden*, 1890, xxxviii. 601, where some observations are recorded on the periods of flowering of the single and double horse-chestnuts, and of *Æsculus carnea*.

[3] *Rev. Horticole*, 1884, p. 98.   [4] *Mitt. Deut. Dendrol. Gesell.* 1905, pp. 13, 14, and 1906, p. 10.

[5] *Handbuch Laubholz-Benennung*, 321 (1903).   [6] *Consp. Fl. Græcæ*, i. 291 (1900).

[7] *Rivista Collez. Botan. in Albania*, 23 (Florence, 1897).

[8] *Verhand. Bot. Vereins Prov. Brandenburg*, 1879, p. 139.   The British Minister at Athens, Sir F. E. H. Elliot, K.C.M.G., who kindly made inquiries, has sent us a letter from Professor Miliarakis of the University of Athens, dated April 2, 1904, which confirms Heldreich's statements.

[9] *Flora Orientalis*, i. 947 (1867).   [10] *Pflanzenverbreitung in Kaukasusländern*, 433, 434 (1899).

in the Caucasus. All the evidence goes to show that it is confined to northern Greece and Albania.

Heldreich states that the horse-chestnut was first found wild in Greece by Dr. Hawkins.[1] In his own travels in Greece in 1897 he observed it in many stations, all lying in the lower fir region, between 3000 and 4000 feet altitude, where it grows in shaded moist gulleys, in company with alder, walnut, plane, ash, several oaks, *Ostrya carpinifolia* and *Abies Apollinis*. These stations, situated in remote uninhabited spots, establish the fact that the tree is really wild. Plants introduced into Greece by the Turks are always found in the neighbourhood of towns. Whether the tree was known to the ancient Greeks is doubtful.

The horse-chestnut was first mentioned[2] by the Flemish doctor Quakleben, who was attached to the embassy of Archduke Ferdinand I. at Constantinople,—in a letter to Matthiolus in 1557. The latter received a fruit-bearing branch, and published the first description[3] of the tree as *Castanea equina*, because the fruits were known to the Turks as *At-Kastane* (horse-chestnut), being useful as a drug for horses suffering from broken wind or a cough.

The tree was introduced into western Europe from Constantinople, the first tree being raised by Clusius at Vienna from seeds sent by the Imperial Ambassador, D. Von Ungnad, in 1576. This tree quickly grew, and was mentioned by Clusius[4] in 1601.

The horse - chestnut was introduced into France[5] in 1615 by Bachelier, who brought the seeds from Constantinople. Gerard mentions it in his *Herbal* of 1579, p. 1254, as a tree growing "in Italy and sundry places of the eastern countries"; and in Johnson's edition of this work, published in 1633, the tree was stated to be growing in Tradescant's garden at South Lambeth. It was probably introduced into England about the same time as into France.   (A. H.)

CULTIVATION

No tree is easier to raise from seed than the horse-chestnut. Its large fleshy fruit are so little hurt by frost and damp that they germinate freely where they fall, and do not seem to be eaten by mice like acorns and beech-mast.

Seeds which have been exposed all winter germinate more readily in spring than those which have been kept dry, and should be sown early and covered with about two inches of soil.

Though it is advised by French writers that the extremity of the radicle should be pinched off before sowing in order to prevent a strong tap-root from forming, as is done in the case of walnuts and chestnuts, I have not observed that they suffer from removal if this is not done; and if transplanted at one or at latest two years after sowing there are abundance of fibrous roots which make the tree an easy one

---

[1] Sibthorp et Smith, *Fl. Græcæ Prodromus*, i. 252 (1806). Hawkins' observation has been disputed, as he records it from Pelion, where the tree does not, so far as we know now, occur wild. Orphanides was the first to establish beyond doubt that the tree is indigenous to the mountains of northern Greece. Cf. Grisebach, *Vegetation der Erde*, French ed. i. 521.

[2] Matthiolus, *Epistol. Medicin. Libri Quinque* (Prague, 1561).

[3] Matthiolus, *Comment. in Dioscorid. Mat. Med.* 211 (Venice, 1565).

[4] Clusius, *Rar. Plant. Hist.* 7 (1601).   [5] Tournefort, *Relation d'un Voyage au Levant*, i. 530 (1717).

to move, even when five or six feet high.   As the tree is liable to form large side branches, the buds should be rubbed off the stem early in order to form a clean trunk, though it bears pruning well as a young tree.

Though somewhat liable to suffer from cold winds and spring frost, which injure the foliage and flowers, the tree is hardier in this respect than many of our native trees, though coming from a warm southern country.

As regards the chemical nature of the soil it is quite indifferent, for though it grows faster on a good loam and does not come to perfection on sandy soil, it attains a large size on dry, rocky, calcareous soils, and even at an elevation of 800 feet and upwards resists wind better than many trees.   I have seldom seen horse-chestnuts blown down, though large heavy branches are often torn off by violent winds.

As an ornamental flowering tree for parks, lawns, and avenues it has no superior, though on account of its branching habit it requires considerable attention in order to form tall shapely trees.   Its principal defect is the tendency of the leaves to become brown and ragged early in the autumn, but they fall quickly, and being easily removed make less litter than the leaves of the beech, oak, or sycamore.

The large branches when allowed to rest on the ground in damp situations frequently take root and become naturally layered, the best instance of this that I have seen being at Mottisfont Abbey, Hants.

For town planting, on account of its beautiful flowers and dense shade during the hottest months, the horse-chestnut is perhaps, next to the plane, one of the best trees we have, and does not seem to suffer much from smoke.   In parks it is valuable for its fruit, which are so much liked by deer that they are eaten as fast as they fall, and would perhaps be worth collecting for winter food.

The extraordinary hardiness of this southern tree is proved by the fact that it will grow to a large size as far north as Trondhjem in Norway, lat. 63° 26′, a tree figured by Schubeler near this place being 37 feet by 8 feet 9 inches.   Another in the Botanic Garden at Christiania, which is considered the largest in Norway, measured in 1861, 16.62 metres by 2.45 metres, and when I saw it in 1903 had increased to no less than 28 metres high by 3 in girth, though it has been exposed to as low a temperature as − 18° to − 20° Réaumur.

As regards the age which the horse-chestnut attains we have few exact records, but it does not seem a very long-lived tree.   J. Smith states[1] that an avenue running south-east from the front of Broadlands House, near Romsey, Hants, was planted in 1735; but in 1887 only two trees remained, which were 11 feet and 12 feet 4 inches in girth.

REMARKABLE TREES

There are so many fine trees in almost every part of Great Britain that I need not go into great detail as to their dimensions, but though it is possible that in Bushy Park, or other places near London, taller trees exist, I have only at

[1] *Trans. Scot. Arb. Soc.* xi. 540 (1887).

Petworth measured one which exceeds in height the group of three which grow near my own house at Colesborne, of which I give an illustration (Plate 63). The height of these as measured in 1902 by Sir Hugh Beevor and myself was 105 feet, and the girth of the largest 11 feet. They grow in a sheltered situation, on damp, cold soil. One of these trees being inclined to split at the base, owing to the great weight and length of one of its principal limbs, was chained up many years ago, and though the iron band which was put round it has become buried in the wood the limb has not broken off.

At Dynevor Castle, Carmarthenshire, the seat of Lord Dynevor, where the park contains a greater number of fine trees than any I have seen in South Wales, there is a very large tree which the Hon. W. Rice measured in 1906 and found to be 109 feet by 17 feet 9 inches. For height and girth combined this seems to be the largest tree in Great Britain.

The tallest tree I have seen is in a grove of beech, chestnut, oak, and silver fir, which grows near the house at Petworth Park, the seat of Lord Leconfield in Sussex, on a deep greensand formation. This tree, though forked at six feet from the ground, has been drawn up to a great height by the trees surrounding it, and though difficult to measure exactly, probably exceeds 115, and may be 120 feet. The two stems are 9 feet 8 inches and 8 feet respectively in girth.

In Bushy Park most of the horse-chestnuts are past their prime; many of the old trees are dead and have been replaced by young ones. The largest, seen in 1906, was growing near the gate; it had a bole of 20 feet giving off four great stems, and measured 100 feet high by 16 feet 5 inches in girth. Another near the pond was 101 feet by 16 feet 1 inch.

At Birchanger Place, near Bishop Stortford, the seat of T. Harrison, Esq., there is one of the largest and finest trees in England, which measures about 80 feet by 20 feet, with a bole about 15 feet high and a spread of 32 yards; a beautiful photograph was taken in 1864 when the tree was in flower, but it is now partially decayed on the north side, and has lost some large branches.

At West Dean Park, Sussex, the seat of W. D. James, Esq., there is a large tree about 70 feet by 16 feet, with branches spreading over an area no less than 36 yards in diameter.

At Hampton Court, Herefordshire, the seat of John Arkwright, Esq., there is a very fine tree growing on deep alluvial soil in the big meadow south of the house. Measured by T. Hogg in 1881[1] it was 93 feet by 16 feet 6 inches. When I saw it in 1905 it had increased about three feet in height and was 18 feet 7 inches in girth, and still handsome and vigorous.

The largest trees I have seen as regards girth and spread of branches are in Ashridge Park, on a bank near the lodge on the Berkhampstead road. The largest of these is about 80 feet high and 20 feet in girth, with extremely wide-spreading branches, and there are several others of 16 to 17 feet girth in the row. These trees are growing on a dry, flinty, calcareous loam.

[1] *Trans. Scot. Arb. Soc.* ix. p. 151 (1886).

There is a fine tree at Syon, which in 1905 was 93 feet high by 15 feet 4 inches in girth; and at Broom House, Fulham, there is a tree 95 feet high.

In the courtyard at Burleigh, near Stamford, the seat of the Marquess of Exeter, there is a large and very beautiful tree, figured by Strutt, plate 37, which was in 1822 60 feet high by 10 feet in girth, with a spread of 61 feet diameter. When I saw it in 1903 it was still in perfect health, and was about 80 feet by 12 feet 6 inches. It had remarkably spiny fruit, and its trunk was covered with small twigs.

At Trebartha Hall, near Launceston in Cornwall, Mr. Enys reports in 1904 a tree 15 feet 6 inches in girth, with an estimated height of 70 feet.

In Scotland the horse-chestnut seems as much at home as in England, and thrives in most places as far north as Gordon Castle, where there is a tree, measured in 1881 by Mr. Webster, 65 feet high by 13 feet 4 inches in girth, and 274 feet in circumference of its branches.

At Newton Don, Kelso, the seat of Mr. C. B. Balfour, there is a tree which was in 1906, 13½ feet in girth with a spread of branches of 165 feet in circumference.

In Perthshire there is a very beautiful tree, remarkable for its weeping habit, in the park at Dunkeld, which measures 80 feet in height by 17 feet 6 inches in girth (Plate 64). At Kilkerran, Ayrshire, Mr. J. Renwick has measured a fine tree 84 feet high by 14 feet in girth, with a bole 22 feet high. At Pollok, near Glasgow, a tree measured, in 1904, 63 feet high by 13 feet 6 inches girth at 2½ feet from the ground, with a bole of 5 feet, giving off four great stems.

None of these are equal to a tree in a group of seven standing at the west end of Moncreiffe House in Perthshire, which Hunter[1] describes as the largest in Scotland, and which then measured 19 feet in girth at five feet from the ground. At ten feet it divides into three great limbs, one of which has become firmly rooted in the ground, and extends so far from the trunk that the total spread of the tree is 90 feet in diameter.

The remarkable hardiness of this tree is shown by the existence of one, reported by Mr. Farquharson of Invercauld, as growing at an elevation of 1110 feet, which was supposed to be 177 years old in 1864, when it was 8 feet 7 inches in girth.[2]

In Ireland the horse-chestnut attains a great size, the largest we know of occurring at Woodstock in Co. Kilkenny, on an island in the River Nore. One tree measured in 1904, 93 feet in height by 18 feet 1 inch in girth, and according to the careful records which have been kept of the growth of the many fine trees on this property, measured in 1825, 10 feet 2 inches in girth; in 1846, 13 feet 2 inches; in 1901, 17 feet 9 inches. Another about the same height, in a meadow near the river, measured in 1825, 11 feet in girth; in 1834, 12 feet; in 1846, 12 feet 11 inches; in 1901, 14 feet 4 inches.

---

[1] *Woods and Forests of Perthshire*, 1883.   [2] *Old and Remarkable Trees of Scotland*, p. 115.

Timber

The wood of the horse-chestnut is one of the poorest and least valuable we have, on account of its softness and want of strength and durability. Though it has a fine close and even grain, white or yellowish-white colour, and is not liable to twist or warp so much as most woods, it does not cut cleanly, decays rapidly, and is only used as a rule for such purposes as cheap packing-cases and linings.

It burns so badly that it is of little use as firewood, and though occasionally cut into veneers or used as a cheap substitute for sycamore, poplar, and lime, in making dairy utensils, platters, and brush backs, it cannot be said to have a regular market. From 4d. to 8d. a foot is about the usual value in most parts of England, though Webster says that it was worth a shilling in Banffshire some years ago.

Holtzapffel says that it is one of the white woods of the Tunbridge turner, a useful wood for brush backs and turnery, preferable to holly for large varnished and painted works on account of its great size.

I am not aware whether it has been tried for pulp-making, but it would seem to be a suitable wood for that purpose on account of its softness, and could, if required, be produced in quantity at a low price. (H. J. E.)

ÆSCULUS CARNEA, Red Horse-Chestnut

Æsculus carnea, Hayne, *Dendrol. Flora*, 43 (1822).
Æsculus rubicunda, Loiseleur, *Herb. Amat.* vi. t. 357 (1822); Loudon, *Arb. et Frut. Brit.* i. 467
    (1838); Carrière, *Rev. Horticole*, 1878, p. 370, coloured figure of var. *Briotii*.
Æsculus Hippocastanum, L. × *Æsculus Pavia*, L., Koch, *Dendrologie*, i. 507 (1869).

A small tree, occasionally 50 feet, but rarely exceeding 30 feet in height. Leaves resembling those of the common horse-chestnut, but darker green with an uneven surface, the leaflets being shortly stalked and more or less curved and twisted. Flowers red, showing as they open an orange-coloured blotch at the base of the petals, which afterwards becomes deep red. Petals five, standing nearly erect, their limbs not spreading horizontally at right angles to the claws, as occurs in the common horse-chestnut; edges of the petals furnished with minute glands, like those present in *Æsculus Pavia*. Fruits with slender prickles.

Identification

In winter, the species is distinguished as follows:—Twigs rather stout, grey, shortly pubescent; leaf-scars as in *Æsculus Hippocastanum*. Buds slightly viscid and smaller than in that species; scales brown, edged with a dry membranous dark-coloured rim. Lateral buds small, oval, pointed, arising from the twig at an acute angle.

II   D

VARIETIES

1. Var. *Briotii*.   Flowers in larger panicles and more brilliantly coloured, the filaments, calyx, and style being red.   Fruits never developing fully, falling soon after the flowers.   This variety [1] was obtained in 1858, by M. Briot at the State Nurseries of the Trianon, Versailles, as a seedling of *Æsculus carnea*.

2. Several variegated forms are known, as var. *aureo-maculata* and *aureo-marginata*.   Var. *alba* is a form with white flowers.   Var. *pendula* is pendulous in habit.

3. *Æsculus plantierensis*, André, *Rev. Horticole*, 1894, p. 246, is supposed to be a cross between *A. carnea* and the common horse-chestnut, as it is intermediate in character.   This variety arose in the nursery of Messrs. Simon-Louis Frères at Plantières-lès-Metz, from a seed of *Æsculus Hippocastanum*.   Other intermediate forms, named by André *Æsculus intermedia* and *Æsculus balgiana*, were derived from seeds of *Æsculus carnea*.

HISTORY

Nothing is known for certain concerning the origin of *Æsculus carnea*.   Loiseleur received the plant from Germany in 1818, and there are no earlier accounts of it.   Its parentage, however, is undoubted : it possesses characters of both the supposed parents.   The leaves and slightly spiny fruit are derived from the common horse-chestnut.   The colour of the petals and the glands on their margins come from *Æsculus Pavia*.   According to André [2] the seeds when sown usually produce plants which bear whitish flowers and are of no horticultural value.   The species is accordingly always propagated by grafting.   Koch,[3] however, reports that while some seedlings are like those of the common horse-chestnut, others produce smooth fruits.   At Kew, according to Mr. Bean, it has come true from seed.

The largest specimen of this tree that we have seen occurs at Barton in Suffolk. It was 50 feet high in 1904, with a bole, however, of only 2 feet, girthing 7 feet 9 inches at a foot above the ground, and dividing into three stems.

It does not seem to live long or to attain any great size in England, and is often supposed to be a red-flowered form of the common horse-chestnut.   (A. H.)

---

[1] *Rev. Hort.* loc. cit.   [2] André, *Rev. Hort.* loc. cit.
[3] *Verhand. Ver. Beförd. Gart. König. Preuss. Staat*, 1855.

ÆSCULUS INDICA, Indian Horse-Chestnut

Esculus indica, Colebrooke, Wallich, *List* 1188 (1828); *Bot. Mag.* t. 5117 (1859); Hiern, in *Flora British India*, i. 675 (1875); Bean, in *Gard. Chron.* 1897, xxii. 155 and 1903, xxxiii. 139, *Suppl. Illust.*; Collett, *Flora Simla*, 97 (1902); Gamble, *Man. Indian Timbers*, 193 (1902); Brandis, *Indian Trees*, 185, 705 (1906).
Pavia indica, Wallich, *ex* Jacquemont, *Voyage dans l'Inde*, iv. 31, t. 35 (1844).

A large tree, attaining in India 150 feet in height and 40 feet in girth of stem. Bark in old trees peeling off in long strips. Leaves large, glabrous, dark green above, pale, almost glaucous beneath; leaflets five to nine, stalked, obovate-lanceolate, acuminate, finely and sharply serrate, with about twenty pairs of nerves in the terminal leaflet. Panicles 12 to 15 inches long, loose, narrow, erect. Flowers large, about 1 inch long; calyx ⅓ inch long, irregularly lobed, often splitting so as to appear two-lipped. Petals four, white, of two unequal pairs; the upper pair narrow and long with a red and yellow blotch at the base, the lower pair flushed with pink. Stamens seven or eight, scarcely longer than the petals, spreading. Fruit brown, rough, without spines, irregularly ovoid, one to two inches long, containing one to three dark brown shining seeds.

IDENTIFICATION

In summer the viscid buds and the large stalked leaflets with finely serrate margins distinguish it from the other species in cultivation. In winter the twigs show the following characters :— Branchlets coarse, shortly pubescent; lenticels like brown raised warts, numerous; pith circular, white; leaf-scars on slightly prominent cushions, each pair wide apart and joined by a raised linear ridge, obovate or semicircular with a raised rim and three groups of bundle-dots. Buds viscid, greenish, the lower scales only being brown; terminal buds ovoid, pointed, the two lowest scales having projecting beaks; scales not ciliate, the outermost four pubescent; lateral buds small, arising at an acute angle.

DISTRIBUTION

It is a common tree in the north-west Himalayas from the Indus to Nepal, occurring at elevations of from 4000 to 10,000 feet, and also occurs in Afghanistan. Sir George Watt informs me that he has measured many trees 150 feet in height with trunks of enormous size, a girth of 40 feet not being uncommon. The wood is used in building and for making water-troughs, platters, vases, cups, packing-cases, and tea-boxes. The twigs and leaves are lopped for use as fodder. The fruit is given as food to cattle and goats; ground and mixed with ordinary flour, it is part of the dietary of the hill tribes. The bark of old trees is very remarkable in appearance, exfoliating in long flakes, which remain attached at their upper ends and hang downwards and outwards. (A. H.)

CULTIVATION

Colonel Henry Bunbury brought seeds from India in 1851, from which plants were raised by Sir Charles J. F. Bunbury[1] at Barton in Suffolk. The large tree[2] now flourishing on the lawn at Barton (Plate 65) is one of the original seedlings, and measured, in 1904, 66 feet high by 7 feet 9 inches in girth. Another tree in the arboretum at Barton measured 65 feet high by 7 feet 2 inches in girth; and divides into two main stems at 7 feet above the ground. This tree flowered for the first time in 1858, producing twelve panicles, being then only seven years old from seed, and 16 feet in height. It did not suffer in the least from the terrible winter of 1860, and flowered as usual in the summer following. In 1868 it ripened fruit, and four thriving plants were raised from its seed. There are no records of the tree on the lawn, which is now the finer of the two. Other trees were planted apparently at Mildenhall,[3] which is about fifteen miles distant from Barton; but these never throve, and none remain. The soil at Mildenhall is a light loam on chalk, and probably did not suit the tree.

I saw the beautiful tree at Barton in full flower on June 24, 1905, when it did not seem to have received the least injury from the severe frosts and cold north-east winds which had occurred a month previously, and which ruined the flowers and destroyed the fruit of the common horse-chestnut in many places.

It seems incredible that this species should be so rare and have remained so little known in England, where it ought to be planted generally in the south and west. Mr. Bean says that the seeds soon lose their vitality if kept dry, and that of some scores received in ordinary paper packets from India in recent years, not one has germinated at Kew. He recommends that the seeds should be gathered as soon as ripe, and be sent packed in fairly moist soil. Mr. Walker, the gardener at Barton, informed me that it ripens seed in good years, and showed me several seedlings raised from them which appeared to grow as well as the common horse-chestnut.

The only other place except Kew, however, where we have seen it, is at Tortworth, where the Earl of Ducie planted in 1890 a few seeds which were sent to him by the late Duke of Bedford. The seedlings were planted at first in sunny places in the open, but did not thrive until moved to a sheltered dell in 1900, where they are now growing well, the best being about 12 feet high.

At Kew there are two or three small trees which have flowered a few times. It seems, therefore, that it only requires a good deep soil and a sheltered situation to succeed as well as it has done at Barton. The late Lord Morley informed me that there was a tree recently planted, but growing very well at Saltram, his place in Devonshire.

According to Jouin,[4] this tree is quite hardy at Metz. (H. J. E.)

---

[1] *Arboretum Notes*, 73 (1889).
[3] *Gard. Chron.* 1903, xxxiii. 188.
[2] Figured in *Gard. Chron.* 1904, xxxvi. 206, *Suppl. Illust.*
[4] *Mitt. Deut. Dendrol. Gesell.* 1905, p. 12.

ÆSCULUS TURBINATA, Japanese Horse-Chestnut

Æsculus turbinata, Blume, *Rumphia*, iii. 195 (1847); André, *Revue Horticole*, 1888, p. 496, figs. 120-124; Bean, *Gard. Chron.* 1897, xxii. 156, and 1902, xxxi. 187, fig. 58; Shirasawa, *Icon. Essences Forestières du Japon*, text 113, t. 71, ff. 16-28.
Æsculus chinensis, Masters (*non* Bunge), *Gard. Chron.* 1889, v. 716. fig. 116.

A tree attaining in Japan, according to Shirasawa, 100 feet in height and 20 feet in girth of stem. Bark thick and scaly. Leaves resembling those of the common horse-chestnut, but much larger, mainly differing in the serration, which is finely crenate. Leaflets five to seven, sessile, obovate-cuneate, occasionally as much as 15 inches long, abruptly acuminate, pubescent beneath. The terminal leaflet has fifteen to twenty-two pairs of nerves. Petiole remaining pubescent towards the tip. Panicles 6 to 10 inches long, dense, somewhat narrow. Flowers yellowish-white, smaller than those of *Æsculus Hippocastanum*. Fruit slightly pear-shaped, $1\frac{1}{2}$ to 2 inches in diameter, four to five on a verrucose rhachis, brown, warty, without spines; valves three, thick; seeds usually two.

Identification

In summer only liable to be confused with the European species, from which it is distinguished by the character of the serration of the leaflets. In winter the twigs closely resemble those of that species, but are not so stout; they are similarly pubescent towards the tip, and are marked with smaller but similar five to seven dotted leaf-scars. Buds smaller, equally viscid, the scales, however, not being uniform in colour, but partly light chestnut brown and partly dark brown. Pith large, irregularly circular in cross-section, and yellowish in tint.

Distribution

The tree is known in Japan as *Tochinoki*, and is common in the forests at 1500 to 5500 feet elevation in the mountains of the main island, descending to lower levels in Yezo. It is recorded by Debeaux, *Fl. Shanghai*, 22, from the provinces of Kiangsu and Chekiang; but no one else has seen the tree in China, and Debeaux's identification is probably incorrect.

The exact date of the introduction of the tree into Europe is uncertain, but it is supposed to be about thirty years ago. It has often passed under the name of *Æsculus chinensis*, an entirely different species. It first produced fruit in 1888 in the arboretum at Segrez in France. It flowered in 1901 at Coombe Wood. As only small trees are known to exist in England, the hardiness of the tree and its suitability for garden decoration are as yet unproved; but at Tortworth it is growing vigorously, and has ripened its buds well whilst still quite small;

and the great size of the leaves on the young trees give it a striking and distinctive appearance.   (A. H.)

In Japan I saw this tree planted in gardens and parks near Tokyo, where it does not seem to grow so large as in its native forests and in higher, colder situations. Sargent says[1] that in the forests of the interior of Hondo, at 2000 to 3000 feet, it attains 80 to 100 feet high, with trunks 3 to 4 feet in diameter, and that these were perhaps the largest deciduous trees that he saw growing wild in the forest. It reaches its most northern point of distribution near Mororan in Hokkaido at sea-level, and I did not see it near Sapporo, in the Aomori district, or near Nikko. At a tea-house called Hideshira, near the village of Sooga on the Nakasendo road, Central Japan, I saw the largest trees of this species growing in a dense grove with *Zelkova acuminata*. They attained over 80 feet high, with clean trunks 40 to 50 feet long, and a girth of 14 feet.

On the Torii-toge Pass, between Wada and Yabuhara, at about 3300 feet, there were many fine trees growing by the side of the road, of one of which I give an illustration from a photograph taken for me by Masuhara of Tokyo in November (Plate 66).

TIMBER

The timber of this tree, though not highly valued in Japan on account of its softness and want of strength, is used for boat and bridge building, furniture making, house-fittings, and for the groundwork of lacquer. It often shows a waved figure, and when old assumes a pale reddish-brown colour, which makes it very ornamental. Such wood, which I procured at Aomori, has been used with good effect in my Japanese wardrobe, and takes a good polish. It is also much used for trays, and from the large burrs and swellings near the root very handsome trays, as much as 18 or 20 inches square, are carved by the Japanese and sold in the villages at a low price. Its value in Tokyo is given at 60 to 100 yen per 100 cubic feet. I saw a plank of this wood in a timber merchant's shop in Osaka measuring 15 feet long and 58 inches wide, showing wavy figure all through. For this plank 90 yen, equal to about £9, was asked, these immense planks being much valued by Japanese connoisseurs for house decoration.   (H. J. E.)

---

[1] *Forest Flora of Japan*, 28.

ÆSCULUS GLABRA, Ohio Buckeye

Æsculus glabra, Willdenow, *Enum. Pl. Hort. Berol.* 405 (1809); Loudon, *Arb. et Frut. Brit.* i. 467 (1838), Sargent, *Silva N. America*, ii. 55, tt. 67, 68 (1892), *Man. Trees N. America*, 644 (1905).
Æsculus pallida, Willdenow, *loc. cit.* 406 (1809).

A tree attaining 70 feet in height and 6 feet in girth in America. Bark dark brown and scaly, becoming in old trees ¾ inch thick, ashy-grey, densely furrowed and broken into thick plates roughened on the surface by numerous small scales. Leaves with long slender stalks; leaflets five, oval or obovate-cuneate, long-acuminate, finely serrate in margin, with tufts of hairs in the bases of the serrations, glabrous underneath except for a few hairs along the midrib and tufts in the axils; petiolules short. Terminal leaflet with about fifteen pairs of nerves. Flowers in pubescent panicles, 5 to 6 inches long; calyx campanulate; petals four, pale yellow; claws shorter than the calyx; limbs twice as long as the claws, broadly ovate or oblong in the lateral pair, oblong-spathulate, much narrower and sometimes red-striped in the upper pair. Stamens usually seven, long, exserted, pubescent. Ovary pubescent. Fruit ovate or obovate, brown, 1 to 2 inches long, roughened by prickles.

The species is distinguished in summer by the glabrous leaves, which always show some cilia in the bases of the serrations. In winter the following characters of the twigs and buds may be recognised :—Twigs glabrous, shining, with orange-coloured lenticels. Leaf-scars slightly oblique on obscure leaf-cushions, crescentic or semicircular, with three groups of bundle-dots, the opposite scars wide apart and often not joined by any linear ridge. Pith large, circular, greenish. Buds not viscid; terminal much larger than the lateral, the latter arising from the twig at an angle of 45°; ovoid, acuminate; scales keeled on the back, ciliate in margin, acuminate, the pointed tips being raised outwardly, dark brown.

Var. *Buckleyi*, Sargent (*Æsculus arguta*, Buckley, *Proc. Acad. Phil.* 1860, p. 448), is a geographical form, occurring in Iowa, Missouri, Kansas, and Texas, and characterised by six to seven leaflets, which are sharply and unequally serrate.

No well-marked horticultural varieties are known.

The type occurs in alluvial soil in Atlantic North America, from Pennsylvania to N. Alabama, and westward to S. Iowa, Central Kansas, Indian Territory, and S. Nebraska. Sargent says that it is nowhere very common and from an ornamental point of view very inferior to *Æsculus octandra*.

This species was introduced, according to Loudon, in 1812, but appears to be very rare in this country. At Devonshurst, Chiswick, a tree cut down in 1905 was 60 feet in height by 6 feet in girth, but though the tree probably exists in some nurseries and old gardens, where it is mistaken for *Æsculus octandra*, more commonly than is supposed, we cannot mention any which are remarkable.

(A. H.)

ÆSCULUS OCTANDRA, Sweet Buckeye

Æsculus octandra, Marshall, *Arbust. Am.* 4 (1785); Sargent, *Silva N. America*, ii. 59, tt. 69, 70
    (1892), and *Man. Trees N. America*, 646 (1905).
Æsculus lutea, Wangenheim, *Schrift. Gesell. Nat. Fr. Berlin*, viii. 133, t. 6 (1788).
Æsculus flava, Aiton, *Hort. Kew*, i. 403 (1789).
Æsculus neglecta, Lindley, *Bot. Reg.* xii. t. 1009 (1826).
Pavia flava, Moench, *Method.* 66 (1794); Loudon, *Arb. et Frut. Brit.* i. 471 (1838).

A tree attaining in America 90 feet in height and 9 feet in girth of stem. Bark of trunk ¾ inch thick, dark brown, slightly fissured, separating on the surface into thin small scales. Leaves with long slender petioles. Leaflets five, occasionally seven, elliptical or obovate-oblong, cuneate at the base, acuminate, finely serrate, pubescent beneath; petiolules short. Terminal leaflet with twenty or more pairs of nerves. Flowers in pubescent panicles, 4 to 6 inches long; calyx campanulate; petals four, yellow, coming into contact at the tips, very unequal, the upper pair much longer than the lateral pair, claws villose within and much exceeding the calyx, limb of lateral pair obovate or round with a subcordate base, limb of upper pair spathulate, minute. Stamens usually seven, shorter than the petals, villose. Ovary pubescent. Fruit 2 to 3 inches long, brown, smooth or slightly pitted.

IDENTIFICATION

In summer distinguished from *Æsculus glabra* by the leaflets being pubescent beneath and devoid of cilia in the serrations; from *Æsculus Pavia*, by the larger leaves, which have petioles with smooth ridges on their upper surface. In winter the twigs show the following characters :—Branchlets glabrous, shining, with a few scattered lenticels. Leaf-scars flat on the twigs (there being no cushion), obovate, with usually three groups of bundle-dots; opposite scars joined by a linear ridge. Pith large, circular, green or white. Buds not viscid, terminal much larger than the lateral, the latter arising at an angle of 45°, long-oval, pointed at the apex; scales brown, the cilia on the exposed margins minute or absent, upper scales rounded at the apex and on the back, lower pair pointed at the apex and keeled on the back.

VARIETIES

1. Var. *hybrida*, Sargent (Var. *purpurascens*, A. Gray; *Æsculus discolor*,[1] Pursh). This is a form occurring wild in the Alleghany mountains. The flowers are purple or red in colour, and the under surfaces of the leaves, as well as the petioles and panicles, are clothed with a dense pale pubescence.

2. *Æsculus versicolor*, Dippel. This is a hybrid between *Æsculus octandra* and *Æsculus Pavia*, and is intermediate in character, the flowers varying in

---

[1] Figured in *Bot. Reg.* iv. 310 (1818).

colour from yellowish to pink. The edges of the petals show a few glands and are tufted ciliate.

A considerable number of forms of this variety are known in cultivation in which slight differences occur in the length and shape of the petals. *Æsculus Lyoni* and *Æsculus Whitleyi* are apparently sub-varieties of this hybrid. The forms with red flowers are often known in gardens as *Pavia rubra*, a name which belongs properly to *Æsculus Pavia*.

DISTRIBUTION

This tree occurs in alluvial soil of river valleys and on moist mountain slopes, from Pennsylvania southward to Georgia and N. Alabama; and westward to S. Iowa, Indian Territory, and W. Texas. Sargent says that when at its best on the slopes of the Tennessee and Carolina mountains, it sends up a straight shaft sometimes free of branches for 60 to 70 feet, and reaches a total height of 90 feet.

(A. H.)

CULTIVATION

According to Loudon this species was introduced into England in 1764, but though more common in cultivation than any *Æsculus* except *A. Hippocastanum*, and apparently not particular about soil, it does not attain any great size. It is perfectly hardy at Colesborne, and ripens fruit in most years, from which I have raised seedlings, which, however, do not grow so fast or well as those of the common horse-chestnut. A seedling raised from a tree at Tortworth in 1905 was 6 inches high in the first year, and some raised from seed which I gathered in the Arnold arboretum, which germinated earlier, were much injured by the frost of May 21-22.

At Syon there are two trees, probably of a great age, both grafted on the common horse-chestnut. One is 65 feet high by 4 feet 4 inches in girth; the other is 56 feet high by 6 feet 4 inches in girth, with a bole of 7 feet, dividing into three stems, which form a wide-spreading crown. A tree at Belton Park, Lincolnshire, was, in 1904, 50 feet high by 3 feet 4 inches in girth, with a fine straight stem, drawn up in a wood. Another, crowded by other trees near the Broad Water at Fairford Park, Gloucestershire, measures about 60 feet by 4 feet 5 inches. A self-sown seedling was growing near it in 1903. There is also a tree, measuring about 50 feet by 5 feet 6 inches, at Charlton Kings, near Cheltenham.

(H. J. E.)

ÆSCULUS CALIFORNICA, CALIFORNIAN BUCKEYE

Æsculus californica, Nuttall, in Torrey and Gray, *Fl. N. America*, i. 251 (1839); *Bot. Mag.* t. 5077 (1858); Sargent, *Silva N. America*, ii. 61, tt. 71, 72, and *Man. Trees. N. America*, 648 (1905); Bean, in *Gard. Chron.* 1902, xxxi. 187, fig. 57.

A tree, attaining in America 40 feet in height, with a short trunk occasionally 9 feet in girth. Bark smooth, grey or white. Leaves with slender grooved petioles. Leaflets five to seven, stalked, oblong lanceolate, acuminate at the apex, cuneate or obtuse at the base, shallowly and crenately serrate, pale glabrescent beneath. Terminal leaflet, with ten to twelve pairs of nerves. Flowers in dense pubescent panicles, 3 to 8 inches long. Calyx two-lipped, upper lip with three teeth, lower lip with two teeth much shorter than the four narrow oblong petals, which are white or pale rose in colour. Stamens five to seven, long, erect, exserted. Ovary pubescent. Fruit pear-shaped, two to three inches long, smooth, pale brown.

In summer it is readily distinguished from the other species with viscid buds by the small leaves, pale beneath. In winter the twigs are slender, grey, glabrous, with numerous lenticels. Leaf-scars wide apart, joined by a linear ridge, flat on the twig, without a leaf-cushion, crescentic or semicircular, with a row of five to seven bundle-dots. Pith large, circular, white. Terminal buds, larger than the lateral buds, which arise at an acute angle, oval, pointed, glistening with white resin; scales gaping at the apex of the bud, broadly ridged on the back, ciliate in margin, with a tuft of hairs at the apex.

The species is a native of California, where it grows on the banks of streams. A very striking picture of a tree, at San Mateo, California, is given in *Garden and Forest*, iv. 523. It shows a very short forked bole, nearly 20 feet in girth at 2 feet from the ground, and an immense umbrella-shaped head only 32 feet high and 60 feet in diameter, densely covered all over with flowers.

It was introduced in 1855 by Messrs. Veitch, and flowered in their nursery at Exeter in 1858. It fruited[1] at the Bath Botanic Gardens in 1901, and again in 1905, though it remains a shrub. It is perfectly hardy in the south of England, and is remarkable for the beauty of its flowers, which appear in June and July. The best specimen we know of in the country is one which Elwes found growing in a shrubbery at Hutley Towers near Ryde, Isle of Wight. It is about 30 feet high, and was in flower on June 22, 1906.   (A. H.)

[1] *Gard. Chron.* 1902, xxxi. 187.

TSUGA

Tsuga, Carrière, *Traité Conif.* 185 (1855); Bentham et Hooker, *Gen. Pl.* iii. 440 (1880); Masters, *Journ. Linn. Soc. (Bot.)* xxx. 28 (1893).
Hesperopeuce, Lemmon, *Rep. Calif. State Board Forestry*, iii. 111 (1890).

EVERGREEN trees belonging to the natural order Coniferæ. Branches horizontal or pendulous, pinnately and irregularly ramified. Buds, one terminal and a few lateral, arising irregularly in the axils of some of the leaves of the current year's shoot, most of the leaves being without buds in their axils. Leaves linear, arising from the branchlets in spiral order, and usually thrown by a twisting of their petioles into a pectinate arrangement, or in one species spreading radially. Petioles short, arising from prominent leaf-bases on the branchlets, appressed against the twigs, a sharp angle being formed by the leaf with the stalk at the point of junction. The leaf has one resin-canal, lying in the middle line between the vascular bundle and the epidermis of the lower surface. The leaves persist for several years; and all the species have in consequence of this and their numerous and fine branchlets very dense foliage.

Flowers monœcious. Male flowers in the axils of the leaves of the previous year's shoot near its apex, composed of numerous spirally arranged, short-stalked, two-celled anthers, with glandular-tipped connectives. Female flowers terminal on lateral shoots of the previous year, short-stalked or sub-sessile, erect, composed of spirally arranged, nearly circular scales, and membranous, usually shorter bracts. Ovules, two on each scale. Cones solitary, small, composed of concave woody imbricated scales, which persist on the axis of the cone after the escape of the seeds, and of inconspicuous bracts, which, except in one species, are concealed between the scales. The cones, ripening in one season, allow the seeds to fall out in the first autumn or winter, but remain on the tree until the summer or autumn of the second year. The seeds, two on each scale, are minute, furnished with resin vesicles and winged. The seedling has three to six cotyledons, which bear stomata on their upper surface.

Tsuga is confined to temperate North America, Japan, China, and the Himalayas. The genus consists of nine species, and is divided into two sections :—

I. *Hesperopeuce*, Engelmann, in Brewer and Watson, *Bot. California*, ii. 121 (1880).

Leaves rounded or keeled above, bearing stomata on both surfaces, and radially arranged; the shorter and lateral branchlets standing in a plane at right angles to the longer and terminal ones.   Cones oblong-cylindrical, large, composed of numerous (about seventy) scales.

This section includes one species :—

1. *Tsuga Pattoniana*, Sénéclauze.   Western North America.

II. *Micropeuce*, Spach, *Hist. Vég.* xi. 424 (1842), identical with *Eutsuga*, Engelmann, *loc. cit.* 120 (1880).

Leaves flat, grooved above, bearing stomata on the lower surface only, pectinately arranged on the branchlets, which are all in one plane.   Cones ovoid, small, composed of few scales, rarely more than twenty-five.

This section comprises the remaining species, of which six are in cultivation in this country.   These may be conveniently arranged as follows :—

A. *Leaves serrulate in margin.   Shoots pubescent.*

2. *Tsuga Canadensis*, Carrière.   Eastern North America.

Leaves, $\frac{1}{3}$ to $\frac{2}{3}$ inch long, usually tapering from the base to the acute or rounded apex; lower surface marked with two narrow well-defined white stomatic bands, the part of the leaf external to them being pure green in colour.   Buds brown, ovoid, pointed, composed of pubescent, keeled acute scales.

3. *Tsuga Albertiana*, Sénéclauze.   Western North America.

Leaves, $\frac{1}{4}$ to $\frac{3}{4}$ inch long, usually rounded at the apex and uniform in breadth; lower surface with two ill-defined broad white stomatic bands, which are indistinctly continued to the margins, there being no distinct bands of pure green.   Buds greyish, ovoid, apex obtuse and flattened; scales keeled, pubescent.

4. *Tsuga Brunoniana*, Carrière.   Himalayas.

Leaves, 1 to $1\frac{1}{4}$ inch long, gradually tapering from the base to the acute apex; lower surface silvery white, stomatic bands well-defined and extending almost to the margins.   Buds globose, flattened on the top, surrounded at the base by a ring of modified leafy scales, the other scales ovate, acute, pubescent.

B. *Leaves entire in margin.   Shoots glabrous.*

5. *Tsuga Sieboldii*, Carrière.   Japan.

Leaves, $\frac{1}{4}$ to 1 inch long, oblong, rounded and notched at the apex, shining above; lower surface with two narrow well-defined white stomatic bands.   Buds red, ovoid, slightly acute at the apex; scales glabrous and ciliate.

C. *Leaves entire in margin.   Shoots pubescent.*

6. *Tsuga diversifolia*, Maximowicz.   Japan.

Shoots pubescent, both on the leaf-bases and in the furrows between them. Leaves, $\frac{1}{4}$ to $\frac{1}{2}$ inch long, oblong, rounded and notched at the apex; lower surface with two narrow well-defined white stomatic bands.   Buds red, pyriform, flattened above; scales obtuse, minutely pubescent.

7. *Tsuga Caroliniana*, Engelmann.   Southern Alleghany Mountains.

Shoots pubescent in the furrows between the leaf-bases, which are glabrous.

Leaves, ¼ to ¾ inch long, oblong, rounded at the apex, which is entire, minutely notched or mucronate; lower surface with two narrow well-defined white stomatic bands. Buds reddish, ovoid, sharp-pointed; scales indistinctly keeled.

In addition to the preceding, two species of Tsuga, belonging to this section, occur in China. They are as yet imperfectly known. *Tsuga chinensis*, Masters,[1] a native of the high mountains of Szechuan, is closely allied to *Tsuga diversifolia*, and, like it, has pubescent young shoots. It differs in the cones, which are quite sessile, and have very lustrous scales. The leaves are described as being green beneath; but this is probably an inconstant character.

Tsuga yunnanensis, Masters,[2] which was discovered by Père Delavay in the mountains near Likiang in Yunnan, is unknown to me. Franchet considers it to be closely allied to *T. Sieboldii*.

TSUGA PATTONIANA, HOOKER'S HEMLOCK

Tsuga Pattoniana, Sénéclauze, *Conif.* 21 (1867); Engelmann, in Brewer and Watson, *Bot. California*, ii. 121 (1880); Masters, *Gard. Chron.* xii. 10, fig. 1 (1892).
Tsuga Hookeriana, Carrière, *Traité Conif.* 252 (1867); and Lemmon, *Erythea*, vi. 78 (1898).
Tsuga Mertensiana, Sargent, *Silva N. Amer.* xii. 77, t. 606 (1898), and *Trees N. Amer.* 51 (1905); Kent, Veitch's *Man. Coniferæ*, 468 (1900).
Pinus Mertensiana, Bongard, *Végét. de Sitcha*, 54 (1832).
Pinus Pattoniana, Parlatore, *D. C. Prod.* xvi. 2, p. 429 (1864).
Abies Pattoniana, Balfour, *Rep. Oregon Assoc.* 1 (1853); Murray in Lawson, *Pin. Brit.* ii. 157 (1884).
Abies Hookeriana, Murray, *Edin. New Phil. Journ.* 289 (1855); and in Lawson, *loc. cit.* 153.
Abies Williamsonii, Newberry, *Pacific R. R. Report*, vi. pt. iii. 53, t. 7, fig. 19 (1857).
Hesperopeuce Pattoniana, Lemmon, *Rep. Calif. State Board Forestry*, iii. 128 (1890).

A tree, occasionally attaining in America 150 feet in height, with a girth of 15 feet. Bark dark cinnamon in colour, deeply divided into rounded connected scaly ridges. Shoots brownish-grey, and densely pubescent. Branchlets in different planes, the shorter and lateral ones usually arising on the upper side of the longer and terminal ones, and disposed at right angles to them, giving a tufted appearance to the branch. Leaves radially arranged on the branchlets, not markedly different in size, ¾ to 1 inch long, curved, linear; apex usually rounded and obtuse, rarely acute; upper surface convex and keeled towards the apex; lower surface rounded with a median groove; both surfaces with about eight lines of stomata, which are sparse and do not form conspicuous white bands; margin entire. Buds brownish, ovoid, acute at the apex, composed of a few closely imbricated, strongly keeled scales.

Cones sessile, about two inches long, oblong cylindrical, tapering at the apex and slightly narrowed at the base, composed of five series of scales, each series with

---

[1] *Journ. Linn. Soc.* (*Bot.*) xxvi. 556 (1902); *Abies chinensis*, Franchet, *Journ. de Bot.* 1899, p. 259.

[2] *Journ. Linn. Soc.* (*Bot.*) loc. cit.; *Abies yunnanensis*, Franchet, *loc. cit.* p. 258; and *cf.* also Masters, *Journ. Linn. Soc.* (*Bot.*) xxxvii. 421 (1906), who identifies the specimens from Szechuan with this species; but judging from Franchet's description, they are the other species.

fourteen to fifteen scales. Scales thin, broader than long, semicircular with a wedge-shaped base, convex, margin irregularly denticulate, pubescent on both surfaces. Bract oblong, abruptly tapering at the apex, which is visible between the scales. Seed with terminal asymmetrical wing, and two resin-vesicles on the side next the scale.

The name Pattoniana is adopted as being the first published under the correct genus Tsuga. The tree is known to American botanists as *Tsuga Mertensiana*, which is unfortunate, as this name was for many years in use for the western hemlock. There is no confusion possible if *Pattoniana* be selected, as no other hemlock has been known at any time by this name.

VARIETIES

The preceding description is drawn up from living specimens of the form with bluish entire leaves, cultivated in this country, and applies, in all essential characters, to dried specimens from trees growing wild in America. I have examined the material in the Kew herbarium and also specimens collected by Elwes on Mount Shasta at 7500 feet elevation; and there do not appear to be two distinct varieties of the tree in the wild state, as the presumed alpine form is only a stunted shrub which agrees in botanical characters with the trees from lower levels.

In England, however, there is a form in cultivation, distinguished by its green serrulate leaves, which differs in many respects from the other form. Concerning its origin, we only know, on the authority of Murray,[1] that it was raised at Edinburgh from seeds collected by Jeffrey in 1851 on the Mount Baker range in British Columbia. Jeffrey found trees growing there from 5000 feet elevation to the snow line, varying in size from 150 feet in height and $13\frac{1}{2}$ feet in girth at lower levels to a stunted shrub not more than 4 feet high close to the timber line. Specimens at Kew from Mount Baker gathered by Jeffrey all have entire leaves and belong to the ordinary wild form.

Engelmann,[2] who visited the Mount Baker range, states that the trees growing there are the ordinary forms of *Tsuga Pattoniana* and *Tsuga Albertiana*. He suggests that the plants raised from Jeffrey's seed may be a mountain form of the latter species; but this cannot be admitted, as they do not resemble that species in botanical characters (buds, leaves, etc.). It is possible that these plants are only a seedling variation of *Tsuga Pattoniana*, and do not correspond with any distinct species or geographical form in the wild state.

Murray,[3] believing that he had two species to deal with, named the bluish form *Abies Hookeriana*, and assigned the name *Abies Pattoniana*, Balfour, to the other form. The original figure of Balfour's species represents, however, the same plant as *Abies Hookeriana* of Murray; and much confusion has resulted in consequence in the use of the two names *Hookeriana* and *Pattoniana*. It is most convenient to

---

[1] *Edin. New. Phil. Jour.* 289 (1855) and *Proc. Hort. Soc.* ii. 202 (1863).　　　[2] *Gard. Chron.* xvii. 145 (1882).
[3] The distinctions relied on by Murray in the cones are trifling; and in the Kew Herbarium there are wild specimens showing these differences, but all belonging to the form with blue entire leaves. I have not seen cones belonging to the other form.

apply the name *Pattoniana* to the bluish form, as it is the earliest name of the wild plant, and to consider the green-foliaged plant to be a variety of it, which may be called var. *Jeffreyi*.

The two forms are distinguished as follows :—

1. Var. *typica*. The form distinguished in cultivation by its bluish foliage. Introduced in 1854 by William Murray, who found the tree on Scots Mountain, in California.

Leaves, though radially arranged, tending on the lower side of the shoot to be in the plane of the branch and not spreading; those on the upper side of the shoot curved and directed outwards and forwards. They are long and narrow, $\frac{1}{2}$ to $\frac{7}{8}$ inch long, and $\frac{1}{20}$ inch wide, entire in margin, convex on both surfaces, the groove in the median line above being very short or absent and never continued to the apex of the leaf, which is rounded or acute; both surfaces marked with conspicuous lines of stomata extending from the base to the apex of the leaf.

2. Var. *Jeffreyi*. Only known in cultivation, distinguished by its greenish foliage.

Leaves spreading radially and directed outwards (never forwards) on all sides of the shoot; straight, short, and broad, less than $\frac{1}{2}$ inch long and about $\frac{1}{16}$ inch in width, serrulate in margin; upper surface flattened and distinctly grooved, the groove continued to the rounded apex; lower surface convex, with lines of stomata the whole length of the leaf. On the upper surface the stomata only occur in four to six broken lines towards the apex.

This form agrees with the typical form in the character of the buds and pubescence of the branchlets; the shoots, however, are not so slender.

(A. H.)

Mr. Gorman gives the following account[1] of the supposed Alpine form, alluded to above :—" Among the hardy alpine trees Hooker's hemlock stands pre-eminent, having a northern range far beyond that of even the white-barked pine. It is a small, dwarfed and stunted tree compared with the type, and seldom exceeds 12 inches diameter or 30 feet in height. It usually ranges in altitude from 5500 to 6400 feet, but is occasionally found up to and beyond 7000 feet where it can find sufficient moisture. Though generally favouring the heads of moist valleys it is sometimes to be found on the leeward side of peaks and slopes, where snowbanks of sufficient size have formed in winter to maintain an adequate supply of moisture during the rest of the year. It is in the latter situations where the tree reaches its highest altitude. In addition to its smaller size and more alpine habit it further differs from its nearest congener in having thinner bark and *small erect* cones, all the other hemlocks having pendent cones. The tree is too small and inaccessible to have any economic value."

This seems to be distinguished principally by its erect cones. Sargent,[2] who alludes to Gorman's account, does not consider this variation to be worthy of distinc-

---

[1] *Survey E. Part Washington Forest Reserve*, p. 336 (19*th Ann. Report of the Survey*, Part v. 1899).
[2] *Silva N. Amer.* xii. 78, note 1.

tion, and explains it by saying that the position of the cones "is evidently due to the thickness of the short lateral branchlets, on which they are terminal and which are sometimes so rigid that the weight of the cones does not make them pendent."

DISTRIBUTION

This tree is only found at high elevations, where it has much the same geographical range as the western hemlock, but it extends farther south in California and reaches its southern limit at 9000 to 10,000 feet on the south fork of King River in the Sierra Nevada.

In the north it descends to sea level on Baranoff Island, and on the shores of Yes Bay in Alaska, lat. 55° 54′ N., where Mr. Martin Gorman collected it. As a rule it is a tree of high altitudes, growing on exposed ridges and slopes near the upper limit of the forest, in company with *Abies lasiocarpa*, *Picea Engelmanni*, and *Pinus albicaulis*. In the Rocky Mountains of British Columbia Mrs. Nicholl found it as a good-sized tree near Glacier up to 7000 feet, though Wilcox,[1] in his excellent account of the trees of that region, pp. 61-65, does not mention it.

Though usually a more or less stunted and ragged tree, it attains a large size on the Cascade Mountains, where I saw it in perfection on the road from Longmire Springs to Paradise Valley, on the south side of Mount Tacoma,[2] in August 1904, first at about 4000 feet, where it was only a scattered tree, and higher up it mixed with the western hemlock in a splendid forest. I was not able to distinguish the two species by their bark, though when not crowded, the habit of Hooker's hemlock is very distinct; but they could be identified by the fallen cones under the trees. The largest that I measured here was about 150 feet by 13 feet 8 inches. Higher up, where the forest[3] opened out into glades at the bottom of the Paradise Valley, which is, in Professor Sargent's opinion, one of the most interesting in America for its alpine flora, it assumed a different and more flat-topped habit; the largest here that I measured was 108 feet by 13 feet 3 inches. It grew in company with *Abies lasiocarpa*, and seedlings of both were numerous on rotten logs on the shady sides of the clumps in which they always grew.

The tree in a very stunted state reaches the timber line—about 7500 feet—in company with *Abies lasiocarpa* and *Cupressus nootkatensis*; but in California, J. Muir[4] measured a specimen at 9500 feet, near the margin of Lake Hollow, which was 19 feet 7 inches in girth at 4 feet from the ground.

Mr. Gorman gives an excellent account of the tree in his *Survey of the Eastern Part of the Washington Forest Reserve*, pp. 335-336, from which I quote as follows :—

"This hemlock is confined to the moist valleys and vicinity of the passes. It is the prevailing tree in Cascade Pass, 5421 feet, and is quite common about the

[1] *The Rockies of Canada*, 61-65 (1900).

[2] The local name is Mount Tacoma, but in maps and writings it is usually called Mount Rainier.

[3] An account of this forest, with two beautiful illustrations of "Patton's spruce," is given in *Garden and Forest*, x. 1, figs. 1, 2 (1897).

[4] *Mountains of California*, p. 20.

sources of the Stehekin, where it attains a very fair size for this region, ranging from 50 to 90 feet in height and from 12 to 27 inches in diameter. The altitudinal range is greater than was expected, from 3100 feet to 5800 feet, and a tree supposed to be of this species was found as low as 2100 feet in the Stehekin Valley.

" The tree is sometimes taken for the western hemlock, but may be distinguished by the erect top of the sapling, the cones long, purple, and more or less massed about the top of the tree ; and the mature tree has an unusually thick, roughly corrugated bark : while in the western hemlock the top is generally drooping, the cones small, oval, and brown, and well distributed over the branches, and the mature tree has a comparatively thin bark. The wood is close grained and of fine texture, and is quite suitable for lumber or fuel, but is not much used on account of its growing usually in inaccessible situations."

Near Crater Lake in Southern Oregon, Mr. Leiberg (*Cascade Forest Reserve Report*, pp. 245, 259), says :—" A few scattered groves of Patton hemlock occur in the southern tracts, some of which are of large size, occasional individuals reaching six to seven feet in diameter. Occasional stands of Patton hemlock 200 to 300 years old exhibit fine proportions at this elevation, 6000 feet ; the species usually grows in close groups, composed of ten or twenty individuals, collected together on what appears to be a common root ; such close growth develops clear trunks, though not commonly of large diameter. Stands of this character sometimes run as high as 25,000 feet per acre."

REMARKABLE TREES

Though now introduced for about fifty-five years this tree has made but little show in our gardens, as the climate of most parts of England is probably too warm for it. I have seen flourishing specimens of no great size in several places, the best, perhaps, being one at Tyberton Court, Herefordshire, the seat of Chandos Lee Warner, Esq., where there is a tree of the typical form 43 feet high by about $3\frac{1}{2}$ feet in girth, said to be fifty years old, and perhaps one of those introduced by William Murray, and sent out by Lawson.

In Scotland it seems to thrive even better, especially at Murthly Castle, where there is a fine group of trees on a lawn (Plate 67). When measured for the Conifer Conference in 1892 the best of these was 35 feet by 3 feet 10 inches, another 30 feet by 4 feet. When I last saw them in September 1906 the tallest tree on the left of the row was 47 feet by 3 feet 8 inches, the tree in the middle with weeping branches 43 feet by 4 feet 2 inches, and the thickest between these two was 6 feet 7 inches in girth. The difference in the habit of these three is well shown in the plate. They produced seed in 1887, from which a number were raised and planted at Murthly. These have grown slowly, and the tallest in 1906 were six or seven feet high, though quite healthy ; and the growth of seedlings which I raised from seed gathered on Mount Rainier is extremely slow.

At Keillour, Henry measured, in 1904, a specimen which was 40 feet by 3 feet 9 inches ; and at the Cairnies, near Perth, the seat of Major R. M. Patton, there were in 1892 two specimens little inferior to those at Murthly.　　(H. J. E.)

TSUGA ALBERTIANA, WESTERN HEMLOCK

Tsuga Albertiana, Sénéclauze, *Conif.* 18 (1867); Kent, Veitch's *Man. Coniferæ*, 459 (1900).

Tsuga Mertensiana, Carrière, *Traité Conif.* 250 (1867); Masters, *Gard. Chron.* xxiii. 179, fig. 35 (1885).

Tsuga heterophylla, Sargent, *Silva N. Amer.* xii. 73, t. 605 (1898), and *Trees N. Amer.* 50 (1905).

Abies heterophylla, Rafinesque, *Atlantic Jour.* i. 119 (1832).

Abies Mertensiana, Gordon, *Pinetum*, 18 (1858).

Abies Albertiana, A. Murray, *Proc. Roy. Hort. Soc.* iii. 149 (1863).

A large tree, attaining in America 200 to 250 feet in height and 20 feet or more in girth, narrowly pyramidal in habit. Bark of old trees reddish brown, and deeply divided into broad, flat, connected scaly ridges. Young shoots whitish grey, and covered with short pubescence, intermixed with scattered long straggling hairs. Leaves pectinately arranged, the shorter leaves on the upper side of the branchlets, those in the median line above often parallel to the twig and directed forwards, exposing their stomatic surfaces. The leaves are $\frac{1}{4}$ to $\frac{3}{4}$ inch long, linear-oblong, uniform in width, serrulate in margin, dark green above, with a median groove continued up to the rounded apex; under surface with inconspicuous midrib and two broad white stomatic bands, which are ill defined on the outer side, there being no distinct marginal green bands. Buds greyish brown, ovoid, with an obtuse and flattened apex; scales keeled and pubescent.

Cones sessile, about one inch long, ovoid, composed of five series of scales, each series with six to seven scales. Scales spathulate, nearly twice as long as broad, wider in the upper half, abruptly narrowed below, rounded with a slightly acute apex, entire and slightly bevelled in margin, striate and slightly pubescent on the outer surface. Bract small, concealed, lozenge-shaped, pubescent and keeled. Seed with a very long wing, decurrent on the outer side of the seed to the base; seed with wing about three-fourths the length of the scale.

The young seedling has three to four cotyledons, which are a little more than $\frac{1}{4}$ inch in length, gradually tapering to an acute apex, sessile, flattened beneath, the upper surface two-sided and bearing stomata, margin entire. The young stem is pubescent and bears first two to three whorls of true leaves (three in each whorl), which are serrulate, shortly stalked, and bearing stomata on their upper surface. These are succeeded by leaves borne spirally. The cotyledons are supported by a caulicle, reddish and glabrous, about an inch in length, which terminates in a very slender flexuose tap-root.

The name *Albertiana* has been chosen, as it appears to have been published as early as that of *Mertensiana* under the correct genus Tsuga. *Tsuga Mertensiana* is now the name given by American botanists to *Tsuga Pattoniana*, and its adoption would cause considerable confusion. *Albertiana*, never having been applied to any other species, is correct on the grounds of common sense as well as of priority.   (A. H.)

DISTRIBUTION

On the west coast of North America it extends southwards from south-eastern Alaska, where it forms the greater part of the great coast forest, which reaches from sea-level up to about 2000 feet, and is associated with Menzies's spruce.

In British Columbia it is very abundant on the coast, and extends as far inland as the heavy rainfall reaches up the valley of the Frazer, on the Gold and Selkirk ranges, and east of the Columbia valley nearly up to the continental divide.[1] In Vancouver's Island it forms with the Douglas fir and red cedar a large though not economically important part of the forest. In Washington and Oregon it is also one of the principal elements of the forest, of which, in the Cascade Forest Reserve, it forms about nine per cent of the timber,[2] and extends up to 5000 feet, crossing the watershed of the coast range in lat. 45°.

In the drier parts of southern Oregon it becomes rare, and though it occurs in the redwood forests of northern California as far south as Cape Mendocino, I did not see it on the Siskyou mountains or on Mount Shasta. In the interior it is found in the wetter parts of northern Montana, Idaho, and in southern British Columbia, where, in company with Douglas spruce, *Picea Engelmanni*, *Abies grandis*, and *Larix occidentalis*, it sometimes forms a considerable part of the forest, and reaches up to 6000 feet in the Cœur d'Alène mountains, though I did not see it in the valley of the Blackfoot river, near Missoula, where the climate is drier.

It attains its finest development on the coasts of Washington and Oregon, where Sargent says that it attains 200 feet in height, with a stem 20 to 30 feet in girth. Plummer, in his Report on the Mount Rainier Forest Reserve,[3] says (p. 101) that it attains an extreme diameter of 6 feet, with a height of 250 feet, of which half to two-thirds is crown. The largest that I actually measured, however, on my visit to Mount Rainier in August 1904, were under 200 feet, with a girth of 12 to 14 feet, and these were growing mixed with *Tsuga Pattoniana* at an elevation of 4000 to 5000 feet.

In the Cascade Reserve Forest of northern Oregon, near Bridal Veil, at about 3500 feet elevation, I measured and Mr. Kiser photographed a tree 175 feet high and 16 feet 6 inches in girth, with a clean bole of about 60 feet, but I am unable to reproduce this, as the negative has not arrived.

The growth of seedlings in all the forests that I saw was exceptionally good. Mr. H. D. Langille says,[4] p. 36 :—

"Certain cone-bearers are better adapted for restocking than others, though the reasons are not apparent. For example, young lovely firs (*A. amabilis*) are abundant everywhere within the zone of that species, whilst noble fir (*A. nobilis*), having a cone and seed of very similar size and nature, seldom germinates, and a seedling of that species is rarely seen.

[1] Mrs. Nicholl, who explored the Rocky Mountains in 1904 and 1905, tells me that it is a large tree at Glacier, on the Canadian Pacific Railway, and grows up to about 5000 feet.
[2] *Forest Conditions of Cascade Reserve*, p. 25, Washington, 1903.
[3] *Twenty-first Annual Report of the U.S. Geological Survey*, part v. Washington, 1900.
[4] *Forest Conditions in Cascade Reserve*, U.S. Geological Survey, Washington, 1903.

"From many observations made in the zone of the hemlock and lovely fir, it is apparent that these trees, from their ability to thrive under the most adverse conditions, are rapidly superseding the others, and will, under natural conditions, be the sole components of the alpine forests. It is a striking fact that, upon many areas where from 50 to 100 per cent of the present forest is red fir (Douglas), the reproduction is entirely hemlock and lovely fir. Should these forests be destroyed by fire it is probable that red fir would rival these species in restocking the burn ; but under natural conditions it is evident that the red fir will be displaced, and the limits of the alpine trees become much lower than at present.

"The yellow pine (*P. ponderosa*), in some instances, does good work in stocking open spots in the timber, but seldom extends far beyond the parent tree. In the yellow pine forests most of the young growth is red or white fir (*A. grandis*), which, taking advantage of the shade and moisture afforded by the yellow pine cover, is growing rapidly, and will in time form a larger percentage of the forest than it has in the past."

I can confirm this from my own observation both in the Cascade Forest and in Vancouver's Island. The seedlings germinate most freely when they fall on the moss-covered rotting trunk of a fallen tree, along which a complete row of young trees often grows ; and Plate 59, vol. i. shows a tree of this species, probably 150 years old, whose roots had completely enclosed the still sound trunk of a red cedar (*Thuya plicata*). A valuable paper[1] by Mr. E. T. Allen, dealing with the western hemlock from a forestry point of view, has been published by the U.S. Bureau of Forestry.

CULTIVATION

It was introduced in 1851 by Jeffrey, and named in 1863 by Murray, at the request of Queen Victoria, in memory of the late Prince Consort, who was a patron of the Oregon Association, and President of the Royal Horticultural Society.[2]

In grace, freedom of growth, and adaptability to varied conditions of culture, in England this, as an ornamental tree, is second to none, and much superior to any other hemlock. Though it has been in cultivation little over fifty years it has already attained a height of about 90 feet in such widely distant counties as Kent, Devonshire, and Perthshire.

The only soils on which it will not thrive are chalk, limestone, and heavy clay, and though it enjoys all the moisture that the wettest parts of England afford, it wants, like all its congeners, a well-drained soil and a sheltered situation.

It ripens seed abundantly in England, and has sown itself in several localities, especially at Blackmoor, the seat of the Earl of Selborne, where there are several self-sown trees, of which the best, growing on the lower greensand formation, is, at about fifteen years old, 10 to 12 feet high, though the parent trees do not exceed about 65 feet.

In Fulmodestone Wood, on Lord Leicester's estate in Norfolk, I have also seen self-sown seedlings ; and though they are very slow in growth for the first four or five

---

[1] "The Western Hemlock," *U.S. Dept. Agric. Forestry Bulletin*, No. 33 (1902).
[2] Hunter, *Woods of Perthshire*, p. 359.

years, yet if kept moist and shaded in a mixture of sand and leaf-mould they may be planted out at five to six years old, with every hope of success.

So far as my experience goes, trees grown from cuttings are not so satisfactory, and there is no excuse for this practice except the saving of trouble, as seedlings are raised in quantity at a very low cost from home-grown seed in Scotland, as I have seen in the nursery at Murthly Castle.

REMARKABLE TREES

Among so many fine trees of this species, all of about the same age, it is hard to choose, but perhaps the largest[1] which we have measured is at Hafodunos, in Denbighshire, which in 1904 was found by Henry to be 94 feet 6 inches by 8 feet 5 inches, and this tree has also produced self-sown seedlings.

At Dropmore there is a very beautiful tree of the spreading type (Plate 68), about 70 feet by 6 feet. At Hemsted, in Kent, I was shown by Lord Cranbrook, in 1905, a tree which is perhaps as tall as any in England, but which, growing in a hole and surrounded by other trees, it was not possible to measure accurately. It is, however, about 90 feet by 4 feet 11 inches, well shaped and growing fast.

At Penllergare, near Swansea, the seat of Sir J. T. D. Llewellyn, Bt., are several fine trees growing in a sheltered valley, which were planted about fifty years ago in company with *Tsuga canadensis*. They are now from 70 to 80 feet high, whilst the best of the eastern hemlock is only 50 feet, and the difference in habit of the two trees is very well shown.

A very large tree, reported[2] to be 110 feet high, is growing at Singleton Abbey, near Swansea, the residence of Lord Swansea, but I have been unable as yet to get confirmation of the height stated. At Castlehill, N. Devon, are several fine trees, the best of which, on a steep bank above a waterfall, where it is somewhat drawn up by beeches, is 90 feet by 6 feet 7 inches. At Carclew, Cornwall, is a fine tree, which in 1902 was 80 feet by 6 feet 3 inches, and in 1905, 82 feet by 6 feet 6 inches, both measurements taken by myself.

At Barton, Suffolk, a young and very thriving tree, shut in by tall beeches and conifers, in 1905 was 80 feet by 4 feet 3 inches, a remarkable instance of height as compared with girth.

In Scotland the tree flourishes exceedingly, and has been planted in many places. Perhaps the tallest is one at Castle Menzies, which in 1904 I made about 90 feet by 7 feet 8 inches, though the gardener thinks it is taller; but one of the most beautiful for its shape, graceful habit, and situation, grows by a deep shady burn on the road from Dunkeld to Murthly Castle, and is about 70 feet by 5 feet (Plate 69), and there are many other fine trees in the grounds there. A tree at Riccarton, near Edinburgh, planted in 1855, measured in 1905, 73 feet by 7 feet 1 inch. A very large tree, measuring in 1907, 10 feet in girth, is reported by Major P. J. Waldron, to be growing at Hallyburton, Coupar-Angus, the seat of Mr. W. Graham Menzies.

---

[1] This tree was in 1868, 28½ feet high by 2 feet 3 inches in girth at the base. In 1883 it measured 65 feet by 4 feet 11 inches at 3 feet from the ground (*Gard. Chron.* 1868, p. 657, and 1885, xxiii. 179). According to the owner, Colonel Sandbach, it was planted probably in 1856. [2] *Gard. Chron.* xxxvii. 136 (1905).

The only place where the tree is reported to have been killed by frost is in the plantations at the Cairnies, Perthshire, where Hunter says (p. 364) that in the severe winter of 1880-81 many were injured and some killed.   Two of the finest specimens in Scotland are, however, growing in the grounds at this place.[1]

In Ireland the best specimen we know of is one at Glenstal, Co. Limerick, which measured in 1903, 78 feet high by 7½ feet in girth.   One of exactly the same height by 6 feet in girth is growing at Kilmacurragh, Co. Wicklow; and around it are several self-sown seedlings.   At Mount Usher, in the same county, there is a fine specimen, 28 years old, from seed, which was 57 feet high by 4 feet 5 inches in 1903.

TIMBER

The timber of the western hemlock has not until recently been much valued, or cut for lumber, on account of its supposed inferiority to that of the Douglas spruce, and is often left standing by loggers, but the increasing scarcity of lumber in some districts has led to its being converted into boards, and it is now largely used for the construction of buildings.   Sargent says that it is light, hard, and tough, stronger, more durable, and more easily worked than the other American hemlocks.   Allen[2] says that in strength it cannot be classed with oak, red fir, or longleaf pine, nor is it suitable for heavy construction, especially where exposed to the weather; but it possesses all the strength requisite for ordinary building material.   It is largely used in Washington for mill frames.

At Mr. Bradley's sawmill at Bridal Veil, Oregon, I saw it being manufactured, and brought away a sample which quite bears out Sargent's high opinion of it.   If such timber existed in Japan or in Europe, I am sure it would be highly valued for joinery, but so far as I can learn none has yet been shipped to Europe.   Hemlock timber[2] has been exported to Manila, and is likely to prove of considerable value in the tropics for housebuilding and indoor finish, as it appears to be free from the attacks of white ants.   The wood is distasteful to rodents, and is used on that account by farmers for the construction of oat-bins.

The bark, according to Sargent, forms the most valuable tanning material produced on the west coast of North America, and the inner bark is eaten by the Indians of Alaska.

James M. Macoun[3] says of it—"The abundance of other wood of better quality has prevented the hemlock from coming into general use, and the same prejudice exists in British Columbia against the western tree that prevailed until very recently against hemlock in eastern Canada.   Though its grain is coarse, western hemlock is for many purposes just as serviceable as other woods which cost more.   The bark is rich in tannin, but is too thin to be extensively used while there is such an abundance of Douglas fir in the same region."   (H. J. E.)

---

[1] These are trees growing in peat soil at 635 feet altitude.   The seeds were sown in 1853, and in 1868 one tree was 29 feet by 1 ft. 11 in., and the other 26 feet by 2 feet at three feet from the ground (*Gard. Chron.* 1868, p. 518).

[2] Allen, "Western Hemlock," 20, 21 (*U.S. Forestry Bulletin*, No. 33, 1902).

[3] *Forest Wealth of Canada*, 82 (1904).

TSUGA CANADENSIS, HEMLOCK OR HEMLOCK SPRUCE

Tsuga canadensis, Carrière, *Traité Conif.* 189 (1855); Sargent, *Silva N. Amer.* xii. 63, t. 603 (1898), and *Trees N. Amer.* 48 (1905); Kent, Veitch's *Man. Coniferæ*, 463 (1900).
Pinus canadensis, Linnæus, *Sp. Pl.* 1421 (1763); Lambert, *Genus Pinus*, i. t. 32 (1803).
Abies canadensis, Michaux, *Fl. Bor. Am.* ii. 206 (1803), and *Hist. Arb. Amer.* i. 137, t. 13 (1810); Loudon, *Arb. et Frut. Brit.* iv. 2322 (1838).
Picea canadensis, Link, *Linnæa*, xv. 523 (1841).

A tree attaining in America over 100 feet in height, but usually only 60 to 70 feet, with a girth of 12 feet as a maximum. Bark of old trees brownish and deeply divided into narrow rounded ridges, covered with appressed scales.

Young shoots greyish in colour and covered with short stiff pubescence. Leaves pectinately arranged, the shorter ones on the upper side of the shoot; those on the median line above pointing forwards, appressed to the twig, and displaying their white under surfaces. They are $\frac{1}{3}$ to $\frac{2}{3}$ inch long, linear, usually broadest towards the base and tapering to the apex, which is rounded or acute; distinctly and sharply serrulate in margin; dark green above with a median groove often not continued to the apex; lower surface with distinct midrib and two narrow well-defined white stomatic bands, the edges being pure green in colour. Buds brown, ovoid, pointed; scales ciliate, pubescent, keeled, acute.

Cones, $\frac{1}{2}$ to $\frac{3}{4}$ inch long, ovoid, on slender puberulous stalks nearly $\frac{1}{4}$ inch long, composed of five series of scales, with about five scales in each series. Scales orbicular oblong, nearly as broad as long, entire and slightly bevelled in margin, striate, glabrescent in the exposed part. Bract small, concealed, lozenge-shaped. Seed with an oblong wing, decurrent half-way on its outer side. The seed with wing about two-thirds the length of the scale.

VARIETIES

A considerable number of horticultural varieties are known, no less than fourteen being described by Beissner. Some of these are variegated forms, as var. *argentea* or *albo-spica*, in which the tips of the young shoots are whitish. Others differ in habit and stature, as var. *pendula*, with pendulous branches, and var. *Sargentii*,[1] a flat-topped bushy form of compact habit with short pendulous branches. The latter was found about forty years ago on the Fishkill Mountains in New York, and was first cultivated and made known by Mr. H. W. Sargent. One of the original plants, growing on the Howland estate, in Matteawan, New York, is now about 25 feet across. Grafted plants of this variety form in a few years an erect stem, and lose the dense low habit which is the charm of the original seedlings.[1]

Var. *parvifolia*, as cultivated at Kew, is a shrub, with stout branchlets, and very short leaves, about $\frac{1}{4}$ inch long, which spread radially outwards from the shoot.

(A. H.)

[1] Sargent, *Garden and Forest*, x. 490 (1897).

DISTRIBUTION

In the colder parts of New England and Canada the hemlock is one of the most characteristic trees of the virgin forest, and extends, according to Sargent, from Nova Scotia and New Brunswick westward through Ontario to eastern Minnesota, southwards through Delaware, southern Michigan, and central Wisconsin, and along the Appalachian Mountains to north-western Alabama. He says that it attains its largest size in the south, in the mountain valleys of North Carolina and Tennessee, and gives its size as usually 60 or 70 and occasionally 100 feet in height, with a trunk 2 to 4 feet in diameter ; but Pinchot and Ashe (*loc. cit.* p. 134) give 110 feet with a diameter of 6 feet as its extreme size, with a beautiful picture of it (pl. xix.). When, however, I was at Ottawa in September 1904 I visited, in company with Mr. James M. Macoun of the Geological Survey, a forest near Chelsea, in the Gatineau valley, where several hemlocks of nearly 100 feet were standing, mixed with birches, maples, and other hardwoods, and found a fallen tree which must have been at least 125 feet, and perhaps 135 feet long, though the top was too rotten to follow it out to the end. Mr. Macoun, however, said he had never seen one so large before.

It often grows on rocky ridges, where it forms dense groves on the north side, and loves the steep banks of river gorges. Henry visited in 1906 Pisgah Mountain, near Hinsdale, in New Hampshire, where there remain on the estate of Mr. Ansell Dickinson about 700 acres of virgin forest. This mainly consists of a mixture of hemlock and hardwoods, with white pine occurring here and there singly and in small groups ; though on one or two areas of a few acres the white pine and hemlock form a pure coniferous stand. The largest hemlock seen measured 113 feet by 7 feet 10 inches, with a clean stem of only 30 feet, being much branched though densely crowded by other trees. A great many small hemlocks throughout the forest formed an undergrowth, and had been suppressed in growth, one which was $\frac{3}{4}$ inch in diameter and 10 feet high showing 65 annual rings.

In the Arnold Arboretum, near Boston, is a fine natural grove of this tree, called Hemlock Hill, which gives a very good idea of its normal growth in New England. The average height here is 60 to 70 feet by 3 to 4 feet, and the best that I measured at the bottom of the hill was 80 feet by 4 feet 6 inches. These trees were rather crowded, and had clean boles for 15 to 30 feet up.

The growth of the tree is very slow, and Sargent says that the specimen of its timber in the Jessup Collection in the American Museum of Natural History at New York (which is the most complete that has ever been formed of the woods of any country) is only $13\frac{1}{2}$ inches in diameter inside the bark, though it shows 164 annual rings, of which the sapwood, 2 inches thick, has twenty-nine.

It seeds freely, but the seedlings do not germinate well in the open or on land which has been recently burned over, and seem to succeed best on a mossy stump or fallen log, where they must often remain eight to ten years before their roots reach the earth. According to Sargent they are only three or four inches high at four years old, under favourable conditions, and are easily destroyed.

CULTIVATION

Though introduced by Peter Collinson about 1736,[1] and at one time planted in almost every garden as an ornamental tree, the hemlock is rarely seen in Europe in a condition to remind the American of it as he knows it at home. Of late years it has been superseded by more modern and faster growing introductions.

I cannot exactly say what are the conditions which suit it best in this country, because I have not seen it planted in the shady, damp, and rocky gorges which it likes at home; but a deep light soil, free from lime and well drained, and a northern aspect, seem to suit it best in gardens. Its graceful habit and perfect hardiness should recommend it to all lovers of trees. It has a general tendency to fork near the ground, and this can only be checked by crowding it when young, or perhaps to some extent by careful pruning, as Loudon says that it bears the knife well, and is used for hedges in American nurseries; though I should consider either common spruce or arbor vitæ much better suited for the purpose here.

It ripens seed freely, but the plants I have raised were so small that frost and March winds destroyed them before I learned the necessity of protecting them; and in future I would imitate nature, and sow them on a mossy piece of half-rotten wood, or in a mixture of sand and leaf mould in a shaded frame.

REMARKABLE TREES

By far the most remarkable specimens of this tree which exist in England, or, as I believe, in Europe, are at Foxley, Herefordshire, the seat of the Rev. G. H. Davenport, which are believed to have been planted by Sir Uvedale Price, who was once the owner of this place. He was born in 1747, and died in 1828. In Nash wood, about half a mile from the house, on a rich soil of old red sandstone formation, in a dell facing south-west, a number of these trees are growing, which, though not quite so large as the tree at Studley, average about 55 feet high by 8 to 10 in girth, and although their trunks are not so straight and clean as in an American forest, are nearly all sound and healthy. I measured twenty of these trees in July 1906 and found the largest, the only one which was forked near the ground, to be 10 feet in girth. Another was 9 ft. 10 in., and had a trunk which would contain from 120 to 130 cubic feet. The others ranged from 7 to 9½ feet at 5 feet from the ground, averaging over 8 feet, and were mostly clear of branches, or nearly so, for 15 to 30 feet from the ground. The dense shade of these trees keeps the soil quite free from vegetation below them, but I saw no seedlings in the grove. Though Mr. Davenport was good enough to have a considerable clearing made in order to get a better view of the trees, and Mr. Foster went to Foxley on purpose to photograph them, the difficulty of the subject was so great that the prints taken (Plate 70) do not show them as well as I could wish.

The largest tree which I have seen in England is at Studley Royal, not far below

[1] A tree said to be the original one planted by him at Mill Hill still survives, but was, when I saw it in 1906, in poor condition, the soil being too dry for it.

Fountains Abbey, and close to two very tall spruce.   This, though hard to measure correctly owing to its crowded position, which makes a satisfactory illustration impossible, is over 80 feet high and 11 feet in girth, but is forked at about 7 feet from the ground.

The next best is at Strathfieldsaye, a very spreading tree in damp soil, also forking near the ground.   The two stems measure 9 feet 6 inches and 8 feet 3 inches, and the height in 1903 was about 75 feet, the branches weeping to the ground on all sides (Plate 71).   At Althorp there is a fine old specimen on the lawn, of a more upright type, which in 1903 was 63 feet by 8 feet 10 inches.   At Walcot, in Shropshire, the seat of the Earl of Powis, is one of the best grown trees I have seen, with a bole about 25 feet high, and measuring 60 feet by 8 feet 8 inches.   At Mr. Heelas' residence, near Reading, part of the old White Knights estate, is a tree, probably planted 150 years ago, which Henry in 1904 found to be 67 feet by 8 feet.   At Arley Castle there is a fine tree dividing into three stems, of which the largest is 6 feet 7 inches in girth and nearly 70 feet high.

At Hardwick, Bury St. Edmunds, there is a tree, forked at 30 feet up, 60 feet by 5 feet 10 inches.   At Beauport, Sussex, a tree measured in 1904, 65 feet by 7 feet. At Osberton, Notts, the seat of Mr. F. Savile Foljambe, there is a remarkably spreading old tree about 42 feet high, and dividing near the ground into three stems, each about 6 feet in girth.   It has some layered branches which are over 20 feet high, and the total circumference is no less than 80 paces.   Bunbury, *Arboretum Notes*, p. 140, mentions as the largest hemlock in the country one growing at Bowood, Wiltshire, the seat of the Marquess of Lansdowne, which, however, cannot now be found.

In Scotland, where the tree should succeed well, I have seen none of great size, except the tree at Dunkeld, which is growing in a thick wood of conifers mixed with beech on rocky ground, close to the Hermitage bridge.   This is mentioned by Hunter as being 80 feet high by 10 feet in girth.   Mr. D. Keir twenty years later made it 85 feet by 11 feet, and when he showed it to me in 1906 I found that, though the top is not easy to see, it is probably as much as 90 feet, and looks as if it would grow taller.   It divides at about 12 feet into several stems, and is believed to be 140 to 150 years old.

At Dalkeith there was in 1891 a tree 42 feet high by 10 feet 6 inches in girth ; and at Buchanan Castle, Stirlingshire, the seat of the Duke of Montrose, one measuring 45 feet by 6 feet 10 inches.[1]

In Ireland the largest known to us is one at Carton, the seat of the Duke of Leinster, which in 1903 was 45 feet by 6½ feet.

TIMBER

Opinions as to the value of this wood differ a good deal, and I have no personal experience in the matter.   Sargent says that it is light, soft, not strong, brittle, coarse, crooked-grained, difficult to work, liable to wind-shake and splinter, and not

[1] *Journ. R. Hort. Soc.* xiv. 520, 544 (1892).

durable when exposed to the air; but that it is now largely manufactured into coarse lumber for the outside finish of buildings, and is also used for railway ties and water-pipes. James M. Macoun, in *The Forest Wealth of Canada*, p. 82, says: "Though little inferior to white pine as rough lumber, a prejudice has for a long time existed against this wood, which is only now dying out. As a coarse lumber it to-day commands almost as high a price as pine. It is one of our best woods for wharves and docks, and great quantities are used annually for piles." It is not, so far as I can learn, imported into Europe. The value of its bark, however, for tanning heavy leather has long been known, and it is used more largely than any other in Canada and the Eastern States of America, often mixed with oak bark in order to modify the red colour of the leather tanned with it alone.[1]

Canada pitch, made from the resin of this tree, and oil of hemlock, distilled from its twigs, were formerly used to some extent in medicine, but are not now of any commercial importance. (H. J. E.)

TSUGA CAROLINIANA, Carolina Hemlock

Tsuga Caroliniana, Engelmann, Coulter's *Bot. Gazette*, vi. 223 (1881); Sargent, *Gard. Chron.* xxvi. 780, fig. 153 (1886), *Silva N. Amer.* xii. 69, t. 604 (1898), and *Trees N. Amer.* 49 (1905); Kent, Veitch's *Man. Coniferæ*, 466 (1900).

A tree attaining in America 70 feet in height with a girth of 6 feet. Bark reddish brown, and deeply divided into broad, flat, connected scaly ridges. Young shoots shining grey, with scattered short pubescence in the furrows between the glabrous leaf-bases. Leaves pectinately arranged, those on the upper side of the branchlet shorter than the others, $\frac{1}{4}$ to $\frac{3}{4}$ inch long, linear-oblong, uniform in breadth or slightly narrowed towards the rounded apex, which is occasionally minutely emarginate; dark green and shining above, with a median groove either continued up to the apex or falling short of it; lower surface with distinct midrib and two narrow, well-defined white stomatic bands, the edges being green; margin entire. Buds reddish brown, ovoid, sharp-pointed; scales indistinctly keeled and pubescent.

Cones on short stout stalks, pendulous or deflected, cylindrical-oblong, 1 to $1\frac{1}{2}$ inch long, consisting of five series of scales, five scales in each series. Scales oblong-orbicular, rounded and slightly narrowed at the apex, pubescent externally, edge thin and bevelled. Bract concealed, wedge-shaped at the base, rounded at the apex. Seed with a long wing, which is decurrent half-way down its outer side.

Tsuga Caroliniana appears to be the American representative of *Tsuga*

---

[1] Prof. H. R. Procter of the Leather Industries Department of Leeds University, tells me that though the bark is still the principal tanning material of North America, it has been cut so recklessly that in many districts the supply is now insufficient, and is supplemented by extracts of other materials, especially that of Quebracho wood (Loxopterygium). In England its use was at one time considerable, but it is no longer a specially cheap material, and its colour has now to a large extent prevented its employment. The bark appears to contain from 8 to 12 per cent of a catechol tannin, yielding large quantities of insoluble "reds," and in this respect it is very inferior to the bark of the common spruce fir, which is largely employed in Austria, though it does not seem to be used in England.

diversifolia, and is remarkable for its limited distribution. It occurs at elevations of 2500 to 3000 feet, usually on dry rocky banks of mountain streams along the Blue Ridge, extending from south-western Virginia through South Carolina to northern Georgia. Sargent states that it occurs either in small groves or mingled with other species, and describes it as a beautiful tree of compact pyramidal habit, with dense dark-green lustrous foliage. Elwes saw it on the Blue Ridge in 1893, and brought home young plants, which, however, died in a year or two.

This tree was discovered in 1850 by Professor L. R. Gibbes. It was first raised in the Arnold Arboretum in 1881, and has proved there quite hardy. It was introduced from thence to England in 1886. There are two or three small specimens in the collection at Kew which are three or four feet in height and have a bushy, spreading habit. This species, judging from the slow rate of growth at Kew, is not likely to attain to timber size in England, and we know of no trees of any size living in this country. (A. H.)

TSUGA BRUNONIANA, Himalayan Hemlock

Tsuga Brunoniana, Carrière, *Traité Conif.* 188 (1855); Hook. f., *Gard. Chron.* xxvi. 72, fig. 14 (1886), and *Flora Brit. India,* v. 654 (1888); Masters, *Gard. Chron.* xxvi. 500, fig. 101 (1886); Kent, Veitch's *Man. Coniferæ,* 462 (1900); Gamble, *Man. Indian Timbers,* 718 (1902); Brandis, *Indian Trees,* 693 (1906).
Tsuga dumosa, Sargent, *Silva N. Amer.* xii. 60 (1898).
Pinus dumosa, D. Don, *Prod. Fl. Nepal.* 55 (1825).
Pinus Brunoniana, Wallich, *Pl. Asiat. Rar.* iii. 24, t. 247 (1832).
Abies Brunoniana, Lindley, *Penny Cyclop.* i. 31 (1833).
Abies dumosa, Loudon, *Arb. et Frut. Brit.* iv. 2325 (1838), and Brandis, *Forest Flor. N.W. India,* 527 (1874).

A tree forming in the Himalayas, according to Hooker, a stately blunt pyramid, with branches spreading like the cedar, but not so stiff, and drooping gracefully on all sides, attaining 120 feet in height and 28 feet in girth. In cultivation in England it assumes a bushy habit, and never makes a clean stem, the trunk being concealed by the dense pendulous branches.

Bark thick and rough. Branchlets light brown in colour with a short and not very dense pubescence. Leaves long, 1 to $1\frac{1}{4}$ inch, narrow linear, gradually tapering towards the acute and recurved apex, serrulate in margin; upper surface dark green and deeply grooved; lower surface silvery white, the bands of stomata extending almost to the margins. Buds globose, flattened on the top; scales ovate, acute, pubescent.

Cones sessile, ovoid, an inch long, composed of about twenty-five woody scales, which are nearly orbicular, vertically striate, shining, showing externally a thickened ridge a little distance from and parallel to the thin entire margin; bract concealed. Seed two-thirds the length of the scale, with an oblong-ovate wing, which is decurrent on the outer side of the seed to its base.

Tsuga Brunoniana occurs in the Himalayas, from Kumaon to Bhotan, at altitudes varying from 8000 to 10,500 feet. Franchet considers that certain Chinese specimens constitute a distinct variety of the species, which he has named var. *chinensis*.[1] These were collected in N.E. Szechuan by Père Farges, and in the mountains of western Yunnan at 9000 feet altitude by Père Delavay. Diels[2] also identifies with this variety specimens collected by Von Rosthorn in Szechuan. I have seen no Chinese examples, and Mr. E. H. Wilson considers that there is only one species of Tsuga in the mountains of Szechuan, which is *Tsuga chinensis*, Masters. Small plants of the Chinese Tsuga are now in cultivation at Coombe Wood; and are as yet too young to entitle us to speak definitely concerning its affinities. (A. H.)

In the interior of Sikkim I saw this beautiful tree in great perfection in the same forests where Sir Joseph Hooker so well describes it,[3] during my journey with the late W. E. Blanford to the Tibetan frontier in 1870. It occurs first in the Lachen valley at about 8000 feet in an extremely moist summer climate, where snow lies for two or three months in winter, growing in company with *Picea Morindoides*, *Abies Webbiana*, and, higher up, with *Larix Griffithii*, in a forest unrivalled in the temperate region for its botanical and zoological wealth; where it commonly attains a height of 100 to 120 feet. Afterwards, on the path from Lachoong to the Tunkralah, I saw even grander specimens, one of which, as measured by Sir J. Hooker, was over 120 feet high by 28 feet in girth. In these almost pathless forests it is covered with ferns and lichens and forms a graceful pyramidal tree with very drooping branches, and reaches an elevation of about 10,000 feet. On the outer ranges it is not so large, but extends into Bhotan, where Griffith found it from 6500 to 9500 feet. It probably occurs throughout Nepal and in the N.W. Himalaya, as far west as Kumaon, where it is a smaller tree and of little economic value, though in Sikkim the bark is used for roofing huts.

The Himalayan hemlock was introduced into England in 1838, according to Loudon,[4] but is rarely seen, except in a stunted state, with several branching stems, and suffering from the absence of sufficient moisture. Like most of the Himalayan conifers, it grows too early and is injured by spring frosts; but in a few favoured districts of Cornwall and Ireland it seems more at home and has attained considerable size and beauty.

The best specimen that I have seen is at Boconnoc in Cornwall, the seat of J. B. Fortescue, Esq. (Plate 72). This tree measures about 53 feet high by 12 feet in girth near the ground, where it branches into several stems, which spread to about 70 feet in diameter. When I saw it in April 1905 it was covered with cones, from which I have raised many young plants.

There is a rather fine tree at Dropmore, planted in 1847, but not so large or healthy as the one described above; and at Beauport, near Battle, Sussex, there is also a fair specimen.

[1] *Jour. de Bot.* 1899, p. 258.
[2] *Flora von Central China,* 217 (1901).
[3] *Himalayan Journals,* i. 209, ii. 108, etc.
[4] *Encycl. Trees and Shrubs,* 1036 (1842).

At Southampton,[1] in the Red Lodge nursery belonging to Mr. W. H. Rogers, there was a tree twenty-five years old in 1884, about 20 feet high, which bore cones in profusion.   At Kew a specimen planted in a sheltered position lived for many years, but ultimately succumbed.   Sir Joseph Hooker[2] knew of no good specimen nearer London than one on a south slope near Leith Hill in a very sheltered and well-watered valley.

At Fota, in the S.W. of Ireland, the seat of Lord Barrymore, Henry measured a tree about 40 feet by 4 feet 10 inches in 1904; and there are trees at Kilmacurragh and Powerscourt, in Co. Wicklow, which are about 30 feet high, all of very branching bushy habit, and with several main stems.

Sargent[3] has never seen a specimen in the United States.   (H. J. E.)

TSUGA SIEBOLDII, SIEBOLD'S HEMLOCK

Tsuga Sieboldii, Carrière, *Traité Conif.* 186 (1855); Masters, *Jour. Linn. Soc. (Bot.)*, xviii. 512 (1881); Mayr, *Abiet. des Jap. Reiches*, 59, t. iv. fig. 12 (1890); Kent, Veitch's *Man. Coniferæ*, 472 (1900).

Tsuga Tsuja, A. Murray, *Proc. R. Hort. Soc.* ii. 508, ff. 141-153 (1862).

Tsuga Araragi, Koehne, *Deutsche Dendrologie*, 10 (1893), and Sargent, *Garden and Forest*, x. 491, fig. 62 (1897).

Pinus Araragi, Siebold, *Verhandl. Batav. Genoot. Konst. Wet.* xii. 12 (1830).

Abies Tsuga, Siebold et Zuccarini, *Fl. Jap.* ii. 14, t. 106 (1842).

Abies Araragi, Loudon, *Trees and Shrubs*, 1036 (1842).

A tree attaining in Japan about 100 feet in height and 12 feet in girth, forming in England a small tree with a short bole and a dense crown of foliage, with numerous branches and pendulous branchlets.

Young shoots greyish in colour and quite glabrous.   Leaves pectinately arranged, variable in size, the smaller on the upper side of the shoot, some of these being directed outwards at right angles to the general plane of the foliage. They are oblong, uniform in width, $\frac{1}{4}$ to 1 inch long, shining and dark green above with a median furrow continued to the rounded and emarginate apex; lower surface with green midrib and two narrow well-defined white bands of stomata; margin quite entire.   Buds reddish, ovoid, slightly acute at the apex: scales glabrous on the surface, ciliate in margin.

Cones elongated ovoid, on a stalk about $\frac{1}{4}$ inch long, pendulous or deflected, composed of five series of orbicular scales, which are rounded at the apex and at the base and have a slightly bevelled margin.   Bract included, very short and bifid.   Seed with a long wing decurrent half-way along its outer side.

This tree has been much confused with the other Japanese species, from which it is very distinct in botanical characters.   Koehne's proposed name, *Tsuga Araragi*, is not adopted by us, the name *Sieboldii* being the first one under the correct genus Tsuga.   (A. H.)

---

[1] Note in Kew herbarium, and Nicholson in *Woods and Forests*, 1884, p. 243.
[2] *Gard. Chron.* xxvi. 72 (1886).   [3] *Garden and Forest*, x. 491 (1897).

TSUGA DIVERSIFOLIA, Japanese Hemlock

Tsuga diversifolia, Masters, *Jour. Linn. Soc. (Bot.)* xviii. 514 (1881); Mayr, *Abiet. des Jap. Reiches* 61, t. xiv. fig. 13 (1890); Sargent, *Garden and Forest*, vi. 495, fig. 73 (1893), and x. 491, fig. 63 (1897); Kent, Veitch's *Man. Coniferæ*, 467 (1900).
Abies diversifolia, Maximowicz, *Mél. Biol.* vi. 373 (1867).

A smaller tree than Siebold's hemlock, which it resembles in habit.

Young shoots pubescent, the pubescence occurring on both the leaf-bases and the intervening furrows. Leaves arranged as in *Tsuga Sieboldii*, but considerably shorter, scarcely exceeding ½ inch in length, oblong, uniform in breadth, shining and dark green above with a median furrow continued to the rounded and emarginate apex; lower surface with green midrib and two narrow well-defined white bands of stomata; margin entire. Buds red, pyriform, flattened above; scales rounded at the apex, minutely pubescent and ciliate.

Cones subsessile, pendent or deflected, ovoid; scales shining, orbicular-oblong, truncate at the base, with edge slightly bevelled and thickened. Bract minute, concealed, rhomboid. Seed with a short terminal wing, which is not decurrent along its side. (A. H.)

Distribution of the Japanese Tsugas

In Japan I saw both species in their native forests; but so far as I could learn they are not distinguished by the foresters and are both called Tsuga (pronounced *tsunga*). By the Japanese botanists *Tsuga Sieboldii* is termed *Tsuga*, the other species being named *Kuro-tsuga* or *Kome-tsuga*. Of the two, the latter apparently has a more northern range than *Tsuga Sieboldii*. I saw it in the forest round Lake Yumoto at 4000 to 5000 feet elevation, where it is a picturesque and graceful tree of no great size. Both species, however, according to Shirasawa, are found in this district. *Tsuga diversifolia* also occurred high up in the Atera valley. Further south in the Kisogawa valley and at Koyasan I saw *Tsuga Sieboldii*, which at 2000 to 3000 feet attains a large size, growing scattered in mixed forests and not gregariously, like the other species at Lake Yumoto. I measured a tree at Koyasan, which had been felled; it was over 100 feet in height, of which half was free from branches, the butt being about 3 feet in diameter. I estimated it as 250 to 300 years old, though the growth had been so slow that I could not count the rings beyond 150. The wood of this tree, as I was told by the chief priest of the Gemyo-in temple, who was my host at Koyasan, is even better than that of Hinoki (*Cupressus obtusa*); and much of the wood used in building the temple had been Tsuga. Old trees, however, are now so scarce that the timber cannot be obtained in quantity. I bought some beautiful boards cut from it at Osaka, which have a pale yellow colour and very fine wavy figure. The wood is also made into shingles, which are said to last about forty years, and it has lately been used for paper-making. The bark is used for tanning fishing-nets, and the timber sells in Tokyo at thirty-five to forty yen per 100 cubic feet.[1]   The growth of the tree from seed is very slow at first as in the allied species.

HISTORY AND CULTIVATION

Tsuga Sieboldii was introduced into Europe by Siebold in 1850.   Cones both of this species and of *Tsuga diversifolia* were brought from Japan by John Gould Vetch in 1861, and the latter species was sent out under the name *Abies Tsuga*, var. *nana*.   Specimens cultivated at Kew as *Tsuga Sieboldii*, var. *nana*, belong to *Tsuga diversifolia*.

Though both species have been introduced long enough to prove their hardiness in favoured parts of the South of England, we have never seen even a moderately large tree, and doubt much if either species will attain timber size in this country. The Japanese hemlocks seem to prefer a light moist rich soil, free from lime, with shade and shelter from cold winds.   They will not grow at all on the limestone soil of Colesborne.   The best specimen we know is in the garden of Mr. W. H. Griffiths at Campden, Gloucestershire, and is about 15 feet high.   It bore cones in 1905.

Sargent[2] says that *Tsuga Sieboldii* is one of the most graceful and satisfactory of the exotic conifers cultivated in American gardens, where it promises to grow to a large size; but in the garden of Mr. Hunnewell at Wellesley, Massachusetts, which I visited in May 1904, I noted that it had been almost killed to the snow line by the exceptionally severe winter of 1903-1904, though it had produced cones in the preceding year.[3]   (H. J. E.)

[1] In *Industries of Japan*, 236 (1889), Rein, who did not distinguish between the two species, probably speaking of *Tsuga Sieboldii*, says that the finest specimens seen by him were in the forests of Kin-shima-yama in Southern Kiu-siu, where it grows with *Picea polita*, and equals it in size, attaining 4 to 5 metres in girth.   This goes to show that the tree enjoys a warm moist climate.

[2] *Silva North America*, xii. 60.

[3] Beissner states in *Mitt. D. D. Ges.* 1905, pp. 165, 167, that *T. diversifolia* is hardier than *T. Sieboldii*, but both of them grow well in East Friesland, and Mayr says that *T. diversifolia* is hardy at Munich.

JUGLANS

Juglans, Linnæus, *Gen. Pl.* 291 (1737); Bentham et Hooker, *Gen. Pl.* iii. 398 (1880).

DECIDUOUS trees with furrowed bark. Twigs with chambered pith. Buds scaly, the lateral buds often extra-axillary or accompanied by superposed accessory buds. Leaf-scars large with three groups of bundle-traces. Leaves large, alternate, compound, imparipinnate; leaflets opposite, entire or serrate. Stipules absent.

Flowers monœcious. Male flowers numerous in pendulous catkins, which arise singly or in pairs above the leaf-scars of the preceding year's shoot, appearing in autumn and then visible as short cones covered by imbricated scales. Stamens eight to forty, in several series on the axis of a scale, which is five- to seven-lobed, the lobes representing a bract, two bracteoles and two to four perianth-lobes. Connective of the anthers clavate or dilated. Pistillate flowers few, in an erect spike terminating the current year's shoot; each flower with a three- to five-lobed or toothed involucre, composed of a bract and two bracteoles, adnate to the ovary. Inside the involucre is an epigynous and adherent four-lobed or toothed perianth. Ovary one-celled with one basal straight ovule. Style divided into two linear or lanceolate recurved spreading fimbriated plumose stigmas.

Fruit a large ovoid, globose, or pear-shaped drupe, with a fleshy, irregularly splitting husk, formed by the accrescent involucre and perianth. Nut ovoid or globose, thick-walled, longitudinally and irregularly wrinkled, two- to four-celled at the base, indehiscent or separating at last into two valves. Seed two- to four-lobed at the base, with fleshy cotyledons, which remain within the shell in germination.

About thirteen species of Juglans have been described; and there are two or three unnamed and little-known species in tropical South America. Of the described species three[1] confined to Mexico, one[2] a native of the Antilles, and the Californian walnut[3] have not yet been introduced, and will not be dealt with in the following account.[4]

Plate 73 illustrates the leaves, branchlets, and leaf-scars of the species in cultivation.

---

[1] *Juglans mollis*, Engelmann; *J. pyriformis*, Liebmann; and *J. mexicana*, Watson.

[2] *Juglans insularis*, Grisebach. Concerning the walnut reputed to occur in Jamaica, *J. iamaicensis*, C. DC., cf. *Kew. Bull.* 1894, p. 371.

[3] *Juglans californica*, Watson.

[4] Since the above was written, Mr. Dode has published a paper containing descriptions of several new species in *Bull. Soc. Dendr. France*, i. 67 (1906); but these seem to us to be founded on variable characters, and to be rather forms due to cultivation.

II 249 H

KEY TO THE SPECIES OF JUGLANS IN CULTIVATION

I. *Leaflets not serrate; usually entire or sinuate* (Plate 73).

1. *Juglans regia*, Linnæus.   Bosnia and Greece, through W. Asia and Himalayas to N. China.

Leaf-scars deeply notched without a pubescent band on their upper edge Leaflets 7 to 9, glabrous beneath except for inconspicuous axil tufts.

II. *Leaflets serrate.   Leaf-scars without a pubescent band on their upper edge.*

* *Leaflets glabrous beneath, except for the axil tufts.*

2. *Juglans regia* x *nigra*.   Two forms: *Juglans Vilmoriniana*, Carrière, and *Juglans pyriformis*, Carrière.

Leaflets 11 to 13, with fine shallow serrations.

** *Leaflets pubescent beneath.*

3. *Juglans rupestris*, Engelmann.   Arizona, Texas, New Mexico, Mexico.

Leaflets small, 7 to 15, ovate or lanceolate, never oblong, green beneath. Young shoots glandular-pubescent.

4. *Juglans nigra*, Linnæus.   Canada and United States, east of the Rocky Mountains.

Leaflets large, 15 to 19, ovate-oblong with long-acuminate apex, pale beneath. Young shoots glandular-pubescent.

5. *Juglans stenocarpa*, Maximowicz.   Manchuria.

Leaflets large, 11 to 13; all oblong, except the terminal one which is broadly obovate, pale beneath.   Young shoots glabrous.

III. *Leaflets serrate.   Leaf-scars with a transverse pubescent band on their upper edge.*

6. *Juglans cinerea*, Linnæus.   Canada and United States, east of the Rocky Mountains.

Leaf-scars semicircular, the upper edge straight and scarcely notched.   Leaflets, 11 to 13, oblong; serrations fine and directed outwards.

7. *Juglans Sieboldiana*,[1] Maximowicz.   Japan, Saghalien.

Leaf-scars obcordate, 3-lobed, notched above.   Leaflets, 13 to 15, oblong; serrations shallow, irregular, directed forwards; base rounded and unequal.

8. *Juglans mandshurica*,[1] Maximowicz.   Manchuria, Korea, China.

Leaflets and leaf-scars practically indistinguishable from those of the last species, though the leaflets are usually longer-acuminate.   Fruit, however, remarkably distinct.   See detailed description.

9. *Juglans cordiformis*,[1] Maximowicz.   Japan.

Leaf-scars and leaflets closely resembling those of *J. Sieboldiana*, the leaflets, however, fewer (11 to 13) and with a cordate base.

---

[1] These three species, though differing remarkably in fruit, are very similar in leaves and shoots.

JUGLANS REGIA, Common Walnut

Juglans regia, Linnæus, *Sp. Pl.* 997 (1753); Loudon, *Arb. et Frut. Brit.* iii. 1421 (1838).

A deciduous tree, attaining 100 feet in height and 15 to 18 feet in girth. Bark smooth and silvery grey in young trees, becoming ultimately more or less deeply fissured.

Leaves large, up to 10 inches long, coriaceous, of five to nine (rarely as many as thirteen) leaflets, sub-opposite or opposite, the terminal leaflet stalked, the others subsessile; elliptic, long-ovate or obovate, shortly acuminate at the apex, tapering and unequal at the base, glabrous on both surfaces, except for inconspicuous tufts of pubescence in the axils of the nerves on the lower surface; dark green above, paler beneath, entire or slightly sinuate in margin; exhaling an aromatic odour. Venation pinnate, with ten to fourteen pairs of lateral nerves, which run nearly straight to near the margin, where they curve forwards and join with the next vein. The leaflets diminish in size from the apex to the base of the leaf. Rachis glabrous, terminal leaflet not articulated. Young shoots glabrous, with yellow sessile glands and white inconspicuous lenticels.

Male catkins arising singly or in pairs (one above the other) above the leaf-scars of the previous year's shoots, green, two to five inches long, sessile, pendulous, thickly cylindrical and densely flowered; flowers with stalked bracts, two to five perianth leaves and two bracteoles; stamens ten to twenty; anthers oblong, apiculate. Female flowers, one to four, at the apex of the young shoots, green, with usually purple stigmas; involucre minute, indistinctly four-toothed; perianth green, with four linear-lanceolate divisions.

Fruit globular, about two inches in diameter; pericarp green, smooth, glandular-dotted, coriaceous, and very aromatic, splitting irregularly when mature. Nut very variable in shape, wrinkled and irregularly furrowed, thin- or thick-shelled; divided interiorly by two thin dissepiments into four incomplete cells; one dissepiment separating the two cotyledons, the other dissepiment dividing them into two lobes. The structure of the fruit of the walnut is very complicated, and the reader is referred for further details to Lubbock's paper[1] on the fruit and seed of the Juglandeæ.

The common walnut, according to Kerner,[2] is truly monœcious, the stigmas, however, ripening several days before the pollen is shed from the anthers.[3] The unripe male catkins have the flowers crowded together in a short thick spike directed upwards. As soon as the pollen develops the spike elongates to three or four times its former length and becomes loose and pendulous, the flowers

---

[1] *Jour. Linn. Soc. (Bot.)*, xxviii. 247 (1890). Cf. also Lubbock, *Seedlings*, ii. 506 *seq.* (1902). Malformed walnuts are occasionally produced, which are very curious. Cf. *Gard. Chron.* 1858, p. 5, and 1890, viii. 758, fig. 154.

[2] Cf. Kerner, *Nat. Hist. Plants*, Eng. trans. i. 742, fig. 184 (1898).

[3] This is not invariable. Delpino observed that while certain trees of the common walnut were protogynous, *i.e.* the stigmas ripening first, other trees were protandrous, the stigmas ripening after the anthers. In such cases the trees behave as if they were diœcious. Cf. Darwin, *Diff. Forms of Flowers*, 10 (1877), and Trelease, *Missouri Bot. Garden Report*, vii. 27 (1896).

separating from one another. The pollen then falls into a depression on the side of the neighbouring flower below, from which it is shaken out by the wind and carried to neighbouring branches of the tree, where it alights on the stigmas of the female flowers.

Seedling [1]

The cotyledons are large, fleshy, obovate, bi-lobed and crumpled, filling the cavity of the seed, from which they do not emerge on germination, but remain underground. The primary root makes its exit by the apex of the nut, and becomes stout and flexuose, giving off a few lateral fibres. The caulicle is very short, stout, and woody. Young stem, erect, compressed, glabrous, greenish, and covered with lenticels. The first four pairs of leaves are mere scales, opposite or sub-opposite on the stem. The ninth leaf is foliaceous, and consists of three leaflets, the terminal one large, obovate or elliptical, and cuspidate, the lateral ones small, oblong and alternate. The next leaf is five-foliolate; the terminal leaflet, oblong-obovate; the middle pair ovate, acuminate, oblique at the base, unequal, and sub-opposite; the basal pair small, ovate, oblique, and unequal. The last leaf is like it, or bears only four leaflets. All these primary leaflets are serrate in margin, and more acuminate than those of the adult plant, which are entire. In these respects they resemble the adult leaves of Carya or other species of Juglans.[2]

Identification

The common walnut is distinguishable in summer from all the other species by its glabrous, entire, few leaflets. In winter the following characters are available :—
Twigs stout, glabrous,[3] shining, greenish or grey, with scattered longitudinal lenticels. Leaf-scars on prominent pulvini, broadly obcordate, the upper margin deeply notched in the centre and not surmounted by a band of pubescence; bundle-dots in three groups. Pith large, white or buff in colour, with wide chambers. Terminal bud ovoid, obtuse at the apex, with four external grey tomentose scales in two valvate pairs, the scales not lobed at their apex and merely representing leaf-bases. In many cases, as in slow-growing old trees, the true terminal bud is aborted on most of the branchlets, and its scar marks the end of the twigs. Lateral buds small, arising at an angle of 45°, globose, the two outer scales usually concealing the inner ones, pubescent at first, but ultimately becoming glabrous. Superposed lateral buds occur only rarely.

Varieties

Two distinct geographical forms are known :—
(a) *typica*, in Europe, Asia Minor, Persia, and the Himalayas. Leaves elliptic; nuts ovoid-globose with thin septa.
(b) *sinensis*, C. DC. in *Ann. Sc. Nat.* 4 Sér. xviii. 33, figs. 38, 39. North

---

[1] Cf. Lubbock, *Seedlings*, ii. 516, fig. 661 (1902).   [2] Cf. Fliche, *Bull. Soc. des Sciences*, Nancy (1886).
[3] Some varieties of cultivated walnuts have the twigs covered with a minute pubescence.

China and Japan. Leaves oval or ovate. Nut globose, scarcely apiculate at the apex, sparingly wrinkled; septa thick and bony.

A large number of varieties have arisen in cultivation.

1. Var. *pendula*. Tree, pendulous in habit.

2. Var. *præparturiens*. A bushy shrub, producing fruit at an early period, some-times when only two or three years old. According to Carrière[1] it was obtained from seed by Louis Chatenay, a nurseryman at Doué-la-Fontaine, about the year 1830, the first mention of it being in *Ann. Soc. d'Hort. Paris*, 1840, p. 741. M. Chatenay found in the midst of a number of seedlings of walnuts three years old a single individual which bore fruit. This variety was put into commerce by M. Janin of Paris. According to Carrière, when the seeds of it are sown, different forms are produced, from young plants which bear fruit in their second year up to others which only produce fruit at an advanced age. The plants are also variable in size. The nuts are generally thin-shelled and small, but good in quality.

3. Var. *præcox*. Comes into flower and fruit a fortnight earlier than the common kind.

4. Var. *serotina*, Desfontaines. This variety flowers very late, and is recom-mended in localities liable to spring frosts. It is said[2] that of this variety, when sown, only three per cent came true, and flowered late in the season.

5. Var. *monophylla*. Leaves simple or trifoliolate. A small tree of this kind, which bears both simple and trifoliolate leaves, the basal pair of leaflets being very small, is growing at Bayfordbury, the residence of Mr. H. Clinton Baker.

6. Var. *rotundifolia*. Leaflets oval.

7. Var. *serratifolia*.[3] Leaves serrate. There is a specimen in the Kew herbarium from a tree in Germany, all the leaves of which were distantly serrate in margin. The leaves of young seedlings are always serrate; and this juvenile character is often retained in some walnut trees up to a considerable age.

8. Var. *laciniata*, Loudon. Leaves very deeply cut. The foliage of this variety is light and feathery, much more so than that of the common walnut, and is retained till late in the autumn. A fine specimen was reported in 1884 to be growing at Bicton.[4] Elwes has seen only three trees of this form, of which the largest, growing on a lawn at Westonbirt, was 30 to 40 feet high. Another was at Melbury, and a third, of no great size, at Poltalloch in Argyllshire.

9. Var. *heterophylla*. Leaflets variable, some of the ordinary form, others irregularly cut.

10. Var. *variegata*.[5] Leaflets with white margins.

11. A tree was growing in 1890 at Chawton Park, Alton, Hampshire, of which specimens with extremely narrow leaflets were sent to Kew.

The number of varieties of the walnut in cultivation, as regards the shape,

---

[1] *Rev. Hort.* 1882, p. 419.

[2] *Gard. Chron.* 1883, xx. 114. See *Rev. Hort.* 1861, p. 430, fig. 108. Called St. John's Walnut, as it does not put forth leaves till Midsummer or St. John's Day, in Parkinson's *Theatrum Botanicum*, 1414 (1640).

[3] The serrate-leaved walnut is mentioned by Parkinson, *loc. cit.* 1413.

[4] *Woods and Forests*, 1884, pp. 164 and 512. See also concerning this variety *L'Horticulteur Français*, 1862, p. 47.

[5] *Rev. Hort.* 1861, p. 429, fig. 104.

colour, and other qualities of the fruit, is very great; but a detailed description of these does not come within the scope of our work. The most remarkable is the huskless walnut [1] of North China, which is cultivated in the mountains to the north-west of Peking. In this curious form the husk is almost wanting, being very thin and irregular. In var. *racemosa* the fruits are numerous, fifteen to twenty-four, and are set close together on the peduncle. In var. *maxima*, Loudon (var. *macrocarpa*), the fruits are very large. The nuts are elongated and very narrow in var. *elongata* (var. *Bartheriana* [2]); very sharp-pointed at both ends in var. *rostrata*; and have very thin shells in var. *tenera*,[3] Loudon (var. *fragilis*). The kernel of the nut is bright red in var. *rubra* (var. *rubrocarpa*).[4]

HYBRIDS

I. *Juglans regia* × *nigra*. Two forms of this are well known in cultivation; they differ mainly in the character of the fruit.

1. *Juglans Vilmoriniana*, Carrière, *Rev. Hort.* 1863, p. 30. Young shoots glabrous. Leaf-scars obcordate, three-lobed, deeply notched above. Leaflets eleven to thirteen, ovate-lanceolate, sub-sessile, apex acuminate, base rounded or tapering; serrations fine and shallow, directed forwards; lower surface green and glabrous, except for conspicuous tufts of pubescence in the axils of the main veins. Rachis glabrous in the upper leaves of the shoot, pubescent towards its base in the lower leaves. Fruit with the thick husk of *J. nigra*. Nut smooth, globose, thicker shelled and more deeply furrowed than that of the common walnut.

In *Garden and Forest*, iv. 51 (1891), M. M. de Vilmorin gives particulars of the original tree in his garden at Verrières les Buisson, near Paris, and an excellent illustration of it in winter. He says that it was planted about 86 years previously as a young seedling by his grandfather as a memorial of the birth of his eldest son. Nothing certain is known of its origin, though it was supposed by Dr. Engelmann to be a hybrid, between the European and the black walnut. The characters of the bark, branchlets, and buds are intermediate; the leaves resemble those of *J. regia* more than those of *J. nigra*. The fruit, which is not produced every year, and never in quantity, is figured, and resembles most that of the black walnut. Of the few seedlings which have been raised from it one is growing beautifully in the Arboretum at Segrez, and produces fertile nuts. All the seedlings have grown well when planted in deep sandy soil mixed with clay. The tree at Verrières was seen by Elwes in 1905, and measured 95 feet high by 10 feet in girth, with a bole about 16 feet long. The habit of the tree was considered by him to resemble the black walnut rather than the common species.

There are young trees of *J. Vilmoriniana* growing at Kew, and one has been recently sent to Colesborne by M. de Vilmorin.

2. *Juglans pyriformis*, Carrière, *loc. cit.* 28, figs. 4 to 9. *Garden*, L. 478, fig. (1896).

[1] See Hance, in *Journ. Bot.* 1876, p. 50.
[2] Figured in *Garden*, L. 478 (1896); and *Rev. Hort.* 1859, p. 147, and 1861, p. 427.
[3] The thin-shelled walnut is mentioned in Parkinson, *Theatrum Botanicum*, 1413 (1640).
[4] See *Gard. Chron.* xxiii. 346 (1898). This variety is figured in *Wien. Illust. Gart. Zeitung*, 1898, p. 165.

Carrière states that this tree arose from a cross between *J. regia* and *J. nigra*. The leaves are identical with those of *J. Vilmoriniana*. The young shoots differ in having a glandular pubescence. The fruits are long-stalked and pear-shaped, but otherwise closely resemble those of *J. nigra*. Young trees of this kind are in cultivation at Kew.

3. Other hybrids between these species have been described. One mentioned by Sargent was an immense tree, found in 1888 by Prof. Rothrock on the Rowe Farm on the north bank of the Lower James River, Virginia. It is described as having the habit, foliage, and general appearance of *J. regia*, but producing a nut not unlike that of the black walnut, though longer and less deeply sculptured. The nut is exactly like that of *Juglans regia gibbosa*, Carrière,[1] which was raised by a nurseryman at Fontenay-aux-Roses in 1848.

De Candolle also described,[2] as *Juglans regia intermedia*, a tree which was found at the Trianon, and supposed to be a cross between the common and black walnuts. M. C. de Candolle informed Elwes that a similar hybrid exists at Geneva, and that its seedlings have characters intermediate between the two parents.

There are specimens at Kew, which were sent by Mr. E. Lyon in 1901 from Hurley, Marlow, where there is a fine old tree of *Juglans nigra*, from the seed of which plants were raised, which are apparently intermediate between that species and the common walnut.

II. *Juglans regia × cinerea*. *Juglans alata*, Carrière,[3] *Rev. Hort.* 1865, p. 447. This is described as having young shoots pubescent : leaflets seven to nine, with the end leaflet stalked, the others subsessile ; all oval or elliptic-lanceolate, abruptly acuminate, obscurely and remotely serrate, pubescent on both surfaces : rachis shortly pubescent. Three trees, presumably of this hybrid, have been observed near Boston in the United States ; and a description and figure of them are given in *Garden and Forest*, 1894, p. 435, fig. 69.

III. *Juglans regia × californica*. A remarkable hybrid between the common walnut and the Californian wild species, has been obtained by Luther Burbank, who names it " paradox."[4]

DISTRIBUTION

The common walnut has a very wide distribution, occurring wild in Europe in Greece, Bosnia, Servia, Herzegovina, Albania, and Bulgaria ; and extending eastward through Asia Minor, the Caucasus, Persia, and the Himalayas to Burma and North China and Japan. Its occurrence as an indigenous plant in Greece was first demonstrated by Heldreich,[5] who found it growing wild in Ætolia at Korax, in Phthiotis on the Œta and Kukkos mountains, and in Eurytania on Veluchi, Chelidoni, etc. It grows wild in Greece in mixture with oaks and chestnuts in great quantity, especially

---

[1] *Rev. Hort.* 1860, p. 99, figs. 21-23, and 1861, p. 428, figs. 101-103. Rehder considers this hybrid to be the same as *J. Vilmoriniana*.

[2] *Ann. Sc. Nat. Sér.* iv. xviii. t. 4.

[3] This is probably the same as *Juglans intermedia quadrangulata*, Carrière, *Rev. Hort.* 1870, p. 493, figs. 66-68.

[4] *Garden and Forest*, 1894, p. 436.   [5] *Verhand. Bot. Vereins Prov. Brandenburg*, 150 (1879).

in the moister valleys and ravines up to the region of the silver fir, at altitudes varying between 2200 and 4300 feet. Small woods of walnut, undoubtedly wild,[1] occur in Bosnia and Servia, especially on the north slopes of mountains rich in springs. It ascends in Herzegovina to 2400 feet, in southern Servia to 1400 feet, and in Albania to 2200 feet. Velenovsky[2] considers it to be truly wild in the Rhodope mountains. According to Radde[3] it occurs in the Caucasus, from the sea-level to 4500 feet altitude; also in Ghilan in North Persia. According to Meakin,[4] it is met with wild in the mountains not far from Bokhara. There are wild specimens at Kew from Armenia. According to Aitchison it is wild in Afghanistan, at 7000 to 9000 feet, and also in the Kuram valley. It occurs in the temperate Himalayas and Ladak, at altitudes of 3000 to 10,000 feet from Kashmir and Nubra eastward. Kurz met with it in the Shan Hills in Burma. It is cultivated throughout China, and appears to be indigenous in North China and Japan;[5] but other species of Juglans are much commoner in the wild state throughout China and Japan.

We are indebted to Sir W. Thiselton Dyer for the following :—

"The walnut found its western natural limit in Greece, but early made its way into Italy. Its classical name Juglans is *Jovis glans*, but in poetry it is always *Nux*. Virgil's *ramos curvabit olentes* hits off the acrid smell of the foliage. The nuts were thrown at weddings, as Virgil tells us, *sparge marite nuces*, because, amongst other reasons, Pliny says, they made the maximum of noise.

"*Relinquere nuces* was to put away childish things: so Catullus, *da nuces pueris iners*. The green rind enclosing the nut contains a dye used to darken the hair, the *viridi tincta cortice nucis* of Tibullus, in modern times more often the skin."

The walnut is extensively cultivated in France, Germany (except in the north where it ripens fruit rarely), and throughout southern Europe. It is cultivated chiefly in the region of the beech, as in Hungary up to 2160 feet, on the southern slopes of the Alps up to 3800 feet, in the Vosges up to 2200 feet. In Norway it is grown on the west coast as far north as Trondhjem, where it has reached a height of 30 feet, and in very favourable summers ripens fruit. Many other localities are mentioned by Schubeler, vol. ii. pp. 429-431. In Sweden it exists near Stockholm, and in Scania, at Cimbrishamn (55° 30′), Linnæus measured, in 1749, a tree 60 feet high.

(A. H.)

PROPAGATION AND CULTIVATION

If the walnut is wanted as a fruit-bearing tree it is better to procure from a nurseryman grafted or budded trees of some of the large-fruited, thin-shelled sorts, which have been raised in France; and which grow best in the south and east of

---

[1] Beck von Mannagetta, *Vegetationsverhält. Illyrischen Ländern*, 219 (1901).

[2] *Flora Bulgarica*, 512 (1891).

[3] *Pflanzenverbreitung in Kaukasusländern*, 170, 182.

[4] *Russian Turkestan*, 23 (1903).

[5] It is included as a wild plant in Japan by Matsumura in *Shokubutsu Mei-I*, 155 (1895); but Sargent in his *Forest Flora of Japan*, p. 60, says, "It is occasionally cultivated in the neighbourhood of temples and as a fruit tree; but we saw no evidence of its being anywhere indigenous, and it is probable that it was introduced from Northern China, where one form of this tree apparently grows naturally."

England. The process of budding or grafting them is fully described by Loudon, p. 1431, and need not be repeated here.

If, however, walnuts are to be planted for timber or ornament, it is far better to raise them from nuts, which may be sown as soon as they are ripe, if they can be protected from mice and vermin; or kept in sand until February, when they should be sown two to three inches deep in rich light soil, which will encourage the production of fibrous roots at an early period. As the large strong tap-root makes the tree difficult to transplant, it should be undercut with a spade about six inches below the soil in the first year, or the nut may be allowed to germinate before sowing and the end of the root pinched off. If this is not done they must be carefully transplanted in March, and protected from late spring frost as much as possible until they have made stems four to six feet high. For though the walnut is one of the latest trees to come into leaf, none is more tender as regards spring frost, and as it does not bear pruning well and has a natural tendency to form branches rather than a clean stem, it is important that the trees should be carefully trained when young.

It is now much less planted than formerly, and the wood is not so much valued by country timber merchants as it ought to be, but there is no reason why it should not be treated as a forest tree on suitable soils, and drawn up among other trees with the object of growing clean timber; though I consider it inferior to the black walnut in this respect. It is evidently a lover of a warm soil and climate, and though on good limestone soil or deep loam resting on chalk it grows fast and to a great size, it should not be planted on heavy clay, on poor sand, or in exposed windy situations.

The walnut is very seldom blown down on account of its strong roots, and I have never seen one struck by lightning. It does not reach a very great age; so far as I know, 200 years is about the limit of its life, and many trees become hollow or decayed before attaining as much as this.

The only place where I have seen walnuts self-sown in England is at Holkham, where, in the Triangle plantation, are several trees, one 17 feet high, in a fairly thick plantation of larch and Scots pine on light sandy soil. They are 100 to 150 yards distant from the parent tree, the nuts having probably been carried by squirrels or rooks. On the sandhills at the same place I saw a self-sown tree five to six feet high, and on the roadside near Colesborne a young tree has sprung up from a nut dropped by a passer-by.

Mr. E. Kay Robinson[1] mentions the occurrence of young walnut trees amidst clumps of other large trees, due to the carrying away by rooks of the fruit from an old walnut tree in a garden near by. He has kindly sent us a photograph of a walnut tree growing in a field at Warham, near Wells, Norfolk, which had evidently been deposited by a rook, as the young tree in its growth had thrust up the roots of an old willow tree, amongst which it had grown.

[1] *Garden*, lxvi. 412 (1904).

REMARKABLE TREES

Though there are many very fine walnuts scattered through the southern half of England I cannot say where the largest tree actually is. Nothing that I know of now living equals a tree recorded by Mr. W. Forbes,[1] which grew on the estate of Sir Charles Isham at Lamport Hall, Northamptonshire, and was sold to Messrs. Westley Richards, gunmakers of Birmingham. According to the measurements given, this tree contained 816 cubic feet of sound wood, of which the butt, measuring 12 feet by 18 feet in girth, contained 243 feet and one limb 108 feet.

A magnificent tree, said to have been the largest in England, grew at Cothelstone, near Bishops Lydeard, Somersetshire, which Loudon records as being 64 feet high and $6\frac{1}{2}$ feet in diameter,[2] but I am informed by Mr. E. V. Trepplin, agent to Viscount Portman, that it was blown down some years ago.

No tree mentioned by Loudon equals the one of which I give a figure (Plate 74), which grows in front of the house at Barrington Park, near Burford, Oxfordshire, the property of Mr. E. C. Wingfield, on an oolite formation. This tree measured in 1903, 80 to 85 feet in height by 17 feet in girth, and has a fine bole and a very burry trunk. There are two other splendid walnuts in this park nearly as tall and over 15 feet in girth, and others have been cut down of which the timber, when cut up in London, was considered by Mr. A. Howard equal in colour and figure to Italian walnut. At the Moot, Downton, Wilts, the residence of my old friend Mr. Elias P. Squarey, are four fine walnut trees, one of which was said by Mr. D. Watney to be the largest he had seen during his long experience as a valuer, and estimated to contain over 400 feet. It measures 17 feet 2 inches in girth, with a short butt dividing into four big limbs which run up to about 80 feet in height. Another is the tallest walnut I have ever seen or heard of, and measured in 1903 about 100 (perhaps more) feet high by 13 feet in girth.

In the village street of Bossington, Somersetshire, I was shown by Mr. S. F. Luttrell of Dunster Castle, a very picturesque old gnarled walnut tree which at 5 feet is 17 feet in girth, but the roots are so spreading that the trunk, measured close to the ground and following the sinuosities, is 35 feet round. A walnut of apparently no great age in a field at Cobham village in Kent measured in 1905 about 70 feet by 13 feet, and the branches spread over a circumference of 99 paces.

An avenue of walnuts is seldom planted in England, but at Moor Court, Herefordshire, there is a short one which from an illustration in the *Gardeners' Chronicle* of February 6, 1875, seems very effective. They are 60 to 70 feet high and 10 to 12 in girth.

At Sudeley, Gloucestershire, the seat of Mr. H. Dent Brocklehurst, there are in a line before the Castle four beautiful trees of great age, the largest measuring 90 feet by 14 feet, and in Rendcombe Park near the Temple there is a fine old tree about 80 feet by 15 feet whose branches cover an area 105 paces in circumference.

---

[1] *Trans. Eng. Arb. Soc.* v. 155.

[2] In *Trans. Eng. Arb. Soc.* ii. 225, measurements of this tree made in 1888 by Dr. Prior are given as follows :— height, 94 feet 6 inches ; girth, 18 feet ; spread, 22 yards by 27.

At Laverstoke Park, Whitchurch, Hants, the residence of Mr. W. W. Portal, there is a fine well-shaped walnut, which was measured by Henry in August 1905, as 80 feet high by 13 feet 8 inches in girth, with a bole of 12 feet, dividing into two stems above.

In the eastern counties there must be many fine walnuts, but the only one of which I have any exact record is a tree which was figured by Grigor[1] at Ketteringham Park, Wymondham, Norfolk, the seat of Sir M. Boileau, Bart., and is said to have been planted at the restoration of Charles II. This tree was one of the best shaped as regards its branches that has been figured, and measured in 1841 68 feet high with a girth of 12 feet.

At Rickmansworth, Herts, Sir Hugh Beevor measured in 1901 a tree 98 feet high by 11 feet 9 inches in girth, the first limb coming off at 18 feet up, the second limb at 36 feet from the ground. At Gayhurst, near Newport Pagnell, Mr. W. W. Carlile showed me a tree growing on a clayey limestone, which, though of great age, is absolutely sound and has lost hardly a branch. It measures no less than 80 feet by 17 feet, and is very perfect in shape. At Ware Park, Herts, Mr. Baker tells us of a tree 16 feet 4 inches in girth, and this seems to be a district where the walnut comes to great perfection. He showed me another of the thin-shelled French variety growing close to the bank of the Lea at Roxford, which, though much cut by frost, was 16 feet in girth.

At Castle Howard, Yorkshire, there is a large tree in the park near the stables, growing among beech and oak which have drawn it up to a height of 80 or 90 feet, though it leans very much to one side. It has a clean bole about 20 feet long by 11 feet 8 inches, dividing into two long straight clean stems, a very unusual form in this tree.

In Scotland the walnut is not so much at home as in England, but in the warmer parts of the east and in Perthshire it attains considerable dimensions. The best that I have seen myself is a tree at Gordon Castle (Plate 75) which in 1904 measured 60 feet by 10 feet, and is, considering the exposed position and latitude, a remarkable tree. But there must have been a still finer one here in 1881, when Mr. J. Webster, father of the present gardener, recorded in the *Trans. Scot. Arb. Soc.* ix. 63, a tree of equal height and 13 feet 4 inches in girth at 5 feet. Col. Thynne has given me a photograph of a fine tree at Cawdor Castle, Nairnshire, which measures 65 feet by 15 feet 7 inches.

Hunter records several very fine trees in Perthshire as follows : "At Gask the largest tree in the policies is a walnut, a little west of 'The Auld House.' It measures 17 feet 5 inches at 5 feet and then swells to a girth of 21 feet at 8 feet from the ground, and at Blair Drummond there is a fine tree," which Mr. A. Morton, the gardener, informs me is now about 80 feet by 13.

Though the walnut is not uncommonly planted in Ireland, we have seen none remarkable for size. The largest one is reported to be growing at Kilkea Castle in Kildare.

[1] *Eastern Arboretum*, 279 (1841).

TIMBER

Until mahogany became common in England about the middle of the eighteenth century, walnut was considered the most valuable wood for furniture, carving, and inside work, and on the Continent most of the best old furniture was made from it. Later it became very valuable for gun-stocks, and is still almost the only wood used, for all except cheap guns.   Loudon states that during the long wars at the beginning of the last century in France no less than 12,000 trees were cut annually for gun-stocks, which caused it to become very scarce, and in England as much as £600 was paid for the wood of one tree.

Sir W. Thiselton Dyer informs me that when for political reasons the War Office thought it no longer desirable to depend on walnut, which was mostly imported from the Black Sea, he was consulted as to what other wood might be found as a substitute ; but though some twenty sorts of colonial woods were sent for trial from the Museum at Kew to the Small Arms Factory at Enfield, none except the black walnut was found to be at all suitable.

The reason for this is that walnut wood does not warp, and can be cut cleanly in any direction to fit the locks and mechanism of the magazine rifle, and is not liable to swell and bind the lock when wet.   But it requires a good deal of care in selection and in cutting out the stocks, so that they are not liable to break at the grip ; and the best gunmakers in England obtain their stocks ready cut to specified sizes from French merchants who make a *spécialité* of this trade.

Maple wood has been found suitable in Japan, for when I was there during the late war, I saw numbers of roughly shaped gun-stocks of that wood being cut in the forest near Koyasan, and carried out on men's backs to supply the immense demand of the arsenal.   But in England it was found to make a rifle stock 4 ounces heavier than walnut, and is also liable to warp.

The late Mr. J. East told me that, in the year 1838, at Missenden in Bucks, four walnut trees were sold in one lot for £200, and about the same time two other trees were sold for £100 each, but the demand is now so much lessened by foreign importations, and by the substitution of other woods, such as mahogany and American walnut, that its average price now is not more than from 1s. 6d. to 3s. per foot.

The wood requires a long time to season thoroughly, and should not be used for good work until three to six years after felling, as it is liable to shrink considerably.   It is also liable to be ring shaken, and has another great defect in the fact that the sapwood, which forms a large proportion of most trees, is pale in colour and very liable to be attacked by wood-eating beetles.   Almost all the old Italian furniture which I have seen is more or less damaged in this way, and though the sapwood is often stained so as to look like the heartwood, it is better in first-class work only to use the latter.

As a rule English walnut does not show so much of the dark markings as is found in the logs imported from Italy and the Black Sea, and Italian walnut is usually specified by English architects.   But I have seen such fine panelling made from English wood alone that I have no hesitation in saying that with careful selection and seasoning, the best effect can be obtained from old trees grown on dry soil in this country; and in a small work on English timber by "Acorn"[1] it is stated that the home-grown timber is harder and more durable than the foreign.

The finest wood as regards colour and pattern comes from near the root, and from the forks in the tree, which, however, are liable to twist if used in the solid, and in order to obtain as much as possible of these figured pieces the tree, if old, should be grubbed, and great care taken in cutting it up into suitable thicknesses for the purpose for which it is wanted. The forked parts should be cut into thin veneers and matched as well as possible. For panelling, walnut comes only second to oak, and is found in some of the best houses in England. As a fine example of Italian walnut panelling I may mention the billiard room at Edgworth, near Cirencester, which was designed for my friend Mr. Arthur James, by Mr. Ernest George. Of modern English walnut panelling I have seen a good example put up in Mr. Franklin's beautiful old house, Yarnton Manor, near Oxford, which he has recently restored, and in which the panelling both of oak and walnut is admirable. The late Mr. Holford of Westonbirt, Tetbury, had his large music-room entirely fitted with walnut cut on his own estate.

A newer system of using walnut wood in large knife-cut unpolished veneers is now adopted by modern decorators, of which a fine example may be seen in the board room of the Royal Insurance Company at Liverpool.

One of the most valuable woods in the world is produced by the burrs or excrescences which are produced on the walnut tree, rarely in England, but more commonly in its native country, and which are sought for by agents travelling for French firms at Marseilles, who seem to have a monopoly of this wood. Sometimes they are very large, measuring two to three feet in diameter, but more usually smaller, and are sold at very high prices, as much as £50 to £60 per ton, according to Laslett. They are called *loupes* in France, and are cut into very thin sheets to cover the very finest pianoforte cases, and much used for cabinet-making. These burrs are said to grow on trees in mountainous and inaccessible regions in Circassia, Georgia, North Persia, and Afghanistan; and I am told by Mr. C. W. Collard that those now imported are not so fine as they used to be some years ago.

I can find no records or measurements of walnuts abroad which show that it ever exceeds in warmer climates the size it attains here; but the largest foreign log which I have ever seen was shown by Messrs. Riva and Massara of Milan at the Exhibition held there in 1906. This log measured about 28 feet long by 15 feet in girth, and was said to contain about 16 cubic metres of timber, equal to about 560 feet. Its weight was 14,800 kilogrammes, and I was informed by the owners that they paid 5800 francs (about £232) for it in Switzerland. But Correvon[2] quotes *La Patrie Suisse* to the effect that a walnut was cut at Bois-de-Vaux, near Lausanne, which required twenty-four horses to haul it. The lower part of its trunk measured about 24 feet, the diameter was 6 feet 4 inches, and the total contents about 700 cubic feet. This butt was sold for £150 to make gun-stocks.

(H. J. E.)

[1] Published by W. Rider and Son, London, 1903.　　　　[2] *Nos Arbres*, 267 (1906).

JUGLANS NIGRA, BLACK WALNUT

Juglans nigra, Linnæus, *Sp. Pl.* 997 (1753); Loudon, *Arb. et Frut. Brit.* iii. 1435 (1838); Sargent,
Silva N. America, vii. 121, tt. 333, 334 (1895), and *Manual Trees N. America*, 128 (1905).

A tree attaining 150 feet in height, with a girth of about 15 to 20 feet, forming in the forest a narrow round-topped head, but with spreading branches when isolated. Bark of old trees dark brown, deeply furrowed with broad ridges, which are scaly on the surface.

Leaves up to 3 feet in length, of fifteen to twenty-three leaflets, which are ovate or ovate-lanceolate, long-acuminate at the apex, rounded at the base, sub-sessile, with coarse sharp irregular serrations; upper surface with a very minute and very scattered pubescence; lower surface with numerous glandular and simple hairs. Rachis with yellow glands and scattered glandular hairs. Young shoots with sessile yellow glands and numerous glandular hairs; older shoots pubescent. Leaf-scars obcordate, deeply notched at the apex, without any band of pubescence on their upper edge.

Staminate catkins three to five inches long; scales with six orbicular concave pubescent lobes, and a bract $\frac{1}{4}$ inch long, which is triangular and tomentose; stamens twenty to thirty. Pistillate flowers, two to five in a spike; involucre laciniate in margin or reduced to an obscure ring below the apex of the ovary; perianth lobes ovate, acute.

Fruit solitary or in pairs,[1] globose or slightly pear-shaped, pubescent, not viscid, yellowish green, $1\frac{1}{2}$ to 2 inches in diameter; nut oval or oblong, $1\frac{1}{8}$ to $1\frac{1}{2}$ inch, deeply ridged irregularly, four-celled interiorly at the base, and slightly two-celled at the apex.[2]

IDENTIFICATION

In summer it is readily distinguishable from *J. cinerea* and the Eastern Asiatic species, which have serrate leaflets, by the character of the leaf-scar, which is deeply notched at the apex and without the transverse band above its upper margin, which characterises those species. The long acuminate pubescent leaflets distinguish it from the hybrids *pyriformis* and *Vilmoriniana*. It has much larger leaflets than *J. rupestris*, and cannot be confused with *J. stenocarpa*, which has a broadly obovate terminal leaflet.

In winter the following characters are available:—Twigs stout, reddish brown, glandular-pubescent; lenticels small. Leaf-scars on prominent pulvini, obcordate, deeply notched above, without pubescent band, with three groups of bundle-dots. Pith large, buff-coloured, with wide open chambers. Terminal bud ovoid or conical, grey-

---

[1] A tree at Albury, Surrey, has, however, borne fruit in clusters of three, four, and six, of which specimens are preserved at Kew.

[2] For a detailed account of the fruit, seed, and cotyledons of the species, see Lubbock, *Seedlings*, ii. 517 (1902).

tomentose, usually with four external scales visible in two valvate pairs, the scales not lobed at the apex. Lateral buds, arising at an angle of 45°, small, globose, pubescent, with two to three scales visible externally ; there are often two buds superposed, the lower one minute and embedded in the notch of the leaf-scar. The reddish-brown pubescent twigs and superposed pubescent lateral buds will distinguish this species from the common walnut.

Varieties and Hybrids

No varieties are known. The Black Walnut forms hybrids with the common walnut, which have been dealt with under the latter species. Burbank has raised a hybrid, which he calls "Royal," between *J. nigra* and *J. californica.*[1]

Juglans nigra × *cinerea.* A tree supposed to be this hybrid grew in the Botanic Garden at Marburg, and was described by Wender as *Juglans cinerea-nigra* in *Linnæa*, xxix. 728 (1857). (A. H.)

Distribution

According to Sargent the black walnut occurs in rich bottom lands and fertile hillsides, from western Massachusetts to southern Ontario, southern Michigan and Minnesota, central and northern Nebraska, eastern Kansas, and southward to western Florida, central Alabama, and Mississippi, and the valley of the San Antonio River, Texas ; most abundant in the region west of the Alleghany Mountains, and of its largest size on the western slopes of the mountains of North Carolina and Tennessee, and on the fertile bottom lands of southern Illinois and Indiana, south-western Arkansas, and the Indian Territory.

The black walnut is not only one of the largest deciduous trees throughout a great part of the Middle States, but also one which, until it became too scarce, furnished a great part of the most valuable hardwood. It reached its maximum of size and abundance in the western foothills of the Southern Alleghany Mountains and in the rich, fertile alluvial river bottoms through which the great rivers of Ohio, Indiana, Tennessee, and Kentucky flow, and which were the first homes of the settlers who crossed the mountains towards the end of the eighteenth century, and for a quarter of a century carried on an unceasing warfare with the Indians. These pioneers also waged war against the forest whenever they could spare time, and for many years used the black walnut for fencing and for house-building, because it was an easy wood to split and to work; but they did not foresee that the trees which they destroyed would become one of the most valuable products of their farms, and would in a century be almost extinct as timber trees in many places where they were formerly the commonest.[2]

When I was travelling in the mountains of North Carolina in 1895, I saw but

---

[1] *Garden and Forest*, 1894, p. 436.

[2] An interesting account of the war waged against the black walnut by pioneers in Indiana in 1834 is given in *Woods and Forests*, 1884, p. 633.

few black walnuts of large size, and met with men who were travelling about purposely to find and buy them in all accessible places.   In the North Carolina forestry exhibit at the St. Louis Exhibition in 1904, I saw a walnut log from a tree in Jackson County, Kentucky, over 12 feet long and 52 inches in diameter which had evidently been lying long in the forest, and had been repeatedly burnt over, which produced over 800 cubic feet of timber, and was sold, as I was told, for $800.   I heard of another still standing in Kentucky which was valued at $1000.

These great trees are now hardly to be seen except in remote regions where it is impossible to get them out, and when I visited the Lower Wabash Valley in southern Illinois, where Prof. R. Ridgway[1] found the largest deciduous trees in the United States, I did not see one of great size.   Dr. J. Schneck, who was my guide and who knows the flora of this region better than anyone, gives in his *Catalogue of the Flora of the Lower Wabash*, the measurements of a tree taken by himself as follows:—Circumference, at 3 feet above the swell of the root, 22 feet; height of trunk to first branch, 74 feet; total height, 155 feet.   Prof. Ridgway measured another 15 feet in girth at 3 feet, and 71 feet to the first branch, where the trunk was 3 feet in diameter.   Assuming such trees to have measured 12 feet in girth in the middle they would contain 600 to 700 feet of clean timber in the first length alone, and now be worth as much as many acres of the land they grew on would fetch when cleared for agriculture.

But in regions which have colder summers and poorer soil, the black walnut does not attain anything like these dimensions, and I have seen none in New England which equal the best trees in Britain.   Emerson[2] speaks of one in the Botanic Garden at Cambridge, Mass., as measuring 6 feet 3 inches at 3 feet from the ground, and the tree which he figures growing near Roslyn was a poor specimen of small size.

In Canada it was once abundant in the rich forests of Southern Ontario, but almost all the old trees have been cut down, and plantations are now being made in various parts of Ontario and Western Quebec, and in Alberta and British Columbia, as well as in many parts of the United States from Kansas to California.

Black walnuts of great size are indeed now so rare that I have been unable to procure a really good photograph of the tree in its native forest, and there is none in Pinchot and Ashe's *Timber Trees of N. Carolina*.   These authors say that it bears seed abundantly only every three or four years, and that young seedlings are not common except in low fertile, rather open lands, or in meadows which border streams.   The growth is very rapid until the tree has reached a large size; only small trees send up shoots from the stump.

The tree, however, has been so largely planted in many parts of the States and in Canada, and succeeds so well, even so far west as British Columbia, that it may again become generally useful as a timber tree.

[1] *Proc. U.S. Nat. Museum*, 1882, p. 49.   [2] *Trees of Massachusetts*, i. 213.

Cultivation

The black walnut was first described by Parkinson,[1] and was introduced into England by the younger Tradescant before 1656, as it is mentioned in the list[2] of the plants growing in his garden at that time. A tree was growing in Bishop Compton's garden at Fulham in 1688, according to Ray.[3]

The black walnut is easy to grow from seed, but, except the hickories, none is more difficult to transplant, on account of the long fleshy tap-roots which it forms at an early age, and which, when grown in the good deep soil which it likes, are at a year old often three or four times as long as the seedling itself. For this reason, unless special care is given to its treatment, it is not likely to become so fine a tree as when sown *in situ*, and, though I have successfully transplanted many at one or two years old, I would much prefer the other method.

Though the nuts ripen in England in hot summers, they are not so large, and do not, I think, produce such strong plants as those imported from North America, and, if possible, I should prefer to get them from trees growing in Canada or New England than from farther south.[4] The nuts are best sown when ripe, as if kept dry for some time, they either lose their germinating power or come up so late that they make weak plants. In any locality which is subject to late frosts it would be better to sow them in boxes at least two feet deep and plant them out when a year old, as like many exotic trees they do not ripen their young wood well, and are liable to be frozen back in winter or spring, which induces a bushy instead of a straight habit of growth.

As this tree requires to be well sheltered and drawn up by surrounding trees in order to form a tall and valuable trunk, it should be sown or planted in small deeply-dug patches in a rich wood, kept free from weeds and protected from mice, rabbits, and boys, until the trees are six to eight feet in height, which they should be under favourable circumstances at four to six years after sowing.

All these difficulties have made the tree unpopular with nurserymen, who rarely care to grow trees for which there is little regular demand. But the great value of the timber, its rapid growth on suitable places, and its perfect hardiness when once established, give it, in my opinion, so much importance, that, however troublesome it may be in its early stages, it should be tried at least on a small scale as a timber tree in the warmest and best soils of the southern, eastern, and west midland counties. For further particulars of the nursery treatment of this tree see Cobbett's *Woodlands*, Art. 553; or *Arboriculture*,[5] iv. 7, July 1905. Cobbett,

---

[1] *Theatrum Botanicum*, 1414 (1640). [2] *Museum Tradescantianum*, 147 (1656).

[3] *Historia Plantarum*, ii. 1798 (1688)—no doubt the tree mentioned by Loudon as existing in 1835 (see p. 268).

[4] But the question as to whether the seeds of trees grown in a comparatively cold climate produce hardier plants than seed from a warm one, is as yet unsolved; and Prof. H. Mayr of Munich, than whom there is no better authority, is inclined to believe that the differences which are observed in the comparative resistance to frost depend on the variable constitution of the individual plant rather than on inherited power. — Cf. H. Mayr, *Fremdl. Wald. u. Park-bäume* (Berlin, 1906).

[5] A magazine of the International Society of Arboriculture; J. P. Brown, Connersville, Ind., U.S.A.

who knew the tree well, considered as I do that it was a hardier and better timber tree than the common walnut.

The black walnut cannot be expected to attain great size except on deep soil in a warm situation. A tree grown from a nut, brought by my father from America over 60 years ago, is now only about 60 feet high and 3 feet in girth, on the dry oolite of the Cotswolds; whilst in Kent, on good loam, it has attained 100 feet by 12 feet in about 100 years, and probably contains as much timber, and that of twice the value, as any oak of its age in Great Britain. It seems indifferent to the chemical nature of the soil, if it is deep, light, and well drained, and should have a southern or western aspect.

It is stated in *Woods and Forests* that the tree is almost if not entirely rabbit-proof, for when nearly everything else is barked it is left untouched, even in a young state.

I have no certain knowledge as to the age which this tree attains, but from the fact that the old ones at Fulham Palace and Syon are dead or dying, I should suppose that, like the common walnut, it is not a very long-lived tree.

CULTIVATION IN GERMANY AND FRANCE

The high value of the timber of the black walnut has led to experiments with the tree in Continental forests. These trials have, however, hitherto been only on a small scale.

In the State forests of Prussia the black walnut has been planted in twenty-two different stations, the whole area under cultivation being thirty-one acres, the separate plots varying in size from seven acres to a rood. Schwappach[1] draws the following conclusions from the results obtained in these experimental plots :— Of all the exotic species which have been tried in Prussia, *Juglans nigra* is the most exacting as regards soil and climate. It only thrives on deep moist rich soils, such as loamy sand rich in humus or pure loam, and never succeeds on shallow soils of any kind, or on wet clay or sand. It requires for its good development a considerable amount of warmth and a long season of vegetation. It has only succeeded on the best oak soils, such as the alluvial lands by the rivers Oder, Mulde, and Elster, and in certain restricted areas of the hilly land of central and western Germany.

Schwappach gives a description of the long tap-root of the seedling, and the consequent difficulty in transplanting; but he lays stress upon the fact that in Germany the seedlings normally make their appearance very late, and believes that this is one of the main causes of failure, as the young plants do not then ripen their wood, and are destroyed by late frosts. He advocates the early germination of the seeds by artificial means, such as by placing them in pits covered with straw, soil, and horse-dung. These speedily germinate, and are then planted in the forest in gaps of about a rood in extent, which are the result of clear felling, or under

[1] *Ergebnisse Anbauversuche Fremländischen Holzarten,* 37 (1901).

the existing canopy of an old wood where the trees will soon be removed. The black walnut requires strong sunlight for its successful growth, yet lateral protection is necessary during the first few years. Heavy shade is hurtful, as it hinders the ripening of the wood of the shoots. The black walnut, after it has successfully passed the dangerous period of youth, becomes perfectly hardy; and older plants resist both spring and winter frosts. Schwappach advocates close planting, with beech or hornbeam as nurses, and recommends thinning at 15 to 20 years old, to remove badly-shaped trees, and to give more light to those which are destined to remain.[1]

In France Henry has seen a small plantation of black walnut near Annecy; but the results obtained were unsatisfactory, as the young plants had suffered much from frost. M. Pardé,[2] however, strongly recommends its cultivation, and points out that, unlike the common walnut, it can be grown as a forest tree; and states that at Les Barres it sows itself regularly and abundantly.

REMARKABLE TREES

The largest tree which we know of in England is growing in the London County Council public park of Marble Hill, Twickenham, in rich alluvial soil close to the Thames. It was measured by Sir Hugh Beevor and Dr. Henry in August 1905, and the height was found to be 98 feet, the stem girthing at 5 feet up 14 feet 3 inches. The bole is about 18 feet long, dividing into two great limbs, with large spreading branches, forming a beautifully symmetrical crown. The diameter of the greatest spread of the branches was 93 feet (Plate 76).

Perhaps the next finest tree now standing in England is the one which I figure (Plate 77), and which grows on a bank at The Mote, near Maidstone, the property of Sir Marcus Samuel, Bart. I have twice measured this tree, first in 1902, when I made it 103 feet by 12 feet in girth, and again in 1905, when I made it 101 feet by 12 feet 6 inches. I am informed by Mr. Bunyard of Maidstone that it was probably planted about 100 years ago by his grandfather. The tree is so healthy, and apparently growing so fast, that it may become very much larger than it now is. At Gatton Park, Surrey, the seat of J. Colman, Esq., there was, in 1904, a tree about 100 feet by 9 feet 6 inches in girth, with a very tall, handsome trunk. Another at the same place, which, when I saw it, was lying on the ground, was about 95 feet by 9 feet, with a bole 10 feet long, and, according to the measurement given me by the late Mr. Cragg, agent for the estate, contained 315 cubic feet of timber.

At Highclere, Berks, there is a fine tree 90 feet by 9 feet 6 inches; and at Bute House, Petersham, Henry measured one 78 feet by 11 feet 10 inches in 1903. At Burwood House, Surrey, Col. Thynne has measured a tree, which I have not seen,

[1] Mr. John Booth of Berlin, who has for many years been one of the best advocates for the planting of exotic trees for timber, tells me that the black walnut has been largely planted near Strassburg under the direction of Forstmeister Rebmann, and the results are extremely successful.

[2] *Les Principaux Végétaux Ligneux Exotiques*, p. 21.

79 feet by 12 feet. At Syon House there was a fine tree mentioned by Loudon, as then 79 feet high and 2 feet 11 inches in diameter. In 1849, according to the manuscript catalogue of trees at Syon, it was 90 feet by 7 feet 3 inches; when I saw it in 1903 its top was gone, the tree fast decaying, and the girth about 10 feet.

At Youngsbury, near Ware, Herts, there are two fine trees which Mr. H. Clinton Baker measured in March 1906. The larger was 90 feet high by 11 feet 10 inches in girth; the smaller 80 feet by 11 feet 3 inches. At Albury, Surrey, near the gardener's cottage, there is a tree which measured in 1904, 90 feet by 9 feet 2 inches. At Arley Castle a black walnut is bearing mistletoe. At Barton, near Bury St. Edmunds, there is one which is about 75 feet by 7 feet, which cannot be more than about 60 years old.

Sir Hugh Beevor reports a fine tree, 80 feet high by 12 feet girth, at Spixworth Hall, Norfolk. In the rooms of the Hall there is some flooring made of locally-grown black walnut. At Wimpole, he measured another tree 78 feet by 12 feet 8 inches.

At Strathfieldsaye there is a plantation of eighteen young black walnuts in a group on the lawn, which, though about eighteen years old when I saw them in 1903, were only 8 to 10 feet high. Three others raised at the same time but planted out younger are twice as high. This seems to me to prove the importance of not keeping this tree long in the nursery. A fine tree on the other side of the house at the same place is about 80 feet by 7 feet, and had a few nuts on it even in the wet season of 1903.

At Fulham Palace there was a tree, which, according to Loudon, was 150 years old in 1835, being then about 70 feet high and 5 feet in diameter. In 1879[1] this tree was 16 feet in girth breast-high, and had passed its prime; and has been quite dead for ten years. This is the largest girth of any black walnut recorded in England.

At Bisham Abbey, near Marlow, the property of Sir H. J. Vansittart Neale, growing in a grove near the garden, where they have been drawn up by other trees, are four fine black walnuts, of the age of which there is no record. The tallest is nearly if not quite 100 feet high, with a clean bole about half as long, and a girth of 8 feet 2 inches; the others have shorter trunks, the biggest being 10 feet 3 inches in girth, and another 8 feet 6 inches, but are nearly as tall.

At Corsham Court, Wilts, the seat of Lord Methuen, is one of the finest specimens in England, with a clean trunk about 35 feet without a branch and 11 feet 5 inches in girth. It is 75 to 80 feet high, and has a very spreading crown of drooping branches, which cover a space 30 yards across. At Lacock Abbey, near Corsham, the seat of Mr. C. H. Talbot, are some good trees planted by the grandfather of the present owner between 1780 and 1800, of which the largest is about 100 feet by 11 feet 5 inches, with a bole of 8 feet, but this has ceased to bear nuts. The others were planted subsequent to 1828, and the best of them is 60 to 70 feet high by 7 feet girth, and bears nuts profusely.

[1] Figured in *Gard. Chron.* 1879, xi. 372, t. 52. *Cf.* p. 265.

At Walsingham Abbey, Norfolk, the seat of H. Lee Warner, Esq., there was a specimen figured in Grigor's *Eastern Arboretum*, p. 300, as a tree clothed to the ground with foliage, and of which the spreading branches were propped up. In 1840 it was 8 feet in girth, with a spread of branches 165 feet in circumference, but is now much decayed.

At Brightwell Park, Oxon, the property of R. Lowndes Norton, Esq., there are three or four well-grown trees about 50 years old, the largest of which measures 68 feet by 5 feet 10 inches, and bears fruit abundantly. The leaves of these trees were conspicuous by their yellow colour in the first week of October.

At all the four last-named places these trees have been known as hickories, and it is probable that others of the so-called hickories in England are really black walnuts.

Two trees[1] growing close together at The Firs, Manor Lane, London, S.E., both measured, in 1886, 10 feet 9 inches at 4 feet above the ground, and were estimated to be 90 feet high. They were then in excellent health, and bore good crops of nuts, which, however, were rarely perfectly developed.

Many other trees no doubt exist in old places south of the Thames; but we have never seen or heard of any large ones in the midland or western counties. Sir Charles Strickland, however, tells us that the black walnut is quite hardy in Yorkshire; and that he has trees at Hildenley, 15 to 20 feet high and ripening seed, whilst at Housham, another place of his in the same county, they thrive even better in the woods, where they look like becoming fine timber trees.

In Ireland, the largest tree seen by Henry is at Ballykilcavan, Queen's County: it measured in 1907, 68 feet high by 9½ feet in girth. We know of no trees of any size in Scotland.

The largest which we have heard of in Europe is a tree growing at Schloss Dyck, the seat of Fürst Salm-Dyck in Germany, which was planted in 1809, and in 1904 measured 35 metres high by 3.58 metres in girth, with a crown diameter of 35 metres.

TIMBER

It is very strange that though this timber has been imported on a large scale from North America for many years, both to England and the Continent, where it commands a very high price, its value is quite unknown to the English country timber merchant, and none of the writers on wood seem to know much about it. Even Marshall Ward, in his edition of Laslett (1894), says (p. 181) that it will not bear comparison with the quality of either Black Sea or Italian walnut wood. Boulger, in *Wood* (p. 339), says that it is "more uniform in colour, darker, less liable to insect attack, and thus more durable than European walnut." Stone says (p. 211), "This wood is readily confused with *J. regia.*"

I can only say that I have seen four different trees felled in England, of which the wood was perfectly distinct by its purplish colour from that of any European walnut; and though I have not been able to get any definite proof of the truth of

---

[1] *Gard. Chron.* 1886, xxvi. 616, fig. 120.

Boulger's statement as to its freedom from insect attack, yet the furniture which I have had made from three of the trees in question is distinctly superior to that of common walnut, and as good as imported black walnut, in colour; and when properly seasoned, for which three or four years should be allowed, as good cabinetmaker's wood as the best Circassian or Italian walnut: and Unwin,[1] quoting Nordlinger and Mayr, says that the timber of trees grown in Germany has the same specific gravity and the same beautifully coloured heartwood as in America. I am informed by experienced cabinetmakers and timber merchants that the colour and quality of the wood now imported is, either on account of its being younger or grown in different localities, inferior to what it used to be when first introduced to this market, and Mr. A. Howard told me that he could not buy American timber of better quality than that of a tree blown down at Albury which was given me by the Duke of Northumberland. It takes a beautiful polish either with oil or French polish, has not warped in the least degree, and has in many cases a beautifully variegated figure. The sapwood is thick and of a paler colour, and should not be used in first-class work any more than that of the common walnut, which is always attacked sooner or later by the larvæ of a woodboring beetle.

From what I saw of it in America, I believe it to be extremely durable when exposed to the weather, as it lasts long in fences, and large canoes were made from it, whilst it was the favourite wood for furniture until its increasing scarcity and price caused it to be superseded by oak and mahogany.

Old trees often show a beautiful wavy grain, as well as a variety of markings, and from the forks and burrs veneers are cut, which, though of a different colour, are equal in beauty and pattern to mahogany or satinwood.

Cobbett[2] states, though he does not appear to have seen it himself, that there was at New York part of a black walnut trunk, which measured 36 feet round at the base, and had been scooped out and used as a bar-room, and afterwards as a grocer's shop, and that this tree, if it had been sawed into inch boards, would have yielded 50,000 feet, worth at that time $1500, but this seems exaggerated; though Loudon states (p. 1438) that a tree, perhaps the same as the one Cobbett speaks of, and grown on the south side of Lake Erie, was exhibited in London in 1827, which was 12 feet in diameter, hollowed out and furnished as a sitting-room.

(H. J. E.)

[1] *Future Forest Trees*, 38 (1905).   [2] *Woodlands*, Art. 553.

JUGLANS CINEREA, Butternut

Juglans cinerea, Linnæus, *Sp. Pl.* 1415 (1763); Loudon, *Arb. et Frut. Brit.* iii. 1439 (1838); Bentley and Trimen, *Medicinal Plants*, iv. t. 247 (1880); Sargent, *Silva N. America*, vii. 118, tt. 331, 332 (1895); and *Manual Trees N. America*, 126 (1905).

A tree attaining in America occasionally 100 feet in height, with a girth of 9 feet, but usually smaller in size, dividing at 20 or 30 feet above the ground into many stout horizontal limbs, and forming a broad, low, round-topped head. Bark of young trees smooth and light grey, becoming in older trees deeply fissured, with broad scaly ridges.

Leaves with eleven to seventeen leaflets, which are sub-opposite, sessile, oblong, unequal-sided, rounded, and slightly unequal at the base, acuminate at the apex; margin finely serrate, the tips of the serrations being directed outwards, ciliate; upper surface finely pubescent with stellate and long simple hairs; lower surface pale, with numerous fine stellate hairs, there being some glandular hairs on the midrib towards its base. Rachis with numerous short glandular hairs. Young shoots with white sessile glands and numerous short white glandular hairs; old shoots pubescent. The leaf-scars are semicircular or triangular, with the upper edge a straight line, and furnished with a transverse band of pubescence.

Flowers: staminate catkins 2 to 3 inches long; scales with six puberulous lobes; bract rusty-pubescent with acute apex; stamens eight to twelve. Pistillate flowers in six- to eight-flowered spikes; involucre with viscid glandular hairs, and slightly shorter than the linear-lanceolate perianth lobes.

Fruit: in drooping clusters of three to five, ovate oblong with two or rarely four obscure ridges, coated with rusty clammy hairs, $1\frac{1}{2}$ to $2\frac{1}{2}$ inches in diameter. Nut ovate, abruptly acuminate and contracted at the apex, with eight ribs, internally two-celled at the base and one-celled above the middle with a narrow pointed apical cavity.

Varieties and Hybrids

No varieties are known. A hybrid between it and *Juglans nigra* has been observed. See *Juglans nigra*.

Identification

The best mark of distinction of this species at all seasons is the leaf-scar, which has a transverse raised band of pubescence above its upper margin, which is never notched, and is straight or slightly convex. In winter the following characters are observable in the twigs and buds. Twigs stout, reddish brown, covered with dense glandular pubescence. Leaf-scars, as described above, obovate, on prominent pulvini, with three groups of bundle-dots. Terminal buds oblong, greyish, densely pubescent, the two outer scales conspicuously lobed at the apex, the two inner scales scarcely lobed.   Lateral buds, directed outwards at an angle of 45°, small, ovoid, pubescent ; frequently two superposed.   Pith dark brown, with narrow chambers.   (A. H.)

DISTRIBUTION

According to Sargent, it occurs in rich moist soil near the banks of streams and on low rocky hills from southern New Brunswick and the valley of the Saint Lawrence in Ontario to eastern Dakota, south-eastern Nebraska, central Kansas, and northern Arkansas, and on the Appalachian Mountains to northern Georgia and northern Alabama ; most abundant and of its largest size northward.   The grey walnut or butternut, as it is commonly called, is a common tree over the same region as that which produces the black walnut, but never attains the same size, and, as a rule, unless drawn up in the forest is a much more spreading and less valuable tree.   It does not in New England usually exceed 30 to 50 feet in height, with a trunk 1 to 4 feet in diameter, but sometimes in the rich forests of the Wabash valley attains greater dimensions.   Ridgway says, *loc. cit.* 76, that two trees felled in the "Timber Settlement," Wabash county, measured 97 feet and 117 feet in length, with clear trunks 50 feet and 32 feet long, and 1 foot 10 inches in diameter. Pinchot and Ashe, *loc. cit.* 82, say that in North Carolina it is nowhere common, but in cool rich mountain valleys it attains 70 feet high with a diameter of 3 feet. In New England Emerson, *loc. cit.* 210, mentions a tree in Richmond, Mass., which was 13 feet 3 inches in girth at the smallest place below the branches. I never saw any such trees as these ; and near Ottawa, where the tree is approaching its northern limit of distribution, it was a small branchy tree bearing little fruit.

INTRODUCTION

The butternut was first described by Parkinson,[1] and was apparently introduced into England at the same time as the black walnut, *i.e.* sometime before 1656, as it is probably one of the species mentioned by Tradescant[2] as growing in his garden. Loudon states that it was introduced into cultivation by the Duchess of Bedford in 1699 ; but the tree referred to by him was *Carya alba*.[3]

CULTIVATION

Though it must have been planted in many places in this country the butternut seems to be now a very scarce tree.   The only one I have seen of any size grows in the grounds of Mr. C. S. Dickens at Coolhurst, near Horsham, and was in 1902 52 feet high and 4 feet 2 inches in girth.   This produced fruit in 1900 from which I raised two seedlings, one of which is now growing at Colesborne.   I noticed that the roots of these seedlings instead of being long, fusiform, and free from rootlets, as in *J. regia* and *J. nigra*, formed a thick, fibrous mass, which made the tree

---

[1] *Theatrum Botanicum*, 1414 (1640).   [2] *Museum Tradescantianum*, 146, 147 (1656).
[3] Aiton, *Hort. Kew*, iii. 360 (1789), *ex Brit. Museum Sloane MSS.* 525 and 3349.

very easy to transplant. I have since then raised numerous seedlings from imported seed, by sowing them both in pots and in the open ground. If allowed to become dry they sometimes lie over a year, and should therefore be sown as soon as ripe. The young trees are distinguishable from those of *J. nigra* by having fewer pairs of leaflets, but they grow quite as fast, and are quite as hardy as the latter. Both *nigra* and *cinerea*, though liable to injury from late spring frosts, are much hardier as regards winter frost when old enough to ripen their wood, but as, like other walnuts, they do not bear pruning well, they require careful attention when young in order to become shapely trees. Sir Charles Strickland has raised from seed plants at Boynton in Yorkshire which grew to five or six feet high, but all ultimately died.

Mr. J. H. Bonny, Ratcliffe Cottage, Forton, Garstang, sent specimens to Kew in 1900 from a tree 60 years old, which fruited for the first time in that year. It had only attained 22 feet high by 2½ feet in girth at 5 feet from the ground. There is a tree at Bayfordbury which produced a few nuts in 1905. It is 35 feet high by 3 feet 2 inches in girth, and is as large as a black walnut planted beside it. At Tredethy in Cornwall, the seat of F. T. Hext, Esq., I am told by Mr. Bartlett, that there was in 1905 a tree 35 feet by 2 feet 2 inches.

At Riccarton near Edinburgh, the seat of Sir James Gibson Craig, Bart., there is a butternut growing in a sheltered spot which Henry measured in 1905, and though its position makes it difficult to measure accurately, he believes it to be about 50 feet by 3 feet 3 inches.

In Ireland Henry measured in 1904 at Kilmacurragh, Co. Wicklow, a tree 32 feet high by 3 feet 4 inches; while at Charleville in the same county, the seat of Lord Monck, a tree, planted probably in 1869, was 25 feet high by 2 feet in girth.

TIMBER

The timber of this tree, though it resembles that of other walnuts in texture and grain, is much inferior in colour to that of the black walnut, but Hough[1] says that though not so high-priced it is nevertheless of great value for interior finish and wainscoting. In Prof. Sargent's house at Brookline, near Boston, I saw a very handsome mantelpiece and some panelling made from it, and it is occasionally used for furniture. It is pale brown in colour, with whitish-grey sapwood, and the burrs are sometimes cut into handsome veneers. Mr. John Booth[2] states that he cut down some exotic trees planted by his father at the celebrated Flottbeck nurseries near Hamburg when about 50 years old; and from the wood of a butternut wainscoted a room; "the polish was even finer than that of *J. nigra*, with a splendid glossy hue."

Emerson says, *loc. cit.* 209, that from the bark a mild purgative is made, and that the Shakers at Lebanon obtain a rich purple dye from it. The common dye used by the early settlers for their homespun cloth was from the husk of the

[1] *American Woods*, p. 61.   [2] *Gard. Chron.* xxx. 407 (1901).

butternut, which gives a fawn colour. The young half-grown nuts make excellent pickles if gathered early in June, but the ripe nuts, though eaten by boys and Indians, are oily and soon become acrid.

According to L. B. Case, who wrote an interesting article[1] on this tree, if an incision is made in the trunk early in spring before the unfolding of the leaves, it yields a rich saccharine sap, nearly if not quite equal to that obtained from the sugar maple. The medicinal uses of the bark are fully explained in Bentley and Trimen's work cited above.　　　　　　　　　　　　　　　　　　　　　　　　　(H. J. E.)

JUGLANS RUPESTRIS, Texan Walnut

Juglans rupestris,[2] Engelmann, *Sitgreave's Report*, 171, t. 15 (1853); Sargent, *Silva N. America*, vii. 125, tt. 335, 336 (1895), and *Manual Trees N. America*, 129 (1905).

The typical form, with small leaflets, which has been introduced into cultivation in Europe, is a shrub or small tree; bark of young trunks smooth, pale, whitish, becoming in older trees deeply furrowed and scaly. Leaflets, seven to fifteen or more, small, one to three inches long, sub-sessile, ovate or lanceolate, never oblong, apex acuminate, base rounded and unequal-sided, crenulate-serrate and non-ciliate in margin; upper surface with scattered minute pubescence; lower surface green with scattered minute brown hairs and axil tufts. Rachis with numerous sessile yellow glands and glandular hairs. Young shoots with numerous sessile yellow glands, interspersed with glandular hairs and obcordate leaf-scars, which are notched above. Older shoots shortly pubescent.

Flowers: staminate, catkins slender, two to four inches long, scales three- to five-lobed, with ovate-lanceolate tomentose bracts; stamens twenty. Pistillate flowers few in a spike, tomentose, involucre irregularly divided into a laciniate border, slightly shorter than the ovate acute calyx-lobes.

Fruit: globose or rarely oblong, very variable in size, $\frac{1}{2}$ to $1\frac{1}{2}$ inch in diameter; husk glabrate or coated with rufous hairs; nut globose without ridges, often compressed at the ends, dark brown or black, grooved with longitudinal simple or forked grooves, four-celled at the base, two-celled at the apex.

Var. *major*, Torrey, *Sitgreave's Report*, 171, t. 16 (1853): usually a tree, attaining 50 feet in height with a trunk 15 feet in girth. In this form the leaflets are large, reaching 6 inches in length; the fruit is also larger. It would appear that this variety is the western form, the typical form being characteristic of the eastern part of the area of distribution of the species.

[1] *Woods and Forests*, 1884, p. 200.

[2] It is probable, as Rehder points out in *Cycl. Am. Hort.* ii. 846, that *Juglans longirostris*, Carrière, in *Rev. Horticole*, 1878, p. 53, fig. 10, belongs to this species.

Identification

The form of this species usually cultivated in England is distinguished in summer by its small leaves, bushy habit, and the other characters given above. In winter the following characters are available:— Twigs very slender, olive-green or brown, densely pubescent. Leaf-scars set obliquely on prominent pulvini, small, obcordate, notched above, without pubescent band above the upper margin; bundle-dots in three groups. Terminal bud elongated, slender, densely and minutely pubescent, the tips of the two outer scales slightly lobed. Lateral buds, arising at an angle of 45°, minute, ovoid, pubescent, usually solitary. Pith small, brownish, with wide chambers.

Distribution

According to Sargent this species occurs on the limestone banks of the streams of central and western Texas, shrubby or rarely more than 30 feet high (var. *typica*); common and of larger size in the cañons of the mountains of New Mexico and Arizona south of the Colorado plateau. It is also met with in northern Mexico,[1] where it frequently leaves the mountain cañons, following the water-courses which are dry throughout most of the year. In such situations its average diameter is 12 to 18 inches, and its height 20 to 30 feet; the nuts, less than an inch in diameter, are scarcely edible.

Cultivation

This species was discovered in western Texas in 1835 by Berlandier. It was growing in 1868 in the Botanic Garden at Berlin, according to a note in Engelmann's Herbarium.[2] It does not seem to have been known in England till 1894, when seeds from Fort Huancha in Arizona were sent to Kew by Sargent. A tree grown from this seed has attained now (1905) about 12 feet in height. There is one nearly as large at Tortworth, and a seedling from Kew is planted at Colesborne, where it seems at least as hardy as the common species and ripens its wood earlier. A tree planted at Mount Edgcumbe, near Plymouth, in 1898 is 9 feet 4 inches high, with a spread of 10 feet. It has been cut back twice, and looks better as a bush than as a tree. (A. H.)

---

[1] *Garden and Forest*, 1888, p. 106.   [2] Sargent, *Silva N. America, loc. cit.* 126.

JUGLANS MANDSHURICA, MANCHURIAN WALNUT

Juglans mandshurica, Maximowicz, *Prim. Fl. Amur.* 76 (1859); and *Mél. Biol.* viii. 630, fig.
  (1872); C. De Candolle, in *D.C. Prod.* xvi. 2, 138 (1864); *Gard. Chron.* 1888, iv. 384, fig. 53.
Juglans regia octagona, in *Revue Horticole*, 1861, p. 429, fig. 106.
Juglans regia cordata, in *Garden*, 1896, p. 478, fig.

A tree attaining 60 feet in height and 5 feet in girth.   Bark dark ashy in colour, furrowed in old trees.   Judging from herbarium specimens, as I have not been able to examine living trees in England, this species differs little in character of leaves and branchlets from *Juglans Sieboldiana*.   Maximowicz, who observed both species growing wild, states that he was unable to find any good distinctions between the two species except in the characters of the nut.

The fruit occurs in short racemes, six to thirteen in a cluster, and is globular-ovate to oblong, viscid, and stellate pubescent.   The nut resembles that of *Juglans cinerea*, but is less sharply ridged, globose or ovate, rounded at the base, abruptly and shortly acuminate at the apex, eight-ribbed, with the intervals much wrinkled.

This species occurs in mountain woods in eastern Manchuria, between the Bureia range and the Sea of Japan, from lat. 50° to the Korean frontier.   It is frequent along the river Amur in its lower part and on its tributaries.   This species is also widely spread throughout Northern and Western China, where it is common in mountain woods at low altitudes, from Chihli through Hupeh and Szechwan to Yunnan.   So far as I have seen it, both in Hupeh and Yunnan, it never makes a large tree, and rarely exceeds 40 feet in height, but Komarov informed us that in Mandshuria it attains 80 feet high by 19 to 20 in girth.

This plant was introduced[1] into the Botanic Garden of St. Petersburg by Maximowicz from seeds sent from the Amur.   A tree[2] from seed planted in 1879 in the Arnold Arboretum bore fruit in 1883, which was large, more nearly spherical and less rough than the butternut, and of good flavour.   The tree is described as being compact and handsome in habit, and likely to become of value as a fruit tree in the northern parts of the United States, where the common walnut cannot be grown successfully.

Specimens were sent to Dr. Masters[3] in 1888 from a tree which had fruited in the nursery of Mr. J. van Volxem at Brussels, where the fruit ripens some weeks before that of the common walnut, and the tree seems less injured by spring frosts.

(A. H.)

[1] Bretschneider, *Hist. Europ. Bot. Discoveries in China*, i. 609 (1898).
[2] *Garden and Forest*, 1888, pp. 396, 443.
[3] *Gard. Chron.*, *loc. cit.*

JUGLANS CORDIFORMIS, Cordate Walnut

Juglans cordiformis, Maximowicz, *Mél. Biol.* viii. 635, *cum fig.* (1872); Shirasawa, *Icon. Ess. Forest Jap.*, text 35, t. 17 (1899); Rehder, *Mittheil. Deut. Dendrol. Gesell.* 1903, p. 117; *Gardeners' Chronicle*, 1901, xxx. 292, Supplementary Illustration.
Juglans Sieboldiana, var. *cordiformis*, Makino, in *Tokyo Bot. Mag.* 1895, p. 313.

A tree attaining 50 feet in height and 6 feet in girth. Bark, according to Shirasawa, remaining smooth for a long time, becoming fissured with age.

Leaves with eleven to thirteen leaflets, which are sub-opposite, oblong with unequal sides, acute or acuminate at the apex, cordate at the base, sessile or sub-sessile, the petiolule not exceeding $\frac{1}{16}$ inch, the base of the leaflet extending over the rachis so that the leaflet appears to be more sessile than is the case in *J. Sieboldiana*; serrations fine, shallow, irregular, directed forwards and ciliate; upper surface finely pubescent, with only tufted hairs; lower surface pale in colour, pubescent, with numerous stellate hairs, dense along the midrib on which the hairs are glandular; rachis with densely glandular long reddish hairs, sessile glands being absent. Young shoots covered with long white hairs, which are tipped with red glands and are much denser than in *J. Sieboldiana*; no sessile glands visible. Leaf-scar as in that species.

Flowers: male catkins twelve inches long or more; female catkins about five inches long, bearing seven to twelve flowers.

Fruit globose; nut heart-shaped, much flattened, sharply two-edged, with a shallow longitudinal groove in the middle of each flattened side, smooth over the surface, rather thin-shelled.

Identification

Readily distinguished in summer by the cordate leaflets and the young shoots densely covered with long white hairs, which bear red glands at the tips. See under *Juglans Sieboldiana*.

In winter the following characters are available:—Twigs stout, brown, covered with long glandular hairs, which tend, however, to fall off from the lower part of the shoot. Leaf-scar large, set slightly obliquely on pulvini which are scarcely elevated, obovate with two lateral lobes and notched above; the upper margin with a transverse raised band of pubescence; bundle-dots in three groups. Terminal bud conical, but compressed laterally, brown, densely pubescent, the two outer scales lobed at the apex. Lateral buds often two superposed, small, brown, ovoid, arising from the twigs at an angle of 60°, densely pubescent. Pith large, brown, with wide chambers.

Distribution

According to Maximowicz, this species occurs in Nippon. Shirasawa says that it is spread along the banks of rivers in the temperate regions of Japan, being rare in the mountains.   The wood, according to this author, is light, with little difference between the sapwood and heartwood, and when well seasoned does not warp or split, and on this account it is much esteemed for making gun-stocks.   Sargent[1] did not find this tree in Japan, and says that its peculiar nuts are considered by Japanese botanists to be merely extreme varieties of *Juglans Sieboldiana*.   However, the species is kept up as distinct by Matsumura,[2] and cultivated specimens at Kew of the two species can be readily distinguished.

Rehder states in 1903 that a tree in the Arnold Arboretum raised from seed of true *Juglans cordiformis* fruited some years ago.   The fruits, however, did not show the characteristic form of this species, and he doubted whether the tree in question was true *cordiformis*, or only a variety of *Sieboldiana* with aberrant fruit.

Nuts were obtained in 1862 by Albrecht,[3] physician to the Russian Consulate at Hakodate, which were sown in the Botanic Garden at St. Petersburg, and produced healthy plants, which were about four feet high, in 1872.   Maximowicz also found the nuts in the market at Yokohama.   Sargent, who found them offered for sale by the Nurserymen's Association of Yokohama, was informed that they were collected on the slopes of Fujisan.

The tree has been recently sent out by Continental nurserymen, and is hardy in this country.   A specimen at Kew, which was raised in 1899 from seed procured from Harvard, is now about twenty feet high.   The male catkins, which are produced freely and expand in May, give the tree a striking appearance, but the fruit has not yet matured.   (A. H.)

---

[1] *Forest Flora of Japan*, 60 (1894).   [2] *Shokubutsu Mei-I*, 155 (1895).
[3] See Maximowicz, *loc. cit.*, and Bretschneider, *European Bot. Discoveries in China*, i. 622 (1898).

JUGLANS STENOCARPA, NARROW-FRUITED WALNUT

Juglans stenocarpa, Maximowicz, *Prim. Fl. Amurensis*, 78 (1859); and *Mél. Biol.* viii. 632, *cum fig.* (1872); Rehder, *Mittheil. Deut. Dendrol. Gesell.* 1903, p. 117.

A tree of which only the fruits are known in the wild state. The following description of the foliage is taken from a specimen cultivated at Kew.

Leaves with eleven to thirteen leaflets, of which the terminal one in well-developed specimens is much broader than the others, being obovate with a short acuminate apex (4 inches broad by 6 inches long). The lateral leaflets ($2\frac{1}{4}$ inches broad by 6 inches long) are oblong, acuminate at the apex, rounded and unequal at the base, subsessile, the petiolule being less than $\frac{1}{16}$ inch; upper surface with scattered stellate pubescence; lower surface pale in colour, with similar pubescence; all the leaflets coarsely and almost crenately (not sharply) serrate and ciliate in margin. Rachis with very scattered stellate hairs and white sessile glands, there being no glandular hairs. Young shoots glabrous with numerous yellow glands, there being, however, a slight pubescence towards the base of the shoot. Older shoots glabrous, grey, shining, smooth. Leaf-scar broadly obcordate, notched at the summit, three-lobed, and without any band of pubescence on the upper margin.

The nuts, on which Maximowicz founded the species, are described by him as being shining, cylindrical or oblong-oval, slightly narrowed at the base, acuminate at the apex, eight-ribbed, with the intervals between the ribs deeply and obtusely wrinkled. The nuts are cinnamon brown in colour and are two-celled.

This species, having serrate pubescent leaflets and non-bearded leaf-scars, can only be confused with *Juglans nigra* and *J. rupestris.* It is readily distinguished in summer from these and all other species of walnut in cultivation by the broad terminal leaflet, which is always well marked in fully developed leaves.

In winter the following characters are available :—Twigs stout, yellowish brown, shining, minutely pubescent towards the apex, glabrous elsewhere. Leaf-scars large, on pulvini which are only slightly elevated, broadly obcordate, notched above and without any pubescent band along their upper margin; bundle dots in three groups. Terminal bud conical, brown, tomentose, the two outer scales slightly lobed at the apex. Lateral buds small, ovoid, tomentose, arising at an angle of 45°. Pith large, buff in colour, with narrow chambers.

The nuts of the tree were found in Russian Manchuria by Maximowicz. Nothing is known about the tree itself.

Specimens are cultivated in the Arnold Arboretum which were obtained from Regel and Keiselring's nursery at St. Petersburg. There are two small plants at Kew which were obtained under the name *Juglans mandshurica* from a Continental nursery.

(A. H.)

JUGLANS SIEBOLDIANA, Siebold's Walnut

Juglans Sieboldiana, Maximowicz, *Mél. Biol.* viii. 633 fig. (1872); Lavallée, *Arbor. Segrezianum*,
  p. 1, tab. I. et II. (1885); *Garden*, 1895, xlvii. 442.
Juglans ailantifolia, Hort. Sieb. *ex* Lavallée, *loc. cit.*; and Carrière in *Revue Horticole*, 1878, p. 414,
  figs. 85 and 86.

A tree attaining 50 feet in height and 5 feet in girth.

Leaves with thirteen to fifteen leaflets, which are sub-opposite, oblong, acuminate at the apex, with base rounded and unequal, sub-sessile, the petiolule being less than $\frac{1}{16}$ inch; serrations fine, shallow, and irregular, directed forwards, ciliate between the teeth; upper surface finely pubescent, with both single and tufted hairs; lower surface pale in colour, covered with numerous stellate hairs, denser close to the midrib on which there are glandular hairs; rachis with long brown glandular hairs and a few small glands near its base. Young shoots green, with long white glandular hairs and white sessile glands; lenticels at first white, becoming brown, conspicuous. Leaf-scars obcordate, three-lobed, deeply notched above, and with a transverse band of pubescence along the upper edge.

Flowers: staminate catkins very long, up to 12 inches, with bracts obtuse at the apex and very villous, scale five-lobed. Pistillate spikes, five to twenty flowered, the rachis and flowers covered with rufous tomentum.

Fruit in long racemes which are ten to twenty inches long; globose to ovate-oblong, shortly acuminate at the apex, viscid and covered with stellate hairs. Nuts ovoid or globose, rounded at the base and acuminate at the apex, with thick wing-like sutures, very slightly wrinkled and pitted, not ribbed, rather thick-shelled.

IDENTIFICATION

This species seems to be practically identical in leaves and shoots with *Juglans mandshurica*, and differs little in these respects from *Juglans cordiformis*, except that the leaflets of the latter are distinctly cordate at the base. All three species differ, however, remarkably in fruit, and must be kept distinct on that account. They belong to the section of walnuts with bearded leaf-scars, and are readily distinguished from *Juglans cinerea*, the other species of this group, by having the leaf-scars deeply notched above.

In winter the following characters are available:—Twigs stout, brown, glabrous except near the tip, where the pubescence of summer is retained. Leaf-scars large, on very slightly raised pulvini, obovate, two-lobed above; upper margin convex, with a central notch, and surmounted by a raised band of pubescence; bundle-dots in three groups. Terminal bud brownish, elongated, covered with a dense minute pubescence; outer pair of scales lobed at the apex. Lateral buds arising at an acute angle, small, ovoid, pubescent; frequently two superposed. Pith large, brown, with narrow chambers and thick plates.

DISTRIBUTION

According to Maximowicz it occurs throughout the whole of Japan, there being large trees around temples at Hakodate. At Miadzi, in Kiusiu, it is wild on the sides of mountain streams, being a tree of about eighteen inches in diameter. It is also supposed to occur in the island of Saghalien, as nuts cast up by the sea were found there by F. Schmidt.

Sargent[1] says that *Juglans Sieboldiana* is a common forest tree in Yezo and the mountainous regions of the other islands of Japan. Specimens more than 50 feet high are uncommon. It is a wide-branched tree, resembling the butternut in habit and in the colour of its pale furrowed bark. The walnuts of this species are an important article of food in Japan, as the nuts are exposed for sale in great quantities in the markets of all the northern towns.

Elwes collected specimens at Asahigawa in central Yezo, and noted that it was always a small tree, 20 to 30 feet in height by a foot in girth. He also saw it near Nikko, but never of any size. It is called *Kurumi*. The wood, though used to some extent in Japan for gun-stocks and ornamental work, does not take a high place among the valuable timbers of the country. It was not included in the collection of woods exhibited at St. Louis.

CULTIVATION

Juglans Sieboldiana was introduced from Japan into Leyden about the year 1860 by Siebold, and was sent from there to Segrez in 1866, under the name of *Juglans ailantifolia*. At Segrez it passed unscathed through the severe winter of 1879-1880, which proved fatal there to the common walnut.

According to Sargent this species is perfectly hardy in New England, where it ripens its fruit. It is not worth growing there as an ornamental tree; but it will produce fruit in regions of greater winter cold than the common walnut can support, and may find some place in planting as a fruit tree.

The largest specimen we know of in these islands is at Belgrove, Queenstown, Ireland, the residence of W. E. Gumbleton, Esq. It was, in 1903, 24 feet in height by 2 feet 9 inches in girth. There are specimens at Kew about 12 feet high, which were grown from seed received in 1894. There is also a small plant at Gunnersbury House, Middlesex, which has borne fruit.

(A. H.)

[1] *Forest Flora of Japan*, 60 (1894).

COMMON OAK

THE following is an account of the three species into which the *Quercus Robur* of Linnæus has been divided :—*Quercus pedunculata*, *Quercus sessiliflora*, and *Quercus lanuginosa*. Brief notes are given also of certain Mediterranean and Oriental forms which are in cultivation. The generic character will be given in another part, with our description of the exotic oaks in cultivation in these islands. Plates 78 and 79 show the twigs and buds of the pedunculate and sessile oaks, as well as those of some other species which will be described in a later volume, and the leaves of the three species now treated and of some of their varieties.

Those wishing to have the latest information on the oak from a physiological point of view are referred to the late Prof. Marshall Ward's work,[1] which contains many details on points with which we do not propose to deal.

Loudon's account of the oak, covering over 100 closely printed pages, is also well worth study, especially with regard to the numerous historical trees, the quality of the timber, and the fungi, galls, and insects which live on or attack the tree.

QUERCUS PEDUNCULATA, COMMON or STALKED-CUPPED OAK

Quercus pedunculata, Ehrhart, *Beiträge*, v. 161 (1790); Loudon, *Arb. et Frut. Brit.* iii. 1731 (1838),
    Boswell Syme, *Eng. Bot.* viii. 145, tab. 1288 (1868).
Quercus Robur, Linnæus, *Sp. Pl.* 996 (*ex parte*) (1753).
Quercus Robur, L., sub-species *pedunculata*, DC. *Prod.* xvi. 2, p. 4 (1864).
Quercus Robur, L., var. *pedunculata*, Hooker, *Student's Flora of the British Isles*, ed. 2, 364 (1878).

A large tree, attaining a height of over 100 feet and a girth of stem of 20 to 30 feet, with the main branches large, long, and irregularly bent.

Bark, when old, irregularly fissured, and gradually increasing to a thickness of two inches or more. Branchlets in winter stout, glabrous, angled, grey, with a five-angled pith and small semicircular leaf-scars, which are set obliquely on prominent leaf-cushions and show three irregular groups of leaf-traces. Buds brown, clustered at the ends of the twigs, and arranged alternately (in 2/5 order) lower on the twigs, arising at an acute angle; blunt-oval, five-angled, with numerous imbricated scales (in five rows), which are glabrous on the surface and

---

[1] *The Oak*, by H. Marshall Ward, F.R.S. (1892).

shortly ciliate on the margin. The bud-scales are stipules, which fall off as soon as the leaves expand.

Leaves deciduous, sessile or with very short stalks, extremely variable in shape and size, but never symmetrical, generally with four to five pairs of entire, irregular, rounded lobes; obovate-oblong, diminishing in size to the base, which has always *two small emarginate auricles*; slightly silky pubescent when young, quite glabrous when adult; coriaceous in texture; dark green above, bluish green beneath. *Some of the lateral nerves run to the sinuses between the lobes.* The leaves from suckers which are very rare are usually entire or only slightly lobed, and are not auricled at the base. The fall of the leaves is very slow, often continuing for weeks, and frequently a part of the leaves remain on the tree till the close of winter.

Flowers appearing with the leaves; the male catkins being pendulous spikes (each bearing about a dozen flowers) arising from the preceding year's shoot and the lower part of the current year's shoot; the female inflorescences being long, obliquely erect spikes (each bearing one to five flowers at the upper end) arising in the axils of the two or three uppermost leaves. Male flower: calyx five- to seven-lobed, enclosing five to twelve stamens. Female flower: calyx six-partite, surrounded by a scaly cupule and enclosing an inferior ovary, surmounted by a cylindrical style terminating in a trifid stigma. Ovary three-celled, each cell containing two pendulous ovules.

Fruits: one to five, sessile on an elongated glabrous peduncle (1 to 6 inches long). Cup hemispheric, composed of many appressed, triangular, obtuse, glabrous or slightly tomentose imbricating scales. Acorn: variable in size and shape, flattened at the base where attached to the cup, and bearing the remains of the style at the apex, smooth and shining, containing one seed in one cell, five ovules and two cells being aborted and only visible as shrivelled remains at the base.

Seedling

The cotyledons remain enclosed in the coats of the acorn, and are not lifted above ground. The caulicle, stout and dark-coloured, gives off a long woody primary root. The plumule arises between the petioles of the two cotyledons, and develops into the young shoot, which at first bears only a few scattered scales, the first green leaf, small and obovate-oblong, coming afterwards; those succeeding are larger, obovate, and lobed. By the end of the first season the seedling has a long primary root with spreading lateral rootlets and a glabrous stem, averaging 6 to 8 inches high, bearing five or six sub-sessile glabrous leaves spirally arranged and ending in an ovoid glabrous bud. Each of these leaves has a minute stalk, with a pair of tiny linear stipules.

Seedlings,[1] according to Brenner, who made many observations, vary considerably in appearance, according to the soil in which they are grown, those in dry ground having leaves with deeper lobes, ending in sharp points; those in moist earth having shallow undulating round lobes.

[1] Brenner, *Flora* (1902), Band 90, p. 122.

Varieties of *Quercus pedunculata*

1. Var. *fastigiata*, Spach, *Hist. Vég.* xi. 151 (1842), Fastigiate or Cypress Oak.

Quercus fastigiata, Lamarck, *Encyc.* i. 725 (1783).
Quercus pyramidalis, Gmelin, *Fl. Bad.* iii. 699 (1808).
Quercus cupressoides, Hort.

The Cypress Oak has the branches pointing upwards, which gives the tree an irregular fastigiate shape; but in foliage and fruit it does not differ from the common oak. It has been found wild in the south-west of France, in the Landes and Pyrenees, in the provinces of Galicia and Navarre in Spain, and in Calabria. A famous tree of this variety stood in 1876 near the village of Haareshausen, close to Babenhausen in Hesse, which was supposed to be 280 years old, and it then measured 100 feet high and 10 feet in girth.[1] It had been celebrated in Germany since the middle of the eighteenth century, and stood originally in the forest, now cleared away. From this tree nearly all the German trees, and possibly many English and French trees of this variety, have been derived. This variety comes true from seed to some extent; of thirty acorns sown at Nancy, twelve produced pyramidal oaks, the remainder reverting to the ordinary type. At White Knights, of several hundred acorns sown by the gardener, only five came true to the fastigiate type. Elwes has raised plants from seed which in youth at least are more or less fastigiate. The tree at White Knights is a remarkably good specimen, being 81 feet high and 8 feet in girth, and is beautifully symmetrical in shape. Sir Herbert Maxwell tells us that there are two trees at Dawick, Peeblesshire. Other fine specimens are at Knole Park, Kent, where Elwes measured one 66 feet by 5 feet; and at Hardwick, Suffolk, where he saw one 61 feet by 4 feet 10 inches. A very well shaped tree of this variety at Melbury Park (Plate 80) measures 65 feet by 3 feet 8 inches, and has the form of a well-grown Lombardy poplar. But none of these are equal to a tree growing at the Trianon at Versailles, which Elwes saw in 1905, and which measures about 90 feet by 10 feet.[2] Several sub-varieties have appeared in various nurseries, and have received names, but as we have seen none of these in cultivation we do not think them worth recording.

2. Var. *pendula*, Loudon, *Arb. et Frut. Brit.* iii. 1732 (1838), Weeping Oak. —In this variety the branches are pendulous. The most famous tree of this kind is at Moccas Court in Herefordshire; but it has now almost ceased to weep, and Elwes would not have been able to distinguish it if it had not been pointed out to him. The present owner, the Rev. Sir George Cornewall, writes that "weeping oaks are far from uncommon in Herefordshire," and showed

[1] Petzold, *Deutschen Reichsanzeiger*, quoted in *Gard. Chron.* v. 51 (1876). See also *Gard. Chron.* xix. 179, fig. 26 (1883), where Mr. Wissenbach states that the oldest and finest specimens in Germany occur in the royal park at Wilhelmshöhe, near Cassel, the best measuring 100 feet high and 8 feet 6 inches in girth. It is 100 years old, being a graft of the original tree in the forest near Babenhausen. An earlier account of the latter tree is given by a correspondent in *Gard. Chron.* 1842, p. 36.

[2] A group of fine trees of this variety, said to be more than 100 feet in height, is reported to be growing in the park of Verdais in Haute Garonne. *Woods and Forests*, 105 (1884).

him a very striking one on the road from Moccas to Bredwardine, from the acorns of which seedlings have been raised. In 1884 there was a weeping oak at the King's Acre nurseries, Hereford, grafted at 3 feet up, which was planted by Cranston in 1785.[1] It bears acorns every year; but none of the seedlings, it is said, show a tendency to droop. The top of this tree is not pendulous; the weeping only occurs on the outer parts of the lower branches.

3. Var. *filicifolia*, Lemaire, *Illust. Hort.* i. t. 32, *verso* (1854), Fern-leaved Oak, also known as *asplenifolia, pectinata, pinnata, taraxacifolia*, etc. The leaves are stalked and cuneate at the base, long and narrow in outline, deeply and irregularly pinnatifid. This was found wild in the mountains of southern Germany; and was sent out by Messrs. Booth and Sons, Hamburg.

4. Var. *heterophylla*, Loudon, *Arb. et Frut. Brit.* iii. 1732 (1844), Various-leaved Oak. This variety has leaves varying greatly in shape; some are lanceolate and entire, others are cut at the edges or deeply laciniate; but all are cuneate at the base. It has received a variety of names, as *comptonæfolia, incisa, dissecta, laciniata, salicifolia, Fennessi, Fenzleyi, diversifolia, cucullata*, etc. Loudon's figure represents a branch from an accidental seedling, raised in 1820 in the nursery of Messrs. Fennessey, Waterford. There is a free-growing tree of this variety at Smeaton-Hepburn, East Lothian, which measured in 1905, 56 feet by 4 feet 8 inches.

5. Var. *hyemalis*, Bechstein, *Forstbot*, 333 (1810). In this variety the fruit stalk is very long, at least as long as the leaf itself. This is also known as *Quercus longipes*, Steven, *Bull. Soc. Nat. Mosc.* i. 385 (1857).

6. Var. *scolopendrifolia*, Hort. This form has leaves with short stalks and cordate bases, somewhat variable in shape. Most of the leaves are long and narrow, with short lobes; but others more angular in form have swollen bladder-like projections on their upper surface. Certain sub-varieties are distinguished as *bullata, cochleata, crispa*, etc.; all having leaves variously deformed and presenting bladder- or blister-like projections on their surfaces.

7. Var. *Concordia*, Lemaire, *Illust. Hort.* xiv. t. 537 (1867). Leaves yellow, much more brightly coloured than in the variety commonly cultivated under the name *aurea*, the colour persisting during the summer. This beautiful form, the *Golden Oak*, originated in the nursery of Messrs. van Geert at Ghent in 1843. The late Mr. Charles Ellis wrote in 1894 to Kew that some golden oaks occur at Inglewood, Hungerford, Berkshire, as bright as the golden elder when seen in May. Mr. Clarke, gardener to H. J. Walmesley, Esq., the owner, informs me that the trees are now in vigorous health, and measure at 6 feet from the ground 45 feet by 6 feet 2 inches and 40 feet by 4 feet 9 inches respectively.

8. Var. *purpurascens*, A.DC., *Flore Française*, vi. 351 (1815), Purple Oak. —This was found wild near Le Mans by De Candolle; and another wild tree was subsequently found in Thuringia. The young leaves, petioles, and branchlets are purple, the colour fading away later in the season. This form

---

[1] *Woods and Forests*, 794 (1884), with a full-page engraving of the tree, which was reported to be 72 feet high and 8½ feet in girth.

has received many names, as var. *purpurea*, Loudon, and var. *sanguinea*, Spach. There are slight sub-varieties which are known as *atropurpurea*, *atrosanguinea*, *nigra*, *nigricans*, etc.   The purple oak was first described by Bechstein (*Forst. Bot.* 333) in 1810 as *Quercus sanguinea*.

9. Var. *variegata*, Endlicher.   Oaks with variegated leaves are· not uncommon in the wild state.   There is a specimen at Kew of a curious form sent by Mr. J. Lindsay Johnston from Eastlodge, Crondall, Hants, in 1882.   The Rev. W. Wilks has sent leaves of an oak at Shirley, which were of a beautiful pink colour in November 1902.   There are many forms of variegated oaks in catalogues ; but it must be remembered that there is often a tendency in them to revert to the green form in a short space of time.   Some of these sub-varieties may be distinguished as follows :—*argenteo-marginata*, margin of leaves white ; *argenteo-picta*, leaves with white streaks ; *aureo-variegata*, leaves with yellow streaks ; *rubrinervia*, veins red in the young leaves ; *aureo-bicolor* and *tricolor*, leaves variously coloured yellow, white, and green.

Elwes has seen a very fine variegated-leaved oak at Haldon near Exeter, the seat of J. F. G. Bannatyne, Esq., and I measured one 57 feet high and 7 feet in girth, at The Grove, Teddington, which, according to Loudon,[1] was 37 feet high in 1837.   This tree bears leaves, which come out variegated green, white, and pink, changing in autumn to a pure pink colour.   The present owner, Charles E. Howard, Esq., informed me that it fruited only once to his knowledge, in 1887.

An account is given in the *Gardeners' Chronicle* of 14th September 1861 of a common oak which became variegated, the result of having been struck by lightning.   This tree grew near Mawley, the seat of Sir Edward Blount, and contained about thirty feet of timber.   It was struck by lightning on 26th June 1838, and did not appear to suffer at the time ; but shortly afterwards the foliage, which was previously green, became beautifully variegated, and continued to produce variegated leaves and remained healthy.

10. Var. *cuprea*, Hort.   This variety has bronze-coloured leaves when young, and is said to be a very distinct and vigorous form.

11. Var. *tardissima*, Simonkai, *Le chêne de juin*.[2]—This variety has more regular branching and denser foliage than the common form ; but is chiefly remarkable for the lateness of its leafing, which occurs five to eight weeks after the common oak.   It was discovered in France in the valley of the Saône, from Pontailler to Saint-Amour ; and has since been found at various places in the departments of Loir-et-Cher and Cher, and also in Hungary.   It appears from experiments made at Nancy to come true from seed ;[3] and the delay in the putting forth of the leaf is as marked in seedlings as in old trees.   It grows vigorously ; and apparently, in spite of the short period each season that it

---

[1] Loudon, *Gard. Mag.* 1837, p. 10.

[2] For interesting accounts of this variety, the following papers may be consulted :—Gilardoni, *Le chêne de juin* (1875) ; Jolyet, *Bull. de la Soc. des Sciences*, 1899.

[3] But seedlings raised by Elwes at Colesborne from acorns sent from France by M. L. Pardé do not seem to retain the late-leafing habit.

carries foliage, it produces as much timber as the common form.[1] The variety is considered of some importance in France, as owing to the lateness of leafing it is never affected by spring frosts; and it is recommended for cold, damp situations where the common oak is injured by this cause.

Many other varieties doubtless occur, both in cultivation and in the wild state. Specimens were sent to Kew from an old oak tree at Springfield, West Wickham, Kent, which bore extremely large leaves all over the tree, measuring as much as 8 inches long and 6 inches wide, and similar leaves occur on a tree at Colesborne. At Tortworth there is an oak about fifty years old, which bears fruit on very long peduncles, and has remarkably glossy coriaceous leaves[2] somewhat variable in shape, but generally obovate-lanceolate, with quite entire or only slightly lobed margin. This is almost identical with a specimen at Kew, gathered near Arcachon in France by Mdme. de Vilmorin. Specimens collected in Wistman's Wood, Dartmoor, are also remarkable for their irregularly shaped and very slightly lobed leaves, which have a cuneate base.

The variation in the size and shape of the leaves in natural wild seedlings growing side by side is often remarkable. Elwes gathered from three trees growing on the rocks above Minard Castle, Lochfyne, leaves varying from about 2 to 8 inches long. Meehan[3] narrates that when he settled in Germantown, near Philadelphia, he found a single *Quercus Robur* on the grounds of Mr. J. Hacker, from the acorns of which he raised hundreds of young seedlings, and has from these a second generation. He found amongst the seedlings numerous varieties, *e.g.* trees with leaves quite sessile, others with a petiole ¼ inch long, others with leaves as entire as those of *Quercus Prinus*, others with pinnatifid lobes; while in some cases the acorns were only a little longer than broad, in other cases cylindrical and twice as long as broad. Evidently here there was no possibility of hybridisation, as there was only one tree. This experiment of Meehan's, however, only goes to show the extreme variability of *Q. pedunculata*; and there is no evidence brought forward that any of the varieties became in the least like *Q. sessiliflora*.

In all the preceding varieties we are treading on safe ground, as there is no doubt that they are all derived from *Q. pedunculata*; but the case is different with certain forms from the Orient and southern Europe, which were considered by De Candolle to be varieties of *Q. pedunculata*, but by other authorities are treated as distinct species. A brief account of such of these as are in cultivation in England follows :—

Quercus Haas, Kotschy, *Die Eiche. Eur. u. Or.* t. 2 (1862) ; *Q. Robur, pedunculata*, var. *Haas*, DC. *Prod.*—This oak occurs in Cilicia and the Taurus, and in habit and size resembles the common oak; it differs in the following respects :— Young shoots white pubescent, puberulous when adult. Buds finely pubescent. Leaves on very short pubescent stalks, obovate, with cordate base, and four or five

---

[1] Mathey, *Exploitation Commerciale des Bois*, 95 (1906), speaks of its timber as being excellent, with very little sapwood, and scarcely any defects.

[2] Figured in Plate 79, fig. 2. [3] *Bull. of the Torrey Bot. Club*, ix. 55 (1882).

pairs of rounded lobes, the lateral nerves reaching to the sinuses as well as to the lobes; coriaceous; under surface bluish green, with a stellate pubescence, often discernible only with a lens. Fruit: 2 to 6 on a long stalk, very large, the acorns being $\frac{4}{5}$ inch in diameter. The cups look very distinct from those of the common oak.

This species is considered by Zabel[1] to be a hybrid between *Q. pedunculata* and *Q. lanuginosa*, but it seems rather to be a geographical form of *Quercus pedunculata*. Elwes saw two stunted trees which may be this at Orton Hall, Peterborough, said to have been raised from acorns sent by the late Sir H. Layard from Kurdistan.

The following three species or geographical forms were considered to be varieties of *Quercus pedunculata* by De Candolle.

Quercus Brutia, Tenore, *Sem. Ann. Hort. Neap.* (1825), p. 12.—Occurs in southern Italy. The difference between it and some northern forms of *Q. pedunculata* is very slight, as the leaves are glabrous. The fruit is large and somewhat peculiar.

Quercus Thomasii, Tenore, *loc. cit.* This also occurs in southern Italy, and is a form with large acorns, having leaves pubescent on the under surface, and standing on short pubescent petioles.[2]

Quercus apennina, Lamarck, *Encyc.* i. 725 (1783).—This is a small oak which occurs on dry situations in the south of France, and is said to form considerable forests in the Apennines in Italy. It has hoary, tomentose shoots and small leaves, with the under surface pale pubescent, and shorter stalks than *Quercus lanuginosa*, which it otherwise much resembles. The fruit is crowded on thick grey tomentose axes, and the cupules are greyish tomentose with appressed scales.

Hybrid or Intermediate Forms.—Hybrids between *Quercus sessiliflora* and *Q. pedunculata* occur; but they seem to be rare in the wild state in England, and I have only seen two or three specimens which could not at a glance be referred to one or other species without doubt. The best name for the hybrid is *Quercus intermedia*, Boenn, in Rchb. *Fl. Germ.* 177 (1830). The type specimen of *Q. intermedia*, Don, obtained by Leighton in Wyre Forest, Shropshire, is true *sessiliflora*. Another specimen in the British Museum labelled *intermedia*, gathered in 1843 in Surrey, is *pedunculata*; in this some of the peduncles are rather short, but there is one fully developed peduncle of the usual length, and the leaves in no way differ from ordinary *pedunculata*. What is often supposed to be *intermedia* is, however, the common oak, bearing leaves with stalks of a moderate length. The word *pedunculata* is apparently a trap to deceive all but the practised botanist. In *Q. pedunculata* the acorns are sessile on a long peduncle, which is distinct from a shoot, as it bears only acorns, never buds or leaves. I have received specimens from professional foresters, labelled "*sessiliflora*, intermediate form," in which the

---

[1] *Laubholz-Benennung*, 78 (1903).

[2] Elwes has received seedlings of both these forms from Herr Sprenger of Naples, and has sent some of them to Kew; but they do not at present show any appreciable difference, which was the case also in the oaks which he saw growing in the Sila mountains in Calabria.

peduncle bearing the acorns overtopped the end of the shoot, and was mistaken for it, and the acorns in consequence were considered to be sessile on the shoot. I think that the alleged occurrence of numerous 'intermediate forms is due to an imperfect appreciation of the real distinctions between the two species; and specimens to support the common occurrence of hybridity are not as yet forthcoming. The first writer who tried to break down the distinctions between the two species in England—Greville [1]—was not at all sure that he had succeeded; and in view of the important sylvicultural differences between the two trees the subject is one of more than academic interest.

Certain cultivated forms may be hybrids, as, *e.g. Quercus falkenbergensis*; and *Q. armeniaca*, Kotschy, from Armenia, is an undoubted hybrid. (A. H.)

The question of the distinctness of the sessile and pedunculate oaks in England has been discussed at great length on many occasions, but is one on which opinions, even among careful observers, always have differed, and differ still. Perhaps the best account of their peculiarities and merits is given by Loudon, pp. 1737-46, and in the *Gardeners' Chronicle* (1900), when a discussion was opened by Prof. Fisher, and continued by other well-known authorities.

Prof. Fisher describes the physiological difference, and maintains the opinion, which, largely based on French experience, is confirmed in some parts of England, that the pedunculate oak is naturally adapted to a wet soil, while the sessile will thrive in comparatively dry situations, and says that these peculiarities are of great importance to planters in selecting seed. As nurserymen rarely distinguish them and are, as a rule, careless of the source from which their seed comes, provided it will produce good nursery plants, I should strongly advise all oak planters to select and grow their own oaks from the trees which thrive best on similar soil in their own district, or in places with similar soil and climate.

Mr. A. C. Forbes says that in many localities the sessile oak is quite rare, and in Wilts "probably the rarest indigenous tree that we have." He accounts for this by the fact quoted from a paper [2] by Mr. J. Smith of Romsey, that at the time when oak timber was in demand for the navy, the durmast oak was not considered fit for that purpose, being, as it was said by the purveyors for the navy, more liable to dry rot, and this tradition still lurks in the minds of the older woodmen, tales being told of how they deceived those worthy gentlemen into passing the durmast oak for the dockyards.

There is a great deal of very interesting information in this paper both on the rate of growth and effects of transplanting of oaks, on their insect enemies and fungoid diseases, and a list is given, with many particulars and measurements of many of the most celebrated oaks of England. No one who is interested in oaks should fail to read it, but it is too long to quote from as freely as I should wish.

Sir Herbert Maxwell, in *Gardeners' Chronicle*, Nov. 10, 1900, says: "The long correspondence in your columns relative to the merits of the durmast or sessile-flowered oak will probably leave most people of the same mind as they were when it

---

[1] *Trans. Bot. Soc. Edin.* i. 65, tt. 4, 5 (1841).   [2] *Trans. Scot. Arbor. Soc.* xiii. 21 (1891).

began "; and goes on to say, "What is important is the fact that the durmast will thrive and ripen its season's growth in moist northern and western latitudes, which are unfavourable to the development of the pedunculate kind. In our salt-laden atmosphere upon the western seaboard much of the growth made by the pedunculate oak during one season fails to ripen before it is nipped by frost, and the tree is much more subject than the durmast to galls—a sure sign of debility; and it never carries with it the wealth of glossy foliage that never fails to distinguish the latter."

He then speaks of the fine oaks at Merevale Park, which are described on p. 318, as being of the sessile variety, and says that at Knole Park, Kent, on the other hand, the general growth is pedunculate; but there is a magnificent avenue of durmast oaks, leading to the house from the direction of the Wilderness, and these tower far and straight above the gnarled and twisted veterans in the rest of the park.

Another peculiarity of the sessile oak is referred to in a letter from the Hon. Gerald Lascelles to Mr. Stafford Howard, in which he says : " I doubt whether there is much difference between the timber of the sessile and pedunculate oaks, but I think that the sessile is straighter and cleaner in growth, and one thing is certain— that it is almost immune from the attacks of the caterpillar (*Tortrix viridana*) which so often destroys every leaf on the pedunculate oak in early summer. Whether this does any real harm or not is a moot point, but I think it must be a check to growth, and that the trees would be better without it. I have seen a sessile oak standing out in brilliant foliage when every tree in the wood around was as bare of leaf as in winter." [1]

Mr. J. Smith, in the paper above referred to, pp. 29-30, confirms Mr. Lascelles' observations, and says that in 1888, which was the worst year for these caterpillars that he remembered, he passed through a wood composed of *Q. sessiliflora* in which, though it had been attacked by the caterpillars, they had left off, evidently either poisoned or starved. He also quotes a resident in the Forest of Dean who, writing in 1881, says : " It was strikingly evident last summer that the *Q. Robur pedunculata*, or old English oak, was attacked by blight (? caterpillars) more severely than *Q. R. sessiliflora* "; and Mr. Baylis, who now has charge of Dean Forest, writes to me on the subject as follows :—" I can confirm the statement that the larva of the green oak moth defoliates *Q. pedunculata*, very much more than *Q. sessiliflora*, and I think the reason is this : the latter is the first to come into leaf, and the leaf has time to get fairly tough before the caterpillar has reached its most destructive stage, which is about the time that *Q. pedunculata* is coming into leaf.[2] I have very frequently noticed this fact that the oak with more decided pedunculate characters is almost invariably attacked rather than the other."

The only published exact observation that I know of with regard to the relative rate of growth of the two forms on the same soil is by Mr. H. Clinton Baker of Bayfordbury.[3] Near his house are growing on sandy loam, close to each other, a pedunculate oak raised in 1811 from the celebrated tree at Panshanger, and a sessile

---

[1] Sir Herbert Maxwell remarks, *in litt.*, that though visitations of *Tortrix* are not common in Scotland, yet in June 1905 the oaks on the shore of Loch Awe and Loch Lomond, which are sessile, were stripped of their leaves by this pest.

[2] Usually *Q. pedunculata* is the first to come into leaf. Cf. p. 292.   [3] *Gard. Chron.* xxxvii. 132 (1905).

oak raised in 1840 from a tree at Woburn Abbey. Measurements show that the former was 6 ft. 7 in. in girth in 1865, and is now 9 ft. 4 in. ; whilst the latter, only 1 ft. 8 in. in 1865, is now 8 ft. 7 in. Mr. J. Hopkinson in *Trans. Hertfordshire Nat. Hist. Soc.* xii. pp. 249, 250, gives diagrams showing the comparative annual increase during two periods of these trees. I may add that the habit of the two trees differs but little, and the soil is more suitable to the sessile oak.

Mr. Sharpe,[1] forester at Monreith, where Sir Herbert Maxwell planted in 1898 a quantity of oaks of the two species, on a fairly deep loam soil, measured ten of each sort in 1905, and informs us that the sessile oak averaged $13\frac{1}{4}$ feet in height, and the pedunculate oak only $10\frac{1}{2}$ feet.   (H. J. E.)

QUERCUS SESSILIFLORA, Sessile or Durmast Oak

Quercus sessiliflora, Salisbury, *Prod. Stirp. Hort. Chap. Allerton*, 392 (1796); Loudon, *Arb. et Frut. Brit.* iii. 1736 (1838); Boswell Syme, *Eng. Bot.* viii. 157, tab. 1289 (1868).

Quercus sessilis, Ehrhart, *Beiträge*, v. 161 (1790).

Quercus Robur, Miller, *Gard. Dict.*, vii. 1 (1759).

Quercus Robur, Linnæus, var. β; *Mantissa*, 496 (1771).

Quercus Robur, L., sub-species *sessiliflora*, DC. *Prod.* xvi. 2, p. 6 (1864).

Quercus Robur, L., var. *sessiliflora*, Hooker, *Stud. Flora Brit. Isles*, ed. 2, 364 (1878).

A tree resembling *Q. pedunculata*, but with more regular branching, resulting in a denser crown of foliage. It differs somewhat in the characters of the branchlets, buds, leaves, pistillate flowers, and fruit, as follows :—

Branchlets pubescent, especially near the top. Buds more sharply pointed, with scales pubescent on the outer surface, especially near the apex, and having long marginal cilia.

Leaves with a long petiole ; symmetrical, obovate-oblong, widest at the middle and gradually diminishing to the base, which is cuneate and generally without auricles ; firm, almost coriaceous in texture ; sinuately lobed or pinnatipartite, the lobes being oblong or triangular, entire, occasionally apiculate ; upper surface glabrous and shining, dark green ; lower surface brighter even glaucous green and always more or less pubescent. *Lateral nerves running to the sinuses are very seldom present.* Pistillate flowers with stigmas almost sessile. Fruit solitary or crowded, inserted on the branchlets, or borne sessile on an erect, stout, short pubescent peduncle. Cups pubescent, with scales more numerous and more closely crowded together than in *Q. pedunculata*.

This species is quite distinct from *Q. pedunculata*, and the characters given above are very constant. The pubescence, which is visible in this species throughout, on the top of the twigs, buds, stalks, peduncles, cups, and under surface of the leaves, is not so pronounced in specimens occurring in rainy districts ; but it can always be made out by a lens. The physiological differences are well marked. The sessile

---

[1] Cf. Sir Herbert Maxwell's account in *Gard. Chron.* xxxvii. 82 (1905).

oak comes into flower and leaf later by some days than the other species, and it is less liable to attacks of the roller moth.　It bears shade better, and on this account can be grown closer as a forest tree.　It grows naturally on drier soils, and on the Continent ascends to higher altitudes than *Quercus pedunculata*.　It is different in habit, the terminal bud being stronger than the others, so that the shoot is continued in the same direction, and the branches keep straight; whereas in *Q. pedunculata* the lateral buds at the apex often develop more vigorously and a crooked branch results, with the leaves much more tufted.

SEEDLING

At first the seedling differs little from that of *Q. pedunculata*, though the young leaves are more distinctly stalked; but towards the end of the first year, the characters shown in the adult stage are well marked, namely :—the stem, leaves, and terminal bud are pubescent, and the leaves have a cuneate base and short but distinct stalks.

VARIETIES OF *QUERCUS SESSILIFLORA*

1. Var. *longifolia*, Dippel, *Laubh.* ii. 67 (1892).—This is also known as *macrophylla*.　The leaves are variable, but are as a rule very long, as much as eight inches, and narrow in proportion to their length, the lobing being never constant. The base of the leaf is always cuneate.

2. Var. *laciniata*, Koehne, *Dendrol.* 130 (1893).—Leaves small with deeply-cut segments, which are directed forwards; base cuneate.

3. Var. *mespilifolia*, Wallroth, *Sched. Crit.* 494 (1822).—Leaves, with a petiole of one inch, lanceolate, long, and narrowed at both ends, averaging five inches long by one inch broad at the widest part; quite entire in margin or very slightly lobed.　This form has been found wild at Nordhausen in the Harz mountains, at Wolgast in Pomerania, and in various places in Austria and Hungary.　Var. *Louetti*, is a somewhat pendulous sub-variety, which is considered by most authors to be identical with var. *mespilifolia*.

4. Var. *sublobata*, Koch, *Dendrol.* ii. 2, 32 (1873).　*Quercus sublobata*, Kitaibel, in *Schult. Oest. Fl.* i. 619 (1814).—This is nearly the same as the last variety, but the leaves are slightly and regularly lobed.　It came into commerce from the Royal nursery at Geltow near Potsdam, and hence is often known as var. *geltoviana*.

5. Var. *cochleata*, Petzold et Kirchner, *Arb. Musc.* 630 (1864).—This resembles the common form, except that the edges of the leaf are curved upwards, so that the centre of it is rendered concave.　It is said to be a free-growing variety.

6. Var. *afghanistanensis*, Hort.—This variety, as cultivated at Kew, has obovate leaves very similar to the common form, except that the lobes of the leaf are more shallow and more numerous, and its bluish under surface is covered with a fine pubescence which extends to the petioles.　It is considered by Zabel[1] to be a hybrid between *Q. lanuginosa* and *Q. sessiliflora*.　It was sent out by Messrs. Booth of

---

[1] *Laubholz-Benennung*, 77 (1903).

Hamburg, who stated in their catalogue that it came from Afghanistan; but Petzold and Kirchner, *loc. cit.*, consider this origin to be improbable. What is sold under this name in some nurseries is *sessiliflora* or *Mirbeckii*.

7. Var. *iberica*, Hort.—This variety, as cultivated at Kew, has small oblong-ovate leaves, broad and cordate at the base, acute at the apex, with numerous small deltoid lobes, each terminating in a callous acute tip, the margins of the lobes being often turned downwards and inwards.

8. Var. *falkenbergensis*, Hort.—This has small dark-green leaves, broadest in diameter in their upper third, lobes few and broad, and the base generally cordate and auricled. The fruit is sessile or on short peduncles. It is very probably a hybrid between *Quercus pedunculata* and *Q. sessiliflora*.[1] This variety was found in 1832 in a wood at Falkenberg in Hanover, and was put into commerce by Messrs. T. Booth and Sons in 1837.

9. Var. *alnoides*, Hort.—This variety, as cultivated at Kew, has small leaves, not exceeding 2 inches in length, with about eight pairs of small lobes, the apex of the leaf being generally acute, the base cordate or cuneate.

10. Var. *pinnata*, Hort.—Leaves deeply pinnate, the sinuses extending almost to the midrib.

11. Var. *rubicunda*, Hort.—Leaves deep red, more especially in the early part of summer.

12. Var. *purpurea*, Hort.—Leaves purple, becoming green with reddish nerves in early autumn. This variety, according to Mr. Nicholson, is a thoroughly distinct and valuable ornamental tree.

13. Var. *variegata*, Hort.—Leaves variegated either with white or yellow tints.

14. Var. *aurea*, DC. *Prod.* xvi. 2, p. 9 (1864). *Quercus aurea*, Kitaibel, in *Reichb. Icon.* xii. 8, t. 645 (1850).—The leaf has generally six pairs of deeply cut lobes, rounded at the top. The young shoots bear yellowish leaves, and are themselves deep yellow. This occurs wild in Austria, and is considered by Zabel[2] to be a hybrid between *Q. conferta* and *Q. lanuginosa*; but a type specimen at Kew does not show evidence of *Q. conferta* parentage.

15. Var. *dschorochensis*, Hort.—The variety which is cultivated under this name does not seem to be the species[3] found by Koch on the Dschoroch range of mountains near Trebizond in Asia Minor; and at Kew is apparently a form of *sessiliflora* with oblong-oval leaves, which have eight or nine pairs of very shallow sinuate lobes.

---

[1] Zabel, *loc. cit.* 79.

[2] *Ibid.* 77. It is *Quercus aurea*, Wierzbicki, of Kotschy, *Eichen*, t. 4 (1862).

[3] *Quercus dschorochensis*, C. Koch in *Linnæa*, xxii. 328 (1849); *Quercus sessiliflora*, var. *dschorochensis*, DC. *Prod.* xvi. 2, p. 9 (1864).

QUERCUS LANUGINOSA, Pubescent Oak

Quercus lanuginosa, Thuillier, *Flora Envir. Paris*, ed. 2, 502 (1799).
Quercus pubescens, Willd. *Sp. Pl.* iv. 450 (1805).
Quercus Robur sessiliflora, var. *lanuginosa*, DC. *Prod.* xvi. 2, p. 10 (1864).
Quercus sessiliflora, Salisbury, var. *pubescens*, Loudon, *Arb. et Frut. Brit.* iii. 1736 (1838).

A small tree, rarely attaining 60 feet in height, and often, in the wild state, a dense shrub with a twisted stem. Bark rather rougher and more scaly than that of the common oak. Twigs and buds densely pubescent, the scales of the latter being ciliate on the margin and pubescent all over their surface. Leaves small, about 3 inches long, variable in shape, wrinkled in margin, cuneate or cordate at the base, with four to eight pairs of rounded lobes variable in depth; always densely pubescent underneath; petiole tomentose, ½ to 1 inch long. Axis of male flowers pubescent. Female flowers with sessile stigmas and tomentose ovary. Fruits, one to four, crowded on a short thick stalk, or sessile; cups tomentose and often tubercular.

This oak occurs on dry soils, especially those of limestone formation, in the south of France, Corsica, Spain, Portugal, Italy, Alsace, south Baden, Thuringia, Austria, Hungary, southern and western Switzerland, Turkey, Greece, Crimea, Caucasus, and Asia Minor. In Provence it forms dense, low thickets covering extensive areas of the very dry lower parts of the limestone mountains. In Corsica it appears to be the only deciduous species of oak; and was seen by me forming scattered groves in the mountains below the zone of *Pinus Laricio*, at about 2000 feet elevation. I observed no trees larger than a foot in diameter; and it is evident that it is very distinct from *Q. sessiliflora*, which, if it occurred, would grow to a large size in the Corsican humid climate. The tree is of no importance in Corsica as a source of timber; and Mr. Rotgès of the forest service considered that it should always be treated as coppice.

It produces hybrids with both *Q. sessiliflora* and *Q. pedunculata*, and differs markedly from both these species in its habit of producing root-suckers, and moreover the bark is different.

Loudon incorrectly states that it occurs in the New Forest, and Sussex. There is a tree of this form growing at Syon with a remarkably curved bole of about 18 feet long and 5 feet 10 inches in girth. If upright this tree might have been 50 feet high. Elwes has seen this species growing wild in the forest of Fontainebleau, which Hickel informed him was about its northern limit as a wild tree; here it is usually small and stunted, so far as he saw, and of no economic value.

Varieties of *Quercus lanuginosa*

1. Var. *Hartwissiana*, Hort.[1] Leaves with six or seven pairs of lobes, which are mucronate at the tips.

---

[1] According to Schneider, *Laubholzkunde*, 194 (1904), the plant so named by Steven in *Bull. Soc. Nat. Mosc.* 1857,

2. Var. *dissecta*, Hort.   Leaves deeply cut.

3. Var. *Dalechampii*, Koch, *Dendrol.* ii. 2, p. 38 (1873); *Quercus Dalechampii*, Tenore, *Ind. Sem. Hort. Neap.* 1850, p. 15; *Quercus sessiliflora*, var. *Tenorei*, DC. *Prod.* xvi. 2, p. 7 (1864).

This form, which is considered by some to be a distinct species, occurs in southern Italy.   It is in cultivation at Kew, and has leaves 3 to 4 inches long on short stalks.   The leaves are oblong-oval, with bases cuneate or truncate, often auricled, coriaceous in texture, shining green above, bluish and only slightly pubescent beneath, with six to eight pairs of acute shallow lobes, which have their margins curved inwards and backwards.   The bark of the tree is very rough and scaly.

i. 387, is either a form of *Q. macranthera* or a hybrid of that species.   The plant, however, usually cultivated as *Hartwissiana* is probably a variety of *Q. lanuginosa*, which Steven collected and described as *Q. crispata* (*Bull. Soc. Nat. Mosc.* 1857, i. 386).

Distribution of the Common Oak

Owing to the general opinion of English botanists that there is only one indigenous species of oak, with two inconstant varieties, there are few accurate records of the distribution of the two species, and in the majority of cases it is impossible to say whether the specimens in our great herbaria are from wild or cultivated trees.   Moreover, owing to the great changes caused by the spread of cultivation and the cutting down of most of the original woodland, the correct distribution of the two species can scarcely be made out.   It is probable, however, that in ancient times the pedunculate oak occupied the alluvial lands and the better soils, now almost entirely devoted to agriculture and pasture.   Hedgerow trees are invariably of this species.   The sessile oak occupied the hilly land and the poorer soils ; and in existing oak-woods occurring in such situations, which have never been touched by the plough, it is always the species met with, as in the Wyre Forest, the Forest of Dean, in the district about Burnham Beeches, in Lord Cowper's woods near Welwyn, Herts, which are on high-lying poor gravel soil, etc.   In Scotland, judging from a specimen at Kew, the famous Birnam wood consisted of *Quercus sessiliflora*.[1]   In Ireland, the ancient wood of Shillelagh, in Wicklow, of which a remnant still exists, was the same species.   The Cratloe wood near Limerick is of pure sessile oak ; and it is the only species in the wilder parts of Kerry.   All the specimens of *Q. pedunculata* which I have received from Ireland, are from planted trees.

In England the oak ascends to 1200 feet in Yorkshire.   In an interesting paper by H. B. Watt on the "Altitude of Forest Trees in the Cairngorm Mountains"[2] in Scotland, 700 to 800 feet is given as the highest level at which the oak was observed ; but Mr. Watt says, in a MS. note, that he found in July 1903 a small oak at Corriemulzie at an elevation of 1200 feet.   The same author gives many interesting particulars of the oak in Scotland, in a paper published in the *Annals of the Andersonian Naturalists' Society*, ii. 89 (1900).   In Ireland the oak ascends in Derry to 1480 feet.   There are remains of virgin forest in Donegal, on Sir Arthur Wallace's property near Lough Esk ; and a very large oak wood, which is of great antiquity, occurs at Clonbrock, the seat of Lord Clonbrock, in Co. Galway, on the limestone formation.   There are smaller woods in many of the mountain glens, and Mr. Welch of Belfast says that where these primitive bits of forest have never been touched by tillage, peculiar and local forms of land-shells occur, and the Clonbrock oak forest contains rare plants, moths, etc., unknown elsewhere.   The oak was in early times much more widely spread ; it has been found, *e.g.*, in a peat moss in the Orkneys.   Mr. T. T. Armistead[3] found a young oak growing in a sheltered ravine on the coast

---

[1] Mr. Steuart Fothringham of Murthly confirms this by leaves from the large oak behind the Birnam Hotel at Dunkeld, which Hunter says is one of the few survivors of the Great Birnam Wood.

[2] *Cairngorm Club Journal*, iv. 111 (1903).

[3] *Zoologist*, 1891, p. 19.

of Hoy, Orkney, and the acorn from which it sprang must have been brought from the mainland by a rock-dove or rook.

Remains of oak are found in all the later geological deposits; in the pre-glacial deposits in the Cromer forest-bed; in inter-glacial deposits in Hampshire, Sussex, Hertford, Middlesex, and Suffolk; in neolithic deposits; common in "submerged forests" everywhere; at the base of peat-mosses in many localities (ascending in them up to 1000 feet in Yorkshire).[1] Mr. S. B. J. Skertchley describes[2] the growth of five successive oak forests in the valley of the Ouse, and considers the oldest of them to be some 70,000 years old. These forests spread downwards towards the fen till checked by water and peat moss, the latter eventually burying and preserving them. The trees in thousands lie to the north-east, having been blown down by the south-west, which is still the prevailing wind. The word oak occurs in place-names both of Celtic and Saxon origin, the Saxon forms in names being ac, oak, wok, and auch. These forms are illustrated by names like Auchley, Auckland, Acworth, Wokingham, Oakingham, Oakham, Oakfield, Oakley, Martock, Holyoak, and Selly-oak. The Gaelic name is *dair*, as in Derry, Edenderry, Ballinderry, Kildare, Adare, Darnock, Kildarragh, Auchindarroch, Craigandarroch.

Quercus pedunculata, according to Willkomm, occurs throughout the greater part of Europe, Asia Minor, and the Caucasus. Its northern limit reaches, on the west coast of Norway, 62° 55′, on the eastern side of Norway 60° 45′, in Sweden 60°, in Finland 61° 30′ at Björneborg and 60° at Helsingfors, then passes along the coast of Esthonia to St. Petersburg, and crosses Russia south of Jaroslav and Perm, then descends southwards, reaching the Ural river between Orenberg and Orsk, and descends along that river to Iletzkoi. Its distribution in the Caucasus and Asia Minor is not known with exactness, owing to the conflicting opinions about the oaks of these regions. In Europe it occurs as far south as Greece, Sicily, and in the Peninsula reaches its southern limit in the Sierra Morena range. The western limit, beginning at the western part of this range, includes the northern part of Portugal and Galicia, and continues up along the coast of France, ending in Ireland and Scotland. It is essentially a tree of the plains and low hills, but it ascends in Southern Scandinavia to 993 feet, in the Berne Oberland to 2530 feet, in the Tirol to 3160 feet, in the Jura to 2216 feet, and in the Pyrenees to 3300 feet.

It is, according to Max von Sivers,[3] a much scarcer tree than it formerly was in the Baltic Provinces of Russia, and only exists in pure forests of any extent in Kurland, where it attains in river valleys and loamy soil very large dimensions, as much as 9 metres (about 30 feet) in girth. Some of the best trees produce logs free from branches over 60 feet long and 5 feet in girth at the top. He attributes its comparative scarcity at present to over-felling during the last two centuries, but states that replanting has been recently carried on to some extent.

Quercus sessiliflora occupies a more restricted area than the other species. Its northern limit is 60° 11′ in Norway, 58° 30′ in Sweden; it then passes through east Prussia, Lithuania, and crosses the central provinces of Russia, Minsk, Mohilev,

[1] C. Reid, *Origin of the British Flora*, 145 (1899).   [2] *Fenland, Past and Present*, chap. xv. (1878).
[3] *Die Forstlichen Verhältnisse der Baltischen Provinzen*, 1903.

Tula, and Penza, to Sergievsk near the southern Ural, in lat. 54°. The eastern limit commencing here, extends southwards, taking in the Crimea and Cilicia in Asia Minor. The southern limit extends through Greece, southern Italy, Sardinia, Catalonia, and the northern provinces of Spain to Asturias.

As a wild tree it does not occur in low-lying plains and alluvial ground; but is met with on the hills and lower ranges of the great mountain chains of Europe. It ascends in Hanover to 1900 feet, in the Alps to 3900 feet, in the Carpathians to 3300 feet, and in the Pyrenees to 5300 feet. In all these localities it ascends considerably higher than the pedunculate oak, reaching, *e.g.*, in the Alps 1500 feet higher than that species. (A. H.)

PROPAGATION AND CULTURE

The oak produces acorns in great abundance in some seasons,[1] generally about one year in three; but this varies very much in different parts of the country; and, so far as I have noticed, fruit occurs oftener and more abundantly in the south and west of England. It begins to bear at a very early age in some cases; and I received, in 1906, a packet of acorns from Miss Woolward, which she assured me were taken from oaks only ten years old from seed. Mr. Emerton, the head gardener at Belton Park, Notts, where they grow, confirms this. In the same season I saw acorns on the Billy Wilkin's Oak, which must be 700 to 800 years old; and was told that the Cowthorpe Oak, which is possibly much older, still bore a few. Acorns are greedily eaten by all domestic animals, but are injurious to cattle if taken in very large quantities.[2] Pheasants and pigeons also consume a great many, and rooks are credited with dropping most of the acorns which so often spring up as seedlings in places far from their parent tree.

The raising of oaks from seed is so easy, and the plants obtained are, as a rule, so much superior to what one can buy, that no one who wishes to plant them should fail to try the experiment by selecting acorns from the best oaks in the neighbourhood. These ripen in October, and should be gathered from the ground as soon as they fall, as dry as possible. They will not keep if stored damp, and my own experience is that they make stronger growth the first year if sown as soon as gathered, because the radicle will then bury itself deep in the ground before winter, and the germination will take place earlier. But if it is desired to sow the acorns where the tree is to grow, they must be protected against mice, rooks, pheasants, and wood-pigeons, all of which are very fond of them. Red lead or paraffin is sometimes used, but the latter is liable to injure the acorn, and it is said that chopped furze placed over the acorn is the best means of protecting them against mice. They should be covered with at least an inch of soil, and, if dibbled, care must be taken that they do not fall in the hole end downwards, but lie on their side in their natural position.

In 1901 I made experiments on the growth of oaks from acorns produced by

---

[1] I saw a large oak on the lawn at Marks Hall, Essex, which produced no less than 31½ bushels of acorns in 1906.
[2] Mr. T. P. Price, of Marks Hall, told me that in 1904 ten bullocks died there from this cause.

many trees in different parts of England, in order to learn whether the size of the acorns and the vigour of the parent tree had much influence on their strength. I have now watched the growth of these young trees for six seasons, and have arrived at no definite conclusion, though I am much surprised by two facts which have become evident. Lord Ducie has an oak in his park which usually produces acorns of unusual size, some that he has weighed being only 36 to the pound. The plants from these were no stronger than those of normal acorns; and some of the very finest plants that I raised were produced by the small acorns of a very stunted grafted tree with variegated leaves, which I only sowed to see whether any variegation would appear in their leaves. I found, however, that on the average the acorns gathered on my own place on similar soil gave the best results, and that those from Hants and Kent did not produce such good seedlings as those from Nottinghamshire.

The shoot appears above ground about the time the oak comes into leaf, or rather sooner, and the first growth is completed in three weeks or a month. A second growth, corresponding to the summer shoots of the parent tree, is produced in July or August, and sometimes even a third shoot. If sown in a nursery-bed they will be 4 to 12 inches high at the end of the first season, and should be transplanted in the following spring before they are a year old. For if the tap root is not cut early it will become so long and strong in good soil that the transplantation is a severe check to the young tree.

When lined out in the nursery they must remain two years longer, in good soil kept clean, after which the best of them should be 2 to 3 feet high and fit to plant out permanently, except where the herbage is long and coarse. They are sometimes left three years, but this is too long, though, where the land they are to go to is good and not too heavy, liberties may be taken with oaks which could not be risked on poor soil. If not planted out at three years they should be transplanted once more in the nursery, and at five or six years old ought to be 4 or 5 feet high, whilst oaks sown *in situ* in land covered with herbage or weeds will at the same age often be not more than a foot high and much less strong. In the long run, however, those which have never been transplanted will probably pass the others when once they have established a good root system, which in poor soil is a very slow process. Transplanted oaks, if they do not come away with good straight leaders, are best cut down to the ground the second or third spring after they are planted, when their roots are sufficiently established to throw up a strong leader. Some say [1] that this should not be done until the beginning of June when the sap is running strongly, but experiments which I have made seem to prove that April or May is better. Mice are the worst enemies of young seedling oaks, and where they are numerous cause an immense deal of damage by barking and biting them off close to the ground.

---

[1] Hayes states, *Planting*, 160 (1794), that from long observation he can aver that the root of an oak never produces a growth of finer young wood than when the tree is felled about the first week in June, when the sap is flowing most freely, and refers to Marshall's *Minutes of Agriculture and Planting in the Midland Shires of England* for evidence in support of this opinion.

Billington's account of the immense losses which were caused by mice to the oaks sown in the Forest of Dean, which is quoted at length by Loudon, pp. 1805-7, shows that in places where mice are numerous it is more economical to plant than to sow; and I have on my own property failed to get anything like a good stand of young oaks by sowing, on account of the ravages of mice and rooks, though every precaution which experience could suggest was taken. I tried dibbling in wheat, and sowing in lines and patches, both on cultivated and uncultivated ground, and have only partial or complete failures to record. In better and lighter soils, and especially in woods of large size where rabbits are kept down, I have seen splendid results from self-sown acorns; and Mr. A. C. Forbes's prize essay on the natural reproduction of woods from seed, published in the *Transactions of the English Arboricultural Society*, v. 239, should be consulted, as well as Loudon's remarks on the same subject, pp. 1804-5.

Mr. Stafford Howard, C.B., who probably knows more about forestry and has done more to improve the management of the Royal Forests than any Commissioner who preceded him, except, perhaps, Lord Glenbervie, has sent me an excellent photograph of a grove of self-sown oaks on his property at Thornbury Castle, Gloucestershire, which has originated from acorns, self sown, in what used to be an osier bed, and which are now about thirty to forty years old. Plate 81 shows their present appearance. On December 29, 1904, Mr. Howard showed me this grove, of which about an acre, containing 139 trees, has been wired in and under-planted with beech at about 6 feet apart. Six trees have been measured and marked with the object of showing whether the future increase of the oaks will pay for the cost of under-planting. As I am not aware that this practice, which in Germany and France is considered good forestry, has ever been properly tested in England, I hope that the results of this experiment will be recorded.

The best illustration of the possibility of converting coppice with standards, into pure oak wood, was shown me in 1900 by Mr. A. C. Forbes in a wood called Derry Hill, on the property of the Marquess of Lansdowne, three miles from Chippenham. In this case the coppice was cut early in the winter, after a good crop of acorns, and completely cleared before the following May. The constant presence of workmen faggoting and cleaning the coppice, not only kept away pheasants and pigeons, but also buried a good many of the acorns; and the soil being suitable for oaks, their growth was so good in the next three years that by cutting away the shoots of the coppice wherever it crowded and overgrew the young oaks, a stand was obtained far thicker, cleaner, and more vigorous than I have ever seen from planted trees. If carefully attended to until the seedlings overtop and smother the remains of the underwood, and provided also the remaining standards are cut and removed before they damage the seedlings, I should expect this wood to become one of the best of its sort in England.[1]

On the property of Dr. Watney, at Buckhold, Berks, I have also seen some

---

[1] On revisiting the place seven years later I found that the growth had not been so good as it promised to be, owing perhaps to the underwood being cut too hard, and the soil having become overgrown with grass.

admirable illustrations of the growth of young oaks from seed, and of the result of converting oak coppice wood into standards, by leaving all the best poles uncut, and carefully thinning out the weakest at intervals. This process, owing to the great fall in the value of oak bark, to the production of which large areas of oak coppice in the west and south-west of England were mainly devoted, has become very generally desirable ; but if the stools are old, it is best to grub them, and replant the ground with seedlings mixed with other trees, as has been largely done on the estates of the Duke of Bedford near Tavistock.

With regard to the effect of transplanting oaks on their future growth and height, opinions differ as much as on any subject. The late Sir James Campbell, who managed Dean Forest for many years, often told me that the oftener you transplanted an oak the better it grew, and he communicated a paper with measurements of some trees in Dean Forest to the International Forestry Exhibition at Edinburgh in 1884 in proof of this ; but Mr. Smith, who quotes and refers to these measurements in the paper on oaks above referred to, agrees with me that they do not prove the case ; and Mr. Philip Baylis,[1] who succeeded Sir J. Campbell at Dean Forest, writes me as follows :—

" At one time I was of the opinion, founded on the above measurements, that trees were benefited by being transplanted, but have long ago given up that opinion. It is true that for a time after the tree has recovered from the shock of moving, you may, in consequence of the greater number of fibrous roots produced by the moving, get a stimulated growth ; but I am convinced that the tree which eventually produces the finest timber tree is the one which is never moved from the place where the seed first germinated."

In this opinion I entirely agree, and believe that though oaks, like other trees, may be drawn up to a considerable height when surrounded closely by other trees, especially the beech, yet that their straight upward growth largely depends on the depth to which the main roots can descend. I do not know that it has ever been proved at what age the tap root decays, and this no doubt depends very largely on the subsoil ; but though one may see very large spreading oaks on a thin soil, I never saw a very tall and straight one except on deep land.

In an appendix to the First Report of the Commissioners of Woods and Forests, published as a blue-book in 1812, will be found (p. 143) some very interesting and valuable observations on the sowing and transplanting of oaks, in which instances are quoted from several places which go to show that oaks on some soils at least, as at Moccas Court, in Bere Forest, and in the Forest of Dean, will grow as fast or faster when transplanted at 8 to 10 feet high, or even more, than when sown *in situ.*

In another appendix to the same report, on page 141, are some further observations, made by men of great experience on the growth of oaks from the stool, which prove that when the stools are young and sound and the land good, sound oak trees of as much as 160 cubic feet may be so produced ; but that when the stool has

---

[1] Mr. Baylis sends me a very interesting photograph showing the difference between the roots of transplanted and untransplanted oaks.

become old and partially decayed, or when the land is poor, such shoots are not likely to attain any size. The best example I know of an oak wood produced entirely from stools is one below the approach to Carclew in Cornwall, which the late Colonel Tremayne showed me in 1902. Here the trees average 15 to 20 feet apart, and have clean boles 25 to 30 feet high, and are about 4 feet in girth.

Marsham's opinion on the growth of oaks, taken from a paper printed in the *Philosophical Transactions*, are so much to the point, and his personal experience was spread over such a very long period (from 1719 to 1795) that I quote him as follows : [1]—

" In 1719 I had about two acres sowed with acorns, and from 1729 to 1770 I planted oaks from this grove, always leaving the best plants standing for the future grove ; but most of the transplanted trees are already larger than those that were not removed ; the largest of which is now (1795) but 5 feet 6 inches 8 tenths in circumference ; and the largest transplanted tree (which was planted in 1735) is 8 feet 8 inches 7 tenths, viz., near 38 inches gained by transplanting in 60 years. And in beeches from seed, in 1733, the largest is now (1795) but 6 feet 9 inches ; and the largest transplanted beech is 7 feet 5 inches 1 tenth, viz., 8 inches larger, although the transplanted beech is eight years younger than that from the seed. This proves that it is better to plant a grove than to raise one from the seed. The expense of planting is inconsiderable, and the planted trees are full as good and handsome, and many years are saved, besides the extra growth of planted trees. But this extra growth will not prove near so great in groves as in single trees. The first grove I planted from these acorns of 1719, was in 1731. In 1732 I made another grove from them, and in 1735 I planted a third grove from them, and in 1753 the last considerable number of plants were taken from the grove, and these are very good trees : so thirty-four years may be saved. But I would by no means advise the planting trees so large, as the trouble and expense will be too much, unless where a shelter or screen is wanted.

" Whether a grove is to be raised from seeds or planted, it is advisable to shelter it round ; if from the seed, with such sorts as will grow quicker ; and if by planting, with larger and taller trees. The soil in Norfolk is unfavourable to elms ; therefore in planting I will venture to recommend hornbeams, as they may be planted large trees. I planted some hornbeams (rather large) in 1757, and, disliking their situation, in 1792 I removed them when they were about three feet in circumference, and did not lose one tree ; and they made shoots of near half a yard that year ; but I ought to say I cut off their heads.

" Before I quit this subject, I will presume to recommend, if young oaks are unthriving, there is reason to hope they may be helped by cutting them down to a foot or six inches ; for in 1750 I planted some oaks from my grove of 1719 into a poorer soil, and although they lived they were sickly ; so in 1761 I cut most of them down to one foot, and then by cutting off the side shoots, in three or four years led them into a single stem, and most of them are now thriving and handsome trees, and you

[1] *Phil. Trans.* 1797, pp. 128-152.

can hardly see where they were cut off, and some are four feet round ; and I have used the same method with unhealthy chestnuts, beech, hornbeam, and wych elm, and with the same success."

RATE OF GROWTH

The rate of growth in the oak is principally governed by the soil and situation, and varies so much that any estimates of its possible increase are of little value unless based on local experience. We often read calculations of the profits of planting, drawn from Continental experience or from exceptionally favourable cases in England, which are very misleading and greatly in excess of reasonable expectations, and there is no tree to which these remarks apply more strongly than to the oak.

Few plantations give more ample proof of this than those made by the Government in the woods at Alice Holt, which were planted between 1810 and 1830, with the object of providing timber for the navy, and which were no doubt done by experienced planters. But the growth has been so poor that, when I visited them in 1905, in company with Mr. Stafford Howard and Mr. Lascelles, we saw but few oaks which looked as if they would ever be fine trees, and their average value was not much over 10s. per tree. In one place, called Willow Green, oaks of seventy years old were not over 30 or 40 feet high and not thick enough for gate-posts.[1]

In many parts of the Forest of Dean the results are not much better, and are largely attributed to over-thinning, and to the fact of the ground being thrown open to grazing too soon ; but the soil and spring frosts must also have had a good deal to do with it.

In the New Forest the results are better, but not at all equal to what might have been expected. I am indebted to Mr. Stafford Howard for the following information on some of these plantations and the way in which they were made :—

Planting in the New Forest.—In order to make provision for the future needs of the navy, in view of the fact that planting had been greatly neglected in the New Forest, an Act was passed, 9 & 10 Will. III., for that purpose. Under this Act it was provided that 2000 acres should forthwith be enclosed and planted with timber for the use of the navy only, underwood and all other produce being excluded ; that 200 acres should be enclosed annually for twenty years following, and that as soon as any of the land thus enclosed was safe from damage from cattle, it should be thrown open and a like area enclosed in its stead. The plantations described were made under the powers of this Act.

The precise form of cultivation employed was as follows :—

" Pits or beds of three spits of ground each were dug a yard apart, and three acorns planted triangularly in each bed. Half a bushel of acorns was allotted for each person to plant in one day. Two regarders attended every day during

---

[1] Mr. Howard says that in the lower part of the Goose Green enclosure, and in the Straights, there is much better timber, and that in Dr. Schlich's report on these woods over 300 acres were classified as good, where the trees attain a mean height of 60 feet.

the time of planting to see that it was properly done; and after the ground was fully planted with acorns it was sown with haws, holly berries, sloes, and hazel nuts, drains were cut where necessary, and traps were set to catch mice, and persons attended daily to reset the traps and to keep off crows and other vermin."

Whether from subsequent neglect or not, the plantations thus formed were never thinned at all, but allowed to grow up like a nursery quarter. Although contrary to every theory of plantation management, it cannot be denied that they were in this bad soil successful in growing a heavy crop of oak timber on moderate land.

Denny Enclosure.—There are some very good examples of natural regeneration in places in this wood, which was reinclosed in 1870. A photograph was sent which contrasts the young growth inside the fence of the enclosure with the bareness of the outside where the cattle graze.

Salisbury Trench.—This plantation was made in or about the year 1700, and measures about 100 acres. It was thrown open under an order dated 20th August 1807. It is calculated that there are now left after frequent thinning about sixty trees to the acre. Two years ago it was reinclosed with a view to its gradual regeneration, and there is already a large number of young oak and beech coming up in the open spaces.

North and South Bentley.—These plantations were made about the same time, probably just before that of Salisbury Trench, and are of the same character, except that there is some beech here and there in North Bentley. During the past twenty years the trees felled in Salisbury Trench, being for the most part the poorest ones, have averaged $23\frac{1}{2}$ cubic feet; and there now remain about sixty to the acre. In North Bentley they have averaged about 25 cubic feet, in South Bentley 29 cubic feet, and about sixty trees to the acre remain standing.

One of the best private oak plantations of which the exact age is known is on the property of Lord Kesteven at Banthorpe, near Casewick, Lincolnshire. It was made by Sir John Trollope, grandfather of the present owner, in 1800, with acorns which had to be sown a second time, as they were eaten by mice in 1799. It is on good soil, and, as near as I could judge by the eye, contains about sixty trees to the acre, straight for the most part, and clean up to 30 to 40 feet. In 1905 twelve average trees in the plantation had an average timber length of 34 feet, an average quarter girth of 18 inches, and contained 903 cubic feet without tops or branches, which would make my rough estimate of 5000 feet to the acre very nearly correct, and if profit alone were considered I should say that these trees had now reached the proper age for felling.

The late Mr. John Clutton, who valued timber for the Crown for many years, gave,[1] in 1873, particulars of the size of oaks.

---

[1] *Transactions of the Surveyors' Institution*, 1873-74, vol. vi.

In New Forest, Aldridge Hill, planted 1813 :—

	Number.	Contents.	Value.
1st acre	75	742	£90
2nd „	79	559	67
3rd „	77	641	78
4th „	72	683	84

In Alice Holt Woods :—

	Number.	Contents.	Value.
Lodge Enclosure	40	837	100
Goose Green	50	812	97
Berewoods, planted 1816	54	771	93
„ „ „	70	618	74

In Dean Forest :—

	Number.	Contents.	Value.
Blakeney Hill, South, planted 1814	72	720	87
Nag's Head Plantation „ „	97	425	57
Bromley Hill Plantation „ 1812	67	700	84
High Meadow Woods (no date stated), 1st acre	30	1528	214
High Meadow Woods (no date stated), 2nd acre	50	1480	207

In Richmond Park :—

	Number.	Contents.	Value.
Upper Pond, planted 1824	60	672	81
Kingston Hill, „ 1826	46	628	75
Isabella, „ 1831	68	450	54
Isabella, „ 1845	110	406	49

In the same volume Mr. Ralph Clutton, in an excellent paper on the self-sown oak woods of Sussex, gives many exact details of the growth of oak without under-wood, with measurements and valuations, which should be consulted by all land-owners in that part of England.

Under more favourable circumstances, however, oak plantations may yield a good profit, as shown by the following extract from the *Norfolk Chronicle*, sent me by Sir Hugh Beevor, and printed in Grigor's *Eastern Arboretum*, p. 360.

" Being enabled from old memoranda of undoubted authority, and from information received several years ago from different persons, who remembered or who assisted in the work, to give you, perhaps, an unusually accurate account of the produce of a piece of land measuring eight acres, planted with acorns in the year 1729, I take the liberty of so doing, and of requesting your insertion of it in your paper whenever you may have the best opportunity. The piece was under the plough at that time, cold and unprofitable, from the practice of underdraining not being then introduced; at Michaelmas 1729 it was sown with wheat, and acorns dibbled in; when reaped, the stubble was left very long, which is supposed to have caused the plants to run up very straight.

"Besides a great many used on the ground, from 1729 to 1763, plants were drawn out and sold to the amount of £100   0   0

In the year 1764 by 1500 poles sold .	.	.	50	0 0
,, 1765 by 1374 ,,	.	.	50	0 0
,, 1767 by 468 ,,	.	.	30	0 0
,, 1770 by 501 ,,	.	.	39 18	0
,, 1771 by 440 ,,	.	.	21	0 0
,, 1777 by 280 ,,	.	.	21	0 0
,, 1781 by 150 ,,	.	.	80	0 0
,, 1793 by 101 ,,	.	.	21	0 0
,, 1794 by 150 ,,	.	.	105	0 0
,, 1797 by 30 trees sold	.	.	20	0 0
,, 1799 by 100 ,,	.	.	60	0 0
From the year 1800 to 1810 by 307 trees sold	.	389 12	0
,, 1811 to the year 1821 by 94 ,,	.	219	0 0
,, 1821 ,, 1833 by 36 ,,	.	108	0 0

£1314 10   0

The underwood never came to perfection, but was stubbed up in the year 1767, and the feed of the ground let for 10s. an acre for thirty years .	120	0 0
Value of the feed at the same price to the present time	144	0 0
There are now 320 trees standing, worth if now felled	1200	0 0

£2778 10   0

"The expenses of felling cannot be now correctly ascertained, but the topwood is not included in the above account of receipts, nor a great many trees which have been used on the premises from the year 1763 to the present time, and at a moderate estimate must have much more than paid for the expenses of the labour.—THOS. HOWES, Morningthorpe, April 22nd, 1834."[1]

The Earl of Darnley showed me an oak in "Mount Meadow," near Cobham, planted by Lady Elizabeth Brownlow, who was born in the year 1800, which therefore could not be much over 100 years old. It has a straight clean bole measuring about 40 feet by 12 feet 10 inches, and a small spreading top.

The following extract from a letter of Robert Marsham to Gilbert White is worth quoting, though I could not identify the tree when I visited the place recently.

"*Stratton, 24th July* 1790.—I early began planting, and an oake which I

[1] Sir Hugh Beevor in 1902 measured eleven of the oaks remaining in this grove, which was nearly all felled in 1885, and found that they averaged 80 to 90 feet high by 8 feet 2 inches in girth at 6 feet, the cubic contents being about 145 feet each.

planted in 1720 is at one foot from the earth 12 feet 6 inches round, and at 14 feet (the half of the timber length) is 8 feet 2 inches. So measuring the bark as timber gives 116½ feet buyer's measure. Perhaps you never heard of a larger oak, and the planter living. I flatter myself that I increased the growth by washing the stem, and digging a circle as far as I supposed the roots to extend, and spreading sawdust, etc., as related in the *Phil. Trans.* vol. lxvii. p. 12."

Blenkam[1] mentions a remarkable instance of rapid growth :—" Three thriving oaks, growing on a hard gravelly and poor soil, were felled in Nottinghamshire, which on an average girthed 15 feet at three feet from the ground, and each tree contained about 430 cubic feet. The trees were planted in 1692 or 1693, and were about 149 years old when felled. They were perfectly sound and yearly increasing in size."

In a paper by Mr. Clayton[2] a photograph is given of a section across the butt of an oak felled at Ravenfield Park between Doncaster and Sheffield in 1885, which had a butt 36 feet long without a branch, and an average diameter of 5 feet, and which showed only 212 annual rings on a radius of $27\frac{3}{4}$ inches. If the actual age of this tree was only 212 years, its growth must have been unusually rapid, and a comparison of this with the section of the oak from Wistman's Wood (cf. p. 326) shows how remarkably the growth of trees depends on their situation.

As an illustration of the possible value of a hardwood plantation about forty acres in area in the Sherwood Forest district, I am able to give the following particulars, for which Mr. Doig, forester to Earl Manvers is my authority. In White's *History of Sherwood Forest* the land in question is called " Robert Fitzorth's land." It now goes by the name of Osland. It had been in cultivation previous to 1730, about which time it was planted, or perhaps sown, with beech, oak, ash, chestnut, larch, and spruce. The conifers had mostly been cut previous to 1846, before which time there are no records of the value of the thinnings taken from it. Since then the following have been cut or blown down :—

	Number.	Cubic Contents.	Value.		
Oak . . .	1801	38,735	£3,732	2	9
Oak poles .	1628	8,696	371	5	1
Beech . .	2054	74,213	3,756	0	8
Ash, elm, etc. .	63	2,215	123	2	5
Chestnut . .	43	1,289	102	4	4
Larch and spruce .	117	660	26	9	1
	5706	125,808	£8,111	4	4

Standing in 1903 :—

	Number.	Cubic Contents.	Value.		
Oak . . .	182	18,200	1,820	0	0
Beech . .	701	63,090	3,154	10	0
Chestnut . .	14	1,336	88	10	0
Larch . .	5	400	23	6	8
Total . .	6608	208,834	£13,197	11	0

[1] *British Timber Trees*, 42 (1862).   [2] *Trans. Bot. Soc. Edin.* xxii. 396.

This shows an average number of trees per acre (omitting the oak poles) of about 125, and a value of £320 per acre.

Perhaps the greatest increase of girth on record in the oak is cited by Gadeau de Kerville[1] of three oaks which were felled at Neauphe-sur-Dives (Orne) in Normandy in 1894. Their exact age was not possible to decide, as they were already trimmed and barked and part of the sapwood taken off, but the rings counted by M. de Kerville were 115 to 120, and the girths 6.16, 4.98, and 4.28 metres respectively. He thought that they might be from 150 to 200 years at most, and this would make the average annual increase of the largest, on the section measured, over 5 centimetres per annum.

Remarkable Trees

The mass of information on the oak which exists in English literature, is so great, so scattered, and often so impossible to verify, that I have had great difficulty in making a selection of what is really valuable and authentic, and have preferred rather to speak of trees and woods that we have seen ourselves, and to quote from the letters of living correspondents, than to repeat what has been written by Evelyn, Hunter, Strutt, Selby, Loudon, and other writers, whose works can always be consulted by those desirous of more detailed particulars than our space will allow.

Some of the most wonderful oaks of England, which we have seen and now figure, must be described more particularly, and among these I think the oaks of Powis Castle come first. Robert Marsham, in a letter communicated by Sir T. Beevor to the *Bath and West of England Societies' Transactions*, i. 78 (1783), says :—" The handsomest oak I ever saw was in the Earl of Powis' noble park by Ludlow in 1757, though it was but 16 feet 3 inches. But it ran straight and clear of arms, I believe, near full 60 feet, and had a large and fine head."

In April 1904 the Earl of Powis showed me some trees growing in his ancient park at Powis Castle, near Welshpool, Montgomeryshire, which I believe to be actually the champion oaks of Great Britain at the present time. They grow on a Silurian formation at about 300 to 400 feet elevation, with an east aspect, and are, as far as one can judge, perfectly sound in the butt, though one of them lost several branches during the dry seasons between 1893 and 1903, and another has a large decayed limb which, if not taken off, may cause the butt to decay.

The measurements which I give were made most carefully by Mr. W. F. Addie, agent for the Powis estates, who used a long ladder and a man to climb nearly all over them and take the length and girth of the principal branches down to 6 inches quarter-girth. I checked the height and girth of the trunks myself as carefully as possible, and believe that the following is a very accurate estimate.

---

[1] *Les vieux arbres de la Normandie*, iii. 373 (1895).

	Girth at Ground.	Height of Bole.	Girth at 4 Feet 6 inches.	Height of Tree.	Cubic Contents.
	Feet. Inches.	Feet.	Feet. Inches.	Feet.	Feet.
No. 1. The Champion Tree, by middle gate (Plate 82) [1]	31 7	25	23 6	105	2026
No. 2. Near the Park Plain (Plate 83) [1]	40 0	12	29 7	95	1925
The girth of this tree at the top of the trunk, where the tall straight branches begin is 38 feet 3 inches.					
No. 3. By Pochfield gate (Plate 84) [2]	32	20	22 6	95	1617
No. 4. In Gwen Morgan Wood	30	31	19 4	93	1432

Of the extraordinary size to which oaks have attained in this district we have a record which is without parallel in this or any country. My attention was called to it by the Earl of Powis, who, knowing the locality, believes it to be true. It is taken from a work called *Collections Relating to Montgomeryshire*, xiii. 424-425 (1880), published by the Powysland Club at Welshpool, and runs as follows :—

"In 1793 and 1796 a large fall of oak timber took place at Vaynor park in the parish of Berriew, when some trees of enormous dimensions were cut down. Major Corbett Winder has kindly favoured us with a copy of the following memorandum of the particulars of the contents of some of the largest trees :—

"Dimensions of twenty-six of the largest oaks cut down in Vaynor Park in 1793 and 1796.

No. of Tree.	Feet.	No. of Tree.	Feet.	No. of Tree.	Feet.
1.	1127	10.	1523	19.	1516
2.	1121	11.	1859	20.	1428
3.	2501	12.	1328	21.	1298
4.	2202	13.	1808	22.	1077
5.	1713	14.	1793	23.	1161
6.	1106	15.	1289	24.	1018
7.	1453	16.	1101	25.	1170
8.	1953	17.	1467	26.	1322
9.	1192	18.	1246		

Total : 37,772 cubic feet, averaging 1452½ cubic feet per tree."

The counties of Hereford, Worcester, Shropshire, and Stafford have produced and perhaps still contain the largest oaks in England, next to those I have just mentioned, but the long years of agricultural depression which have impoverished so many of the squires of England, have caused the felling of many of the finest. Among these the most celebrated was the Hereford Monarch which grew at Tyberton, near the house of Chandos Lee Warner, Esq., to whom I am indebted for two copies of a print taken from drawings which were made by G. L. Lewis, and published in a scarce work called *Portraits of British Forest Trees*.[3] One of

[1] The photographs from which these plates are reproduced were taken in June 1904 by Mr. R. G. Foster of Burford.
[2] This plate is from a photograph taken in 1906 by Lord Powis.   [3] Vale, Hereford, 1837.

these shows the tree in summer, the other in winter, and prove it to have been a tree of faultless shape and beauty, if not quite equal in bulk to the Champion Oak at Powis Castle. I visited the site of this tree in 1905, but the stump was no longer visible, and the soil, though a good deep red loam, did not show in the other trees any striking evidence of unusual fertility.

Its measurements, as given me by Messrs. Openshaw of Woofferton Court, to whom I am indebted for many particulars about trees in their district, were as follows :—

Butt . . 30 feet by 55½ inches quarter-girth	}	923 feet.
Second length . 60 ,, ,, 26 ,, ,,		
One branch . 18 ,, ,, 42 ,, ,,		220 ,,
Other branches more or less damaged by lightning, about .		400 ,,
		1543 feet.

A record of the tree was sent me by Messrs. Stooke and Sons of Palace Yard, Hereford, as follows :—" *The Hereford Monarch.*—An Oak tree, containing 1200 cubic feet, felled in Tyberton Park, ten miles from Hereford, April 1877. Length of tree, cut off at 18 inches diameter, 88 feet. Length of butt only 29½ feet. Height of tree when growing 130 feet. Circumference at 5 feet from the ground 22 feet 8 inches. Photograph taken of tree as felled, and showing the larger bough as shattered by lightning. Purchased by Messrs. R. and T. Groom and Sons, Wellington, Salop."

Mr. T. E. Groom of Hereford, whose firm bought it, informed me that though the tree would have been worth about £300 before it was struck, it did not actually cost them more than £200. It was felled in consequence of its having been disfigured by a stroke of lightning. Before this it was a perfectly sound tree with over 1500 feet of timber in it. It was still growing and might have become much larger. The butt was quartered and sold to a vat maker who cut it all into thin rims. At the end of the 30 feet of butt were two parallel spires each containing several hundred feet. The larger one was so much broken that it had but little useful timber left in it. The smaller was 60 feet long and about 2 feet in diameter at the top end. This was cut up into railway planking. The tree also made several thousand keys and trenails used on the railway.

Another immense tree was felled in Staffordshire on May 29, 1786, of which Messrs. Openshaw give me the following particulars :—" It grew in the middle of the Grove field on Bath farm, Chillington estate, and measured as follows :—

	£	s.	d.
Butt, 30 feet by 60 inches = 750 feet at 5s. . .	£187	10	0
Limbs (22), 560 at 1s. 8d.	46	13	4
Thirteen cords of wood at 10s. 6d. . . .	7	7	0
The root	2	10	0
2½ tons bark	8	8	0
	£252	8	4

" No branches under 9 inches quarter-girth were included in the above. Twelve men worked twelve hours each in felling this tree."

One of the tallest oaks which I have ever measured in England is a comparatively young tree in perfect health and vigour, which, though not shut in by other trees, appears to be still growing, and may even attain a greater height. It stands on the edge of a plantation at the bottom of a steep slope facing north-east in Whitfield Park, Herefordshire, the seat of Capt. Percy Clive, who showed it me in 1906. A careful measurement from both sides made it 130 feet high, or perhaps a little more, by 11 feet 10 inches in girth, with a straight bole of 55 feet free from branches, though two or three small ones had been cut off four years ago. For symmetry and height combined I have not seen its equal in England, and the photograph of it taken by Mr. Foster, though under the circumstances a very good one, fails to give a correct idea of its great height (Plate 85). The soil is old red sandstone, and the tree is of the sessile type.

At Foxley, near Hereford, the seat of the Rev. G. H. Davenport, are many fine oaks, all of which, so far as I saw, are sessile. The best is about 104 feet high by 20 feet girth, with a bole of 20 feet. In the Nash Wood there is a superb lot of young oaks with the tallest and cleanest stems in proportion to their thickness I have seen in England. They may average 90 feet high, and one which I measured was clean and straight to 62 feet and only 3 feet 4 inches in girth. Mr. Davenport believes them to be sixty to seventy years old, and if well taken care of they should in a hundred years be some of the finest of their type in England.

The largest oaks now standing in Herefordshire that I know of are at Holm Lacy, one of which, a short-boled spreading tree now much decayed, was in 1905 75 feet by 30 feet 2 inches, and 125 yards in circumference of the branches. The other, 90 to 95 feet high, with a bole 25 feet by 23 feet 9 inches, is vigorous and healthy, though perhaps not quite sound.

In Lord Leigh's park at Stoneleigh Abbey, Warwickshire, are many fine old oaks, relics of the Forest of Arden, which grow on a red sandstone soil, and are in many cases long past their prime. The largest stands near the Abbey, and is 28 feet 3 inches in girth ; though the top is much broken and decayed, the butt seems sound. Another, just outside the Tantarra Lodge, is a vigorous tree of later date, and measures 22 feet 10 inches in girth, with a fine spreading crown ; a third, near the river, is 27 feet 5 inches in girth. The most interesting, however—of which I hope to give an illustration later—is Shakespeare's Oak, so called from the tradition that Shakespeare used to sit and write under it. It grows on the top of a low sandstone cliff, over which at least half the thickness of its trunk projects, and is supported entirely by the roots on the other side to which it leans ; it measures no less than 25 feet in girth, and though deeply cleft on one side and hollow, has vigorous branches.

The oak grove at Kyre Park, Worcestershire, the property of Mrs. Baldwyn Childe, was first noticed by the Woolhope Club, who visited it in 1893, and described later by Sir Hugh Beevor, who published a short account of it.[1] I had the pleasure of visiting this wonderful grove in March 1904, when some

---

[1] *Trans. of the English Arboricultural Society,* v. 473.

photographs were taken (Plates 86 and 87), which give a good idea of the remarkable size and height of the trees.   The soil is a good deep loam on the red sandstone formation.   The grove is unfenced and has been open to cattle for many years, and there is no visible evidence of the trees having been drawn up by beech.   The majority of them are of the sessile variety, though some are pedunculate oaks, as proved by specimens kindly sent me by Mrs. Baldwyn Childe and by the observation of her very obliging agent, Mr. J. W. Openshaw, who found six trees of the pedunculate to about twenty-four of the sessile form. Sir Hugh Beevor speaks of them as sessile, and at the time I was there it was difficult to distinguish one from the other.   As to their age, Mr. Openshaw writes that he could not count the rings because they were so minute, but from the evidence of Habingdon's *History of Worcestershire*, written in the time of Queen Elizabeth, they must be very old.   Habingdon says:—" The Parcke of Cure Wyard is not to be shutt up in silence, for it is adorned with so many tall and mightie oakes as scarce any ground in England within that quantity of akers can showe so many."   Most of these trees do not show decay in their tops like so many of our great park oaks, and may thrive for centuries to come.

Sir Hugh Beevor's measurements of their height agree very fairly with my own, but exact measurements of the heights of such trees are difficult to obtain, and they are not so remarkable for their girth as for the way in which they run up with clean stems to a great height.   The two tallest are certainly over 130 feet by my own measurements in 1907.   Sir Hugh Beevor gives 78 and 79 feet as the first length of two, and one which was blown down in 1897 was 82 feet to the first limb, though only 16 inches in quarter-girth, and with no measurable tops.   These trees show very few burrs, but some have large buttresses at the base.[1]   The largest, according to Mr. Openshaw, has a stem 83 feet long by 17 feet 8 inches in girth at 5 feet, and contains 1031 cubic feet of timber. Fourteen of them contain over 600 feet, and the smallest tree in the grove has 97 feet, which is considered a big oak in many districts.   The tree I have figured (Plate 86), with Kyre House in the background, is on the outside of the grove, and of different type from most of them.   It is the third largest tree in contents, having 694 cubic feet in the butt and 150 cubic feet in the tops.   I made it 115 feet high by 18 feet 6 inches at 5 feet, and it looks vigorous and is growing fast.   The other tree figured (Plate 87) is 85 feet to the first limb, 13 feet 6 inches in girth at 5 feet, and contains 604 feet in the butt, and 112 in the tops. The measurements given below, taken by Mr. Openshaw, may be thoroughly relied on.   They were taken in the usual way by strap, and good allowance made for taper.   The heights were taken with the help of a long pole; and both Mr. Openshaw and his father, who has probably as much experience in measuring big oaks for sale as anyone in England, are confident that the grove contains more than they have estimated, though no doubt a quantity of the timber would be broken in falling if cut.   Of this, however, there is not the least risk in the lifetime of the present owner, who is much interested in, and very proud of her trees.

[1] One of these measures no less than 44 feet round the base, and at five feet from the ground is 20 feet in girth.

"*Kyre Park.*—Measure of oak trees in Woodpatch grove made by John W. Openshaw, November 1904. The tape girths are over bark taken at 5 feet. The quarter-girth is the middle of first length taken under bark. Eleven trees removed (1883, 1887, 1897) contained 2990 feet, average 272 feet. Ninety-seven trees now standing contain 38,365 feet, growing on 5 acres, 2 roods, 19 poles of land; an average of 395⅝ feet per tree. A hundred and eight trees contained 41,365 cubic feet, an average of 383 feet per tree. There remain distinct traces of sixty and indistinct traces of ten trees having been removed, including the eleven referred to above."

Number (in Mr. Openshaw's Table).	Girth at 5 Feet High.		Length of Stem.	Quarter-Girth.	Contents.		Total Content of Tree.	Remarks.
					Trunk.	Tops.		
No.	Feet.	Inches.	Feet.	Inches.	Feet.	Feet.	Feet.	
10	21	6	65	54	644	129	773	Very large spurs at base.
13	15	0	61	40	538	83	621	
16	17	6	49	45	447	156	603	
23	19	6	72	30	450	200	650	Blown down, 1904.
25	17	0	28	45	393	258	651	Forks.
57	18	0	73	45	421	249	670	Forks.
60	13	6	85	32	604	112	716	Tree in group.
62	17	5	60	40	666	224	890	
63	19	6	45	46	447	184	631	
72	15	9	88	42	573	60	633	
78	15	2	62	36	558	145	703	
79	18	9	75	50	694	150	844	Single tree in photo., facing Kyre House.
81	16	6	69	41	532	160	692	Leans, large top.
82	14	0	82	35	522	110	632	
91	17	8	83	47	851	180	1031	Ivy growing, largest tree.
97	14	5	95	34	633	100	733	By holly tree.
							11,473	

There is an oak of remarkable size in another part of the Kyre estate called the Hannings, growing on high ground exposed to the north, in a rough pasture overgrown with trees, which no doubt have drawn it up in youth. It is 113 feet in total height, with a trunk nearly straight to about 90 feet high, where the head begins, and 15 feet 10 inches in girth. Mr. Openshaw and I estimated its contents as follows :—

1st length	. . .	18 feet by 48 inches	= 288 feet.
2nd ,,	. . .	20 ,, 40 ,,	= 222 ,,
3rd ,,	. . .	50 ,, 24 ,,	= 200 ,,
			710 feet.

£100 was refused for this tree a few years ago.

There is also in the deer park a circle with a diameter of fifty yards formed by ten (formerly twelve) oaks of great age and very spreading in habit, and a very ancient oak near by, called the Gibbet Oak, on which tradition says that criminals were formerly hung in chains.

Of the difficulty and risk of removing some of these immense trees when steam traction engines were not in use by timber merchants, Mr. Openshaw gave me an excellent instance which he actually saw himself. A very large oak was felled in a field near Woofferton and sold to a naval timber buyer at Exeter. It was so long and heavy that two of the largest timber carriages were fastened together, and 28 horses brought to get it away. In rolling it up on to the carriage one of the chains got round a horse's leg, but they dared not stop to clear it, and the horse was killed. Mr. Openshaw saw the carriage coming down the road with the log on it, and, believing that it could not pass through the turnpike gate, warned the woman who kept it, to get out of the house, as if the log touched it the house would certainly come down. The man in charge of the team, however, ran on in front and steered the leaders so accurately through the gate that, with an inch to spare, it got past in safety.

It seems probable that many of the great oaks in England which are now decayed, owe their lives to the cost and risk of converting and removing them in the days when there were no railways, and good roads were scarce or absent.

The Nunupton Oak.—The remains of a very large fallen oak, not, however, so big as the one at Croft Castle, is described in the *Transactions of the Woolhope Naturalists' Field Club*, 1870, p. 307. It had long been hollow, and was large enough to contain forty-two sheep at once. It was alive and covered with leaves up till about 1851, when it was set on fire by accident, and was felled soon afterwards, with what object I do not know. In 1870 it was 60 feet long and 26 feet 8 inches in girth, and was still lying in much the same condition when I visited it in 1904.

According to the late Mr. Edwin Lees, whose knowledge of the botany of Worcestershire was very accurate, and whose sketches of old trees, some of which I have, through the kindness of his widow, been allowed to copy, the finest old oak in the county known to him in 1867 stood in a field near the Severn, below Holt, and was known as the Boar Stag Oak. It measured about 34 feet in girth at 3 feet from the base, and might be roughly calculated at 800 years old.

Other remarkable oaks in Worcestershire were described and figured by W. G. Smith, in the *Gardeners' Chronicle*, 1873, p. 1497. They grew in the Lug Meadows, near Moreton, and were known as Adam and Eve. When the Shrewsbury and Hereford Railway was made, Eve, which measured 25 feet in girth, and was quite hollow, was converted by the navvies into a residence: the top was thatched in, a brick fireplace built, and a door fitted, and for months after the line was opened this tree was the only residence of the stationmaster, and was afterwards converted into a lamp-room and so used for fourteen years.

The finest oaks that I know of in Somersetshire are at Nettlecombe Court, the seat of Sir Walter Trevelyan, Bart. When staying at Dunster Castle, in March 1904, Mr. Luttrell was good enough to give me an opportunity of seeing them. He told me that at a previous time, which, from the information received from the agents for the property, I gather to have been about 1847, but Mr. Luttrell thinks it was earlier, £40,000 was offered for about forty acres of oak timber on this property; and an old man at Nettlecombe said that the tools were actually brought to the place ready to fell them, when the owner changed his mind and they were allowed to stand. A considerable part of these oaks have been since felled, but a magnificent grove still remains on the slopes of a combe, at an elevation of five to six hundred feet on the south-west side of Nettlecombe Park, facing to the north and east, and on a soil locally called "shiletty," which is a reddish rocky formation, overlaid by a thin layer of rubbly stone, probably old red sandstone, which would appear too thin and dry to produce big oak timber. The age of these trees, so far as I could judge by counting the rings of one which had been blown down, is not more than 200 to 250 years, but some may possibly be much older.[1] The majority are very clean and free from limbs to from 40 to 60 feet up, and average 10 to 12 feet in girth. One, about 210 years old and over 100 feet long, was 3 feet in diameter at the butt, and had fifty annual rings in a radius of 9 inches near the heart, but outside of this the growth had been much slower. I had not time to measure them carefully, or estimate the number now standing on an acre; but two of the finest trees on the steep banks of the combe were 116 by 14 feet, with a bole 65 feet long; another was 116 by 16 feet, with a bole of 50 feet by 36 inches quarter-girth. The thickest trees, which I did not measure, are on the outside of the grove. Assuming the price of £1000 per acre to have been based on 4s. per foot for the butts, which for trees of this size and character would, sixty years ago, have been about the value, and the trees to have averaged 200 cubic feet, there would have been perhaps forty trees to the acre, averaging £25 each, and though the cubic contents do not come up to what we are told is produced in some of the picked areas of oak forest in France and Germany, I have never heard of an actual sale of any timber in England at so high a price.

At Hazlegrove, Somersetshire, the property of the Rev. A. St. John Mildmay, is a remarkably fine oak, reported to be the largest in the county. It is about 75 feet high by 29 feet 9 inches at 5 feet from the ground, and at ground level spreads out to no less than about 18 yards in circumference. Though it seems sound, yet it has a rent on the north-east side, as though struck by lightning, and many of the largest limbs have been broken by wind, and are mended with lead. A drawing of it, made in 1833 when it seems to have been in full vigour, is in Hazlegrove House.

In Melbury Park, Dorsetshire, the seat of the Earl of Ilchester, there is an extraordinary oak, known as Billy Wilkin's Oak (Plate 88), which swells into an immense burry trunk, 38 feet in girth at the ground, and 35 feet at 5 feet up. Above this it falls away a good deal, and is only about 50 feet high. Like all the trees I have seen of this type, of which perhaps it is the largest in England, it is of the pedunculate variety, and bears acorns abundantly.

At Longleat, Wilts, which has a most beautifully timbered park, and is one of the finest places in England, there is an extremely fine tall oak growing in the

---

[1] The Rev. Mr. Hancock, who is a connection of the Trevelyans of Nettlecombe, says that he has always heard that they were planted about 1600, when part of the existing house was built.

grove of limes which I shall describe later, in a position which makes it difficult to photograph. This tree measures about 100 feet high by 23 feet in girth, and has a fine clean bole of 40 feet. It contains, according to Mr. A. C. Forbes's estimate, about 950 feet of timber.

The finest oak I have seen in Devonshire is in the park of the Hon. Mark Rolle at Bicton, a place long celebrated for its arboretum and for its avenue of Araucarias, which I have elsewhere described. It measures about 78 feet high by 24 feet 8 inches girth at 3 feet, and has a spread of branches of 103 feet in diameter. There are some fine but not extraordinary oaks at Powderham Castle and at Poltimore in the same county.

Near Mottisfont Abbey, Hants, there is a very thick but short pollard oak on the banks of the Test, of which a photograph, by Mr. J. Bailey, Southampton, has been kindly sent me by Mrs. Meinertzhagen, who long resided at Mottisfont. It measures 32 feet in girth and spreads considerably, and, though evidently of very great age, is full of healthy foliage. It must have been frequently flooded, as it stands close to the river.

Near Bramley, Hants, by the road leading to " The Vine," is an oak, which Henry measured in 1905, 100 feet by 22 feet, and which seems quite sound. There are, so far as I know, no oaks now living in the New Forest which are remarkable for their size as compared with the trees I have mentioned.

Of the historical parks of England I know none which contains so many fine oaks as Bagot's Park, near Rugeley, Staffordshire. This must be one of the oldest parks in England, for though Lord Bagot cannot tell me the exact date of its enclosure, he states that it belonged to his family long before 1367, and that in the " Peregrinations of Dr. Boarde, *temp.* Henry VIII.," printed at the end of Hearne's *Benedictus Abba*, p. 795, " Baggotte's Park " is mentioned in the list of Staffordshire parks. It is generally said to contain 1500 acres within the pale, but varies from time to time, as land has been added in some places and taken out in others for planting, to be again restored when the woods are grown.

This practice seems to be well worthy of more general adoption, for no one who is acquainted with the condition of the trees in many of our oldest parks can have failed to notice, that they are as a rule going back; and as trees cannot be successfully raised to a great height if deer are not excluded—unless enclosures of considerable size are made about once in a generation, in which trees can be properly drawn up to a sufficient height, before they are thinned and the deer admitted—the time must come, and in some cases already has come, when nothing but wrecks are left, and the singly planted trees, though protected by iron or wooden guards at great cost, are a mere mockery of their predecessors.

The soil in Bagot's Park is poor and cold, being a moist gravelly loam upon a clay or marl bottom, and Lord Bagot says it is not worth 10s. per acre at the present time. It affords, however, an excellent proof of the fact that land which is not valuable from an agricultural point of view, may often be of great value for planting. The woods extend over many hundred acres and consist almost wholly of oak, mostly, I believe, of the pedunculate variety. Many of the trees are of great age, being mentioned by Dr. Plot in 1686 as full-grown timber.

I visited it in March 1904, and, though the weather was dull, Mr. Foster was able to secure some excellent photographs, of which I reproduce the following :—

Plate 89 represents the Beggar's Oak, which has been well figured by Strutt in his plate No. 2, and though eighty years have elapsed since that picture was taken, a comparison with my plate shows that very little change has taken place in the tree—thanks to the care with which it has been treated by successive owners, who have worthily kept up the spirit described by Strutt in his account of this tree. It now measures, as nearly as I could estimate, 62 feet high, with a bole of about 33 feet long, and a girth of 24 feet. The roots measure 25 paces round, and the branches cover an area of 114 paces round (according to Lord Bagot's measurement 7850 square feet). It is one of the finest and best-preserved oaks of its type that I know, for though the Major Oak in Sherwood Forest (Plate 95) is bigger, it is not nearly so sound; and the Bourton Oak (Plate 93), which is taller and in better condition, is not so large in girth or so spreading at the base.

Another very fine tree in this park is the Squitch Bank Oak, also figured by Strutt (Plate 34), who gives its measurements as follows :—height, 61 feet; girth, 21 feet 9 inches; contents, 1012 feet. When I saw it in 1905 its top was dead, and the butt seemed to be decaying at the base internally. I measured it as about 60 feet by 24 feet 10 inches, so that it has increased three feet in girth in eighty years. The Beggar's Oak, in the same time, has increased rather more, but in measuring the girth of such trees as this a few inches higher or lower will often make a great difference, and therefore these rates of increase cannot be considered exact.

Other great trees in this park mentioned by Strutt were the Rakeswood Oak, the Long Coppice Oak, and the twisted oak on the Squitch Bank, which, though I did not see them, still survive. In the Horsepool grove are a number of younger but very tall and straight trees, which have been grown close together, and which Lord Bagot's old woodman, W. Jackson (now dead), said he "could remember so thick that you could hardly swing an axe amongst them." Of these, one, which was called Lord Bagot's Walking-Stick, is the straightest and cleanest oak I ever saw in England, though recently struck by lightning; another was 95 feet by 8 feet 6 inches, with a clean stem 65 feet high. On the other side of the park, at the west end of the grove called the Cliffs, are a number of splendid trees of great size. Two of them, standing near each other, are figured in Plate 90. Of these, the one in the foreground measures about 112 feet by 16 feet 8 inches, with a bole 35 feet high and four great erect limbs. The other, about the same height and a foot less in girth, has a clean bole 45 feet high. One hundred pounds was offered and refused for it. In the same grove, farther east, is an oak with a bole about 40 feet by 15 feet 3 inches, twisted from right to left, and another called the King Oak, which, though now partly hollow, has been perhaps the finest timber oak in the park (Plate 91). It is now about 100 feet high, but has been taller, as the topmost branches are dead, with a straight clean bole 21 feet 3 inches in girth, and must have contained over 1000 feet of timber. It is stated[1] that in 1812 £200 was offered for the first length of this tree, estimated at 12s. per foot, and £93 for the

[1] *Gard. Chron.* xvi. 230 (1881).

remainder, including the bark, estimated at £14 per ton.   Near it is a tree of great height, leaning at an angle of about one in four to one side, though quite firm in the ground ; and it seemed to me that all the trees in this grove owed their great height and clean stems to their having been drawn up by beech trees, many of which are now dead or dying.   Close to the Park Lodge are three very curious and picturesque old trees, one of which is called the Venison Oak, because King John is supposed to have dined under it ; another, which we christened the Beer-barrel, is an immense burry shell 10 or 12 feet high and 28 feet round, with hardly any branches ; a third we called Gouty Toes, because of a huge swollen root, like a gouty foot, on one side of it.

Dr. Plot, in his *Natural History of Staffordshire*, p. 213, after speaking of different species of trees growing together, among which were an oak and an ash near Chartley, hollies and oaks at Bagot's Park, and an oak and thorn at Drayton Basset, goes on to speak of trees "that grow so conjoynd that they seem (after the manner of some sort of animals) to prey upon one another," and says : "But the most signal example of this kind is the large fair birch, about the bigness of one's thigh, that grows on the bole of an oak in the lane leading south from Adbaston Church, which has sent down its roots in six branches perpendicularly through the whole length of its trunk and fastened them in the ground, which might be seen in a hole cut in the bottom of the oak ; having eaten out the bowells of the old tree (as all the rest will doe) that first gave it life and then support.   All which are occasioned, no doubt, by the seeds of those trees dropt by birds in the mould on the boles of the others that lyes commonly there, and is made of the annual rottings of their own leaves."

He goes on to speak of another great oak, "lying near the Lodge house in Ellen Hall Park, of so vast a bulk that my man upon a horse of 15 hands high, standing on one side of it, and I also on horseback on the other could see no part of each other " ; and also of an oak that " was felled about twenty years since in Wrottesley Park which, as the worthy Sir Walter Wrottesley (a man far from vanity of imposition) seriously told me, was 15 yards in girth."—" How much less in bigness and number of tuns the oak might be that grew in the New Park at Dudley, and made the table now lying in the old hall at Dudley Castle, is not remembered, but certainly it must be a tree of prodigious height and magnitude out of which a table all of one plank could be cut, 25 yards 3 inches long and wanting but 2 inches of a yard in breadth for the whole length, from which they were forced (it being so much too long for the hall at Dudley) to cut off 7 yards 9 inches, which is the table in the hall at Corbins Hall hard by, the ancient seat of the Corbins."

In the park at Merevale Hall, Warwickshire, the seat of W. F. S. Dugdale, Esq., are a quantity of very fine and tall oaks, which rival those at Bagot's Park, and are, according to Sir H. Maxwell, of the sessile variety, though when I saw them they were not in leaf.   They stand at a considerable elevation, on a dry and seemingly rather shallow red sandstone.   Many of them are 100 feet and more in height, with clean trunks of 40 to 60 feet long.

The best that I could find measured as follows :—112 feet by 13 feet, with a straight bole 65 feet long; 107 feet by 15 feet, with a clean bole of 70 feet, and probably containing about 600 feet of timber; 107 feet by 17 feet 3 inches, with a bole 48 feet long, and about the same cubic contents as the last; 114 feet by 15½ feet, bole about 60. This last is, I believe, the same tree which Mr. Dugdale had measured some years ago, when it was thought to be 133 feet high; but I do not think it can be nearly so much, the sloping ground on which it stands making a base line difficult to get. He tells me that these trees are believed to have been planted by the monks who lived at Merevale Abbey at the foot of the hill, which would make them at least 370 years old, and that most of them have now passed their best. The timber being very straight in the grain is largely used for cleaving spokes.

Chirk Castle in Denbighshire, the seat of R. Myddleton, Esq., and one of the most ancient inhabited castles in England, is in a park full of oaks, most of which I believe to be of the sessile variety. They are not of great age, having been planted, as Mr. Parker, agent for the property, told me, after the Commonwealth, but are remarkable on account of their uniformly straight boles 30 to 60 feet high. They grow on millstone-grit, where the rock comes very near the surface, on land where the pedunculate variety would not, I think, make nearly such fine trees. I only measured two, one just below the castle which was 100 feet by 11 feet 8 inches, with a straight clean bole of 60 feet; another, probably of greater age, about 90 feet by 18 feet 2 inches, was beginning to decay at the base. A curious growth is seen on an oak in this drive, a branch having grown out of one stem into another, somewhat in the same style as the beech in Plate 4 of this work.

The trees in the Great Park of Windsor have been described by many writers, and especially by the late Mr. William Menzies in a rare folio published by Longmans in 1864,[1] which gives photographs of some of the finest trees, these being, so far as I know, the first large photographic plates of trees published, and, considering the imperfect development of the art forty years ago, wonderfully good.

They show Queen Victoria's Favourite Oak, which was chosen by her late Majesty shortly after her accession, and which stands with the three other royal trees between High Standing Hill and New Lodge. This is a very well shaped tree of fair size, 70 feet high and 11 feet in girth when Menzies measured it in 1864. Now, as I am informed by Mr. Simmonds, it has increased only 9 inches in girth. Queen Anne's Oak, a very handsome tree in shape, but past its prime, though supposed to be only 400 years old, measured 60 feet in height by 15 feet 3 inches in girth. Queen Charlotte's Oak, a tree of no special beauty, was 65 feet high by 17 feet 3 inches in girth. The great Pollard Oak at Forest Gate, known as William the Conqueror's Oak, and figured in the Supplement to *Gardeners' Chronicle*, 31st October 1874, supposed by Menzies to be 800 years old, though about 37 feet in girth, and the largest in the forest, is now a wreck; but there are near the Prince Consort's chapel, and in the Cowpond grove, many beautiful tall and straight-grown oaks, one of which, growing near the culvert of the pond, measured by me in March 1904, was from 114 to 118 feet high and 10 feet

[1] *History of Windsor Great Park and Windsor Forest.*

10 inches in girth.　For this tree Mr. Simmons told me, £100 was offered to make the keel of a ship forty or fifty years ago.　It should live for many years, and may perhaps become the finest timber oak in Windsor Forest.

Mr. Menzies gives[1] an excellent explanation of the old custom of pollarding oaks and beeches, which has produced the picturesque veterans which are so common in most of our really old parks.　For the support of the deer in winter it was customary to lop off the boughs of the oak and beech.　The law required that no bough should be cut larger than a buck could turn over with its horns, and after they had been stripped by the deer these branches became the perquisite of the keepers, under the name of "fireboote," or "houseboote."　Any timber fit for the navy could not be cut without the sign manual of the King, a rule yet extant; but in times of civil war, and in royal forests which were granted to favourites in the times of the Stuarts, the keepers often cut and sold as timber or firewood a great deal more than the deer needed; and notwithstanding that these matters were investigated by James I. with his national and personal thriftiness, and that the surveyors whom he employed were spoken of by the country people as "shroade and terrible men,"[2] these abuses increased to such a point that the growing scarcity of naval timber was a common complaint for centuries.

There is no doubt that browse or lop, being the natural winter food of deer in hard weather, is more suitable for them than beans and maize, which is now given in so many places probably to save trouble.　I find in my own park that ash and elm are the favourites, and beech the next best lop for deer, and only give hay when the ground is frozen or covered with snow; but many parks are so overstocked with deer and with cattle in summer that in February and March the former must have some extra food, or a heavy death-rate follows.

Gloucestershire is not famous for fine oaks, though the Boddington Oak, near Tewkesbury, now gone, must have been an exceptionally large tree.　The Newland Oak, near Coleford, is an immense pollard, with a short burry trunk no less than 43 feet in girth.　An excellent photograph of it has been published as a postcard by Mr. J. W. Porter of Coleford.　There are some fine ones in the Winchcombe Valley, near Sudeley Castle, one of which is 25½ feet in girth; but in the Vale of Gloucester elms are commoner than oaks, and I know none of special note, though Mr. J. R. Yorke tells me of a large tree still standing near Forthampton Court.

The largest I have seen are in Witcombe Park, the seat of W. H. Hicks-Beach, Esq., a small but picturesque park lying under the steep Birdlip Hill.　Here on fertile clay soil, facing north and west, are a number of very fine trees, which, judging from the rings counted on one of the largest which has recently been felled, are not so old as they appear to be.　This tree, which measured about 90 feet by 17½ feet, and contained 400 to 500 cubic feet, was only about 210 years old, and beginning to fail in the upper branches, which were dying off.　The largest tree, in a very exposed position, has lost some of its biggest limbs, and measures 25 feet in girth at about 5 feet from the ground, and 50 feet round the roots at the base.　A very tall, well-shaped, handsome tree, with its bole clean and straight

---

[1] *Op. cit.* 7.　　　　　　　　　　[2] Arthur Standish, *The Common Good.*

for 30 to 40 feet, stands on high ground in the centre of the park ; and at the bottom of the hill near the house is a pollard which seems sound, and is 24½ feet in girth at the smallest part of its trunk.

In a grove near Campden, close to Norton House, which has been lately restored by the Earl of Harrowby, I was shown a remarkably tall and clean oak over 100 feet high with a straight bole clean for 60 feet, but only 7 feet 5 inches in girth.

Near Bourton-on-the-Water, on the east side of the road to Stow, stands a pedunculate oak which, of its type, is almost equal in size to any I have seen, and which is specially remarkable on account of the perfect condition of all its branches, which, as Plate 93 shows, are growing to the very tips, and which spread over an area of 115 paces in circumference, equal to that of the Beggar's Oak. This tree grows in a grass field on the property of Mrs. Butler of Wick Hill.[1] It measures about 85 feet high by 22½ in girth, and has the appearance of having been pollarded at about 12 feet up very early in life. There are some fine tall oaks at Wick Hill, not far off, measuring 85 feet by 14 feet and 80 feet by 13 feet, and there are still some big ones in the cow pastures at Sherborne Park in the same district. But the best of these were felled fifty years ago by the father of the present Lord Sherborne, who has never ceased to lament their loss.

There are many superb oaks in Earl Spencer's park at Althorp, Northants, which were carefully measured by the former forester, Mr. Mitchell, now at Woburn. Lord Spencer's ancestors were evidently great lovers of trees, and followed a practice which is much to be admired. In Althorp Park are several inscribed stones, giving the date of planting and the name of the planter. The earliest of these is in the Heronry, and is dated 1568.

Of the others one reads as follows—   Another has the legend—

<table>
<tr><td>

This Wood was planted by
Robert Lord Spencer
In the year of our Lord,
1602-1603

</td><td>

This Wood was planted
by Sir William Spencer, Knight of the Bath
in the year of our Lord
1624
Up and be doing, and God will prosper

</td></tr>
</table>

When one sees how small are the trees planted about 300 years ago, when compared with the older trees, one realises the immense time it takes for such oaks to grow. The finest at Althorp is shown on Plate 94. It grows near a farmyard, and is No. 8 in Mitchell's list.[2] It measures about 90 feet in height, and carries a thick straight stem up to about 45 feet high, and girths 19 feet 6 inches at 5 feet. It must contain at least 1000 feet of timber, and is apparently sound, healthy, and growing, with no signs of decay in the top.[3]

There are some very fine oaks in Burleigh Park, Stamford, the seat of the

---

[1] In 1906 I saw this tree again, and found that a large fungus had attacked its trunk, and that some of the branches were showing signs of decay at the ends. Steps are being taken to preserve it as far as possible.

[2] A description of some of the finest trees at this place is given in *Trans. Scottish Arb. Soc.* xiii. 83.

[3] Sir Hugh Beevor measured fifteen oaks standing on one acre in a grove planted at Althorp in 1561-1568, and found them from 100 to 115 feet high, with an average girth of 11 feet 8 inches, and the average cubic contents of the first length of 54 feet was 330 feet. In another plantation, made in 1589 on stiffer soil than the last, there were more trees per acre, but their size was less, the average being 90 feet by 9 feet 7 inches.

Marquess of Exeter.   The best, known as the King Oak, is 100 feet high by 16 feet 6 inches in girth.   At Ashridge the oaks are not so fine as the beeches, but the King Oak in that park is a splendid tree, measuring 98 feet by 21 feet 8 inches.

Sherwood Forest, in Nottinghamshire, contains an immense number of very ancient, picturesque, and curious oaks, many of them now mere wrecks, but preserved with care by Earl Manvers, who is the owner of a large area of the unenclosed part of what was formerly a royal forest.   I have seen no other place where so many of the trees are covered with immense burrs, and where they assume such extraordinary shapes, as in that part of Sherwood Forest between Edwinstowe and the Buck gate entrance to Thoresby Park.   The soil in this district is mostly a poor-looking sand on which the birch thrives remarkably.   About seventy years ago the open forest which up to that time had been grazed by sheep, came into the possession of Lord Manvers.   An immense quantity of seedling birch then sprang up, and large quantities of acorns were sown to fill up the vacant spaces caused by the decay of the old oaks, most of which are now stag-headed, and dead at the top.

The finest oak now standing in Sherwood Forest is the Queen or Major Oak (Plate 95).   This tree, though hollow, and having its branches partly supported by iron stays, is still healthy and vigorous.   It measures about 60 feet in height by 30 feet 5 inches in girth, and the spreading roots are about 18 paces round at the ground. The spread of the branches is 30 yards in diameter.   It is about three-quarters of a mile from Edwinstowe, and is not far from another tree known as Simon Foster's Oak, which is about 44 feet high and 25 feet in girth.

At Welbeck, the seat of the Duke of Portland, in the same beautiful and well-wooded district, known as the Dukeries, on heavier soil than that at Thoresby, are a number of magnificent oaks which were described and figured in 1790 in a scarce pamphlet by Major Rooke.   Of these I saw the Porter Oaks, so called because they stand opposite each other on each side of a gate in the park.   When measured by Rooke about 1779 they were as follows :—No. 1. 98 feet high, 23 feet girth at 6 feet ; contents, 840 feet.   In October 1903, 25½ feet ; the top having been dead for many years it is now much less in height.   No. 2. 88 feet high, 20 feet girth at 6 feet ; contents, 744 feet.   Now it is 23 feet and rapidly decaying.

Another tree, called by Rooke the "Duke's Walking Stick," of which there is a small figure in Loudon, p. 1766, was in 1779, 111 feet 6 inches high, and 70 feet 6 inches to the first branches ; at 6 feet it measured 12 feet in girth, and was estimated to contain 440 cubic feet.   A very celebrated oak at Welbeck is the Greendale Oak, which has often been figured and described.   In my copy of Strutt there is a good plate of this tree, without number or description, bound at the end of the volume.   Tradition says that a bet was made by a former Duke of Portland, that he had an oak so large, that a coach and four could be driven through its trunk, and the hole having been cut, he won his bet.   When measured by Rooke it was, above the arch of the hole, 35 feet 3 inches in girth, the hole being 10 feet 3 inches high and 6 feet wide.   Even at that time Rooke's figure shows it to have been a mutilated wreck, but the tree is still alive.

Near the Greendale Oak there is a magnificent though dead specimen of burr oak, about 50 feet high and 28 feet 9 inches in girth, and though all the veterans are long past their prime, there are still healthy growing oaks at Welbeck on the south side of the road to Norton, of which I measured one with a butt 32 feet high and 19 feet in girth, which Mr. Michie, the forester, considered would contain 500 feet in the butt alone. Such oaks have actually been cut and sold here in recent times ; and I have a photograph, given me by Mr. G. Miles of Stamford, of a tree which he bought at auction for £40, and whose trunk measured 38 feet 6 inches long by 43¼ inches quarter-girth—equal to 511 feet 8 inches. It was so heavy that the weight on the wheels of the timber carriage broke through the road, and when brought to the station after much risk and trouble, the railway company refused to take it to Peterborough except on a special train by itself.

In Rockingham Park, Northants, the seat of the Rev. Wentworth Watson, there are a number of wonderful oaks, many of which are brown, and I had the opportunity, through the kindness of Mr. C. Richardson of Stamford, of seeing several of these felled in September 1903. He told me that, in the whole course of his long experience, he had never seen so many fine brown oaks together as these. The park lies high, on land which looks like oolitic limestone, the rock in some places coming near the surface ; but where these oaks grow there is a good depth of loamy soil. Some of the trees which I saw lying were more or less hollow, and required no saw to bring them down. I was anxious to photograph one in the act of falling, and as the fellers were at work on one of the best, I asked them to let me know how long it would take ; the roots only being then cut all round the tree. I expected that some hours would be required, but before the camera was fixed to take the tree as it stood, they suddenly called out, " stand clear," and down it came.

Plate 96 shows what the roots of these brown oaks are usually like, but if there is a foot or two of sound wood in the lower part, and the brown colour extends a good way up the trunk, they are still very valuable. I asked the fellers if they could tell a brown oak standing without boring it, and they said they could make a good guess at the colour, though they could not be sure. Probably long experience in a district where brown oak seems to be commoner than elsewhere, is the only guide, if there is one ; but stories are told of men going in the night to bore such trees with an auger before trying to buy them, in the hopes of getting a bargain. From a statement sent me by Mr. Richardson, it appears that twenty-six of these trees were sold for £1100, five of them for £100 each, and contained about 8030 feet, all measured over bark, and nothing allowed for defects.

The best of this lot were eventually sold to Messrs. J. T. Williams of New York, and afterwards bought by the Pullman Company at a very high price. Mr. Richard Dean, of that Company, informs me that he considered the wood superior to any that they had previously used, and was good enough to send me some samples of the veneer made from them, which has been used in decorating their palace railway cars. The largest of these specimens measures 6 feet 1 inch by 2 feet 8 inches without a flaw, and is throughout of a uniform chestnut-brown colour, mottled with silvery patches, formed by the medullary rays, showing that it has been cut from a quartered plank.

The sandy and gravelly tracts in Essex have extensive woodlands, in which the oak is the principal timber tree. Sound oak trees with boles measuring from 16 to 20 feet in girth are scattered through the county. Oak trees of larger dimensions, many in a more or less decayed condition, have been measured and described by Mr. J. C. Shenstone.[1] Some of these I visited under his guidance in March 1907, and I think the following are worthy of notice :—At Thorrington are four trees from 27 to 31 feet in girth, decayed ; at Danbury Park two trees of 31 feet in girth, decayed ; at Hatfield Broad Oak the Doodle Oak, 42 feet in girth, decayed ; at Havering-atte-Bower Bedford's Oak, 27 feet in girth, decayed ; in Easton Park the finest tree is 80 feet by 23 feet, sound and vigorous, and there are many old pollards of great size. One of these, covered with burry growths, is 29 feet in girth ; and another, on which the burr is very peculiar from its kidney-shaped lobes, is 33½ feet, of which the burr takes up 14 feet. At Marks Hall, near Coggleshall, the property of T. P. Price, Esq., there are very large sound oaks, as well as some relics of the ancient forest ; the largest, which is perhaps the finest tree of its kind now standing in the county, is 90 feet by 24 feet 3 inches, and though some large branches are gone on one side it seems sound and vigorous. The only very large oak now left in Epping Forest is the Fairmead Oak, 30 feet in girth, and much decayed. At Thorndon Park, the ancient seat of Lord Petre, are many picturesque relics of the ancient forest ; and at Wealside House, Brentwood, is an oak 27 feet in circumference of bole.

Mr. E. R. Pratt of Ryston Hall kindly sends me the following account of—

Kett's Oak at Ryston, Norfolk.—In the year 1547 this tree was the trysting-place of the West Norfolk rebels under the brothers Robert and William Kett. The former and the other "Governors" selected large oak trees under which their Courts sat to administer justice and regulate disorders. The Court in this case did not seem to look upon sheep-stealing as other than a necessary evil, since they left on the tree the following inscription :—

Mr. Prat, your shepe are very fat
and we thank you for that
we have left you the skinnes
to buy your ladye pinnes
and you must thank us for that.

	Dimensions in 1840.	In 1906.
On the ground level,	46 feet 6 inches.	49 feet 6 inches.
Three feet from the ground,	27 ,, 4 ,,	26 ,, 6 ,,
Five feet do.	24 ,, 3 ,,	23 ,, 11 ,,

From the photograph which accompanied this account it seems that the old tree is still fairly sound and vigorous. In an old map of the seventeenth century Kett's Oak is marked, showing that it was then known as a landmark.

Other remarkable oaks in Norfolk which I have seen are at Merton Hall, the

[1] *Essex Naturalist,* June 1904.

seat of Lord Walsingham, where the largest, now much decayed, is about 27 feet in girth ; at Blickling, where an oak in the kitchen garden 95 feet high, said by Grigor to have been planted by the Earl of Buckinghamshire, has a straight clean trunk 32 feet high and 15½ in girth ; and at Stratton Strawless, where there is a beautiful straight-stemmed oak close to the house clean to 40 feet high and over 10 in girth.

Cowthorpe Oak.—No oak in England has probably been the subject of so much writing as the Cowthorpe Oak, near Wetherby, which perhaps never was such a great tree as has been supposed, and is now a mere wreck. It has been figured several times, so that I need only refer those who wish to know more of it to a paper with illustrations by Mr. John Clayton, published in the *Transactions of the Botanical Society of Edinburgh*, 1903, p. 396. A comparison of the various measurements taken at different times shows great discrepancies. Mr. Clayton attempts to prove by a diagram that the decay of its roots have allowed it to settle into the ground, and thus explains the diminution in its girth, but the discrepancy between measurements taken by different people is considerable. The girth at 5 feet given by Marsham as 36½ feet in 1768, when no hollow or cavity is mentioned as existing in the tree, and the girth given by Mr. Clayton of 36 feet 10½ inches, at 5 feet 3 inches in 1893, are so nearly identical that I do not think Mr. Clayton proves his argument. Whether trees ever subside owing to the decay of their roots is to me a very doubtful point, and I have certainly seen oaks felled which, though of great age and completely hollow, were supported in their original position by a mere shell. I visited the Cowthorpe Oak in July 1906, and found that in its present condition no accurate measurement of it could be taken, a large part of one side having fallen in. I could see no evidence to support Mr. Clayton's idea that the base of the tree had sunk into the ground. The few living branches still bear acorns, from which some seedlings were raised in 1905 by Messrs. Kent and Brydon, nurserymen of Darlington.

The finest oaks in Yorkshire that I have seen or heard of are in the park at Studley Royal, which were described and figured by Loudon from drawings which I have seen in the Marquess of Ripon's library. Though I could not identify the drawings with trees now standing, Loudon gives the dimensions of the largest pedunculate tree as 80 feet by 24 feet 4 inches, and the largest sessile oak, which he says was then the largest in England, as 118 feet by 33½ feet. The best that were shown me were a pedunculate oak 80 feet by 23 feet, a good deal past its prime, and a sessile oak which I made 114 feet by 12 feet 2 inches, a vigorous and healthy tree.

One of the most remarkable oaks in England on account of its shape is the Umbrella Oak at Castlehill, North Devon (Plate 97). This tree had not altered materially during the recollection of the late Earl Fortescue, who lived to be over eighty, though it does not give the impression of very great age. It grows on a slope called Eggesford Bank, near the house, and has a clean bole about 8 feet by 6 feet 8 inches. The branches spread horizontally from one point, and form a close flat surface, of which the twigs are interlaced, and spread to a diameter of about 25 yards. Seedlings have been raised from its acorns, which do not produce this very curious habit, and attempts to reproduce it by grafts have not succeeded.

Another freak of nature is the Marriage Oak in Eridge Park, Kent, which Lord H. Nevill was good enough to show me.   Here a yew and an oak have grown up together, though the two trunks, which measure 16 feet 3 inches in girth, have not combined, the yew having spread its branches widely over the top of the bole of the oak.   A similar case is recorded by Mr. A. D. Webster[1] in the South Park at Holwood, Kent.   Here the two trees have combined their stems into a normally shaped trunk, which girths 7 feet 10 inches at 5 feet, the yew being only 15 feet high, and spreading 36 feet, while the oak is 35 feet high with a spread of 54 feet.

Pollard oaks, when they are hollow at the top, sometimes support other trees of considerable size, which originate from seeds dropped by birds or brought by the wind. The best living instance of this that I know, is on an oak of no great size at Orwell Park, the seat of E. G. Pretyman, Esq., in Suffolk.   This grows in a wood near the Decoy Pond, which is full of large self-sown hollies mixed with oaks, and looks as if it might be part of an original forest.   Here a birch about 30 feet high, 20 inches in girth, is growing on the top of the oak, and has formed inside its hollow trunk what on one side seems to be a woody stem, whilst on the other side the roots are still in process of formation within the bole of the oak, which on that side is dead, but on the other has living branches.[2]   Henry has seen a similar example on the road between Byfleet and Cobham, on Lady Buxton's property, where the birch, growing out of a stout oak bole, is 49 feet high and 8 inches in diameter.

Wistman's Wood.—After having said so much of big oaks, I must now mention one of the most remarkable oak woods in Britain, called Wistman's or Welshman's Wood, which is on Crockern Tor, Dartmoor, at an elevation of about 1400 feet.   It contains a number, perhaps a thousand, of the most stunted and dwarf oaks in existence, growing among granite boulders in a very exposed and windy situation.

Wistman's Wood was described by Burt in his Notes to the second edition of Carrington's *Dartmoor*, p. 56, and also by Mr. W. Borrer.[3]   I am indebted to Mr. E. Squarey of Downton, Wilts, for information in a letter to him by Mr. P. F. S. Amory of Druid, Ashburton, which brings our knowledge up to date, with photographs showing the curious appearance of these trees.   The *Journal of Forestry*, v. 421, in a description of them, says that no acorns are produced; but Mr. J. B. Rowe, editor of the *Perambulations of Dartmoor* (ed. 1896), in 1895 found two acorns after a long search, one of which, planted at Druid on 9th November 1902, is over 4 feet high.

In September 1868 Mr. Wentworth Buller obtained leave from the Prince of Wales to cut down one of these trees in order to find out its age.   One section was sent to Kew; and another now in Mr. Amory's possession is 9 inches by 7 in diameter, and shows 163 years' growth, with distinctly marked medullary rays and several deep shakes.   The bark is extremely thin, probably owing to the thick coat of moss and lichen which covered it.   The slowness of growth in this tree is remarkable, no less than forty years to the inch.

[1] *Trans. Scot. Arb. Soc.* xii. 313 (1889).
[2] Compare Plot's account of a similar case quoted on p. 318 *supra.*   [3] Loudon, *loc. cit.* 1757.

Strutt in *Sylva Britannica*, published in 1822, figured no less than twenty-one oak trees, and as I have seen a good many of these myself, it may be interesting to notice their present condition after a lapse of over eighty years.

Plate 1. The Swilcar lawn Oak in Needwood Forest was then supposed to be about 600 years old, and was 21 feet 4½ inches at 6 feet, having increased 2 feet 4 inches in 54 years. When I saw it in 1904 it was about 25 feet in girth, but nearly dead at the top.

Plate 2. The Beggar's Oak in Bagot's Park, fully described above. It measured in 1822, 20 feet; in 1904, 24 feet 1 inch.

Plate 3. The Great Oak at Fredville was in 1822, at 8 feet from the ground, more than 28 feet in girth, and contained above 1400 feet of timber. Now, I am informed by the Rev. S. Sargent, who sends me a photograph, showing that it is in good health, it measures at 3 feet, which seems to be about its smallest girth, 33 feet 6 inches.

Plate 4. The Panshanger Oak, near Earl Cowper's house in Herts, seemed to Strutt to have scarcely reached its prime, though his plate shows that the spire was already dead. It measured in 1822, 19 feet at a yard from the ground, and was supposed to contain 1000 feet of timber. When I saw it in 1905 the topmost limbs were dead or dying, and there was a large rift in the trunk on one side. The girth was 21 feet 4 inches at 5 feet.

Plate 9. The Salcey Forest Oak was a mere wreck in 1822. I know not if it still exists.

Plate 10. The Abbot's Oak at Woburn Abbey was never a very large tree, but if it is the same that I saw in 1905 it remains sound.

Plate 11. The Chandos Oak at Michendon House, Southgate, was also not a first-class oak, though a very handsome one. It was then only 60 feet by 15 feet 9 inches. Henry's measurements in 1904 were 80 feet in height and 18 feet in girth, with a spread of branches 143 feet in diameter.

Plate 12. The oak called Beauty at Fredville, not a first-class tree among great oaks and figured with a dead top, measured only 16 feet in girth.

Plate 17. The Shelton Oak near Shrewsbury I have not seen. It was a hollow tree of great age, 26 feet in girth, in 1822, and I am told that it is now a mere wreck.

Plate 18. The Bounds Park Oak, near Tonbridge Wells, was a tree in perfect health and vigour when figured by Strutt, and measured 69 feet by 17 feet 9 inches at 12 feet. It is still standing, and as I am informed by Mr. H. J. Wood, has not much changed in appearance.

Plate 19. The Moccas Park Oak was much decayed in 1822, when it measured 36 feet in girth; it still survives, but is fast going to ruin.

Plate 20. The Wotton Oak was never a first-class tree, judging from the plate, and I do not know what is its present condition.

Plate 25. The Cowthorpe Oak has been already discussed.

Plate 26. Queen Elizabeth's Oak at Huntingfield in Suffolk is, I believe, the same tree which I saw on Lord Huntingfield's property, near the Hill Farm, Strutt quotes from Davy's letters but gives no measurement. It was quite hollow in 1773, and is now divided into three great sections which lean outwards and measure in all 39 feet 8 inches in girth. It has a few green and healthy branches, and the sound parts of the shell are about a foot thick.

Plate 27. Sir Philip Sidney's Oak at Penshurst, Kent, was, in 1822, a very old tree measuring 22 feet in girth.

Plate 28. The King Oak in Savernake Forest was quite hollow when figured by Strutt, and measured 24 feet in girth.

Plate 33. The Twelve Apostles at Burley Lodge, New Forest.

Plate 34. The Squitch Bank Oak in Bagot's Park was in 1822, and still is, one of the finest in England, and is now considered by Lord Bagot the best oak in his park.

Plate 35. Two trees called Gog and Magog near Castle Ashby still survive, and, judging from photographs of them sent me by Mr. Scriven, have not changed much in appearance, though Gog has apparently lost its bark on one side. Though very picturesque, they are not well-shaped trees. The former is now 58 feet by 28 feet, at 3 feet, with contents 1668 feet; the latter is 49 feet by $28\frac{1}{2}$ feet.

Plate 36. The Tall Oak at Fredville is to my eye the best shaped of Strutt's oaks, though not of extraordinary size. He says the stem went up straight and clean to about 78 feet, and the girth at 4 feet was 18 feet.

Among the trees figured in *Sylva Scotica*, a continuation of the work just cited, there is only one oak, namely, Wallace's Oak at Elderslee or Ellerslie, near Paisley. Many larger and finer oaks than this occur in Scotland. Judging from the figure it is not remarkable except from its historic interest, which seems rather mythical.

THE OAK IN SCOTLAND

The oak rarely attains in Scotland the size and vigour so commonly met with in England.[1] Mr. Hutchinson[2] has catalogued 151 Scottish oaks, remarkable for size; and of these only six exceeded 20 feet in girth at 5 feet above the ground; the largest recorded by him, at Lee, Lanarkshire, was 23 feet girth at 3 feet up, the total height being 68 feet. The tallest oak recorded by Hutchinson was one at Hopetoun, Linlithgowshire, 110 feet high, with a bole of 93 feet, girthing 8 feet 8 inches, but I saw no tree approaching this height at Hopetoun in 1904. In Dr. Christison's[3] paper on the " Rate of Girth Increase in Trees," the average rate of increase is given for trees at the Edinburgh Botanic Garden; Craigiehall, Linlithgowshire; Pollok, Renfrewshire; and Methven Castle, Perthshire. The rate of course depends on the age of the trees, and is very variable even in the same locality. At

---

[1] Sir Herbert Maxwell thinks that this is not owing to soil or climate, but to the fact that Scotland was denuded of trees before the seventeenth century. Planting was carried on slowly and sporadically after the Union, and there are few planted oaks in Scotland over 200 years old.

[2] *Trans. Highl. and Agric. Soc. Scot.* xiii. 218 (1881).   [3] *Trans. Bot. Soc. Edin.* xix. 461 (1892).

Methven, an oak planted in 1811 had attained, in 1893, 16 feet in girth, and during the last sixteen years had increased as much as 18 inches in girth.

The "Capon Tree,"[1] near Jedburgh, in 1893 was 22 feet 7 inches in girth at the narrowest part of the trunk. It divides at 6 feet into two stems, girthing 16 feet 2 inches and 10 feet 9 inches. The "Pease Tree" at Lee,[1] Lanark, measured, in 1890, 23 feet 7 inches in girth.

There is a fine oak at Methven Castle called the Pepperwell Oak, which Henry measured in 1904, 85 feet by 20 feet 4 inches. Colonel Smythe informed him that when his ancestor Peter Smythe was imprisoned in the Tower of London in 1722, his wife, though in sore straits for money, refused 100 marks for this tree.

In the shrubbery of Scone Palace, Perth, in ground which was formerly gardens belonging to a village, there is an oak, planted in 1805 (growing in black loam $4\frac{1}{2}$ feet deep, resting on sand of unknown depth), which in 1904 was 102 feet high, 36 feet to the first branch, and 11 feet 4 inches in girth at 5 feet up. This shows unusually rapid growth. Near it is another oak, probably of the same age, 98 feet in height, 10 feet 8 inches in girth, with a bole of 25 feet.

The finest oak seen by Henry in Scotland is growing in front of the house at Blair Drummond, Perthshire, the seat of H. S. Home Drummond, Esq. It is 118 feet in height and 17 feet in girth, the first bough coming off at 24 feet up. This oak and a number of others near it probably date from some time after 1730, the year in which the house was built. At Drumlanrig, Dumfriesshire, there is an oak 16 feet in girth, with a bole of $22\frac{1}{2}$ feet, which is estimated to contain 361 cubic feet of timber. At Dalswinton, Dumfriesshire, there are two remarkable oak stools, standing close together. The larger is $28\frac{1}{2}$ feet in girth near the base; and gives off five great stems, 81 feet in height, which average 8 feet in girth.

(H. J. E.)

THE OAK IN IRELAND

The most famous oak wood in Ireland was that of Shillelagh in Wicklow, from which is derived the name formerly given to an oak stick, but now erroneously transferred to the blackthorn. From the wood of Shillelagh, according to tradition, were derived the timbers which roof Westminster Hall, and also those on the roof of the chapel of King's College, Cambridge. There is said[2] to be a record in St. Michan's Church, verified by "Hanmer's Chronicle" in the library of Trinity College, Dublin, which states: "The faire greene or commune, now called Ostomontoune Greene, was all wood, and hee that diggeth at this day to any depth shall find the grounde full of great rootes. From thence, anno 1098, King William Rufus, by license of Murchard, had that frame which made up the roofes of Westminster Hall, where no English spider webbeth or breedeth to this day." According to Hayes,[3] the finest trees in Shillelagh were cut down in the time of Charles II. and exported to Holland for the use of the Stadt House, under which hundreds of thousands of piles were driven. In 1692 iron forges were introduced into Shillelagh; and the ruin of the wood

---

[1] *Trans. Nat. Hist. Soc. Glasg.* 4th Sept. 1894.   [2] *Woods and Forests*, Jan. 28, 1885, Suppl. p. iii.
[3] *Practical Treatise on Planting*, 111 (1794).

speedily followed.   Great trees, however, still remained till near the end of the eighteenth century.   At that time Mr. Sisson, who had purchased large quantities of timber, was given an oak tree of his own choice as a present, and this tree was so large that though forked at the base, each stem was big enough for a mill shaft at more than 50 feet from the butt.   Two pieces being appropriated to this use, he sawed the remainder into panels for coach-building, which were sold for £250.   In the MSS. of Thomas, Marquess of Rockingham, it is recorded that in 1731 there were standing in the deer park of Shillelagh 2150 oak trees, then valued at £8317, the timber being rated at 1s. 6d. a foot, and the bark at 7s. a barrel.   In 1780 there remained of the old reserves 38 trees, which contained 2588 feet of timber.   In the adjoining woods of Coolattin, in Hayes' time, there was a considerable number of young healthy oaks, several being 7½ feet in girth.

I visited Coolattin in 1906 and was shown many fine trees, though none were of great thickness, the best tree seen being 118 feet high with a clean bole to 40 feet and a girth of 13 feet.   All the trees were *Quercus sessiliflora.*

The largest oak wood in Ireland is in Viscount de Vesci's park at Abbeyleix, Queen's County, where there are several hundred acres of trees of the pedunculate species, growing very close together, especially on the alluvial flats along the river Nore.   The trees are of no great height, and have usually short boles with wide-spreading, stout branches, the largest tree measured being 21 feet in girth.

Hayes gives several instances of the remarkable growth of oak in Ireland.   At Ballybeg in Wicklow, a tree growing in alluvial soil, eighty years old, was 12 feet in girth at 8 feet from the ground.   At Muckross, Killarney, six trees sown in 1760 measured in 1794, from 3 feet to 4 feet 11 inches in girth at 5 feet from the ground. Ireland, renowned in ancient days for its oak timber, which was valued abroad, is now singularly wanting in even good specimens of solitary oak trees; and Loudon gave in 1838 no examples of fine oak trees growing in Ireland.   The finest which have been seen by me are :—At Dartrey, Cootehill, the seat of the Earl of Dartrey, a beautiful symmetrical pedunculate oak, 100 feet high with a girth of 14 feet 4 inches ; at Kilmacurragh, Wicklow, a sessile oak 14½ feet in girth ; at Glenstal, Limerick, a tree of the same species 16½ feet in girth ; and at Shane's Castle, Antrim, a pedunculate oak 19 feet in girth.   There are also many fine trees with good boles at Doneraile Court, Co. Cork, the largest about 13 feet in girth.   (A. H.)

REMARKABLE TREES ABROAD

As the oak is one of the most characteristic British trees we give only a few details of the remarkable oaks which we have seen on the Continent.   A good account of the trees in the forests of Retz, Compiegne, and St. Amand was written by Prof. Fisher in the *Trans. Eng. Arb. Soc.* v. 205.   I took part in the excursion which this paper records, and saw the splendid sessile oaks at Compiegne, of which the one called the Czarina's Oak is the finest.   This is as well-grown, but not a finer tree than some of those which I have described and figured in England, though in cubic contents inferior to several of them.   The French measurements given on the trunk of the tree are—height 36 metres = 118 feet; girth at 1.30 metres, 5.20 metres = 17 feet; volume 32 cubic metres = 1130 feet; value £100. Mr. George Marshall, Past-President of the English Arboricultural Society, who is a timber valuer of great experience, estimated the butt of the tree to contain (46 feet by 42 inches quarter-girth) 550 cubic feet; plus 150 cubic feet for the top, making a total of 700 cubic feet; which, with the addition of an unknown quantity for the branches, always reckoned in France, plus 20 per cent for the difference between the total volume and the English quarter-girth measurement, will come near the French estimate. A photograph of an oak in the Forêt de Bellême was reproduced in this report. Its total height was 119½ feet, and its girth at 4 feet 6 inches was 9 feet 9 inches. It is impossible to imagine a tree containing more useful timber and less waste than this tree, which has rather the appearance of a gigantic mop than of an oak as we know them. Prof. Fisher considers Bellême as the finest oak forest in France, and in the *Gardeners' Chronicle*, xxviii. 220 (1900), speaks of a sessile oak which he measured there 146 feet high, with a clean bole 113 feet by 9 feet 10 inches girth, and a volume of about 500 cubic feet.[1]

Another renowned forest in France is that of Bercé near Chateau du Loir (Sarthe), visited by Henry in 1903 and in 1906, which covers 13,350 acres; and is made up of about 90 per cent of sessile oak and 10 per cent of beech. It is situated on a plateau; the soil being a deep loamy sand, poor in lime. There is not a single pedunculate oak in the forest itself, yet, curiously enough, all those in the hedgerows of the surrounding country are of this form. The sessile oak, owing to its ability to bear shade, is grown densely in the forest, and attains an astonishing height, though it is slow in growth, as far as regards diameter of stem, which averages at 200 years old only 20 inches. The best individual tree, the *Chêne Boppe*,[2] in 1905 measured 115 feet high, 75 feet to the first branch, and 14 feet in girth. Another tree, measured in 1906, was 125 feet total height, 92 feet to the first branch, and 8 feet in girth. In one section, containing a little less than twenty acres, there stood in 1903, aged 211 years, 1314 oaks and 268 beeches; the oaks averaging 28 inches in diameter. The total amount of the timber[3] was estimated by an accurate survey in 1895 at 275,000 cubic feet, valued at £14,720, or about £740 an acre. The yield of the first and second series in this forest, 2700 acres in extent, over which felling is done in sections once every 216 years, works out at 66 cubic feet of timber per acre per annum, equivalent to a net annual revenue per acre of £2 : 3s. A photograph taken by Henry, shows the shape of these forest oaks, all beautiful, clean, cylindrical stems, and illustrates the way in which the

---

[1] Henry visited Bellême in 1906, and does not consider it to be quite as fine a forest as Bercé. The best tree seen, possibly the same as the one measured by Prof. Fisher, was 125 feet total height, 95 feet to the first branch, 10 feet 4 inches in girth, and about 425 cubic feet in volume. On referring to Prof. Fisher as to this measurement, he sends me two photographs given him by M. Granger, then Garde Général at Bellême, representing (1) the Chêne de Brigonnais, which is 37 metres = about 120 feet high; girth at 4 feet 6 inches, 3 metres = 9 feet 10 inches; height to the first branch, 23 metres; (2) the Chêne Lorentz, which is 40 metres in height = about 130 feet, girth 4½ metres = about 14¾ feet, and 18 metres long to the point where it divides into two nearly equal stems. It therefore appears that we have in England a few oaks at least as tall, and many larger in bulk than any recorded in France.

[2] Near this tree Henry observed an oak bearing misletoe on a branch at 60 feet up.

[3] Huffel, *Economie Forestière*, i. pp. 370, 372 (1904). The capital or volume of wood in the forest is not diminished by its felling, but is steadily increasing slightly all the time, owing to careful management.

woodcutter, to save the seedlings beneath from damage, lops off the crown of the tree with an axe at a point below the first branch before felling the trunk.

The oaks in the German forest of Spessart have been so frequently mentioned by recent writers on Forestry that I need not say anything of them, but doubt whether they equal the oaks in some of the few remaining virgin forests of Slavonia. In 1900 I saw a splendid lot of clean straight logs 3 to 5 feet in diameter, which had been felled in these forests, and floated down the Save to Bosnabrod.

We are indebted to Dr. Simonyi Semadam Sandor, a member of the Hungarian Parliament, for an account of the oaks of Slavonia in the Forest of Brod Petevarad, which is published in a Hungarian journal called *Erdzesti Lapok*, at Buda Pesth, June 1889, with photographs showing splendid clean-stemmed trees, 30 to 40 metres high, and 2 to 3 metres in diameter.

The European oak seems able to grow well in temperate parts of the southern hemisphere. In Chile it seems as much at home as in Europe, and not only grows much faster, but reproduces itself with such ease from seed on land capable of being irrigated that I saw no reason why it should not be cultivated for its timber.[1]

In South Africa the original Dutch settlers planted oaks near Cape Town, and under one of these trees the Convention was signed by which the Colony was transferred to Great Britain in 1814. On 5th April 1905 my brother posted me a few acorns from this tree, the trunk of which is now hollow and bricked up. I sowed them in May to see whether they would at once revert to their proper season of growth; and out of twenty acorns, three germinated in June, and are now nice young trees, the others never coming up at all.

In North America I have seen no European oaks of any great size, though there is one in Prof. Sargent's grounds near Boston, which has puzzled several good botanists as to its origin.

INJURIES TO OAKS

The liability of the oak to be struck by lightning was noted by Shakespeare, who, in *King Lear*, Act III. Scene ii., wrote—

> You sulphurous and thought-executing fires,
> Vaunt-couriers to oak-cleaving thunderbolts.

Mr. Menzies says,[2] " Of all forest trees oaks are, in my experience, the most dangerous. If they have a large spreading head, they are shivered into shreds when struck. If they have long tapering stems, and thus can act almost as conductors, they are not so dangerous, and the lightning will run down the side, ploughing out a deep furrow. I have once seen a beech struck, an ash once, an elm once, a cedar of Lebanon once, but never any other trees, except the oak. And while the others stood comparatively singly in an open space, the oaks have been selected and struck in the midst of a thick wood."

[1] Sir W. Thiselton Dyer informs me that on the Blue Mountains of Jamaica Sir Daniel Morris found a characteristic English oak.
[2] *History of Windsor Great Park*, p. 8.

Several interesting particulars of the effect of lightning on oaks are given by Loudon, who also states that the oak, owing to its roots not being so liable to rot in the ground as those of most trees, is not often blown down. He describes the effect of a hurricane in October 1831, on the splendid oaks growing in Lord Petre's park at Thorndon Hall in Essex, which reminds me of a similar case in April 1890, when I saw, at Narford, in Norfolk, oaks of 2 to 3 feet in diameter broken off at 4 to 10 feet from the ground by the force of the wind, which tore up many plantations of spruce and other shallow-rooting trees by the roots.

Sir Charles Strickland tells me that a very tall young oak tree 54 feet to the first branch, and quite straight, growing at Housham in Yorkshire, nearly on a level with the river Derwent, was, in the severe winter of 1860-61, completely killed by a frost which was the severest in his recollection. Though he has no record of the temperature at Housham, yet he believes that at Appleby, in Lincolnshire, it was as low as $15\frac{1}{2}$ below zero,[1] and generally in the northern counties the thermometer went below zero. Many other oaks were killed in the woods and in the hedgerows between Malton and Pickering by the same frost.

The various insects which attack the oak are too numerous to be mentioned in detail, but are described at length by Loudon and by many other authors.

The galls, which are so common on the leaves, are produced by several species of Cynips, and the so-called oak-apples are the result of an injury by an insect of the same family.[2]

Mistletoe on the Oak

Since the time of Pliny, who describes the worship of the oak, and especially of the mistletoe-bearing oak by the Druids, the occurrence of this parasite on the oak has always been looked on as a rarity. Loudon only mentions two trees known to him, of which one near Ledbury was cut down in 1831, and another at Eastnor Castle is still living; but we have now been able to collect many more authentic records. A paper on the subject by the late Dr. Bull of Hereford[3] gives particulars of several, and states that it is considered a dangerous practice to interfere with a mistletoe-bearing oak. One at St. Diels, near Monmouth, was cut down by the bailiff about 1853, and the owner of the estate immediately dismissed him. A woodman who climbed the Eastnor tree to get some mistletoe, fell down and broke his leg, and other similar stories are quoted. The finest mistletoe oak I have seen was shown me by Sir George Cornewall, at Bredwardine, in 1902. When described by Dr. Bull, mistletoe was growing on it in no less than fifteen different places, and it measured 78 feet by 11 feet 6 inches in girth. Sir George has lately found another in his park, and has a third on his estate in Woodbury Wood.

This part of England seems to be, for some reason, the most prolific in England

---

[1] This is a little lower than any temperature recorded by the Meteorological Office, but the subject of meteorology as affecting trees will be discussed fully later.

[2] An article in the *Kew Bulletin of Miscellaneous Information*, Additional Series, v. 1906, on Oak Galls, by R. A. Rolfe, gives much information on the subject, but is too long to quote. Nearly one hundred different kinds have been described which occur on the roots, buds, leaves, stamens, ovaries, and fruit.

[3] *Trans. Woolhope Nat. Field Club*, 1870, p. 68.

in mistletoe oaks; and it will be observed in the list which follows that there are none reported in the northern half of Great Britain.

The subject has been recently studied by M. H. Gadeau de Kerville,[1] who records in Normandy alone no less than 26 mistletoe-bearing oaks, living or recently felled, of which a list with exact particulars of their locality is given, pp. 298-301. An excellent illustration of one of the finest of these growing on the farm of Bois, at Isigny-le-Buat, Department of Manche, shows a large and well-shaped tree, about 60 feet by 16 feet, of the pedunculate variety, which is covered with tufts of mistletoe, some of them growing on the trunk, and of very large size. M. de Kerville estimates the age of this tree at 200 to 300 years, and says that it has begun to deteriorate, as the dead branches show. M. Eugène Ormont states that a tuft of mistletoe of about a foot in length, which he examined on an oak, was eleven years old and seemed slower in its growth and yellower in colour than mistletoe growing on the apple.

LIST OF REPORTED MISTLETOE-BEARING OAKS IN ENGLAND [2]

Locality.	Authority.	Date.	Particulars.
Bredwardine, Hereford,	Dr. Bull,	1870	
	H. J. E.	1902	
Moccas Park, do.,	Rev. Sir G. Cornewall,	1904	
Woodbury Wood, do.,	do.	do.	
Tedstone Delamere, do.,	Dr. Bull,	1870	
Haven in the forest of Deerfold, do.,	do.,	do.	
Badham's Court, near Chepstow, Monmouth,	do.,	do.	
Near the Hendre, Llangattock, do.,	do.	do.	This tree is not known to exist now, so far as I can learn, at the Hendre.
Eastnor Castle, Worcestershire,	do.,	do.	
	H. J. E.	1903	
Lindridge, Worcestershire,	*Leisure Hour*,	1873	
Frampton-on-Severn, Gloucestershire,	H. Clifford, Esq.,	1904	Mentioned by Lees in 1857, and still living.
Knightwick Church, Worcester,	*Leisure Hour*,	1873	
Plasnewydd, Anglesea, in Marquis of Anglesea's Park,	Lees,	1857	
Hackwood Park, Basingstoke, Hants,	*Leisure Hour*,	1873	
Lee Court, Kent,	do.,	do.	
Burningfold Farm, Dunsfold, Surrey,	do.,	do.	
Bodlam's Court, Sunbury Park, do.,	do.,	do.	
Shottesham, Norfolk,	Francis, in Trimmer's *Flora of Norfolk*,	1866	
Alderley, Norfolk,	Winter, in do.,	do.	
Not far from Plymouth, by side of S. Devon railway,	Britten,	1884	
Near Cheltenham,	*Leisure Hour*,	1873	I can get no confirmation of this.
Seven miles from Godalming,	Menzies,	1860	

[1] *Les Vieux arbres de la Normandie*, pt. iv. (1905).

[2] Sir Herbert Maxwell in *Memories of the Months*, p. 285, mentions the existence of mistletoe-bearing oaks at Stoulton in Worcestershire, in Sherwood Forest, Windsor Forest, and Richmond Park.

BARK

The bark of the oak was until recently a valuable source of revenue in England, but, owing to the introduction of other materials for tanning, has now fallen so much in price that in some districts it hardly pays to take off, and large areas of coppice oak in the western counties have become almost worthless in consequence. Whether the leather made by these modern substitutes is as durable as that produced under the old system is doubtful, but the comparative slowness of the process of tanning by oak bark seems to be one of the chief reasons for the change.

Professor H. R. Procter of the Leather Industries Department of the Leeds University, whom I consulted on this question, tells me that though he agrees with me that no tree at present grown in England is worth growing for the sake of its bark alone, yet he thinks that it will be long before the demand for oak bark entirely disappears. He considers that though leather tanned with oak bark alone is perhaps the best for boots and shoes, the cost of the slow process is so much greater in proportion to quality, that the leather so tanned is practically an article of luxury.

In the *Land Agents' Record* for October 29, 1904, there is an article on the price of oak bark, which is stated to have fallen from £8 a ton in the writer's experience to 47s. 6d. ; and when the cost of peeling, which averages about 25s. per ton, the cost of loading and delivering to the station, and the cost of railway carriage is added, little or nothing is left for bark grown at any distance from its market. Since then the price in some districts has risen a little, but in this case, as in others, it is clear that chemically prepared substitutes are killing an industry of much importance to landowners and labourers.

TIMBER

With regard to the difference in the timber of the two varieties of oak, we have, strange to say, little or no certain experience in England. Laslett says that though he agrees generally with the opinion then prevalent, that *Q. sessiliflora* was slightly inferior to *pedunculata*, he feels bound to admit that during a long experience in working them, he has not been able to discover any important difference between the two varieties. He says that very fine specimens of long clean oaks of the sessile form were found in the Forest of Dean, which, however, were liable both to cup and star shake, and that he is inclined to believe that these defects are less common in *Q. pedunculata*.

Though little attention is now paid to the difference of winter- and spring-felled oak timber, and it seems as if most users of wood will pay as much for the latter as for the former, yet, considering the low price of bark and the importance of durability, I should strongly advise the former being used for all first-class work.

Laslett,[1] who, as timber inspector to the Admiralty, had probably as much experience as any man of his day, and more than any one at the present time,

[1] By far the best account that I know of is in Laslett's *Timber and Timber Trees*, of which a new edition, revised by the late Prof. Marshall Ward, was published in 1894.

gives in chapter xi. many proofs in support of his opinion that winter-felled oak is better than spring-felled ; though the practice he recommends was to bark the trees standing, and fell them in the succeeding winter, a custom which is still followed in some parts in England.   He also states, on page 73, that having carefully examined and compared many pieces of both winter- and spring- or summer-felled logs, he found almost invariably that the winter-cut timber, after being a few years in store, was in better condition than that which had been cut in the spring.   "The winter-felled logs were sounder, less rent by shakes, and the centres or early growths generally showed less of incipient decay than the spring-felled."

So much has been written about the timber of the oak that it seems unnecessary to go into very great detail with regard to this subject, especially as this timber, of which little is now required for the navy, is being ousted by iron and by cheaper imported timber from many of its former uses, and is of far less value than formerly ; but though at the present time English oak is out of fashion, there is no doubt that such durable and beautiful wood must always have a considerable value to those who do not sacrifice durability to cheapness, and who have patience to wait until it has been properly seasoned, which requires from two to six years according to the thickness of the plank.

There are so many proofs of its everlasting character in the form of roofs and in the old timbered buildings which are common in Cheshire, and of which so many beautiful illustrations are given in *Country Life*, that I need not repeat them, but an extraordinary instance of its longevity when exposed to the weather was pointed out to me by the late Lord Arundell of Wardour in the ruins of Wardour Castle.   This building, according to an account of it published in *The Antiquary*, November 23, 1873, was inhabited before the reign of Edward III., and was besieged and sacked by the Parliamentary army in the reign of Charles I., and blown up by its owner, Lord Arundell, in 1644, rather than allow it to remain in the hands of the enemy. An oak lintel, which must therefore have been exposed to the weather for 260 years, still remained *in situ* in 1903, and as far as I could see from below was not much decayed.

In a paper by W. Atkinson[1] it is said that during the last thirty years he had taken every opportunity of procuring specimens of wood from old buildings, and particularly what the carpenters called chestnut, but never in a single instance had he seen a piece of chestnut, the wood so called being always that of *Q. sessiliflora*, mistaken for chestnut from a deficiency of the flower or silver grain.   He goes on to say : "The roof of Westminster Hall has been said to be chestnut ; while it was under repair I procured specimens from different parts of the roof, the whole of them were oak, and chiefly the *Q. sessiliflora*.   Most of the black oak from trees dug out of the ground I have found to be of the same kind.   From finding the wood of the oldest buildings about London to be chiefly of the *Q. sessiliflora*, I should suppose that some centuries ago the chief part of the natural woods were of that kind ; at present the greater part of the oak grown in the south of England is *Q. pedunculata*.   Specimens of oaks that I have procured from different parts of

---

[1] *Trans. Hort. Soc.* Second Series, i. p. 336 (1835).

Yorkshire and the county of Durham have all been *Q. sessiliflora*, which is very scarce in the south. There are some trees of it at Kenwood, the Earl of Mansfield's, near Highgate, which I believe to be one of the oldest woods near London, and a greater part of the *Q. sessiliflora* appears to be trees from old stools." To this the Secretary, Mr. G. Bentham, adds a note, as follows :—" Mr. Atkinson's opinion on this subject is confirmed in a remarkable manner by the discovery that the oak in an extensive submarine forest near Hastings is *Q. sessiliflora.*"

BROWN OAK

In a paper on British timber which I read before the Surveyors' Institution in February 1904,[1] I called attention to a form of oak timber, known as "brown oak," which does not appear to have been much noticed by any previous writer.[2] Though after very careful investigation I have failed to ascertain with certainty the causes which produce it, I am inclined to believe that it is not, as some have thought, caused by a fungus ; though spores of some fungoid mycelium are often found running through it ; but that the change of colour is produced, especially on certain soils and in certain localities, by age. And though I have evidence that in exceptional cases the heartwood of quite young oaks is brown,[3] the majority of the trees which produce this beautiful and valuable wood are in an incipient stage of decay, and often hollow, leaving only a shell of more or less sound wood. The change of colour in some trees commences at the ground and extends upwards, or less commonly begins in the upper part and extends downwards. No one can be certain, without boring or felling the tree, whether the wood is brown or how far the colour may extend ; but if the tree is allowed to stand too long after it has become brown it loses its "nature," to use a carpenter's expression, and is often so shaky and full of cracks that it is of little use. The sapwood always remains of the normal colour. But when a brown oak of good rich colour contains sound and solid timber it is superior to any wood I know for the interior decoration of houses, and for the making of sideboards and other heavy furniture.

Until about fifty years ago this wood was little valued in England, and I am told that on the Duke of Bedford's estate its use was prohibited in building contracts because it was supposed to be unsound. Even now it is hardly known or recognised as valuable except in certain parts of England, and is often sold far below its real value by inexperienced persons. But the Americans have created such a demand

---

[1] *Trans. Surveyors' Institution*, vol. xxxvi. pt. vii.

[2] Laslett, ed. 2, p. 96, only says of it, "and even when in a state of decay or in its worst stage of 'foxiness,' the cabinetmaker prizes it for its deep red colour, and works it up in a variety of ways."

[3] Mr. Alexander Howard tells me that he has seen a group of young oaks felled in Essex, which were not more than 12 to 18 inches in diameter, all perfectly sound, in which the wood was of a rich brown all through the trunk up to and beyond the first main branch. He purchased near Chelmsford a very fine oak which had no less than five secondary trunks growing out of the butt, all of a very rich brown colour, and a number of younger trees growing near it in the same park also proved to be of the same colour. Thus it seems that though the conditions of the soil have some influence, yet the colour may in some cases be inherited. Mr. Howard has inquired for many years but never heard of a brown oak on the continent, and believes it to be only found in this country. Some woodmen in Essex have thought that the trees which carry their leaves longest in winter produced "red oak," which is the local term for brown oak, but I could get no definite proof of the truth of this idea.

for it, that most timber merchants are now quick to appreciate the difference between brown and common oak, and the best qualities of it are sometimes sold for as much as 10s. per cubic foot.

When the wood shows the blackish streaks running through it, which is known as tortoise-shell grain, it is most valued for cutting veneers. These are laid in thin sheets on other wood, partly to make it go farther, and also because this wood is so difficult to season properly, and so wasteful in conversion that it is not safe to use in the form of thick boards.

My friend, Dr. Weld of Boston, U.S.A., who is a great connoisseur and admirer of fine woods, and especially of brown oak, showed me at his house the most magnificent specimens of panelling and wainscoting, done under his own supervision by Messrs. Noyes and Whitcombe of Boston, with oak which he selected and purchased himself in England. In their works I saw a quantity of carved brown oak pews, and a very large brown oak organ front designed by Mr. C. Brigham, architect, of 12 Bosworth Street, Boston, for a memorial church at Fairhaven, Mass. Mr. Whitcombe was good enough to show me the manner in which the boards are seasoned after they are cut from the logs, which are imported in the rough as an unmanufactured product to escape the heavy duty. Dry white pine boards fresh from the hot-air kiln are laid on each side of the oak boards, and properly stripped in an open covered shed. When the moisture has been partially absorbed, they are all turned over and again sandwiched between fresh dry pine boards; thus saving a great deal of time, which is rarely given to season timber properly in America, and preparing the wood to stand the conditions of dryness, which are more trying to furniture in American than in English houses.

Mr. C. M'Kim, a distinguished American architect, writes me as follows respecting English oak :—"We regard it as the most beautiful oak in the world, costly because of its scarcity and the duties imposed upon it; requiring the best workmanship in putting it together; but preferred above all others for its finer quality, richer colour, and endurance. The most important and dignified panelled rooms in this country are furnished in English oak." I also was pleased to find that the great dining-room in the White House at Washington is completely panelled with English brown oak.

Mr. F. H. Bacon of the A. H. Davenport Company of Boston, one of the best firms for cabinet work in the United States, writes :—"Mr. Davenport has been using it in his business for at least thirty years, and we think it is a wood which will always be in demand, as a room furnished with English oak has a richness and depth of tone which is impossible to get with any other oak. The wood is becoming more expensive, but I think it will always be used by people who can afford it. It is difficult to work; the plain surfaces are generally veneered. It stands perfectly well without warping and twisting, and is not attacked by worms as walnut wood is."

The best example that I have seen of fine brown oak work in England is at Rockhurst, the residence of the late Sir Richard Farrant, in Sussex. This was done by Marsh, Cribb, and Company of Leeds, with brown pollard oak, showing very varied figure, and superior in this respect even to that of Dr. Weld's house.

This wood requires no varnish, but when simply polished with wax and shellac only, in the manner adopted by Dr. Weld, is as rich as any mahogany. It is to some extent imitated by a practice called fuming, which is now very commonly used to give a darker colour to foreign oak, and thus make it resemble old oak, which has become so fashionable; but fumed oak can easily be distinguished from, and is far inferior to real brown oak, which also varies a good deal in colour when new.

POLLARD OAK

There is another form of oak wood usually called pollard, which is produced from the burrs or swellings which often appear on old oaks, especially in very dry and in wet ground. The real cause of these excrescences is not yet fully explained; but in some places, and especially in Sherwood Forest, they are very common, and when cut, show a twisted and contorted grain, sometimes full of little eyes which resemble those of the so-called "birds'-eye maple," a variety of the wood of American maple, of which we shall speak later.

Pollard oak is usually full of little cracks, and is best cut into thin slices or "plating" $\frac{1}{8}$ inch thick or less. When polished the little cracks are filled up, and when the wood is mottled with brown, yellow, and pink in various shades it is very beautiful. An oak of this type, which was only about 10 feet high and 9 feet in girth, grew on Chedworth Downs, Gloucestershire, and was given to me by the Earl of Eldon. Its wood, when cut into veneer, was throughout the whole thickness of the tree more like that of birds'-eye maple than oak, and has served to make the front of a very handsome bookcase.

OAK PANELLING

I cannot pass from this subject without alluding to the use of English oak for panelling walls, a practice which was almost universal in houses built in the sixteenth and seventeenth centuries, and of which many beautiful examples still exist. Modern architects, however, do not seem to have properly appreciated, that the beauty and fitness of oak for such work depends on the extent to which the "figure," "flower," "silver grain," or "flash" is shown—all these terms are used to express the bright glossy patches and lines which the medullary rays of oak show when cut "on the true quarter."

In our ancestors' time, when roads were bad or non-existent, and when sawmills were unknown, it was necessary to cut up large oak trees where they fell, either by digging a saw pit near or under them, or by cross-cutting them into suitable lengths, and then "rending," cleaving, or splitting them into slabs. This practice is now adopted principally for making oak palings and for wheel spokes, which are much stronger when rent than when sawn ; but it will be found on examination of the back of old panelling that it was usually rent, and as you can only cleave oak on the line of the medullary rays, the figure shown by rent oak is much better and more abundant than when sawn on the quarter, and though the practice is more wasteful and is only possible in the case of straight-grained trees, yet it should certainly be tried by those who admire finely figured oak.

Strange to say, the importance of selecting and matching the figured pieces, and of placing them in the most conspicuous positions, does not seem to be noticed, for I have seen in modern houses, and in old castles on whose restoration no expense has been spared, panelling in which new and plain pieces have been introduced amongst splendid old panels, and finely figured new and old panels put in dark corners where they were unseen.   When one considers how small a proportion the cost of the wood bears to the workmanship, it is extraordinary that this should be allowed, or that American oak should be used, as I have seen sometimes done, in restoring ancient houses, when infinitely better and more beautiful wood was growing, and often rotting on its roots, within a very short distance.

Experienced cleavers are not to be found in every county, but in Kent, Surrey, Sussex, and Hants, and where rent oak palings are much used, as in the neighbourhood of London, such men may be found, who with a tool called a " break-axe " or "flammer," will convert straight-grained oak into slabs of suitable dimensions for panelling, which, when properly seasoned, show better figure than sawn timber.   For this purpose logs of not less than three feet diameter should be selected, as straight as possible in the grain, and cut into the lengths of which the panels are required. The slabs come out rather irregular in size, and are, of course, much thicker on the outside.   They should be carefully piled for about twelve months in a dry, airy place, when they can be reduced by a thin circular saw and by planing to the proper thickness, choosing whichever side shows the best figure for the face. Longer and narrower pieces, either rent or sawn, must be selected for the stiles and rails, and if put together by a competent joiner, I can say from experience that the effect will be much superior to work done by the best London firms with foreign timber, especially when brown oak can be found fit for rending.

The diagram, Fig. 1, on the following page shows the best method of rending oak to show its fine figure.

For quartering by the saw different methods are adopted, the best being that shown on the following page, Figs. 2 and 3, taken, by permission of Messrs. Rider and Son, from a very useful little book.[1]   By this method, which, though rather wasteful, produces the best results, only the central boards of each cut are on the true quarter, and the others are narrower, and more or less across the natural line of cleavage.

Of the different styles of oak panelling it is not my intention to speak, but it seems to me that elaborate carving is out of place in such wood as this, which wants no extraneous adornment.   Many beautiful specimens of ancient panelling in various

[1] *English Timber and its Economical Conversion*, London, 1904.

styles may be seen in the galleries of the South Kensington Museum, among which that taken out of Sizergh Castle, Westmoreland, is, though rough in workmanship, a good example of ornamentation with native wood.

One of the most elaborate instances of room-decoration in woodwork of old times is seen in the dining-room at Gilling Castle, near York, formerly the property of the Fairfax family, now belonging to W. S. Hunter, Esq. It is a room about 30 by 20 feet, and is panelled with large panels of oak, in oblongs 2 feet 4 inches wide and 3 feet deep, surrounded by heavy carved mouldings. Each panel is inlaid with highly intricate and varied geometrical patterns in narrow lines of black and white wood, which I believe to be bog oak and holly, inlaid in narrow lines, and forming an elongated diamond in the middle of the panel. The four corners of each

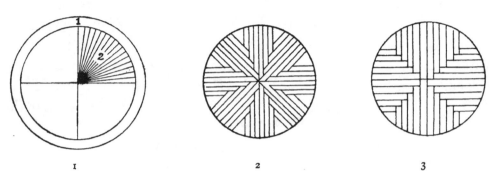

1   2   3

FIG. 1. (1) Sapwood; best taken off. (2) Feather-edged boards somewhat variable in width and thickness, but following the natural line of cleavage on the medullary rays of the wood.

FIGS. 2 and 3. Methods of quartering by the saw.

panel are also inlaid with flowers done in similar wood. This work runs from the ground up to about 10 feet high, above which an elaborate decoration in colour, containing many family trees and coats of arms, reaches to the ceiling. Some good judges think this is the most beautiful room in England, but without resorting to such minute and fanciful patterns, I may safely say that good plain oak panelling, in which the stiles and rails are duly proportioned, and the silver grain well matched in each panel, gives not only the handsomest and richest effect of any wall covering I know, but is also the most durable, improving in colour with age, and if done with one's own timber, affords an interest which no Italian frescoes or plaster work can give.

In the chapel, in the hall, and in the Earl's study at Powderham Castle, Devonshire, are very good examples of pews and panelling, both of the linen pattern and carved panels, but though the linen pattern was once a favourite one, and is still copied by some decorators, it seems to me a mistaken notion to imitate the folds of a textile material in wood, and especially in oak.

Wainscot Oak

What is usually known under this name was for many years imported from the Baltic seaports of Dantzic, Riga, and Libau, and was the produce of forests in the interior of the Russian Baltic Provinces, and of Russian Poland, from whence it was brought to the coast by water, until railways were made. According to Laslett, the Riga timber, though of moderate dimensions, had the medullary rays more numerous and better marked than the Dantzic oak, and came to market in the form of hewn billets of about 18 feet.

But as the supplies of this oak became less, and the demand greater, a fresh source of supply was found in Slavonia and South Hungary, which for many years has furnished about half the total import through the ports of Trieste and Fiume. Mr. A. Howard tells me that the size and quality of this was better than the Baltic oak, but owing to the Austrian Government having recently diminished their cuttings in consequence of the rapid diminution of mature timber, a large quantity of billets are now exported from Odessa, which are believed to come from the forests of Podolia and Volhynia, and other provinces of South-West Russia.

All this imported oak is milder and more easily worked than English oak, and as only selected logs free from knots are shipped, it can be converted into boards with less waste and risk than home-grown timber. We have no certain evidence as to the existence of a sufficient quantity in Russia to keep up the supply either from the Baltic or Odessa, and though the more scientific foresters of Austria are taking steps to restore their oak forests by natural regeneration, it is probable that the French, who consume an immense quantity of oak from this region, will take all they can get, and this, coupled with the approaching disappearance of American oak large enough for quartering, must, sooner or later, cause our own timber when long and clean to be much more valuable than it is at present.

A note in Holinshed's *Chronicles*,[1] vol. i. p. 357 (ed. 1807), seems to show that wainscot oak was already exported from the Baltic as long ago as Queen Elizabeth's reign, but whether " Danske " means that it came in Danish ships or from the port of Dantzig I cannot ascertain, though Colonel Brookfield, H.B.M. Consul at that port, has made inquiry on the subject.

Laslett is the only practical English writer I know of who was personally acquainted with the oak in its native forests in the east of Europe, having been employed by the Admiralty to survey the forests near Brussa, in Asia Minor, as well as in Bosnia, Herzegovina, Croatia, Styria, and Hungary. He states that in the

---

[1] According to Mr. J. C. Shenstone, Harrison of Redwinter in Essex, who lived in the reign of Henry VIII., was the author of this note. " Of all oke growing in England the parke oke is the softest, and far more spalt and Prickle than the hedge oke. And of all in Essex, that growing in Bardfield parke is the finest for joiner's craft ; for often times have I seene of their workes made of that oke so fine and faire as most of the wanescot that is brought hither out of Danske, for our wanescot is not made in England. Yet diverse have assaied to deale with our okes to that end, but not with so good suceesse as they have hoped, because the ab or juice will not so soone be removed and cleane drawne out, which some attribute to want of time in the salt water."

forests south-east of Brussa he found oaks resembling the English *Q. Robur* on the upper ranges of the mountains, while in the valleys *Q. Cerris* or the mossy-cupped oak was found. In Austria, he says, that in the Kogarate mountains, and in the district between the rivers Verbas and Okvina they were chiefly of the sessile variety, mixed occasionally with *Q. Cerris*, and all of straight growth with long clean stems, generally of good quality, but at that time no attempt had been made to utilise them except for cleaving cask staves.

Of all the oaks of which trials were made in our Government dockyards during the period at which British oak became scarce, Laslett says that the white oak of North America compared very favourably with all the foreign oaks, but proved to be slightly inferior in strength to English oaks.

Bog Oak

This is obtained from trees which have been buried in peat bogs for centuries, and which has become blackened by the peat water. It is very commonly found in Ireland, and in some parts of England and Scotland. When large and sound enough it is used for furniture, picture frames, and for small ornamental work, but as a rule is so full of shakes, and cracks so much in drying after it is dug up, that it is of no use for cabinet-making except in the form of inlay, or marqueterie. Occasionally, however, fine sound logs are dug out, which if slowly seasoned in an airy cellar may be used for larger work. One of the best examples I have seen of black oak was a door exhibited by Mr. E. R. Pratt of Ryston at the Royal Agricultural Society's Show at Park Royal in 1905, made from oak found on his property in Norfolk. He tells me that the planks after being sawn are dressed two or three times with "fuel" or "dead" oil which replaces the evaporated water by the refuse of petroleum, a substance theoretically similar to that lost by age. The result is certainly very successful.

Many cases have been recorded and published of the great durability of the timber of the oak under ground and under water ; but I have come across no relic of the past so interesting in this respect as the prehistoric boat which was dug up at Brigg, in Lincolnshire, in 1884, when digging a foundation for a gasometer. This has been well described by the Rev. D. Cary Elwes in a lecture, which was published in 1903,[1] and a photograph of it is published in a recent pamphlet by the Rev. A. N. Claye,[2] for which I am indebted to Miss Woolward. This wonderfully preserved dug-out was hollowed out of one huge oak log 48½ feet long, and approximately 6 feet in diameter, which showed no signs of branches, a log which must have contained nearly 1000 feet of timber, and which could not be matched now in England, or, so far as we know, in Europe or North America. The boat is 4 feet 3 inches wide by 2 feet 8 inches deep at the bows, and 4 feet 6 inches by 3 feet 4 inches at the stern, which was the root end of the tree. The sides are about 2 inches thick, the bottom 4 inches at the bows, and as much as 16 inches at the stern. The stern piece was

[1] *A Prehistoric Boat.* Stanton and Son, Northampton.
[2] *Brigg Church and Town.* Jackson, Brigg.

ingeniously fitted in, though not found *in situ*, and a large rift on one side had been still more cleverly repaired with wooden patches caulked with moss. No metal had been used in any part of it. The boat was found embedded in the blue and brown clay which underlies the peat, and is considered on geological evidence, which is given with great detail, to be from 2600 to 3000 years old. It was offered by Mr. Cary Elwes to the British Museum, but was declined as being too large ; it is, however, now suitably housed at Brigg.

Many similar oaken boats of smaller dimensions have been discovered in various parts of England, and I saw one myself which had been just dug out of a peat bed close to Shapwick Station, in Somersetshire, in September 1906, which was 20 feet long by 2 feet 10 inches wide.

At Brigg an ancient causeway was discovered, which is described by Mr. Claye in the same pamphlet, and a photograph given. This roadway was found in a brickyard lying between the two branches of the river, under a deposit of blue alluvial clay, and above the forest bed which lies on the top of the glacial drift, and was probably made by the early Britons to secure a safe passage across the valley when it was little more than a swamp. Small trees and branches of yew were laid lengthwise, and across them rough planks of oak, which were fixed in their place by long wooden pins driven through holes at each end. From the photograph the wood appears to have been well preserved, but having been covered up again shortly after the excavation was made, I can give no further details of its condition. In the same place was discovered a sort of raft or flat-bottomed boat, 40 feet long and 6 feet wide, which was also covered up again. From the illustration given, this seems to have been made of five logs placed side by side, and held together by cross ties passing through holes in projections on the upper side of the logs.

In the foundations of Winchester Cathedral, oak piles had been used to form a solid foundation in the wet peaty soil on which part of the structure rested. When the Cathedral was under restoration in 1906, samples of these piles sent me by Mr. Jackson, the architect of the work, who said that they were put down in the time of William Rufus, were in places decayed. Some logs of beech laid horizontally under the same building, which Mr. Jackson attributes to Bishop de Lucy, about A.D. 1206, remained comparatively sound, and, though the wood has changed from its natural colour to a grey, is fit to use as boards for book-binding.

With regard to the foundations of the Campanile at Venice, it has been stated that they were laid on larch piles, which are still used in that city for the same purpose ; but when I was at Venice in 1905 I inquired into this, and was given a section of an oak pile only about 6 inches in diameter, but perfectly sound and very hard, which was cut from one of the piles taken from the foundation of the Campanile after it fell.   (H. J. E.)

LARIX

Larix, Adanson, *Fam. Pl.* ii. 480 (1763); Bentham et Hooker, *Gen. Pl.* iii. 442 (1880); Masters,
    Journ. Linn. Soc. (Bot.) xxx. 31 (1893).
Pinus, Linnæus, *Gen. Pl.* 293 (in part) (1737).
Abies, A. L. de Jussieu, *Gen.* 414 (in part) (1789).

TREES belonging to the order Coniferæ, with thick scaly bark, irregular and not whorled branches, and deciduous foliage. Branchlets of two kinds, long shoots bearing solitary leaves spirally arranged, and short shoots bearing numerous leaves in tufts at their extremities, these leaves being of unequal lengths and arising each in the axil of a bud-scale. Leaves linear, either flattened or keeled above, always strongly keeled beneath, with a single fibro-vascular bundle and two resin-canals close to the epidermis of the outer angles. Buds of three kinds: (1) terminal on the long branchlets and developing either into long or short shoots; (2) axillary on the long branchlets, scattered, solitary in the axils of the leaves, and developing occasionally into long shoots, or more commonly producing short shoots with apical tufts of leaves; and (3) apical buds on the short shoots, which usually on developing slightly prolong the short shoot and produce again a tuft of leaves, this process being repeated for several years; or occasionally suddenly elongate into long shoots with solitary leaves, or produce flowers. In this way a complicated and irregular system of branching results, very different from that produced by the regular whorled buds of pines, silver firs, and spruces.

Flowers monœcious, fertilised by the wind, arising solitary on the apices of short shoots of two to six years old. Male flowers always much more numerous than the females, directed downwards; globose, ovoid or oblong; sessile or stalked, surrounded at the base by scales, and composed of numerous stamens with short stalks spirally arranged on a central axis; anthers two-celled, dehiscing longitudinally; connective rounded. Female flowers always erect, subglobose, girt at the base by a bundle of leaves, and consisting of a series of orbicular, stalked, ovular scales, each in the axil of a much longer mucronate, oblong bract. The scales, each bearing two ovules, increase in size, as the flower develops into the fruit, while the bracts do not increase.

Fruit a cone, short-stalked and always erect, composed of concave imbricated woody scales, which are persistent and are either longer or shorter than the bracts; cones ripening at the end of the first season, the scales opening and letting out the seeds, which are distributed by the wind in autumn or in the following spring, the empty cones remaining on the branches for several years. Seeds, two on each scale, with a translucent wing, which remains coalesced with the seed, covering it entirely on the upper side, and extending for some distance along its outer edge.

The genus is confined to the temperate and colder regions of the northern hemisphere, and comprises about fourteen described species. Four of these, which we have not seen either growing wild or in cultivation, will now be briefly alluded to.

Larix Cajanderi, Mayr, *Fremdländ. Wald- u. Parkbäume*, 297, fig. 88 (1906). Discovered by Dr. Cajander in eastern Siberia, where it occurs along the banks of the river Lena from the mouth of the Aldan at 68° N. lat. northwards to 72° N. lat., becoming here a stunted tree only 10 to 20 feet in height. It usually forms mixed woods with the Siberian spruce or *Betula odorata*, assuming in wet soil the same appearance as is presented by *L. americana* in the swamps of Wisconsin; or on unflooded land growing pure to a height of about 70 feet. Judging from the description it is closely allied to, if not a mere variety of, *L. dahurica*. The young branchlets are yellowish brown with scattered hairs, older branchlets becoming ashy grey. The leaves are very long, up to 2 inches in length; and are accompanied on the opening of the bud by a tuft of dense whitish pubescence, which is absent in *L. dahurica*. The cones are small, with about twenty scales, which gape widely when ripe, and are broad and concave on the upper margin.

Larix Principis Rupprechtii, Mayr, *op. cit.* 309, figs. 87, 94, 95 (1906). This species was discovered by Mayr on the Wu Tai mountain in the province of Shansi in northern China; and appears to resemble strongly the European larch, from which it differs in the cone-scales being finely denticulate and glabrous, with bracts short and only visible towards the base of the cone. This species has been introduced into Europe by Mayr, who brought a living plant to Grafrath, near Munich, which is growing there very vigorously.

Larix kamtschatika, Carrière, *Conif.* 279 (1855); *Abies kamtschatika*, Ruprecht, *Beit. Pflanzenkund. Russ. Reich.* ii. 57; *Pinus kamtschatika*, Endlicher, *Conif.* 135 (1847). This species, which occurs in Kamtschatka, is said to differ from *L. dahurica* in having larger cones. It is imperfectly known, and has not been introduced.

Larix chinensis, Beissner, *Mitteil. Deutsch. Dendrol. Gesell.* 1896, p. 68, and 1901, p. 76; and *Nuov. Giorn. Bot. Ital.* iv. 183, t. 5 (1897). A tree, dimensions of which are not stated. Branchlets yellow, glabrous. Leaves up to $1\frac{1}{4}$ inch long, triangular in section, stomatose on the under surface. Cones ovoid-cylindrical, $1\frac{1}{2}$ to 2 inches long; scales numerous, orbicular, entire, coriaceous, furrowed and tomentose on the outer surface, standing horizontally in the opened cones; bracts lanceolate, truncate at the narrowed apex, with a short mucro, extending considerably beyond the upper margin of the scale, and appressed and not recurved in the unripe cone. Seeds about $\frac{1}{6}$ inch in length with a broad wing slightly exceeding the seeds in length.

This species, specimens of which I have recently seen in the Museum at Florence, was discovered at 10,000 feet altitude in the Peling mountains of Shensi in China by Père Giraldi in 1893. Beissner has raised seedlings from seeds sent in

1899, and some of these have been grafted on the common larch and are now growing in the Arnold Arboretum, Massachusetts.

This larch in botanical characters stands nearest to *L. occidentalis*. Occurring at a high elevation in Shensi at about lat. 38°, it should prove perfectly hardy in this country ; but must not be expected to be of much importance as a forest tree.

The remaining species, ten in number, are tolerably well known, and are readily distinguishable by the characters of the cones and flowers. In the absence of cones, the following arrangement will give a good clue to the species :—

A. *Leaves deeply keeled on both surfaces.*
  1. *Larix Lyallii*, Parlatore. Western N. America.
    Young branchlets completely covered with a dense greyish tomentum, which persists in part in the second year.
  2. *Larix Potanini*, Batalin. Western China.
    Young branchlets bright yellow in colour, with a scattered pubescence.

B. *Leaves keeled only on the lower surface, the upper surface being flattened or rounded.*
  * *Young branchlets pubescent.*
  † *Leaves glaucous, bluish, with two conspicuous bands of stomata, each of five lines, on the lower surface.*
  3. *Larix leptolepis*, Endlicher. Japan.
    Branchlets of the second year reddish, with a glaucous tinge. Leaves numerous in the bundle, at least forty, long and slender, arranged in an erect cone-like pencil.
  4. *Larix kurilensis*, Mayr. Kurile Islands.
    Branchlets of the second year shining reddish brown, pubescent, not glaucous. Leaves few in the bundle, often only twenty to thirty, short and very broad, spreading so as to form an open cup around the bud.
  †† *Leaves greenish, with two inconspicuous bands of stomata, each of two to three lines, on the lower surface.*
  5. *Larix Griffithii*, Hooker. Himalayas.
    Branchlets of the second year very stout, dull reddish brown, pubescent. Short shoots broad and fringed above by very large loose reflected pubescent membranous bud-scales.
  6. *Larix occidentalis*, Nuttall. Western N. America.
    Branchlets of the second year slender, light brown, shining, pubescent. Short shoots slender, with narrow inconspicuous fringe of bud-scales.
  7. *Larix sibirica*, Ledebour. Russia, Siberia.
    Branchlets of the second year slender, shining, greyish yellow, glabrous, the long hairs present in the furrows between the pulvini of the first year's shoot having fallen off. Leaves very long and slender, up to 2 inches in length.

**   *Young branchlets glabrous.*

† *Branchlets yellowish grey in colour.*

8. *Larix europæa*, De Candolle.   Europe.

Branchlets of the second year shining, glabrous, yellowish grey.

8A. *Larix sibirica*, Ledebour, *var.*   Russia.

In certain specimens of this species the branchlets are indistinguishable from those of *Larix europæa*, and in the absence of cones only show a difference in the leaves, which are very long and slender in *L. sibirica*.

†† *Branchlets brown in colour.*

9. *Larix americana*, Michaux.   North America.

Young branchlets often glaucous.   Branchlets of the second year shining brown.   Short shoots blackish.   Leaves short, not exceeding $1\frac{1}{4}$ inch in length.

10. *Larix dahurica*, Turczaninow.   Siberia.

Young branchlets never glaucous.   Branchlets of the second year shining brown.   Short shoots blackish.   Leaves long, exceeding $1\frac{1}{4}$ inch.

These two species strongly resemble each other in technical characters, but are readily distinguished, as seen in cultivation in this country, by the appearance of the branchlets, which in *L. dahurica* are vigorous, long, and straight, whereas in *L. americana*, which makes slow growth, they are short, curved, and twisted.

10A. *Larix occidentalis*, Nuttall, *var.*   In glabrous specimens of this species the chestnut-brown coloured short shoots will readily distinguish them from either of the two preceding species.

Mayr says that though the various species of larch seem very different at the first sight, yet that they all have the same biological character, and are all inhabitants of the coldest limits of the forest, whether produced by latitude or altitude, and that when introduced into warmer regions or zones, they lose their economic usefulness through premature fruitfulness or fungoid attacks.   This opinion, though so often expressed in various forms by foresters of continental experience, is not strictly applicable to Great Britain, as the pages of this work will prove; and though the liability to spring frost is greater with the more northern and alpine species, yet in their native countries larches are also subject to frosts during almost every month in the year, and though the young shoots in spring and the unripened wood in autumn are often much injured by frost, yet no trees have a greater power of recovering from injuries produced by climatic influences, provided the soil is suitable; and Mayr truly says that the warmer the climate in which the larch is cultivated the better the soil it requires.   He considers that the timber of all larches is practically of equal value, its quality depending on the slowness at which it is grown, rather than on the species or origin of the parent tree.

LARIX EUROPÆA, Common Larch

Larix europæa, De Candolle[1] in Lamarck, *Fl. Franç.* 3rd ed. iii. 277 (1805); Loudon, *Arb. et Frut. Brit.* iv. 2350 (1838); Willkomm, *Forstliche Flora*, 140 (1887); Mathieu, *Flore Forestière*, 555 (1897); Kent, *Veitch's Man. Coniferæ*, 391 (1900).

Larix decidua, Miller, *Dict.* ed. 8, No. 1 (1768); Kirchner, Loew, u. Schröter, *Lebengesch. Blütenpfl. Mitteleuropas*, 155 (1904).

Larix pyramidalis, Salisbury, *Trans. Linn. Soc.* viii. 314 (1807).

Larix vulgaris, Fischer, *ex* Spach, *Hist. Vég.* xi. 432 (1842).

Larix Larix, Karsten, *Pharm. Med. Bot.* 326 f. 157 (1882).

Pinus Larix, Linnæus, *Sp. Pl.* 1001 (1753).

Pinus læta, Salisbury, *Prod.* 399 (1796).

Abies Larix, Poiret in Lamarck, *Dict.* vi. 511 (1804).

A tree attaining 100 to 150 feet in height[2] and 10 to 15 feet in girth. Bark of young stems and branches smooth and grey; on older stems (twenty years and upwards) fissuring and scaling off in thin irregular plates, exposing the reddish cortex below; at the base of old trunks in the Alps becoming extraordinarily thick, a foot or more. Young branchlets slender, glabrous, greyish yellow, with linear pulvini separated by narrow grooves; in the second and third year shining yellow with more elevated pulvini, at the apices of which are the scars of the fallen solitary leaves; base of the shoot girt by a sheath of the bud-scales of the previous season, within which is visible a ring of pubescence. Short shoots dark brown, with rings of pubescence marking each year's growth. Terminal buds small, globose, resinous, with glabrous scales, the lowermost of which are subulately pointed. Lateral buds hemispherical, glabrous, broadly conical, surrounded at the base by a dense ring of hairs.

Leaves light green, soft in texture; those solitary on the long shoots shorter, broader, and more acuminate than those in the tufts, the latter differing in length, the longest about $1\frac{1}{2}$ inch long, and rounded at the apex; upper surface flat or rounded, with one line of stomata on each side; lower surface deeply keeled, with two to three lines of stomata on each side.

Male flowers sessile, ovoid, $\frac{1}{5}$ to $\frac{2}{5}$ inch long. Pistillate flowers, reddish or occasionally whitish, ovoid, about $\frac{1}{2}$ inch long; bracts, with their mucronate apices pointing upwards and outwards and not reflected or recurved, about $\frac{1}{4}$ inch long, oblong, widest at the base, deeply notched above between two pointed projections; mucro about $\frac{1}{12}$ inch long.

Cones ovoid, with the tips of the bracts slightly exserted, $1\frac{1}{4}$ to $1\frac{1}{2}$ inch long,[3] the terminal scales small and not gaping but closing the rounded or flattened apex of the

---

[1] We adopt the name *Larix europæa*, although it is not the oldest one, because it has been in general use for over a century. According to a note at Kew of Alph. de Candolle the *Flore Française*, 3rd ed., was published in reality in 1805, and not in 1815, as it is printed in the volume at Kew.

[2] Kerner, *Nat. Hist. Plants*, Eng. trans., i. 722 (1898), gives the greatest certified height of the larch as 53.7 metres, equal to 176 feet; and this refers to a tree growing in Silesia, mentioned by Mathieu, *loc. cit.* 556.

[3] In the Museum at Florence there are specimens from Courmeyeur, in the Piedmontese Alps, with cones two inches in length, the largest which I have seen, and remarkable for the dense velvety pubescence of their scales.

cone. Scales in four to five spiral rows, nine to ten scales in each row, about $\frac{1}{2}$ inch broad and long, convex from side to side but flattened longitudinally, with the apex usually retuse, often emarginate or rounded; margin thin, entire, not bevelled, and neither inflexed nor recurved; outer surface light brown pubescent, the pubescence most marked towards the base. Bracts oblong, widest at the base, truncate or rounded at the apex, with a short mucro extending about half-way up the scale. Seeds in shallow depressions on the scale, with wings narrowly divergent and extending almost quite to its upper and outer margin; body of the seed $\frac{1}{6}$ inch long, wing short and broad, widest near the base; seed with wing less than $\frac{1}{2}$ inch long; wing $\frac{1}{6}$ inch broad.

VARIETIES

The flowers of the common larch are occasionally white in colour.[1] This occurs both in the wild state and in cultivated trees, as at Arley Castle.

Various kinds of weeping larch have been found wild or have originated in cultivation: and some have been propagated by grafting. *Var. pendula*, Lawson, is noted by Loudon, who states that there were large trees of this kind at Dunkeld, which had been raised from Tyrolese seed. In this form there is an erect leader, and the branches are spreading or even ascending, the branchlets being very slender, elongated, and quite pendulous. In another form of weeping larch the habit is quite different, as it has no tendency to form an erect leader, the trunk remaining short and often divided near the top into several secondary stems that are bent downwards, as are the branches and branchlets. A remarkable example[2] of the latter form, with extremely long slender pendulous branchlets, was growing in 1888 in Mr. Maurice Young's nursery at Milford. *Var. pendulina*,[3] Regel, *Gartenflora*, 1871, p. 101, does not seem to be essentially different in habit from this. Loudon[4] mentions a remarkable pendulous larch at Henham Hall, Suffolk, which was planted in 1800, and was supported on pillars, the main branches forming a covered alley 80 feet long and 16 feet wide in 1841. I am informed by Mr. Simpson, gardener at Henham Hall, that this tree is now in good health, the tall shrubs which surrounded it having been cleared away on one side some three years ago, since when it has made surprising growth. At three feet from the ground it measures 8 feet 2 inches in girth, and at about eight feet forms an angle, and extends laterally for a great distance, being supported on pillars and cross pieces which form a pergola 140 feet long, 8 feet high, and 10 to 14 feet wide, which is almost completely covered by its branches, and will shortly require extension. In a note at Kew, dated 1882, Sir J. Hooker mentions a weeping European larch at Waterer's nursery, Bagshot, which was 50 feet high and had the habit of *Larix Griffithii*.

---

[1] Referred to in *London Catalogue of Trees*, 43 (1730).
[2] Well figured in *Gard. Chron.* iii. 430, 531, Supplementary Illustration (1888). Var. *pendula*, Lawson, is figured in *Gard. Chron.* ii. 684, fig. 132 (1887).
[3] Cf. Beissner, *Nadelholzkunde*, 325, fig. 89 (1891).
[4] *Gard. Mag.* 1841, p. 353. Another weeping larch is figured in the same journal, 1839, p. 574.

DISTRIBUTION

The most recent account of the distribution of the European larch is by Cieslar,[1] the distinguished Austrian forester, who points out that the tree in the wild state occupies four distinct and separate regions, namely, the Alps, the Silesia-Moravia boundary, Russian Poland, and the Tatra mountains in the Carpathians. Cieslar strongly disputes the commonly accepted view that the larch is everywhere an alpine tree, occurring at high elevations; and holds that the Silesian and Alpine larches are two distinct climatic varieties, differing in habit and mode of growth, in period of vegetation and in the altitude at which they naturally grow. He has not apparently studied the Polish tree, of which I have seen no specimens, nor the Carpathian larch.

In the Alps, the larch is widely distributed, occurring in French territory in Savoy, Provence, and Dauphiné; and in the Maritime Alps it reaches about 44° 30′ N. lat., its most southerly and at the same time its most westerly limit. In Switzerland the larch, while generally found, does not occur in the Jura and in the cantons of Glarus, Schwyz, Upper and Lower Unterwald; it reaches its most northerly point in Switzerland on the Gäbris in Appenzell. Extending eastwards it occurs in Vorarlberg, in the Alps of Bavaria and Salzburg, in the Tyrol and in Carinthia. According to Cieslar it is wild in the provinces of Upper and Lower Austria only south of the Danube, but is found near Vienna as a planted tree. It is absent from lower Styria and nearly the whole of the Karst; and in Carniola does not occur wild south of the Sannthaler Alps; from Idria the southern limit of distribution runs westward into Italy through the Isonzo valley. In Italy the larch is confined strictly to the Alps and is not wild in the Apennines, where it has been occasionally planted with unfavourable results, as the tree, after growing rapidly for twenty years, slackens in growth and becomes decrepit at 40 to 45 years old.[2] Elwes saw it planted in the Sila mountains of Calabria, where it was producing seed at 10 to 15 years old.

In the Alps the larch is certainly an alpine tree, often reaching the timber line in company with *Pinus Cembra* and *Pinus montana*; while lower down, but above the zone of the beech, it is usually met with either pure or in company with the spruce and silver fir. It occurs, mixed with the beech, at low elevations, according to Cieslar in certain valleys of the Tyrol. M. Coaz,[3] Inspector-General of Forests of Switzerland, is of opinion that the forests of pure larch which now exist in the Alps are not natural, but have been produced artificially by cutting the ancient mixed woods. The larch has taken possession of the felled areas and has succeeded well as regards growth; but the pure forests are liable to insect attack and possibly also to disease; so that he thinks that it is necessary to restore artificially the ancient and natural condition of the forest. The highest elevation recorded for the larch is 8200 feet in the Dauphiné. The upper limit in the Central

---

[1] *Waldbauliche Studien über die Lärche*, 4 (1904).   [2] Borzi, *Flora Forestale Italiana*, 25 (1879).

[3] See *Garden and Forest*, 1895, p. 238, for a résumé of M. Coaz's monograph on "Insect Ravages in the Forests of Larch on the High Alps."

Alps varies from 6500 to 8000 feet, and in the Engadine is 7622 feet; on Mont Blanc 7218 feet; at Zermatt 7874 feet; in Northern Switzerland, Salzburg, and the Bavarian Alps, 6400 feet; in the Venetian Alps 6700 feet. The lower limit to which the larch descends in the Alps is 1400 feet at Martigny, 2300 feet at Castasegna, 2000 feet at Chur, 3000 feet in the Bavarian Alps, 2000 to 2300 feet in the South Tyrol, 1300 feet in Lower Carinthia, and 1600 feet in Lower Austria.

The Larch occupies on the boundary between Silesia and Moravia a small area, about 30 German square miles, lying between the rivers Mohra and Oppa, and occupying a zone on the mountains between 1170 and 2840 feet elevation; but only occurring in a very scattered condition above 2600 feet. It grows here in mixture with spruce, silver fir, and beech; and appears to be indifferent to soil, as it is met with on primitive schists, grauwacke, and basalt: it occurs also on all aspects. It is absent from the adjacent high mountain of Altvater, which rises to 4900 feet, and is clothed with spruce and mountain ash. According to Cieslar, the Alpine larch has been unadvisedly introduced into Silesia, and it will be difficult in the future to obtain pure seed of the Silesian variety. Cieslar considers this form to be entirely distinct from the larch of the Alps, as it has a cylindrical stem, with slender branches and twigs which are directed upwards, and form a very narrow slender crown. The Alpine larch has stouter branches and twigs, which are directed horizontally, and form a much more spreading crown of foliage, the stem being much more tapering. Introduced into cultivation at low elevations, the Silesian larch is later to come into leaf, and sheds its leaves earlier in autumn, grows much faster, is less liable to damage from snow, and can, on account of its narrow pyramidal form, be planted much more densely. The Alpine larch will not bear crowding, according to Cieslar, and is an inferior tree for planting in every respect.

In Russian Poland, the larch is mainly met with on the hilly land of Lysa Gora, where it forms large forests on sandy soil between Konskie and Szydlowice, near Samsonow. It also extends over the right bank of the Weichsel into Galicia. According to Vrzozowski, the larch at one time was spread over the governments of Piotrkow and Warsaw, as churches and manor houses built 300 to 500 years ago of larch wood are still standing. The distribution of the larch in ancient times must also have extended considerably to the eastward, as a church built of larch in 1419 is reported to exist at Slucz in the government of Minsk in West Russia. Count Dzieduszycki's forester at Poturzyca, near Sokal, in Galicia, reports that larch occurs there between 600 and 800 feet elevation.

The larch occurs also, but not extensively, in the Tatra mountains, between Hungary and Galicia, where it grows on southwest slopes up to 5200 feet, reaching a somewhat higher altitude than the spruce and not ascending as high as the Cembra pine. Cieslar finds no reliable evidence for the larch being wild in the Carpathians east of the Tatra mountains; and does not credit its occurrence in Transylvania.

Prof. Huffel[1] of Nancy states that the larch occurs, but is very rare, in Roumania, where he saw it in the mountains which separate the valleys of the Ialomitza and Prahova at 6300 to 6600 feet elevation, Here it was growing either in mixture

---

[1] *Forêts de la Roumanie*, 6.

with the spruce or higher than it. In Moldavia he reports it on the Ceahlaù, where it rises on a southern slope to 5550 feet. The larch in Moldavia and Roumania has been considered to be *Larix sibirica*; but Huffel doubts this.

Herr F. Mack, forest administrator at Azuga in Roumania, states[1] that larch is common at Bucecii above the beech region, at from 1300 to 1600 metres, mixed with spruce. It attains 60 to 65 centimetres, or about 2 feet in diameter, and is often clear of branches to a considerable height. The wood is hard, red, and durable, and was used in the construction of the Royal Palace of Sinaia.

INTRODUCTION

There is little doubt that the larch was introduced into England about the beginning of the seventeenth century, as Parkinson, who published his *Paradisus* in 1629, speaks of the tree as rare. Evelyn,[2] writing in 1664, mentions "a tree of good stature not long since to be seen about Chelmsford in Essex," and urged its cultivation as a useful timber tree. The earliest trees in Scotland are supposed to be those at Dunkeld, the history of which is given below; but we have no reliable evidence as to the exact date and locality where it was first planted. Loudon's account is very full and should be consulted. The very useful little book by C. Y. Michie on the larch, published in 1885 by Blackwood, must not be overlooked, as it gives a very good résumé by a practical forester whose experience in Scotland was considerable. A. H.

PROPAGATION

Ever since it was realised by landowners that the larch was the tree which before all others could be looked on as profitable to plant, its propagation has been one of the most important branches of the nurseryman's business, especially in Scotland, where by far the larger part of the trees grown in England are raised; and until the disease spread all over the country, and it became evident that precautions must be taken, which in the palmy days of larch-growing were not considered necessary, the majority of raisers were not very careful as to the source from which their supplies of seed were obtained. It was generally supposed that Scottish seed was best, though in years when it could not be obtained in sufficient quantity foreign seed was used.

So far as I have been able to ascertain from very numerous inquiries, the reason for this idea was, that foreign seed usually germinated more quickly, and that the seedlings were therefore more liable to be killed by severe spring frost just as they were germinating. But as all the old larches in England and Scotland must necessarily have been raised from foreign seed, it seems obvious that though Scottish seedlings may have been most profitable to the nurseryman, yet that unless the seed was gathered from carefully selected trees, they were liable in after-life to show weakness of constitution, and succumb, as they often did, to the attacks of *Peziza Willkommii*.

---

[1] *Zeit. für Forst. und Jagdwesen*, Oct. 1904, p. 644. [2] *Silva*, Hunter's ed., 1776, p. 297.

Another reason has been assigned, with some probability, to the apparently greater liability to disease of larch now than formerly, namely, that the cones are often gathered too early, and exposed to too much heat in the kilns in order to extract them. The cone of the larch does not open of its own accord usually until spring; often in this climate so late that the seedlings make little growth the first year, and the seed cannot be extracted without heat, or by breaking up the cones in a mill, which bruises and destroys many seeds; and in the climate of Scotland they do not often ripen so early or thoroughly as in the drier, colder, and sunnier climate of its native Alps: therefore it seemed to me desirable to make experiments with larch seed from abroad, in order to find out whether there was any real difference in the vigour of foreign and home-grown seedlings; and though my experience in this way now extends over fifteen years, I cannot say that I have solved the question.

On many occasions I have sown seed from Scotland, the French Alps, and the Tyrol, and have found that on my poor calcareous soil, which, though it grows larch very well, is not at all suitable for raising it, a large proportion of these seedlings from all sources either perished in infancy, or grew so slowly in the first two years that they were far inferior to seedlings raised in Scotland on a better soil and climate, and probably on manured land. But many of these weaklings have afterwards grown into robust young trees, and the difference in their liability to suffer from spring frost, which is their greatest enemy, is not sufficiently marked to enable me to form a sound opinion as to which are best.

What I have learnt, however, is that, though seedlings cannot be raised as cheaply or as rapidly at Colesborne as in a Scotch nursery, they are more satisfactory in other ways, because it is better to eliminate the weaklings before they are planted out than to have to replant them afterwards; and I believe that the greater the risk of disease the more careful one must be, not only in the selection of seed, but also in their nursery management. Another point in favour of home raising is that the seedlings are not exposed in their younger stages to the extreme drying of the roots which arises from the careless way in which they are often lifted and packed by nurserymen, and from the long delays in transit on the railway; and, finally, the transplantation in a private nursery is more carefully done, and the roots are better developed and more able to endure the severe check of final transplantation to a soil which is less favourable to their growth than that of the nursery.

Mr. J. P. Robertson, forester to the Duke of Devonshire at Chatsworth, writes me as follows with regard to the comparative hardiness of larch raised from native and foreign seed in 1903 :—

"We have two nurseries, one at an elevation of 900 feet, the other at about 600. In both a large quantity of larch from home seed have been put in this spring, while in each a break of the Tyrolese, 10,000 in number of similar size, has been placed. These last were from two different nurserymen. In both nurseries the home variety has suffered severely from the strong white frosts that we had in Easter week, while the Tyrolese in each case is practically untouched."

But on inquiry in 1905 whether this apparent superiority was still the case, he wrote :—

" I did not require to wait long to see the results reversed, as severe frosts in June of that same year, I think on 21st, 22nd, and 23rd, when we had 9°, 10°, and 8° of frost respectively, nearly put the Tyrolese bed out of existence, while those that had been cut earlier in the season (the home variety) did not suffer to anything like the same extent. I am now so thoroughly convinced in my own mind of the superiority of larch from home or British seed, that I have entirely discarded the Tyrolese."

Though the seed is ripened in ordinary seasons in all parts of the country, and a few self-sown trees may be found on most estates where rabbits are kept down, yet our conditions of soil and climate are so unlike those of the natural larch forests of the Alps, that it is useless to attempt natural reproduction with any economic advantage. The only cases in which I have seen any number of self-sown seedlings in the southern half of England are where a clean felling has been made of the larch, and the ground more or less broken up by hauling out the logs immediately afterwards. Of the seedlings which germinate, so large a proportion are destroyed by frost, drought, or vermin in the first season, that the number remaining is not worth consideration, and their growth is so slow for five or six years that planted trees of half the age will usually be stronger. On sandy land, however, or at high elevations, and especially in Scotland, it may sometimes be worth while to encourage self-sown trees, but I cannot say that I have ever seen even a small area which is either sufficiently or regularly stocked by self-sown larch.[1] In the Alps, on the other hand, where the soil is covered with snow for three months or more, natural regeneration is both easy and regular, and I have, both in the French and Italian Alps, seen the ground covered with larch seedlings, which, taken up as late as May, when just uncovered by the snow, I have brought to England when a few inches high, with success. Indeed, it is wonderful how long seedlings will live if taken up when vegetation is just commencing, and sent by post in small tin boxes, tightly packed with a little damp moss or soil, and such trees are my most agreeable souvenirs of many visits to distant countries.

The manner in which the seed is collected in the French Alps is described to me as follows by M. Surel, Inspector of Forests at Briançon, a district which is celebrated for its larch forests :—" In February, before the season when the cones are ripe, we choose trees of which the cones are still closed, and spread large cloths round their trunks at about 10 feet from their base. When the cones open, the seed falls on the cloths. It is then dried in the sun, or preferably, in order to avoid excessive drying, under an open shed. The collection takes place at a minimum altitude of 5500 feet, where the snow is still frozen, and the drying of the seed by the sun, which in this district is remarkably strong, the thermometer rising in the sun in February to 30° to 32° Centigrade, is therefore carried on under very favourable conditions. Drying by the stove would give deplorable results. If I were obliged to work in a climate where the climatic conditions made our practice impossible, I should use closed rooms, slightly heated, but of which the air was freed from moisture by chloride of calcium."

[1] Prof. Fisher tells me that on old roads, and other places where the soil has been exposed, on the shores of Lake Vyrnwy in Wales, and also on old pit banks in Dean Forest, he has seen numerous self-sown larches spring up.

I was informed by M. Mougin, Conservator of Forests at Chambery, that in the Modane district the cones are collected at the end of November by men who climb the trees, with a long hooked pole, and gather the cones by hand into a bag which they carry. The cones are received twice a week at the drying-place, where they are spread out in an airy shed, and turned over every day to dry them and prevent them from heating. When the fine weather returns, they are spread out on a cemented floor, exposed to the sun, which opens them, after which the seed is collected and cleaned, and put in boxes, which are shaken frequently to prevent the attacks of insects. Sometimes seed can be collected on the snow under the trees in January by shaking the trees. But in no case is a stove used to extract them, as seems to be the usual practice in Scotland.[1]

From Prof. A. Fron,[2] of the Forestry School of Les Barres, I have received valuable information on the germination of larch seed, which I summarise as follows : He considers that the process usually adopted of grinding the cones in a mill is very inferior to either of those which I have described as in vogue in the Alps, because the seeds of good quality which come from the central portion of the cone are mixed with those from its upper and lower ends, which are usually empty or imperfect. In 1905 he made experiments on the germination of larch seed obtained at Modane, which had been extracted by the heat of the sun, and obtained the following result :—

Purity	.	.	.	.	.	98 per cent.
Germinative power	.	.	.	61.3	,,
Cultural value	.	.	.	.	60.1	,,

whilst the average of the seeds obtained from seedsmen only gave the following result :—

Purity	.	.	.	.	80 to 85 per cent.
Germinative power	.	.	.	45 ,, 50	,,
Cultural value	.	.	.	40	,,

I may say that the seeds I have gathered from my own trees late in March, and extracted by exposing them to the sun under glass in a garden frame, have germinated quicker and grown better than any which I have purchased.

An ideal way of raising larch would be as follows : To gather cones in the month of March or April from the best and healthiest mature trees in one's own district, or, failing this, from trees known to be healthy on a similar soil; or to purchase seed of known origin direct from a reliable firm abroad, among whom I can highly recommend Messrs. Vilmorin of Paris and Messrs. Jenewein of Innsbruck.

The seed-bed should be in an elevated position, where spring frosts are not likely to be severe, and sheltered as much as possible from the morning sun by trees

[1] Prof. Fisher tells me that in Germany larch seed is extracted from the cones by a toothed axis rotating in a drum, also lined with shorter teeth, and driven by water or steam power.

[2] For further particulars concerning the purity and germinative power of larch seed from different sources, cf. Fron, *Analyse et Contrôle des Semences Forestières*, 92 (1906).

or by clipped hedges or high walls. The soil, light and friable, and, if naturally poor, enriched by manuring with leaf mould and road scrapings, and as free from weed seeds as possible, should be laid up into beds 4 feet wide, with perfect drainage, in the previous autumn, so as to have a fine mould on the surface.

About the middle of April, but earlier or later according to the climate, the seed should be steeped in warm water for a day or two and dusted with red lead when damp, in order to keep mice and birds from attacking it.[1]

On a dry day the seed should be sown broadcast as evenly as possible, and thick enough to have about one plant to the square inch, or less if the seedlings are to remain two years before transplanting. If the soil becomes dry the beds must be watered and shaded as the seed begins to germinate, which should be in fourteen to twenty days after sowing. At this time great care must be taken to prevent the seedlings from damping off, and it is better to keep the bed rather dry than wet. If weeds appear they must be carefully pulled up when quite small. If the soil and season are favourable the seedlings will in the first season be 4 to 6 inches high. If they should be too thick to stand a second season in the seed-bed, the strongest should be lifted and pricked off in lines about the end of March in the second year, and if there is much risk of a severe frost it is wise to transplant them all, as this check will retard their too early growth.

After transplanting they should remain two years in the nursery lines, except in the case of strong one-year seedlings, which may be fit to plant out one year after transplanting, but this must depend on the soil and the nature of the ground where they are to be planted out. Except for planting in woods or in places overgrown with coarse grass or fern, seedlings of one year old plus two years transplanted, or two years old plus one year transplanted, are, in my experience, large enough ; and any which, from overcrowding or other causes, are not then strong enough may be rejected and have another year or two in the nursery. There will always be a considerable proportion of young trees which are inferior to the rest in size and vigour, and these are better separated when transplanting ; whilst all those whose leaders are frosted or immature should be rejected, no more than forty to fifty per cent of the whole being usually fit to plant out at three years old.

The raising of such trees may cost from 20s. to 30s. per thousand in a private nursery, and though they can often be bought cheaper are, in my opinion, worth the extra cost.

Mr. Robertson writes on the same subject as follows :—

" Here we are now, I am glad to say, as little troubled with larch disease as most people, and the reason is simply that we endeavour to keep the plants strong and healthy *at all stages of their growth*, so as to be better able to resist attack. It must not be forgotten in these days of continental forestry that larch is a light-demanding tree, and ought not to be grown on the same principle of density as

---

[1] It is a regular practice in some nurseries where large quantities of larch are raised from seed to soak it for a day or more in water, and then spread it out on a floor where it is daily turned over and sprinkled with water until it seems ready to germinate. By adopting this practice the germination is quicker and more regular.

advocated for shade-bearing species ; but it is equally dangerous to over-thin, and thus bring about starvation, and consequently weakness, by cold winds."

CULTIVATION

Whether the system of notching or pitting is adopted must depend on the local conditions and the size of the trees to be planted. If not more than 18 inches to 2 feet in height, notching, when carefully done, is sometimes as successful as pitting, but in very dry summers a large proportion of the trees will die whichever way the planting is done. In my own experience allowance must be made, in calculating the cost of planting, for a loss of about 20 per cent on an average, though this is often much exceeded when the trees are planted after Christmas, or when their roots have become dry, or when careless workmen have been employed without very close and constant supervision. This is allowing nothing for damage done by hares and rabbits, which, unless thoroughly killed down before planting and kept out by a really effectual wire fence, will soon destroy a great many of the young trees.

Having once planted the trees, the success of the plantation will depend more on soil and climate than on the skill of the planter. For though larch will, owing to their extraordinary vigour, grow almost anywhere up to fifteen to twenty years old, they will not attain a large size unless the soil is moderately fertile and well drained and the situation open and airy. If large trees are desired, I should always advise a mixture of beech or birch being planted with them or three to four years later ; but where the crop can be profitably realised as small poles, or where the soil and climate are really favourable for larch, they may be planted at four feet apart without mixture. The distance apart and the mixture of other trees can only be decided by local experience, the object in view being to keep the trees thick enough to suppress the grass without depriving them of enough light and air to keep their lateral branches alive until they are fifteen to twenty years old. All thinnings should be based on these considerations, and the poorer the soil the more distance is required between the trees to keep them growing. On my own soil I have repeatedly noticed that if grass already exists when the trees are planted, it is impossible to keep the larch thick enough to smother the grass, without crowding each other to the point of starvation and disease, and in such land a mixture of beech, at the rate of one beech to two or three larch, is essential. The result of this mixture is that the larch, instead of beginning to decay at forty to sixty years old, as it often does when on soil deficient in natural fertility, at which period it may be worth 5s. to 15s. per tree, will live and increase in girth till at least 100 years, when they may be worth from £1 to £3 or £4 each. After they have been cut the beech may remain, or if not thick enough to stand with advantage, the land will be left in a very much better condition for replanting than after a crop of pure larch. [1]

---

[1] Prof. H. M. Ward gave in *Nature*, xxxvii. 207 (1887), the following account of an experiment conducted by Prof. Hartig :—"There is a plantation of larches at Freising, near Munich, with young beeches growing under the shade of the larches. The latter are seventy years old, and are excellent trees in every way. About twenty years

With regard to the probable profit arising from a crop of larch planted pure, and realised at 30 to 50 years as compared with a crop mixed with hardwood and realised at 80 to 100 years, I have, with the assistance of Sir Hugh Beevor and Dr. Schlich, made several calculations, but it depends so much on local conditions, on the price realised for thinnings, and on other circumstances which cannot be foreseen, that it seems impossible to estimate it with any certainty.

I have, however, arrived at the conclusion that the short rotation is, as a general rule, the more profitable, especially where a sporting rent varying from 2s. 6d. to 5s. per acre can be realised from pure larch plantations after the age of 15 to 20 years, when rabbits can be admitted freely without risk of serious damage, or where, as in many parts of Scotland, larch plantations are thrown open to sheep grazing.

What is undoubtedly the best system of forestry is not always the most profitable to the landowner, and every one must decide from his own experience which system will suit his own circumstances best.

When mixed with Spruce or Scots or Corsican pine, as is often done, the larch on suitable soil will usually far exceed the other conifers in value at the same age; and I see no advantage, but rather a loss in such a mixture.

In woods which have been treated as coppice-with-standards the larch is a more profitable tree than beech or oak, and may be introduced to the number of thirty to forty per acre immediately after each cutting of the coppice. If left till sixty to eighty years old there would thus be eventually about 100 trees per acre, which will pay much better in these times than the underwood; for if only ten trees, worth say 30s. each, be taken at each rotation, the value will amount to £15 per acre, and there are not many districts in England where underwood is now worth half as much. In Earl Bathurst's extensive woodlands near Cirencester this system has been adopted for many years with great success; but if rabbits exist it is necessary to protect each tree by a wire cage until it is old enough to be safe from their attacks, which it is in this district after twenty to thirty years of age.

The produce per acre of larch in plantations on really good land has in many instances been surprising, and so profitable to the owner that some writers have greatly exaggerated the average returns that may be expected. Prof. Charles E. Curtis,[1] assuming that 300 trees per acre may be grown to maturity, which I greatly doubt, states as a reasonable possibility of production for the larch, no less than 10,000 to 12,000 feet per acre, and says that it will be found possible to bring 1000 to 1200 poles per acre to a useful and profitable size in thirty to forty years. I have

ago these larches were deteriorating seriously, and were subsequently underplanted with beech, as foresters say, *i.e.* beech plants were introduced under the shade of the larches. The recovery of the latter is remarkable, and dates from the period when the underplanting was made. The explanation is based on the observation that the fallen beech-leaves keep the soil covered, and protect it from being warmed too early in the spring by the heat of the sun's rays. This delays the spring growth of the larches; their cambium is not awakened into renewed activity until three weeks or a month later than was previously the case, and hence they are not severely tried by the spring frosts, and the cambium is vigorously and continuously active from the first. But this is not all. The timber is much improved; the annual rings contain a smaller proportion of soft, light spring wood, and more of the desirable summer and autumn wood consisting of closely-packed, thick-walled elements. The explanation of this is that the spring growth is delayed until the weather and soil are warmer, and the young leaves in full activity; whence the cambium is better nourished from the first, and forms better tracheides throughout its whole active period."

[1] *Journ. Roy. Agr. Soc.* lxiv. 36 (1903).

never seen or heard of such an instance, and on writing to Prof. Curtis he could not tell me of anything at all near it.

The best estimate I have is from Sir Hugh Beevor of a plantation at Petworth belonging to Lord Leconfield. It is growing on a steep hill facing east, on sandy loam overlying sandstone, and at thirty-two years after planting contained about 300 trees or more per acre, averaging 15 feet each, which makes about 4500 feet per acre. At Mailscott Lodge, in High Meadow Wood, Forest of Dean, he saw a small plantation, thirty-four years planted, in which the trees on an area of half an acre numbered 214, averaging 9 feet per tree, equal to about 3800 feet per acre.

The best example on my own property is a plantation at Hilcot, now about fifty-four years old, in which there are 2500 trees on an area of about twenty acres. The trees average about 25 feet on the better parts of the land, and 10 to 15 feet on the worst, or about 18 feet over the whole area, equal to about 2200 feet per acre. There are some beech, wych elm, and other hard woods amongst them, which might make up a total of 3000 cubic feet per acre, and though the larch might stand ten to twenty years longer they are not now making a profitable increment.

Mr. J. E. Hellyar Stooke of Hereford sends me the following particulars of a sale in 1907, of larch sixty to seventy-five years old, growing on a hill 400 to 500 feet high, the soil being stiff clay overlying limestone facing east to south. There was no disease except on some of the smaller branches; the trees were all sound, and would probably have continued to grow for many years.

Lot.	Acreage.			Number of trees per acre.	Number of poles per acre.	Estimated contents per tree.	Estimated contents per pole.	Total estimated contents per acre.	Prices realised for larch on estimated contents.		
									Per cubic foot.	Per acre.	
	a.	r.	p.			cubic feet.	cubic feet.	cubic feet.	pence.	£ s.	d.
1	2	0	5	138.0	32.5	26.00	5.6	3769.6	8.89	139 18	0
2	2	2	18	134.7	54.3	26.27	6.26	3881.0	8.85	143 3	1
3	1	3	20	155.7	24.0	26.00	6.25	4197.8	9.01	157 14	8
	6	2	3								
Average of the above.				142.8	36.9	26.09	6.03	3949.4		146 18	7

These trees were sold standing, by auction, at such a distance from a railway station that the hauliers could only make one journey daily.

At what age it pays best to fell a crop of larch is a question which depends entirely on the growth of the trees and the local value : in some cases thirty to forty years may be the most profitable age, in most fifty to sixty ; and where the trees are planted with a good mixture of beech, and continue to grow well after this period, it may pay to let them stand to 100 years old, beyond which they will seldom if ever continue to make a profitable increase.

I should say that £100 per acre was a very fair average valuation of a clean larch crop at forty to sixty years old, and though it is often exceeded, yet in many more instances I believe that at present prices the return will be less, even when disease has not seriously deteriorated the value of the trees by the scars and cankers which disfigure the trunks.

DISEASES OF THE LARCH

Though it is not within the scope of this work to describe the diseases of trees, yet an exception must be made in the case of the larch, because it is a subject of such vast economic importance that it may truly be said, that the losses of all other trees, from all kinds of diseases, whether induced by climatic causes, by insects, or by fungi, do not collectively approach the loss caused to English landowners by larch disease. In using this term without qualification I mean the disease caused by the fungus usually known as *Peziza Willkommii*, but which is now named by mycologists *Dasyscypha calycina*, and which is perhaps best described in English by the name "Canker," or "Blister." This began to attract attention in this country about 1859, when the Rev. M. T. Berkeley[1] made known its existence in England, and Charles M'Intosh in 1860 wrote a small book on larch disease, though what he described more especially was heart-rot, a very different thing from canker.

Hartig and de Bary were the first to describe the fungus. Prof. Marshall Ward in his *Timber and some of its Diseases*, published in 1889, described it more fully; and since then Mr. Carruthers, Dr. Somerville, and other scientific writers have written largely on the subject. In the *Gardeners' Chronicle*, 1896, are many interesting articles respecting the larch disease by J. S. W., Sir Charles Strickland, A. C. Forbes, and C. Y. Michie; and an excellent paper on it with coloured illustrations, by Mr. Geo. Massee, appeared in the *Journal of the Board of Agriculture* for September 1902.

The most practical observations on the larch disease I know of, are in Mr. A. C. Forbes's excellent work on *English Estate Forestry*, pp. 289-307 (1904). These should be studied carefully by every one who is in any degree interested in the subject. After giving a summary of the more important opinions and facts noticed in connection with this disease, he says—and I entirely agree with him—that the disease is as much the result as the cause of the bad health and unthrifty condition of many plantations throughout the country; and that the temporary debility which is induced by the conditions under which planting is conducted is largely responsible for a great deal of disease. He goes on to say that the practically permanent nature of the blister, when once established, renders the result of this temporary debility a much more serious matter than it otherwise would be. If the return to normal health and growth were accompanied by the disappearance of the disease, little harm would be done, but the existence of a blister, once established, is perpetuated indefinitely, and in most cases only ceases with that of its host, so that the occurrence of a blister on the stem of a young tree is much more serious than it would be on a branch or older stem. Cases commonly occur of the disappearance of

---

[1] *Gard. Chron.* 1859, p. 1015.

the blisters when the trees recover health and vigour; and he mentions a plantation over twenty years old, more or less mixed with beech, on greensand, where a number of old blisters are gradually becoming occluded. That they were genuine blisters is evident from the remains of the Peziza cups still present, and the only possible theory respecting their disappearance must be found in the improved health of their host. I have frequently observed similar cases, both on my own land, where in some of the worst diseased plantations, individual trees, which on account of their greater vigour have taken a lead from the first, remain almost untouched and growing vigorously, when most of the surrounding trees are killed or severely injured; and also in Hertfordshire, where tall slender larch trees growing amongst beech showed at various points, from near the ground up to 50 or 60 feet, signs of repeated attacks, which had neither killed them nor apparently checked their growth materially. Forbes says that one may pick up dead twigs or branches under the largest, finest, and most isolated larches that can be found, and the fructifications of Peziza are invariably present on them. This fact he thinks is sufficient to prove that the mere existence of the fungus does not necessarily lead to diseased trees, using the term diseased in its practical sense.

The year 1879 will long be remembered by all gardeners, farmers, and land-owners in the southern half of England as the most disastrous in its effects on plants, farm crops, and trees generally. There was practically no summer, and the rainfall was so continuous, that in late districts much of the corn never ripened at all, and being followed by two severe winters, the disease spread to a degree which ruined hundreds of acres of young larch on my own estate, and caused a loss which must have amounted to millions of pounds throughout the whole country. Though after bad seasons and in smaller areas there had been disease before, it was generally assumed by planters that larch might be successfully grown on almost any kind of land without mixture, and without any special precautions, and there is little doubt in my mind that a large percentage of the worst cases originated in that season, and may be directly traced to the exceptionally bad climatic conditions which prevailed.

Mr. A. M'Dougal, forester to the Earl of Feversham at Helmsley in Yorkshire, who has charge of something like 10,000 acres of woodland, and, having been brought up on the Duke of Atholl's estate, has had unusual experience of the larch, tells me that in Yorkshire the disease first began to be prevalent about 1862 when two plantations died clean off. Since then it has been very prevalent on thin red loam overlying limestone rock; and this applies as much to localities which have previously been under oak wood, or cultivation, as to those where larch has been replanted after larch. He considers that severe spring frosts, together with low-lying situations and heavy soil, are the conditions which bring on the disease most severely.

Sir W. Thiselton Dyer, who has been good enough to read this article, does not quite agree with me with regard to the disease, which he considers due to physical causes alone, and not influenced by heredity. He says that the fungus is a wound parasite, whose spores can only develop in lesions where the bark is injured either by frost, weight of snow, insect punctures, or otherwise, and that it is usually worst in sheltered hollows, where damp air lies and spring frosts are severely felt, and that on high situations facing north and east the disease rarely causes much injury.

All this I admit in full, but I am also convinced that, as the spores of the fungus are now so generally present everywhere, it is impossible to eradicate it, the only way by which it can be combated is by planting only on soils and in situations which experience has shown enable the tree to grow vigorously, and on poor and dry soils mixing it with hardwoods, the fall of whose leaves enriches the soil and keeps it cool and free from grass.

Heart-Rot in Larch.—Though sometimes confused with Peziza by careless observers, this is a totally different disease. C. M'Intosh[1] describes it very fully, and Hartig refers to it under the name of root-rot. Forbes believes, as I do, that it is the direct result of unsuitable soil, either too wet or too dry. It is most common on very poor limestone, sand, and chalk, but also occurs on clay and gravelly soils. In my experience it is especially noticeable where larch follows larch on soils containing insufficient nourishment, and can only be avoided by not planting larch where it is found to be prevalent. It usually attacks trees of about twenty years old, when they have got over their first period of vigorous growth and have practically exhausted the available sources of nutrition. According to Mr. Simmonds, late Deputy Surveyor of Windsor Forest, larch grown on what is called iron pan in that district gets red rot at the heart and is then said to be "pumped."

Larch Bug.—What is commonly known as larch aphis or larch bug is an insect called *Chermes laricis.* The life-history of this insect is at present somewhat obscure, some continental observers believing that it passes through an intermediate stage of existence on the spruce, as no males have yet been found on the larch, in which case it is evident that the insect cannot spread or become numerous unless spruce exists in the neighbourhood. But this is contested by Dreyfus, and I have observed that in England at any rate it multiplies exceedingly where no spruce are near. The females pass the winter under the bark, and are wingless, oval, of a purplish-black colour, and have a long bristle-like sucker through which they feed on the sap of the leaves. In spring they lay eggs which produce young, which grow rapidly, and are covered later by a whitish woolly down, and when numerous give the trees a whitish appearance. They increase rapidly by successive broods, and seriously weaken the constitution of the trees when young, rendering them especially liable to succumb to the attacks of Peziza, which often accompany and succeed them. Whenever I have seen bad attacks of the bug I have noticed that the Peziza is more than usually destructive, and it seems as though the climatic conditions which favour the one also favour the other. In the autumn the bark of the trees in a badly attacked plantation appeared quite black ; and though this plantation was in a high situation, exposed to the east, and was heavily thinned the year afterwards, the greater part of the trees, which were thirteen years planted from Tyrolese seed, and had been growing vigorously at first, were so sickly on the thinner and drier parts of the land

[1] *The Larch Disease* (1860).

that many will not survive.   In the nursery I have observed that some trees were practically immune, though growing side by side with others whose branches touched them and were covered with the Chermes ; but having marked these trees and watched them after they were planted out, I have not as yet been able to assure myself that this immunity is permanent.   Though I have not found this insect attack Japanese, American, and Siberian larches at Colesborne as severely as the common species, yet I have seen it upon them all, both there and elsewhere. Blandford[1] says that washing the trees in April with a soft soap and paraffin mixture in hot water may prove effective, and suggests other forms of wash ; yet it is evident that such remedies cannot be economically employed in plantations, and I know of no means of preventing the ravages of this insect ; though thin planting, mixing with hardwoods, and the avoidance of thin dry soils and damp shady situations are undoubtedly the best means of avoiding severe injury from this pest, as well as Peziza.

Leaf-Miner of the Larch.[2]—The only other insect that I know of which causes serious injury to larch in this country is a small tineid moth, *Coleophora laricella*. This is extremely prevalent almost every year in some of my own plantations having a south-west aspect, and has been supposed by some authors to be directly connected with the attacks of Peziza, which usually accompany or succeed it. According to Stainton the larva is hatched in the autumn, and at first feeds as a miner inside the leaves, and at the approach of winter retires to the stem of the tree, where it passes the winter without feeding.   In the spring as soon as the leaves appear it begins to work, and frequently becomes so numerous that most of the buds have several leaves injured.   In May it is fully fed, and attaches the case which it has formed for itself from the leaves of the tree to a twig, and appears as a perfect insect in July.   The tree is undoubtedly very much weakened by severe and repeated attacks, which render it more liable to die from Peziza, but as far as I know there is no practicable remedy for it in plantations.

A new enemy to the larch which has recently appeared in the north of England was described in the *Journal of the Board of Agriculture* in 1906, p. 375, and more fully in a paper by Mr. J. Smith Hill.[3]   This is the larva of a sawfly, *Nematus Erichsonii*, Hartig, which was first noticed about 1904 by Mr. Cyril F. Watson, of Cockermouth, and which has done considerable damage in the Lake district of Cumberland by defoliating the larch.   Mr. Gillanders has recently found the larva near Rothbury, and Mr. Forbes in Chopwell woods, but I have not heard of its appearance in the south of England.

I am informed by Mr. R. D. Marshall, of Castlerigg Manor, that he has known periodical visitations of the same insect for several years, and that, owing to the late period of the season at which the larva appears, the trees have not suffered as seriously as they would if attacked earlier.   He states that the plantation alluded to by Mr. Smith Hill first suffered from this cause as much as forty years ago, and has survived the attack in three consecutive years recently.   It was noticed that during these

[1] *Journ. Roy. Hort. Soc.* 1892, p. 170.   [2] Cf. *Gard. Chron.* xxxvi. 181, figs. 70, 71 (1904).
[3] *Quarterly Journal of Forestry*, i. p. 67 (1907).

seasons the plantation was full of small birds, which were apparently feeding on the larvæ.

REMARKABLE TREES

To enumerate all the larches which are remarkable for their size and age in Great Britain would be impossible, as in almost all places of sufficient age or importance this was one of the first exotic conifers to be planted, but it will suffice to say that many still exist in a sound condition which are 150 years or more old and exceed 100 feet in height. The tallest trees I have ever heard of were felled about the year 1890 in a deep valley near Croft Castle, Herefordshire, the seat of Capt. H. Kevill Davies, which I visited in 1904 under the guidance of Mr. Openshaw, who assured me that some trees there were 135 feet long at the point at which the tops were cut off, with a diameter of 6 inches. This was confirmed by the woodman on the estate, H. Prince, who estimated the tops to have been 10 to 15 feet long, making the trees nearly if not quite 150 feet high.[1] The soil is Old Red Sandstone and the situation very sheltered. I have a record of a tree measuring 134 feet by 10 feet 8 inches which grew in Yorkshire on Lord Masham's estate, and at Penrhyn Castle, North Wales, Henry measured a tree 118 feet by 7 feet 10 inches, and I saw another at the same place growing in a low, very wet, almost swampy situation very near the sea among hardwoods which was about 90 feet by 12 feet, and judging from the rings of felled trees lying near it was about 130 years old. This is remarkable from the fact of the conditions of growth being so extremely unlike those which are usually considered natural to and suitable for the larch, and I can only explain them by the fact that the natural drainage was better than it seemed. Certainly I would not expect larch now planted in such situations to escape disease.

At Ombersley Court, Worcestershire, the seat of Lord Sandys, a tree is growing on the lawn in deep red loam, which exceeds in girth any larch that I know of in England. It is no less than 15 feet 7 inches at five feet from the ground, though it falls away rapidly higher up, and is only about 80 feet high, and has very large and wide-spreading branches.

At Stoneleigh Abbey, Warwickshire, the seat of Lord Leigh, there are some very large and picturesque larches, near the park-keeper's house, which look as old as any in England. One of them, measuring 14 feet 8 inches in girth, has a mass of rugged branches, some of which touch the ground, where they seem to have taken root. Another is about 80 feet by 14 feet. In the grounds of Warwick Castle there is a group of seven ancient larches, as well as one in the castle yard whose top curves into a drooping form.

In Gloucestershire there are many fine trees of this species on the Cotswold hills, among which may be mentioned two near the Woodhouse in Earl Bathurst's woods (Plate 98). These are growing on dry and rather shallow soil, overlying Oolite rock, and are over 100 feet high by 11 feet and 12 feet in girth respectively.

---

[1] Mr. T. E. Groom of Hereford writes to me that he measured several of these trees himself, and has a clear recollection that two of them were over 140 feet long as topped for sale, where they would be 5 or 6 inches in diameter. The quarter-girth under bark half-way up was, however, only about 14 inches, which gives their cubic content as about 190 feet.

They seem to have been drawn up by surrounding trees, though now open on one side ; and the fine trees beside them are Lawson Cypress, about fifty years old.   At Sherborne House, the seat of Lord Sherborne, is a fine group of six old larches on the lawn, planted in a circle of very small diameter, which seed freely, and from which I have raised good plants.   They are remarkable for their symmetry and equality rather than for their great size (Plate 99).   At Mickleton Manor I measured in 1903 a very curious larch, 10½ feet in girth, but of no great height, whose branches spread to a distance of over twenty yards from the trunk. Plate 100 shows the tallest larches which I have measured myself in England, growing on a very dry, stony bank, composed of Oolite gravel, at Lyde, near Colesborne.   These have no doubt been drawn up by the surrounding beeches to their great height, which exceeds 120 feet, the tallest, whose top is now dying, was, when measured in 1903, about 125 feet ; but their girth is only 7 to 8 feet. They are remarkable from the fact that a part at least of their roots is under water, and must derive some part of their nourishment from the decaying beech leaves which accumulate there, as the trees higher up the bank are not nearly so large.

The tallest larch mentioned by Loudon in England was at Strathfieldsaye, where one was recorded as being 130 feet high by 3 feet 6 inches in diameter ; but none over 80 feet were reported at the Conifer Conference in 1891.   At Eridge Park, Kent, are some very fine larch trees growing on sandy soil, in what seems a damp situation below sandstone rocks, which average well over 100 feet in height, and one which I measured was 115 feet by only 5 feet 3 inches, a very unusual proportion of height to girth.   Mr. R. Anderson has heard of a tree which was felled near Moorhampton which contained 356 cubic feet as measured over bark on the railway, and trees of over 200 cubic feet were not uncommon near this place. At Savernake House, Wilts, he has measured a tree 12 feet in girth, and tells me that the growth on this estate is sometimes so rapid that eight or nine rings may be found together with an average width of half an inch.

In the north-western counties there are, or have been, many very fine larches. Sir Maurice Bromley Wilson tells me of two on the shores of Windermere, which he• thinks are the largest in the Lake district ; but the best I have seen myself are at Greystoke Castle, the seat of the Howards of Greystoke, where Lady Mabel Howard showed me a tree in a plantation near the castle called John-by-Park, which is believed to have been planted by Charles, eleventh Duke of Norfolk, about 130 years ago, and which measured 11 feet 10 inches at 5 feet from the ground, and contains about 230 cubic feet.   There are also two trees, taller but not so thick as the one at Greystoke, in the sunken garden at Lowther Castle in the same district.

In Wales the larch has been planted as extensively as in England on most of the large estates, and as a rule grows as well as, or better than, in England up to 800 or 1000 feet above the sea.   Among the most remarkable trees are two at Chirk Castle, Denbighshire, the seat of R. Myddelton, Esq., one of which measures 74 feet by 13 feet 5 inches, and has very wide-spreading branches.   The other forks low down and is 12½ feet in girth.   At Maesllwch Castle, Radnorshire, the seat of

W. de Winton, Esq., there is a very fine group of twelve old larches 90 to 100 feet high, the largest of which measured 11 feet 10 inches, 11. feet 1 inch, and 10 feet 6 inches in girth when I saw them in 1906. At Dynevor Castle, in a low-lying damp spot, there is a very fine larch about 100 feet high and 9 feet 10 inches in girth, which may contain as much as 300 feet of timber. At Hafod, in Cardiganshire, the seat of T. J. Waddingham, Esq., there were planted in the year 1800 400,000 larch trees on a surface of 44 acres, for which the then proprietor, J. Jones, Esq., obtained a gold medal from the Society for the Encouragement of Arts.[1] Of these, I am informed by M. D. Barkley, Esq., many still remain, and a section of one which he sent me shows that they have grown to magnificent trees. As a rule, however, the large plantations in Wales are not allowed to stand to any great age, being more valuable when large enough to make pit timber.

In Scotland the number of larches remarkable for their size is so great that it is not easy to make a selection, almost every large estate, especially in the Highlands, having splendid trees of great age. So far as I can learn, the trees on Drummond Hill, near Taymouth Castle, the Perthshire seat of the Marquis of Breadalbane, are actually the largest in Great Britain. I visited this place in April 1904 and carefully measured the best trees myself. They are growing on the slope of a hill facing south in good open loamy soil, overlying rock, from which, in some places, springs of water rise; and seem to owe their immense size in part to the fact of their having been mixed with beech and oak, which were planted at or about the same time, and which they have far surpassed in height. The finest tree is figured in Plate 101, and is about 115 feet in height by 17 in girth. It carries its bulk very well up to at least fifty feet, where some large branches go off, and contains, according to Mr. Peter Mackay, the forester, over 500 feet of timber. I estimated the first length alone at 450 feet, the next at 100 feet, and the top and branches at about 50 feet more, so that this tree must contain nearer 600 than 500 cubic feet. In November 1893 a tree near it on the same hill was blown down, and the butt, which was sold, weighed ten tons on the railway, or about 500 cubic feet, besides which three tons more were cut up on the estate. Near it is a tree (Plate 102) remarkable for being divided at about 20 feet up into four large upright stems, a rare occurrence in this species. It is nearly the same height and girth as the first, and may contain as much timber. A third, as measured by the forester, has a bole of only 6 feet long, girthing at 1 foot from the ground no less than 24 feet, and at 5 feet 17 feet 9 inches; it divides into two huge trunks over 100 feet high. These trees are believed to be from 160 to 180 years old, and were probably planted as early as those at Dunkeld.

The next largest and probably the best known larches in Scotland are the so-called Mother Larches, which stand close to the ruins of the Cathedral at Dunkeld (Plate 103), and which were planted, according to the inscription on a stone slab in the wall close by, in 1738 by James, third Duke of Atholl, who, according to Hunter, obtained them from Mr. Menzies of Culdares, who brought a few small plants from the Tyrol in his portmanteau; but in an account of the larch plantations on the estates of Atholl and Dunkeld, published in the *Transactions of the Highland Society*

---

[1] Michie, *The Larch*, 63 (1885).

(vol. v.), which is largely quoted by Loudon, it is said that Mr. Menzies of Migenny was the introducer, and Walker[1] gives 1727 as the date of their introduction. When measured in 1831, Loudon says that the largest was 100 feet by 10 feet 6 inches at 5 feet from the ground. In 1888, according to the tablet mentioned above, it was 102 feet high, and girthed at 3 feet 17 feet 2 inches, at 5 feet 15 feet 1 inch, at 17 feet 12 feet 10½ inches, at 51 feet 8 feet 8 inches, and at 68 feet 6 feet 1 inch, the estimated contents being .532 cubic feet. When measured by Mr. Keir, forester to the Duke, in 1899 it was 15 feet 6 inches in girth; and I made it in 1904 15 feet 8 inches and 100 feet high; so it is still growing and vigorous, though the smaller tree beside it has lost most of its top and many of its branches.[2] There are many other fine larches on this estate, of which the largest perhaps, on the Kennel Bank at Dunkeld, is 120 to 125 feet high by 11 feet 10 inches in girth, with sound top and clean bole to 50 to 60 feet, containing about 350 feet of timber. Three trees of the same age as the Mother Larches are growing near the Castle at Blair Atholl, but are not nearly as large or well shaped.

At Gordon Castle there are some fine larches, one of which, growing in a plantation called Cotton Hill, exposed to the full blast of the North Sea, is figured on account of its remarkable trunk (Plate 104). An immense limb comes off close to the ground, where the trunk girths 20 feet 6 inches, and at about 5 feet, where the tape is seen in the plate, it is 11 feet in diameter. The branches spread to at least 15 yards on each side and measured 198 feet in circumference, and the tree in April 1904 was covered with fine cones, which were beginning to shed their seed, and from which I have raised some plants.

At Monzie Castle, near Crieff, are some splendid larches of the same age and origin as those at Dunkeld, of which the largest, according to Hunter, was 100 feet by 16 feet 3 inches, and contained about 380 feet of timber when he wrote in 1883. I have not seen these trees myself, but Henry measured the largest in 1904 as 109 feet by 17 feet 4 inches, and describes them as very beautiful trees with immense pendent branches in full health and vigour. Hunter says that John, fourth Duke of Atholl, called "the Planting Duke," because he is said to have planted over 10,000 acres of larch, considered them to be the only rivals to the Mother Larches at Dunkeld, and sent his gardener every year to report on their progress. They are figured by Michie on p. 205.

At Inveraray there are some very fine larches on the level ground near the Castle. The best that I measured was about 110 feet by 11 feet, but there may be taller ones;[3] none approached the silver firs in the same locality in height or girth. They serve to show, however, that the larch will succeed well in a climate as unlike that of its native mountains as it is possible to find in Scotland, provided the soil is good and there is shelter from the west wind.

---

[1] *Economic History of the Hebrides and Highlands*, ii. 214 (1812).

[2] I was told by Mr. Keir in 1906 that the largest tree had lately been struck by lightning and was now quite dead.

[3] In *Old and Remarkable Trees of Scotland*, p. 64, it is stated that a larch at Ben-an, in the parish of Inveraray, was 130 feet high and 10 feet in girth at 3 feet, and others are reported at Glenarbuck, in the county of Dumbarton, and at Auchintorlie, in the same county, 143 feet and 140 feet high, but these latter measurements are not reliable, and have never been confirmed. Mr. Renwick has recently measured the Auchintorlie tree, and finds it only 95 feet high.

In the *Scottish Arb. Soc. Trans.*, viii. 233, J. Hutton states that at Keppoch, in Inverness-shire, there were in 1878, 124 larch trees, said to have been brought home as two-year seedlings by Ranald Macdonald of Keppoch in 1753. They grew on an area of about eight acres, and had an average height of about 90 feet, and were then estimated to contain altogether 18,848 cubic feet of timber. The two largest, close to the banks of the Roy, were 108 feet by 12 feet 2 inches, contents 355 feet, and 86 feet by 14 feet 7 inches, contents 358 feet; and he mentions that upwards of forty similar trees were blown down in 1860, so that the timber on this area would have exceeded 3000 feet per acre. This property now belongs to the Mackintosh of Mackintosh, whose forester, Mr. A. Rose, tells me that at the present time there are only seventy-seven trees left, of which twenty-five are small ones which have suffered from various causes; the remaining fifty-two are fine trees with an average content of 120 feet, making, together with the smaller ones, only 7192 cubic feet in all. The largest now standing, which is about twenty-five yards from the banks of the Spean, is 74 feet by 18 feet 6 inches at 3 feet from the ground, and contains 395 cubic feet. The tallest is 108 feet by 11 feet 2 inches. The two largest in 1878 have both been since cut on account of decay, but the rings counted on the stump were 123 and 131 only, which does not agree with their reputed history.

There are very tall and large larches at Brahan Castle and elsewhere in East Ross, one of which was reported by Mr. Pitcaithley[1] as being 115 feet by 11 feet. Mr. Munro-Ferguson tells me that a very large larch was recently felled on his property at Novar; and his factor, Mr. Meiklejohn, sends me the following measurements:—at 5 feet from the ground 12 feet 8 inches, at 25 feet 10 feet, at 40 feet 9 feet 4 inches. The cubic contents of the trunk were 400 feet, and the branches probably contained 50 more.

The highest elevation which I found recorded for the larch in Scotland is in the Ballochbuie forest, where three larches of great size were reported, in 1860, to be in a sound condition at 117 years old and 1110 feet above the sea.

Michie[2] gives a long account of some fine larches growing in the Paradise at Monymusk, in Aberdeenshire, with details of their measurements; the largest in 1881 was 100 feet by 10 feet 5 inches at 20 feet from the ground, and was supposed to contain 416 cubic feet.

A remarkable instance of the manner in which the roots of the larch may continue to grow after the tree has been cut is described and figured in *Gardeners' Chronicle*[3] from a specimen submitted by the late Mr. Webster, head gardener at Gordon Castle. The figure shows the felled stump, rotten in the centre, and with the new wood surging over the edges of the wound, and also two roots of the foster tree, inosculating by means of various branches with those of the stump.

The larch has been extensively planted in Ireland, and has given, when grown on ordinary soils, excellent results, as it has usually remained free from disease. As an instance of good growth, Mr. Mitchell, land-agent at Doneraile and an experienced forester, told Henry that many trees cut in 1891 in a plantation on the Kilworth

[1] *Trans. Scot. Arb. Soc.* xi. 505.　　[2] *The Larch* (1885).　　[3] *Op. cit.* 31st Aug. 1872, p. 1161.

property in Co. Cork must have been 135 feet in height, as he measured them lying on the ground 120 feet to the small end, where they had been cut off at 6 inches diameter. There are still trees as large growing on the same property. Attempts have been made to plant pure larch on peat-bogs; but even when the bogs have been well-drained and good soil has been added to the pits at the time of planting, the trees have not grown. In such cases a preliminary plantation of Scots pine, or in localities with a mild climate the maritime pine, will prepare the bog for larch, which after a few years can be planted in amongst the pines. The conditions for success in bog-planting are delicate, depending apparently on moderate drainage, as when the bogs are quite dry the trees are starved for want of water, and when they are too wet, trees will hardly grow at all. Mr. Richards, forester at Penrhyn, who has had great experience, is confident that good larch can be grown on peat-bogs; and isolated trees doing well on peat have been seen by Henry in various parts of Ireland. Experiments with larch and various mixtures of trees that will grow easily on bogs should be attempted. The American larch has never been tried, and possibly might succeed better than the common species, as it is a swamp-loving tree.

The most remarkable old larches in Ireland are at Doneraile Court in Cork, the seat of Lord Castletown. The history of these trees, which were seen by Henry in February 1907, is obscure, but there is a tradition that they were sent in the eighteenth century to Doneraile by the Duke of Atholl. Five trees out of six originally planted now remain, all of peculiar habit, with numerous more or less weeping branches, the lowermost of which spread over the ground to a great distance, and in one tree are layering. This tree is about 70 feet high, and is 12 feet 7 inches in girth at 5 feet from the ground, the base of the tree below 4 feet being much swollen and covered with very thick bark, like that of old trees in the Alps. On one side the branches spread to 70 feet distance, and on the other side, where there was less room on account of other trees, to 30 feet. Another tree, 10 feet 10 inches in girth, has a spread of 91 feet in diameter. None of them attain more than a moderate height, which is difficult to explain, as ordinary larch grows very tall in the neighbourhood. From the seed of the old trees, sown in 1890, plants were raised, which were put out in 1893 on a hillside, seven acres in extent, and with good soil. This small plantation is now remarkably healthy, though the trees are very dense on the ground, and, at seventeen years old from seed, they average 37 feet in height and 20 inches in girth.

At Carton Park, the seat of the Duke of Leinster, there is a curious tree with the trunk inclined and pendulous branches, which was in 1903 60 feet high and 9 feet in girth. It is considered to be one of the original importations from Scotland in the 18th century. A fine tree in the same place with a straight stem measured 98 feet by $10\frac{1}{2}$ feet. At Abbeyleix House, the seat of Viscount de Vesci, a tree is growing on the lawn similar to those at Doneraile in having weeping branches, some of which are layering. At Dartrey Castle, Co. Monaghan, the seat of the Earl of Dartrey, there are three very old trees, also with more or less pendent branches, which were in 1903 13 feet 10 inches, 13 feet 8 inches, and 11 feet 7 inches in girth respectively. At Emo Park, Queen's

County, the seat of the Earl of Portarlington, there are about twenty fine trees in the pleasure ground, one of which measured in 1907 105 feet by 7 feet 9 inches, another being 92 by 10 feet.

LARCH IN THE ALPS

In its native home the larch loves a dry cold winter climate, where the snow lies from December to April or May, and at the higher elevations does not begin to vegetate before the end of the latter month. It is not very particular as to the geological character of the soil provided that the rock is sufficiently disintegrated for the roots to penetrate and there is a fair amount of soil in which the seeds can germinate, and as a rule natural reproduction is fairly regular and abundant. It is not often allowed to attain its full age, which may be 150 to 300 years or more, on account of the value of its timber for building and other purposes.

As to the size it attains in its native home I have few exact particulars. The largest that I have measured myself was near Modane, in the forest de Villarodin, at 4500 feet elevation, growing on schist with a north aspect. This tree, said to be the largest in the district, was about 90 feet high by 16 feet in girth, but tapered rapidly, and would not contain more than about 200 feet of timber.

By far the finest specimen of the larch in the Alps is figured in Plate 105, made from a negative which was very kindly lent me by M. Coaz, Chief Forest Inspector of the Swiss Forest Department, and which is described in *Les Arbres de la Suisse*[1] as follows :—

"The larch of Blitzlingen grows opposite the little village of this name in the district of Conches in the upper Valais at an elevation of 1350 metres. At the foot of a slope facing north-west, on a narrow terrace this tree grows in a deep and fresh loam, rich in humus, and overlying gneiss rock. There it has become one of the largest in Switzerland, and measures at its base 8 metres 70 cent., and at $1\frac{1}{2}$ metre is still $7\frac{1}{2}$ metres in girth. Its branches extend 10 metres from the trunk. Its top is dead, and thus it is only 29 metres high. Strongly attacked by decay, its trunk does not allow its age to be exactly determined, but no one can accuse us of exaggeration if we estimate it at about five centuries."

According to Dr. L. Klein, who gives an excellent account of the larch,[2] it sometimes attains in the Alps an age of 600 to 700 years. Some stumps which he saw in the so-called Park of Saas-Fee, in the canton of Valais, showed that number of rings, but these trees did not exceed from 1 to $1\frac{1}{2}$ metre in diameter. Dr. Klein counted on a sawn stump near the Findelen glacier 417 annual rings in a diameter of 85 centimetres. He gives several excellent illustrations of Alpine larches taken near the Riffel Alp, one of which shows a tree forking close to the ground into four stems, and another a so-called Candelabra larch with branches rising parallel to the main stem.

---

[1] Schmid u. Francke, *Baum Album der Schweiz* (1900).
[2] Karsten u. Schenck, *Vegetationsbilder*, ii. tt. 25-28 (1905).

LARCH IN OTHER COUNTRIES

In Norway, so far as I have seen, the larch does not grow well on the coast, though there are fine trees 70 to 80 feet high at a farm called Kjostad near Trondheim, and in the interior and farther south.   Schübeler tells us that it has been successfully grown as a forest tree, especially at Brandvold, in the Glommen valley, where trees planted in 1803 had attained in 1878, according to Forstmeister Mejdele, from 70 to 95 feet high, the largest having a diameter of 14 inches at 58 feet from the ground. A very large tree said to be 150 years old existed in 1866 near Gothenburg in Sweden.

The larch is one of the few European trees which appears to grow really well in New England.   The following instances of its success are recorded in *Garden and Forest*:—vol. ii. p. 9, an acre of larch planted in 1877 by Mr. T. H. Lawrence of Falmouth, Mass., on gravelly soil, in an exposed situation, a mile from the coast, was awarded a prize in 1888, when the trees formed a regular and complete cover on the ground, and many of them were over 25 feet high; vol. iv. p. 538, records the success of a plantation made by Mr. J. Russell at East Greenwich, Rhode Island, with 100 small seedlings costing one dollar, which were planted in 1879, and in 1891 were 20 to 27 feet high.   Here the larch has been planted alternately with the native *Pinus Strobus*, to which they form an excellent nurse.   In 1896 Sargent (vol. ix. p. 491) speaks of it as a tree likely to produce valuable timber in the northern states; but in Virginia, on the lower Chesapeake river, the climate is too wet and hot for it, and the trees did not thrive (vol. i. p. 500).

European larch has been tried in various places in the Himalaya, but not with much success, those at Manáli, in Kulu, being apparently the most successful; in 1881 young trees four years old were 6 feet high.

TIMBER

The value of larch timber for all purposes where durability and strength are required has been so well known for so many years past and is so fully dealt with by Loudon, Michie, "Acorn," and many other writers that I need not say very much about it.   There is no home-grown timber so generally used on estates for building and fencing, and though its price has fallen considerably of late years on account of the increasing competition of foreign timber, it is likely to remain in demand, and is easier to market at all ages than almost any timber except ash.

The only country from which larch timber is at present imported or from which any possible supplies can come in future is the north of Russia, and this at present is not used to any great extent; but shipbuilders, collieries, and railway companies are not buying home-grown larch so freely as they used to do except in districts where it can be procured close at hand.

For long telephone poles, for bridge-building and other purposes where lengths of 50 feet and upwards are required, heavy larch poles exceeding 50 cubic feet fetch prices of from 1s. 2d. to 1s. 4d. a foot standing, and cannot always be procured when wanted. But the greater strength and durability of the red heartwood in trees of great age does not command the increased price which it ought to be worth, and it is often best to keep this for private use and sell the smaller and younger trees, whose timber cannot be expected to last as long. For trees of 30 to 50 cubic feet 1s. per foot and upwards, if not too far from a railway, is about the present price. For trees of 15 to 30 cubic feet 9d. to. 1s. should be realised, and for small thinnings the price fluctuates according to the local demand for fencing, hop-poles, and pit-timber.

On account of the durability of larch wood under water, it is specially adapted for piles, wharves, and groins ; but owing to its propensity to warp and twist and the difficulty of sawing, planing, and jointing it in comparison with most other coniferous woods, it is seldom used for inside work. It makes very handsome panelling, however, if the red heartwood is carefully selected and seasoned, and is preferred to all other woods in its native Alps for building log-houses, which in some cases are known to have remained sound for 400 years.

The Duke of Atholl informs me that the larch used in the construction of the stables at Dunkeld in 1809 appears to be still quite sound ; and I saw at Blair Castle a handsome table 5 feet in diameter made from a transverse section, laid as veneer, of a larch grown on the property, which shows eighty-seven annual rings. In the museum at Innsbruck I saw a very handsome antique chest made from very dark-coloured larch wood, which had been dug out of the ground, akin to bog oak in character ; and the wood is used in conjunction with that of *Pinus Cembra* for making artistic furniture by Messrs. Colli Brothers of Innsbruck.

For ship- and boat-building it was at one time much more used than at present, and knees cut from its roots are at least as strong and durable, if not more so, than oak knees.

The bark, though used to some extent for tanning, is now seldom worth stripping except in the case of large trees felled in the spring, when, if taken off in large slabs, it makes a very durable covering for summer-houses, sheds, and other rustic buildings.

Venice turpentine is a resinous product of the larch formerly much valued in medicine and surgery, and for making varnish, of the production of which Loudon gives ample details ; but like so many similar products, it has gone out of use in this country at least, but is still sold in Venice, where I procured a sample of it. Manna of Briançon is a saccharine exudation from the leaves of the tree in the form of small white opaque grains which formerly had some repute in medicine.

(H. J. E.)

LARIX SIBIRICA, RUSSIAN LARCH

Larix sibirica, Ledebour, *Fl. Alt.* iv. 204 (1833); Willkomm, *Forstliche Flora*, 153 (1887); Kent, *Veitch's Man. Coniferæ*, 402 (1900); Mayr, *Fremdländ. Wald- u. Parkbäume*, 311 (1906).

Larix intermedia, Lawson, *Agric. Man.* 389 (1836); Turczaninow, *Bull. Soc. Nat. Mosc.* xi. 101 (1838).

Larix archangelica, Lawson, *loc. cit.*

Larix europæa, De Candolle, var. *sibirica*, Loudon, *Arb. et Frut. Brit.* iv. 2352 (1838).

Larix rossica, Sabine, *ex* Loudon, *Encyl. Trees*, 1054 (1842); Trautvetter, *Act. Hort. Petrop.* ix. 211 (1884).

Larix altaica, Nelson (Senilis), *Pinaceæ*, 84 (1866).

Larix decidua, Miller, vars. *sibirica* and *rossica*; Regel, *Gartenflora*, xx. 101, t. 684, ff. 1, 2, and 4 (1871).

Pinus intermedia, Fischer, *Scht. Anz. Entdeck. Phys. Chem. Nat. et Techn.* viii. 3. Heft. (1831). (Not Wangenheim.)

Pinus Ledebourii, Endlicher, *Syn. Conif.* 131 (1847).

Abies Ledebourii, Ruprecht,[1] *Beit. Pflanz. Russ. Reich.* ii. 56 (1845).

A tree attaining in Siberia over 100 feet in height and 9 to 12 feet in girth. Bark resembling that of the European larch. Young branchlets slender; in specimens from the Ural mountains and Tobolsk, pubescent with long hairs in the furrows between the pulvini; in specimens from the Altai, glabrous; girt at the base by a sheath of the previous season's bud-scales, within which a ring of pubescence is visible. Branchlets of the second year glabrous, greyish-yellow, shining. Terminal buds broadly conical, resinous, with ciliate scales. Lateral buds hemispherical, dark brown, resinous. Apical buds of the short shoots broadly conical, girt at the base by a dense ring of pubescence. Leaves soft in texture, very long and slender, up to 2 inches in length, narrower than in *L. europæa*, sharp-pointed, agreeing with that species in the arrangement of the stomata, but more deeply keeled on the lower surface. Staminate flowers as in the European larch. Pistillate flowers according to Willkomm, ovoid, pale green. Cones, when unopened, cylindrical, with the terminal scales not gaping and the bracts quite concealed; variable in size, up to $1\frac{1}{2}$ inch long, composed of five spiral rows of scales, five to six scales in each row. Scales convex from side to side and also from the base to the apex, quadrangular, about as long as broad ($\frac{1}{2}$ inch); upper margin rounded or truncate, thin, entire, not bevelled, inflected; outer surface finely striate, covered with a reddish-brown pubescence, which is most marked towards the base of the scale. Bract ovate or oblong with a cuspidate point, extending about one-third the height of the scale. Seeds lying on the scale in shallow depressions, with their wings widely divergent and not extending to its upper and outer margin. Seed $\frac{1}{6}$ inch long; with its wing $\frac{1}{2}$ to $\frac{5}{8}$ inch long; wing about $\frac{1}{5}$ inch in width, broadest about the middle.

This species is amply distinct from *L. europæa*, differing in the long and slender leaves, which appear about ten days earlier in the spring; and in the

---

[1] This name is quoted wrongly as *Larix Ledebourii*, Ruprecht, in *Index Kewensis*, ii. 31, and in Sargent, *Silva N. Amer.* xii. 4.

cones, which have fewer and differently shaped scales and short concealed bracts. In the Siberian larch the scales are convex both laterally and longitudinally, whereas in the European larch they are flattened longitudinally. The seeds, moreover, of the former have longer and differently shaped wings, and do not cover the scales of the cone up to their margin as is the case in the latter.

VARIETIES

In wild specimens both pubescent and glabrous branchlets occur. Cones from a tree, cultivated in the Botanic Garden at St. Petersburg, differ in being narrowly cylindrical, with oblong scales only half the width of wild specimens ; and the bracts are also much narrower. The seeds, however, lie on the scales as in wild specimens ; and the scales have the convex form and inflected upper margin of typical *L. sibirica*.

A supposed variety, *rossica*, occurring in northern Russia, was distinguished by Regel as having small cones ; but as Beissner informs me in a letter, it was subsequently abandoned by Regel, and is now not noticed by Willkomm or by any Russian botanist. Sir C. Wolseley, Bart., vice-consul at Archangel, has kindly sent me excellent fruiting specimens from Archangel, which differ in no respect from the Ural larch.

DISTRIBUTION

The Siberian larch has an extremely wide distribution, occurring in north-eastern Russia and throughout a great part of Siberia.

In European Russia it occurs wild in the governments of Archangel, Vologda, Viatka, Perm, and Orenburg. According to Korshinsky,[1] it grows rather sparingly in the plains of northern Russia, as isolated trees in the pine forests ; whereas on the mountains of the Ural chain and its branches it forms extremely large forests, sometimes pure, and sometimes mixed with pine and spruce. Its exact distribution is differently stated by various Russian authorities. Herder[2] adds to the preceding provinces Ufa, Olonetz, eastern Finland, and the northern parts of Kostroma and of Nijni-Novgorod. Ruprecht[3] states that it commences to grow in the northern part of the government of Olonetz beyond the city of Kargopol, from whence extensive woods of it stretch to the Ness river in the Kanin peninsula. In this peninsula it attains its most northerly point in Europe, on the Arctic circle. Further east its distribution sinks to the southward, and its most northerly point on the Ural range is about 58° latitude.

Its distribution in Siberia is not yet clearly known, as it has been confused with *Larix dahurica*. It would appear to be the species common in middle and southern Siberia west of Lake Baikal, while *Larix dahurica* apparently occupies eastern Siberia and Manchuria, a close ally of it, *Larix Cajanderi*, occurring in the extreme north in the lower part of the valley of the Lena, north of lat. 63°. *Larix sibirica* is reported from Olga Bay in Manchuria, but this requires confirmation ; and it has

---

[1] *Tent. Fl. Rossiæ Orientalis*, 493 (1898).   [2] *Act. Hort. Petrop.* xii. 101 (1892).   [3] *Loc. cit.*

been supposed to occur in Mongolia and north China; but Mayr has recently described the North China larch as a new species—*Larix Principis Rupprechtii.* In Siberia its most northerly limit is lat. 69° on the Yenesei and Kolyma, its southern limit extending from the Ural at lat. 54° to the Altai in lat. 52°.

The Siberian larch was reported by Kanitz[1] to occur as a shrub in upper and middle Moldavia at about 6000 feet elevation. He identified it on the authority of Parlatore in a letter. I have seen no specimens from this locality, and consider the identification very doubtful.[2]   (A. H.)

An excellent account is given by Mayr[3] of a plantation of this tree which was made in 1750-1760 for the Czarina Elizabeth at Raivola on the Russian-Finnish frontier north of St. Petersburg. The seed was procured from Ufa, and the trees have on the better land grown remarkably straight and clean without branches for 20 metres up, and attain 40 metres in height with a diameter of 70 centimetres. The wood of these trees, which was shown at the Paris exhibition of 1900, was of remarkably good quality, and Prof. Mayr recommends this tree strongly for cultivation. But as summer does not commence in Finland until June, and the trees had already turned yellow on September 18th, it is probable that the species is not unlikely to succeed in Great Britain except perhaps in elevated districts in the north and east of Scotland.

On my journey to Siberia in 1897 I saw larches in the Ural mountains near Zlataoust, but only after passing the watershed into Asia, and these were of no great size. In the Altai they first appeared at about 3000 feet, and at 4000 feet they became more numerous and larger, some of them 3 feet to 4 feet in diameter and about 100 feet high, but nearly all were dead at the top, and not yet in full leaf on 7th June. They grow scattered in open forest on the drier hillsides as well as on marshy flats, and where the soil is damper are often mixed with *Picea obovata.*

Farther to the south in the upper valleys of the Katuna and Tchuya the larch became the prevalent tree, and extends to a higher elevation than any other, following the banks of the mountain streams on the Mongolian frontier up to about 7500 feet. At this elevation I saw a grove of young larches from 8 to 15 feet high, and cut one of the smallest to count the rings, of which there were twenty-five in a diameter of only 1½ inch. Some of the old trees were remarkably stunted, only 10 to 12 feet high and 5 feet to 6 feet in girth. In this region the climate is extremely severe, frost and snow occurring even in July. The bark of the tree is used all over the region where it grows for covering the winter huts of the nomad Tartars, which are in shape and construction very like the lodges of the Indians in Montana.

CULTIVATION

It was introduced by the Duke of Atholl in 1806 from Archangel, as stated in the fourth volume of the *Transactions of the Horticultural Society,* p. 416, and

---

[1] *Plant. Romaniæ,* 139 (1881). Cf. also Janka, *Flora Siebenbergens,* xvi. 366 (1866).
[2] See distribution of the European larch.
[3] *Loc. cit.* and *Naturw. Stud. Nordw. Russl. Allg. Forst. u. J.-w.* 1900.

was described as follows :—" The bark quite cinereous, not of a yellowish-brown colour, and not distinctly scarred as in the common larch, but, on the contrary, the vestiges of the scars are scarcely visible ; the leaves come out so soon that they are liable to be injured by spring frosts, and what is remarkable, the female flowers are not produced till some time after those of the European larch appear ; they are like those of *Pinus* (*Larix*) *microcarpa*. Mr. Sabine has a plant of this sort in his garden at North Mimms, which he received under the name of *Larix sibirica* from Messrs. Loddiges, who obtained the seed originally from Professor Pallas, whose *Pinus Larix* it probably is. He contrasts the cinereous bark of his plant with the pale brown colour of the common larch ; it may probably prove to be a distinct species." So far as I can learn no trees of this introduction are now living at Dunkeld.

Large quantities of seed were procured by Messrs. Little and Ballantyne of Carlisle, and raised in their nurseries about eight years ago, but the trees from them have generally been a complete failure owing to the very early bursting of their leaf-buds.

I received in 1902, from the Tula Government, through Professor Fischer de Waldheim, some seed of the Siberian larch, and a few of the seedlings look rather more promising than those from North Russia ; but we are not aware that any fair-sized tree of this species now exists in England.

In December 1902 I received seed of this tree from Herr E. Rodd, which was gathered in the Ouimon valley in the Altai mountains early in September, but he tells me that it is not naturally shed there until spring. This seed germinated, but the plants raised from it are small and unhealthy, and vegetate very early in the spring, so that they seem likely to grow as badly in this climate as the larch from the Ural.

In England, as a forest tree this species seems likely to be worthless, for it opens its leaves so early, and suffers so much from spring frost, that with few exceptions the young trees I have grown are unhealthy, and many have already died, though planted in a very cold and exposed situation.

In the north of Norway I saw it growing at the Government nurseries in Saltdalen in 1903 from Russian seed sown in 1882. Trees only 15 feet high were already bearing cones, but were much healthier and more vigorous than the common larch ; and in the Botanic Garden at Christiania I noticed that though growing at the rate of a foot annually, the leaves were attacked by a Chermes like *C. laricis*.

TIMBER

The tree is common in the north of Russia, where it forms a large part of the forests on the east side of the White Sea ; and in the valley of the Petchora, seems to attain very large dimensions. Seebohm[1] says that Alexievka at the mouth of this river is the shipping port of the Petchora Timber Company, where ships are loaded with larch for Cronstadt. " The larch is felled in the forests 500 or 600 miles up the

---

[1] *Siberia in Europe*, 174 (1886).

river, and roughly squared into logs varying from 2 to 3 feet in diameter. It is floated down in enormous rafts, the logs being bound together with willows and hazel boughs. These rafts are manned by a large crew, many of whom bring their wives with them to cook for the party, sleeping huts are erected on the raft, and it becomes to all intents and purposes a little floating village, which is frequently three months in making the voyage down the river."

This larch is now shipped to London in some quantity for various purposes, and has been considerably used for piles in the Dover harbour works, and elsewhere. Mr. D. J. Morgan of Morgan Gellibrand and Company informs me that it is one of the most durable timbers that can be used, but so hard that when it is being sawn water is poured on the saw to keep it from heating, and this is probably the reason why it is not much used in England. He informs me that all the lighters at Onega were built of larch timbers, which lasted a very long time, and that when an old house at Archangel, which had been built on a foundation of larch logs, was pulled down, they were found to be quite sound after lying on the ground for possibly a hundred years. The experiments which have been made with it in the quays at the Surrey Commercial Docks, where the wood was continually wet and dry, have proved the lasting power of this wood, which, from what I have seen of it, is much closer in the grain than English-grown larch. But Mr. G. Cartwright, engineer of the Grimsby Docks, tells me that though he has no actual personal experience of its use, it is considered inferior to the best English larch, as indeed its lower price would imply, and inferior in strength and durability under water to English oak, greenheart, jarrah, or even to Danzig red fir, and that for constructional purposes he would consider its value less than half that of large oak.

Messrs. Crundall and Company of Dover inform me that Messrs. Pearson and Sons have used a large quantity of larch deals for their block moulds, and for other purposes where much wear and rough usage is entailed, and the wood has given entire satisfaction. I purchased from Messrs. Howard Bros. and Company of London a long clean log of this tree, from north Russia, in order to compare it with that of home-grown larch, and find the wood is very slowly grown, there being fifteen rings in an inch of radius. The heartwood is less red and apparently much less resinous than that of the European larch. My carpenter reports that when free from knots it works as well as some red deal, and he considers it very well suited for the roofs of plant houses. Its present value is from £11 to £13 per standard.

(H. J. E.)

LARIX DAHURICA

Larix dahurica, Turczaninow, *Bull. Soc. Nat. Mosc.* xi. 101 (1838); Trautvetter, *Pl. Imag. Fl. Russ.* 48, t. 32 (1844); Regel, *Gartenflora*, xx. 105, t. 684 (1871); Kent, *Veitch's Man. Coniferæ*, 390 (1900).

Larix pendula, Salisbury,[1] *Trans. Linn. Soc.* viii. 314 (1807); Lawson, *Agric. Man.* 387 (1836); Forbes, *Pinet. Woburnense*, 137, t. 46 (1839).

Larix europæa, De Candolle, var. *dahurica*, Loudon, *Arb. et Frut. Brit.* iv. 2352 (1838).

Larix americana, Michaux, var. *pendula*, Loudon, *op. cit.* 2400.

Pinus pendula, Aiton, *Hort. Kew.* iii. 369 (1789); Lambert, *Pinus*, i. 56, t. 36 (1803).

Pinus dahurica, Fischer, *ex* Turczaninow, *loc. cit.*

Abies pendula, Poiret, *Lamarck's Dict.* vi. 514 (1804).

Abies Gmelini, Ruprecht, *Beit. Pflanz. Russ. Reich.* ii. 56 (1845).

A tree attaining in Saghalien 140 feet to 150 feet in height, but in Siberia usually much smaller. Bark scaling in broad, thin, irregularly quadrangular plates. Young branchlets slender, glabrous, becoming pinkish at the end of the season, shining brown in the second year; older branchlets yellowish grey. Shoots girt at the base by a sheath of the previous season's bud-scales, with no ring of pubescence visible. Short shoots slender, dark brown or blackish, glabrous. Terminal buds globose, glabrous, resinous, with the basal scales subulately pointed. Lateral buds hemispherical, resinous, dark brown, glabrous. Apical buds broadly conical and surrounded by a ring of brown pubescence. Leaves light green, similar to those of *L. europæa* in size and arrangement of the stomata, with the tips usually blunter than in that species.

Staminate flowers sessile, smaller than those of the European larch. Pistillate flowers ovoid, red, with the bracts and scales more closely appressed than in the common larch, making the flower narrower and shorter; bracts slightly recurved, $\frac{1}{5}$ inch long, oblong, with a shallow notch at the upper margin between two pointed projections; mucro short, less than $\frac{1}{12}$ inch long.

Cones variable in size, dependent upon the number of the scales, $\frac{3}{4}$ to $1\frac{1}{4}$ inch long, cylindrical, slightly narrowed at the apex, where the scales gape open in the ripe cone, composed of three to four spiral rows of scales, six to eight in each row, bracts concealed. Scales longer than broad, about $\frac{1}{2}$ inch long; upper margin rounded, truncate, or slightly emarginate, bevelled, slightly denticulate, not recurved; outer surface glabrous,[2] channelled, shining light brown when ripe. Bracts not exserted, about $\frac{1}{5}$ inch long, much shorter than the scales. Seeds lying upon the scale in slight depressions, their wings narrowly divergent and not extending quite to its upper margin. Seed about $\frac{1}{8}$ inch long; together with its wing scarcely $\frac{1}{2}$ inch long; wing broadest just above the seed.

The Dahurian larch is a native of eastern Siberia, Manchuria, Corea, and

[1] Though this is the oldest correct name under the genus, I have not adopted it, as it has been erroneously applied to the American larch, and its use now would cause considerable confusion.

[2] Cultivated specimens, as those from Boynton and Murthly Castle, occasionally have slightly pubescent scales; but the cones and seeds in all other respects are typical of *L. dahurica*.

Saghalien. According to Herder[1] it occurs in the northern Ural range at lat. 68°, and at Nijni Kolymsk in north-eastern Siberia at the same latitude; but it is probable that in the former locality he may be referring to *Larix sibirica*, and in the latter case to the form now distinguished by Mayr as *Larix Cajanderi*. It is uncertain whether the larch which occurs in Kamtchatka is *L. dahurica* or a distinct species.[2]

Larix dahurica is very plentiful on the Stanovoi mountains, and along the southern half of the coast of the sea of Ochotsk. Middendorff found it on the Aldan mountains up to 4000 feet elevation. According to Komarov[3] it forms woods in moist situations in the mountain valleys throughout the Amur, Ussuri, S. Ussuri, and Kirin provinces of Manchuria and in northern Corea. Korshinksy[4] states that it is frequent in the whole Amur region, forming forests in the mountains of the upper Amur and Bureja, but that it does not occur in the plain between the Zeja and Bureja.

It occurs in Saghalien, in the northern half of which it grows mixed with common birch and attains a great size, a fallen tree in the forest having been measured by Hawes[5] as 145 feet in length. Elsewhere it forms part of the coniferous forest of the island, being mixed with *Abies sachalinensis*, *Picea ajanensis*, and *Picea Glehnii*. It also occurs on the island of Shintar.

Elwes saw at Wellesley, Mass., a young larch raised in the Arnold Arboretum from seed received at Petersburg as *L. dahurica*, which had a peculiar growth of the branches, which, according to Prof. Sargent, is seen in all the trees of the same origin. At the commencement of each season's growth the new wood made a distinct angle, turning upwards a little, so that in four years' growth it became erect. Prof. Sargent states that he saw many larches in eastern Siberia which he considered to be *L. dahurica*, and that they all had the same habit. The young trees at Boston have not yet borne cones, but the main stems were making annual growths about 2 feet long, and the tree seemed more at home in that climate than in England.

History

Pinus pendula was first described by Aiton in 1789; and Solander's[6] MS., on which the description was founded, states that the tree is a native of Newfoundland, with leaves longer and cones shorter than the European larch. A sheet of specimens preserved in the British Museum bears in Salisbury's handwriting "*Pinus pendula*"; three specimens are unmistakable *L. dahurica*; the fourth, a small branch, is *L. americana*.

Lambert's figure of *P. pendula*, published in 1803, is certainly *L. dahurica*, the drawing being made from specimens obtained from a tree in Collinson's garden at Mill Hill which was planted in 1739, the supposed first introduction of the species. Lambert also figures and describes, as a distinct species, *P. microcarpa*, identical

---

[1] *Act. Hort. Petrop.* xii. 98 (1892).   [2] *Larix kamtschatica*, Carr.   [3] *Floræ Manshuriæ*, i. 190 (1901).
[4] *Act. Hort. Petrop.* xii. 424 (1892).   [5] *Uttermost East*, 105 (1903).
[6] According to Loudon, *op. cit.* 2401, Solander's description was taken from the tree at Mill Hill, which, according to Lambert's figure, must have been *L. dahurica*.

with *L. americana.* He states that cones of both species were sent annually from America to Loddiges, *P. pendula* under the name of black larch, and *P. microcarpa* as red larch; and that both kinds were growing in Loddiges's nursery.

Lawson's *Manual*, published in 1836, gives a careful description of both species, and repeats the information that they are natives of North America.

So far as we know *Larix dahurica* does not grow in N. America; and no traveller or botanist except Pursh ever claimed to have seen in the eastern part of the continent any species but *L. americana.* Pursh[1] asserts that *L. pendula* and *L. microcarpa* are distinct species, and were seen by him, the former growing in low cedar swamps from Canada to Jersey, the latter occurring about Hudson's Bay and on the high mountains of New York and Pennsylvania. As *L. americana* varies in the size of the cone, it seems certain that Pursh only saw forms of *L. americana.* It is very difficult to understand how seeds of *L. dahurica* from eastern Siberia could have been introduced so early.

Until about 1840 the American origin of *L. pendula* was unquestioned; and a tree planted in that year at Bayfordbury, and recorded in the planting book as *L. pendula*, is still living, and is undoubtedly *L. dahurica.* *Larix dahurica* was noticed first in Lawson's *Manual* as a stunted bushy tree, growing poorly, as it was propagated from cuttings or layers; and is stated to have been introduced in 1827. (A. H.)

REMARKABLE TREES

The finest specimen we know is figured in Plate 106, and is growing on the edge of a grassy drive at Woburn Abbey, where I first noticed its peculiar bark on the occasion of the visit of the Scottish Arboricultural Society to that place in July 1903. None of the members present could name the tree, and on comparing the foliage with the specimens at Kew I came to the conclusion that it must be a tree which is mentioned in the *Pinetum Woburnense* as *Larix pendula.* I went to Woburn again on purpose to see it in flower, on 31st March 1905, when the difference in the flowers from those of a pendulous form of the common larch growing close by was evident. But the less rugged bark, which resembles that of a cedar, is the best distinction, and is clearly shown in our illustration. It measured 86 feet high by 6 feet 7 inches in girth in 1905. I have raised a seedling from this tree.

A very similar tree is growing by the side of the entrance drive at Beauport, which from its bark and habit we believe to be of the same origin.

At Bayfordbury the tree planted in 1840 as *Larix pendula* is now 56 feet high and 5 feet in girth, with a conical stem, and bark scaling in large thin plates. European larches planted near it at the same time are 70 feet high and 5 to 6½ in girth. A tree at Denbies, near Dorking, the seat of Lord Ashcombe, was in 1903 40 feet high and 2 feet in girth. It is said to have been sent to Denbies as *Larix Griffithii* by Sir Joseph Hooker, but some mistake had evidently been made in the plant that was forwarded from Kew some forty years ago.

---

[1] *Fl. Amer. Sept.* ii. 645 (1814).

In the Cambridge Botanic Garden there are two trees of this species, one 56 feet high by 5 feet in girth, in 1906.   The bark scales off in smaller plates than the common larch, and shows more red-coloured cortex below.   The second tree, labelled *L. pendula*, is grafted at 6 feet up on the common larch, and has its stem bent over at a right angle a few feet higher up.

At Ribston Park, Yorkshire, there is a well-grown tree of *L. dahurica* which cannot be more than about forty years old, as Major Dent remembers its being planted, though its origin is unknown.   It has somewhat pendulous branches and smooth bark without ridges, and measures 71 feet by 5 feet 2 inches.   It had both new and old cones on it in 1906.

There are some larches at Boynton, near Bridlington, Yorkshire, which Sir Charles Strickland has always known as red larches, and supposed to have been of American origin, but which I believe, on account of their smoother bark, to be *L. dahurica*.   The best of them is 75 feet by 7 feet 8 inches ; another, with a very spreading top, was 9 feet 4 inches in girth ; and both had cones from which seedlings have been raised.   Sir Charles Strickland has written of these in the *Gardeners' Chronicle*, 1896, pp. 399 and 494.   He says that the trees which have been grown at Boynton for eighty or ninety years under the name of red and black larch are the two trees described in Loudon as varieties of *Larix americana* ; and that the red larch is more like the European larch, and in loose, rather wet, sandy soil grows at Boynton as fast and to as large a size, but he does not consider the wood quite as good as that of the common larch ; it is more liable to twist and warp, though probably as durable.   On drier soils the red larch is much less healthy and vigorous than the common one.

At Murthly Castle there is a row of fifteen trees which were planted about 1881 by Mr. D. F. Mackenzie, who informs me that they were probably from the nursery of Messrs. B. Reid of Aberdeen, but their origin cannot now be traced with certainty.   Their habit varies very much, the first one, coming from the Castle, having very pendulous branches and a weeping top, which none of the others possess.   The cones also vary somewhat in size and colour, but with one exception— which I believe to be a common larch planted subsequently to replace a dead tree of the original lot—are characteristic of *L. dahurica*.   The trees average 40 to 45 feet high and 3 to 4 feet in girth, and have the bark distinctly smoother and less corrugated than the bark of common larch growing under similar conditions. They are fairly healthy in appearance, with no evidence of having suffered from Peziza, but are bearing cones so freely that I do not expect they will become large trees.   Mr. Mackenzie attributes this to their growing on dry, gravelly soil.

(H. J. E.)

LARIX KURILENSIS

Larix kurilensis, Mayr, *Abiet. Jap. Reiches*, 66, t. 5, f. 15 (1890), and *Fremdländ. Wald- u. Parkbäume*, 300 (1906).
Larix dahurica, Turczaninow, var. *japonica*, Maximowicz, in Regel, *Rev. Sp. Gen. Larix*, p. 59, and *Gartenflora*, xx. 105, t. 685 (1871); Miyabe, *Mem. Boston Soc. Nat. Hist.* iv. 261 (1890).

A tree, attaining in the Kurile Islands a height of 70 feet and a girth of 7 to 8 feet. Bark, according to Mayr, scarcely distinguishable from that of the Japanese larch. Young branchlets covered with a moderately dense, wavy, irregular pubescence. Branchlets of the second year shining reddish brown, pubescent. Base of the shoot girt by a ring of the previous season's bud-scales, the uppermost of which are loose and reflected, no ring of pubescence being visible; short shoots dark red, or almost black, shining. Terminal buds dark red, ovoid, with comparatively few scales, which are acuminate, non-resinous, ciliate with brown silky hairs. Lateral buds ovoid, dark red, with ciliate scales. Apical buds of the short shoots hemispherical, dark red, with no ring of pubescence at the base.

Leaves glaucous, short, broad, and curved, about an inch long, rounded at the apex, few in a bundle, usually twenty to thirty, spreading so as to form a wide open cup around the bud; upper surface flattened, green without stomata; lower surface deeply keeled, with two bands of stomata, each of five lines.

Flowers not seen. Cones small, cylindrical, about $\frac{3}{4}$ inch long, composed of few scales, less than twenty, with the bracts conspicuous at the base of the cone, but concealed elsewhere by the upper scales. Scales oval, longer than broad, about $\frac{1}{3}$ inch long; upper margin thin, emarginate, slightly bevelled, not reflected; outer surface minutely pubescent towards the base. Bract panduriform, about half the length of the scale, terminated by a very short mucro. Seeds lying on the scale in two depressions which are separated by a membranous ridge, with the wings slightly divergent and extending up to the margin of the scale. Seed about $\frac{1}{8}$ inch long; seed with wing about $\frac{1}{3}$ inch long; wing broadest just above the seed.

(A. H.)

This tree was first distinguished as a species by Dr. Mayr, the distinguished dendrologist and traveller, who found it in the Kurile Islands, especially on Iturupp,[1] where it forms forests of some extent. Sargent gives an excellent illustration, plate xxvi. in the *Forest Flora of Japan*, from a photograph taken by Dr. Mayr, and I am able to show its aspect in the same island from two photographs kindly given me by the Imperial Japanese Forest Department (Plate 107). The upper shows a forest of larch on Iturupp; the lower a scattered group near the shore on the same island.

The tree was commonly planted in the neighbourhood of Sapporo, and it was introduced into Europe in 1888 by Dr. Mayr, and seems to grow almost as well as the Japanese larch, at least when young. There is a tree 15 feet high at Grafrath,

[1] We adopt this spelling on Dr. Mayr's authority, as the correct Aino name for the island. Eterofu is the Japanese form of the word, and Eterop a corrupt combination of both forms of spelling.

the experimental forestry station near Munich, where the thermometer goes down to 15° Fahr. below zero, and seedlings only four years old are already 5½ feet high. They resembled *Larix americana* more than *L. leptolepis* in the blackish colour of their young shoots. Dr. Mayr says that it is the first larch to become green in Europe, though in my nursery seedlings of the Altai and north Russian larches are both earlier. He says that its dark shoots have gained it the name of black larch from visitors to his nursery, and that in the park of The Duke of Inn- and Knyphausen at Lütetsburg in east Friesland it grows faster than any other species of larch, being 6 metres high at the age of seven years.[1]

So far as our very short experience of this tree in England enables us to judge, it is likely to thrive well, at any rate in its youth. Several young trees which are in my nursery grow fast, and ripen their growths earlier than common larch. Some seed received from Japan in June 1906 germinated very quickly, and made healthy little plants the same season. It should be tried especially in the wetter parts of Great Britain.

(H. J. E.)

LARIX LEPTOLEPIS

Larix leptolepis, Endlicher,[2] *Syn. Conif.* 130 (1847); Gordon, *Pinetum*, 128 (1858); Mayr, *Abiet. Jap. Reiches*, 63, t. 5, f. 14 (1890), and *Fremdländ. Wald- u. Parkbäume*, 302 (1906); Kent, *Veitch's Man. Coniferæ*, 397 (1900).

Larix japonica, Carrière, *Conif.* 272 (1855).

Larix Kaempferi, Sargent, *Silva N. Amer.* xii. 2, adnot. 2 (1898).

Pinus Larix, Thunberg, *Fl. Jap.* 275 (1784) (not Linnæus).

Pinus Kaempferi, Lambert, *Pinus*, ii. preface, p. v (1824).

Abies Kaempferi, Lindley, *Penny Cycl.* i. 34 (1833).

Abies leptolepis, Siebold et Zuccarini, *Fl. Jap.* ii. 12, t. 105 (1842).

Pinus leptolepis, Endlicher, *Syn. Conif.* 130 (1847).

A tree attaining in Japan a height of 100 feet and a girth of 12 feet. Bark of native trees, according to Mayr, similar to that of the European larch, the freshly ex-foliating scales being more brownish than red; but in cultivated trees in England the bark begins to scale very early, peeling off usually in large long strips and giving a red appearance to the trunk. Young branchlets glaucous, usually covered with a dense, erect, brown pubescence, but occasionally almost glabrous, only a few brown hairs being present. Branchlets of the second year reddish with a glaucous tinge, retaining some pubescence or quite glabrous. Base of the shoots girt by a sheath of the previous season's bud-scales, the uppermost of which are loose and reflected, with no ring of pubescence visible. Short shoots stouter than in the common larch,

---

[1] In *Mitt. Deutsche Dendr. Ges.* 1906, p. 27, the age of this tree is stated erroneously as twenty-five to thirty years. Its height in 1906 is given as 9 metres.

[2] *Pinus leptolepis* was the name preferred by Endlicher; but he quotes *Larix leptolepis*, Hort., as a synonym; and as this is the first publication of *Larix leptolepis*, Endlicher is responsible for the name, and it is credited to him; and being the first published name under the correct genus is adopted by us. Moreover, it is the name by which this species is universally known; and the adoption of Sargent's name, *Larix Kaempferi*, would cause great confusion, as this has been used for *Pseudolarix Kaempferi*, the golden larch of China. The Japanese larch, though known to Kaempfer and Thunberg in the eighteenth century and mentioned by Lambert, was first described by Lindley in 1833.

reddish, glabrous. Terminal buds sharply conical, resinous, glabrous, the lowermost scales subulately pointed. Lateral buds ovoid, glabrous, resinous, directed slightly forwards. Apical buds of the short shoots conical, with loose scales, surrounded at the base by a ring of pubescence.

Leaves glaucous, about $1\frac{1}{4}$ inch long, rounded at the apex; upper surface flattened, with two bands of stomata, variable in the number of lines, often two to four in each band on leaves of the long shoots, usually one to two irregular lines on leaves of the short shoots; lower surface deeply keeled, with two conspicuous bands of stomata, each of five lines.

Staminate flowers ovoid, sessile, smaller than in *L. europæa*. Pistillate flowers ovoid, pinkish; bracts all recurved, about $\frac{1}{6}$ inch long, oblong, broadest at the base, truncate, and scarcely emarginate at the apex, brownish with pink margins, mucro about $\frac{1}{20}$ inch long. Cones shortly ovoid, broad in proportion to their length, 1 to $1\frac{1}{4}$ inch long, readily distinguished by the thin reflected upper margins of the scales, of which there are four to five spiral rows of eight to nine in each row. Scales almost orbicular, about $\frac{2}{5}$ inch long and wide; upper margin very thin, reflected, truncate or slightly emarginate; outer surface furrowed, slightly pubescent. Seeds in very shallow depressions on the scale, their wings slightly divergent and extending to its upper margin; seed about $\frac{1}{6}$ inch long, with wing $\frac{2}{5}$ inch long.

A stunted form, growing on the higher parts of Fuji-yama, was collected by John Gould Veitch, and was considered to be a new species by A. Murray;[1] and is recognised as a variety by Sargent. According to Mayr, it scarcely deserves to be ranked as a variety, as it only differs in being a low tree, with smaller cones than usual, which are only $\frac{3}{8}$ inch in diameter and globular in shape.   (A. H.)

INTRODUCTION

It was introduced by J. G. Veitch in 1861 from seeds which he procured during his visit to Japan. Nothing is said by Kent as to the number of plants raised and sent out at that time, but probably the number was small, as we know of few trees as old as forty-five years. Larger importations were made later, and the tree grew so well generally that it is now being planted almost everywhere, and some of the older trees have produced good seed for ten years or more.

DISTRIBUTION

In Japan this larch grows naturally on the slopes of volcanic mountains in a sandy soil at 4000 to 6000 feet elevation, in a climate very much warmer and moister in summer, drier in winter, and less liable to late frosts than England.

---

[1] *Larix japonica*, A. Murray, *Pines and Firs of Japan*, 94 (1863).
Larix leptolepis, var. *minor*, A. Murray, *Proc. Roy. Hort. Soc.* ii. 633, f. 155 (1862).
Larix leptolepis, var. *Murrayana*, Maximowicz, *Ind. Sem. Hort. Petrop.* 1866, p. 3.
Larix japonica, var. *microcarpa*, Carrière, *Conif.* 354 (1867).
Larix Kaempferi, var. *minor*, Sargent, *Silva N. Amer.* xii. 2, adnot. 2 (1898).
Abies leptolepis, Lindley, *Gard. Chron.* 1861, p. 23.

Where I first saw it, on a sandy plain above the Lake Chuzenji on the slopes of the volcano of Nantai-san, the trees were of no great size, averaging perhaps 60 to 70 feet in height, with a girth rarely exceeding 6 feet in mature trees, and more often 3 to 4 feet. They were very similar in habit to the larch in the Alps, and had not an excessive development of branches. Higher up above Yumoto in rich forest soil, thinly scattered among deciduous trees of many species, they were larger, sometimes attaining 80 feet high and 10 to 12 feet in girth; but I saw none anywhere which rivalled our larch in height, and am inclined to think it is not nearly such a long-lived tree, though, as I saw none felled, I was unable to count the rings. Prof. Sargent, who saw the tree in the same place as I did, came to a very similar conclusion. Mayr states that he found it wild on the volcanoes of central Hondo, Fuji, Ontake, Asama, Shiranesan, Norikura, and others, always growing near the timber line, with *Abies*, *Tsuga*, and *Picea hondoensis*.

The tree is valued for its timber, which is used for ship- and boat-building, and has lately come into great demand for railway sleepers and telegraph poles. In consequence of this it has been largely planted at elevations of 4000 to 5000 feet in the central and northern provinces, and many plantations that I saw of ten to fifteen years old were very similar to larch plantations in England in growth and habit. I also saw it planted experimentally in Hokkaido, along the lines of railway, where it seemed to grow as well in this rich black soil as in its native mountains.

CULTIVATION

In 1890 I sowed seeds from three different localities—Dunkeld, Hildenley, and Tortworth—and raised plants from each of them, which grew better than seedlings raised at the same time from Japanese seed; but this may have been partly due to the fact that the latter were dressed with paraffin by my forester to protect them from birds and mice in the seed-bed. At six years old these plants are now from four to eight feet high, and though some of them have been more or less checked by severe spring frosts, they are generally growing well.

As a proof of the hardiness of the tree I may mention that the late Sir R. Menzies showed me three young trees which he had planted, at an elevation of about 1250 feet, in the garden of the inn near the top of the pass between Glen Lyon and Loch Rannoch; and in some of his plantations on the north shore of Loch Rannoch they were growing very vigorously in mixture with Douglas fir.

No conifer of recent introduction has attracted so much attention among foresters as the Japanese larch, which, during the last ten years, has been sown very largely by nurserymen (Messrs. Dickson of Chester are said to have sold no less than 750,000 in the year 1905), and is looked upon by many foresters as likely to replace the common larch, because it is, so far as we yet know, less liable to the attacks of *Peziza Willkommii*. But this pest has already in more than one place been certainly identified on the Japanese larch, and I have little doubt that as time goes on we shall hear more of this. Henry visited in 1904 six plantations of Japanese larch of ages from five to sixteen years, and in none could detect any sign of canker. There were plantations of European larch in every case adjoining those of the Japanese tree, and the former were all badly affected by disease. Henry concluded that the Japanese larch was practically immune from disease, though on his return to Kew he received specimens from estates in Perthshire and Dumfriesshire which were undoubtedly suffering from Peziza.[1] As, with the exception of Prof. Sargent and Dr. Mayr, no one had studied this tree in its native climate, I paid particular attention to it during my visit to Japan in 1904, and, as I have stated[2] elsewhere, came away with the impression that it is not likely to supersede the European larch as a forest tree, and am very doubtful whether it can be expected to become a profitable one, to plant under ordinary conditions. Though when young its growth is extremely rapid and vigorous, it has a great tendency to form spreading branches, and even in the much more favourable soil and climate of Japan, rarely, if ever, attains anything like the dimensions which the European larch does in Great Britain.

Mayr's opinion on the suitability of the tree for economic plantations in Europe is the same as my own, and he considers that though it may grow faster than the European larch for the first twenty years, yet that it will eventually be surpassed if planted under precisely similar conditions. He also agrees with me that though in selected positions and under careful cultivation it may not seem so liable as the European larch to the attacks of Peziza, yet that it is not immune, and the figures which he gives of its growth in Germany show that it has the same tendency to produce spreading branches there as in Great Britain. In a note on this tree by K. Kumé, chief of the Forestry Bureau in Japan, in *Trans. Scot. Arb. Soc.* xx. 28, January 1907, a yield table at various ages is given, which shows that on soils of medium quality in Japan the mean basal diameter at 100 years old is about a foot, the mean height 92 feet, and the stem volume per acre 6330 cubic feet. I will only note that what is meant by land of medium quality in Japan is very superior to what it is in this country. In Germany Mayr says that the seed falls in autumn from the cones, which are busily sought for by squirrels, and that self-sown seed has germinated freely at Grafrath under trees twenty-two years old.

REMARKABLE TREES

There are many specimens now of about 40 feet high in various parts of the country, but of those that I have seen the one figured, which is growing at Tortworth (Plate 108), is perhaps the finest. It measured in 1904, 45 feet by 4 feet 7 inches, and was covered with cones. It is growing on red sandy soil, and Lord Ducie thinks it is one of the earliest introductions. At Hollycombe, Sussex, the seat of J. C. Hawkshaw, Esq., Mr. G. Marshall measured a tree 45 feet by 2 feet 4 inches in 1904. At Hildenley, Yorkshire, there is a fine tree about 40 feet high, which produces good seed. A clump of fine trees is reported[3] to be growing at Bothalhaugh, near Morpeth. There is also a fine specimen at Brook House, Haywards Heath, the residence of Mrs. Stephenson Clarke.

[1] See note by Mr. Massee in *Journ. Board Agriculture*, 501 (1904).
[2] *Trans. Scot. Arb. Soc.* xix. 77 (1906). [3] *Gard. Chron.* xxxix. 282 (1906).

At Dunkeld there is a tree planted close to a common larch, from which seedlings were raised at my suggestion by the late D. Keir, which appear to be hybrids between the two species.[1] His son, who succeeded him as forester to the Duke of Atholl, and who has watched the growth of these seedlings, considers them to be intermediate between the two species ; but it is yet too soon to be certain.

At Abercairney, Perthshire, the seat of Col. Drummond Moray, there is a tree, raised from seed brought from Japan in 1883, which, measured by Henry in 1904, was 38 feet by 3 feet 5 inches. At Blair Drummond, in the same county, he measured ten trees planted in 1888, one of which was 44 feet high, and the average girth 2 feet 5 inches. They were all healthy though growing among common larch which was diseased.

At Cullen House, Banffshire, Mr. Campbell tells me that there is a tree 45 feet by $3\frac{1}{2}$ feet. At Kirkennan, near Dalbeattie, Kircudbrightshire, two larches sown in 1885 were in 1904 41 feet by 2 feet and 35 feet by 1 foot 11 inches. We are indebted for this information to the owner Mr. W. Maxwell.

In Germany at Schloss Lütetsburg, it seems to have grown faster than with us, for it is stated[2] that trees thirty-five to forty years old are 17 to 20 metres high, with a girth at 1 metre of 1.80 to 2.70 metres. (H. J. E.)

LARIX GRIFFITHII, Sikkim Larch

Larix Griffithii, J. D. Hooker, *Ill. Himal. Pl.* t. 21 (excl. ff. 1-4) (1855), *Flora Br. India*, v. 655 (1888), and *Gard. Chron.* xxv. 718, f. 157 (1886); Masters, *Gard. Chron.* xxvi. 464, f. 95 (1886) ; Kent, *Veitch's Man. Coniferæ*, 395 (1900) ; Gamble, *Indian Timbers*, 720 (1902).
Larix Griffithiana, Carrière, *Conif.* 278 (1855).
Abies Griffithiana, Lindley and Gordon, *Journ. Hort. Soc.* v. 214 (1850).
Pinus Griffithii, Parlatore, DC. *Prod.* xvi. 2, p. 411 (1864).

A tree, attaining in the Himalayas about 60 feet in height, with thick brown bark, and wide-spreading, long and pendulous branches.

Young branchlets, reddish, covered with a dense wavy, more or less appressed pubescence, and girt at the base by a sheath of the previous season's bud-scales, the uppermost of which are very broad, loose, membranous, and reflected. Branchlets of the second year very stout, dull reddish brown, pubescent. Short shoots broad and stout, fringed above by very large, loose, reflected, pubescent, membranous bud-scales. Terminal buds broadly conical, non-resinous, with pubescent scales. Lateral buds ovoid, pointing outwards and forwards, non - resinous, pubescent. Apical buds of the short shoots conical, with loose pubescent scales.

Leaves light green in colour, about $1\frac{1}{4}$ inch long, ending in a short rounded point ; upper surface rounded or flat, with one or two broken lines of stomata near the apex ; lower surface deeply keeled with two bands of stomata, each of three

[1] Cf. *Trans. Roy. Scot. Arbor. Soc.* xviii. 62 (1905).
[2] *Mitt. Deutsche Dend. Ges.* 1906, p. 29.

(occasionally five) lines. In cultivated specimens, the leaves are fringed on each side with a very thin and narrow membranous translucent border.

Staminate flowers, $\frac{3}{8}$ inch long, cylindrical, raised on short stout stalks, about $\frac{1}{16}$ inch long. Pistillate flowers ovoid; bracts reflected at their bases, with the mucros pointing downwards, oblong, truncate or slightly concave at the apex, the green midrib being prolonged into a mucro about $\frac{1}{8}$ inch long.

Cones 3 to 4 inches long, cylindrical, tapering to a narrow, flattened apex, supported on a short stalk, glaucous green or purplish, with orange-brown bracts before ripening, composed of five spiral rows of scales, eighteen to twenty scales in each row, which, on the opening of the cone, stand almost at right angles to its axis, the bracts being exserted with their mucros directed upwards. Scales quadrangular, with a cuneate base, about $\frac{1}{2}$ inch in width and length; upper margin truncate and slightly emarginate; outer surface radially furrowed, densely pubescent towards the base. Bract lanceolate, nearly as long or quite as long as the scale, the mucro, often incurved, projecting beyond the scale about $\frac{3}{16}$ inch. Seeds lying in slight depressions on the scale, their wings widely divergent and not extending to its upper margin. Seed, white on the inner side, shining dark brown on the outer side, about $\frac{1}{8}$ inch long; seed with wing about $\frac{7}{16}$ inch long; wing brownish, rather opaque, broadest about the middle. Cotyledons[1] five to six, which, in the seedling, are linear, pointed, and much longer than the succeeding leaves.   (A. H.)

The Sikkim larch is confined, so far as we know at present, to a rather narrow area in the Himalaya, from eastern Nepal to Bhutan, but very possibly will be found farther east. It was discovered by Griffith, but not distinguished until Sir Joseph Hooker found it in E. Nepal in December 1848.[2] Here it was only a small tree 20 to 40 feet high, differing from the European larch, in having very long, pensile, whip-like branches. It is called "Saar" by the Lepchas, and "Boargasella" by the Nepalese, who said that it was only found as far west as the sources of the Cosi river. In Sikkim it is common in the interior valleys of the Lachen, Lachoong, and their tributaries from about 8000 to 12,000 feet elevation, and here attains a larger size, but is not found in the forests of British Sikkim. In *Illustrations of Himalayan Plants from Drawings by Cathcart*, where it is beautifully figured, Sir Joseph states that it grows to a height of 60 feet in deep valleys, but prefers the dry rocky ancient moraines formed by glaciers, and also grows on grassy slopes where the drainage is good. On my journey to Tibet in 1870 I saw this tree in the Lachoong valley, but nowhere forming a forest, and usually scattered singly in rather open places, where it seemed to me to have a much less erect and regular growth, with branches more drooping in habit than any other larch. Sir Joseph Hooker says that the wood is soft and white, but a specimen from the Chumbi valley, authenticated by cones, is described by Gamble as having red heart-wood with a slow growth, twenty-one rings to the inch, and a weight of 32 lbs. to the foot.

Though introduced by Sir Joseph Hooker, who sent seeds to Kew in 1848, this tree has, except in a few places in the south-west of England, failed to grow in Europe. He says that the seedlings raised from his seeds were 3 to 4 feet high in

---

[1] Masters, *loc cit.*   [2] *Himalayan Journals*, i. 255.

1855, and that some had withstood the severe winter of 1854-5 without protection, though others were killed, a difference which he attributes to some of the seed having been gathered from trees which grew at 8000 and some from trees at nearly 13,000 feet. Hooker[1] further states that hundreds of plants were raised and widely distributed by Kew, but in every case these succumbed in a few years to virulent attacks of *Coccus laricis*. As the climate of the Chumbi valley is much drier than that of Sikkim, it is quite possible that seed from that locality would give better results; but I have never been able to keep the tree alive at Colesborne for long, as it suffers from the dry climate, and seems to object to lime in the soil. Mr. Barrie, forester to the Hon. Mark Rolle, has been very successful in growing this tree from English-grown seed, and has sent me healthy young plants of it; but the seedlings I have raised at Colesborne both from imported and home-grown seed have always died, though protected by a frame.

REMARKABLE TREES

The largest specimen of the Sikkim larch we know of in this country is one at Coldrinick, near Menheniot, Cornwall, the seat of Major-Gen. Jago-Trelawney. I have not seen this tree, but the gardener, Mr. Skin, informs me that in 1905 it measured no less than 57 feet by 4 feet 6 inches in girth. It has very spreading branches, the width from point to point of the lowermost branches being 43 feet. The cones were admirably figured in the *Gardeners' Chronicle*,[2] and have produced fertile seed. The seedlings require careful treatment, as they easily "damp off."

A tree of the original introduction is growing at Strete Raleigh, Devonshire, the seat of H. M. Imbert Terry, Esq., who showed it to me in 1903, when it measured 40 feet high by 4 feet in girth. It is growing on poorish soil at a considerable elevation, where it is a good deal exposed to the damp south-west winds, and perhaps in consequence of this has thriven very well, and has borne fertile seed for some years past (Plate 109).

Another much smaller tree, which also bears cones, is growing at Leonardslee in Sussex. There is also an old tree at Pencarrow, in Cornwall, which in 1905 was only 12 feet high by 15 inches in girth, stunted and covered with lichen. It also bears cones.

Dr. Masters[3] received flowering specimens in 1896 from The Frythe, Welwyn, Herts; but the tree from which they were obtained could not be found when Henry visited this place in 1906.　　　　　　　　　　　　　　　　　　　　(H. J. E.)

[1] *Gard. Chron.*, *loc. cit.*

[2] After this was printed a good illustration of the tree appeared in the same journal on 2nd March 1907, which shows that it is not only larger, but a better shaped tree than the one I have figured.

[3] *Gard. Chron.* xxvii. 296 (1900).

LARIX POTANINI, Chinese Larch

Larix Potanini, Batalin, *Act. Hort. Petrop.* xiii. 385 (1894); Masters, *Gard. Chron.* xxxix. 178,
   f. 68 (1906).
Larix thibetica, Franchet, *Jour. de Bot.* 1899, p. 262.
Larix Griffithii, Masters, *Jour. Linn. Soc. (Bot.)* xxvi. 558 (1902). (Not Hooker.)

A tree attaining in western China a height of 70 feet and a girth of 6 feet. Young branchlets bright yellow, with a scattered pubescence, densest near the base of the shoot, which is girt by a sheath of the previous season's bud-scales, showing within a ring of pubescence. Buds ovoid, with ciliate scales.

Leaves slender, up to an inch in length, ending in a sharp cartilaginous point, tetragonal in section, keeled above and below, with two bands of stomata, each of two lines, on both the upper and lower surfaces.

Staminate flowers, $\frac{1}{4}$ inch long, on a short but distinct stalk. Pistillate flowers ovoid, narrow and rounded at the apex; bracts closely appressed, on one side of the young cone with their tips pointing towards its apex, on the other side reflected about their middle with their apices pointing towards the base of the cone, ovate or oblong, rounded and entire at the apex, which is prolonged into a short mucro. The bracts in the pistillate flower, described above as seen in herbarium specimens, are probably all reflected at first; and gradually by the growth of the scale assume the erect position.

Cones cylindrical, rounded at the apex, $1\frac{3}{4}$ inch long, with the scales and bracts pointing upwards and outwards, or more or less spreading. Scales small, about $\frac{1}{3}$ inch long, almost orbicular, reddish brown, pubescent on the lower part of the outer surface; upper margin rounded or truncate, entire, thin, slightly inflected, not bevelled. Bract extending beyond the scale, exserted with the mucro about $\frac{1}{4}$ inch. Seeds in slight depressions on the scale, with their wings widely divergent and not reaching to its upper margin. Seed about $\frac{1}{8}$ inch long; seed with wing $\frac{1}{3}$ inch long; wing broadest just above the seed.

Larix Potanini has been collected in western China by Potanin, Prince Henry of Orleans, Pratt, and Wilson, who found it in the neighbourhood of the Szechuan-Thibetan frontier near Tachienlu at 7500 to 11,000 feet above sea-level. The same species, according to Franchet, was probably collected by Père Delavay farther south on the Likiang range in Yunnan at 11,600 feet altitude. Mr. A. Hosie, Consul-General in Szechuan, informs me that forty miles north-east of Tachienlu, there is a pure forest of this larch between 11,000 and 12,000 feet elevation on the southern slope of the mountain range, and extending for about a mile. It consists of fine straight trees, which he estimated to be about 70 feet high. At lower altitudes the larch gives place to silver fir and birch. The tree is known to the Chinese as "hung-sha," red fir, and produces the most valuable coniferous timber in western China.

Seed was collected by Wilson in 1904, and plants have been raised, which are growing well at Veitch's nursery, Coombe Wood.

This species, being a purely alpine tree of no great size, will probably be of no value as a forest tree, resembling in that respect its immediate allies *L. Griffithii* and *L. Lyallii*, between which it occupies an intermediate position as regards botanical characters. (A. H.)

LARIX AMERICANA, Tamarack

Larix americana, Michaux, *Fl. Bor. Am.* ii. 203 (1803); Sargent, *Silva N. Am.* xii. 7, t. 593 (1898), and *Trees N. Am.* 35 (1905); Kent, *Veitch's Man. Conif.* 389 (1900).

Larix americana, Michaux, var. *rubra*, Loudon, *Arb. et Frut. Brit.* iv. 2400 (1838).

Larix tenuifolia, Salisbury, *Trans. Linn. Soc.* viii. 314 (1807).

Larix microcarpa, Desfontaines, *Hist. Arb.* ii. 597 (1809); Lawson, *Agric. Man.* 388 (1836).

Larix laricina, Koch, *Dendrologie*, II. ii. 263 (1873).

Larix pendula, Masters, *Journ. Roy. Hort. Soc.* xiv. 218 (1892). (Not Salisbury.)

Pinus Larix americana nigra, Muenchausen, *Hausv.* v. 226 (1770).

Pinus laricina, Du Roi, *Obs. Bot.* 49 (1771).

Pinus intermedia, Wangenheim, *Beit. Hölz. Forst. Nord Am. Hölz.* 42, t. 16, f. 37 (1787).

Pinus microcarpa, Lambert, *Pinus*, i. 58, t. 37 (1803).

Abies microcarpa, Poiret, *Lamarck's Dict.* vi. 514 (1804).

A tree attaining in America about 80 feet in height and 6 feet in girth. Bark separating in thin small polygonal or roundish scales about an inch in diameter, which are closely appressed, and show when they fall off the reddish cortex beneath. Young branchlets slender, often glaucous, glabrous, or with a few scattered hairs in the grooves between the pulvini; older branchlets glabrous, shining brown. Base of the shoot girt with a short sheath of the previous season's bud-scales, no ring of pubescence being visible. Short shoots small, blackish, glabrous. Terminal buds globose, slightly resinous, glabrous, with the basal scales subulately pointed. Lateral buds hemispherical, resinous, dark brown. Apical buds of the short shoots broadly conical, surrounded at the base by a ring of brown pubescence.

Leaves short and slender, not exceeding $1\frac{1}{4}$ inch in length, rounded at the apex, light green; upper surface flat or rounded, without stomata, except two broken lines near the tip; lower surface deeply keeled with two bands of stomata, each of one to two lines.

Staminate flowers sessile, shorter than in *L. europæa*. Pistillate flowers ovoid, reddish, very small; bracts pointing upwards and outwards, not reflected or recurved, $\frac{1}{8}$ to $\frac{1}{6}$ inch long, oblong, scarcely emarginate at the apex, reddish with a green midrib and mucro, the latter cuspidate and very short, about $\frac{1}{30}$ inch long.

Cones small, globose, consisting of three to four spiral rows of five scales each, reddish brown when ripe, $\frac{1}{2}$ to $\frac{2}{3}$ inch long. Scales gaping widely at the apex of the cone, longer than broad, about $\frac{2}{5}$ inch long; upper margin rounded, bevelled, slightly crenulate, not recurved or reflected. Bract concealed, minute, about $\frac{1}{8}$ inch long.

Seeds lying on the scale in minute depressions, with their wings only slightly divergent and not reaching to its upper margin, ¼ inch long; wing ⅓ inch long, broadest just above the seed.   (A. H.)

DISTRIBUTION

The American larch is found in the United States from North Pennsylvania, Northern Indiana and Illinois, and Central Minnesota through the New England States, where, however, it is only found in cold and swampy places. In Newfoundland, Labrador, and the eastern provinces of Canada it occupies swampy ground, and extends from York Factory on Hudson Bay as far as Fort Churchill, 67° 30′ N., and west to Athabasca and Peace river districts, and in Alberta where it has been found forty miles S.W. of Edmonton.[1] Northwards it extends to the border of the barren lands. Mr. J. M. Macoun informs me that it was found by Mr. Camsell in the angle between the Snake river and the upper part of Peel river. This place is just within the Yukon district. He also states that it extends westward twenty-two miles up the Dease river, and northward along the upper Liard river to lat. 61° 30′. He has heard several people who have been on the Yukon speak of the larch, so that it must be quite common in some parts, though no definite data are as yet given.

The tamarack, as it is called in most parts of N. America, is a tree which I know but little in a state of nature, and which never seems to have received the attention from foresters which it deserves; for though it nowhere attains the size of the common larch, it seems able to thrive in undrained and swampy ground where that would die; and though a slow-growing tree in comparison with the common larch, its timber has the same valuable qualities as others of the genus.

Henry saw this species in Minnesota in 1906. On the Cass Lake Forest Reserve it occurs in the swampy ground between the pine-covered sand-dunes, in company with balsam fir, Thuya, black and white spruce, birch, and willow. The largest that he saw measured 81 feet by 4 feet 7 inches. The trees are remarkable for their buttressed roots, which branch and extend close to the surface and even above ground for as much as 6 feet. Seedlings were numerous in felled areas near Erskine, where the larch remaining uncut, occurs in swamps either pure or mixed only with birch. They grow very rapidly in the wet ground, taking root in mossy elevated patches and not in the water of the swamps; and averaged 10 feet high at seven years old, and were making leaders of 1 to 2 feet annually. He saw no stumps larger than 2 feet in diameter, and the tree in Minnesota rarely attains a greater size than 80 feet by 6 feet. In *Garden and Forest*, 1890, p. 60, there is, however, mention of a tamarack in Minnesota, which measured 7 feet 8 inches in girth and was estimated at 125 feet high.

In most parts of New England and over the greater part of British North America the tamarack is a well-known tree, but rarely attains any great size. The average in the neighbourhood of Ottawa is not over 50 to 60 feet, but when the tree is planted on drier, better land it will grow faster and attain 80 feet or more. I noticed that though it seeds freely the seedlings require more light than those of the spruce, balsam fir, and Thuya, which often grow with it, and it was only where clearings had been made, or in wet places on the edge of the groves, that they seemed able to thrive. Their growth is slow at first, but when established may be as much as two feet annually.

Dr. Bell gives the probable life of the white spruce in Canada as from 100 to 140 years, that of the black spruce 150 to 175 years, and that of tamarack 175 or 200 years. Of the latter he says:[1] About 1893 or 1894 the imported sawfly[2] came up from the direction of New York and got into the forests north of the Ottawa river. In a year or two it reached James bay and killed the tamarack throughout that district, which was only able to live three or four years after it was first attacked by the larva. This destruction continued to spread to the centre of Labrador, and now it has gone pretty well all over that great peninsula. But Mr. J. C. Langelier (loc. cit. p. 65), speaking of the same attack in the northern part of the province of Quebec, says that a great portion of the young trees were spared, and that the dead trees which remain standing are not attacked by rot, and would supply excellent railway ties.

REMARKABLE TREES

In this country there are not many large trees of this species, though it was introduced, according to Loudon,[3] by the Duke of Argyll in 1760 at Whitton, near Hounslow. It has been entirely neglected by modern arboriculturists, and is seldom or never procurable in English nurseries. The largest trees that I know of are at Dropmore, where there is a well-grown tree 78 feet by 5 feet (Plate 110), and at Arley Castle, where there are three trees of nearly the same size standing together, of which the best measures 71 feet by 4 feet 8 inches. A fourth is nearly as large, and differs in having larger cones.

At Boynton, Yorkshire, there are two in a wet situation among other trees, about 50 feet high and sixty years old, which were raised by Sir Charles Strickland from seed produced by trees planted by his grandfather. These again have produced fertile seeds, from which seedlings are growing vigorously in a low frosty situation at Colesborne, and have never suffered from frost or bug, though one of them in 1906 was attacked by Peziza. Sir Charles adds that on dry soil they have grown very badly.

At Beauport there are three rather stunted specimens of American larch, one of which, however, is 5 feet 10 inches in girth, and has the bark very smooth in comparison with the common larch. No specimen seems to have been sent to the Conifer Conference, but one is mentioned as growing in the grounds of Dalkeith Palace,[4] which we have identified with *L. dahurica*. Several trees mentioned by Loudon are either not now in existence or were not correctly named.

(H. J. E.)

[1] *Can. For. Ass. Annual Report*, 1905, p. 59.
[2] According to Sargent this is *Nematus Erichsonii*, Hartig, a European insect which was not much noticed in America before 1880, and which has recently attacked the larch in England. Cf. *supra*, p. 364.
[3] *Op. cit.* 2400, 2401. The original tree at Whitton was between 40 and 50 feet high in 1837: it has long since been cut down.
[4] *Veitch's Man. Coniferæ*, 390 note (1900).

LARIX OCCIDENTALIS, Western Larch

Larix occidentalis, Nuttall, *Sylva*, iii. 143, t. 120 (1849); Lyall, *Journ. Linn. Soc.* vii. 143 (1864); Sargent, *Gard. Chron.* xxv. 652, f. 145 (1886), *Silva N. Amer.* xii. 11, t. 594 (1898), and *Trees N. Amer.* 36 (1905); Kent, *Veitch's Man. Coniferæ*, 400 (1900); Mayr, *Fremdländ. Wald- u. Parkbäume*, 306 (1906).

Pinus Nuttalli, Parlatore, DC. *Prod.* xvi. 2, p. 412 (1868).

A tree attaining in America 200 feet in height and over 20 feet in girth ; narrowly pyramidal in habit, the branches being much shorter than in the other species. Bark of young stems thin, dark-coloured, and scaly ; becoming near the base of old trunks 6 inches thick and breaking into irregularly shaped oblong plates, often 2 feet in length and covered with thin reddish scales. Young branchlets covered with a minute dense pubescence intermixed with longer hairs in the grooves between the pulvini. In certain cultivated specimens the branchlets are glabrous from the first. Branchlets of the second year light brown, shining. Base of the shoot girt with a sheath of the previous season's bud-scales, no ring of pubescence being visible. Short shoots chestnut brown, shining. Terminal buds globose, with pubescent and ciliate scales, the lowermost of which are subulately pointed. Lateral buds hemispherical with pubescent and ciliate scales. Apical buds of the short shoots broadly conical, reddish brown, pubescent.

Leaves light green in colour, up to $1\frac{3}{4}$ inch long, rounded on the back, deeply keeled beneath, with stomatic lines as in *L. europæa*.

Staminate flowers raised on short stalks at maturity. Pistillate flowers ovoid ; the bracts pointing upwards and outwards and not recurved, $\frac{1}{4}$ inch long, brownish in colour with a green midrib and mucro, oblong, emarginate at the apex ; mucro $\frac{1}{10}$ inch long.

Cones ovoid, $1\frac{1}{4}$ to 2 inches long, with the bracts long-exserted and the scales opening early in the season to let out the seeds and then standing at right angles to the axis of the cone. Scales in six spiral rows, each row of nine to ten scales ; orbicular, $\frac{1}{3}$ to $\frac{1}{2}$ inch long ; upper margin entire or emarginate, thin, slightly recurved, not bevelled ; outer surface densely pubescent. Bracts ovate-lanceolate, extending up to near the margin of the scale, beyond which the mucro projects $\frac{1}{8}$ to $\frac{1}{2}$ inch. Seeds lying in two deep depressions on the scale, their wings narrowly divergent and extending up to its upper margin ; body of the seed $\frac{1}{8}$ inch long ; wing pale coloured, short and broad, widest at the base ; seed with wing $\frac{1}{4}$ to $\frac{2}{8}$ inch long.

Varieties

In the wild state the tree shows little variation, except in the pubescence of the branchlets, which in rare cases is entirely absent ; while in other cases, noticed occasionally at high elevations, the amount of pubescence becomes so dense as to be almost similar in character to the tomentum of *Larix Lyallii*. In the few cultivated trees in England, two distinct forms are apparent. Certain trees have pubescent branchlets and bear large cones, up to two inches in length, which have large scales purplish in colour before ripening, long exserted bracts and long-winged seeds. Other trees with glabrous branchlets bear small cones, about $1\frac{1}{4}$ inch in length, with scales green before ripening, shorter exserted bracts and small seeds with short wings. The former trees are more narrowly pyramidal in habit.

HISTORY

This splendid tree is the largest of the genus, and though it has been known to botanists for many years, it was till quite recently, on account of its being neglected by the early explorers of the limited region which it inhabits, one of the rarest exotic conifers in cultivation.

It was first discovered by David Douglas[1] in 1826 near Fort Colville on the Upper Columbia river; but was mistaken by him for the European larch. His specimens in the Kew Herbarium are labelled "in aqueous flats on the mountain valleys near Kettle Falls and in the Rocky Mountains, 1826." The tree was first described in 1849 by Nuttall, who found it on the Blue Mountains of Oregon in 1834.

It was introduced into cultivation in the Arnold Arboretum in 1881, seedlings having been imported from Oregon; but in the climate of New England these have remained small and stunted, though branches grafted on the Japanese larch have grown vigorously. Forty plants were sent from the Arnold Arboretum to Kew in 1881, and one tree survives (the fate of the other plants being unknown), which is remarkable for its beautiful straight stem and narrow, almost columnar habit. This tree bears large purplish cones, and is now (1906) 33 feet in height and 17 inches in girth.

Ten plants were subsequently sent in 1889 from the Arnold Arboretum to Kew, of which two survive. One of these trees is, however, identical in cones and pubescent branchlets with the tree of 1881, and may be erroneously labelled 1889; it has suffered damage at the top. The other tree, which has glabrous branchlets and bears small green cones, is not quite so narrow in habit, and measured in 1906 29 feet in height and $17\frac{1}{2}$ inches in girth.

The only other large tree in Britain with which we are acquainted is growing at Grayswood Hill, Haslemere; and measured in 1906 28 feet high by 19 inches in girth. It has pubescent branchlets, and bears purple cones, which are, however, smaller than those of the Kew tree, labelled 1881. Mr. Chambers informs us that this tree was obtained from Messrs. Dickson of Chester in 1889.

DISTRIBUTION

The western larch is confined to the more humid parts of the region, which extends from the western slope of the Rocky Mountains in British Columbia and

---

[1] *Comp. Bot. Mag.* ii. 109 (1836), where Douglas states that he measured trees 30 feet in girth.

Montana to the eastern slope of the Cascade Mountains in Washington and Oregon.

In British Columbia it is abundant and large in the Kootenay and Columbia river valleys, reaching as far north as the head of Upper Columbia lake, and attaining its most westerly point, where it was found by Prof. Dawson, in long. 124° E., on a tributary of the Blackwater river. It grows sparingly about the Shuswap lake and in the Coldstream valley near the head of Okanagan lake.

The tree, however, attains its greatest development in Montana, where it is abundant and constitutes a great part of the timber of the Flathead, Lewis and Clarke, and Bitter Root Forest Reserves; and is met with east of Missoula on the Big Blackfoot river. The tree can be most conveniently seen by the traveller on different points of the Great Northern Railway between Nyack and Bonner's Ferry. It attains also great perfection in Northern Idaho and North-East Washington, where it constitutes an important part of the timber of the Priest River Forest Reserve. It also occurs in Oregon, in the Blue Mountains, and on the foothills of the eastern side of the Cascade Mountains,[1] as far south as Mount Jefferson.

The western larch occurs between 2500 and 6000 feet altitude; and attains its maximum height and is most abundant in mountain valleys and on alluvial flats, where the average elevation is 3000 to 3500 feet. On the sides of the mountains, owing to the lack of moisture in the soil, it rapidly diminishes in size and vigour. It requires a wetter soil than either *Pinus ponderosa* or Douglas fir, and is restricted in its distribution where the rainfall is slight.

With regard to the opinion, prevalent even in America, that it grows in a semi-arid climate, my experience is entirely different. The meteorological stations are almost invariably in towns in the prairie regions, where the rainfall is small and trees only occur on the banks of streams; and the maps and statistics of the rainfall give on that account an imperfect picture of the climatic conditions which prevail in the forest regions between the Cascades and the Rocky Mountains. At Kalispell in the Flathead country, which is situated in a treeless plain, surrounded by densely forested mountains, the annual rainfall varies from 13 to 19 inches; whereas at Columbia Falls, placed on the edge of the plain and amidst the larch forests, the rainfall increases to from 20 to 29 inches; and in the mountain valleys, as at Lake Macdonald and Swan Lake, where *Thuya plicata* attains a large size, the rainfall must exceed 30 inches. The meteorological data of Columbia Falls, which is at 3100 feet elevation, give a fair idea of the climate in which *Larix occidentalis* thrives, though it is scarcely here at its best. The figures for 1905, which was a dry year, are :—

---

[1] Mr. Cohoon, Forest Assistant in the Northern Division of the Cascade Forest Reserve, wrote to me in 1906 as follows : " The only locality in which larch came under my observation in the reserve was on the east slope of the Cascade Mountains about 15 miles west of Durfur, Oregon. It did not occur abundantly, but was more or less scattered, in mixture with yellow pine, red fir, and lodge-pole pine. It was found on moist but well-drained soil at an altitude of about 2500 to 3000 feet." He adds that he never saw it west of the summit of the Cascades, which he has travelled over from Columbia river to California.

At Bridal Veil, Oregon, and other places on the Pacific slope, the term larch is erroneously applied to *Abies nobilis*.

	Precipitation in Inches.		Min. Temp.	Max. Temp.	Mean Temp.
	Snow.	Rain.	Fahr.	Fahr.	Fahr.
January	2.14	...	−1°	46°	24°
February	0.93	...	−35°	52°	18°
March	0.34	...	14°	63°	38°
April	...	0.45	15°	76°	44°
May	...	3.13	20°	83°	49°
June	2.23	...	28°	89°	56°
July	...	0.38	34°	96°	65°
August	...	0.12	29°	96°	64°
September . . .	...	2.04	24°	83°	55°
October	...	2.54	9°	60°	38°
November . . .	2.47	...	−11°	56°	30°
December . . .	2.79	...	3°	46°	25°
Total precipitation, 1905 .	19.56 inches.				

Average precipitation for ten years 21.70 inches.

Rain or snow fell on 76 days; 91 days were cloudy; 49 days were partially cloudy; and the sky was clear on 149 days.

The above figures show that the climate is an extreme one, the winter season being cold and severe and lasting five months, while in summer a high temperature is often reached.

The western larch grows usually mixed with other conifers; and the number of accompanying species and the proportions of the admixture are very variable, being dependent on the climate and altitude, and on the quantity of moisture in the soil. Douglas fir is the most common companion of the larch, and *Pinus ponderosa* steps in where the soil is dry. Engelmann's spruce and *Abies lasiocarpa* descend into the larch forests, but never constitute any large element of it. *Pinus monticola, Tsuga albertiana*, and *Abies grandis* are often met with in small quantity at low altitudes in the larch forests of Montana; farther west, in the Priest River Forest Reserve, *Pinus monticola* is more abundant than the larch itself between 2400 and 4800 feet. *Thuya plicata*, in regions with a moist climate, forms a notable part of certain larch stands, often to the exclusion of the other species which usually accompany the larch.

The following notes on a few of the larch forests visited by me will illustrate some different types in Montana.

Near Missoula, in Pattie Cañon, which is a very dry valley at 3500 feet elevation in a rather arid climate, the larch only grows on the cool northern aspect, and is mixed with Douglas fir and *Pinus ponderosa*. An acre contained, of trees over a foot in diameter, twenty larches, four firs, and three pines. An average good larch tree measured 143 feet by 9 feet 7 inches; and a tree which we cut down, 14 inches in diameter, showed 211 annual rings, the sapwood being 1¼ inch in thickness and containing thirty-one rings.

On the southern end of Lake Macdonald, at 3500 feet altitude in a humid climate, I saw a fine stand composed almost exclusively of larch and *Thuya plicata*. The soil was glacial clay, very deep, and covered with a thick layer of humus. The

Thuya only attained about 110 by 7 feet, and had been overtopped by the larch, which ran from 140 to 150 feet high, and 7 to 14 in girth. The trees were extremely dense upon the ground, standing often only 12 feet apart, and averaging 200 to the acre. The ground was covered with seedlings of Thuya, 3 to 6 feet high, and more than thirty years old. The Thuya trees were being felled for telegraph and telephone poles, but never had clean stems, being covered with dead branches to 6 to 20 feet above the ground, and with living branches above this, and when of a large size were always decayed at the heart. The larch, as usual, was quite sound.

A wood near Whitefish, on flat land in a moderately rainy district at 3000 feet altitude, was composed of about nine-tenths larch and one-tenth Douglas fir, *Pinus ponderosa*, and Engelmann's spruce. The larch were 160 feet high by 6 to 9 feet in girth, overtopping the other trees, and with clean stems up to 80 or 90 feet. A stump, 40 inches in diameter, showed 585 annual rings, the sapwood with forty-two rings being only an inch in thickness, and the bark two inches.

The largest tree which I saw was growing on a high bank beside the Stillwater Creek, some miles west of Whitefish. It measured 19 feet 4 inches in girth at 5 feet from the ground, but the top was blown off. Near it were many large trees, 12 feet to 15 feet in girth, but the tallest was only 151 feet in height.

With regard to the height attained by the western larch, Sargent in his *Report on the Forest Trees of North America*, 216 (1884), states that it ranges from 100 to 150 feet, but in the *Silva* he gives the maximum height as 250 feet. I could find no confirmation of the latter figure either at the Arnold Arboretum or Washington, and I am of opinion that 180 feet is rarely if ever exceeded. The tallest tree recorded by any accurate observer is, I believe, the one cut down by Ayres[1] in the Whitefish Valley at 3500 feet altitude, which measured 181 feet high, with a diameter of 3 feet on the stump, and scaled 3500 feet board measure. He mentions[1] also another tree growing on the middle fork of the Flathead river, which was 180 feet high by 4 feet in diameter.

J. B. Leiberg states in his account of the Priest River Forest Reserve that the larch in the sub-alpine zone, above 4800 feet elevation, averaged 60 to 100 feet in height, 1 to 2 feet in diameter, and eighty to a hundred years old; while in the white pine zone, from 2400 to 4800 feet, the trees were 150 to 200 feet in height, 2 to 4 feet in diameter, and 175 to 420 years old. Here the heights are evidently estimates, and cannot be relied on implicitly.

The western larch is rarely seen as pure forest, and then only as the result of forest fires. Mr. Langille in his account of the Cascade Forest Reserve, p. 36, says that the larch "has done more than any other species to restock the immense burns that have occurred on a part of the reserve. This is largely due to the fact that the thick bark of this tree resists fire better than any other species, and more trees are left to cast their seed on the clean loose soil and ashes immediately after a fire. The seeds are small and light, and are carried to remote places by the wind and covered deeply by the fall rains. In the spring a dense mass of seedlings covers the

[1] *U.S. Geol. Survey, Flathead Forest Reserve*, 256, 314 (1900).

ground and grows rapidly.   The thickets become so dense that it is impossible to travel through them.   In time only the fittest survive, and there remains a thrifty, vigorous stand of this valuable timber."   In Montana the lodge-pole pine usually takes possession of burnt areas; but I saw near Belton on the Great Northern Railway a hillside which had been swept by a fire, leaving a good number of larch trees unharmed, all the trees of other species being destroyed, and larch seedlings were coming up in profusion.   On the Stillwater Creek farther west I noticed a burnt area on which the lodge-pole pines were about 30 feet high; and amongst them larch seedlings were growing in openings exposed to sunlight during at least a part of the day.   Here in time the lodge-pole pine will be supplanted by the larch.   Sargent's statement,[1] that young seedlings of the western larch are able to grow up under the shade of other trees, which they finally overtop and subdue, requires modification. Seedlings never occur in the shade of the forest, and are most numerous in open places exposed to full sunlight; but on good soil, as on a recently burnt area, they will spring up in the partial shade of small pine trees.   The western larch is not a fast grower in the young stage; at Belton seedlings twelve years old, growing on rather poor rocky ground, were from 7 to 12 feet high.

As the seed of the western larch had never been collected, so far as we knew, by any one except Mr. Carl Purdy's collector in 1903, I visited Montana in 1906, with the object of collecting a large quantity for Sir John Stirling Maxwell and Lord Kesteven.   In the common larch the seeds do not fall out of the cones until spring, and their collection during winter is an easy matter.   The western larch behaves very differently, as will be seen by the following notes of my observations in Montana.   About the middle of August the squirrels begin to throw down cones, a sign that the seeds are nearly ripe.   About the 10th September the leaves, which form a tuft at the base of the cone, begin to turn yellow, and in a day or two become brown and withered, showing that the supply of nutrition to the cone is stopped. The cones, which until now were purplish in colour, become brown, and the scales gape open widely, allowing the seeds to escape.   By the 20th September all the cones on the trees have become quite brown, and have emptied all their seeds.   The empty cones remain on the branches till the autumn of the following year, by which time their peduncles have rotted and the cones are ready to fall.   For collecting seed the larch forests must be visited during the first three weeks of September; and localities where felling is being carried on should be chosen, as the cones occur only at the summit of very tall trees, which are troublesome to cut down, even if permission to do so has been obtained from their owners.   The western larch appears to produce a good crop of seed once every two or three years, and this is general over the whole region.   1906 was a remarkably poor year, scarcely any cones having been formed. In 1905, judging from the old cones of that year still remaining on the trees, the crop of seed was very abundant.   (A. H.)

As I had long been trying to find a larch that would in England be less liable to the attacks of *Peziza Willkommii* than the common larch, I made inquiries as

---

[1] *Garden and Forest*, ix. 491 (1896), where there is an article on the tree, with an illustration of the trunk, fig. 71, showing the very thick bark.

to how seeds could be procured, and Prof. Sargent was good enough to do his best for me. Mr. Leiberg, in 1901, went on purpose to the Flathead Lake country, but found all the seed shed as early as September, and could only send a few seedlings by post. These heated on the way to England, and though I saved a few of them, they were always sickly, and most of them died before coming into leaf. Again I tried through the United States Forestry Bureau, who were also unable to get seed. In 1903, however, I procured a small parcel from Mr. Carl Purdy, and distributed the seed to many arboriculturists in England in 1904. These have germinated fairly well, and I hope that my efforts to make this grand tree better known may succeed.

The seedlings raised in 1904, from the seed which I distributed, have grown in several places, best perhaps at Murthly, under the care of Mr. Lowrie, where in September 1906 I saw some hundreds thriving very well, though not so large as common larch of the same age. At Walcot, in rather dry soil, they were 6 to 9 inches high. At Colesborne they grew slowly, and many were killed or injured in the seedbed by the frost of May 1905; but I have just planted out a number which were raised for me by Messrs. Herd of Penrith, and which are 12 to 18 inches high.

I visited Missoula in June 1904 on purpose to see the tree, and was fortunate enough to do so in company with Prof. Elrod of the Montana University, to whom I am greatly indebted for the excellent photographs of the tree here reproduced (Plate 111). They were taken on the Big Blackfoot river about twenty miles up the valley from Bonner, on the Northern Pacific Railway, where a large sawmill, managed by Mr. Kenneth Ross of the Big Blackfoot Lumber Company, has its headquarters. Guided by this gentleman we reached the logging camp in the Camas prairie and found the larch growing in deep bottom land at about 3500 feet, mixed with *Pinus ponderosa* and Douglas fir, but far exceeding both of them in size. The tree grows on slopes and in ravines where there is a good depth of soil not liable to dry up, and best on slopes with a north and east aspect, and on the rich detritus at their foot, and along the sides of the river. It differs strikingly from other larches in habit when adult, having very short branches, which are not produced singly or at regular intervals but grow in irregular groups of four or five, starting near together on the trunk. It forms a tall, very narrow column, and as it gets old loses many of its branches. It carries its girth to a great height and is, when grown in a thick forest, sometimes clear of branches for over 100 feet. The tallest tree I have heard of was figured in the *Butte Miner* of 29th February 1904, and was said to be the largest in Montana, 233 feet high and 24 feet in girth at or near the ground. This tree grew on the Upper Clearwater between Salmon and Seely lakes. It could be seen for miles above the surrounding trees, and must have contained over 2000 feet of timber. The best I saw, however, were from 150 to 180 feet in height, with a girth at 5 feet of 10 to 15 feet.

Frank Vogel, a timber surveyor who has had much experience with this tree, told me that it grew up to 6000 feet elevation on the hills above the Blackfoot river, and that he saw no difference between these trees and those lower down except in size. The age of those of which I counted the rings, and which would be about the same age as the one photographed, was 330 to 350 years, these trees showing no signs of decay. The bark in dense forest is very thin for such large trees, sometimes only 2 to 3 inches thick, and though in older and more isolated trees it attains a much greater thickness, as much as 9 to 15 inches near the ground, it struck me as not being so thick and rugged as the bark of old European larch.

The undergrowth in the forest was not dense, and was composed of *Berberis aquifolium*, *Cornus canadensis*, *Linnæa borealis*, *Symphoricarpus*, *Thalictrum*, with violets, strawberries, and in some places that lovely little orchid *Calypso boreale*. There were abundant seedlings of larch and Douglas fir springing up wherever there was enough light and moisture, but in the drier parts of the forest pine only was seen. The young cones were already formed on 29th May, and I came away with the impression that though this tree may not rival the European or Japanese larches in rapidity of growth, it will be valuable in the mountains of Central Europe and will probably succeed on the better soils of England and Scotland.

With regard to the timber of the western larch, Prof. Sargent says that " it surpasses that of all other American conifers in hardness and strength, it is very durable, beautifully coloured, and free from knots ; it is adapted to all sorts of construction, and beautiful furniture can be made from it. No other American wood, however, is so little known." Through the kindness of Mr. K. Ross I was able to bring back from the St. Louis Exhibition a door and frame made from this wood which fully bears out Sargent's high opinion of it.

Until a few years ago the timber of the western larch was invariably called tamarack, and was of no great commercial importance. The use of this name, which is properly applied to *Larix americana*, the timber of which is little esteemed, proved prejudicial to the reputation of the western larch in the eastern states. Of late years the timber merchants of Idaho and Montana insist on the use of the term larch ; and large quantities of this lumber are now being exported even as far east as New York. Coarse grades are used for joints, beams, and railway ties. Finer grades are sawn into planks, used for flooring, and are converted into materials for indoor finish, as ceiling, laths, mouldings, panelling, etc. The timber is remarkably free from knots, and is variable in colour, being often nearly white, though it is usually reddish in tint.

(H. J. E.)

LARIX LYALLII, Lyall's Larch

Larix Lyallii, Parlatore, *Enum. Sem. Hort. Reg. Mus. Flor.* 1863, *Journ. Bot.* i. 35 (1863), and *Gard. Chron.* 1863, p. 916; Sargent, *Gard. Chron.* xxv. 653, f. 146 (1886), *Silva N. Amer.* xii. 15, t. 595 (1898), and *Trees N. Amer.* 37 (1905); Kent, *Veitch's Man. Coniferæ*, 399 (1900).

A tree attaining in America 80 feet in height and 12 feet in girth, but usually considerably smaller. Bark of young stems and branches thin and pale grey, on larger stems loose and scaly, on older trunks 2 inches thick and fissuring into irregular plates covered by reddish-brown loose scales. Young branchlets covered with a dense greyish tomentum, concealing the pulvini, and partly persistent on older branchlets, which become greyish black in colour. Short shoots stout and greyish pubescent. Bud-scales fringed with long cilia. Base of the long shoots girt with a sheath of the previous season's bud-scales, the uppermost of which are loose, membranous, and reflected.

Leaves bluish green, rhombic in section, deeply keeled on both surfaces, 1 to 1½ inch long, rigid, ending in a sharp cartilaginous point.

Staminate flowers ovoid, acute at the apex, ⅓ inch long, raised on stalks ⅕ inch long. Pistillate flowers ovoid, with the bracts reflected about their middle, their mucros curving outwards; bract oblong, ⅕ inch long, truncate at the apex, the midrib being prolonged into a rigid mucro about ¼ inch long.

Cones ovoid, acute at the apex, 1½ to 2 inches long, on a short tomentose stalk: scales numerous, loosely imbricated, thin, ovate, of a beautiful pink colour before ripening, ½ inch long, fringed with matted hairs; outer surface sparingly pubescent: bracts extending up to the margin of the scale, with their mucros projecting beyond about ¼ inch and at first directed upwards; when ripe the scales spread at right angles and finally, together with the bracts, become much reflexed. Seeds in slight depressions on the scale, with their wings narrowly divergent and not reaching its upper margin. Seed together with wing about $\frac{7}{16}$ inch long; wing pale pink in colour, broadest near the base.

This species has been supposed to be an alpine form of *L. occidentalis*; but is readily distinguished from it by the structure of the leaves, the tomentum of the branchlets, the beautiful pink cones, which have fringed scales, and the pink-winged seeds. (A. H.)

This tree was discovered by Dr. D. Lyall when surgeon to the International Boundary Commission in British Columbia in 1858, and though I have raised seedlings which I believe to be this species, it has not as yet been introduced into cultivation either in America or Europe, though it is a tree which must have been seen by thousands of travellers while crossing the Rocky Mountains in the Canadian Pacific Railway. Plate 112 shows a typical tree growing near Laggan, and is from a negative which I purchased at Victoria.

It is a strictly alpine tree, of somewhat limited range, its northern limit being about 51° N. on the Rocky Mountains, not extending to the moister climate of the Gold or Cascade ranges in British territory, nor has it as yet been discovered in the more northern parts of British Columbia. Southwards, it extends along the Cascade Mountains of Northern Washington to Mount Stewart on the north fork of the Yakima river, and along the continental divide of the Rocky Mountains to the middle fork of Sun river and to Pend d'Oreille pass in North-Western Montana.[1] In its northern habitat—near Laggan, Alberta—I have seen it from about 5000 up to 7000 feet. Though Mr. J. Macoun reports it on a mountain near Morley as low as 4500 feet, yet Wilcox,[2] who must have seen as much of this tree as any one who has written of it, says it is rarely seen below 6000 feet, and that its extreme range of altitude might be placed between 5600 and 7600 feet.

Lyall's larch is a very beautiful tree of moderate size, from 50 to 70 feet high being about the average, with a girth of 5 to 6 feet, but on Mount Stewart Mr. Brandagee reported that it attained as much as 4 feet in diameter. Its growth is extremely slow, Wilcox having counted 30 rings of growth in a branch only $\frac{3}{4}$ inch in diameter; whilst a tree cut by Brandagee on Mount Stewart which showed 562 annual rings was only $16\frac{1}{2}$ inches in diameter under the bark.

Mr. M. W. Gorman says:[3]—Near Lake Chelan it was not seen at all in the moist valleys, and was generally found to favour the passes and sheltered sides of the crest lines and divides, and here it ranges in altitude from 5800 to 7100 feet. The best grove seen was at about 6700 feet elevation near War Creek pass. The tree ranges in height from 50 to 90 feet, and in diameter from 10 to 25 inches. The mature tree has a rather thick greyish bark, and is well fruited with oval, mostly erect persistent cones. The branches are mostly lateral, very brittle, and quite small in proportion to the tree. The foliage changes colour with the first severe frosts about October 1.

L. Lyallii has to contend with a climate as severe as, and very similar to that of the Altai Mountains, the snow usually lying till late in June or even July, and snow and frost often occurring in July and August. The bark is rough and greyish and the branches short, irregular, brittle, and easily broken by a heavy snowfall. Wilcox says that the trees growing at the highest altitude have a curious development not found on those only a few hundred feet lower. The tufts of leaves spring from a hollow woody sheath, which is sometimes more than an inch long on the trees at high altitudes, whilst elsewhere this is not present.

The seed appears to ripen and shed early like that of the western larch, for though I have made several attempts to procure it from friends visiting the Rockies they have been, like myself, always too early or too late, and though I tried to bring home seedlings in 1893 they died on the journey home.

It is not, however, at all likely to succeed in this country, except possibly on the higher parts of the Grampian Mountains, and even there I fear the climate will be too damp, and the winter too short for it. (H. J. E.)

---

[1] Sheldon, in *Forest Wealth of Oregon*, says that it is " rare on the high peaks of the Wallowa Mountains."
[2] *The Rockies of Canada*, 63 (1900).
[3] *U.S. Geol. Survey, Eastern Part of Washington Forest Reserve* (1899). Mr. Gorman calls the tree *L. occidentalis*; but his specimens, which we have seen, are labelled *L. Lyallii* by himself, and are this species.

Lyall's Larch in Montana

Larix Lyallii occurs in five isolated areas in the mountains of Northern Montana, between 113° and 115° E. long. and 47° 25′ and 49° N. lat.

One of these localities was discovered by Prof. Elrod and myself in our ascent of the unexplored peak of St. Nicholas, which lies just west of the continental divide, about ten miles east of Nyack on the Great Northern Railway. Here about 1000 trees grow on a rocky precipitous slope, with a strictly northern aspect, and extend in scattered groves over about a mile of ground between 6600 and 7500 feet altitude. The tree is, owing to lack of moisture in the soil, unable to exist on the sunny southern slopes, where *Pinus albicaulis* thrives at similar altitudes. Separate groves of Engelmann's spruce accompany the Alpine larch. The largest tree measured 71 feet by 5 feet 2 inches; and another tree, felled by us, which was 8 inches in diameter, showed 220 annual rings, the sapwood with 25 rings being half an inch thick. Younger trees up to 40 feet high are gracefully pyramidal in shape, with wider branches than *L. occidentalis*; older trees have twisted and irregular branches and flattened crowns, the result of age, as is the case in all species of larch. The branches are remarkably brittle. On another part of the mountain, but still on the northern aspect, eighteen trees in two groups were seen at 8250 feet elevation, the tallest of which was only 10 feet high. The trees in Montana bore in 1906 only a few cones, but the crop in the preceding year had been plentiful. I procured only twenty or thirty seeds, which are now being raised at Kew. The cones in this species resemble those of the western larch in the manner in which they quickly cast their seeds in September.

The western larch in this region did not mingle with the Alpine larch, the former ascending, in company with Douglas fir, the northern slope up to 5900 feet; and between this elevation and 6600 feet, where the lowermost Alpine larch was found, no trees were growing.

Two other localities farther south are mentioned by Ayres,[1] who states that on the summit of the continental divide (long. 113°, lat. 47° 25′), between the Sun river and Willow Creek, there is a fine forest of the species, with trees about 70 feet high and 15 inches in diameter. Twenty miles due west on the summit of the range north of Pend d'Oreille pass there are a few scattered trees.

In the Whitefish range and in the mountains between the Kintla and Chief Mountain lakes, the tree is common on northern slopes from the Canadian boundary line to about 15 miles south of it. In the Whitefish range, Ayres[2] reports that the trees attain a maximum size of 80 feet by 6 feet in girth, the largest growing about the heads of basins where the snow lingers late into summer or lies in banks throughout the season. I visited the Whitefish range, which is a few miles from Fortine, on the Great Northern Railway, late in September, in company with Mr. Eastland, forest ranger, and at 7000 feet altitude could distinguish numerous groves of Alpine larch, extending over the mountains for an immense distance, as the foliage,

---

[1] *U.S. Geol. Survey, Lewis and Clarke Forest Reserve*, 42, 43 (1900).

[2] *U.S. Geol. Survey, Flathead Forest Reserve*, 268 (1900).

which had turned yellow at this season, rendered the trees very conspicuous ; but in all cases the groves were confined to strictly northern slopes.   We encamped in a small grove, where the trees did not exceed 40 feet in height, and observed numerous seedlings ; but were forced to descend on account of a heavy fall of snow and to leave the larger and more important forests unvisited.

Further east, in the Kintla lake region, Ayres[1] reports that the mountain slopes are best wooded on the northern slopes, where the Alpine larch reaches a height of 80 feet and a diameter of 30 inches.   It is more vigorous here than in any other locality seen by Ayres, who considers that the tree will produce timber suitable for mining purposes.   (A. H.)

[1] *U.S. Geol. Survey, Flathead Forest Reserve,* 277 (1900).

PINUS LARICIO[1]

Pinus Laricio,[2] Poiret, *Lamarck's Dict.* v. 339 (1804); Lambert, *Genus Pinus*, i. 11, t. 4 (1832); Loudon, *Arb. et Frut. Brit.* iv. 2200 (1838); Forbes, *Pinetum Woburnense*, 23 (1839); Parlatore, DC. *Prod.* xvi. 2, p. 386 (1868); Masters, *Gard. Chron.* xx. 785, fig. 142 (1883); xxi. 18, fig. 1 (1884); iv. 692 (1888), *Journ. Linn. Soc. (Bot.)* xxxv. 624 (1904); Willkomm, *Forstliche Flora*, 226 (1887); Mathieu, *Flore Forestière*, 596 (1897); Kent, *Veitch's Man. Coniferæ*, 338 (1900).

Pinus nigra, Arnold, *Reise nach Mariazell*, 8 (1785); Kirchner, *Lebengesch. Blutenpfl. Mitteleuropas*, 231 (1906).

Pinus austriaca, Höss, *Flora*, viii. *Beiträge*, 113 (1825); *Gard. Chron.* ix. 275, figs. 49, 50 (1878).

Pinus nigricans, Host, in Sauter, *Versuch Geog. Bolan. Schilderung Umgeb. Wiens*, 23 (1826).

Pinus taurica, Loddiges, *Cat.* (1836).

Pinus caramanica, Bosc. *ex* Loudon, *op. cit.* 2201 (1838).

Pinus dalmatica, Visiani, *Fl. Dalmat.* i. 199 (1842).

Pinus monspeliensis, Salzmann, *ex* Dunal, *Mém. Acad. Montpell.* ii. 82 (1851).

Pinus Salzmanni, Dunal, *loc. cit.*

Pinus calabrica, cebennensis, and *poiretiana*, Hort, *ex* Gordon, *Pinetum*, 168 (1858).

Pinus Fenzleyi, Carrière, *Rev. Hort.*, 1864, p. 259.

Pinus Fenzlii, Antoine et Kotschy, *ex* Carrière, *Conif.* 496 (1867).

Pinus pindica, Formanek, *Verhandl. Naturf. Verein Brünn*, xxxiv. 20 (1896); Masters, *Gard. Chron.* xxxi. 302, figs. 95, 96 (1902).

A species very variable in habit, dimensions, and foliage, comprising several different geographical forms, which under cultivation preserve in a great measure their peculiarities. The following description is drawn up from wild specimens of the Corsican tree, which is the finest form.

A tree attaining 150 feet in height and 20 feet in girth. Bark on old trees about an inch thick, deeply fissuring into irregular longitudinal plates, which exfoliate in small rounded scales, leaving exposed pale brown, slight oval depressions where they fall off. Buds $\frac{1}{2}$ to 1 inch long, elongated, abruptly contracted to an acuminate apex, light brown in colour, tinged with white, the lowermost scales loose and reflected, the uppermost bound together by white resin. Branchlets stout, glabrous, brown in colour; leaf-bases very prominent, keeled, and imbricated, persisting for several years on the older leafless branchlets.

Leaves, in pairs, densely covering the whole branchlet on barren shoots, forming an apical cup-like tuft above, directed upwards and forwards below; deciduous in the fourth or fifth year; stout, 4 to 6 inches long, about $\frac{1}{16}$ inch wide, straight or curved, often twisted,[3] serrulate, ending in a short callous point; semi-terete in section, with

---

[1] The generic description of Pinus will be given in a later part. There is no English name in common use for the whole species. The different forms are well known, as the Corsican, Austrian, and Pyrenean Pines.

[2] The oldest name for the species is *Pinus nigra*, Arnold, which has lately been revived by some German writers. We adopt the name *Pinus Laricio*, Poiret, as it has been in general use for more than a century.

Pinus pallasiana, Lambert, *Genus Pinus*, i. 13, t. 5 (1832), is impossible to recognise, being supposed by some to be *Pinus Laricio* and by others to be *Pinus Pinaster*.

Pinus pyrenaiaca, Lapeyrouse, *Hist. Pl. Pyrén., Suppl.* 146 (1818), points, so far as the locality is concerned, to the Pyrenean variety of Laricio; but the description is doubtful. Mr. H. L. de Vilmorin, who gives a history of this name in *Bull. Soc. Bot. France*, xl. p. lxxvii (1893), considers it to refer to *Pinus Brutia*; but M. Calas, in his account of the *Pin Laricio de Salzmann*, p. 22, controverts this opinion, and believes the description to apply to the Pyrenean Laricio.

[3] The twisting of the leaves, supposed to be characteristic of the Corsican variety, is an inconstant character.

twelve lines of stomata on the convex surface and eight lines on the flat surface ; resin canals median, surrounded by stereome cells, meristele elliptic, fibro-vascular bundle branched. Basal sheath about ½ inch long, brown near the base, whitish above, becoming on old leaves short, lacerated, and blackish.

Male flowers clustered, three to ten or more in number, on the lower half of the branchlet of the first year, which grows beyond the inflorescence and bears leaves above ; later, when the flowers drop off, these fertile branches appear to be bare of leaves in their lower half. The male flowers are upright, yellow, cylindric, stalked, about an inch long ; connective crest large, purplish, finely toothed. Female flowers single or two to three at the top of the young branchlets, very shortly stalked and bright red in colour, remaining as small (½ inch diameter) globular cones till the beginning of the second year.

Cones ripe at the end of the second year, solitary or in pairs or threes, sub-terminal, sessile ; variously directed, upwards, horizontally, or even curving down-wards ; shining brown ; ovoid-conic, 2 to 3 inches long by an inch in diameter, straight or curved, symmetrical, ending in a narrow apex. The cones open in the spring or summer of the third year and soon after the escape of the seeds fall off. Scales about an inch long ; concealed part thin, dark reddish brown below and light brown above ; apophysis or visible part shining yellowish brown, raised, rounded at the upper margin, with a transverse keel, curved on each side of the central umbo, which is reddish brown and bears a minute or obsolete prickle. Seeds greyish or brownish, more or less mottled, about ⅙ inch long ; wing three or four times as long, striated light brown, straight on one side and gently curved on the other, about ¼ inch wide at the broadest part, which is at the middle or just below it. Seedling with six or seven cotyledons.

The different geographical forms may be arranged as follows :—

1. Var. *corsicana*, Loudon, *loc. cit.* (var. *poiretiana*, Antoine, *Conif.* 6 : 1840), Corsican Pine. Occurs in south-east Spain, Corsica, southern Italy, Greece, and Crete.

A tall tree with straight stem and slender branches. Leaves light green in colour, not extremely dense upon the branchlets, the whole aspect of the foliage being lighter in colour and sparser in quantity than in the Austrian pine. Buds not very resinous. Cones usually without radiating cracks on the apophyses.

Var. *calabrica*, Loudon, *loc. cit.*, is scarcely distinguishable. As seen under cultivation at Les Barres, it has perhaps slightly denser foliage than the Corsican variety growing beside it.

2. Var. *austriaca*, Loudon, *loc. cit.* (*Pinus nigra*, Arnold ; *Pinus austriaca*, Höss ; *Pinus nigricans*, Host ; *Pinus Laricio*, var. *nigricans*, Parlatore). Austrian Pine. Austria, Balkan Peninsula, Crimea, Caucasus, Asia Minor.

Shorter tree, with numerous stout branches. Leaves dark green in colour, extremely dense upon the branchlets, giving the whole tree a dense dark crown of foliage. Buds resinous, whitish, stouter than in the Corsican pine. Cones usually showing radiating cracks in the apophyses.

Var. *pallasiana*, Endlicher, *Syn. Conif.* (*Pinus pallasiana*, Loudon, *op. cit.* 2206).

This name is given in England to trees with numerous stout branches, the lowermost of which ascend parallel to the trunk ; but in foliage scarcely different from the Austrian pine.[1] The cones are usually larger than in that variety and have the radiating cracks strongly marked. This form is supposed to have come from the Crimea. The Laricio which occurs in the Crimea, Asia Minor, and the Caucasus appears, however, to be identical with the Austrian form.

Var. *caramanica*, Loudon, *loc. cit.* (var. *Karamana*, Masters, *Gard. Chron.* 1884, xxi. 480, fig. 91). This is the Austrian pine as regards the foliage ; but producing extraordinarily large cones, up to four inches or more in length. It is supposed to be identical with a form introduced into Paris by Olivier, who sent seeds in 1798 from Caramania in Asia Minor ; but is perhaps only a mere sport of the common Austrian pine. The only specimens known to us are two trees at Syon, grown on the lawn west of the mansion ; and one of these measured, in 1903, 72 feet by 8 feet 6 inches.

3. Var. *tenuifolia*, Parlatore, *loc. cit.* (vars. *pyrenaiaca et cebennensis*, Grenier et Godron, *Flore de France*, iii. 153 (1856). *Pinus monspeliensis*, Salzmann. *Pinus Salzmanni*, Dunal). Pyrenean Pine. Cevennes and Pyrenees.

Small trees, often stunted in growth, with remarkably slender leaves, only half the thickness of the other forms. Young branchlets orange-coloured. Cones smaller than in the Corsican variety. Owing to its slow growth, the annual shoots are very short, and the older branchlets remain slender and bare of leaves for a great distance behind the short tuft of leaves at their extremities.

Pinus leucodermis, Antoine, treated by us as a distinct species, is considered by many authorities to be only an alpine form of Laricio; and there appear to be similar forms occurring in high regions elsewhere, as *Pinus Fenzlii*, Carrière, which resembles *P. leucodermis* in having short leaves, almost appressed together in the bundles.

Pinus pindica, Formanek, reported as growing in the Pindus and the Thessalian Olympus, is not recognised by Halacsy ;[2] and is probably only a slightly aberrant form of the ordinary Corsican variety. It has been fully described and figured in *Gardeners' Chronicle*, *loc. cit.*, by Dr. Masters.

Horticultural varieties of Laricio are few and unimportant. Beissner[3] mentions pendulous, variegated, and dwarf forms. A golden variety[4] of the Austrian pine, said to have been raised or introduced by Mr. Mongredien of the Heatherside Nursery, has the leaves, especially those on young growths, tipped with gold. Ilsemann[5] saw a tree, in which the leaves were beautifully variegated with yellow, growing wild in a forest in Hungary. A peculiar form of Austrian pine with stout falcate leaves has been observed at Breslau.[6]

---

[1] Probably some trees called *Pallasiana*, on account of their habit, are really of Corsican origin.
[2] *Consp. Fl. Græca*, iii. 452 (1904).
[3] *Nadelholzkunde*, 243 (1891). Masters saw at Moser's Nursery, Versailles, in 1903, a dwarf variety of very compact habit with dense bright green foliage : *Gard. Chron.* xxxiv. 338 (1903).
[4] *Gard. Chron.* xvi. 507 (1881) and ii. 730, 785 (1883).
[5] *Gartenflora*, 1897, p. 643.   [6] Baenitz, *Gartenflora*, 1903, p. 58.

INTRODUCTION

According to Loudon,[1] the Corsican variety was introduced into England, as long ago as 1759, under the name *Pinus sylvestris*, ε *maritima*, which was adopted by Aiton.[2] In France, the tree in the Jardin des Plantes at Paris was planted in 1774; but the date of introduction of the first seed is probably earlier.   The Austrian pine was introduced[1] in 1835 by Lawson of Edinburgh.   Var. *pallasiana* was first raised by Messrs. Lee and Kennedy, Hammersmith, from seeds sent to them about the year 1790 from the Crimea by Professor Pallas.[1]   Captain Cook[1] imported seed in 1834 from the Sierra de Segura in the south of Spain; but the plants raised were probably indistinguishable from the ordinary Corsican variety; and there is no record of the introduction of the Pyrenean or Cevennes variety, of which we know of no large trees in this country.

DISTRIBUTION

The species has a widespread distribution, extending westwards from Spain into the Cevennes in France, finding its northerly limit in Austria, and descending into Corsica, Italy, Sicily, the Balkan peninsula, Greece, Crete, and Cyprus, it re-appears in the Crimea and in Asia Minor, and reaches its most easterly point in the Caucasus.

In Spain, a form considered by Willkomm to be identical with the Corsican variety occurs scattered through the plateaux and mountains of the south-eastern and central provinces, at altitudes between 1000 and 3500 feet.   The largest forests occur in the Serrania de Cuenca, and in the sierras of Segura and Cazorla, the most southerly point reached being in the last-named mountain in N. lat. 37° 40′ and W. long. 3°.

. PYRENEAN LARICIO.—The form[3] which occurs in the Pyrenees and the Cevennes is remarkable for its stunted growth and slender leaves.   It grows on the Spanish side of the Pyrenees in the province of Aragon, not far from Venasque, between the rivers Esera and Cinca.   From this locality, which was visited by Mr. H. L. de Vilmorin in his investigations of the Pyrenean Laricio,[4] seeds were regularly sent to Paris for many years early in the 19th century, by M. Boileau, pharmacist at Bagnères-de-Luchon.

M. Calas, who has written an elaborate memoir[5] on this variety, accompanied by a map of its distribution and numerous illustrations of the forests reproduced from photographs, discovered it in 1890 on the north side of the Pyrenees near Prades. Here it covers a scattered area of about 3600 acres in the hills south of the river Têt and north of Mount Canigou, the district being called Conflent; and grows on glacial clay at elevations between 1880 and 3300 feet.   In most places the original forest has been ruined by sheep-grazing and fires, and usually only small isolated

---

[1] *Op. cit.* 2204, 2206, 2208, 2209.   The date for the Corsican pine is not improbable, as Loudon (viii. t. 315) gives a figure of a tree at Kew, which was 85 feet high in 1838.

[2] *Hort. Kew.* iii. 366 (1789).   [3] Cf. Durand, in *Bull. Soc. Bot. France*, xl. p. ccxxviii (1893).

[4] *Ibid.* p. lxxvii.

[5] *Le Pin Laricio de Salzmann*, pp. 50, tt. 1-19.   Published at Paris by the Minister of Agriculture in 1900.

groups of trees are to be seen, in the ravines and on the precipices. There are, however, two woods of considerable extent ; and one of these, situated in the basin of the stream of Masos, is considered by M. Calas to be the finest which he has seen, as regards the density, regularity, size, and vigour of the trees, which are, however, only about 80 to 90 years old. The best trees in the district are 50 to 60 feet high by 3 to 4 feet in girth.

In the Cevennes, this variety occurs in three localities. In Herault, near Saint-Guilhem-le-Désert,[1] it covers, between 1700 and 2300 feet elevation, about 2400 acres, of which 1900 have lately been purchased by the Government. The soil is dolomite limestone and is extremely poor and shallow ; and the trees growing either on southern arid slopes or on wind-swept plateaux are in a worse condition than elsewhere. They usually have twisted stems and average 15 feet in height ; attaining at their best 30 feet high by 3 feet in girth.

Another locality[2] occurs north of Bessèges, in the valley of the river Gagnières, which forms the boundary line between the departments of Gard and Ardèche. The tree grows here at 650 to 1100 feet elevation on siliceous soil, and covers a scattered area of 2500 acres, half of which belongs to the State. It often attains, on northern slopes and on slightly better soil than usual, 60 feet high by 4 feet in girth. This appears to be the only locality where the tree is regularly felled, the timber being sold for pit-props. The maritime pine has been planted in the district in the open spaces caused by forest fires, and though slightly faster in growth than the native Laricio, has proved to be a poorer tree, on account of the inferior quality of its timber.

M. Fabre discovered in 1897 a third locality in the Cevennes, at the Col d'Uglas, eight miles west of Alais in Gard. The area is only 250 acres ; but is interesting, on account of *Pinus sylvestris* growing wild in company with *Laricio* in the upper part of the forest.

The Pyrenean pine has been planted in a few localities in Ardèche, Herault, Aude, and Pyrénees Orientales ; and has done slightly better than the Austrian pine tried with it. Calas considers it to be a useful tree, on account of its capability of growing on the worst possible soils ; and is of opinion that its meagre growth in the wild state is entirely dependent on the poor conditions of soil and climate to which it is subjected.

CORSICAN PINE.—This species is widely spread in Corsica in the great mountain range and its ramifications, which occupy the centre of the island. On northern slopes it grows between 2700 and 5500 feet elevation, the lower margin of the forest being often contiguous with dense woods of *Quercus Ilex* or with scattered groves of *Quercus lanuginosa*. On southern sunny slopes it only descends to 3700 feet, the zone below that altitude being usually occupied by *Pinus Pinaster*, the two species mingling slightly at the line of junction. The forests of *Laricio*, often of great extent, belong almost entirely to the State and to the Communes, and are all treated by the selection

[1] Here this variety was first discovered in France by Salzmann in 1851.
[2] First mentioned in 1856 by Grenier and Godron, *loc. cit.*

method. The pine usually occurs pure; but in the ravines small and unimportant groups of silver fir are often seen, and the edges of the streams are bordered in many places by *Alnus cordifolia*. The beech in Corsica attains as high an elevation as *Laricio*, and in some cases the two species are mixed, and a struggle occurs for predominance. Birch is occasionally a component of the pine forest, but is comparatively rare. The soil on which *Laricio* grows is usually extremely poor, consisting of debris of granite rocks, and contains very little humus or decayed vegetable matter.

The following observations which were taken in 1906, at 3200 feet altitude, in the midst of the *Laricio* forest at Vizzavona, show the climate in which the tree thrives :—

	Precipitation in inches.	Days of Fall of		Temp. Fahr.	
		Rain.	Snow.	Max.	Min.
January	10.98	4	5	47°	18°
February	3.74	3	13	46°	16°
March	4.05	5	6	64°	21°
April	3.46	12	4	59°	27°
May	5.28	13	1	75°	32°
June	0.31	7	...	75°	43°
July	1.61	8	...	77°	46°
August	0.12	2	...	79°	48°
September	1.73	3	...	77°	39°
October	7.60	5	...	66°	41°
November	14.61	14	1	61°	30°
December	15.44	8	11	59°	16°
Total	68.93 inches.	84 days rain.	41 days snow.		

Snow and low but not extreme temperatures are common during nearly six months of the year, from November to the beginning of May. The sky is generally clouded more or less completely during a greater part of the year; a clear blue sky only being recorded on 77 days out of the whole year.

The *Laricio* forests are easy of access, owing to the railway, which goes through the heart of the mountains from Ajaccio to Bastia; and in spite of a heavy fall of snow I succeeded in seeing some of the most important forests in the last week of December 1906. The finest is Valdoniello, which lies about twenty miles west of Corte railway station, the road to it passing through the magnificent gorge of the Scala di Santa-Regina. This forest occupies the upper basin of the river Golo, which has a north-easterly exposure, and its wooded area covers 6682 acres lying between 3100 and 5100 feet altitude. The soil is very dry and extremely poor, consisting of granite debris; and the few beech and silver fir that were seen could only obtain a footing in the ravines. The forest is divided into two series, one of which, about 4000 acres in extent is being regularly felled, whilst the other series at a greater elevation is left untouched as a zone of protection. In the first series there are 109,000 trees over 16 inches in diameter, 4000 of which are decayed or diseased. Only trees over 9 feet in girth are marked for felling; and these are being cut down gradually, two or three trees in each spot, so that gaps are left in which seedlings may spring up. Though good seed years occur about once every three years, natural regeneration is always difficult on account of the poverty and dryness of the soil, and only occurs in open spaces exposed to sunlight. As a great deal of the best timber has been removed in past years, the number of excessively large trees is limited, there being only thirteen over 14 feet in girth. The largest tree now standing, the " Roi des Laricios," is growing in a dense part of the forest at 3850 feet altitude, and measured 143 feet in height by 18 feet 9 inches in girth, with a clean stem to 100 feet. Plate 113, from photographs taken by me, shows the stem of this tree and a dense stand of pines. Plate 114, from a negative kindly lent us by M. A. André, Inspector of the French Forest Service, shows very well the peculiar habit assumed by the Laricio in old age, the crown becoming remarkably flattened, owing to the bending over of the leading shoot and the increase in size of the upper branches, which become very stout and horizontal or even curve slightly downwards. The frontispiece is reproduced from a sketch taken in Corsica by the late Robert Elwes of Congham, Norfolk.

In this forest the presence of a considerable number of diseased trees is probably explained by the fact that some twenty years previously most of the large trees had been tapped for resin, an operation which was not justified by its financial results, and which exposed the trees to the attacks of fungi. In many parts of the Valdoniello forest, as in *parcelle F*, the trees are very tall, and stand very close together, and have beautifully clean stems, showing that the tree bears crowding without injury. The foliage of the trees in Corsica struck me as being denser than is the case usually in isolated trees growing in England; and I agree with Prof. Fliche that the canopy of Laricio is considerably denser than that of the Scots pine, and as a corollary that plantations should not be over-thinned. In Corsica, as only trees of large size are saleable, no thinning operations are ever attempted.

The railway passes through another fine forest, that of Vizzavona, which is about 3400 acres in extent. The trees here are as a rule younger than those at Valdoniello, and in many parts of the forest are mixed with beech, between 3000 and 4000 feet. In one place it was evident that, owing to an excessive felling of *Laricio* several years ago, the young forest coming up will consist almost entirely of beech. In pure stands of young but tall pines there is usually a slight undergrowth of beech and holly. Near the forester's house I measured a large tree, 145 feet high by 12 feet 3 inches in girth, which was growing at 3200 feet altitude.

With regard to the size attained by *Laricio* in Corsica, a tree in the forest of Pietropiano with a short stem measured 23 feet in girth. In the forest of Marmano trees have been felled which were clean in the stem to 115 feet, and yielded 950 cubic feet of dressed and squared logs. At Aitone there is a fine forest of Laricio which I was unable to visit from Valdoniello, as the pass across the mountain was impassable owing to deep snow. I was informed that the forest of Asco has been practically untouched by the axe, and contains many very old trees of peculiar habit.

The *Laricio* grows with extreme slowness in the mountains of Corsica, trees 40 inches in diameter averaging about 360 years old, and those over 5 feet in diameter are often as much as 700 years.

The timber of young trees is valueless in Corsica, as it contains practically only sapwood, which rapidly decays on exposure to the air. The sapwood is white in colour, and always considerable in thickness, varying on an average from 8 inches in young trees (77 years old) to 2 to 3 inches in old trees (250 years old and upwards). The heartwood, which is reddish brown, only develops in quantity when the trees attain an advanced age, exceptionally at 120 to 150 years, usually at 300 years. At the latter age the trees average 3 feet in diameter, and are considered to be mature and at the most profitable period for felling. Most of the timber is exported in the form of logs to Italy, where it is much esteemed, and is used for shipbuilding purposes generally. The logs are squared in the forest, all the sapwood being chipped off except a little at the four corners. Saleable logs must be at least 23 feet in length, and have a minimum section at the small end of 1 square foot. They fetch at Bastia, after a long haulage by road and railway, 36 to 40 francs per cubic metre, or about 10d. to 11d. per cubic foot. A small proportion of the timber in the forests is cut up into planks and joists for local use. The timber is very strong but heavy, and often contains a great deal of resin; when of the first quality it is considered to be as good as American pitch pine. It is very seldom used in France, and the reasons for this are not very clear.

I could obtain no information as to the collection of the seed of *Laricio* in Corsica, though I made inquiries when visiting the forests and also at the Conservator's office in Ajaccio. Mr. M. L. de Vilmorin, however, kindly informs me in a letter that the annual collection amounts to about three or four tons, of which his firm disposes of about one-half. The main localities for collecting are near Corte and Calacuccia, and at Vivario, which is not far from Vizzavona. The cones are put in the ovens which the villagers use for drying chestnuts, and as the amount of heat is not regulated with any precision, the seed is often over-heated. Though the crop of cones in the forest varies very much in different years, there has been no difficulty so far in procuring always a quantity of seed sufficient to meet the demand.

In Sardinia the Corsican pine is recorded from only one locality, the valley of the Flumini Maggiore, where it was collected by Moris.[1]

CALABRIAN PINE.—In Sicily the Corsican pine is common, according to Schouw,[2] on Mount Etna, where it forms woods between 4000 and 6000 feet. It is, however, in Calabria, in Sila and Aspromonte, that *Laricio* occurs in abundance, and there is little doubt that the tree here is identical with that of Corsica. Schouw,[2] who compared specimens from the botanical garden at Naples with the large Corsican pine growing in the Jardin des Plantes at Paris, is convinced of their absolute identity. Longo, who has recently written an article[3] on the flora

[1] Parlatore, *Fl. Italiana*, iv. 53 (1867). Moris's specimens, though without flowers or fruit, are probably *Laricio*, according to Parlatore. [2] *Ann. Sci. Nat.*, III Ser., iii. 234 (1845). [3] *Annali di Botanica*, iii. 1-17, tt. 1-6 (1905).

of Calabria, gives five plates, reproductions from photographs, of the Calabrian forests, and a plate showing the variation in the cones; but he has added little to our knowledge of these interesting forests in his short description of them. He states that the finest one is the State forest of Gallipano. (A. H.)

As I could find no account of this tree in its native country, and it was then little known in England; from the information I received from Signor Siemoni, chief of the Forest Department at Rome, I visited Cosenza, a town in Calabria, in April 1903. Here I was kindly received by Signor Carlo Pagliano, Inspector of Forests, who directed me to a village called Spezzano Grande, two hours' drive from Cosenza, from where I rode with Signor D. Greco, the sub-inspector, to the Sila Mountains, on which the largest forests of this tree now exist. The snow was still lying on the pass at about 4800 feet, but on the plateau beyond this it had melted except in shaded places. The forest is composed mainly of pine, here called Pino della Sila, Pino Rosso, or Pino Butello, mixed with beech in some places; but the forest has been considerably diminished by felling in former times, when the dockyards of Naples drew a large part of their timber from this district. The inspector told me that the only place he knew of where virgin forest of this tree still remained, was on a mountain called Femina Morte in the forest of Carigleone, in the district of Cattanzaro, 60 to 70 kilometres south-east of Cosenza. The average size of the trees which I saw being cut for the sawmill was not above 80 to 90 feet by 6 to 8 feet in girth, and smaller where they grew densely. These trees were 80 to 90 years old, and the heartwood, 10 inches in diameter, was reddish. In places where fire and cattle had not destroyed them, the natural reproduction was very good, and the seedlings when once established were making 2 to 3 feet of growth every year. The trees grew best in a south aspect on a soil which appeared to be decomposed granite, and, as far as I could learn, there is no limestone in this district. On my way back I visited Potenza in the Basilicata, whence, according to M. de Vilmorin's information, the seeds of the tree originally were introduced; but if the tree ever existed in the district, I could hear nothing of it.

AUSTRIAN PINE.—The Austrian pine has been the subject of a monograph by Prof. A. von Seckendorff[1] which gives very elaborate details of its literature, economy, and distribution in Austria, with maps and illustrations of remarkable trees in various places, which should be consulted by those who wish to know more than the brief résumé which I give. It occurs as a wild tree abundantly only in Lower Austria in an area extending from Mödling, near Vienna, south to near Pitten and south-west to Reichenrau, especially on the Alpine chalk formation, and attains an elevation of about 4000 feet. It attains a very great age, the rings of one felled near Stixenstein showing no less than 584 years, though the tree was only 65 feet high and about 6 feet in girth. In very rocky situations it grows so slowly that a tree near Mehadia was 270 years old, with a trunk only 8 feet high and about a foot in girth at the base.

Among the trees most remarkable for size may be mentioned a splendid tree at Vostenhofer (fig. ii. of Seckendorff) which is about 75 feet high and 21 feet in girth.

---

[1] *Beiträge zur Kenntniss der Schwarzföhre* (Vienna, 1881).

It is divided into 4 stems near the ground and has a diameter of branches of about 25 yards.   A tree called the Broad Pine at Mödling, near Vienna (fig. iii.), has an umbrella shape, very unusual in this species.   It is only about 35 feet high but is no less than 60 feet broad.   A tree called the Cross or Picture Pine in the Grossen Föhrenwalde (fig. v.) is considered the finest tree there.   It measures about 65 feet high, of which two-thirds are clean trunk, and is 9 to 10 feet in girth at about 9 feet from the ground.   The tallest specimen which is mentioned is not much over 90 feet, very much less than those I saw in Bosnia, some of which were considerably over 100 feet and probably over 120 feet, with clean stems to two-thirds of their height.

On good ground, however, in Austria this pine forms very fine timber; an example (shown on fig. viii.) at Gutenstein, near Zellenbach, is said to be 280 years old with an average height of 30 metres.   Another of the same age at Fahrafelde is so like the growth of the tree in Bosnia that the photograph illustrating it (fig. ix.) shows the best form of this tree very well.

A hybrid between this tree and *Pinus sylvestris* was described by Reichhardt[1] as growing in the Forest of Merkenstein.   (H. J. E.)

In Hungary, according to Pax,[2] the Austrian pine is only found at Mehadia on the lower Danube, where there are woods on dry stony mountain slopes.   He noticed it, however, as a mere shrub at Talmacsel in the valley of the river Alt.   In Styria its occurrence as a wild tree is doubtful.   In Carinthia there are limited areas of this species on calcareous soil on the southern slopes of the Dobratsch.   It is also recorded from Istria, Carniola, Croatia, and the island of Cherso.   Ascherson[3] mentions one locality in Galicia.   In Bulgaria[4] it grows in several localities in the Rilo-Dagh, and in the Rhodope Mountains above Stanimaka.

An excellent account of the distribution and forest conditions of this species in the western states of the Balkan peninsula is given by Beck.[5]   The most extensive forests in this region lie in south-eastern Bosnia and extend across into Servia, in the district of *Novibazar*.   Fine pine forests occur at Semec, on the slopes of the Lim valley, and on the hills between the Lim and Ceatina rivers.   Between the middle part of the course of the river Drina in Bosnia and the river Morava in Servia the tree usually grows on palæozoic rocks, though it is occasionally seen on lime-stone.   In Servia the forests of Austrian pine are less extensive, but extend from Ivica to Kapaonik.   In middle Bosnia, where the tree is found growing on serpentine, and in western Bosnia, it is not at all common.

Elwes saw the tree growing abundantly in the valley of the Drina, as already mentioned in our account of *Picea Omorika*, and brought home a quantity of seed from this locality in 1901, which he distributed under the MS. name of *Pinus Laricio*, var. *bosniensis*, believing at the time that it was not the same variety as the common Austrian pine; but he now considers that the difference observed is no more than might be caused by a good soil and a more southerly and warmer climate.

[1] *Verh. Zool. Bot. Ges. Vienna*, xxvi. p. 462.   [2] *Pflanzenverb. in Karpathen*, 104 (1898).
[3] *Syn. Mitteleurop. Flora*, i. 213 (1897).   [4] Velenovsky, *Flora Bulgarica*, 518 (1891).
[5] *Veg. Illyrischen Länder*, 139, 226 (1901).

In Herzegovina, according to Beck, the tree grows down the Neretva valley to the Plasa Planina and the southern slope of the Prenj Planina. In Montenegro it is comparatively rare, *Pinus leucodermis* having been often mistaken for it. It occurs scattered through Albania. In Dalmatia there are peculiar forests of Austrian pine, in which there is a dense undergrowth of evergreen Mediterranean shrubs and *Juniperus Oxycedrus*; and Beck describes the most remarkable of these, which occur at about 2500 feet elevation, on the peninsula of Sabioncello and the island of Brazza. The greatest altitude in these regions at which the Austrian pine was seen growing by Beck was 5300 feet on the west slope of Mount Dinara in south-western Bosnia, on the Dalmatian frontier.

In Greece, *Laricio*, probably of the Corsican variety, occurs in the mountains, often forming extensive woods, and Halacsy[1] mentions various localities in the provinces of Epirus, Thessaly, Eubœa, Ætolia, Peloponnesus, and in Crete. In Cyprus[2] *Laricio* is only met with on the summit of Troodos and on some crests to the west, at 4000 to 5000 feet altitude, just above the zone of *Pinus halepensis*, the two species mingling slightly together at the line of junction, as is the case in Corsica. Mr. Madon, who cut down a hundred trees, says that the timber is of no value, on account of the large amount of sapwood in immature trees, until it has reached the age of 250 years. Hartmann,[3] who has recently visited Cyprus, gives an elaborate account of the *Laricio* forest. He states that pure woods of this species are rarely met, as in its lower zone, from 4000 to 4500 feet, it grows mixed with *Pinus halepensis*; and above this, to the summit of Troodos, it is accompanied by *Juniperus fœtidissima*. It attains a height of 80 feet and a girth of as much as 16 feet.

In Asia Minor, according to Tchihatcheff,[4] it grows mixed with silver fir on Olympus in Bithynia at 2700 to 5000 feet altitude, and in the same province, on Mount Samanly, at 1600 to 2100 feet, and in the island of Thasos, where it forms with *Juniperus excelsa* a wood in the littoral region. He records it near Soma in the mountains of Mysia; in the valley of the Meander in Troas; between Mughla and Eskischer in Caria; in the Antitaurus, where it forms mixed woods with *Juniperus excelsa*, *Abies cilicica*, cedar, and oak; and in various localities in Pisidia, Isauria, and Cilicia.

In the Crimea[5] it grows on dry, poor, calcareous soil, forming woods on the western slopes of the mountain chain which extends along the coast of the Black Sea. The Crimean pine has been made a distinct variety, *pallasiana*, but it is probably identical with the Austrian pine.

According to Radde,[6] *Pinus austriaca*, as he terms it, is rare in the Caucasus. Steven discovered it in 1840 in the neighbourhood of Gelentschik; and Kusnezoff has since found it at a place called Wulanskaja, 35 kilometres south-east of Gelentschik, where there is a small open grove with sound trees attaining 2 metres in girth. Radde adds that it grows near the Black Sea at Bulanka.   (A. H.)

---

[1] *Consp. Fl. Græcæ*, iii. 452 (1904).
[2] *Forests of Cyprus; Parly. Paper, Cyprus*, No. 366 of 1881, *Encl. No. 2*, pp. 28, 34.
[3] *Mitt. Deutsch. Dendrol. Ges.* 1905, p. 172.   [4] *Asie Mineure*, ii. 497 (1860).
[5] Antoine, *Conif.* 6 (1840).   [6] *Pflanzenverb. in Kaukasusländern*, 169, 184 (1899).

Cultivation : Corsican Pine

Of all the conifers introduced into England, of which great expectations have been formed, none except the larch has shown such good results as the Corsican pine, which has proved a hardy and vigorous grower on almost all soils, and in almost all parts of Great Britain and Ireland. It has not, however, been long enough in the country to have established a position in the English timber market, and until it does it is difficult to say much of its economic value in the future. All accounts of this wood for estate purposes, though often used long before it has attained sufficient age to give the best results, agree in saying that though rough and knotty when grown singly, it is at least as good as Scots pine ; probably more durable and stronger when used before maturity. Though it does not grow so fast on very barren and stony soils as the Austrian pine, it is far better from a timber point of view, and occupies less space. Its greatest defect is the difficulty of transplanting it when young on account of its very scanty root system, and as this often, indeed usually, entails considerable loss on both nurserymen and planters, the cost of getting a crop of *Laricio* established is very much higher than in the case of the Scots pine.

I have been most successful in avoiding a high death-rate by purchasing two-year seedlings with as many roots as possible from French nurseries in the spring, not before the middle of March, planting them at once in nursery rows on as sandy a soil as possible, and transplanting them to their permanent habitation in March or April, two years afterwards. But the plants will not then be large enough for the better class of land, and may require another transplantation before finally going out, by which time they will have cost 40s. to 50s. per 1000, and in some cases much more. The seedling has a very long primary root at first with very little fibre. By cutting this tap-root when the plant is only a year old, without lifting it from its seed-bed, it may be induced to make more roots, but if left unprotected for the first winter on wet or heavy soil a great many of the seedlings will be thrown out of the ground altogether. In my own ground I prefer to sow the seed in boxes, as their growth in the open ground is slow in comparison with what are raised in France. In order to overcome this difficulty some nurserymen adopt the practice of lifting all their one-year seedlings before winter sets in, and laying them in until spring, when they are lined out for two seasons' growth before being again transplanted.

I have on two occasions tried sowing the seed in the field where I wished the trees to grow, but with little success. The seedlings remain so small for the first two or three years that they cannot be seen among the grass, which soon covers them, and though this species seems to suffer less than any tree from being planted among coarse grass, it takes five or six years before the seedlings become conspicuous, and it will also be found that in some places they are too thick, and in others have entirely failed.

The Corsican pine is distasteful in the young state to hares and rabbits. An experiment to test this was made some years ago at Tortworth Court, where Lord Ducie planted a young *Laricio* in the centre of a rabbit warren, which, until the ground was covered with snow, the teeming population of the spot did not touch ;

and even then, when starving, after an attempt to consume the young needles of the buds, they abandoned the experiment.[1]

Captain the Hon. R. Coke, a very close observer of trees, sends us the following notes from Holkham :—

" In distinguishing between *P. Laricio* and *P. austriaca*, one must apparently be guided rather by the general appearance and habit of the trees, than by any hard and fast rules. *Laricio* always looks well-bred in comparison with the coarseness of *austriaca*. Even when the former develops great limbs, coarse in themselves, the more delicate foliage will distinguish it from its Austrian relative. A good instance of this may be seen at Wolterton, where a fine specimen of each are growing side by side.

" Though the curved or twisted leaves are usually considered to mark the Corsican, yet this feature has been noticed in trees thoroughly Austrian in every other respect; moreover, some Corsicans have straight leaves. Sometimes the branches being produced in regular whorls up the stem is considered to be the mark of a *Laricio*, but all Corsicans do not follow this rule.

" When planting the sandhills at Holkham at various times between 1855 and 1890, Lord Leicester took the precaution of wiring in *austriaca* against rabbits and omitting to do so in the case of *Laricio*. This was done because it had been found that the *P. Laricio*, which were all raised from the seed of the old trees at Holkham introduced from Corsica in the early part of the 19th century, were unharmed by rabbits, which eagerly devoured *P. austriaca*. At the present time, of the trees growing on the sandhills, namely, *P. Laricio*, *P. austriaca*, *P. sylvestris*, *P. maritima*, practically the only one which reproduces freely is the *Laricio*, as the rabbits, though no longer numerous, seem to be able to distinguish this tree from its congeners, and leave it untouched. On the other hand, some trees bought as *Laricio* from an English nurseryman, which had every appearance of being genuine, were recently planted to fill up gaps in a belt at Holkham, and in this case the rabbits ignored the nurseryman's label, and made short work of the so-called *Laricio*."

Mr. J. D. B. Whyte, agent to Lord Iveagh, confirms the statement that rabbits will eat Austrian, and will not touch Corsican pines when planted together; but though the gamekeeper says that he has never anywhere seen a Corsican damaged by rabbits, Mr. Whyte does not think that the question has been fully tested at Elveden. This tree and the Austrian pine are sometimes planted in the Eastern counties as belts and hedges, but do not form so dense a shelter, or bear clipping so well as the Scots pine.

The Corsican pine is apparently less liable than some other pines to the ravages of insects and fungi. A specimen, however, sent in July 1905 to Kew by Mr. Wellwood Maxwell of Kirkennan, near Dalbeattie, showed a branch attacked by *Peziza Willkommii*, and Sir Herbert Maxwell showed me a similar case on a tree at Monreith.

On the sandhills of the Norfolk coast, near Holkham, are a number of Austrian

---

[1] Hutchison, in *Trans. Scot. Arb. Soc.* vii. 55 (1875).

and Corsican pines, planted on what appears to be pure drift sea sand, but Colonel Feilden suggested to me that their health and vigour may be due to the presence of lime, produced by sea-shells in the underlying beds. These trees were, as I was told by Mr. Donald Munro, forester to the Earl of Leicester, partly raised from seeds produced by the old trees in the garden at Holkham, and planted thirty to forty years ago, together with *Pinus insignis*, *P. Pinaster*, and *P. sylvestris*, to form a shelter belt and bind the loose drifting sand. Though some of the trees had preserved the peculiar leaf, colour, and habit of the Corsican and Austrian varieties, there were many others which could not be identified with certainty. A great number of seedlings have sprung up on the south or landward side of the hills, of which the largest were twelve to thirteen years old and 9 to 10 feet high ; and many smaller ones of various ages were growing freely even in wet spots among tall rushes. Plate 115 shows the appearance of these seedlings. Rabbits and hares do not seem very abundant here, and I saw none of the Corsican seedlings barked, though one or two of the much scarcer *Pinasters* had suffered.

Mr. Richards, forester to Lord Penrhyn, is enthusiastic as to the merits of this tree, and writes to me that in North Wales it will grow where all other trees fail, that it stands wind better than any other conifer, and if planted in March and April few deaths take place. He grows it from seed collected in March and April and sown in May. He says there are many trees on the Penrhyn estate 80 to 90 feet high, but I did not see any quite so large as this. He considers that the timber is very good, better than that of any conifer he knows.

Captain Rutherford, agent to the Earl of Carnarvon at Highclere, also speaks very well of this tree, and sends me the dimensions of two not over seventy years old, one of which contains 201, the other 150 cubic feet, and a plank which he was good enough to give me certainly bears out his good opinion of the timber. It has pale red heartwood and yellowish sapwood, though it seems somewhat coarser in grain, and inferior to the wood of the Calabrian variety which I brought from Italy.

The Corsican pine [1] has not proved hardy in New England. It may be occasionally seen in the middle States, but there is no evidence, in large or old specimens, that this tree will really become a valuable acquisition for American plantations.

CULTIVATION: AUSTRIAN PINE

This tree is often sold as Corsican pine, but should never be planted knowingly except upon land where no better tree will grow, or to form a shelter belt on windy exposed hillsides of chalk or limestone, or on the sea-coast. For though a tree of extraordinary hardiness and rapid growth, it produces such a mass of large branches, and is so much inclined to fork, that its timber is extremely coarse, rough, and knotty, and would be unsaleable except at a very low rate or for pit-props. My father planted many of this tree, and I have found that though they make girth more rapidly than any other pine, they only thrive on sunny situations, where

---

[1] *Garden and Forest*, x. 471 (1897).

they have plenty of light and air; and though the great bulk of timber they produce in a short time may make them worth planting on such soils, yet I doubt the possibility of getting a sale at remunerative prices in most districts. In mixed or pure plantations their lower branches die off and leave large snags which are difficult and costly to remove, and though the very resinous nature of the wood may fit it for some purposes, I have never heard of its being utilised to any extent, except for pitwood. Austrian pine[1] has been planted very successfully as a shelter belt on the southern shore of Belfast Lough, about forty yards from the sea, in heavy clay; and behind it hardwoods and other trees are doing well. The tree has been extensively planted in many provinces of Austria and Hungary, mainly, according to Seckendorff, with the object of improving the soil for other trees; it has been recommended for this purpose on the poorer limestone soils of England, but the cost of so doing would in my opinion make the operation very unprofitable.

Though there is no reason why the Austrian pine should not sow itself in Great Britain, as the seeds ripen in hot years freely, yet I have never seen self-sown plants except near Sarsden Park, Oxfordshire, the property of Lord Moreton, and here only two or three young trees have sprung up on the rough limestone close to some old quarries.

The Austrian pine, according to Schübeler, is hardy in Norway as far north as Stenkjaer, at the upper end of the Throndhjem fjord. A tree in the Botanic Garden at Christiania, which Schübeler says was planted in 1842, is over 40 feet high, but was not a fine specimen when I saw it in 1906.

The Austrian pine[2] has been largely planted in the northern United States as an ornamental tree, and in youth is a handsome tree; but it generally succumbs to the attacks of boring insects before it has lost its bushy juvenile habit, and an Austrian pine in the United States more than fifty feet high is exceptional.

An account of Austrian turpentine,[3] which is derived from *Pinus Laricio*, is given by Georg Schmidt in an inaugural dissertation before the University of Berne in 1903.

CULTIVATION: CALABRIAN PINE

The Calabrian variety of *Laricio* was introduced into France by M. de Vilmorin in 1819-21, and a full account of its development at Les Barres was given in a catalogue of the trees cultivated there, published at Paris in 1878 by the Forest Department.[4] From this it appears that the tree has proved superior to other pines as a forest tree, and is especially recommended for planting in mixture with oak, which it rapidly surpasses in height, but without injuring it, on account of the slight development of its lateral branches. It has attained on this poor sandy soil a considerable size, and the young trees raised from seed grown there have preserved their superiority in the second and third generation. It produces seed abundantly there, but has the same defect as *P. Laricio* of being difficult to transplant. It is not easy to distinguish from the Corsican variety. M. Maurice de Vilmorin tells

[1] *Journal of Forestry*, 1879, p. 165.
[2] *Garden and Forest*, ix. 453 (1896) and x. 470 (1897).
[3] *Harzbalsam von Pinus Laricio* (Bern, 1903).
[4] Cf. Pardé, *Arb. Nat. de Barres*, 61 (1906).

me that "in nearly every place where this variety has been planted in France, it has proved to be in comparison with true Corsican pines the larger and finer of the two."

In Calabria the cones are gathered in December before they open, and kept till the following July, when they are spread out in the sun, and the seed falls out naturally, not being sown till the year after. I brought back in 1903 a sack of this seed which proved very good, and a large quantity of plants were raised from it by Prof. Fisher at Cooper's Hill, where they grew extremely well; better, as it seemed to me, than the Corsican pine, and much better than they did on my limestone soil. A number of these were sent to Culford, the seat of Earl Cadogan, in Suffolk, where his forester, Mr. Hankins, says that they stood the drought of 1906 very well on sandy soil. So far as I can see at present, the tree is quite hardy, and grows as fast or faster than the Corsican variety. It is equally difficult to transplant. Time alone will prove whether this tree has any economic value in England, but its superiority over the Corsican pine will be, I expect, only on soils deficient in lime, which the latter endures; and on granitic sand, in the warmer parts of England, it would certainly be worth a trial, either as a pure plantation, or, as recommended at Les Barres, in mixture with oak or beech.

A tree[1] reputed to be of the variety *calabrica* is growing in the Royal Botanic Garden, Belfast, and was 39 feet high by 3 feet in girth in 1905. It is said to be columnar in habit. A tree at Glasnevin, growing on the side of a hill, measured in 1906 41 feet by 4 feet, and is pyramidal in habit, with branches ascending at an angle of 45°. It is reported to have been planted in 1888, when four years old from seed.

REMARKABLE TREES

CORSICAN PINE.—One of the oldest, if not the oldest tree in England, stands near the entrance gate of Kew Gardens, and in 1903 measured 86 feet by 9 feet 3 inches. It was figured in the *Gardeners' Chronicle*, 1888, iv. 692, fig. 97, and according to J. Smith[2] was brought to England by Salisbury in 1814, when a seedling only 6 inches high.

In the pleasure ground at Holkham are three large trees which the Earl of Leicester believes to have been brought to England by a relative early in the nineteenth century, but the date of planting is somewhat uncertain. In 1907 they measured 85 feet by 11 feet, 80 feet by 9 feet 11 inches, and 80 feet by 9 feet 4 inches. Plate 116 shows two of these trees.

The tallest I have seen is at Brocketts, Herts, the seat of Lord Mountstephen, which, growing in a sandy soil and sheltered situation, was, when I measured it in 1905, no less than 119 feet by 8 feet 6 inches.

At Arley Castle, six fine trees, all over 100 feet high, measure 10 feet 8 inches, 9 feet 8 inches, 7 feet 9 inches, 8 feet 1 inch, 7 feet 8 inches, and 6 feet in girth respectively. Plate 117 shows the largest of these. Two of them have the habit of var. *Pallasiana*, but are indistinguishable in cones and foliage from

[1] Mentioned in *Gardeners' Chronicle*, 1870, p. 1537, as a prominent sort, distinct from the Caramanian or Corsican varieties.   [2] *Records of R. Bot. Gardens, Kew*, 286.

the others. At Albury, Sussex, there is one over 100 feet high by only 6 feet 9 inches in girth. At Highclere, Berks, in Great Pen wood, on sandy soil, are the best plantation *Laricios* which I have seen. At about 70 years old they measure about 90 feet high by 7 to 8 feet in girth, and have clean boles for about half their height: several of these, however, are forked at some distance from the ground. At Bayfordbury there is a tree which in 1906 was 94 feet by 8 feet 7 inches, and in many other places we have seen specimens 80 to 90 feet high, which need not be specially mentioned.

AUSTRIAN PINE.—Of the Austrian pine we have seen no specimens in England which rival the Corsican in height, though at Wolterton Park, Norfolk, the seat of the Earl of Orford, there are two large trees about 85 by 9½ feet, which show the characteristic difference in habit and in the colour of the leaves very clearly. From Grigor's account of this place in the *Eastern Arboretum*, p. 114, they seem to have been planted before 1840. Among the largest is a large spreading tree of this type at Nuneham Park, the seat of the Right Honourable L. Harcourt. Another at Canford Manor, Dorset, measured 83 feet by 9 feet; and at Williamstrip Park, on rather heavy soil, which this tree by no means seems to dislike, there is one of nearly the same dimensions, the largest I know in Gloucestershire.

Var. *Pallasiana*.—The best authentic specimen I know is a fine tree at Elveden, Suffolk, the property of Lord Iveagh. It is a flourishing tree with the foliage and cones of the Austrian variety, and measured when I saw it in 1907 94 feet by 8 feet 3 inches (Plate 118). Prof. A. Newton of Cambridge informs me that this tree was raised from seed sent by his eldest brother General Newton of the Coldstream Guards from Balaclava in 1854. The parent tree stood in a garden, which was used as a cemetery during the early days of the occupation of the Crimea. In the historic gale of 14th November 1854 the tree was blown down, and the graves covered with rubbish, and a cone was sent home in memoriam.

Other noteworthy trees are as follows :—

At Dropmore	.	. 108 feet by 11 feet 5 inches	*fide* A. Henry, 1904.
,, Beauport .	.	. 85 ,, by 11 ,, 5	,, ,, ,, ,,
,, Penrhyn .	.	. 95 ,, by 11 ,, 4	,, ,, ,, ,,
,, Smeaton-Hepburn .		64 ,, by 6 ,, 5	,, ,, ,, 1905.

At Chiswick House there is a good-sized tree, remarkable for having an immense growth of the character of what is usually called " witches' broom."

M. Gadeau de Kerville has figured[1] a very fine example of this pine, which was considered to be of the Calabrian variety by M. L. Corbiere (though this identification seems to me somewhat uncertain), which measured in 1894 35 metres (about 110 feet) high and 3.84 metres in girth. This tree is growing at Vatimesnil (Eure) in the park of M. de Vatimesnil, who believes it to have been planted by his ancestor about the year 1780. If this is correct, it is the oldest and probably the largest planted tree of the species either in France or England.

[1] *Les vieux arbres de la Normandie*, fasc. iii. p. 317, plate ix.

PINUS LEUCODERMIS, HERZEGOVINIAN PINE

Pinus leucodermis, Antoine, *Oestr. Bot. Zeitung*. xiv. 366 (1864); Beck v. Mannagetta, *Weiner Illust. Gartenzeit*, 1889, p. 136, and *Veg. Illyrischen Länder*, 353 (1901); Ascherson u. Graebner, *Syn. Mitteleurop. Flora*, i. 212 (1897).

Pinus Laricio, Poiret, var. *leucodermis*, Christ, *Flora*, l. 81 (1867); Masters, *Journ. Linn. Soc. (Bot.)* xxxv. 626 (1904).

An alpine tree attaining rarely 90 feet in height and 6 feet in girth. Bark ashy grey, fissuring into irregular plates, averaging 6 inches in length and 3 inches in breadth. Buds like those of *P. Laricio*, but darker brown in colour. Young branchlets glaucous. Leaves in pairs, persisting five or six years, densely covering the branchlets, except at the base of each year's shoot, which is bare for a short distance, forming an apical cup-like tuft, and on the rest of the branchlet directed forwards and slightly outwards; the two leaves in each bundle only slightly divergent; dark green, stiff, short, 2 to 3 inches in length, ending in a sharp cartilaginous point; basal-sheaths as in *P. Laricio*. According to Koehne,[1] the structure of the leaf differs from *P. Laricio* in the resin-canals not being surrounded by stereome cells; and Masters states that the hypoderm projects in wedge-shaped masses into the substance of the leaf, which is not the case generally in forms of *Laricio*.

Cones short-stalked, ovoid-conic, with a flat base, about 3 inches long, resembling generally those of *Laricio*, but differing in the uniform dull brown colour of the whole cone, the umbo being of the same colour as the rest of the apophysis. The lower scales of the cone have very prominent pyramidal apophyses, and the umbo has a well-marked short spine directed backwards. Concealed part of the scales light brown on both surfaces. Seeds as in *P. Laricio*.   (A. H.)

Pinus leucodermis was discovered in 1864 by Maly, who introduced it into cultivation the same year in the Belvidere, Vienna. The best account of the tree is given by Beck, who considers it to be specifically distinct from Laricio, and names it the Panzerföhre or Smré of the Herzegovinians. It is found in four distinct areas in Bosnia, Herzegovina, and Montenegro; and as the most southerly of these is on the Montenegro-Albanian frontier (lat. 42° 25′), it is probable that it also grows on the Peristeri[2] mountain, which lies west of Monastir in Albania. The most northerly locality (lat. 43° 40′), where it was discovered by Beck, is the Prenj Planina in the heart of Herzegovina. Here it occupies an area of about sixty kilometres in diameter, surrounding the western part of the Bjelasnica mountain, and forms a coniferous belt at from 4600 to 5500 feet elevation, rising solitary or in small groups to 5800 feet. Another area is the Bjela Gora, where the political boundaries of Bosnia, Montenegro, and Herzegovina unite around Mount Orjen. Reiser found it also in the Sinjavina Planina in Montenegro. Its occurrence in Servia is not yet established.

---

[1] *Deutsche Dendrologie*, 37 (1893).

[2] This must not be confused with another mountain of the same name, east of Janina in the Pindus range.

It seems to resemble *P. Cembra* in its way of growth, and is confined. to mountains of Triassic and limestone formation, where it forms a zone of scattered forest just below the limit of trees, usually not more than 1000 feet in depth, and finds its lowest level at 1000 metres on the Preslica planina, according to Reiser, near the railway station of Bradina; ascending on the Prenj and Orjen mountains to 1700 or 1800 metres. At the lowest elevation it is mixed with beech; at the highest with *P. montana, Juniperus nana*, and *J. sabina*.

In some places at the upper levels, where the snow lies very deep, it becomes very stunted, not rising more than 2 to 4 metres from the ground, but does not assume the procumbent habit of *P. montana*. It roots itself so firmly on the dry bare rocks of these mountains that no wind can hurt it, and it endures the burning sun and bitter winds of this region without injury. I am indebted to Herr Reiser of Serajevo for the photographs showing the habit of this tree (Plate 119).

In the upper Idbar valley there is a forest where *P. leucodermis* grows mixed with spruce, silver fir, Austrian pine, and yew, as well as with beech, ash, sycamore, *Pyrus torminalis*, and *Acer obtusatum*. Its smooth grey bark,[1] divided into irregular segments, makes it very easy to distinguish from the Austrian pine, but Beck does not think the name of whitebark pine so applicable as that of Panzerföhre or armoured pine. The tree attains under favourable circumstances a height of 90 feet, with a diameter of 6 feet at the age of 294 years.

Of its timber Beck says nothing, but a story which was current in Bosnia when I was there in 1899, and which doubtless has some foundation, leads one to suppose that it is very hard. A Bosnian Turk was said to have bought a lot of trees of this species, which he felled and floated down the Narenta, and sold the timber as that of larch.

With regard to the occurrence of this species elsewhere, Christ described as a new species, *Pinus Heldreichii*,[2] specimens which were collected on Mount Olympus in Thessaly. Afterwards, in a letter to Dr. Masters, he stated that this is only a remarkable alpine variety of *Pinus Laricio*, very reduced, and approaching in some respects *Pinus montana*. Halacsy[3] considers that this tree, which grows on Mount Olympus in company with the ordinary form of *Laricio* and with *Abies Apollonis*, is identical with *Pinus leucodermis*.

A tree referred to this species has been recently found in southern Italy by Dr. Biagio Longo. He mentions[4] two localities, the alpine zone of the Calabrian Apennines from Orsomarso to Mount Montea, and the mountain of La Spina in the province of Basilicata, where it grows in the zone of the beech, and rivals that tree in thickness of trunk; but the foresters in the Sila mountains do not recognise this as a distinct species, or did not know of its discovery when I was there in 1903.

Seeds were sent by Beck to Kew in October 1890; and five plants were raised, which have grown with remarkable slowness, being only 9 to 12 inches high in 1901.

---

[1] The bark is figured in Hempel u. Wilhelm, *Bäume u. Sträucher*, i. 161, fig. 84 (1889).

[2] Christ, in *Verh. Naturf. Ges. Basel*, iii. 549 (1863), but later, in *Flora*, l. 83 (1867), he states that *Pinus Heldreichii* is identical with *P. leucodermis*, which he considers to be only an alpine variety of *P. Laricio*.

[3] *Consp. Fl. Græcæ*, iii. 453 (1904). [4] *Annali di Botanica*, iii. 13, 17 (1905), iv. 55 (1906).

One of these trees, planted out in a bed near the pagoda, is barely 3 feet high at present.   Another which was sent to Colesborne was planted in a high exposed situation in my park, where it grows very vigorously on oolite soil.

When in Bosnia, on my way to collect seeds, I was obliged to return home suddenly, but my companion, Mrs. Nicholl, who visited the Prenj mountain, procured a quantity of seeds which I sowed in 1902, and which have grown as fast as either the Corsican or Austrian pines, and look more healthy and vigorous on my soil than any other pine I have raised.   They form a much better root-system when young than either the Austrian or Corsican pine, and in consequence are much more easy to transplant.   I moved a number in September last just before a period of drought, and they have passed through a severe winter with very few deaths; I therefore believe that the tree will be a good one for planting in dry limestone soils, and may have a greater ornamental if not economic value than the Austrian pine.

(H. J. E.)

GYMNOCLADUS

Gymnocladus, Lamarck, *Dict.* i. 773 (*ex parte*) (1783); Bentham et Hooker, *Gen. Pl.* i. 568 (1865).

Guilandina, Linnæus, *Gen. Pl.* 518 (*ex parte*) (1742).

DECIDUOUS trees, belonging to the division Cæsalpinieæ of the order Leguminosæ. Branches stout and without thorns. Leaves large, alternate, bipinnate, the number of pinnæ being either odd or even; pinnæ and leaflets usually alternate. Stipules foliaceous, early deciduous.

Flowers polygamous or diœcious, terminal or axillary, in racemes or racemose corymbs, on long pedicels. Calyx tubular, lined with a glandular disc, ten-ribbed, five-lobed, the lobes narrow and nearly equal. Petals four to five, slightly unequal, imbricated, inserted on the. margin of the disc, spreading. Stamens ten, free, shorter than the petals and inserted with them, those opposite the calyx lobes longer than the others; anthers oblong. Ovary rudimentary or absent in the staminate flowers, sessile or sub-sessile in the polygamous and pistillate flowers; style short and dilated above obliquely into a two-lobed stigma.[1] Ovules four or numerous.

Pod oblong, thick, coriaceous, dark brown, flattened, beaked at the apex, slightly curved or falcate, on stalks ½ to 2 inches long, pulpy between the seeds. Valves two, narrowly winged on the margins. Seeds on long slender stalklets; seed-coat thick and bony; embryo surrounded by a layer of horny albumen.

Only two species are known, one occurring in China and doubtfully hardy in this country, the other a native of N. America and cultivated in England.

GYMNOCLADUS CHINENSIS, CHINESE SOAP TREE

Gymnocladus chinensis, Baillon, *Compt. Rend. Assoc. Franç. Avanc. Sc.* 1874, p. 418, t. 4, and *Bull. Soc. Linn. Paris*, 1875, p. 33; Oliver, in Hooker, *Icon. Plant.* xv. 9, t. 1412 (1883); Hemsley, *Journ. Linn. Soc (Bot.)* xxiii. 207 (1887).

Dialium sp. ?, Hanbury, *Science Papers*, 238, fig. 5 (1876).

A tree attaining 40 feet in height. Young shoots rusty pubescent. Leaves 1 to 3 feet long; pinnæ alternate or sub-opposite, all composed of numerous (twenty

---

[1] The stigma of *Gymnocladus chinensis* is not correctly shown in Hook. *Ic. Pl.* t. 1412.

427

to twenty-four) leaflets, which are ¾ to 1½ inch long, alternate, oblong, rounded at the base, obtuse or rarely acute at the apex, densely silky appressed pubescent beneath, on short pubescent petiolules; rachis densely pubescent, swollen at the base, and forming a conical sheath enclosing the bud.

Flowers polygamous, in pubescent racemes, those with staminate flowers shorter than the others. Calyx pubescent, with subulate lobes. Petals oval-oblong. Ovary glabrous with four ovules. Pod, 4 inches long by 1½ inch broad, glabrous. Seeds, two to four, black, globose, smooth, ¾ inch in diameter.

This tree is rather rare in China, though specimens have been collected in the provinces of Anhwei, Kiangsi, Chekiang, Hupeh, and Szechuan. Near Ichang it grows at 1000 to 2000 feet altitude. The pods, called *fei-tsao*, after being steeped in water, produce a liquid esteemed for washing the hair and cleansing silk articles.

Plants[1] were raised at Kew from seeds sent by me in 1888; but died in a year or two. Seeds, which could be easily procured from Shanghai, where they are sold in the shops, might be tried in the warmer parts of England and Ireland, as the tree is worth cultivating on account of its beautiful delicate foliage.   (A. H.)

GYMNOCLADUS CANADENSIS, Kentucky Coffee Tree

Gymnocladus canadensis, Lamarck, *Encycl.* i. 733 (1783); Loudon, *Arb. et Frut. Brit.* ii. 656 (1838).
Gymnocladus dioicus, Koch, *Dendrologie*, i. 5 (1869); Sargent, *Silva N. Amer.* iii. 69, tt. 123, 124 (1892), and *Trees N. Amer.* 554 (1905).
Guilandina dioicus, Linnæus, *Sp. Pl.* 381 (1753).

A tree attaining in America over 100 feet in height and 9 feet in girth. Bark fissured, dark grey, and roughened by small persistent scales. Young shoots covered with short pubescence. Leaves (Plate 125, fig. 4) 1 to 3 feet long, with 5 to 11 pinnæ, which are usually alternate but occasionally sub-opposite, the two or rarely the four lower pinnæ simple, the others composed of six to fourteen alternate pinnate leaflets. Leaflets 2 to 3 inches long, on pubescent stalklets, ovate, rounded at the base, acuminate at the apex, entire and ciliate in margin; under surface with scattered long hairs.

Flowers usually diœcious, the inflorescence of the staminate tree a short racemose corymb, that of the pistillate tree a long raceme. Calyx tomentose, with five narrow oblong lobes. Petals five, tomentose, longer and broader than the calyx-lobes. Ovary pubescent; ovules ten or more.

Pod, 6 to 10 inches long by 1½ to 2 inches broad, minutely pubescent. Seeds,

[1] Cf. Nicholson, *Garden and Forest*, 1889, p. 139.

five to ten, surrounded by dark-coloured sweet pulp, ovoid, ¾ inch long, and covered by a hard dark brown shell.

In the young leaf[1] of *Gymnocladus canadensis*, the rachis is prolonged an inch or more above the insertion of the upper pinnæ; and the axes of the pinnæ are similarly prolonged beyond the leaflets. These terminal appendages are very slender and tendril-like, and disappear before the leaf attains its full size. They have been supposed to be rudimentary tendrils, such as occur normally in a developed state in many leguminous plants; but they may represent simply degenerate terminal leaflets.

Sargent states that this species is diœcious; and that in order to obtain fruit male and female trees must be close together. C. M. Hovey,[2] however, writing from Boston, states that he knows a solitary tree, no other being within two miles, which produces fruit and fertile seeds, from which he has raised many plants. The so-called pistillate flowers have stamens, which doubtless are usually not fully developed; but it is possible that in some cases they may produce good pollen.

The flowers[3] in America are visited by bees, which are attracted by the nectar secreted by the inner wall of the calyx tube.

IDENTIFICATION

In summer the foliage of the tree is unmistakable. In winter the fewness of the branches and the stoutness of the branchlets, which are very short in adult trees, are remarkable. The latter show the following characters:—

Twigs coarse, grey, glabrous, with numerous small brown lenticels and wide, circular, orange-coloured pith. Leaf-scars large, obcordate, slightly oblique on prominent pulvini, with a narrow raised yellowish margin and a whitish convex surface, marked by three to five irregular tubercles, which are the scars of the vascular bundles. Buds very small; two to three vertically superposed, in the axil of each leaf-scar, the lower one rarely developing; projecting slightly out of circular depressions in the bark, which form pubescent rings around the buds. Each bud shows two to three minute scales, which become accrescent and green in the spring at the base of the shoots. No true terminal bud is developed, the tip of the branchlet falling off in summer and leaving at the apex of the twig a circular scar.

DISTRIBUTION

The Kentucky Coffee tree, though occupying a wide area in North America, is nowhere common. It is found scattered amongst other trees on hillsides where the soil is rich, and in alluvial land beside rivers. It is met with in central

---

[1] Cf. B. D. Halstead, in *Torreya*, ii. 5 (1902).   [2] *Garden*, xiv. 240 (1878).
[3] Robertson, *Trans. Acad. Sc. St. Louis*, vii. 165 (1897).

New York and western Pennsylvania, through southern Ontario and southern Michigan to the valley of the Minnesota River and to eastern Nebraska, eastern Kansas, south-west Arkansas, the Indian territory, and central Tennessee.

The tree is noted[1] in America for its habit of suckering from the roots when it is cut down. After a tree is felled the ground around to a distance of often 100 feet becomes filled with numerous suckers; and this is one of the ways in which the trees are reproduced in the American forests. The tree never develops any epicormic branches, and is very seldom attacked by any insect or fungus.

(A. H.)

An article by Sargent in *Garden and Forest*, ii. p. 75, gives an excellent account of this tree, and states that by far the largest and handsomest that he has seen was planted in 1804 directly in front of the historical Verplanck mansion at Fishkill-on-Hudson, and was, in 1889, 75 feet high and a little over 10 feet in girth below the point where it divides into three stems at 3 feet from the ground. Though it was struck by lightning in 1887, the tree is an extremely graceful and well-shaped one, as the picture shows.

The tree grows well as far north as Ottawa, where I saw two spreading trees about 40 feet high, planted in front of Rideau Hall, the residence of the Governor-General. The gardener informed me that they were the latest trees to come into leaf, and, though they flowered in good seasons, produced no fruit.

At Mount Carmel, Illinois, I measured a tree in the forest 92 feet by 8 feet, one of the few remaining relics of the splendid trees described by Ridgway, one of which was 109 feet high, with a clear stem 76 feet to the first limb, but only 20 inches across the stump. Dr. Schneck has measured one in the same locality no less than 129 feet high. It is, however, nowhere an abundant tree in this district, but grows scattered through the richer bottoms.

The tree from which a specimen log in the Jessup collection in the American Museum of Natural History was cut, grew not far from St. Louis, and although only 18 inches in diameter was 105 years old. This represents the average rate of increase of the tree growing naturally in the forest, cultivated trees in favourable conditions growing much more rapidly.

CULTIVATION

Gymnocladus canadensis was introduced into England by Archibald, Duke of Argyll, who had a tree in cultivation[2] at Whitton in 1748. This tree was afterwards removed to Kew, on the establishment of the gardens there by the Princess of Wales, mother of George III., who obtained it and many other interesting trees as a present from the Duke of Argyll in 1762. This tree died[3] about 1870; and as old trees reported by Loudon at Syon and elsewhere cannot now be found, it goes to show that the tree lives little over 100 years in England.

---

[1] *Garden and Forest*, vii. 358 (1894).　　　　[2] Aiton, *Hort. Kew.* v. 400 (1813).

[3] J. Smith, *Dict. Econ. Plants*, 235 (1882), mentions this tree as if it was still living in 1882; but according to Nicholson it had died several years previously to that date.

According to Nicholson,[1] it is very easy to transplant, and bears drought well. It is propagated either by seeds or by root-cuttings. Pieces of the roots, 4 to 5 inches long, placed in prepared beds and kept moist, will develop in the first year into plants three or four feet high. Some of the cuttings, however, will not start into growth until the following year.

I have raised seedlings from American seeds, which, being large and hard, should be soaked in warm water for some days before sowing. The seedlings grow slowly, and should be kept under glass for a year or two before planting out.

In spite of Loudon's assertion to the contrary, it appears to flower very rarely in England, the only record being at Claremont, where Mr. Burrell[2] says it produces flowers freely early in summer. Pods have never been produced, so far as we know, in this country.

It is a rare tree in cultivation; but though stiff and peculiar in habit, it is not at all ungainly when well-grown, even when bare of leaves. It comes into leaf very late in the season, and it drops its leaves early in autumn, the stalks, however, often remaining on the tree for weeks. The foliage, like that of many leguminous plants, shows the phenomenon of sleep, the leaflets drooping and closing together soon after sunset in summer.

REMARKABLE TREES

There are two trees at Claremont, which were about 55 feet high in 1888. When I measured them in 1907 the largest was 60 feet by 6 feet 7 inches, and seemed quite healthy; the other was broken.

A tree at Chiswick House measured, in 1903, 53 feet high by 3½ feet in girth. Another at Barton, Suffolk, was in 1904 57 feet high by 5 feet 2 inches in girth at two feet from the ground, and divided above this into two stems. In the Botanic Garden at Cambridge there is a good specimen, which was 45 feet by 3 feet 9 inches in 1906. There are three smaller trees in the Oxford Botanic Garden.

At Kayhough, Kew, in the garden of Mr. Charles Wright, there is a healthy and well-shaped tree, which was in November 1905, 40 feet high by 2 feet 9 inches in girth, with a bole of 6 feet, dividing into two main stems. This tree was purchased from a nurseryman at Kingston in 1878, when it was said to be twenty-two years old, and was then about two-thirds its present height. After transplanting, it made no growth for three years; but since then it has grown steadily though very slowly, and has not been injured in any way by severe winters, though it has never flowered. It has been much surpassed in rate of growth by an Ailanthus in the same garden. There is a tree of about the same size growing close to Mr. Clarke's house at Andover, Hants, which is fifty to sixty years old and measures 43 feet by 2 feet 10 inches. There are several small trees in Kew Gardens, the largest one being near the main entrance.

It seems evident that the tree, to attain a large size, requires a much greater

---

[1] *Garden*, xxiv. 29 (1883).   [2] *Garden*, xxxiii. 229 (1888) and xlv. 404 (1894).

degree of summer heat than it gets in England, for in the south of France it becomes a splendid tree. I saw in the Museum Gardens at Chambery, in the grounds of the Castle formerly belonging to the Dukes of Savoy, a tree which, though forked near the ground, had two tall clean trunks each about 100 feet by 5 to 6 feet. The leaves were only just appearing on 18th May, and many of the large bean-like pods full of greenish pulp, which had fallen in the winter, lay on the ground. Seeds from these pods germinated, but the seedlings, with one exception, withered soon afterwards. It is not uncommon in Savoy, and I saw a fine specimen, 81 feet by 9 feet 6 inches, in the Public Gardens at Aix-les-Bains, which in October 1906 had ripe pods on it. It is known in France by the name of " Bonduc."

In the old Botanic Garden at Padua a splendid tree was in 1895, according to Prof. Saccardo,[1] 135 years old, 21 metres high, and 2.60 metres in girth. When I saw it in 1905 the trunk was broken off at about 12 feet, but long shoots, which were in flower, had been produced from the stump.   (H. J. E.)

[1] *L'Orto Botanico di Padova* (1895).

CEDRELA

Cedrela, Linnæus, *Gen. Pl.* 109 (1764); Bentham et Hooker, *Gen. Pl.* i. 339 (1862).
Toona, Roemer, *Synops.* i. 131 (1846).

Trees, belonging to the order Meliaceæ, with unequally pinnate leaves, without stipules, and composed of numerous opposite or sub-opposite stalked leaflets.

Flowers in panicles, perfect, regular   Calyx short, four- to five-cleft.   Petals, four to five, nearly erect, imbricated, free.   Stamens, four to six, free, inserted at the top of a four- to six-lobed hypogynous disc; filaments subulate, anthers versatile. Ovary sessile on the disc, five-celled, each cell containing in two series eight to twelve pendulous ovules.   Fruit, a coriaceous or woody capsule, composed externally of five valves, and almost filled up internally by a central column, between which and the valves are five thin cells, containing the seeds, which are numerous, compressed, and with one or two wings.

The genus is divided into two sections:—

I. *Eu-Cedrela.*—Seed with a single wing on its lower side.   Nine species in tropical America.

II. *Toona.*—Seed with either two wings, one at each end, or with a single wing above.   Eight species in India, Indo-China, China, and Australia, all in tropical regions except *Cedrela sinensis*.

CEDRELA SINENSIS

Cedrela sinensis, A. Jussieu, *Mém. Mus. Par.* xix. 255, 294 (1830): *Rev. Hort.* 1891, p. 573, figs. 150, 151, 152; Hemsley, *Journ. Linn. Soc. (Bot.)* xxiii. 114 (1886).
Toona sinensis, Roemer, *Synops.* i. 138, 139 (1846); Diels, *Flora von Central China*, 425 (1901).
Ailanthus flavescens, Carrière, *Rev. Hort.* 1865, p. 366.

A tree of moderate size, attaining in China a height of 60 to 70 feet.   Bark scaling off in narrow longitudinal strips, 1 to 2 inches in width, and leaving exposed in parts the reddish inner bark below.   Young shoots covered with minute pubescence.   Leaves (Plate 125, fig. 7), large, 1 to 2 feet in length.   Leaflets, eleven to nineteen, about 4 inches long, on pubescent stalklets (nearly ¼ inch long), opposite or sub-opposite, divided into two unequal parts by the midrib, the upper part larger and rounded at the base, the other part usually cuneate at the base; apex caudate-acuminate; margin repand, minutely ciliate, distantly and minutely serrate or with occasional short teeth; nerves, fifteen to eighteen pairs, usually dividing and forming loops close to the margin; upper surface dark green, glabrous; lower surface pale green, glabrescent.

Flowers fragrant, in pubescent terminal panicles, which are a foot or more in length; pedicels short. Calyx with five short, rounded, ciliate lobes. Petals five, white, oblong, sub-cordate at the base, converging at the apex. Stamens five, alternating with five staminodes. Fruit about an inch long; valves, opening longitudinally from above downwards. Seed with an oblong wing attached to its upper side, the wing two to three times as long as the body of the seed.

In summer the large pinnate leaves give the tree much the appearance of Ailanthus; but the bark is different, and the leaflets of Cedrela are devoid of the glandular teeth near the base, which are so characteristic of Ailanthus. In winter the following characters are available (Plate 126, fig. 2):—

Twigs stout, brown, minutely pubescent; lenticels small, scattered; pith white, circular in section. Leaf-scars large, alternate, slightly raised, obcordate or oval, with five bundle-dots. Terminal bud, much larger than the others, broadly conical, of four to six triangular scales, which are swollen externally and hollowed internally, brown, shining, with acuminate pubescent tips. Lateral buds minute, solitary, inserted immediately above the leaf-scars, hemispherical, showing three to five shining brown scales.

Lubbock,[1] who gives a detailed account of the structure and development of the buds, the scales of which are modified leaves, states that the terminal bud usually dies in winter, but sometimes lives, and then is always later in developing in spring than the lateral buds.

Cedrela sinensis is a native of northern and western China. It is very common in the neighbourhood of Peking, and was found in Kansuh, beyond the Great Wall, by Piasetski. According to von Rosthorn and Wilson, it is wild in the forests of the province of Szechuan. It is commonly cultivated in central China, where it never attains a great size, mainly because the Chinese spoil its growth by lopping off in spring the young shoots, which are much esteemed as food. These are eaten after being chopped and fried in oil. The tree is known to the Chinese as the *hsiang-ch'un*.[2] The timber is good, reddish in colour, and often used in making furniture.

The tree was first made known to Europeans by Père d'Incarville, who sent dried specimens from Peking to Paris in 1743. In China it has been well known from classical times, and references to it occur in the earliest Chinese literature.

Cedrela sinensis was introduced in 1862 by Simon, who sent a living plant from Peking to the Museum at Paris, which was described by Carrière in 1865 as *Ailanthus flavescens*. On the tree flowering in 1875 it was recognised to be *Cedrela sinensis*. This tree, which was planted in the nursery attached to the garden of the Museum, had attained in 1891 a height of 40 feet; and, when Elwes saw it in 1905, it was very little taller, and about 4 feet in girth.

Many trees have been raised in the vicinity of Paris, both by seed and by root-

---

[1] *Journ. Linn. Soc.* (*Bot.*), xxx. 478 (1894).　　　[2] Cf. name given to *Ailanthus*, p. 32.

cuttings; and it appears to be perfectly hardy in the north of France, having sustained without injury the severe winter of 1879-1880. Its large fragrant foliage renders it perhaps more suitable than the Ailanthus for planting in towns. It is said by Nicholson to be now largely used in Holland for that purpose.

The tree is rather rare in England, and we have seen no specimens remarkable for size. There is a tree in Kew Gardens which measured in November 1905 33 feet by 2 feet 4 inches. This is probably of the same age as an Ailanthus of equal height growing beside it. A tree much about the same size is growing and thriving in Messrs. Veitch's Nursery at Coombe Wood. Mr. Cassels informs me that young trees of Cedrela are planted in some of the London County Council parks, as Meath Gardens and Bethnal Green.

Cedrela sinensis is also cultivated in the United States,[1] where a tree flowered at Meehan's nurseries, Germanstown, in 1895. Another only eight years old had attained in the same year 20 feet in height in western Virginia. Professor Sargent thinks it might be used as a street tree in New England, though introduced plants have proved rather tender in that climate. It has frequently flowered in France, but has never produced fruit there. There is no record of its having flowered as yet in England.

Mouillefert[2] speaks of this tree as one which, in his opinion, has a great future in Europe on account of the high quality of its wood, which he compares to that of mahogany and that of the so-called cedar of the West Indies (*Cedrela odorata*). He says that the tree grows fast from seed, attaining 5 feet in the third year, and adds that on calcareous soil of middling quality at Grignon a tree about twenty-five years old measured 10 metres high. (A. H.)

---

[1] *Garden and Forest*, 1896, pp. 260, 279.   [2] *Principales Essences Forestières*, 471, 472 (1903).

PTEROCARYA

Pterocarya, Kunth, *Ann. Sc. Nat.*, sér. I. ii. 345 (1824), Bentham et Hooker, *Gen. Pl.* iii. 399 (1880).

DECIDUOUS trees belonging to the order Juglandeæ, with large, alternate, compound, imparipinnate leaves; leaflets serrate; stipules absent. Buds scaly or naked, the lateral ones often multiple, two to three in a vertical row above the insertion of the leaf. Pith chambered. Flowers monœcious, numerous in long pendent catkins. Male catkins usually several, arising singly in the leaf axils; in some species (*caucasica, stenoptera*) lateral on the preceding year's shoots, with an occasional catkin on the current year's shoot; in other species (*rhoifolia, Paliurus*) all on the new shoots. Stamens nine to eighteen in several series on the axis of a three- to six-lobed scale, to which a bract is adnate on the back, the scale representing two bracteoles and one to four perianth segments. Female catkins solitary, terminating the young shoot. Female flowers with a bract and two bracteoles at the base; perianth four-lobed, adnate to the ovary, which contains one ovule, and is surmounted by a short style, divided above into two papillose stigmatic divisions. Fruit catkins long, with numerous nut-like fruits, which have in most species two lateral wings, in one species a single orbicular wing all round, due to the enlarged bracteoles of the flower, the bract persisting little changed at the base of the fruit. Nutlet, with a thin pericarp and a hardened endocarp, the latter divided below into four imperfect cells, and containing one seed, which is four-lobed below. Cotyledons bi-partite, each division being again deeply divided, forming four linear segments; carried above ground in germination.

Pterocarya and Juglans have similar foliage, and agree in the chambered pith of the twigs. They are readily distinguished when in fruit, that of Pterocarya being always small and winged. When specimens in leaf only are obtainable, the best mark of distinction lies in the buds, which in Pterocarya are either without scales or are enclosed in a long conical beaked funnel-like covering, composed of membranous scales—differing in either case from the short buds of Juglans with two to three external scales.

Seven species of Pterocarya are known, occurring in Persia, the Caucasus, China, Tonking, and Japan. A hybrid species has been obtained in cultivation, which will be described under *P. caucasica*. The seven species which occur in the wild state may be arranged as follows :—

Section I. Cycloptera, Franchet, *Journ. de Bot.*, 1898, p. 318.

Fruit surrounded by an orbicular wing, composed of the connate bracteoles, which cover the nutlet at the base.

1. *Pterocarya Paliurus*, Batalin, *Act. Hort. Petrop.* xiii. 101 (1892); Franchet, *loc. cit.*; J. H. Veitch in *Journ. R. Hort. Soc.* 1903, xxviii. 65, fig. 26. China: mountains of Szechwan, Hupeh, and Chekiang.

Tree 40 feet. Twigs pubescent and glandular. Buds naked. Leaf-rachis villous or pubescent, not winged. Leaflets seven, coriaceous, oblong-ovate, with sub-acute apex, glabrous below except along the midrib. Fruits samara-like, the nutlet in the centre of an orbicular wing, 2 inches across, several on a raceme a foot long.

This species was introduced in 1903 by Mr. E. H. Wilson from the mountains of Central China; and young plants, which seem perfectly hardy, are now growing at Messrs. Veitch's Nursery, Coombe Wood. The tree when in fruit presents a remarkable appearance, and is well worth trial, as it should prove hardier than *P. stenoptera*, which grows at a lower level.

Section II. Diptera (*Sectio nova*).

Fruit with two lateral wings, the developed bracteoles, which do not cover the nutlet at the base.

* *Buds naked, without scales.*

2. *Pterocarya stenoptera*, C. DC. China, Tonking.

Tree 60 feet. Twigs bristly-pubescent. Leaf-rachis winged. Leaflets nine to twenty-five, coriaceous, underneath glabrescent with pubescent tufts in the axils of the nerves. Fruit with long lanceolate upright glabrous wings. In cultivation. See description below.

3. *Pterocarya hupehensis*, Skan, *Journ. Linn. Soc.* (*Bot.*), xxvi. 493 (1899). China: mountains of Hupeh.

Small tree about 30 feet. Twigs glabrous. Leaf rachis not winged, glabrous except for some tomentum near its insertion. Leaflets five to nine, lanceolate; under surface with brown scurfy scales and glabrous except for stellate rusty tomentum in the axils of the nerves. Fruit minutely glandular, with sub-orbicular wings, $\frac{1}{2}$ inch diameter. Introduced by Mr. E. H. Wilson in 1903. Young plants are now growing at Coombe Wood and seem to be perfectly hardy.

4. *Pterocarya Delavayi*, Franchet, *Journ. de Bot.* 1898, p. 317. China: mountains of Yunnan.

This species, which I have not seen, appears closely to resemble the last, differing mainly in the fruits being covered with short hairs. Not introduced.

5. *Pterocarya caucasica*, C. A. Meyer. Persia, the Caucasus.

Tree attaining 100 feet. Twigs glabrous except for some pubescence at the tips.   Rachis of the leaf not winged.   Leaflets fifteen to twenty-seven, membranous ; under surface without glands and glabrous except for stellate pubescence on the nerves and in their axils.   Fruit, ½ inch broad, glabrous ; wings semi-orbicular.   In cultivation.   See description below.

** *Buds long, conical, beaked at the apex, enclosed during summer and autumn by a membranous funnel-like covering, composed of several scales.*

6. *Pterocarya macroptera*, Batalin, *Act. Hort. Petrop.* xiii. 100 (1893).   China : mountains of Kansuh.

Small tree, about 20 feet in height.   Twigs glabrous.   Rachis of the leaf not winged, rusty-tomentose.   Leaflets nine to eleven, acute, rusty-tomentose on the midrib and nerves beneath.   Fruit : nut pubescent, wings broadly ovate, pilose, 1¼ in. long by 1 inch broad.   Not introduced.

7. *Pterocarya rhoifolia*, Siebold et Zuccarini.   Japan.

Tree, rarely attaining 100 feet.   Twigs glabrous.   Rachis of the leaf not winged.   Leaflets fifteen to twenty-one ; under surface glandular with tomentum along the midrib and veins and in their axils.   Fruit, 1 inch wide ; wings rhombic, broader than long, glabrous.   Introduced.   See description below.

PTEROCARYA CAUCASICA

Pterocarya caucasica, C. A. Meyer, *Verz. Pflanzen Caucasus*, 134 (1831) ; Loudon, *Arb. et Frut. Brit.* iii. 1452 (1838).

Pterocarya fraxinifolia, Spach, *Hist. Nat. Veg.* ii. 180 (1834) ; Lavallée, *Arb. Segrez. Icones.* 73, t. 21 (1885).

Pterocarya Spachiana, Lavallée, *op. cit.* 69, t. 20.

Pterocarya sorbifolia, Dippel (*non* S. et Z.), *Laubholzk.* ii. 327 (1892).

Juglans fraxinifolium, Lamarck, *Encyc. Meth.* iv. 502 (1797).

Juglans pterocarpa, Michaux, *Fl. Bor. Am.* ii. 192 (1803).

Rhus obscura, Bieberstein, *Fl. Taur. Cauc.* i. 243 (1808).

A tree attaining 100 feet in height and 10 feet or more in girth, usually however smaller, and tending to branch into several stems at no great height above the ground.   Bark dark grey and furrowed.   Shoots glabrous.   Leaves (Plate 125, fig. 1) 16 to 20 inches long, on a stalk 2 to 3 inches long, only slightly swollen at its base ; rachis not winged.   Leaflets fifteen to twenty-seven, opposite or sub-opposite, sessile or sub-sessile, 3 to 5 inches long ; oblong or oblong-lanceolate ; acute, acuminate, or obtuse at the apex ; unequal and rounded or narrowed at the base ; dark green above ; under surface lighter green, without glands, glabrous except for some stellate pubescence along the nerves and in their axils ; thin in texture ; sharply and finely serrate.   Staminate catkins several, each in the axil of a leaf-scar on the preceding year's shoot, rarely one or more on the current year's shoot ; scale usually five-lobed, stamens twelve to fifteen.   Fruiting catkins up to eighteen inches long.

Fruit $\frac{1}{2}$ inch broad; wings semi-orbicular, concave below, conspicuously veined; nutlet with beaked apex.

Seedling.[1]—The caulicle terete, erect, and about two inches in length, raises the two cotyledons well above the ground. Each cotyledon is shortly stalked, about an inch in width, and deeply bipartite, the two primary divisions being again divided for nearly two-thirds of their length, the whole forming four linear-oblong obtuse diverging segments. The cotyledons are palmately five-nerved at the base, the three middle nerves each ending at the base of a sinus and sending divisions into the segments. The young stem is slightly glandular near the apex. The first five leaves are alternate, simple, lanceolate or ovate, rounded at the base, acute or acuminate at the apex, penni-nerved, serrate, and vary in length from 1 to 2 inches. Succeeding leaves are compound, unequally pinnate, and with many leaflets.

IDENTIFICATION

In summer this tree is only liable to be confused with *Pterocarya rhoifolia*, which has scaly buds. It is distinguished from all species of Juglans by its naked buds.

In winter the following characters are available:—Twigs stout, olive green, glabrous except at the minutely pubescent, glandular tip. Leaf-scars oblique on the twigs, their lower part projecting, large, obcordate, marked by three crescentic prominences, which are the fused cicatrices of the vascular bundles. Pith pentagonal in cross section, chambered in longitudinal section. Buds without covering scales, consisting of a short shoot and three to four undeveloped leaves, which are stalked below, enlarged and lobed above, rusty brown in colour, minutely pubescent and glandular. Lateral buds multiple, two to three superposed vertically above each leaf-scar; the uppermost one like the terminal bud, but smaller and stalked; the lowermost close to the upper margin of the leaf-scar, minute and rudimentary.

VARIETY AND HYBRID

1. Var. *dumosa*, Schneider, *Laubholzkunde*, 94 (1904); *Pterocarya dumosa*, Lavallée, *Arb. Segrez.* 217 (1877). This is a shrubby form, with yellowish brown twigs, and small closely-set leaflets, about $2\frac{1}{2}$ inches long. The fruit and flowers are unknown; but it is probably a horticultural variety of *P. caucasica*.

2. *Pterocarya Rehderiana*, Schneider, *op. cit.* 93. This is a hybrid between *P. caucasica* and *P. stenoptera*, which was described by Rehder in *Mitth. Deut. Dendrol. Gesell.* 1903, p. 116. It grows in the arboretum at Segrez; and plants of it are now cultivated in the Arnold Arboretum,[2] Massachusetts, where it is perfectly hardy. It is intermediate in character between the two species. The leaflets in texture, serration, etc., resemble those of *P. caucasica*, being a trifle smaller; but

[1] Cf. Lubbock, *Seedlings*, ii. 521, fig. 662 (1892)

[2] Two seedlings were raised by Elwes from seeds of this tree, one of which is now about eighteen inches high, and shows evidence of its hybrid origin in the leaves.

the rachis shows here and there a very slight wing, like that of *P. stenoptera*, only never serrate in margin.   The fruits have oval wings, shorter and broader than those of *P. stenoptera*, the nut being more beaked than in that species.   The veining of the fruit-wings resembles *P. caucasica*.

DISTRIBUTION

Pterocarya caucasica has been found in the northern provinces (Astrabad and Ghilan) of Persia, and in Russian Armenia, as well as in the Caucasus. According to Radde,[1] it occurs in the marshy delta of the Rion in company with *Alnus glutinosa*, and along the coast of the Black Sea, mixed with oak, beech, and hornbeam.   It grows sometimes as a tree, but oftener as a tall shrub, on the banks of streams.   It extends up to about 1200 feet only in Kachetia, and is met with as far eastward as Talysch, on the coast of the Caspian Sea, where in damp places it forms the principal underwood.   It is not found wild in the interval between the lower Rion on the west and the lower valley of the Alazan on the south side of the central Caucasus, and is again absent from here to the province of Talysch.

Mr. Younitsky of the Russian Forest Service has kindly sent me the following account of the tree in the Caucasus.   He says it is only found in certain stations, rarely over 1200 feet elevation, and always in moist or very wet places, to which it is better adapted than even the alder.   In the young stage the tree is very delicate and susceptible to spring frosts, requiring shelter when young ; and when older does not bear shade well.   Very large trees occur, of 100 feet in height and 10 feet in girth, and logs of it are obtained bare of branches for 50 feet, with a girth of 5 feet at the smaller end.   It grows very rapidly in youth, making a height of 30 feet in ten years.   The wood is light and soft, resembling much that of the lime-tree, and is chiefly used for making boxes and packing-cases.   The bark is used for sandals and roofing.   The leaves contain a poisonous matter, and when thrown into water intoxicate the fish, which rise to the surface and are easily caught. The tree is rarely cultivated, but is recommended for planting in the wettest situations, where it will thrive better than almost any other tree.

CULTIVATION

Pterocarya caucasica was introduced into France by the elder Michaux on his return from Persia in 1782.   According to Bosc the first tree was planted at Versailles, others a little time after being planted about the Museum in Paris. According to Mouillefert,[2] there are still growing at the Trianon, Versailles, and at the Museum, Paris, two fine specimens which are probably original trees.[3]   The tree flowered and produced fruit in 1826 in the park at Malesherbes, according to a note by Gay in the Kew Herbarium.   There is a tree 80 feet high and 9 feet in girth in the Old Botanic Garden at Geneva, which was seen by Elwes in 1905.

---

[1] Radde, *Pflanzenverbreitung in Kaukasusländern*, 109, 139, 159, 182, 205, etc.

[2] *Traité des Arbres*, ii. 1195 (1898).   [3] I could not find either of these trees in 1905.—H. J. E.

This species was introduced into England some time after 1800, the largest tree mentioned by Loudon in 1838 being one 25 feet high and fifteen years planted at Croome; but it is long since dead. (A. H.)

I have raised numerous plants of Pterocarya from seed sent me from the Caucasus by the late Dr. Radde in 1903, some of which was distributed by the Royal Horticultural Society. The seedlings grow fast, attaining 2 feet or more in height at two years old, but do not ripen their wood well when young, and are extremely liable to be injured by frost if not protected in spring.[1] The leaves appear about the same time as those of Liriodendron. The tree does not seem to dislike lime in the soil, and should be planted out when 3 or 4 feet high, in a situation where the ground is not liable to drought in summer, or near running water.

REMARKABLE TREES

This is one of the most ornamental hardwoods that we have; and is well worth planting in warm and sheltered positions in the south of England, where it thrives from Kent to Devonshire.

By far the largest and finest tree of this species known in England is at Melbury, Dorsetshire, the seat of the Earl of Ilchester. This magnificent tree (Plate 121) is growing on a sheltered bank below the house, on soil which contains lime, close to the finest specimen I know of *Picea Morinda*. It is no less than 90 feet high by 11 feet in girth, and has a straight clean bole about 15 feet long, spreading out into a symmetrical head of branches, and when I saw it in September 1906 had many catkins of fruit hanging on it.

Its spreading habit is shown by a fine tree at Claremont Park, near Esher, Surrey, which grows on deep sandy soil, and is a noble ornament of a lawn. The illustration of this tree (Plate 122) is from a photograph taken in 1903, when it measured about 50 feet in height, with a bole of only 4 feet high but no less than 18 feet in girth. It divides into eight large limbs, each of which is about 4 feet in girth, and the foliage spreads over an area of 30 yards in diameter. The tree is believed by Mr. Burrell, the gardener, to be about eighty years old, and seems to be decaying at the heart. The bark is very rough and deeply furrowed, and the leaves and flower-buds were just appearing, after a very mild winter, on 6th March. A self-sown seedling from it was about 2 feet high.[2]

Another fine tree is growing at Tortworth Court, from which I gathered ripe seed in October 1900, one of which grew in the following spring. The Earl of Ducie has raised several young trees from the same parent in other seasons. At Linton Park, Kent, there is a fine tree, which was about 50 feet high in September 1902, but not so large as the one at Claremont. Ripe fruiting specimens were sent from Devonshire by Sir John Walrond in 1888, which were figured by

---

[1] The severe frost of 20th-22nd May 1905 seriously injured all my young trees, and it is evident that this tree should only be planted in situations where spring frosts are not severe.

[2] Mr. Burrell found a seedling in the summer of 1899. See *Garden*, 1902, lxii. 234, where a figure and description of the tree are given. See also *Garden*, 1894, xlv. 404, fig., and *Gard. Chron.* 1894, xvi. 192. According to a note in the Kew Herbarium, the Claremont tree was, in 1887, 45 feet high by 13½ feet in girth.

Dr. Masters in the *Gardeners' Chronicle*, but I have been unable to procure particulars of the tree from which the specimens were obtained.

In the Botanic Garden, Cambridge, there is an old tree which was 58 feet high in 1903, with eight stems, girthing from 3 feet to 4 feet 3 inches; and from the roots of another tree which was blown down about 1885 a number of strong stems, about twenty, have sprung up, which average about 50 feet in height and 2½ feet in girth. These particulars, which have been kindly sent me by Mr. Lynch, the curator, show the remarkable power of the tree in producing root-suckers (Plate 123).

A tree at Fota, near Queenstown in Ireland, seen by Henry in 1903, measured 42 feet high by 3 feet 9 inches in girth. It produced flowers and fruit in 1902.

Dr. Masters[1] recommends it for planting in towns, and says that there was a good specimen in the Chelsea Botanic Garden (since cut down) in 1891. There are said to be good specimens in some of the towns in Holland.   (H. J. E.)

PTEROCARYA RHOIFOLIA

Pterocarya rhoifolia, Siebold et Zuccarini, *Abh. Bayr. Ak. Wiss. Math. Phys. Kl.* iv. 2, 141 (1845); Maximowicz, *Mél. Biol.* viii. 637 (1872); Shirasawa, *Icon. Ess. For. Japon.* text 35, t. 16 (1900).

Pterocarya sorbifolia, Siebold et Zuccarini, *loc. sit.*; Rehder, *Mitt. Dendrol. Deut. Gesell.* 1903, p. 115.

A tree attaining, according to Shirasawa, 100 feet in height, with a straight stem 10 feet in girth. Bark greyish brown with deep longitudinal fissures. Shoots glabrous. Leaves (Plate 125, fig. 3) 8 to 16 inches long, on a stalk about 2 inches long, which is swollen at its insertion; rachis without wings. Leaflets, fifteen to twenty-one, usually opposite, sessile or sub-sessile, 2½ to 5 inches long, oblong-lanceolate, acuminate at the apex, unequal at the base, which is rounded or somewhat narrowed; dark green above; under surface lighter green, with glandular scales, and some tomentum on the midrib and nerves and in their axils; somewhat thicker in texture than the leaves of *P. caucasica*; margin sharply and finely serrate.

Flowers appearing with the leaves. Staminate catkins two to three at the base of the young shoots; scale three-lobed, pubescent, bearing nine to twelve short-stalked stamens. Pistillate catkins, solitary, terminal at the end of the young shoot, later apparently lateral owing to the growth of the upper axillary bud. Fruiting catkins, 8 to 10 inches long; fruit an inch across; nut with a short, scarcely beaked apex; wings rhombic, broader than long, without any hollow at their base, inconspicuously veined.

The above description applies to the glabrous form, which is in cultivation in England and is common in Japan. In wild specimens from Yezo the leaves appear to be much more pubescent, the rachis and nerves being often covered with dense long hairs.

[1] *Journ. R. Hort. Soc.* 1891, xiii. 86.

This species is readily distinguished by the peculiar buds, which are formed early, and by the scars at the base of the shoot, left by the fall of the bud-scales of the previous year. The buds at first are long, conical, with a curved beak, and are covered by a funnel-shaped membranous sheath, which is composed of two external and two to three internal glabrescent glandular scales. The scales fall off in November, leaving four or five narrow scars at the base of the buds, which in this stage resembles in structure those of *P. caucasica*, but are whitish and densely tomentose. Lateral buds usually solitary at some distance above the leaf scars. Twigs quite glabrous, otherwise as in *P. caucasica*. (A. H.)

In Japan this is a large tree known as *Sawa gurumi*, which I saw in the central provinces of Hondo, where it grows to a height of 50 to 60 feet, old trees attaining a girth of 8 or 10 feet. It generally grows on the banks of streams in mixed forest, and did not seem to be very common or to be valued for its timber, though I got a specimen of the wood from the Government sawmills at Atera, which is now at Kew.

Sargent found it very abundant on the slopes of Mt. Hakkoda, in the north of Hondo, at 2500 to 4000 feet elevation, where it attains as much as 80 feet in height, being next to the beech the largest deciduous tree in the forest. It is a broad-topped tree with stout spreading branches, and when covered with its long hanging slender racemes of fruit, is very handsome. It is hardy at the Arnold Arboretum near Boston and produces seeds there.

Pterocarya rhoifolia is recorded by Diels[1] as having been collected by Von Rosthorn in the province of Szechuan in China.

It seems to have been introduced into cultivation by the Duke of Bedford, to whom seeds were sent from Japan in 1889. Young plants from some of this seed were raised at Kew in 1890; and these have now attained about 12 feet in height. They are the only specimens we have seen in England.

(H. J. E.)

PTEROCARYA STENOPTERA

Pterocarya stenoptera, C. de Candolle, *Ann. Sc. Nat.* sér. IV. xviii. 34 (1862); Lavallée, *Arb. Segrez. Icones*, 65, t. 19 (1885); Franchet, *Journ. de Bot.* 1898, p. 317.

A tree, 50 to 60 feet in height, with a girth of stem of 6 or 8 feet. Bark rough. Leaves (Plate 125, fig. 2) about a foot in length; rachis covered with bristles, slightly swollen at its insertion, and having on each side a conspicuous irregular membranous wing, occasionally slightly serrate in margin. Leaflets nine to twenty-five, opposite or alternate, terminal leaflet often wanting; coriaceous; under surface with a few scattered glands, and some pubescence on the midrib and nerves and in their axils; oblong or oblong-lanceolate; acute at the apex, unequal and rounded or narrowed at the base, finely and sharply serrate in margin, 3 to 5

[1] *Flora von Central China*, 274 (1901).

inches long. Male catkins, arising as in *P. stenoptera*; scale glandular, four-lobed; stamens six to ten. Female catkins 8 inches long; bract minute, bracteoles oblong and longer than the style, perianth with four subulate lobes. Fruit: catkins a foot or more in length; nut with conic beak-like apex; wings linear-oblong and erect.

The above description applies to the form in cultivation, which is also common in the wild state. The species is, however, very variable as regards the amount of pubescence, the twigs being often glabrous and the leaf-rachis only slightly pubescent. In many wild specimens the wing of the rachis is very slight.

This species is readily distinguishable in summer by the winged rachis of the leaf. In winter the twigs are slender and covered with a rusty-red bristly pubescence; but in other respects resemble those of *P. caucasica*. The buds, more slender than in that species, but similar in structure and position, are greyish in colour.

This is a common tree in the central and southern provinces of China, extending in a slightly different form into Tonking.[1] It is usually met with in the plains and low hills, along rivers and water-courses; and never grows to be a large tree. It is recorded from near Moukden in Manchuria, where it was collected by James; but was probably only cultivated there. It is usually called *ma-liu*[2] by the Chinese; and is much planted in the streets of Shanghai, where it is often called "Chinese ash" by the European inhabitants. As the climate of the regions where it grows naturally is very different from that of England, it is liable to be injured by spring frosts, and fails from want of heat in autumn to ripen its wood. The timber is considered in China to be of little value.

The tree was introduced into Europe apparently by Lavallée, who received the seeds from Siebold, about 1860. It supported at Segrez very low temperatures in 1870 and 1871; but succumbed during the severe winter of 1879-1880. Lavallée considered it to be about as hardy as the common walnut.

The only specimen that we have seen in England of any size is at Tortworth, where Elwes measured in 1905 a tree 32 feet high by 2 feet 3 inches in girth, believed by Lord Ducie to have been planted about twenty years. It is in a shady and sheltered valley and produced small racemes of fruit in 1905.   (A. H.)

---

[1] Var. *tonkinensis*, Franchet, *Journ. de Bot.* 1898, p. 318. A geographical form, distinguished by large leaflets, up to 6 inches long, and linear wings to the fruit, which diverge at a wide angle.

[2] Henry, "Chinese Names of Plants," *Journ. China Branch R. Asiat. Soc.* xxii. 256 (1887).

CLADRASTIS

Cladrastis, Rafinesque, *Cincinnati Literary Gazette*, i. 66 (1824); and *Neogeniton*, i. (1825); Bentham
et Hooker, *Gen. Pl.* i. 554 (1865).
Maackia, Ruprecht et Maximowicz, *Mél. Biol.* ii. 440, t. ii. (1856).

DECIDUOUS trees or shrubs belonging to the division Papilionaceæ of the order
Leguminosæ. Leaves alternate, unequally pinnate; leaflets opposite, sub-opposite
or alternate, on stout petiolules, entire in margin, and without stipels. Flowers in
panicles or racemes, on slender pedicels; calyx with four or five short unequal teeth;
corolla papilionaceous, petals unguiculate, standard nearly orbicular, wing and keel-
petals oblong; stamens ten, free or slightly united at the base; anthers versatile;
ovary with numerous ovules; style incurved, subulate; stigma terminal, minute.
Pod linear, flattened, thin, thickened on the upper margin; valves membranous;
seeds four to six.

Four species of Cladrastis have been described, constituting two sections, which
have been considered by Sargent and other botanists to form two distinct genera,
Cladrastis and Maackia. The difference in the buds of the two sections is remarkable; but analogous differences occur in other genera, as Carya and Pterocarya; and
in the absence of important differential characters in the flowers and fruit, it is
advisable to unite the sections into one genus.

SECTION I. EU-CLADRASTIS.

Buds several together, compressed into a cone, and concealed in the base of
the petiole of the leaf. Leaflets usually alternate. Flowers in panicles;
calyx five-toothed.

1. *Cladrastis tinctoria*, Rafinesque. Kentucky, Tennessee, Alabama, and N.
Carolina.

Shoots glabrous. Leaflets seven to eleven, oval or ovate, acuminate, almost
completely glabrous.

2. *Cladrastis sinensis*, Hemsley. Central and western China.

Shoots rusty pubescent towards the base. Leaflets nine to eleven, oblong-
lanceolate, obtuse or sub-acute, rusty pubescent towards the base and along
the midrib.

SECTION II. MAACKIA.

Buds solitary, axillary, not concealed. Leaflets opposite or sub-opposite.
Flowers in racemes; calyx four- or five-toothed.

445

3. *Cladrastis amurensis*, Bentham et Hooker.   Amurland, E. Manchuria, Korea, and Japan.

 Shoots pubescent.   Leaflets nine to eleven; deltoid, ovate or oval; obtuse or acute; densely appressed pubescent; calyx four-toothed.

4. *Cladrastis Tashiroi*, Yatabe.[1]   Loochoo Islands.[2]

 Allied[3] to *C. amurensis*, but always a small shrub; with smaller leaflets, acute and not truncate or rounded at the base as in that species, glaucescent and scarcely pubescent beneath.   Flowers and pods also smaller; calyx five-toothed.

CLADRASTIS TINCTORIA, Yellow-Wood

Cladrastis tinctoria,[4] Rafinesque, *Neogeniton*, i. (1825); J. D. Hooker, *Bot. Mag.* t. 7767 (1901).
Cladrastis fragrans, Rafinesque (name only), *Cincinnati Literary Gazette*, i. 66 (1824).
Cladrastis lutea, Koch, *Dendrologie*, i. 6 (1869); Sargent, *Silva N. America*, iii. 57, tt. 119, 120 (1892), and *Trees N. America*, 568 (1905).
Virgilia lutea, Michaux, *Hist. Arb. L'Amér.* iii. 266, t. 3 (1813); Loudon, *Arb. et Frut. Brit.* ii. 565 (1838).

A tree attaining 60 feet in height, and rarely 12 feet in girth.   Bark smooth and silvery grey.   Branchlets brittle, glabrous.   Leaves (Plate 125, fig. 5) alternate, unequally pinnate, 8 to 12 inches in length.   Leaflets seven to nine, usually alternate; the terminal one largest, articulate and directed to one side, often broadly rhombic; the others gradually diminishing in size towards the base of the leaf, 3 to 4 inches long by $1\frac{1}{2}$ to 2 inches wide, on stout pubescent petiolules, oval or ovate, entire and non-ciliate in margin; base broadly cuneate or rounded, apex acuminate; upper surface light green and glabrous; lower surface pale green with occasional hairs on the midrib and veins.   Rachis of the leaf terete, glabrous, with the base swollen and hollowed out, enclosing the buds, which are usually four, the largest and uppermost one developing, the others minute and rudimentary.

Flowers in nodding terminal panicles, 10 to 20 inches long, white, with a yellow spot at the base of the standard.   Pedicels slender and not grouped in pairs.   Calyx canpanulate, enlarged on its upper side; teeth five, short, obtuse, nearly equal. Corolla papilionaceous with clawed petals; standard nearly orbicular; wings oblong and two-auricled at the base; keel-petals free, oblong, and sub-cordate or two-auricled at the base.   Stamens ten, free.   Ovary linear, stalked, villose; ovules numerous. Pod glabrous, short-stalked, linear, glabrous.   Seeds four to six, attached by slender stalklets, oblong-compressed, without albumen.

---

[1] *Tokyo Bot. Mag.* vi. 345, t. 10 (1892).
[2] Cf. Ito and Matsumura, *Journ. Science College, Imp. Univ. Tokyo*, xii. 436 (1899).
[3] Judging from the description, as I have seen no specimens.   There are specimens in the Kew Herbarium (*Cladrastis, sp.* ? Hemsley, *Journ. Linn. Soc.* (*Bot.*) xxiii. 201 (1887)) which were collected by Millett, probably in the vicinity of Canton, which are very near to the Loochoo species.
[4] This name is adopted as being the first one with a description published under the correct genus.

SEEDLING

A plant, raised from seed sown at Colesborne on 2nd March, showed the following characters on 7th July:—Root white, fleshy, tapering, 3 inches long, giving off numerous lateral fibres. Caulicle striated, glabrous, 1½ inch long. Cotyledons two, sub-sessile, oblong, tapering slightly at the base, broader towards the rounded apex, green above, white beneath, coriaceous, entire. Stem terete, with a few scattered hairs below, densely white pubescent above. Leaves, all with petioles swollen at the base; first pair opposite, on pubescent stalks, simple, ovate, entire, 2 inches long by 1½ inch broad. The third, fourth, and fifth leaves are alternate; the third simple and like the first pair; the fourth and fifth trifoliolate on a stalk 2 inches long, terminal leaflet ovate, lateral leaflets oval and smaller.

IDENTIFICATION

Cladrastis tinctoria is readily distinguishable in summer by the pinnate leaves with alternate leaflets, of which the terminal one is directed to one side of the leaf; and by the swollen base of the petiole, which encloses and conceals the buds.

In winter the following characters are available (Plate 126, fig. 4):—Twigs zig-zag, shining, brown or grey, terete, glabrous; lenticels minute, numerous. Leaf-scars alternate, obliquely set on slightly prominent pulvini, oval, whitish, with five bundle-dots on the outer rim, the centre of the scar being occupied by a projecting cone, which consists of four buds compressed together and superposed one above the other, the uppermost one the largest, all pubescent. Terminal bud not formed, the apex of the twig showing a small circular scar or a short stump, indicating where the top of the branchlet fell off in early summer.

DISTRIBUTION

Cladrastis tinctoria is one of the rarest trees in the American forest, growing only in a few isolated localities in central Kentucky, central and eastern Tennessee, northern Alabama, and the south-western part of N. Carolina. It is met with on limestone ridges and cliffs, usually in rich soil, and frequently overhangs mountain streams. (A. H.)

CULTIVATION

The yellow-wood is a favourite ornamental tree in American gardens, where, according to Sargent,[1] it adapts itself readily to varied conditions of soil and climate, though it requires deep rich soil in order to attain its full size and beauty. It has a tendency, however, which in England is equally marked, to divide into several spreading stems, which are rather brittle and liable to split the trunk. Its long racemes of white fragrant flowers make it a very pretty tree early in June, but in our climate these are not produced as freely as in America, and I have never seen fruit ripened in this country. In autumn the leaves turn a bright yellow.

[1] *Garden and Forest*, i. p. 92.

Sargent[1] gives an illustration of a beautiful specimen in a garden near Boston which, 35 years after planting, was 35 feet high and had a spread of nearly 60 feet. I saw several in this district, but none so large as those which I have seen in England.

Though it germinates quickly, and seems easy to raise from seed, the tree is now seldom planted in England, but may be recommended for warm sheltered situations in good soil in the south and east, though perhaps the damp climate of the west does not suit it; and as most of the trees mentioned by Loudon have disappeared, it seems to be short-lived in this country. The seedlings which I have raised from American seed are fairly hardy, and after the first two years grow better than many American trees on my soil.

This species was introduced into cultivation in England in 1812, by John Lyon, a Scotsman who travelled in Carolina, Georgia, and Florida.

REMARKABLE TREES

The largest tree known to us is at Syon (Plate 124), which in 1904 was no less than 60 feet in height by 7 feet in girth and still a fine tree, though its trunk is decaying inside. There is another in Kew Gardens, near the Director's office, which measures 35 feet high, with a bole of 3 feet girthing 5 feet 4 inches and dividing into six main stems, which sub-divide into numerous upright branches. At the Knaphill Nursery near Woking is a very well grown tree about 45 feet high and 8 feet in girth, the head spreading to 16 yards in diameter.

At Highclere there is a tree which measures 42 feet by 7 feet with a spread of branches of 45 feet. Although there is some decay near the root the tree seems to have become more vigorous recently. At Blenheim there is an old specimen, with a stem divided close to the ground, and forming rather a large bush than a tree. At Cornbury Park there is also a fair-sized tree. At Barton, Suffolk, a tree planted[2] in 1832 was in 1904 25 feet high with a short bole, 5 feet 6 inches in girth, dividing into three wide-spreading main branches.

We have not seen any large enough to mention in Scotland or Ireland.

TIMBER

The wood, according to Sargent, is heavy, hard, strong, and close-grained, and is susceptible of a fine polish. At one time it was used in Kentucky for making gun-stocks; but is too rare to have any commercial importance. It produces a yellow dye.   (H. J. E.)

---

[1] *Garden and Forest*, i. p. 92.   [2] Bunbury, *Arboretum Notes*, p. 1.

CLADRASTIS AMURENSIS

Cladrastis amurensis, Bentham et Hooker, *Gen. Pl.* i. 554 (1865); Maximowicz, *Mél. Biol.* ix. 72
(1873); Franchet et Savatier, *Enum. Pl. Jap.* i. 115 (1875) and ii. 327 (1879); J. D. Hooker,
Bot. Mag. t. 6551 (1881); Shirasawa, *Icon. Ess. Forest. Jap.* text 85, pl. L. figs. 1-12 (1900).
Maackia amurensis, Ruprecht et Maximowicz, *Mél. Biol.* ii. 418, 441 (1856) and 534 (1857);
Maximowicz, *Prim. Fl. Amur.* 87, 390, t. v. (1859); Morren, *Belgique Horticole*, 1890, p. 301,
t. 18; *Gartenflora*, 1875, p. 152.

A small tree, attaining 40 or 50 feet in height, with bark peeling off in old trees
like that of a birch. Young shoots minutely pubescent. Leaflets (Plate 125, fig. 6)
seven to eleven, opposite or rarely sub-opposite, the terminal one articulate, the
lateral ones on short, stout pubescent petiolules; 2 to 3 inches long; deltoid, ovate
or oval; base truncate or rounded; apex obtuse or acute; entire; upper surface
dark green and minutely pubescent; lower surface pale green, densely appressed
pubescent; rachis pubescent, swollen at the base.

Flowers greenish white, on long pedicels, in simple or occasionally branched
erect terminal dense racemes. Calyx teeth four, short, broad, unequal. Petal-
claws long, slender; standard obovate, emarginate; wings oblong, obtuse, two-
auricled at the base; keel petals partially coalesced, one-auricled. Stamens slightly
connate below. Pod, 2 to 3 inches long, oblong, flattened, brown, slightly appressed
pubescent; seeds, one to five, oblong.

In specimens from the Asiatic continent the leaflets are larger and much less
pubescent than in the Japanese tree, which has been distinguished by Maximowicz
as var. *Buergeri*,[1] and is characterised by very dense appressed pubescence on the
lower surface of the leaflets and white tomentose shoots.

In winter the twigs (Plate 126, fig. 5) are shining, glabrous; leaf-scars on prominent pulvini, semicircular, marked by a central large tubercular bundle-scar and
two minute dots close to the upper margin; true terminal bud absent, the top of
the branchlet having fallen off in early summer and leaving a short stump at the
apex of the twig. Buds solitary, dark brown, shining, pubescent towards the apex,
showing two scales visible externally.

Cladrastis amurensis occurs in Amurland as far north as lat. 52° 20′, and grows
throughout Eastern Manchuria and Korea, the largest tree seen by Maack being
only 35 feet high and 1 foot in diameter. According to Shirasawa, it is met with in
Japan on moist rich soils in the temperate parts, ascending to 4300 feet in the central
chain of the main island, and attaining a height of 50 feet and a diameter of 28
inches. It was collected by Elwes in the forest near Asahigawa in central Hokkaido,
where, however, it was not abundant or conspicuous. It is called *Inu-enju* in Japan.

Cladrastis amurensis was introduced from the Amur in 1864 by Maximowicz;
and has been spread throughout Europe by the St. Petersburg Botanical Garden. It
probably came into England about 1870.

[1] *Mél. Biol.* ix. 72 (1873).

It is propagated either by seed or by root-cuttings. At Kew it is rather a shrub than a tree, and produces flowers when quite young, which appear late in the season, in the end of July or the beginning of August. It ripens its fruit in October, the pods remaining on the tree during winter.

The timber, according to Shirasawa, is hard and tenacious, and is used in building and in making furniture. Elwes purchased planks of it at Sapporo, which are of a yellowish-brown colour, and seem to be of good quality for cabinet-making.

(A. H.)

CLADRASTIS SINENSIS

Cladrastis sinensis, Hemsley, *Journ. Linn. Soc.* (*Bot.*) xxix. 304 (1892).

A tree attaining 70 feet in height and 10 feet in girth. Young shoots rusty pubescent towards the base. Leaflets nine to eleven, alternate, entire, oblong-lanceolate, obtuse or acute at the apex; broad and rounded, rarely cuneate, at the base; lower surface with appressed pubescence most marked towards the base and along the midrib. Leaf-rachis pubescent, with swollen base enclosing two or three buds. Leaf-scars on older shoots, oblique on prominent pulvini, orbicular; the raised circular rim, discontinuous above, surrounding a central densely pubescent depression, in which lie two or three buds, the upper one of which is the largest.

Flowers pinkish-white, fragrant, in large terminal, rusty-pubescent panicles. Calyx rusty-pubescent; teeth short, broad, rounded. Petals long-clawed, erect, free; standard broadly obovate, bifid; wings and keel-petals oblong. Stamens slightly connate at the base; ovary pubescent. Pod linear-oblong, flattened, with thickened margins.

This tree, which resembles *Sophora japonica* in habit and foliage, was discovered by Pratt, in 1890, in Western Szechuan, where E. H. Wilson subsequently saw large trees at 7000 feet altitude in the Hsiang Ling range, west of Mt. Omei. It also occurs in the high mountains of the Fang district in Hupeh, from whence seeds were sent home by Wilson in 1901. Plants raised at Coombe Wood were, in 1906, 5 feet high, and for so far have proved perfectly hardy. The tree has beautiful flowers, and, growing at high altitudes in western China, should thrive in this country.

(A. H.)

Printed by R. & R. Clark, Limited, *Edinburgh.*

PLATE 58 A.

WEEPING BEECH AT ENDSLEIGH

(*See Page* 10)

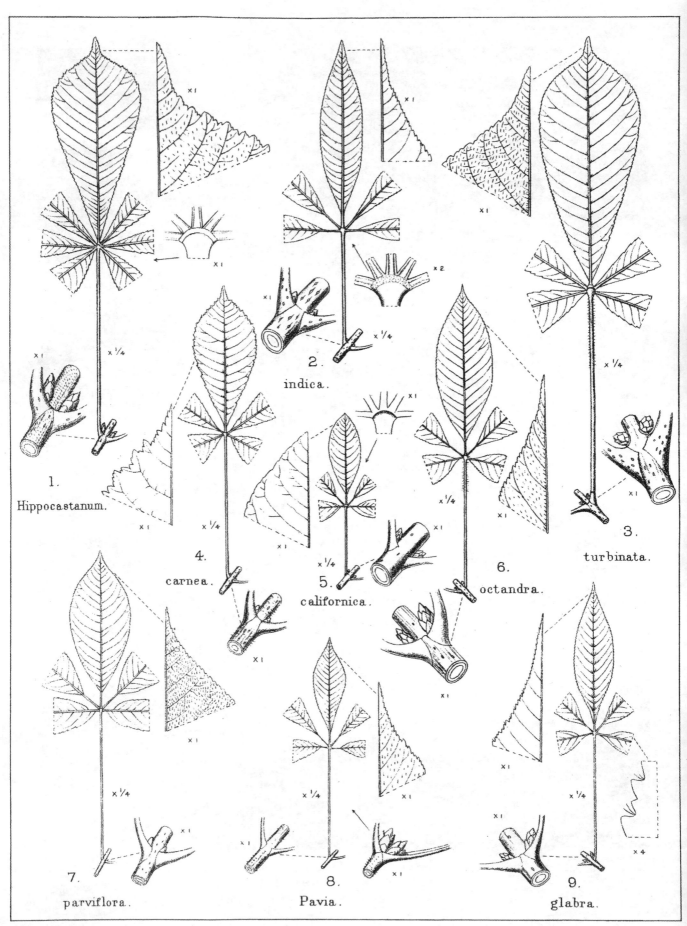

1.
Hippocastanum.

2.
indica.

3.
turbinata.

4.
carnea.

5.
californica.

6.
octandra.

7.
parviflora.

8.
Pavia.

9.
glabra.

Huitt del, Huth lith.

PLATE 61.

AESCULUS.

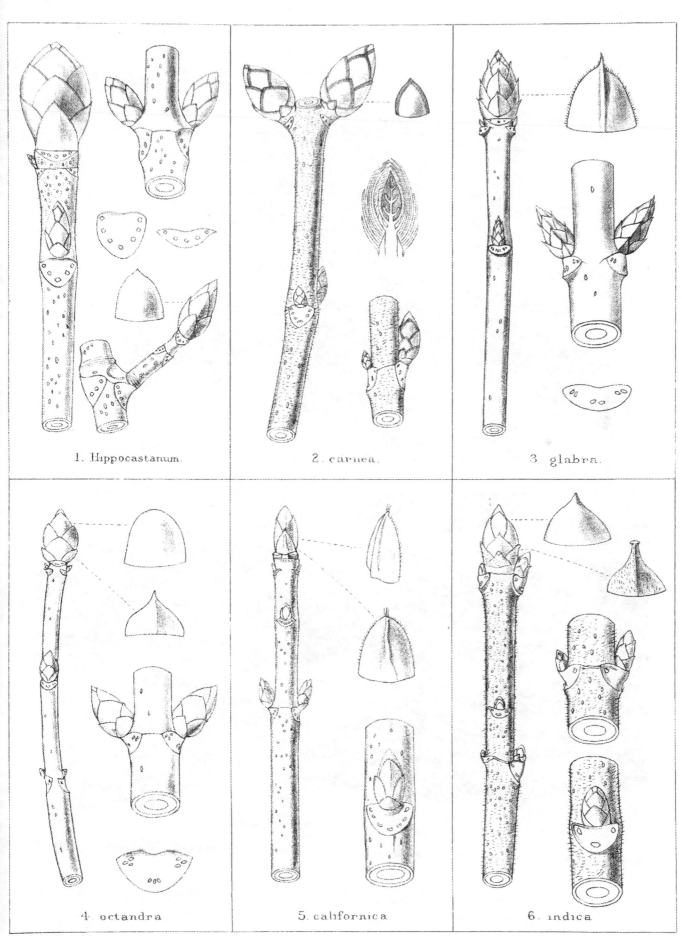

1. Hippocastanum.   2. carnea.   3. glabra.

4. octandra   5. californica   6. indica

PLATE 82.

AESCULUS.

PLATE 63.

HORSE CHESTNUT AT COLESBORNE

WEEPING HORSE CHESTNUT AT DUNKELD

PLATE 64.

PLATE 65

ÆSCULUS INDICA AT BARTON

PLATE 66.

ÆSCULUS TURBINATA IN JAPAN

PLATE 67.

HOOKER'S HEMLOCK AT MURTHLY

PLATE 68.

WESTERN HEMLOCK AT DROPMORE

PLATE 69.

WESTERN HEMLOCK AT MURTHLY

PLATE 70.

HEMLOCK SPRUCE AT FOXLEY

PLATE 71.

PLATE 72.

HIMALAYAN HEMLOCK AT BOCONNOC

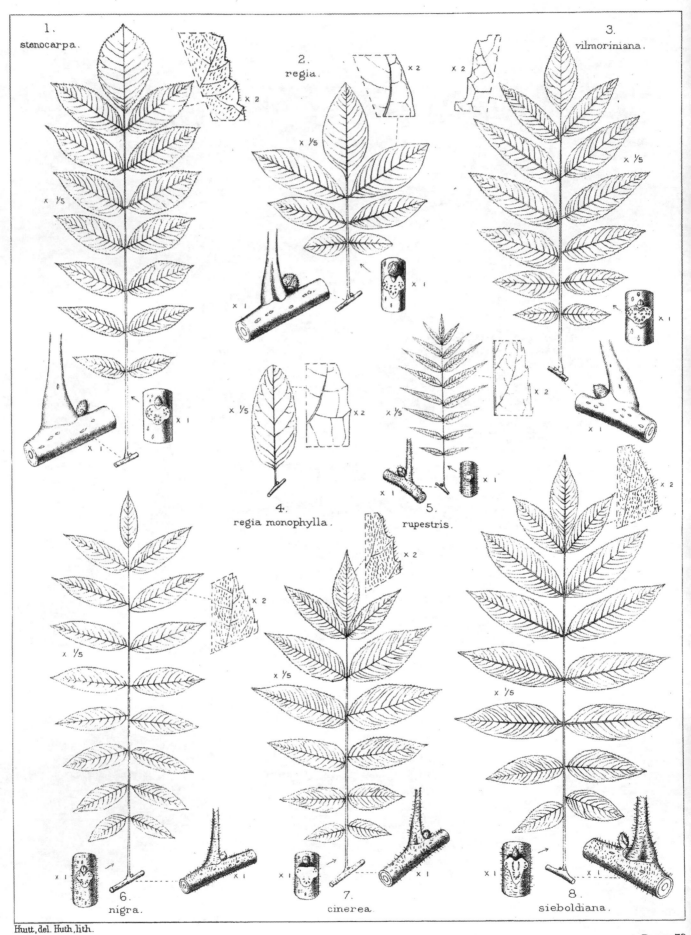

1. stenocarpa.

2. regia.

3. vilmoriniana.

4. regia monophylla.

5. rupestris.

6. nigra.

7. cinerea.

8. sieboldiana.

Huitt, del. Huth, lith.

PLATE 73.

JUGLANS.

PLATE 74.

WALNUT AT BARRINGTON PARK

WALNUT AT GORDON CASTLE

PLATE 75.

BLACK WALNUT AT TWICKENHAM

PLATE 76.

PLATE 77.

BLACK WALNUT AT THE MOTE

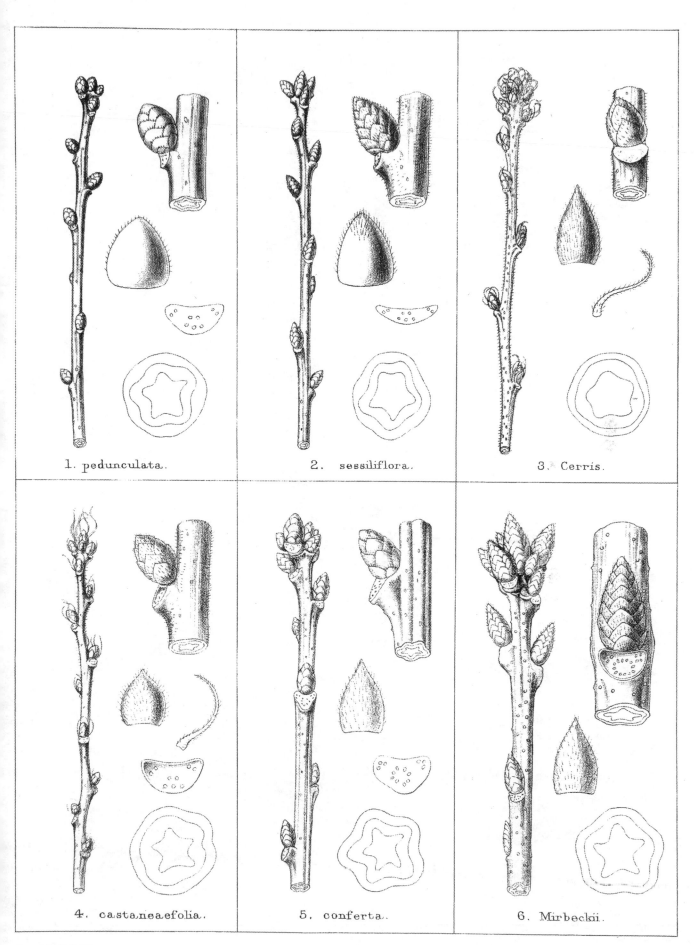

1. pedunculata.

2. sessiliflora.

3. Cerris.

4. castaneaefolia.

5. conferta.

6. Mirbeckii.

Hewitt del. Huth lith.

PLATE 78.

QUERCUS.

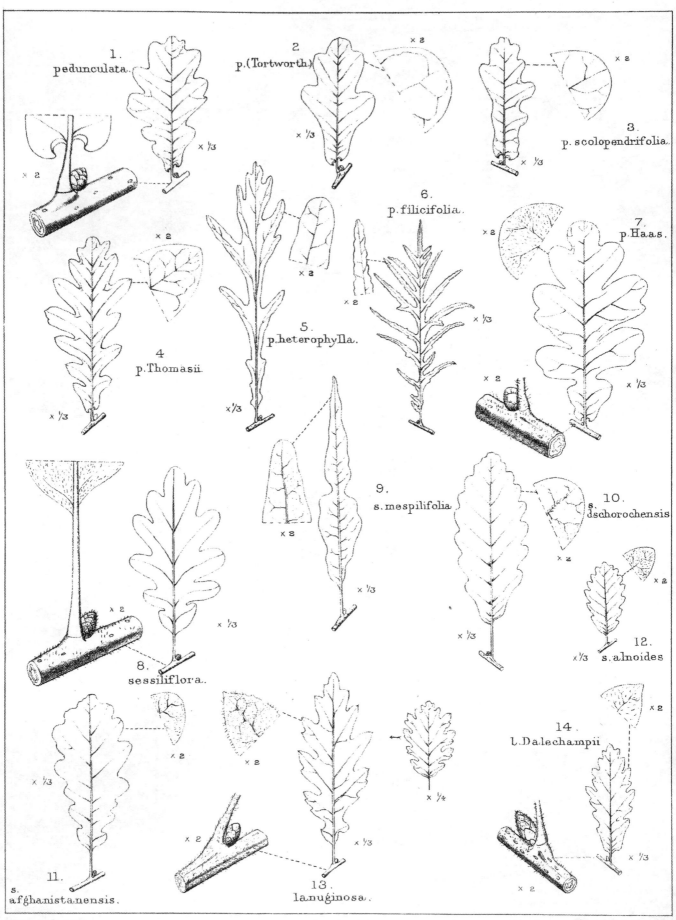

1. pedunculata.
×2

2. p.(Tortworth.) ×2
×⅓

3. p. scolopendrifolia.
×2 ×⅓

4. p.Thomasii.
×2 ×⅓

5. p.heterophylla.
×2 ×⅓

6. p.filicifolia.
×2 ×⅓

7. p.Haas.
×2 ×⅓ ×2

8. sessiliflora.
×2 ×⅓

9. s.mespilifolia
×2 ×⅓

10. s. dschorochensis
×2 ×⅓

11. s. afghanistanensis.
×2 ×⅓

12. s.alnoides
×2 ×⅓

13. lanuginosa.
×2 ×2 ×¼ ×⅓

14. L.Dalechampii
×2 ×2 ×⅓

Britt, del. Huth, lith.

PLATE 79.

QUERCUS PEDUNCULATA, SESSILIFLORA, AND LANUGINOSA.

PLATE 80.

CYPRESS OAK AT MELBURY

PLATE 81.

SELF-SOWN OAKS AT THORNBURY CASTLE

PLATE 82.

CHAMPION OAK AT POWIS CASTLE

PLATE 83.

OAK AT POWIS CASTLE

PLATE 84.

LADY POWIS' OAK AT POWIS CASTLE

PLATE 85.

TALL OAK AT WHITFIELD

PLATE 86.

OAK AT KYRE PARK

PLATE 87.

TALL OAKS AT KYRE PARK

BILLY WILKIN'S OAK AT MELBURY

PLATE 88.

PLATE 89.

BEGGAR'S OAK IN BAGOT'S PARK

PLATE 90.

OAKS AT BAGOT'S PARK

PLATE 91.

KING OAK AT BAGOT'S PARK

PLATE 92.

SESSILE OAK AT MEREVALE PARK

PLATE 93.

PLATE 94.

OAK AT ALTHORP

PLATE 95.

MAJOR OAK IN SHERWOOD FOREST

PLATE 96.

BROWN OAK AT ROCKINGHAM PARK

PLATE 97.

UMBRELLA OAK AT CASTLE HILL

PLATE 98.

LARCHES IN OAKLEY PARK

PLATE 99.

LARCHES AT SHERBORNE, GLOUCESTERSHIRE

PLATE 100.

LARCHES AT COLESBORNE

PLATE 101.

CHAMPION LARCH AT TAYMOUTH

PLATE 102.

FORKED LARCH AT TAYMOUTH

PLATE 103.

MOTHER LARCH AT DUNKELD

PLATE 104.

FORKED LARCH AT GORDON CASTLE

PLATE 105.

LARCH IN THE ALPS

PLATE 106.

DAHURIAN LARCH AT WOBURN

A

B

PLATE 107.

LARCH IN KURILE ISLANDS

PLATE 108.

JAPANESE LARCH AT TORTWORTH

PLATE 109.

SIKKIM LARCH AT STRETE RALEIGH

AMERICAN LARCH AT DROPMORE

PLATE III.

WESTERN LARCH IN MONTANA

PLATE 112.

LYALL'S LARCH IN ALBERTA

PLATE 113.

CORSICAN PINE IN CORSICA

B

A

CORSICAN PINE IN CORSICA

(André)

PLATE 114.

PLATE 115.

PINUS LARICIO ON SANDHILLS AT HOLKHAM

PINUS LARICIO AT HOLKHAM   PLATE 116.

PLATE 117.

PINUS LARICIO AT ARLEY

CRIMEAN PINE AT ELVEDEN

PLATE 118.

PLATE 119.

PINUS LEUCODERMIS IN BOSNIA

Plate 120 *is reserved for Gymnocladus and will be issued with a later Volume.*

PLATE 121.

PTEROCARYA CAUCASICA AT MELBURY

PLATE 122.

PTEROCARYA CAUCASICA AT CLAREMONT

PLATE 123.

PTEROCARYA CAUCASICA AT CAMBRIDGE

PLATE 124

YELLOW-WOOD AT SYON

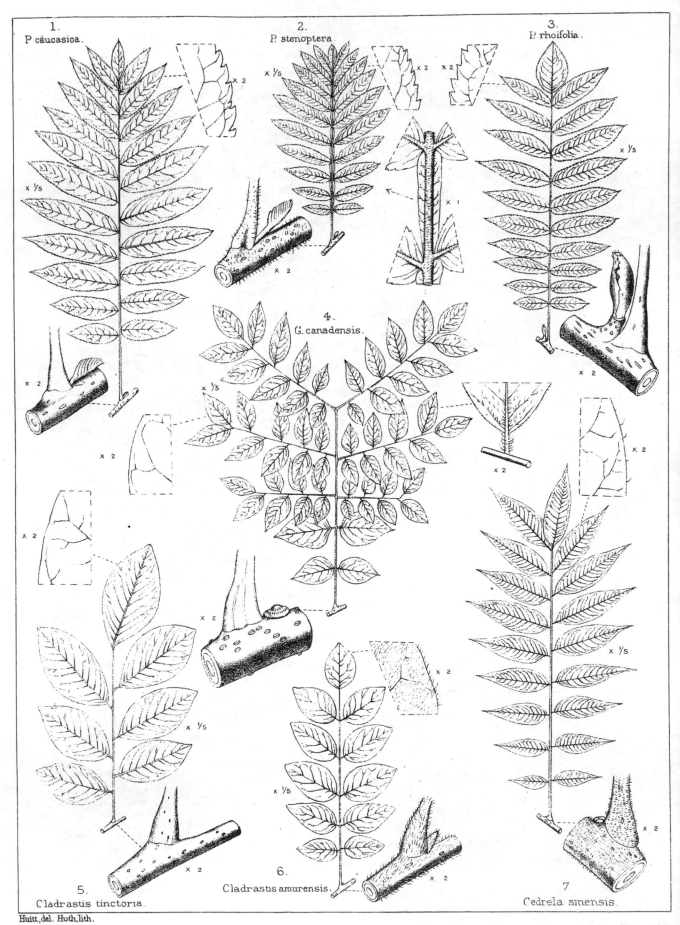

1. P. cáucasica.
2. P. stenoptera.
3. P. rhoifolia.
4. G. canadensis.
5. Cladrastis tinctoria.
6. Cladrastis amurensis.
7. Cedrela smensis.

Huitt, del. Huth, lith.

PLATE 125.

PTEROCARYA, GYMNOCLADUS, CLADRASTIS AND CEDRELA.

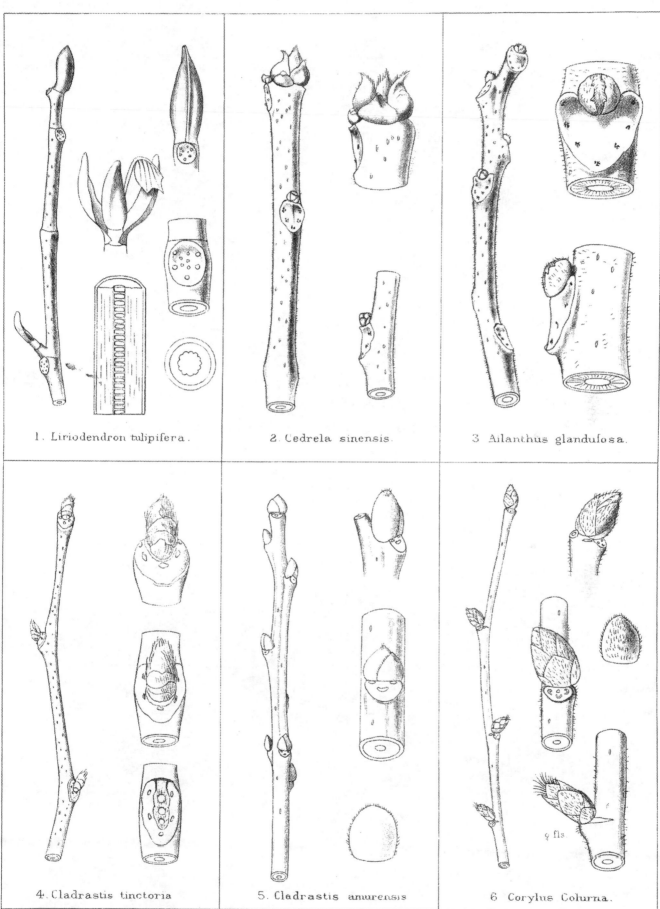

1. Liriodendron tulipifera.

2. Cedrela sinensis.

3 Ailanthus glandulosa.

4. Cladrastis tinctoria

5. Cladrastis amurensis

6 Corylus Colurna.

♀ fls.

PLATE 126.

LIRIODENDRON, CEDRELA, AILANTHUS, CLADRASTIS AND CORYLUS.

Plate 27, which was unavoidably omitted from Volume I., shews the Tulip trees at Leonardslee and Horsham Park mentioned on page 70. We are indebted to Mrs. F. Du Cane Godman for the admirable negatives from which this plate was made.

TULIP TREE AT HORSHAM PARK

PLATE 27.

TULIP TREE AT LEONARDSLEE

Printed in the United States
By Bookmasters